INDUSTRIAL MECHANICS

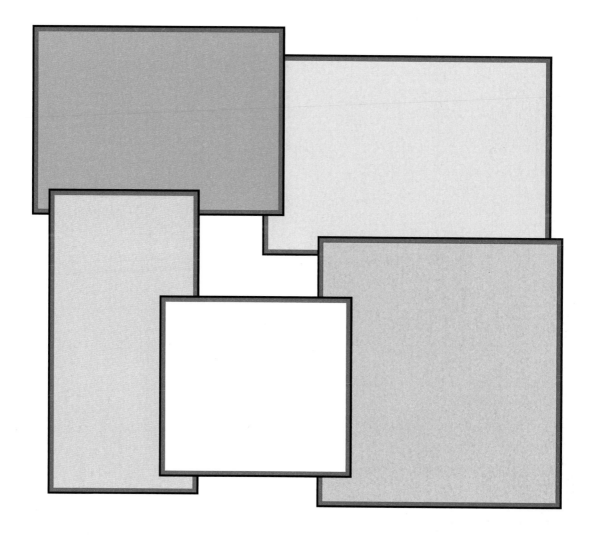

AMERICAN TECHNICAL PUBLISHERS, INC.
HOMEWOOD, ILLINOIS 60430-4600

Albert W. Kemp

1 2 3 4 5 6 7 8 9 – 99 – 9 8 7 6 5 4

Printed in the United States of America

ISBN 978-0-8269-3690-5

Acknowledgments

The author and publisher are grateful to the following companies and organizations for providing technical information and assistance. Companies preceded by an asterisk (*) have provided photographs that were used on the cover.

- Acco Chain & Lifting Products Division
- Advanced Assembly Automation Inc.
- ARO Fluid Products Div., Ingersoll-Rand
- Atlas Technologies, Inc.
- * Ballymore Company, Inc.
- Baldor Electric Co.
- Bimba Manufacturing Company
- Boeing Commercial Airplane Group
- Boston Gear®
- The Caldwell Group, Inc.
- Cincinnati Milacron
- Clippard Instrument Laboratory, Inc.
- Columbus McKinnon Corporation Industrial Products Division
- Cone Drive Operations Inc./Subsidiary of Textron Inc.
- Cone Mounter Company
- Continental Hydraulics
- CooperTools
- The Crosby Group, Inc.
- Crowe Rope Industries LLC
- DoALL Company
- Dow Corning Corporation
- Eaton Corporation
- * Emerson Power Transmission
- Engelhardt Gear Co.
- Exxon Company
- Fenner Drives
- Flow Ezy Filters, Inc.
- * Fluke Corporation
- Gast Manufacturing Company
- The Gates Rubber Company
- GE Motors & Industrial Systems
- Greenlee Textron Inc.
- Guardian Electric Mfg. Co.
- * Harrington Hoists Inc.
- Heidelberg Harris, Inc.
- Honeywell's MICRO SWITCH Division
- Humphrey Products Company
- Ingersoll-Rand Material Handling
- Lift-All Company, Inc.
- Lift-Tech International, Division of Columbus McKinnon
- Lovejoy, Inc.
- LPS Laboratories, Inc.
- Ludeca Inc., representative of PRUEFTECHNIK AG.
- * Manufacturing & Maintenance Systems, Inc.
- Martin Sprocket & Gear Inc.
- Miller Equipment
- * North American Industries, Inc.
- * NTN Bearing Corporation of America
- Oil-Rite® Corporation
- Pacific Bearing Company
- Panduit Corp.
- Parker Hannifin - Pneumatic Division North America
- PLI Incorporated
- Power Team, Division of SPX Corporation
- Precision Brand Products, Inc.
- Predict\DLI
- Ratcliff Hoist Co.
- Ruud Lighting, Inc.
- Saylor-Beall Manufacturing Company
- SEW-Eurodrive, Inc.
- The Sinco Group, Inc.
- Snorkel
- * SPM Instrument, Inc.
- Sprecher + Schuh
- The Timken Company
- Tractel Inc., Griphoist® Division
- Vibration Monitoring Systems, Inc.
- * Werner Ladder Co.
- Wire Rope Technical Board

Contents

1 Calculations 1

2 Rigging 17

3 Lifting 63

4 Ladders and Scaffolds 87

5 Hydraulic Principles 109

6 Practical Hydraulics 133

7 Pneumatic Principles 167

8 Practical Pneumatics 183

9 Lubrication 211

10 Bearings.......... 225

11 Flexible Belt Drives 243

12 Mechanical Drives.......... 257

13 Vibration 271

14 Alignment 289

15 Electricity 319

A Appendix 347

G Glossary.......... 383

I Index 399

Introduction

Industrial Mechanics is a comprehensive introduction to fundamental industrial mechanical concepts, principles, and equipment. The textbook is designed for industrial mechanics, technicians, and maintenance personnel. Photographs from over 60 major manufacturers and organizations have been included to help illustrate the broad range of technical information used in industry today.

Industrial Mechanics covers topics such as rigging and lifting, ladders and scaffolds, hydraulics and pneumatics, lubrication, bearings, flexible belt and mechanical drives, vibration, alignment, and electricity. The comprehensive Appendix contains many useful tables, charts, and other supplemental reference material. The Glossary defines nearly 800 technical terms used throughout the textbook. An extensive Index is also included.

Industrial Mechanics Workbook tests for each of the major concepts presented in the textbook. Answers to the workbook problems are in the *Instructor's Guide*. To obtain information on related training products, visit the American Tech web site at www.americantech.org.

The Publisher

Calculations

Chapter 1

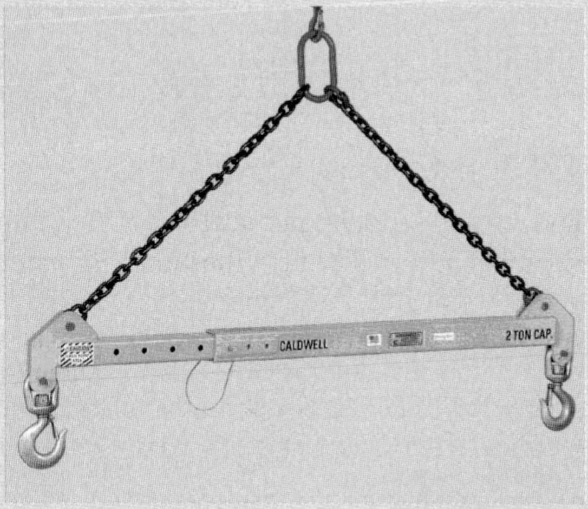

The Caldwell Group, Inc.

Formulas are used daily in calculations to determine the various quantities used in industry. Linear measurements are used to determine the area of plane figures. The size and weight of objects are used to determine fit and weight for lifting. Measurements may be made in the British (U.S.) system or in the metric system. Volume is calculated to determine the capacity of containers used for storage and in hydraulic and pneumatic systems.

FORMULAS

An *equation* is a means of showing that two numbers or two groups of numbers are equal to the same amount. See Figure 1-1.

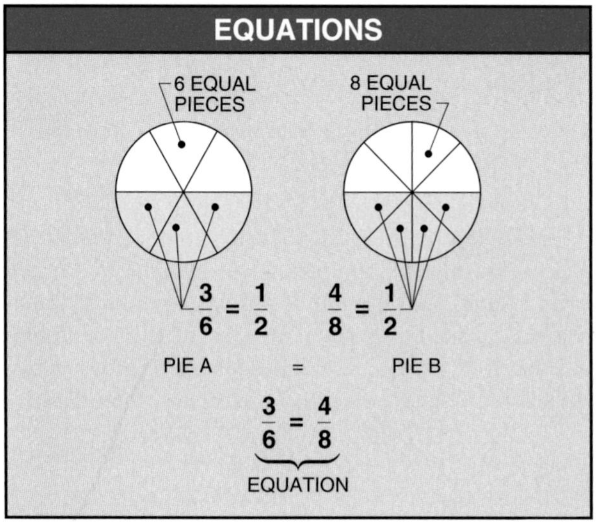

Figure 1-1. An equation is a means of showing that two numbers, or two groups of numbers, are equal to the same amount.

A *formula* is a mathematical equation that contains a fact, rule, or principle. Formulas are used in all trade areas. They are used to find the area of plane figures such as circles, triangles, quadrilaterals, etc. Italic letters are used in formulas to represent values (amounts).

For example, $a + b = c$ is a formula. See Figure 1-2. In a formula, any number or letter may be transposed from left to right or from right to left of the equal sign. When transposed, the sign of the number or letter is changed to the opposite sign.

The sign is always in front of the number or letter of which it is a part. For example, in the formula $a + b = c$, if $b = 4$ and $c = 12$, the value of a is found by changing the formula to $a = c - b$, or $a = 12 - 4$. A formula can be changed to solve for any unknown value if the other values are known. Subscript letters or numbers may be used in formulas to distinguish between similar dimensions of different objects. For example, V_c may indicate the volume of a cylinder and V_t may indicate the volume of a tank.

A motor operating at a standard speed of 1800 rpm rotates 2,592,000 times in a 24-hour period (1800 × 24 × 60 = 2,592,000).

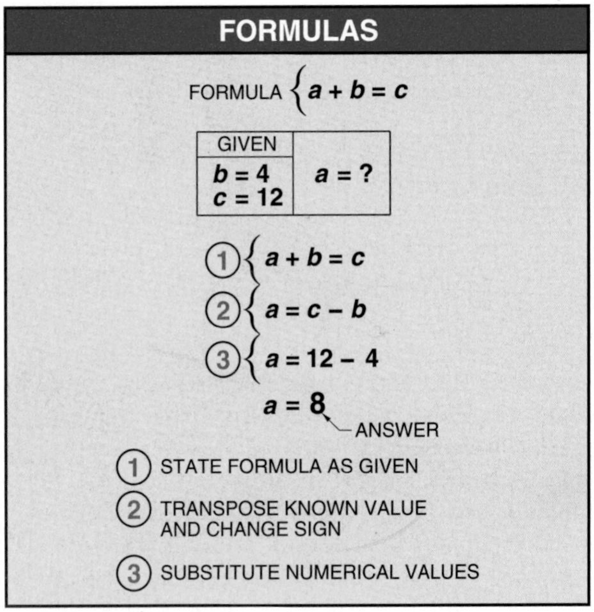
FORMULAS
FORMULA { a + b = c
GIVEN
b = 4
c = 12
a = ?
1 { a + b = c
2 { a = c − b
3 { a = 12 − 4
a = 8
ANSWER
1 STATE FORMULA AS GIVEN
2 TRANSPOSE KNOWN VALUE AND CHANGE SIGN
3 SUBSTITUTE NUMERICAL VALUES

Figure 1-2. A formula is a mathematical equation that contains a fact, rule, or principle.

Common Formulas

In the trades, common formulas related to plane and solid figures are used when laying out jobs. For example, a welder may be required to lay out and build a cylindrical tank to hold a specified number of gallons of liquid. By applying the volume formula for cylinders, the welder can determine the required size of the cylindrical tank.

Units of Measure

The common measurement systems are the British (U.S.) system and the SI metric system (International System of Units). Conversions can be made from one system to the other.

British (U.S.) System. The British (U.S.) system, primarily used in the United States, is also known as the English system. This system uses the inch (in. or ″), foot (ft or ′), pint (pt), quart (qt), gallon (gal.), ounce (oz), pound (lb), etc. as basic units of measure. See Appendix.

Metric System. The metric system is the most common measurement system used in the world. This system is based on the meter (m), liter (l), and gram (g). See Appendix. Prefixes are used in the metric system to represent multipliers. For example, the prefix kilo (k) has a prefix equivalent of 1000, so 1 kilometer (km) = 1000 m = 1 km.

To change from a quantity to another prefix, multiply the quantity by the number of units that equals one of the original metric units (conversion factor). For example, to change 544 m to kilometers (1 m = $\frac{1}{1000}$ k), divide 544 by 1000 (544 ÷ 1000 = .544 km).

It may be necessary to change a quantity to a base unit before changing to a new prefix. For example, to change 2000 mg to decagrams, change 2000 mg to grams (base unit) by multiplying 2000 by .001 (2000 × .001 = 2). Change grams to decagrams by multiplying 2 by .1 (2 × .1 = .2 dg).

Additionally, a conversion table may be used to change a metric quantity to another prefix. For example, to change from liters to kiloliters, move the decimal point three places to the right. See Appendix.

Conversions. Conversions can be performed in all units of measure. English and metric measurements are converted from one system to the other by multiplying the number of units of one system that equals one unit of measurement of the other. The number of units is the conversion factor. Conversions are performed by applying the appropriate conversion factor. Equivalent Tables are used to convert between systems. See Appendix.

To convert a metric measurement to English, multiply the measurement by the number of English units that equals one of the metric units (conversion factor). For example, to convert 50 mm to inches, multiply 50 by .039 (50 × .039 = 1.95″).

To convert an English measurement to metric, multiply the measurement by the number of metric units that equals one of the English units (conversion factor). For example, to convert 3 mi to kilometers, multiply 3 by 1.609 (3 × 1.609 = 4.827 km).

PLANE FIGURES

A *plane figure* is a flat figure with no depth. All plane figures are composed of straight or curved lines. Plane figures include circles, triangles, quadrilaterals, and polygons. The area of a plane figure is measured in square units such as square inches, square feet, square millimeters, square meters, etc.

Lines

A *line* is the boundary of a surface. See Figure 1-3. Lines are measured in linear units such as inches, feet, yards, millimeters, meters, etc.

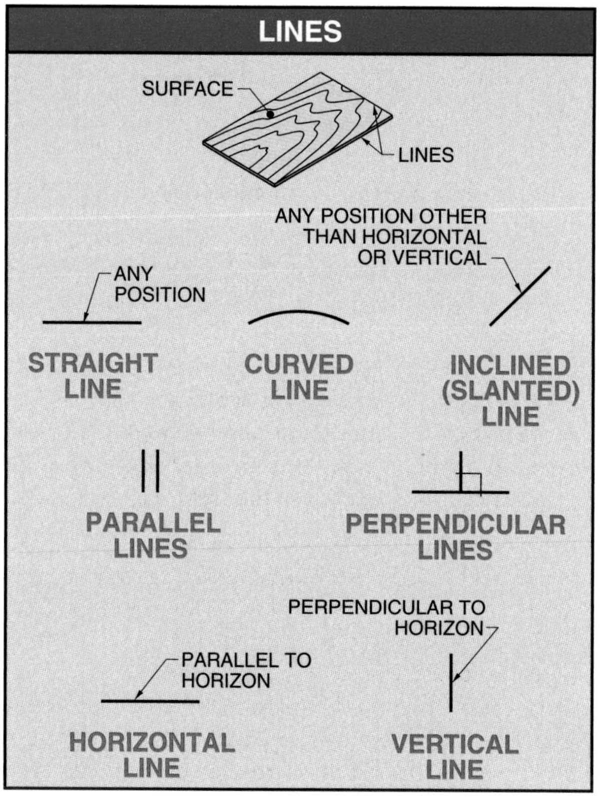

Figure 1-3. A line is the boundary of a surface.

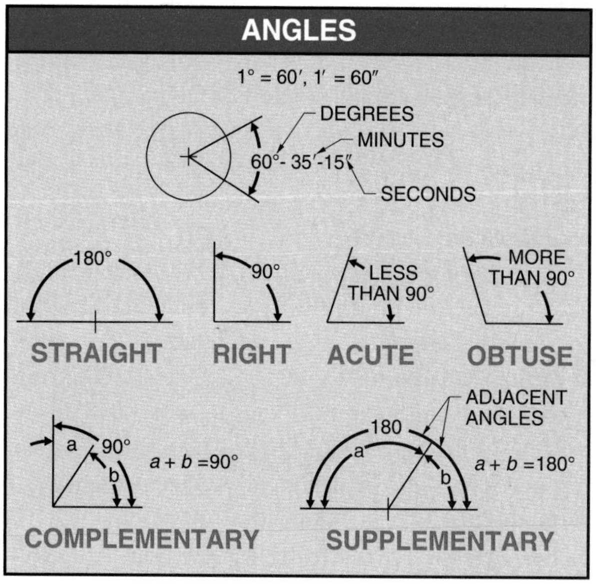

Figure 1-4. An angle is the intersection of two lines.

A *straight line* is the shortest distance between two points. It is commonly referred to as a line. A *curved line* is a line that continually changes direction. It is commonly referred to as a curve.

All lines may be drawn in any position, unless they are horizontal or vertical. An *inclined line* is a line that is slanted. It is neither horizontal nor vertical. *Parallel lines* are two or more lines that remain the same distance apart. The symbol for parallel lines is ||.

A *perpendicular line* is a line that makes a 90° angle with another line. The symbol for perpendicular is ⊥. A *horizontal line* is a line that is parallel to the horizon. It may be referred to as a level line.

A *vertical line* is a line that is perpendicular to the horizon. It is often referred to as a plumb line. *Plumb* is an exact verticality (determined by a plumb bob and line) with the Earth's surface.

Angles

An *angle* is the intersection of two lines or sides. See Figure 1-4. The *vertex* is the point of intersection of the sides of an angle. To identify angles, letters are placed at the end of each side and at the vertex. When referring to an angle, the vertex letter is read second. The angle symbol (∠) is used to indicate an angle. The way the two sides of an angle intersect determines the size and type of angle.

Angles are measured in degrees, minutes, and seconds. The symbol for degrees is °, the symbol for minutes is ′, the symbol for seconds is ″. There are 360° in a circle (one revolution). There are 60′ in one degree and 60″ in one minute. For example, an angle might contain 112°-30′-12″.

A *straight angle* is two lines that intersect to form a straight line. It is one-half of a revolution, or $^{360}/_{2}$ = 180°. A straight angle always contains 180°.

A *right angle* is two lines that intersect perpendicular to each other. It is one-fourth of a revolution, or $^{360}/_{4}$ = 90°. A right angle always contains 90°.

An *acute angle* is an angle that contains less than 90°. An *obtuse angle* is an angle that contains more than 90°. For example, a 45° angle is an acute angle, and a 135° angle is an obtuse angle.

Lines can intersect to create more than one angle. *Complementary angles* are two angles formed by three lines in which the sum of the two angles equals 90°. Each complementary angle is an acute angle. For example, a 30° angle and a 60° angle are acute angles that are complementary angles.

To find the complementary angle of a known acute angle, subtract the known angle from 90. For example, to find the complementary angle of a 40° angle, subtract 40 from 90 (90 − 40 = 50). The complementary angle to a 40° angle is a 50° angle.

Supplementary angles are two angles formed by three lines in which the sum of the two angles equals 180°. For example, a 45° angle and a 135° angle are supplementary angles.

To find the supplementary angle of a known angle, subtract the known angle from 180. For example, to find the supplementary angle of a 70° angle, subtract 70 from 180 (180 − 70 = 110). The supplementary angle to a 70° angle is a 110° angle.

Adjacent angles are angles that have the same vertex and one side in common. Adjacent angles are formed when two or more lines intersect.

For example, when two straight lines intersect, four angles and four sets of adjacent angles are formed. The sum of adjacent angles that form a straight line equals 180°. The two angles opposite each other when two straight lines intersect are equal.

Warning: Always verify quantities when mixing chemicals.

Linear Measure

Linear measure is the measurement of length. It is used to find the one-dimensional length of an object. It measures distances such as how far, how long, etc. The common units used for linear measure are the inch (″) in the English system and the millimeter (mm) in the metric system.

Linear measurement is the measurement of length.

Area

Area is the number of unit squares equal to the surface of an object. For example, a standard size sheet of plywood is 4′ × 8′. It contains an area of 32 sq ft (4 × 8 = 32 sq ft).

Area is expressed in square inches, square feet, and other units of measure. A *square inch* measures 1″ × 1″ or its equivalent. A *square foot* contains 144 sq in. (12″ × 12″ = 144 sq in.). The area of any plane figure can be determined by applying the proper formula. See Figure 1-5.

Circles

A *circle* is a plane figure generated about a centerpoint. See Figure 1-6. All circles contain 360°. The *circumference* is the boundary of a circle.

The *diameter* is the distance from circumference to circumference through the centerpoint. The *centerpoint* is the point a circle or arc is drawn around.

An *arc* is a portion of the circumference. The *radius* is the distance from the centerpoint to the circumference. It is one-half the length of the diameter.

A *chord* is a line from circumference to circumference not through the centerpoint. A *quadrant* is one-fourth of a circle containing 90°.

A *sector* is a pie-shaped piece of a circle. A *segment* is the portion of a circle set off by a chord. A *semicircle* is one-half of a circle containing 180°.

Concentric circles are two or more circles with different diameters but the same centerpoint. *Eccentric circles* are two or more circles with different diameters and different centerpoints.

A *tangent* is a straight line touching the curve of the circumference at only one point. A tangent is perpendicular to the radius. A *secant* is a straight line touching the circumference at two points.

Circumference of a Circle (Diameter). When the diameter is known, the circumference of a circle is found by applying the formula:

$$C = \pi D$$

where

C = circumference

π = 3.1416

D = diameter

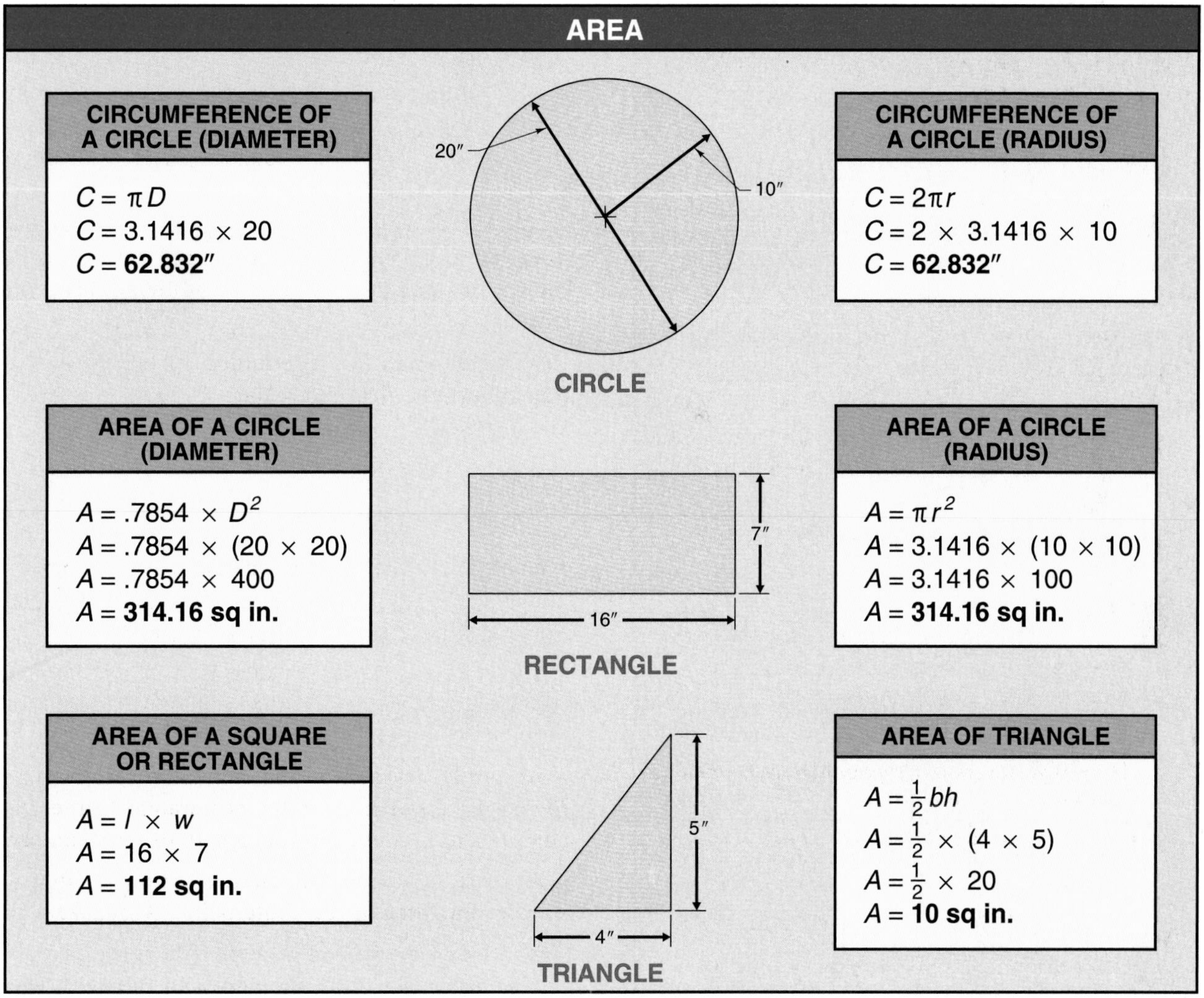

Figure 1-5. Area is the number of unit squares equal to the surface of an object.

For example, what is the circumference of a 20″ diameter circle?

$C = \pi D$

$C = 3.1416 \times 20$

$C = \mathbf{62.832''}$

Circumference of a Circle (Radius). When the radius is known, the circumference of a circle is found by applying the formula:

$C = 2\pi r$

where

C = circumference

2 = constant

π = 3.1416

r = radius

For example, what is the circumference of a 10″ radius circle?

$C = 2\pi r$

$C = 2 \times 3.1416 \times 10$

$C = \mathbf{62.832''}$

Area of a Circle (Diameter). When the diameter is known, the area of a circle is found by applying the formula:

$A = .7854 \times D^2$

where

A = area

.7854 = constant

D^2 = diameter squared

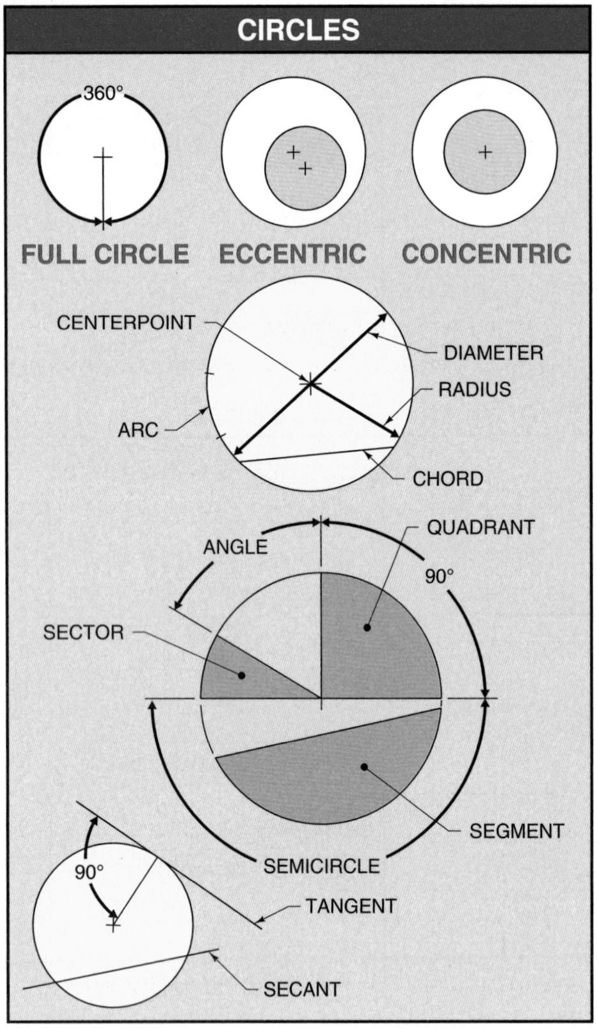

Figure 1-6. A circle is a plane figure generated about a centerpoint.

For example, what is the area of a 28″ diameter circle?

$A = .7854 \times D^2$

$A = .7854 \times (28 \times 28)$

$A = .7854 \times 784$

$A = \mathbf{615.754\ sq\ in.}$

Area of a Circle (Radius). When the radius is known, the area of a circle is found by applying the formula:

$A = \pi r^2$

where

A = area

π = 3.1416

r^2 = radius squared

For example, what is the area of a 14″ radius circle?

$A = \pi r^2$

$A = 3.1416 \times (14 \times 14)$

$A = 3.1416 \times 196$

$A = \mathbf{615.754\ sq\ in.}$

Triangles

A *triangle* is a three-sided polygon with three interior angles. The sum of the three angles of a triangle is always 180°. The sign (Δ) indicates a triangle. See Figure 1-7. The *altitude* of a triangle is the perpendicular dimension from the vertex to the base. The *base* of a triangle is the side upon which the triangle stands. Any side can be taken as the base.

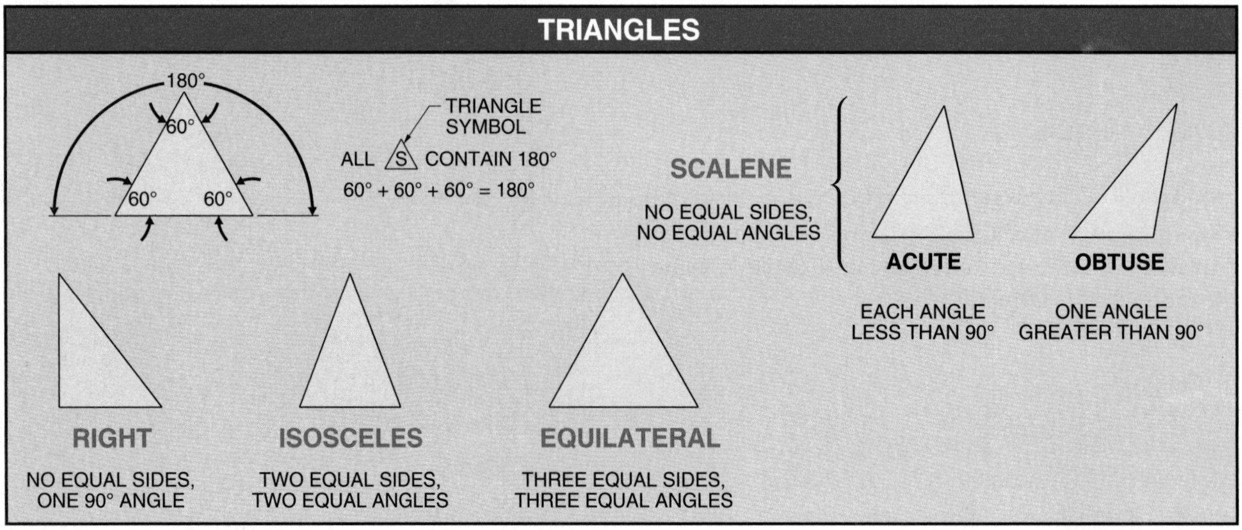

Figure 1-7. A triangle is a three-sided polygon with three interior angles.

The angles of a triangle are named by uppercase letters. The sides of a triangle are named by lowercase letters. For example, a triangle may be named ΔABC and contain sides d, e, and f.

The different kinds of triangles are right triangles, isosceles triangles, equilateral triangles, and scalene triangles. A *right triangle* is a triangle that contains one 90° angle and no equal sides. An *isosceles triangle* is a triangle that contains two equal angles and two equal sides. An *equilateral triangle* is a triangle that has three equal angles and three equal sides. Each angle of an equilateral triangle is 60°. A *scalene triangle* is a triangle that has no equal angles or equal sides. A scalene triangle may be acute or obtuse. An *acute triangle* is a scalene triangle with each angle less than 90°. An *obtuse triangle* is a scalene triangle with one angle greater than 90°.

Area of a Triangle. The area of a triangle is found by applying the formula:

$A = \frac{1}{2}bh$

where

A = area

$\frac{1}{2}$ = constant

b = base

h = height

For example, what is the area of a triangle with a 10″ base and a 12″ height?

$A = \frac{1}{2}bh$

$A = \frac{1}{2} \times (10 \times 12)$

$A = \frac{1}{2} \times 120$

$A = \mathbf{60\ sq\ in.}$

Pythagorean Theorem. The *Pythagorean Theorem* states that the square of the hypotenuse of a right triangle is equal to the sum of the squares of the other two sides. The *hypotenuse* is the side of a right triangle opposite the right angle.

A right triangle is said to have a 3-4-5 relationship and often is used for laying out right angles and checking corners for squareness. To check a corner for squareness, measure 3′ along one side and 4′ along the other side. These two points measure 5′ apart when the corner is square. See Figure 1-8.

The length of the hypotenuse of a right triangle is found by applying the formula:

$c = \sqrt{a^2 + b^2}$

where

c = length of hypotenuse

a^2 = length of one side squared

b^2 = length of other side squared

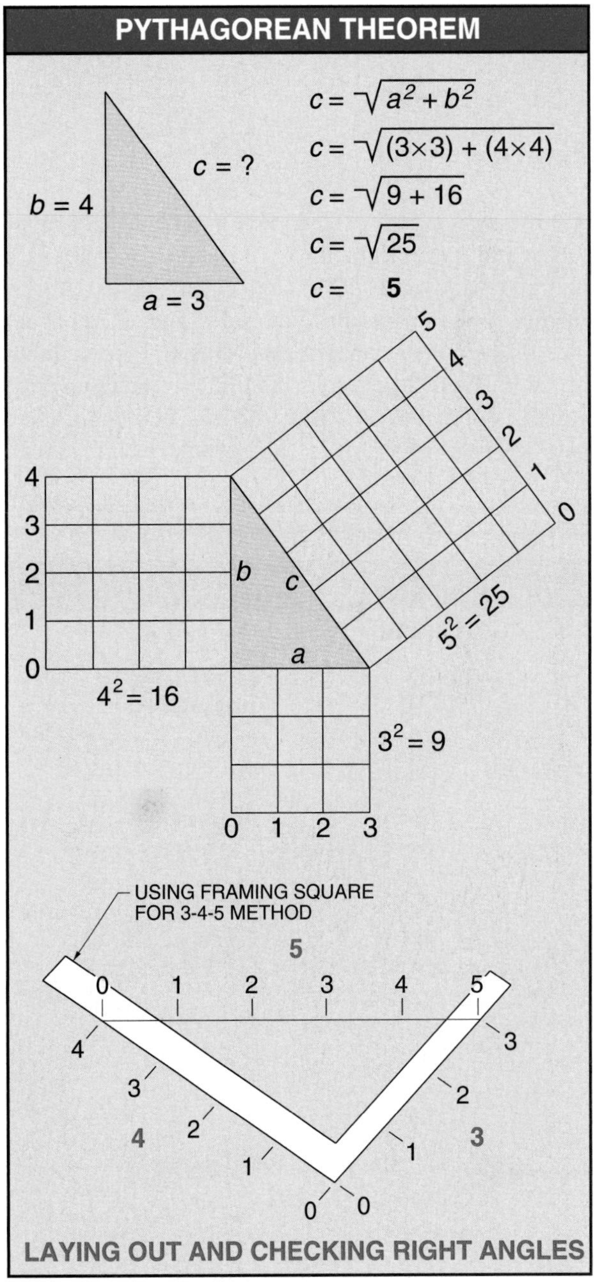

Figure 1-8. The square of the hypotenuse of a right triangle is equal to the sum of the squares of the other two sides of the triangle.

For example, what is the length of the hypotenuse of a triangle having sides of 3′ and 4′?

$$c = \sqrt{a^2 + b^2}$$

$$c = \sqrt{(3 \times 3) + (4 \times 4)}$$

$$c = \sqrt{9 + 16}$$

$$c = \sqrt{25}$$

$$c = \mathbf{5'}$$

Regular Polygons

A *polygon* is a many-sided plane figure. All polygons are bound by straight lines. A *regular polygon* is a polygon with equal sides and equal angles. An *irregular polygon* has unequal sides and unequal angles. Polygons are named according to their number of sides. Typical polygons include the triangle (three sides), quadrilateral (four sides), pentagon (five sides), hexagon (six sides), and octagon (eight sides). See Figure 1-9.

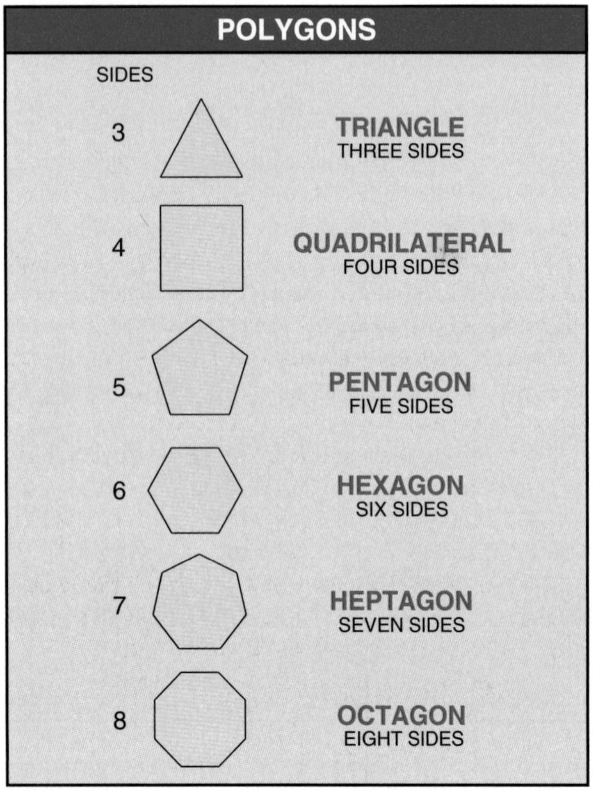

Figure 1-9. A polygon is a many-sided plane figure.

Quadrilaterals

A *quadrilateral* is a four-sided polygon with four interior angles. The sum of the four angles of a quadrilateral is always 360°. The kinds of quadrilaterals are squares, rectangles, rhombuses, rhomboids, trapezoids, and trapeziums. See Figure 1-10.

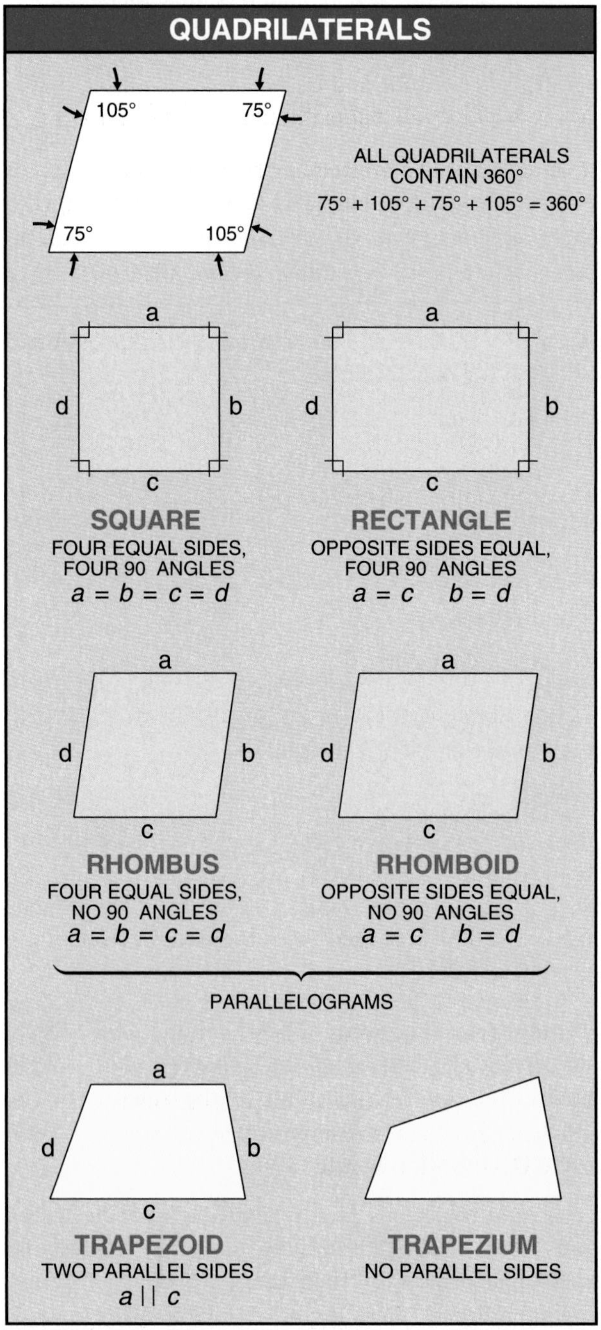

Figure 1-10. A quadrilateral is a four-sided polygon with four interior angles.

A *square* is a quadrilateral with all sides equal and four 90° angles. A *rectangle* is a quadrilateral with opposite sides equal and four 90° angles. A *rhombus* is a quadrilateral with all sides equal and no 90° angles. A *rhomboid* is a quadrilateral with opposite sides equal and no 90° angles.

The square, rectangle, rhombus, and rhomboid are parallelograms. A *parallelogram* is a four-sided plane figure with opposite sides parallel and equal.

A *trapezoid* is a quadrilateral with two sides parallel. A *trapezium* is a quadrilateral with no sides parallel. Trapezoids and trapeziums are not parallelograms because all opposite sides are not parallel.

Area of a Square or Rectangle. The area of a square or the area of a rectangle is found by applying the formula:

$A = l \times w$

where

A = area

l = length

w = width

For example, what is the area of a 22′-0″ × 16′-0″ storage room?

$A = l \times w$

$A = 22 \times 16$

$A = \mathbf{352\ sq\ ft}$

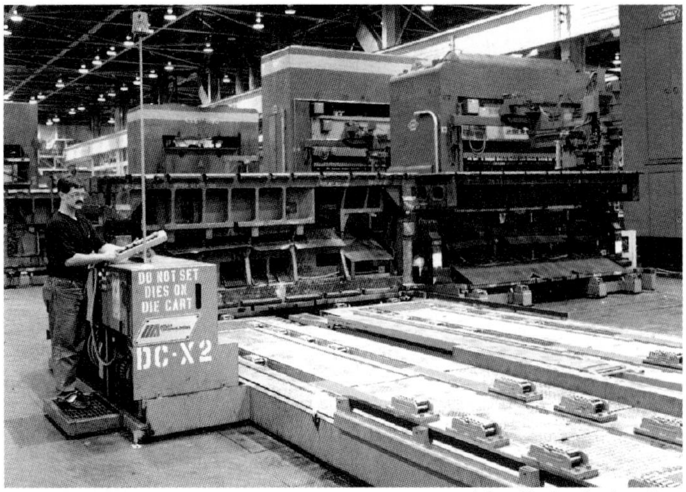

Atlas Technologies, Inc.

Industrial facilities are designed based on square footage for the efficient layout of production line equipment.

SOLIDS

Polyhedra are solids bound by plane surfaces (faces). *Regular solids* (polyhedra) are solids with faces that are regular polygons (equal sides). *Irregular polyhedra* are solids with faces that are irregular polygons (unequal sides).

Solids have length, height, and depth. The five regular solids are the tetrahedron, hexahedron, octahedron, dodecahedron, and icosahedron. Other common solids are prisms, cylinders, pyramids, cones, and spheres. Less common solids include the torus and ellipsoid. See Figure 1-11.

The National Institute of Standards and Technology (NIST) is the federal agency that establishes accurate measurement standards for science, industry, and commerce in the United States.

Regular Solids

A *tetrahedron* is a regular solid of four triangles. A *hexahedron* is a regular solid of six squares. It is commonly referred to as a cube. An *octahedron* is a regular solid of eight triangles. A *dodecahedron* is a regular solid of twelve pentagons. An *icosahedron* is a regular solid of twenty triangles.

Prisms

A *prism* is a solid with two bases that are parallel and identical polygons. *Bases* are the ends of a prism. The three or more sides of a prism are parallelograms. See Figure 1-12. A prism can be triangular, rectangular, pentagonal, hexagonal, octagonal, etc., according to the shape of its bases.

Lateral faces are the sides of a prism. There are as many of these lateral faces as there are sides in one of the bases.

The *altitude* of a prism is the perpendicular distance between the two bases. When the bases are perpendicular to the faces, the altitude equals the edge of a lateral face.

A *right prism* is a prism with lateral faces perpendicular to the bases. An *oblique prism* is a prism with lateral faces not perpendicular to the bases.

A *parallelepiped* is a prism with bases that are parallelograms. A *right parallelepiped* is a prism with all edges perpendicular to the bases. A *rectangular parallelepiped* is a prism with bases and faces that are all rectangles.

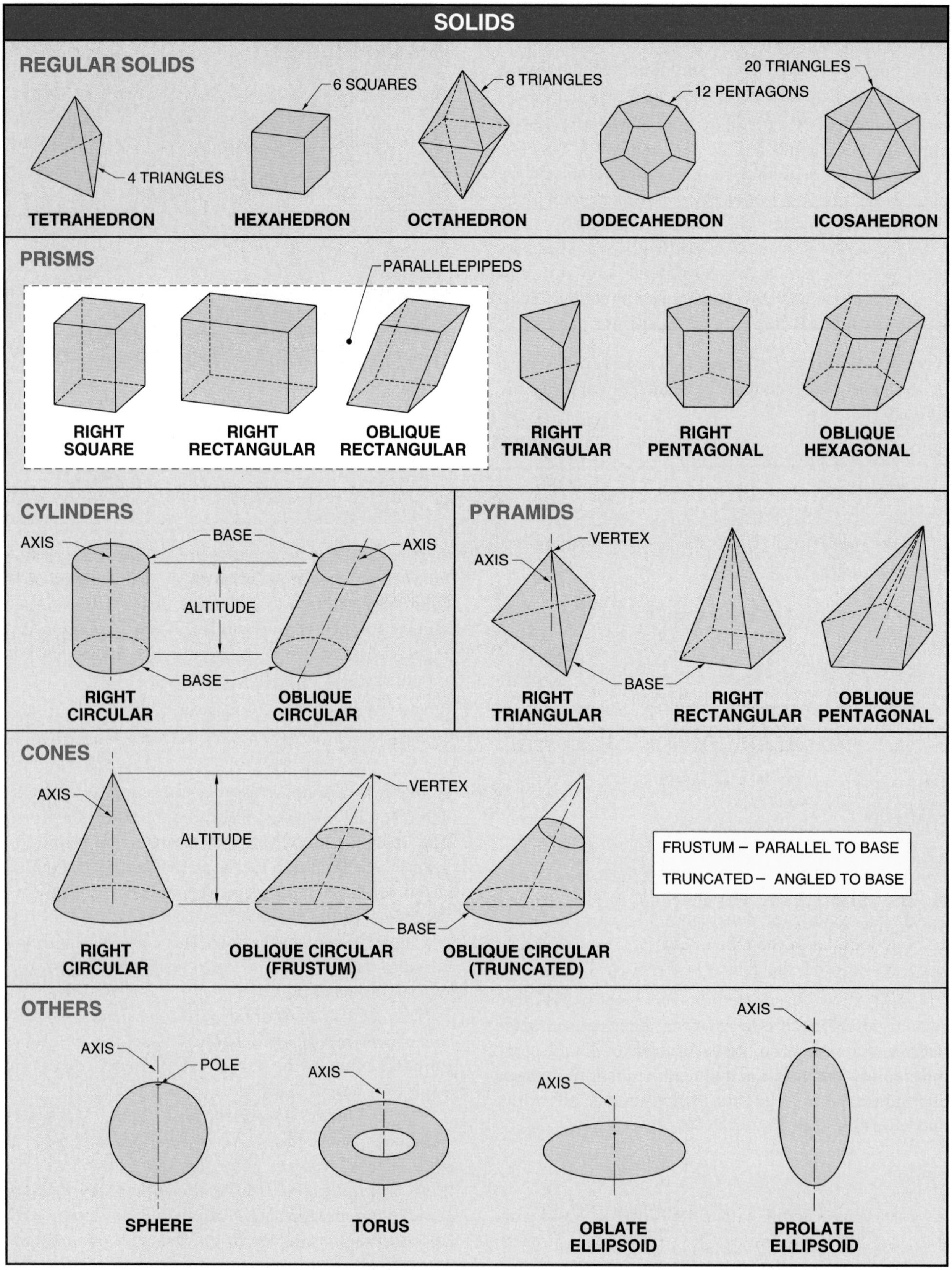

Figure 1-11. Solids have length, height, and depth.

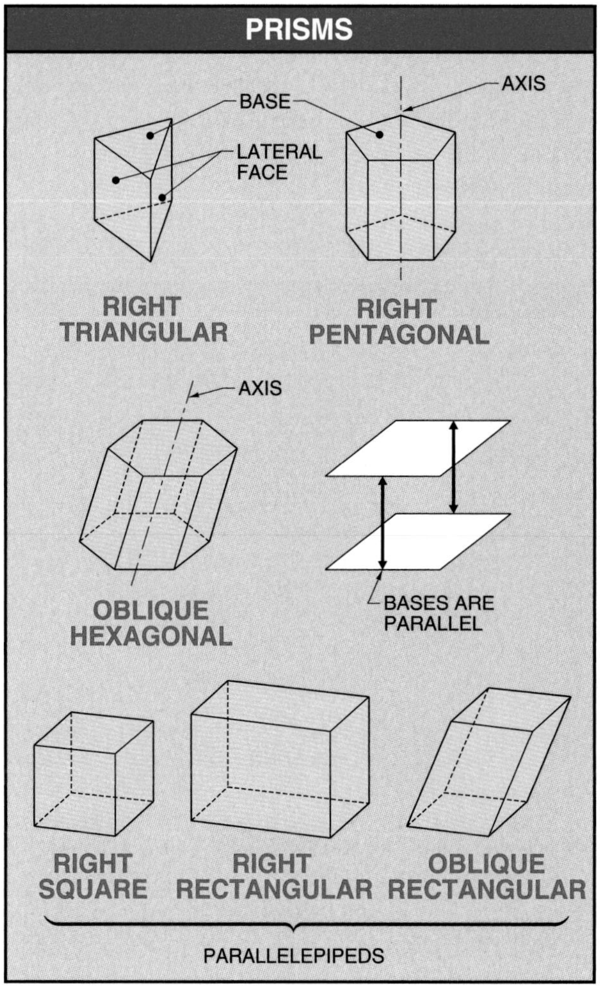

Figure 1-12. A prism is a solid with two bases that are parallel and identical polygons.

Cylinders

A *cylinder* is a solid generated by a straight line (genatrix) moving in contact with a curve and remaining parallel to the axis and its previous position. Each position of the genatrix forms an element of the cylinder.

A *right cylinder* is a cylinder with the axis perpendicular to the base. An *oblique cylinder* is a cylinder with the axis not perpendicular to the base. See Figure 1-13.

Pyramids

A *pyramid* is a solid with a base that is a polygon and sides that are triangles. The *vertex* is the common point of the triangular sides that forms the pyramid. See Figure 1-14.

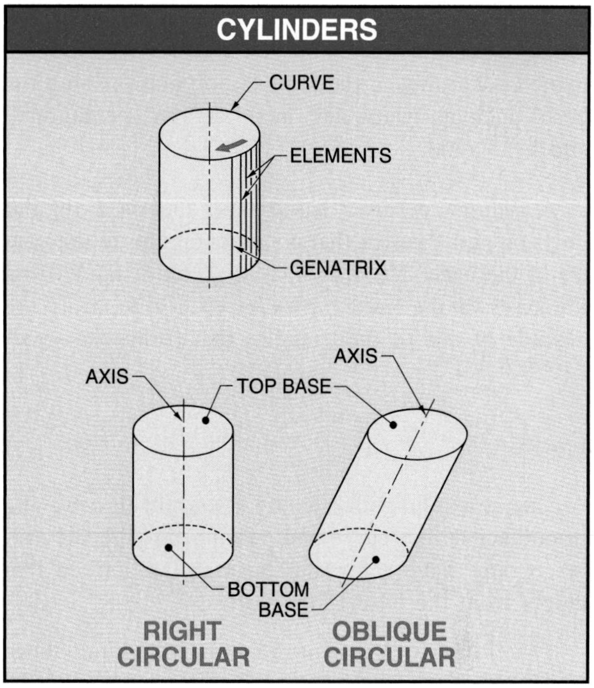

Figure 1-13. A cylinder is a solid generated by a straight line (genatrix) moving in contact with a curve and remaining parallel.

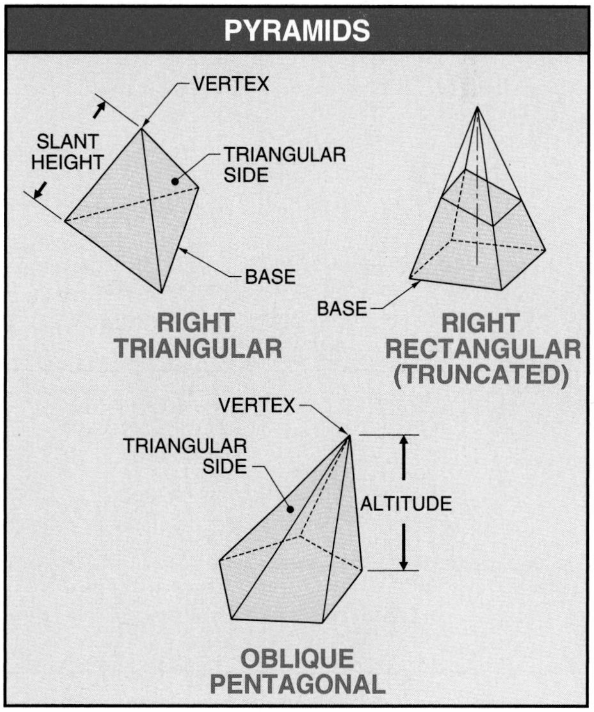

Figure 1-14. A pyramid is a solid with a base that is a polygon and sides that are triangles.

The *altitude* of a pyramid is the perpendicular distance from the vertex to the base. Pyramids are named according to the kind of polygon forming the base, such as triangular, quadrangular, pentagonal, and hexagonal.

A *regular pyramid* has a base that is a regular polygon and a vertex that is perpendicular to the center of the base. The *slant height* is the distance from the base to the vertex parallel to a side. It is the altitude of one of the triangles that forms the sides.

Cones

A *cone* is a solid generated by a straight line moving in contact with a curve and passing through the vertex. Cones have a circular base and a surface that tapers from the base to the vertex.

The altitude of a cone is the perpendicular distance from the vertex to the base. The slant height is the distance from the vertex to any point on the circumference of the base. See Figure 1-15.

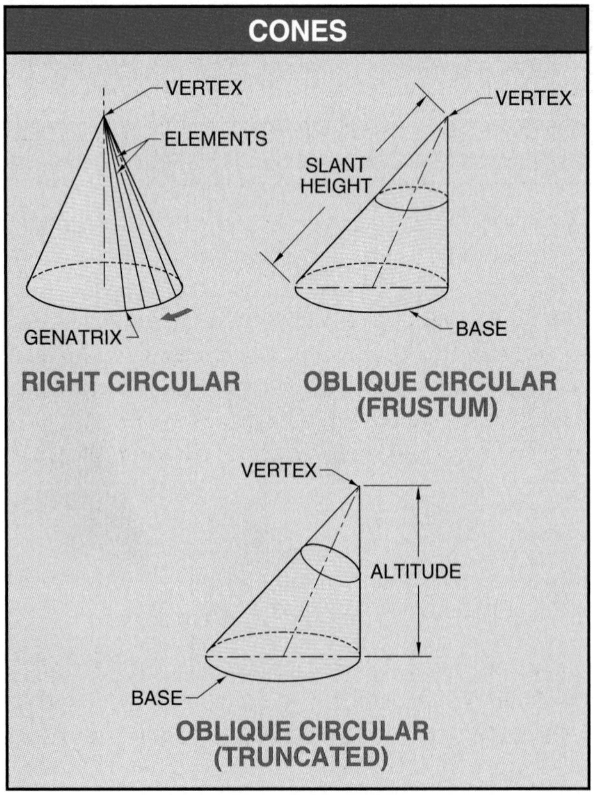

Figure 1-15. A cone is a solid generated by a straight line (genatrix) moving in contact with a circle and passing through the vertex.

Conic Sections. A *conic section* is a curve produced by a plane intersecting a right circular cone. A *right circular cone* is a cone with the axis located at a 90° angle to the circular base. The four conic sections are the circle, ellipse, parabola, and hyperbola. See Figure 1-16.

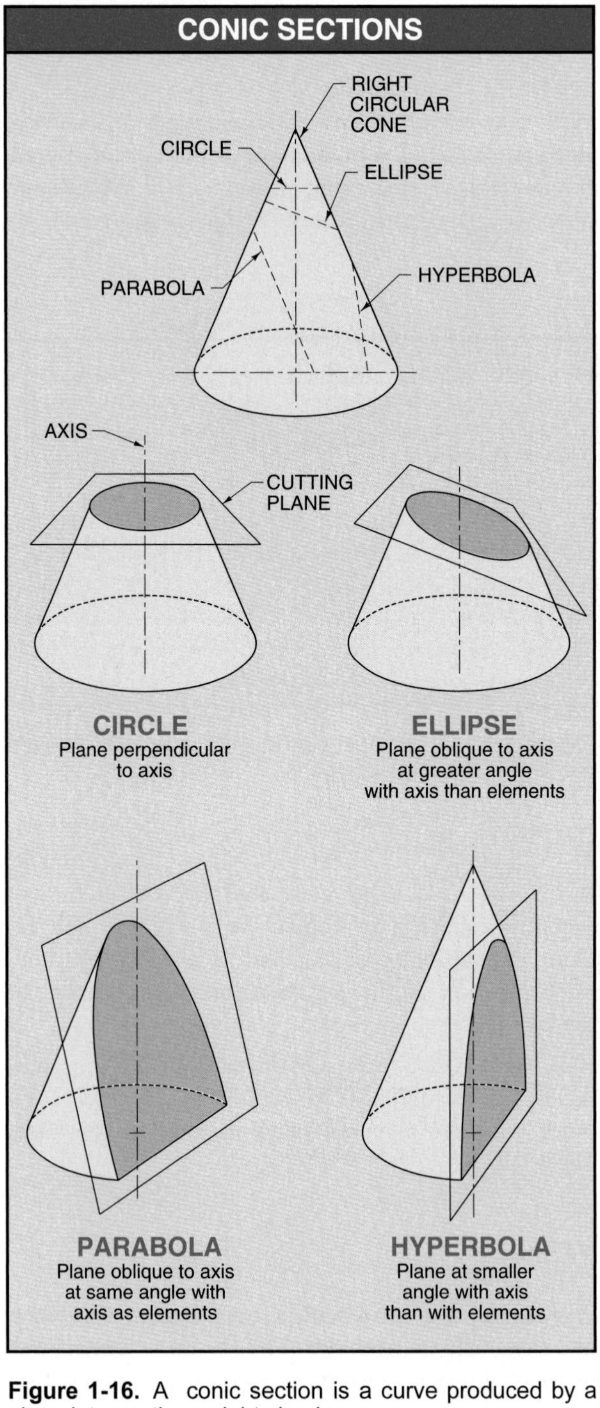

Figure 1-16. A conic section is a curve produced by a plane intersecting a right circular cone.

A *circle* is a plane figure formed by a cutting plane perpendicular to the axis of a cone. An *ellipse* is a plane figure formed by a cutting plane oblique to the axis of a cone, but at a greater angle with the axis than with the elements of the cone.

A *parabola* is a plane figure formed by a cutting plane oblique to the axis and parallel to the elements of the cone. A *hyperbola* is a plane figure formed by a cutting plane that has a smaller angle with the axis than with the elements of the cone.

Frustums. A *frustum* of a pyramid or cone is the remaining portion of a pyramid or cone with a cutting plane passed parallel to the base. A truncated pyramid or cone is the remaining portion of a pyramid or cone with the cutting plane passed not parallel to the base. See Figure 1-17.

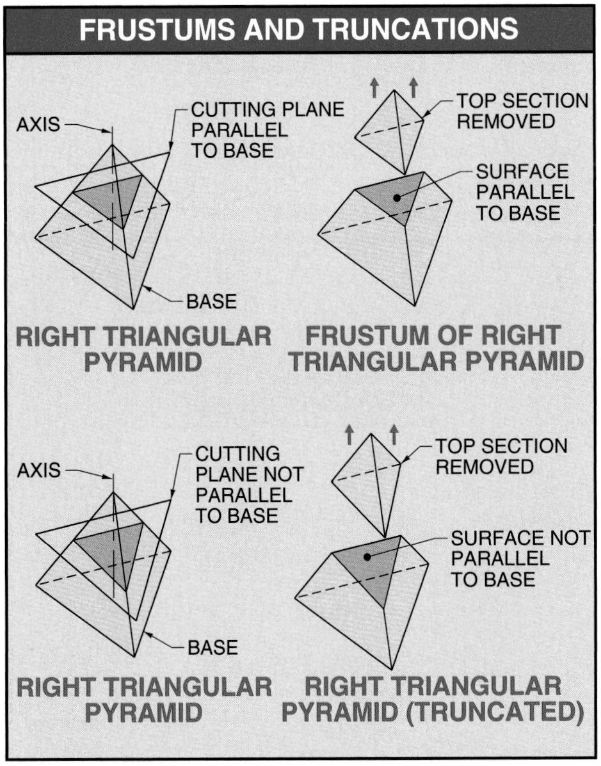

Figure 1-17. A frustum is the remaining piece of a pyramid or cone with a cutting plane passed parallel to the base.

Spheres

A *sphere* is a solid generated by a circle revolving about one of its axes. All points on the surface are an equal distance from the center of the sphere. See Figure 1-18.

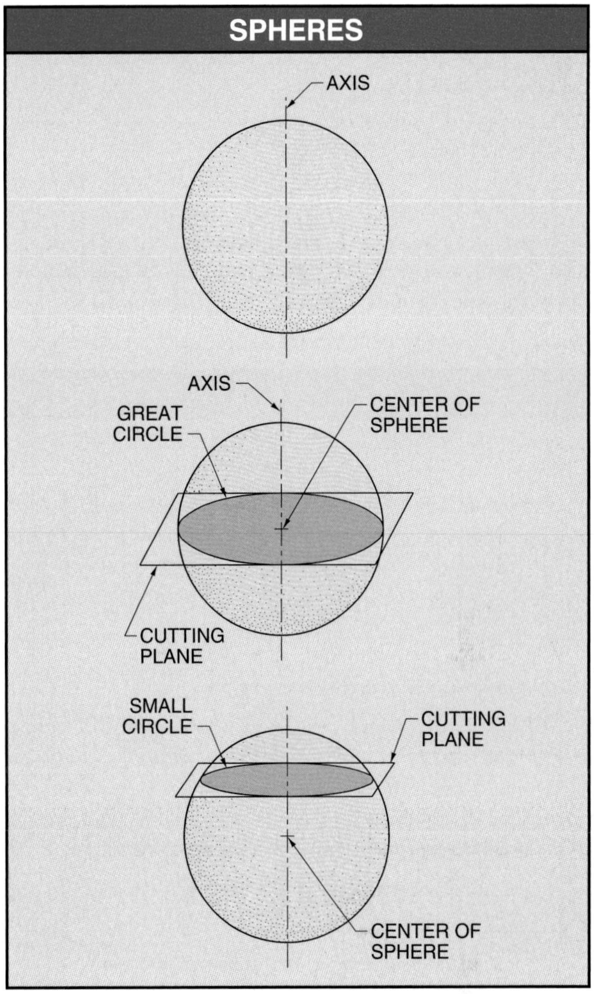

Figure 1-18. A sphere is a solid generated by a circle revolving about the axis.

A *great circle* is the circle formed by passing a cutting plane through the center of a sphere. A *small circle* is the circle formed by passing a cutting plane through a sphere but not through the center. The circumference of a sphere is equal to the circumference of a great circle.

VOLUME

Volume is the three-dimensional size of an object measured in cubic units. For example, the volume of a standard size concrete block is 1024 cu in. (8″ × 8″ × 16″ = 1024 cu in.).

Volume is expressed in cubic inches, cubic feet, cubic yards, and other units of cubic measure. A *cubic inch* measures 1″ × 1″ × 1″ or its equivalent.

A *cubic foot* contains 1728 cu in. (12″ × 12″ × 12″ = 1728 cu in.). A *cubic yard* contains 27 cu ft (3′ × 3′ × 3′ = 27 cu ft).

Finding Volume

The volume of a solid figure can be determined by applying the proper formula. See Figure 1-19.

Volume of a Rectangular Solid. The volume of a rectangular solid is found by applying the formula:

$V = l \times w \times h$

where

V = volume

l = length

w = width

h = height

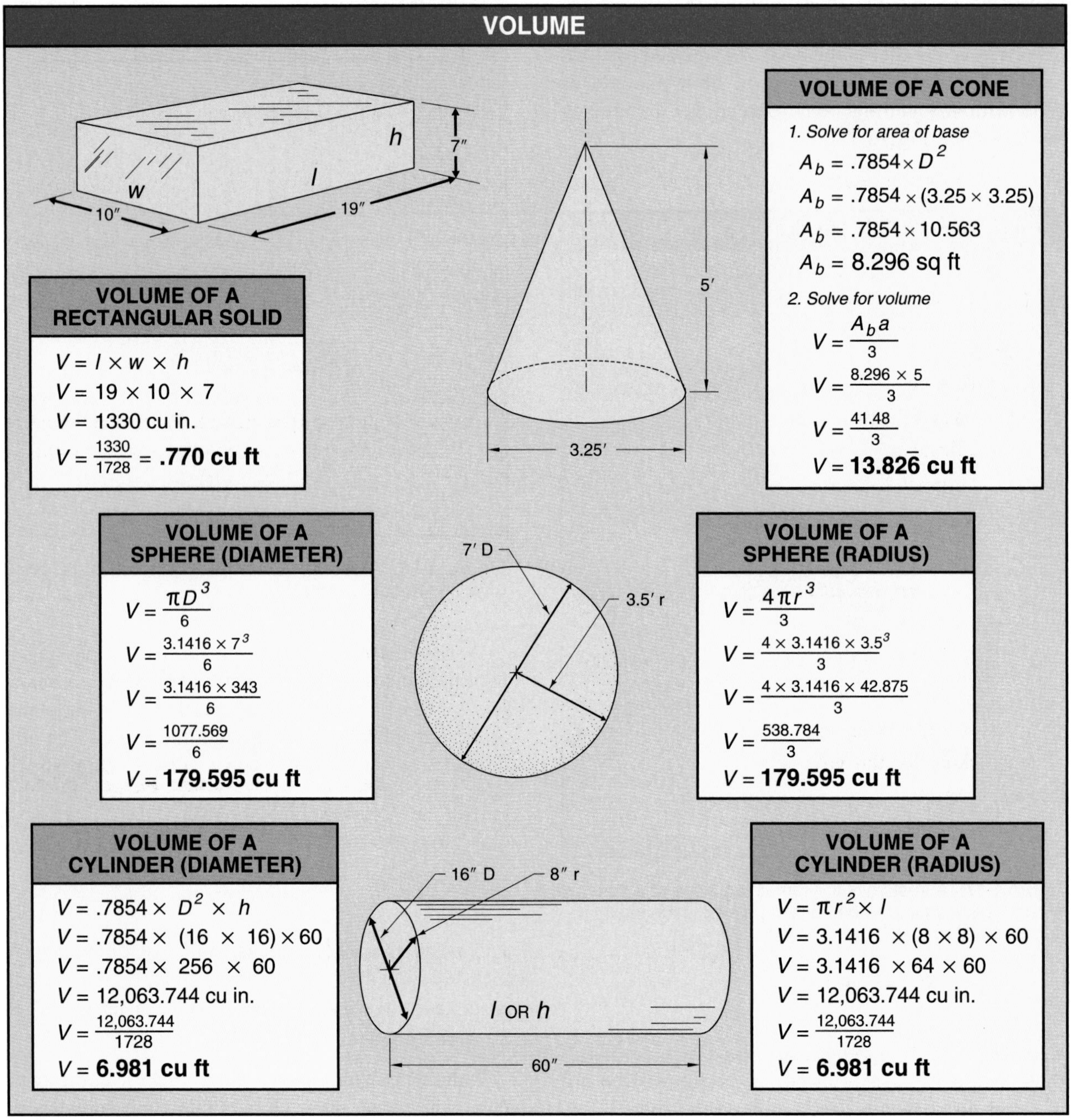

Figure 1-19. Volume is the three-dimensional size of an object.

For example, what is the volume of a 24″ × 12″ × 8″ rectangular solid?

$V = l \times w \times h$

$V = 24 \times 12 \times 8$

$V =$ **2304 cu in.**

Volume of a Cone. The volume of a cone is found by first solving for the area of the base and then solving for volume. The area of the base is found by applying the formula:

$A_b = .7854 \times D^2$

where

A_b = area of base

.7854 = constant

D^2 = diameter squared

The volume of the cone is then found by applying the formula:

$$V = \frac{A_b a}{3}$$

where

V = volume

A_b = area of base

a = altitude

3 = constant

For example, what is the volume of a cone that has a 14″ diameter and a 35″ altitude?

1. Solve for the area of the base.

$A_b = .7854 \times D^2$

$A_b = .7854 \times (14 \times 14)$

$A_b = .7854 \times 196$

$A_b = 153.938$ sq in.

2. Solve for the volume.

$$V = \frac{A_b a}{3}$$

$$V = \frac{153.938 \times 35}{3}$$

$$V = \frac{5387.83}{3}$$

$V =$ **1795.943 cu in.**

The pressure in an enclosed air tank increases when the tank in left in the sun because the volume of air increases when heated.

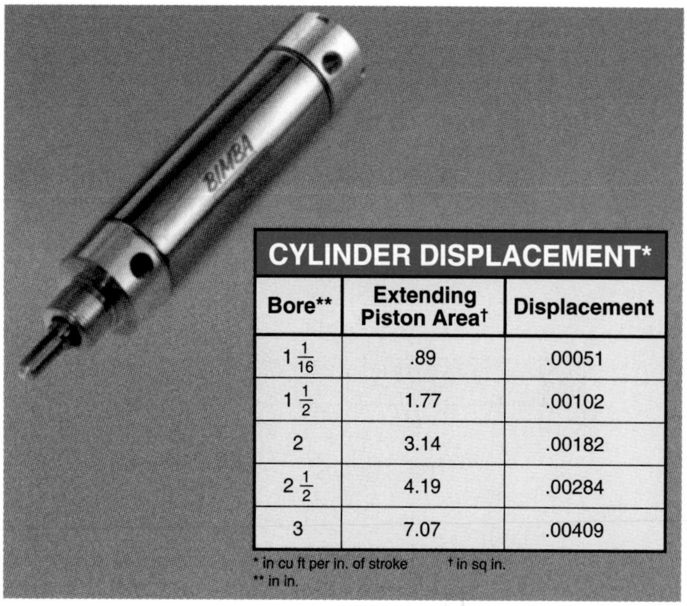

CYLINDER DISPLACEMENT*

Bore**	Extending Piston Area†	Displacement
$1\frac{1}{16}$	.89	.00051
$1\frac{1}{2}$	1.77	.00102
2	3.14	.00182
$2\frac{1}{2}$	4.19	.00284
3	7.07	.00409

* in cu ft per in. of stroke † in sq in.
** in in.

Bimba Manufacturing Company

The volume of fluid power cylinders is calculated to determine required pump or compressor flow rates and pipe sizes.

Volume of a Sphere (Diameter). When the diameter is known, the volume of a sphere is found by applying the formula:

$$V = \frac{\pi D^3}{6}$$

where

V = volume

π = 3.1416

D^3 = diameter cubed

6 = constant

For example, what is the volume of a sphere that is 4′-0″ in diameter?

$$V = \frac{\pi D^3}{6}$$

$$V = \frac{3.1416 \times 4^3}{6}$$

$$V = \frac{3.1416 \times 64}{6}$$

$$V = \frac{201.062}{6}$$

$V =$ **33.510 cu ft**

The volume of air contained in the tank of a compressor is calculated by applying the volume formula for a cylinder.

Volume of a Sphere (Radius). When the radius is known, the volume of a sphere is found by applying the formula:

$$V = \frac{4\pi r^3}{3}$$

where

V = volume

4 = constant

π = 3.1416

r^3 = radius cubed

3 = constant

For example, what is the volume of a sphere that has a 2′-0″ radius?

$$V = \frac{4\pi r^3}{3}$$

$$V = \frac{4 \times 3.1416 \times 2^3}{3}$$

$$V = \frac{4 \times 3.1416 \times 8}{3}$$

$$V = \frac{100.531}{3}$$

V = **33.510 cu ft**

Volume of a Cylinder (Diameter). When the diameter is known, the volume of a cylinder is found by applying the formula:

$V = .7854 \times D^2 \times h$

where

V = volume

.7854 = constant

D^2 = diameter squared

h = height

For example, what is the volume of a tank that is 4′-0″ in diameter and 12′-0″ long?

$V = .7854 \times D^2 \times h$

$V = .7854 \times (4 \times 4) \times 12$

$V = .7854 \times 16 \times 12$

V = **150.797 cu ft**

Volume of a Cylinder (Radius). When the radius is known, the volume of a cylinder is found by applying the formula:

$V = \pi r^2 \times l$

where

V = volume

π = 3.1416

r^2 = radius squared

l = length

For example, what is the volume of a tank that has a 2′-0″ radius and is 12′-0″ long?

$V = \pi r^2 \times l$

$V = 3.1416 \times (2 \times 2) \times 12$

$V = 3.1416 \times 4 \times 12$

V = **150.797 cu ft**

Rigging

2
Chapter

Lift-All Company, Inc.

Rigging is securing equipment in preparation for lifting by means of rope, chain, or webbing. Loads must be balanced and load weights calculated for safe load lifting. Rope is used for lifting because of its length and flexibility. Chain is used in situations in which other materials would be damaged by the load or environment. All rigging components should be inspected initially, frequently, and periodically. Repair and testing of rigging components must be completed by the manufacturer.

RIGGING

The shape, weight, and location where a load is to be moved must be known prior to rigging and lifting. *Rigging* is securing equipment or machinery in preparation for lifting by means of rope, chain, or webbing. *Lifting* is hoisting equipment or machinery by mechanical means.

The weakest component determines the strength of the entire lifting system. For this reason, the limitation of every component used to move a load must be determined. Other considerations include the conditions (indoor or outdoor), travel path, equipment, and the skill level of the workers.

Load Balance

The shape of a load normally determines its center of gravity. The shape of a load may be symmetrical or asymmetrical. See Figure 2-1. A *symmetrical load* is a load in which one-half of the load is a mirror image of the other half. Symmetrical loads include straight pipe sections, motors, paper rolls, and sheet metal.

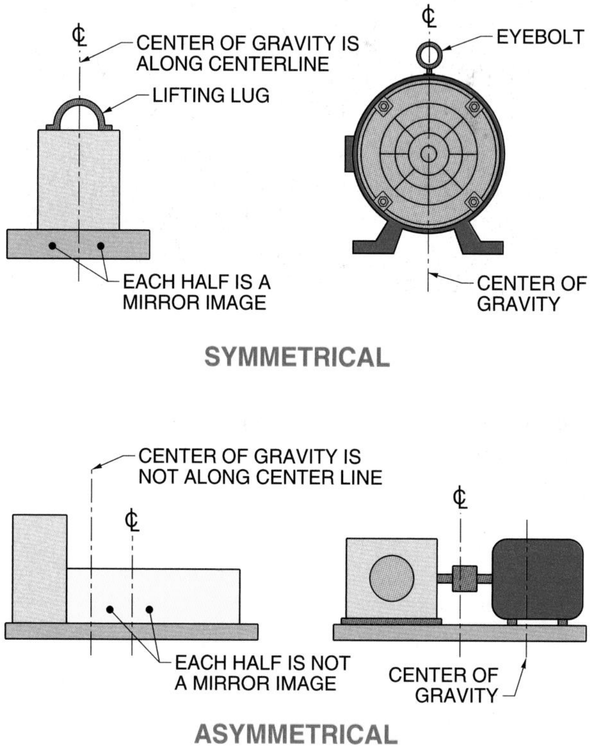

Figure 2-1. The shape of a load normally determines its center of gravity.

An *asymmetrical load* is a load in which one-half of the load is not a mirror image of the other half. Asymmetrical loads include most machinery, motor and pump assemblies, pipe and valve assemblies, and engines.

Some loads are equipped by the manufacturer with lifting lugs for ease in lifting and transporting. A *lifting lug* is a thick metal loop (eyebolt) welded or screwed to a machine to allow balanced lifting. Lifting lugs eliminate the need for additional rigging.

Center of gravity is the balancing point of a load. The center of gravity of a load must be determined before lifting to prevent tipping or toppling of the load. Complex shapes and materials make it difficult to determine the center of gravity for most loads. Some manufacturers mark the center of gravity on their equipment, while others offer specification sheets that include center of gravity information. Calculations determining the center of gravity may be made from weight, shape, and material standard information. An educated guess may be made, placing the center of gravity in an approximate location.

A load may be lifted without chance of tipping or toppling once the center of gravity is determined (balanced load). See Figure 2-2. Load tipping occurs when a load is unsteady, unbalanced, or unstable.

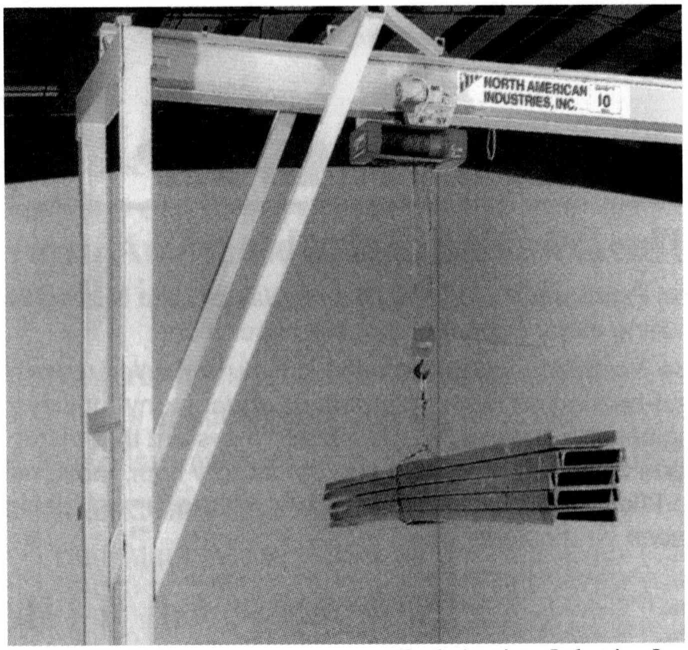

North American Industries, Inc.

Symmetrical loads are lifted using one sling by placing the sling at the center of the load and lifting slightly to observe weight shifting. The sling is adjusted and the process repeated until the load is stable.

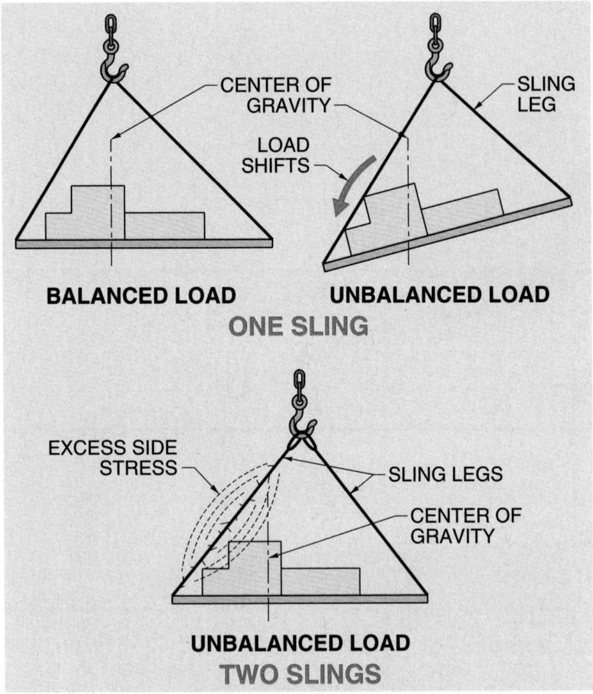

Figure 2-2. A load lifted outside its center of gravity may shift or create excess side stress on one sling.

Load toppling occurs when a load is unbalanced because it is unstable or top heavy. Unbalanced loads may shift toward the center of gravity and create a dangerous condition when lifted. Most of the weight is placed on one sling if two slings are used to lift an unbalanced load.

Lifting equipment is connected after the center of gravity is determined. The load is lifted slightly to observe stability or weight shifting. The lifting equipment is readjusted towards the heavy (dipping) side of the load if an imbalance is observed. This procedure is continued until the load is balanced and stable. *Imbalance* is a lack of balance.

In addition to the vertical center of gravity, the horizontal weight center (horizontal center of gravity) of a load must also be determined. The *horizontal weight center* is a weight mass above a pivot point that causes a load to topple because it is top heavy. The lifting equipment must not be attached to the load at any point lower than the horizontal weight center. A load may be unstable and subject to toppling if lifting equipment is placed below the horizontal weight center of a load. See Figure 2-3.

The sling apex must be above the horizontal weight center if a load is to be lifted from the base of a machine or skid. The *sling apex* is the uppermost point where sling legs meet.

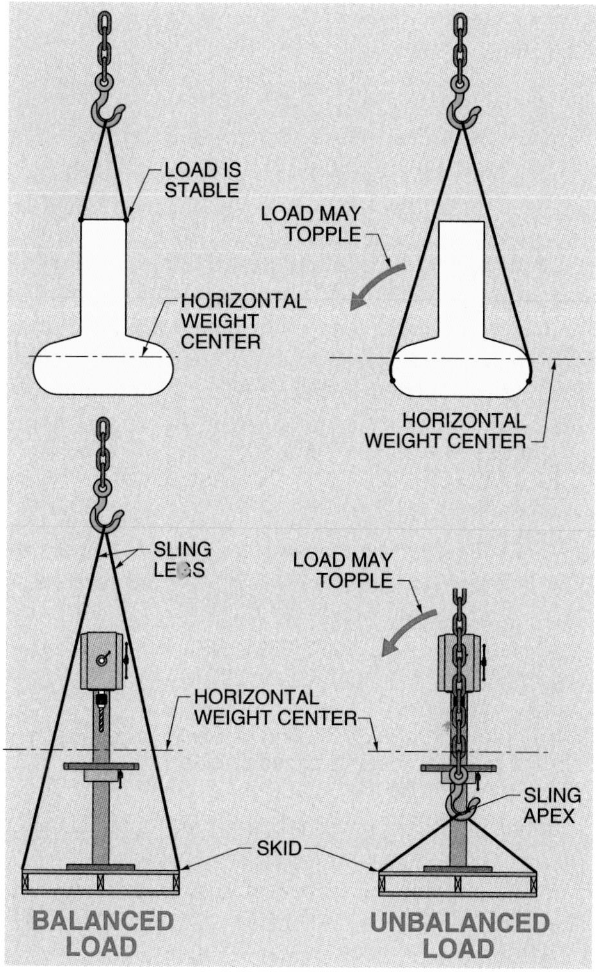

Figure 2-3. A load may be unstable and subject to toppling if lifting equipment is placed below the horizontal weight center of the load.

CALCULATING LOAD WEIGHTS

The weight of a load is calculated after considering the shape and size of the load. The weight of the load may be found on the data plate located on the equipment, on shipping documents, or on the manufacturer's product bulletin. Always ensure that the weight has not changed since the last printing or entry when data is obtained from printed documents. Weight changes often occur when equipment has been modified or when a product is left in a holding vessel.

> ***Per the National Safety Council, a disabling injury occurs on the job every 9 sec and a person is killed on the job every hour and a half.***

Load weights are normally determined by calculation if manufacturer's printed information is not available. Load weights may be calculated using stock material weight tables or the area, volume, and load material weight information. The full-load weight should include the weight of the rigging equipment. To obtain the full-load weight, add the total material weight to the weight of the rigging equipment.

Sprecher + Schuh

Overhead cranes driven by electric motors are used to move heavy metal beams.

Stock Material Weight Tables

Stock material weight tables are used when a load consists of basic stock materials such as steel or brass bar stock. Stock material weight tables are available listing the weight of materials by their linear, square, or cubic measurements in either the English or metric systems. See Appendix. For example, a 1″ diameter round steel bar weighs 2.67 lb/ft (from Weight of Steel and Brass Bar Stock table). See Figure 2-4.

Common stock materials include round and square bar, round and square tubing, I-beam, angle stock, tee stock, channel, and plate. Tables for these common shapes may be located through the American National Standards Institute (ANSI). The weight of stock material is found by referring to the proper stock material weight table and applying the formula:

$W = l \times w/ft$

where

W = weight (in lb)

l = length (in ft)

w/ft = weight (in lb/ft)

WEIGHT OF STEEL AND BRASS BAR STOCK*			
Diameter or Thickness**	Round Steel	Square Steel	Brass
¼	.167	—	.181
½	.667	—	.724
¾	1.50	—	1.63
1	2.67	3.4	2.89
1¼	4.17	—	4.52
1½	6.01	7.7	6.51
1¾	8.18	—	8.86
2	10.68	—	11.57
4	42.7	54.4	—
5	66.8	85.0	—
6	96.1	122.4	—
10	267.0	340.0	—
12	384.5	489.6	—

* in lb/ft

** in in.

WEIGHT OF STEEL PLATE*	
Thickness**	Weight
1⁄16	2.55
⅛	5.1
3⁄16	7.65
¼	10.2
5⁄16	12.75
⅜	15.3
½	20.4
⅝	25.5
¾	30.6
1	40.8
1¼	51.0
1½	61.2
2	81.6

* in lb/sq ft

** in in.

Figure 2-4. Stock material weight tables list the weight of materials by ft, sq ft, or cu ft.

Industrial lifting equipment is not to be used for lifting, supporting, or transporting people.

For example, what is the total material weight of a 10′ long, 2″ diameter brass bar?

$W = l \times w/ft$

$W = 10 \times 11.57$ (from Weight of Steel and Brass Bar Stock table)

$W =$ **115.7 lb**

Load weights of multiple plates or sheets of steel, aluminum, or brass stock are calculated by finding the material weight using stock material weight tables, finding the square footage, and multiplying by the number of plates or sheets. The weight of multiple plates or sheets of stock is found by applying the procedure:

1. Find the weight of one sheet.
2. Find the area (in sq ft) of material. The area of the material is found by applying the formula:

   $A = l \times w$

   where

   A = area (in sq ft)

   l = length (in ft)

   w = width (in ft)
3. Find the total material weight. The total material weight is found by applying the formula:

   $W = w/sq\ ft \times A \times q$

   where

   W = weight (in lb)

   $w/sq\ ft$ = weight per square foot (in lb)

   A = area (in sq ft)

   q = number of plates or sheets

For example, what is the total material weight of a load consisting of 35 4′ × 8′ sheets of steel that are 1⁄16″ thick?

1. Find the weight of one sheet.

   1⁄16″ plate steel weighs 2.55 lb/sq ft (from Weight of Steel Plate table)
2. Find the area of the material.

   $A = l \times w$

   $A = 4 \times 8$

   $A = 32$ sq ft
3. Find the total material weight.

   $W = w/sq\ ft \times A \times q$

   $W = 2.55 \times 32 \times 35$

   $W =$ **2856 lb**

Material Weight Calculations

Material weight calculations are made using the volume of the object and the weight of the material if the load weight cannot be determined using a material weight table. The figures used in this method should be rounded off to allow rapid calculations. *Rounding off* is the process of increasing or decreasing a number to the nearest acceptable number. For material weight calculations, numbers are rounded to the nearest 50. Any numbers in question should be rounded up for added safety. For example, 490, 325, 782, 110, 231, and 506 can be added rapidly if the numbers are rounded off to 500, 350, 800, 100, 250, and 500.

Rounded-off calculations give dependable information for calculating load weights. With experience, rounded-off calculations become rapid and precise enough for safe rigging. The total material weight of loads with multiple regular shapes is estimated by applying the procedure:

1. Find the area of the vessel bottom. The area of the vessel bottom is found by applying the formula:
   $A_b = \pi r^2$
   where
   A_b = area of bottom
   π = 3.1416
   r^2 = radius squared
2. Find the area of the sides. The area of the vessel side is found by applying the formula:
   $A_s = d_s \times \pi \times l_s$
   where
   A_s = area of side
   d_s = diameter of side
   π = 3.1416
   l_s = length of side
3. Find the total area of the vessel. The total area of the vessel is found by applying the formula:
   $A_v = A_b + A_s$
   where
   A_v = total area of the vessel
   A_b = area of bottom
   A_s = area of side
4. Find the total material weight. The total material weight is found by applying the formula:
   $W = A_v \times$ w/sq ft
   where
   W = total material weight (in lb)
   A_v = area of vessel
   w/sq ft = material weight per square foot

For example, what is the total material weight of an uncovered steel tank 4′-8″ in diameter and 5′-11″ deep with a wall thickness of ¼″? See Figure 2-5. *Note:* Dimensions are rounded off to 5′ diameter and 6′ depth. Pi is rounded to 3.

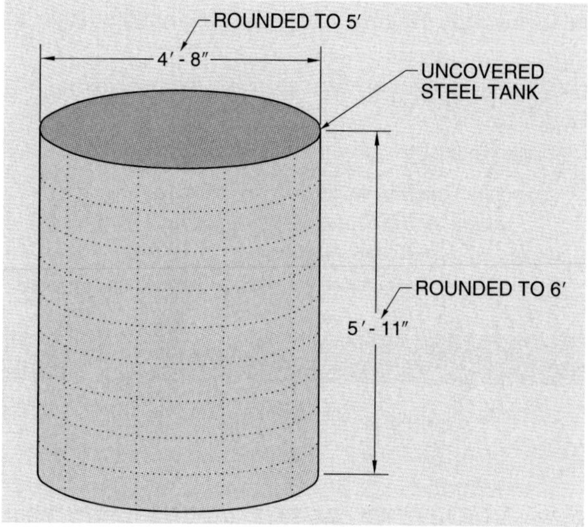

Figure 2-5. Numbers may be rounded off to allow rapid material weight calculations.

1. Find the area of the vessel bottom.
   $A_b = \pi r^2$
   $A_b = 3 \times (2.5 \times 2.5)$
   $A_b = 3 \times 6.25$
   A_b = 18.75 sq ft (round to 19 sq ft)
2. Find the area of the sides.
   $A_s = d_s \times \pi \times l_s$
   $A_s = 5 \times 3 \times 6$
   A_s = 90 sq ft
3. Find the total area of the vessel.
   $A_v = A_b + A_s$
   $A_v = 19 + 90$
   A_v = 109 sq ft (round to 110 sq ft)
4. Find the total material weight.
   $W = Av \times$ w/sq ft
   $W = 110 \times 10.2$ (from Weight of Steel Plate table)
   W = **1122 lb**

SLINGS

A *sling* is a line consisting of a strap, chain, or rope used to lift, lower, or carry a load. Slings used to lift loads are made of various components. The main sling components lift the load. Main sling components include wire rope, fiber rope, chain, webbing, and round sling. See Figure 2-6. Other sling components include rigging hardware attachments such as clips, hooks, eyebolts, shackles, sockets, wedge sockets, triangle choker fittings, and master links. See Figure 2-7.

The American Society of Mechanical Engineers (ASME) helps establish safe structural design of hoists and cranes.

Sling Combinations

A sling consists of a section of the main component (rope, chain, etc.) with a loop at both ends. Basic slings include vertical (single-leg), choker, U, basket, and bridle. Tables are used to determine the safe load capacities of specific rigging components such as rope, webbing, chain, etc., and may also rate their attachments. See Figure 2-8.

Lifting using a vertical sling is a straight vertical pull. The straight connection between the hook and the load offers 100% load capacity of the single sling. A single-leg sling is used to lift loads such as pumps, motors, gear drives, or any device equipped with a single eyebolt or lifting lug that has not been modified.

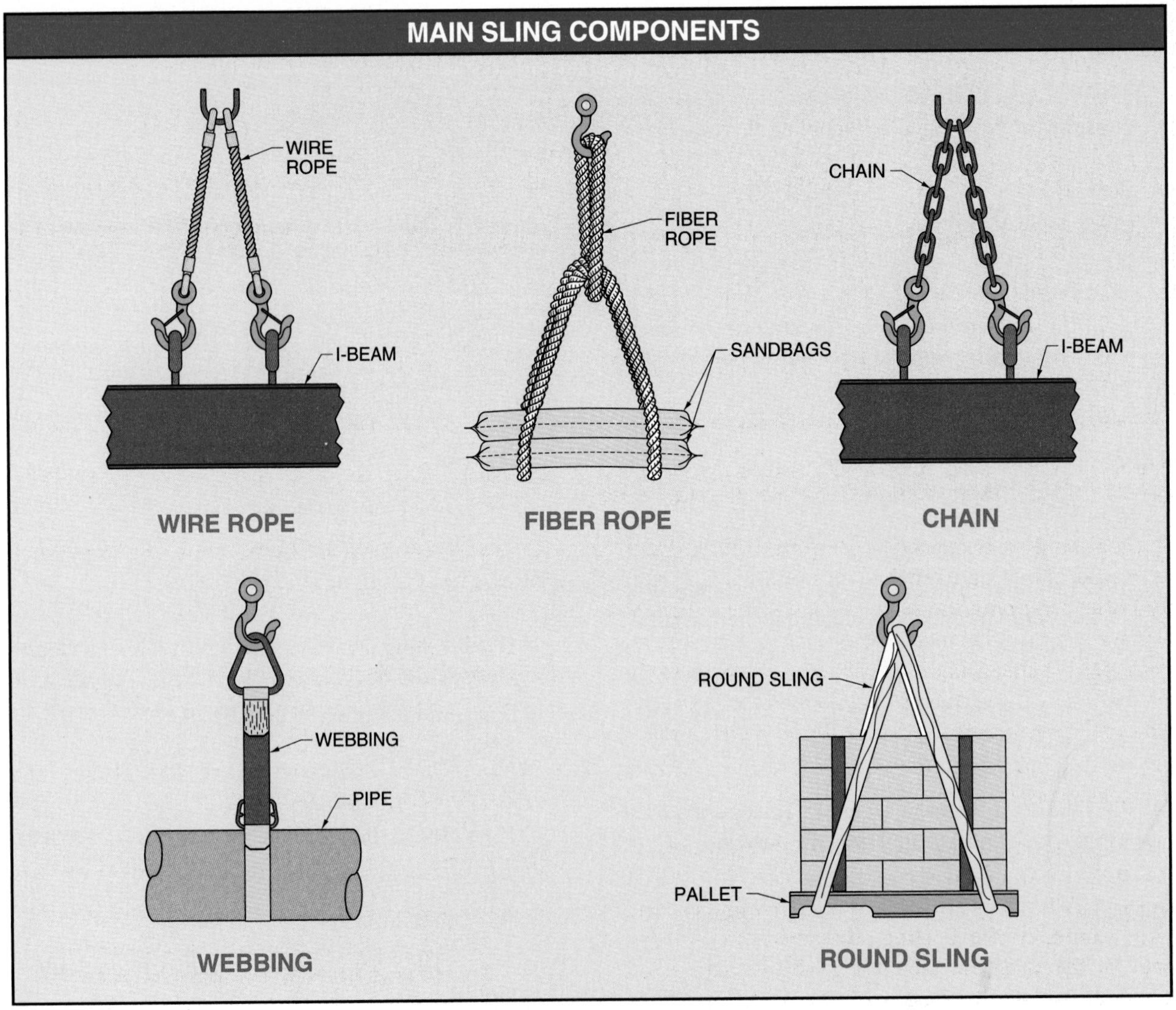

Figure 2-6. Main sling components include wire rope, fiber rope, chain, webbing, and round sling.

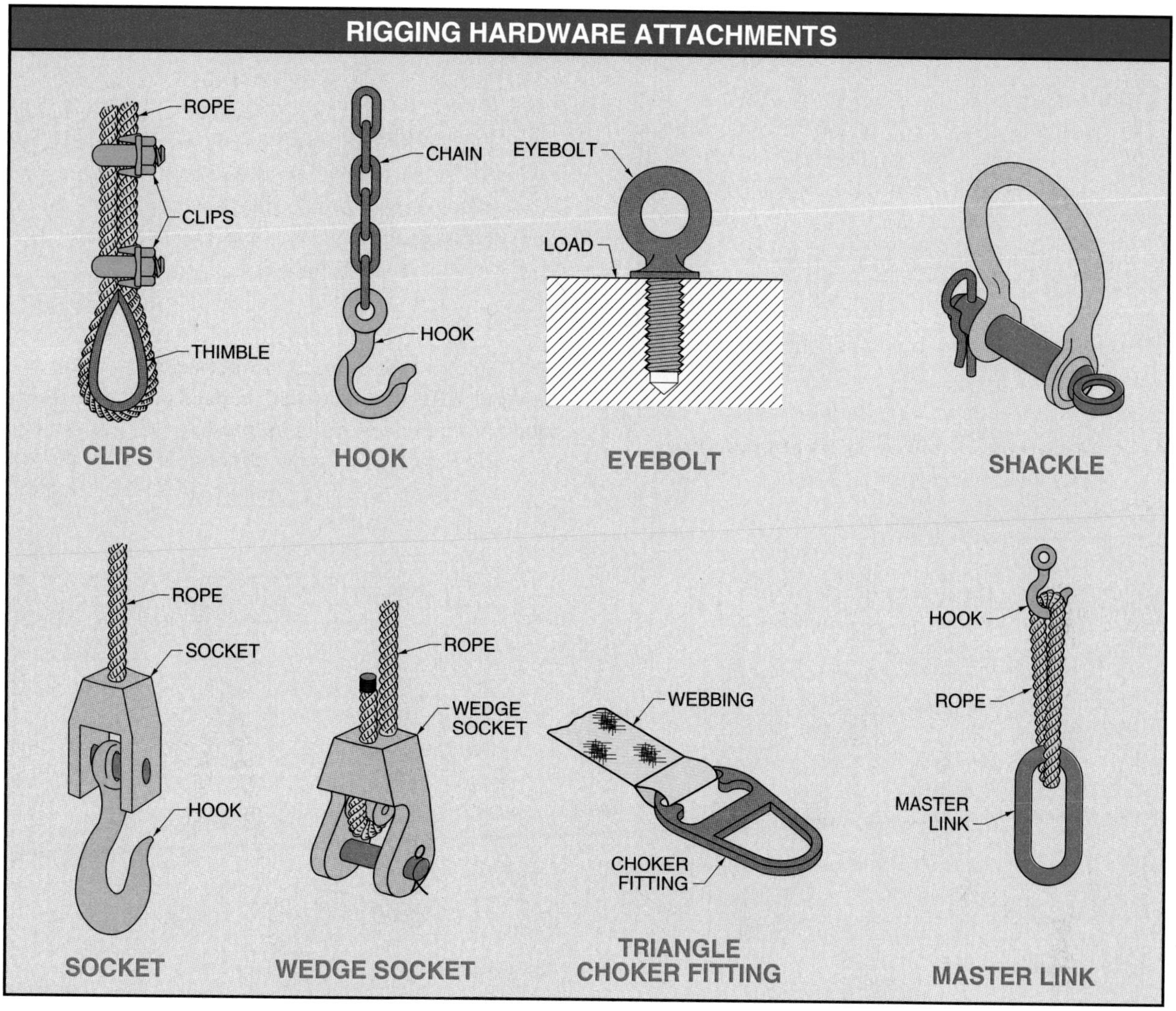

Figure 2-7. Other sling components include rigging hardware attachments such as clips, hooks, eyebolts, shackles, sockets, wedge sockets, triangle choker fittings, and master links.

A choker sling is created by slipping the loop from one end of the sling over the other end (choke junction) after wrapping the load. Choker sling loads are commonly center-balanced loads such as pipes, bars, poles, etc. Choker sling load capacity is considerably less than that of a vertical sling. The reduced capacity is due to the angle of pull created at the choke junction. Each degree from the vertical 0° angle position increases tension on the sling.

A U-sling is a single line looped under the load. The ends of the U-sling are attached to different hooks of the lifting device. A basket sling uses one sling, similar to the U-sling. However, the two eye loops of the sling are attached to the lifting device at a single point. The basket sling has a reduced capacity due to the angles of the sling legs. The rated load must be determined by multiplying the vertical load rate by a sling angle loss factor.

A bridle sling consists of two or more straight slings using identical sling constructions, length, and previous loading experience. Normal stretch must be the same for paired slings to avoid overloading individual legs and unbalancing the load during the lift. Bridle slings are used where more than one straight sling is required to make the lift, such as lifting drums, machinery, or lengths of material.

Welding rigging attachments can be hazardous. Knowledge of materials, heat treatment, and welding procedures are necessary for proper welding.

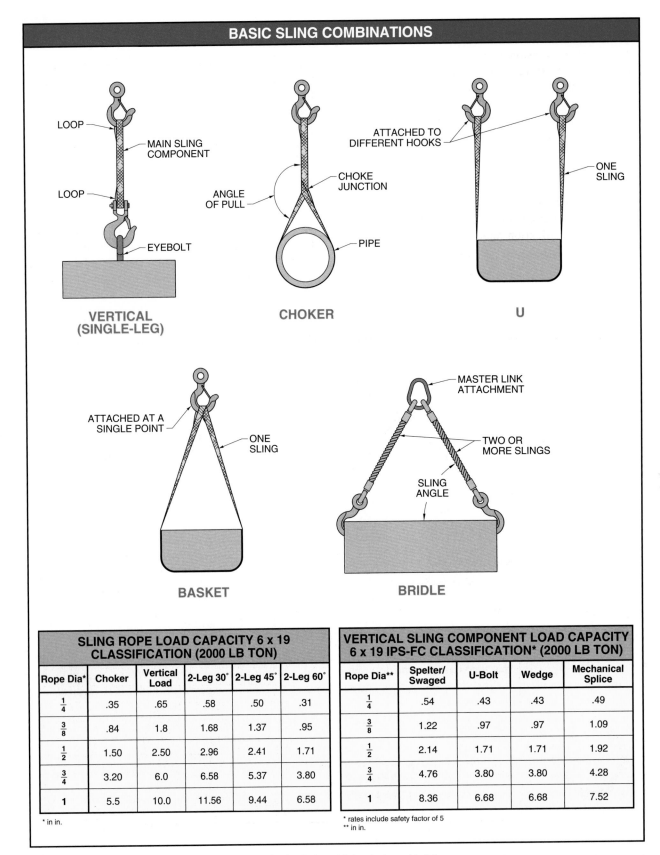

BASIC SLING COMBINATIONS

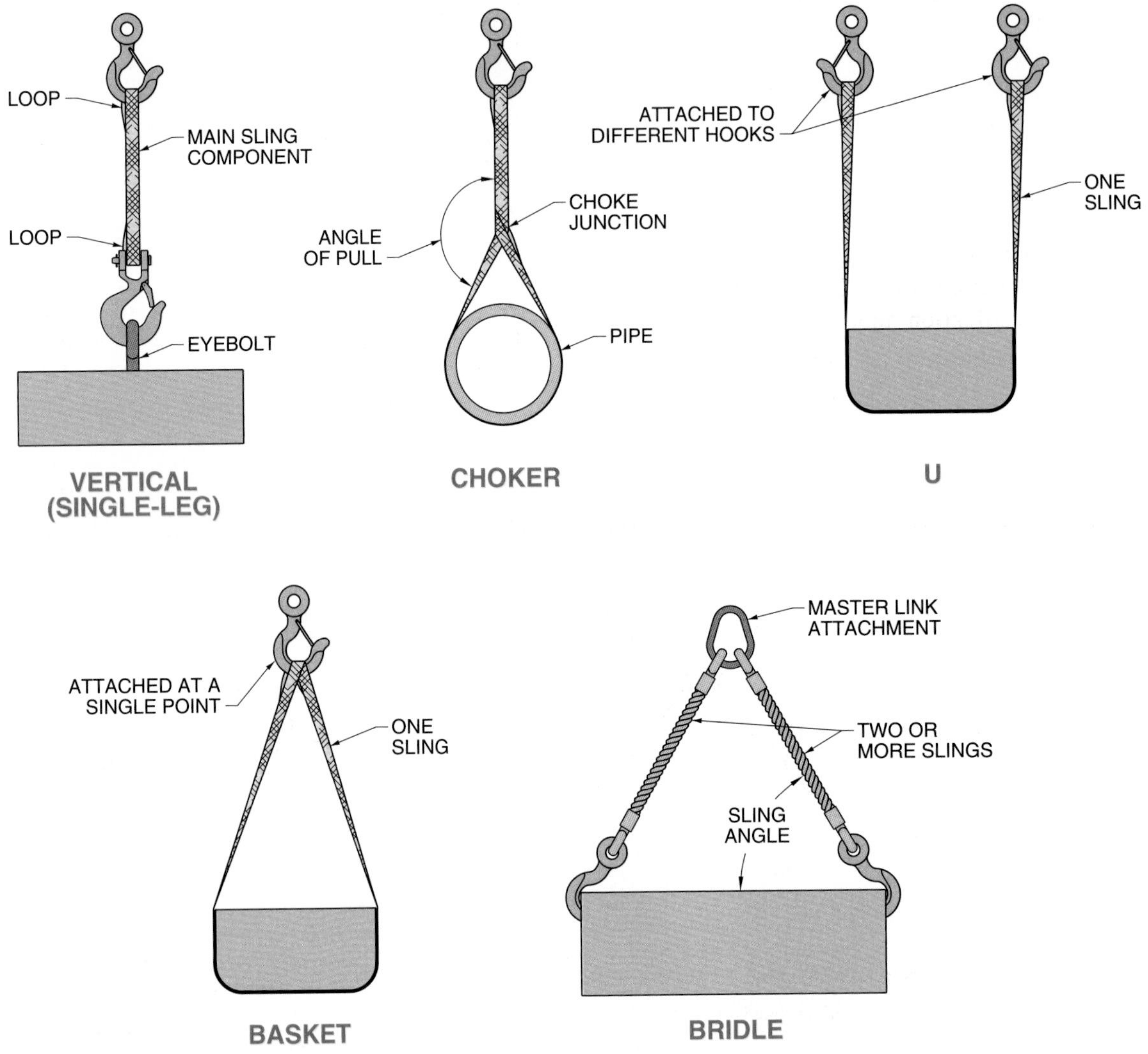

SLING ROPE LOAD CAPACITY 6 x 19 CLASSIFICATION (2000 LB TON)

Rope Dia*	Choker	Vertical Load	2-Leg 30°	2-Leg 45°	2-Leg 60°
$\frac{1}{4}$	.35	.65	.58	.50	.31
$\frac{3}{8}$	.84	1.8	1.68	1.37	.95
$\frac{1}{2}$	1.50	2.50	2.96	2.41	1.71
$\frac{3}{4}$	3.20	6.0	6.58	5.37	3.80
1	5.5	10.0	11.56	9.44	6.58

* in in.

VERTICAL SLING COMPONENT LOAD CAPACITY 6 x 19 IPS-FC CLASSIFICATION* (2000 LB TON)

Rope Dia**	Spelter/ Swaged	U-Bolt	Wedge	Mechanical Splice
$\frac{1}{4}$	.54	.43	.43	.49
$\frac{3}{8}$	1.22	.97	.97	1.09
$\frac{1}{2}$	2.14	1.71	1.71	1.92
$\frac{3}{4}$	4.76	3.80	3.80	4.28
1	8.36	6.68	6.68	7.52

* rates include safety factor of 5
** in in.

Figure 2-8. Basic slings include vertical (single-leg), choker, U, basket, and bridle.

Angular Sling Load Capacity

Angular lifting tension increases and load capacity decreases as the sling angle decreases. The sling angle decreases as the lifted load widens. See Figure 2-9. The load capacity of choker, basket, or bridle slings is calculated by applying the formula:

$LC = vl \times l \times s$

where

LC = load capacity (in t)

vl = vertical load rate (from Vertical Sling Component Load Capacity 6 × 19 IPS-FC Classification table)

l = number of sling legs (not more than two)

s = loss factor (from Sling Angle Loss Factors table)

For example, what is the total lifting capacity of a two-leg sling made of 3⁄8″, 6 × 19, IPS-FC wire rope with sling loops constructed of U-bolt clips and sling angles of 50°? *Note:* The vertical load rate for U-bolt clips is .97.

$LC = vl \times l \times s$

$LC = .97 \times 2 \times .766$

$LC =$ **1.486 t**

ROPE

Rope is used for lifting because of its length and flexibility. Rope is flexible due to its construction of many wires or fibers and small strands. Rope is manufactured from wire, organic fibers, or synthetic fibers, and is widely used for transporting loads.

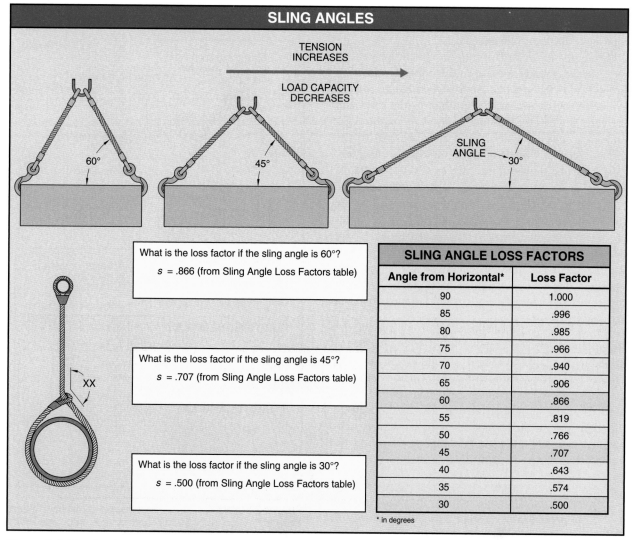

SLING ANGLES

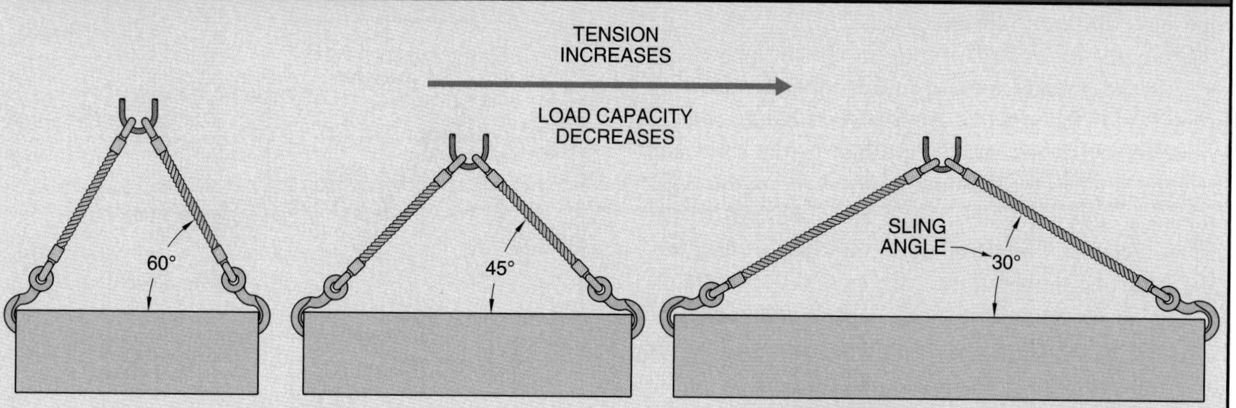

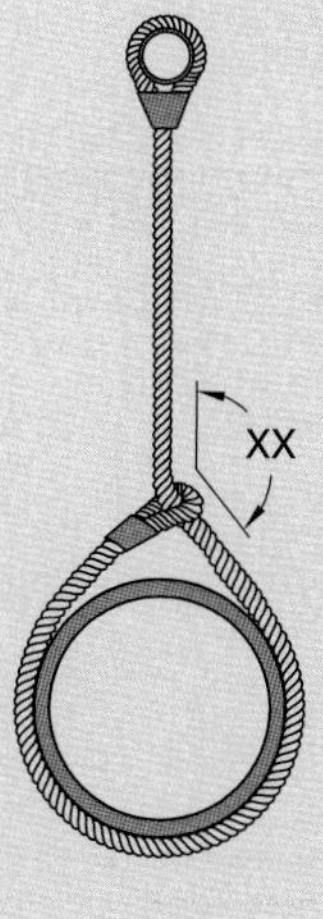

What is the loss factor if the sling angle is 60°?
s = .866 (from Sling Angle Loss Factors table)

What is the loss factor if the sling angle is 45°?
s = .707 (from Sling Angle Loss Factors table)

What is the loss factor if the sling angle is 30°?
s = .500 (from Sling Angle Loss Factors table)

SLING ANGLE LOSS FACTORS

Angle from Horizontal*	Loss Factor
90	1.000
85	.996
80	.985
75	.966
70	.940
65	.906
60	.866
55	.819
50	.766
45	.707
40	.643
35	.574
30	.500

* in degrees

Figure 2-9. Tension increases and load capacity decreases as sling legs widen and sling angles are reduced.

Rope Construction

Fiber rope is constructed by twisting fibers into yarn, yarn into strands, and strands into rope. *Yarn* is a continuous strand of two or more fibers twisted together. A *strand* is several pieces of yarn helically laid about an axis. A *helix* is a spiral or screw shape form. Strands are twisted (laid) to form the rope. Wire rope is constructed by twisting wires into strands around a wire core. The strands are laid to form the wire rope. The strands are often laid around a fiber core. See Figure 2-10.

Lay and Strands. *Rope lay* is the length of rope in which a strand makes a complete helical wrap around the core. A rope lay also designates the direction of the helical path in which the strands are laid. The strands form a spiral to the right in a right lay rope. The strands form a spiral to the left in a left lay rope. The right and left lay rotation produces regular-lay or lang-lay rope. Regular-lay and lang-lay rope have different advantages.

A *regular-lay rope* is a rope in which the yarn or wires in the strands are laid in the opposite direction to the lay of the strands. A *right regular-lay rope* is a rope in which the strands are laid to the right and the yarn or wires are laid to the left. A *left regular-lay rope* is a rope in which the strands are laid to the left and yarn or wires are laid to the right. A *lang-lay rope* is a rope in which the yarn or wires and strands are laid in the same direction. A *right lang-lay rope* is a rope in which the yarn or wires are laid to the right and the strands are laid to the right. A *left lang-lay rope* is a rope in which the yarn or wires are laid to the left and the strands are laid to the left. Right regular-lay rope is generally used for rigging purposes due to its resistance to rotation. Lang-lay ropes are used where flexibility and fatigue resistance are required.

Rope Diameter. The diameter of wire rope is determined by the largest possible outside dimension. See Figure 2-11. The outside dimension (diameter) is the circle that fully encircles the rope. The rope is measured from the high spot on one side of the rope to the high spot on the opposite side using vernier calipers. New ropes are normally slightly larger in diameter than the specifications indicate.

Fiber core wire rope slings of all grades shall be permanently removed from service if they are exposed to temperatures exceeding 200°F.

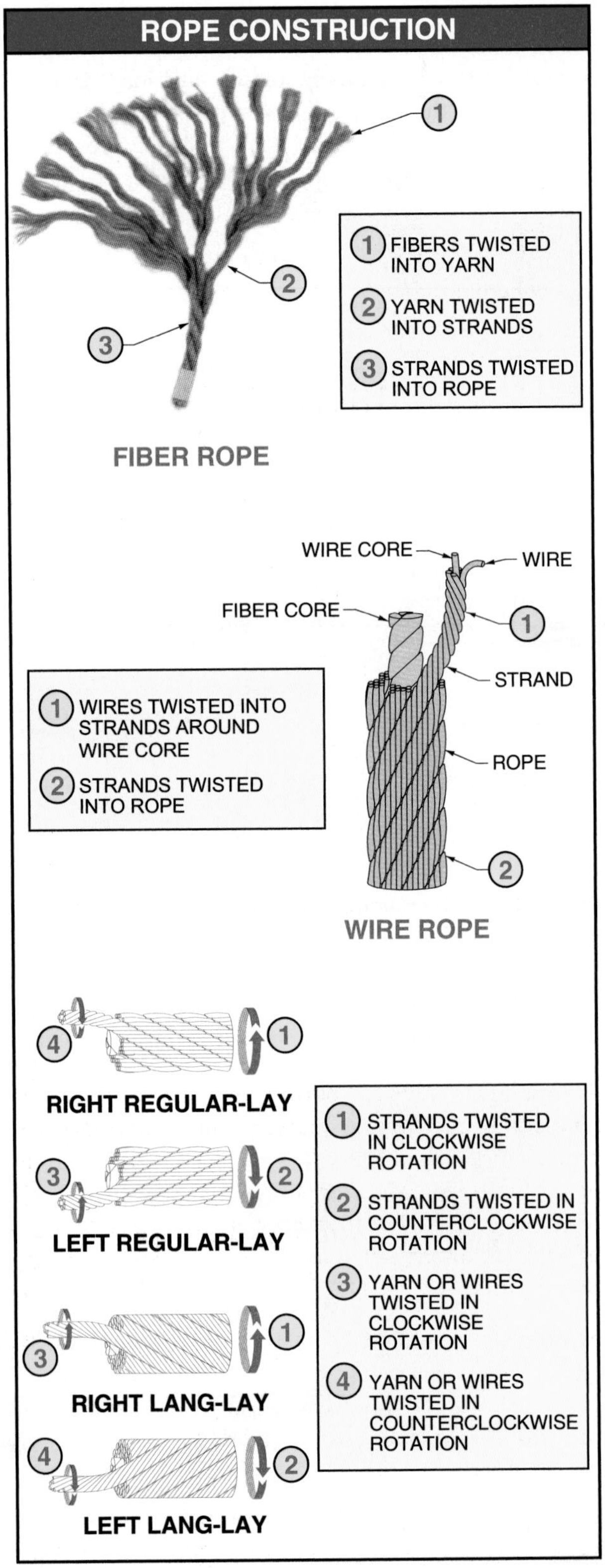

Figure 2-10. Fiber rope is constructed by twisting fibers into yarn, yarn into strands, and strands into rope. Wire rope is constructed by twisting wires into strands around a wire core.

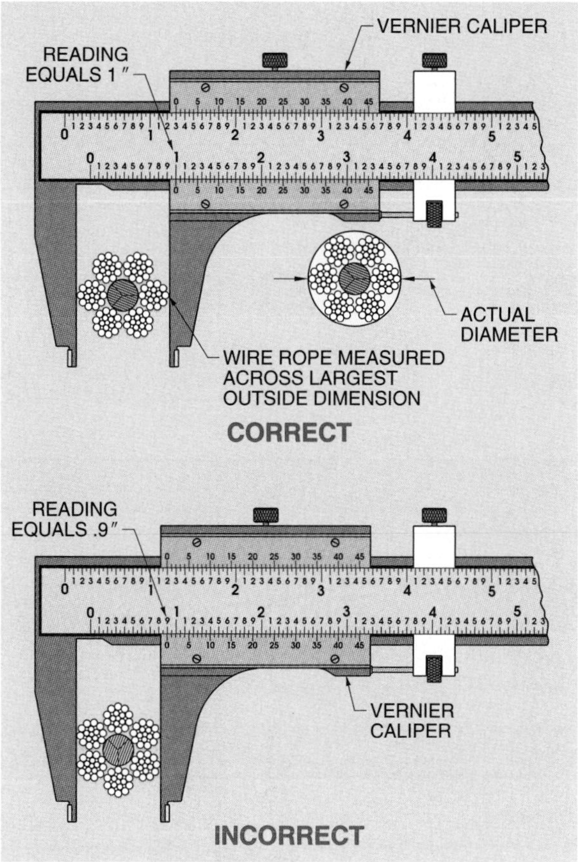

Figure 2-11. The diameter of wire rope is determined by the largest possible outside dimension.

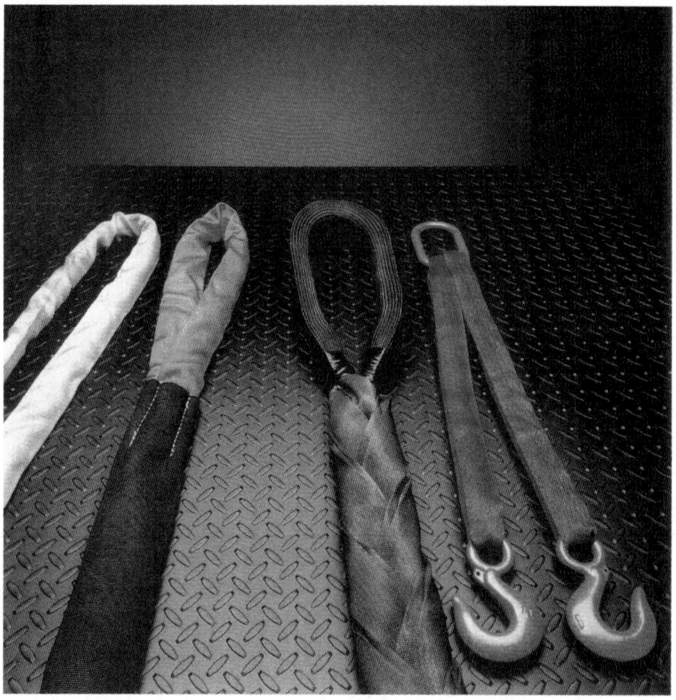

Lift-All Company, Inc.

Synthetic slings, such as web slings and roundslings, are used to lift loads that must be protected from damage.

Rope Strength

A rope loses strength during use due to moisture, temperature, chemical activity, and bending. The typical breaking strength of rope shown in charts or tables is based on new rope. Rope wear is indicated by abrasion marks, stretching or breaking of wire or fibers, or a reduction in rope diameter. See Figure 2-12.

Moisture. The effects of moisture vary between rope types. Wire ropes should be kept lubricated to prevent rusting. Moisture affects wire rope by causing it to rust from the inside out. Weakening from rust may not be indicated until the rope breaks.

Natural fiber rope may absorb moisture and decay or rot. Most rope manufacturers treat natural fiber rope with waterproofing. However, enough moisture may still be absorbed to significantly weaken a natural fiber rope when frozen. Natural fiber rope must be completely thawed before use.

Synthetic fiber rope is normally not affected by moisture because moisture is not absorbed by the fibers. However, the fibers may become brittle and weakened if synthetic rope is coated with ice.

Temperature. Manufacturers supply data on the temperature limits of rope. Wire rope with a fiber core should not be used in temperatures over 180°F (82.2°C). Wire rope with a wire core is used in temperatures up to 400°F (204.4°C). The strength of fiber rope is rated for use in the temperature range of −20°F to 100°F (−28°C to 65°C).

Chemical Activity. Reaction to acids, alkalies, caustic solutions, or fumes cause rapid damage to rope. An *acid* is any of a large class of sour substances with a pH value less than 7. An *alkali* is a bitter substance with a pH value greater than 7. Acids and alkalis turn certain materials to salts by means of corrosion. *Corrosion* is the action or process of eating or wearing away gradually by chemical action. A *caustic solution* is a liquid that creates heat and corrosion. Common caustic solutions include potash (water and potassium hydroxide) and soda (water and sodium hydroxide).

CONDITIONS AFFECTING ROPE

Moisture		
Wire Rope	Fiber Rope	
	Natural	Synthetic
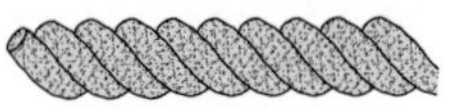 RUSTING (MAINLY FROM WITHIN)	 ROT, MILDEW, AND BRITTLENESS FROM ICING	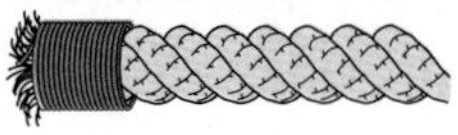 BRITTLENESS FROM ICING
Temperature		
Overheating destroys lubrication applied during manufacturing	High temperature dries out fibers making them brittle	Excess heat softens synthetic fibers allowing them to stretch
Chemicals		
Must be coated with vinyl, nylon, Teflon, or zinc for use in certain chemical environments	Begins to deteriorated almost immediately in chemically-active environments	Designed to withstand many chemical conditions
Bending		

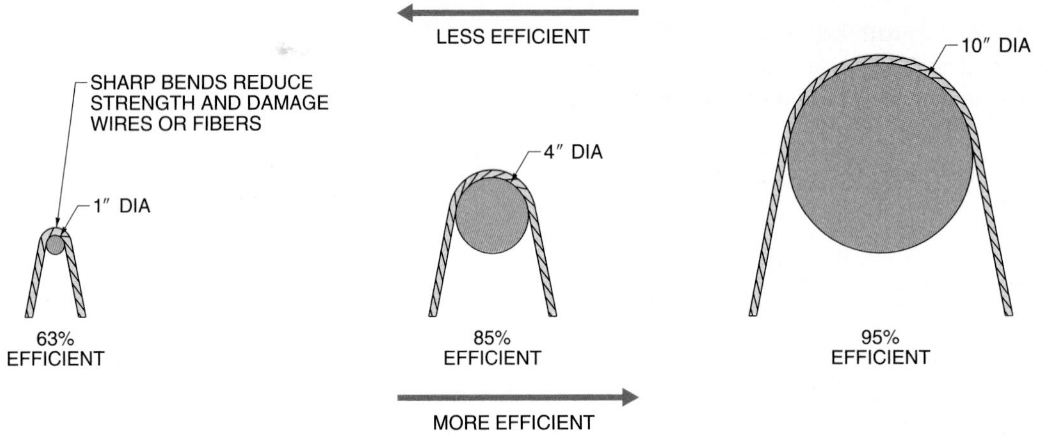

Figure 2-12. Moisture, temperature, chemical activity, and bending reduce rope strength.

Rope used in acidic or alkaline environments must be designed specifically for such use. Wire rope used in certain chemical environments such as battery shops, metal-plating shops, pickling plants, or pulp and paper mills is coated with vinyl, nylon, Teflon, or zinc. Natural fiber rope begins to deteriorate immediately in a chemically-active environment. Synthetic fiber rope materials such as vinyl are manufactured to withstand many chemical conditions. Consult the rope manufacturer before using a rope in a chemical environment.

Bending. Bending subjects rope to stress. Small diameter bends can reduce strength efficiency by more than 50%. Rope efficiency depends on the curve/rope (D/d) ratio. The *curve/rope (D/d) ratio* is the ratio between the diameter of a curved component (D), such as a pulley, and the nominal diameter of the rope (d).

The D/d ratio is compared to a rope bending efficiency chart to determine the rope's efficiency factor. See Figure 2-13. The information obtained for plotting the rope bending efficiency chart is established from static load tests applied to rope bent over stationary diameters.

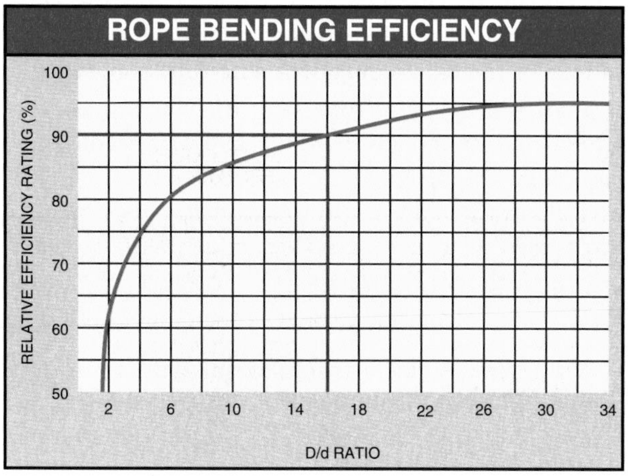

Figure 2-13. Rope bending efficiency rating increases as the diameter of the pulley increases.

A *static load* is a load that remains steady. A static load exerts a straight vertical pull on a rope. Rope bending load rating is determined by applying the procedure:

1. Calculate D/d ratio. D/d ratio is calculated by applying the procedure:

   $$R = \frac{D}{d}$$

   where

   R = D/d ratio

   D = diameter of rope curve (in in.)

   d = diameter of rope (in in.)

2. Determine relative efficiency rating (from Rope Bending Efficiency chart).
3. Calculate rope bending load rating. Rope bending load rating is calculated by applying the formula:

   $$R_{br} = R_{lr} \times R_{eff}$$

   where

   R_{br} = rope bending load rating

   R_{lr} = rope load rating

   R_{eff} = relative efficiency rating

For example, what is the rope bending load rating of a ⅜″ (.375) rope traveling over a 6″ pulley with a load rating of 1350 lb?

1. Calculate D/d ratio.

   $$R = \frac{D}{d}$$

   $$R = \frac{6}{.375}$$

   $$R = 16$$

2. Determine relative efficiency rating.

   A D/d ratio of 16 = 90% (from Rope Bending Efficiency chart)

3. Calculate rope bending load rating.

   $$R_{br} = R_{lr} \times R_{eff}$$

   $$R_{br} = 1350 \times .90$$

   $$R_{br} = \mathbf{1215\ lb}$$

Rope efficiency is increased when thimbles are used in rope ends or rope loops. A *thimble* is a curved piece of metal around which the rope is fitted to form a loop. The thimble increases the radius of the otherwise sharp bend and the rope is also protected from abrasion. Abrasion or cutting is also prevented if protection pads are placed over any rough or sharp corners of the load.

Wire Rope

Wire rope is made of a specific number of strands wound helically (spirally) around a core. Each strand is made of a number of wires. Wire rope is classified according to the number of wires in a strand and the number of strands in the rope. For example, a 6 × 19 rope indicates a six-stranded rope with approximately 19 wires per strand. An 8 × 9 rope is an eight-stranded rope with approximately nine wires per strand. The second figure in the designation is nominal in that the number of wires in a strand may be slightly higher or lower.

A *nominal value* is a designated or theoretical value that may vary from the actual value. A 6 × 7 rope consists of six strands with 3 to 14 wires per strand. A 6 × 19 classification consists of six strands with 15 to 26 wires per strand. The 6 × 19 classification includes constructions such as 6 × 21 filler wire, 6 × 25 filler wire, and 6 × 26 Warrington-Seale. These constructions are classified as 6 × 19 rope, despite the fact that none of these constructions have 19 wires. Wire rope is selected based on the largest load weight, shock from acceleration or deceleration, speed, condition of the rope, attachments used, and temperature.

Wire Rope Construction. Wire rope is also classified by the design of the wire pattern. While many pattern variations exist, the most common wire ropes used for basic rigging are filler wire, Warrington, Seale, and Warrington-Seale. Each type uses wires of different sizes and offers more or less flexibility and wear than the others. See Figure 2-14.

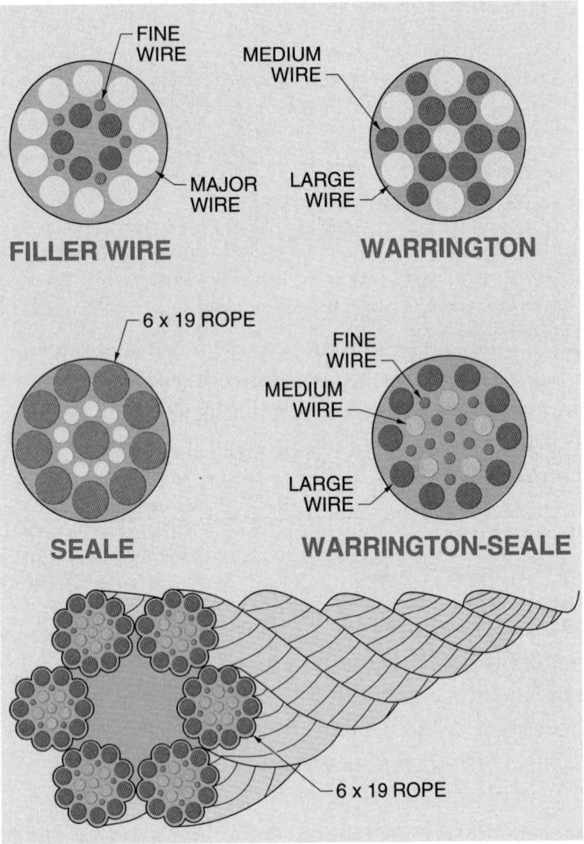

Figure 2-14. The most common wire ropes used for basic rigging are filler wire, Warrington, Seale, and Warrington-Seale.

Filler wire is wire rope that uses fine wires to fill the gaps between the major wires. Filler wire rope construction is the most flexible, but wears more than Warrington or Seale wire rope. *Warrington wire* is wire rope constructed of strands consisting of more than one size wire staggered in layers. Warrington wire rope is less flexible than filler wire rope, but wears better. *Seale wire* is wire rope that uses different size wire in different layers. Seale wire rope is less flexible than Warrington wire rope, but is the least susceptible to wear. A combination of Warrington and Seale wire provides the best wear and flexibility. The outer layer is Seale wire used for wear and the inner layers are Warrington wire used for flexibility.

Most rigging rope is constructed with a center (core) of various materials. Core materials include fiber, polyvinyl, thin wire rope, or a multiple-wire strand. Fiber cores (FCs) are made of sisal or manila fibers and are shaped to keep the strands in order and to act as a protecting cushion. Polyvinyl cores are used in areas of certain chemical or caustic solutions or fumes. Independent wire rope core (IWRC) is a small 6 × 7 wire rope with its own core of wire strands. IWRC resists crushing and offers consistent stretching. Multiple-wire strand core (WSC) is generally constructed of multiple wire strands similar to those used in wire rope construction. WSC is generally less flexible than IWRC. However, WSC and IWRC are rated 7½% stronger than other core types.

Rope wire is made from several types of metals, including steel, iron, stainless steel, monel, and bronze. Wire rope manufacturers select the wire that is most appropriate for general requirements of the finished product. The most widely used material is steel with a high carbon content. This steel is available in a variety of grades, each having properties related to the basic curve for steel rope wire. Grades of wire rope are traction steel (TS), mild plow steel (MPS), plow steel (PS), improved plow steel (IPS), extra improved plow steel (EIPS), and extra extra improved plow steel (EEIPS).

Galvanized (coated with zinc) wire rope may be required for harsh or corrosive environments. Typical rigging and lifting wire rope is uncoated (Bright). Galvanized wire rope is approximately 10% lower in strength than Bright rope.

Wire Rope Strength. The strength of rope used for safely lifting a load is determined by its breaking strength. See Figure 2-15. The breaking strength (ultimate strength) of rope is obtained from actual breakage tests. Safe load limits vary with different lifting applications but are generally established by dividing the breaking strength by a safety factor.

A life sign is a sign that identifies a caution, hazard, danger, instruction, or position of life saving devices. Life signs are generally designed to be used as a standard international safety device. They are pictorial so they can be understood regardless of language, and are color coded to indicate importance.

WIRE ROPE STRENGTH			
Nominal Diameter*	**Classification**	**Nominal Breaking Strength per 2000 lb Ton**	
		IPS**	**EIPS†**
1/4	6 x 19 STANDARD HOISTING FIBER CORE	2.74	–
	6 x 19 STANDARD HOISTING IWRC	–	3.40
	8 x 19 SPECIAL FLEXIBLE HOISTING FIBER CORE	2.35	–
	18 x 7 NONROTATING	2.51	–
3/8	6 x 19 STANDARD HOISTING FIBER CORE	6.10	–
	6 x 19 STANDARD HOISTING IWRC	–	7.55
	8 x 19 SPECIAL FLEXIBLE HOISTING FIBER CORE	5.24	–
	18 x 7 NONROTATING	5.59	–
1/2	6 x 19 STANDARD HOISTING FIBER CORE	10.7	–
	6 x 19 STANDARD HOISTING IWRC	–	13.3
	8 x 19 SPECIAL FLEXIBLE HOISTING FIBER CORE	9.23	–
	18 x 7 NONROTATING	9.85	–

STANDARD HOISTING
6 x 19 SEALE
WITH FIBER CORE

SPECIAL FLEXIBLE HOISTING
8 x 19 WARRINGTON
WITH FIBER CORE

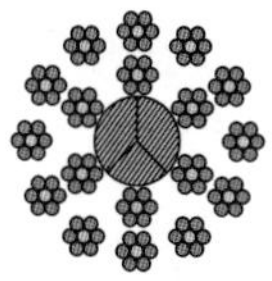

NONROTATING WIRE ROPE
18 x 7
WITH FIBER CORE

* in in.
** IPS - improved plow steel
† EIPS - extra improved plow steel

Figure 2-15. The strength of rope used for safely lifting a load is determined by its breaking strength.

Safety factors used for safe rope limits are normally on a scale of 5 to 8. A safety factor of 5 is used for steady or even loads. A safety factor of 8 is used for shock or uneven loads. Higher values should be used where there is a hazard to life or property. For example, a safety factor of 5 is used for a rigging assembly that is rated to lift a steady load of 2.35 t. The safe lifting capacity of the rigging assembly is .47 t (2.35 ÷ 5 = .47 t). Generally accepted safe rope strength required to lift a load is found by applying the procedure:

1. Convert load weight to tons. Load weight must be converted to tons because the nominal breaking strength of rope is rated in weight per ton. Load weight is converted to tons by applying the formula:

$$t = \frac{w}{2000}$$

where

t = weight (in t)

w = weight (in lb)

2000 = constant (to convert lb to t)

2. Determine rope strength needed to lift the load. The strength of the rope is calculated by applying the formula:

$$R_s = t \times 5$$

where

R_s = rope strength (in t)

t = weight (in t)

5 = constant (safety factor)

3. Determine rope size to be used (from Wire Rope Strength table).

For example, what is the generally accepted safe rope strength required to lift a 4000 lb milling machine? *Note:* A safety factor of 5 is used because the load is a steady lift without shock.

1. Convert load weight to tons.

$$t = \frac{w}{2000}$$

$$t = \frac{4000}{2000}$$

$$t = 2 \text{ t}$$

2. Determine rope strength needed to lift the load.

   $R_s = t \times 5$

   $R_s = 2 \times 5$

   $R_s = 10$ t

3. Determine rope size to be used.

The proper wire rope to be used for lifting the machine is determined as ½″ 6 × 19 standard hoisting fiber core rope (from Wire Rope Strength table).

Wire Rope Technical Board

A tension break is the result of overloading wire rope.

Once the proper wire rope size is determined, the load capacity of choker, basket, or bridle slings used to lift the load are determined. See Figure 2-16. The load capacity of a sling is found by applying the formula:

$$LC = vl \times l \times s$$

where

LC = load capacity (in t)

vl = vertical load rate (from Sling Material Strength Capacities table)

l = number of sling legs (but not more than two)

s = loss factor (from Sling Angle Loss Factors table)

For example, what is the working load capacity of a wire rope sling using a ½″ 6 × 19 IPS/FC rope, a 2-leg sling, and a 70° load angle?

Note: The loss factor of a sling with a 70° load angle equals .940 (from Sling Angle Loss Factors table).

$$LC = vl \times l \times s$$

$$LC = 2.0 \times 2 \times .940$$

LC = **3.76 t**

SLING MATERIAL STRENGTH CAPACITIES*

6 x 19	Rated Capacities (in Tons)†		
ROPE DIA**	VERTICAL	CHOKER	BASKET
1/4	.51	.38	1.0
5/16	.79	.60	1.6
3/8	1.1	.85	2.2
7/16	1.5	1.1	3.0
1/2	2.0	1.5	4.0
9/16	2.5	1.9	5.0
5/8	3.1	2.3	6.2
3/4	4.4	3.3	8.8
7/8	6.0	4.5	12.0
1	7.7	5.9	15.0

* improved plow steel/fc
** in in.
† rates include safety factor of 5

Figure 2-16. Rated strength capacities of 6 × 19 wire rope are based on the rope diameter and sling.

Wire Rope Rotation. Due to the nature of its construction, rope must not be allowed to spin or rotate while being used. A special rotation-resistant rope is used in applications such as a single part line or situations where operating conditions require a rope to resist cabling. *Cabling* is a rope's attempt to rotate and untwist its strand lays while under stress.

Rotation-resistant ropes are available in single-layer and multi-layer strand classifications. A single-layer strand wire rope consists of a single layer of strands, without a core, where each strand supports one another. A multi-layer strand wire rope consists of two or more layers of strand laid in opposing directions.

Cutting Wire Rope. The ends of wire rope must be secured by binding (seizing) to prevent raveling, unsafe loose wires, or strength reduction before cutting a wire rope. *Seizing* is the wrapping placed around all strands of a rope near the area where the rope is cut. See Figure 2-17. Seizing holds the strands firmly in place by the tight turning of seizing wire. This must be done twice to both ends before the cut.

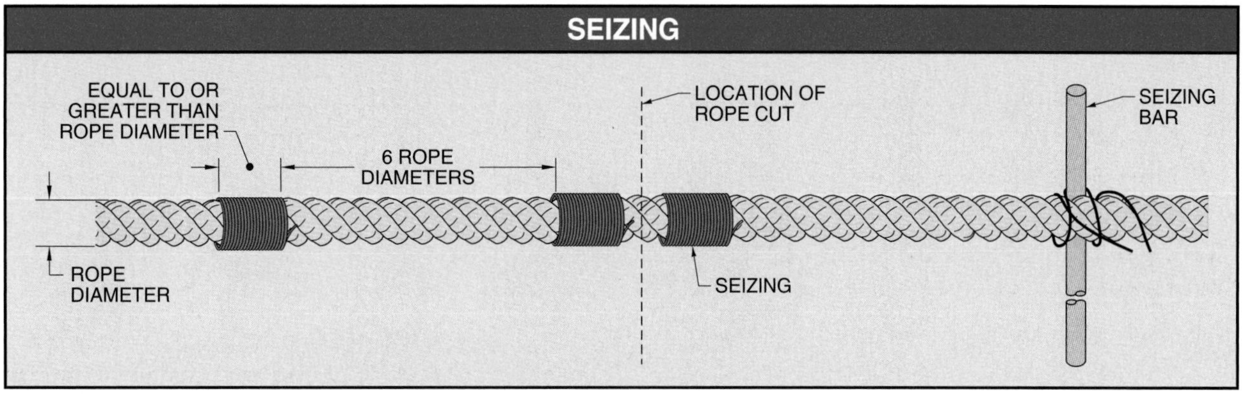

Figure 2-17. Seizing is the wrapping placed around all strands of a rope near the area where the rope is cut.

Normally, one seizing on each side is sufficient for preformed wire rope. Common wire rope (those that are not preformed or are rotation-resistant) normally requires a minimum of two seizings on each side of the cut. *Preformed rope* is wire rope in which the strands are permanently formed into a helical shape during fabrication. Adequately seized ends prevent rope distortion, flattening, or strand loosening. Inadequate seizing shortens rope life by allowing uneven distribution of the strand load during lifting. Seizing is placed six rope diameters apart and the length of each seize should be equal to or greater than the rope diameter.

The recommended method for seizing a wire rope is to lay one end of the seizing wire between two strands of the wire rope. The other end of the seizing wire is wrapped around the rope over the dead end of the seizing wire. In a looped wire or rope, the loose end is the dead end and the load-lifting portion or working end is the live end. A seizing bar is placed at a right angle to the rope. A *seizing bar* is a round bar ½″ to ⅝″ in diameter and about 18″ long used to seize rope. The live end of the seizing wire is brought around the back of the bar and the bar is twisted around the rope to wind the seizing wire around the rope. The wire is twisted in a tight helix, without overlapping, until the required seizing length is obtained. The seize is secured by twisting the ends of the seizing wire together.

Wire rope may be cut using a rope shear, an abrasive cutoff wheel, or an acetylene cutting torch. Shearing or abrasive cutting leaves a sharp edge that should be filed smooth. An acetylene cutting torch is preferred because the heat fuses the strands and strand wires together. Seizing specifications vary based on rope diameter. Always check manufacturer's specifications.

Wire Rope Terminations. Wire rope ends (terminations) are fastened to fittings or spliced into loops when wire rope is attached to a load. Common wire rope terminations include thimbles and sockets. Fittings, loops, or other attachments may not have the strength of the rope and may reduce the sling efficiency. See Figure 2-18.

A thimble is a curved piece of metal around which the rope is fitted to form a loop. A thimble protects the wire rope from sharp bends and abrasion. The length of wire rope that is looped back is determined by the loop style or manufacturer's specifications. Always consult the manufacturer's specifications whenever possible.

In a looped rope, the loose end is the dead end, and the load-lifting portion is the live end. Loops are secured using a U-bolt clip (clamp). A U-bolt clip consists of a saddle, threaded U-section, and two nuts. The clip should be assembled with the threaded U-section contacting the dead-end section of the rope for maximum rope strength and to prevent damage to the live end of the thimble and clip assembly. Clip connections must be arranged, spaced, and assembled properly to maintain the strength of the rope. See Figure 2-19.

Assembling a thimble and the correct number of clips is determined by using manufacturer's specification charts. The first clip is placed approximately 4″ from the end of the rope. The other clips are spaced at a minimum of 6 rope diameters apart.

The preferred method of splicing two wire ropes together is to use interlocking turnback eyes with thimbles that contain the recommended number of clips on each eye.

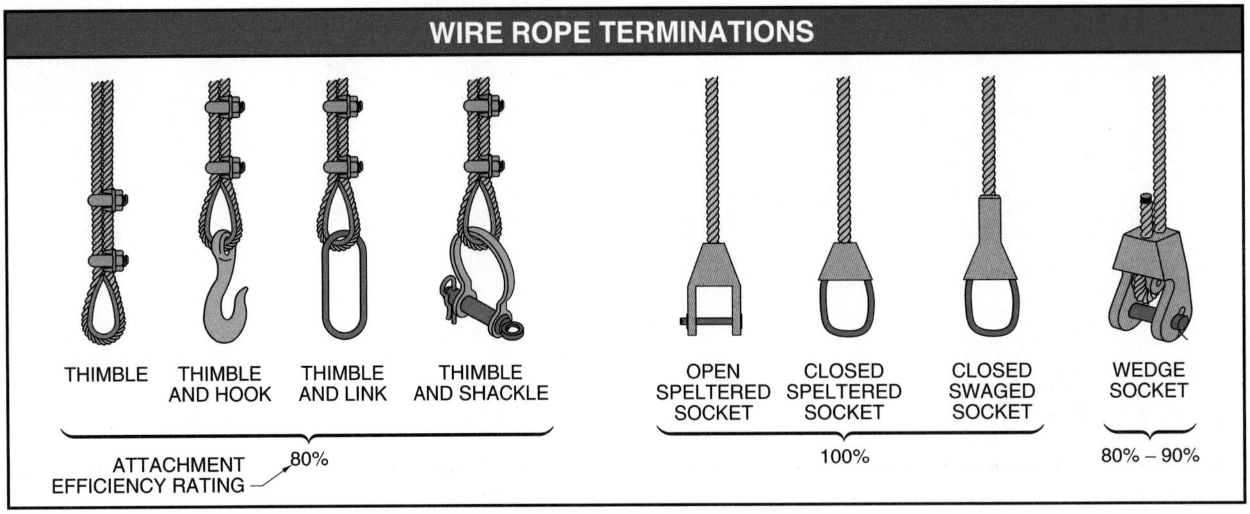

Figure 2-18. Common wire rope terminations include thimbles and sockets.

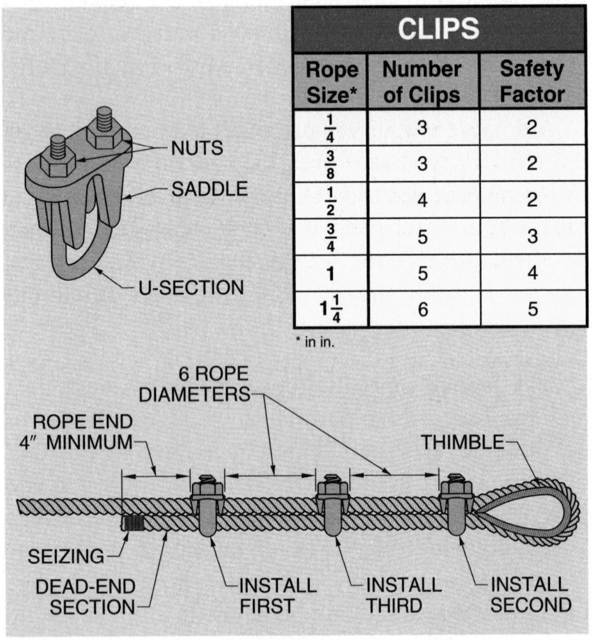

CLIPS

Rope Size*	Number of Clips	Safety Factor
1/4	3	2
3/8	3	2
1/2	4	2
3/4	5	3
1	5	4
1 1/4	6	5

* in in.

Figure 2-19. Clip connections must be arranged, spaced, and assembled properly to maintain the strength of the rope.

The U-section is assembled against the dead-end section. Rope damage occurs if the U-section is assembled against the live end. The first clip is installed 4″ from the dead end and the nuts are tightened. Next, the second clip is installed at the thimble and the nuts are finger-tightened. The remaining clips are assembled finger-tight. Finally, a strain is placed against the rope and the remaining nuts are alternately tightened.

A *socket* is a rope attachment through which a rope end is terminated. Wire rope sockets include swaged, speltered, and wedge designs. Swaged and speltered sockets are permanent wire rope attachments. Permanent wire rope attachments have the highest efficiency rating.

A *swaged socket* is a compressed socket assembled to the wire rope under high pressure. Swaged sockets are compressed in a hydraulic press. A *speltered socket* is a socket assembled by separating the wire rope ends after inserting the rope through the socket collar. Molten zinc or resin is poured into the collar, creating a solid assembly. Swaged and speltered sockets are 100% efficient due to the manufacturing and assembly process.

A *wedge socket* is a socket with the rope looped within the socket body and secured by a wedging action. Wedge sockets are popular because rapid position changes are possible and installation and dismantling processes are fairly easy. However, because of its design, a wedge socket can be incorrectly installed, creating a sharp bend on the live end of the rope. The live end must be in line with the socket. The exposed dead-end section must extend out of the wedge a minimum of eight rope diameters. See Figure 2-20.

Fiber Rope

Rope constructed of fibers is preferred for some applications because fiber is less likely to gouge or mar equipment surfaces than wire rope or chain. Fiber rope is classified by the materials used to construct the rope. Fiber rope can be made from either natural or synthetic fibers.

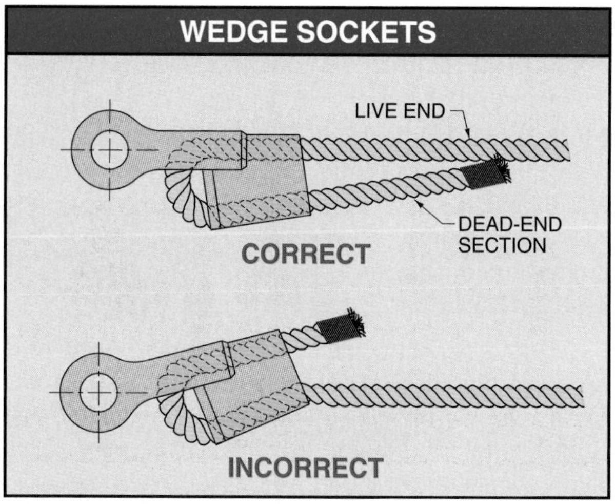

Figure 2-20. Proper installation of a wedge socket has the live end of a rope in line with the socket.

Natural fibers are obtained from plants. Plants used in the manufacture of fiber rope include cotton, hemp, manila, and sisal. Manila fiber is used predominantly for lifting and is derived from the banana plant.

Natural fiber quality varies because the living plant quality varies. This affects the quality of the finished rope. The grade of rope is classified by the quality (grade) of fiber used. Common manila rope classifications include yacht rope, number 1, number 2, and hardware. Only yacht and number 1 class manila rope should be used for lifting. Number 2 and hardware classes of manila rope should not be used for lifting because, in some cases, the quality, type, and grade of fiber is unknown.

Synthetic materials used for rigging and lifting include nylon, polypropylene, and polyester. Synthetic ropes are used more commonly today because of the consistent quality of fiber from one rope to another. Also, the breaking strength of synthetic fibers is approximately twice that of manila fibers. Synthetic fibers are generally stronger than short natural fibers because the synthetic fiber is continuous throughout the length of the rope. See Figure 2-21.

Synthetic materials are used in ropes specifically for their special properties. Nylon is used primarily for strength, polyester for dimensional stability, fiberglass for electrical properties, vinyl for chemical properties, and polypropylene for flotation. Another advantage of synthetic fibers is that they do not mildew, rot, or decay as natural fibers do.

NOMINAL FIBER ROPE STRENGTH*

Rope Dia**	Manila	Poly-Propylene	Polyester	Nylon
1/4	600	1200	1150	1500
5/16	1000	2100	1750	2500
3/8	1350	3100	2450	3500
1/2	2600	4200	4400	6000
3/4	5400	8000	9500	13,500
1	9000	13,500	16,000	23,500
	NATURAL FIBER	SYNTHETIC FIBERS		

* in lb
** in in.

Figure 2-21. Synthetic fibers are generally stronger than natural fibers.

Fiber Rope Construction. Fiber rope is constructed by twisting fibers into yarn, yarn into strands, and strands into rope. The first step in making fiber rope is to create a yarn by twisting the fibers in one direction. The yarn is twisted together in the opposite direction to create a strand. Strands are twisted (laid) in the reverse direction to create the rope. Reversing the twist of each step prevents the rope from unwinding. See Figure 2-22.

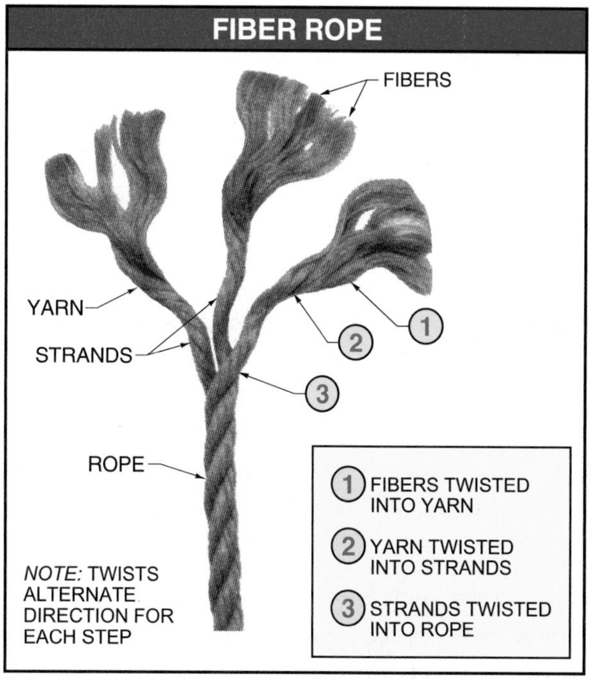

Figure 2-22. Fiber rope is constructed by twisting fibers into yarn, yarn into strands, and strands into rope.

Fiber rope is constructed of three or more strands (normally three) with or without a core. A *lay* is a complete helical wrap of the strands of a rope. The lay is used when inspecting a rope. Remove any rope from service if there is a measurable increase in lay length or reduction of rope diameter. For example, a rope classified as having three rope lays per foot must have a measurement of 4″ per lay.

The rope lay length measurement is used to determine if the rope has stretched any measurable amount due to age or load. The initial measurement is made when the rope is new. This gives a comparison figure for all future measurements. Measurements should be at a length of 1′ or 2′ and at the same area of rope at each measurement. The measurement should include whole lay lengths. For example, if four complete lays of a rope are recorded at 1′-4″, all future measurements would determine if the rope has or has not stretched beyond the 1′-4″ at four lays.

Fiber Rope Strength. Rope strength varies according to the degree of twist of each construction step. A high strength grade has a light degree of twist and is referred to as soft lay grade. Rope with a high resistance to abrasion is referred to as hard lay and has a high degree of twist. Ropes formed into cables use two or more three-strand ropes twisted together. Specific breaking strength of fiber rope varies greatly. Always consult the rope manufacturer's specifications when choosing rope for an application.

Safe rope strengths for fiber rope are calculated using the same method as for wire rope. The vertical breaking strength is divided by a safety factor to determine the maximum allowed load for a straight vertical pull. The safety factor for manila rope is 5, polypropylene 6, and polyester and nylon 9. All rope strengths should be increased where life, limb, or valuable property is involved.

Fiber Rope Applications. Binding (fastening) loads with a rope normally requires some form of hitch or knot. A *hitch* is the interlacing of rope to temporarily secure it without knotting the rope. *Knotting* is fastening a part of a rope to another part of the same rope by interlacing it and drawing it tight.

All knots involve changing direction of the rope axis and pinching against another part of the rope. For a rope, an *axis* is an imaginary straight line that runs lengthwise through the center of the rope. Each sharp change of direction weakens a rope. A knot in a rope can reduce rope efficiency by as much as 55%. For example, straight rope has an efficiency of 100%, an eye splice over a thimble has an efficiency of 90%, a short splice has an efficiency of 80%, and an overhand knot has an efficiency of 45%.

Most rope hitching and knotting terminology was derived from nautical (sailing) terms. Basic elements used when working with rigging knots include bight, loop, whipping, working end, working part, standing part, standing end, kinks, nips, and eye loops. See Figure 2-23.

A *bight* is a loose or slack part of a rope between two fixed ends. A *loop* is the folding or doubling of a line, leaving an opening through which another line may pass. The *working end* is the end of the working part of a rope. The *working part* is the portion of the rope where the knot is formed. The *standing end* is the end of the rope that is normally fixed to a permanent apparatus or drum, or is rolled into a coil. The *standing part* is the portion of the rope that is not active in the knot-making process.

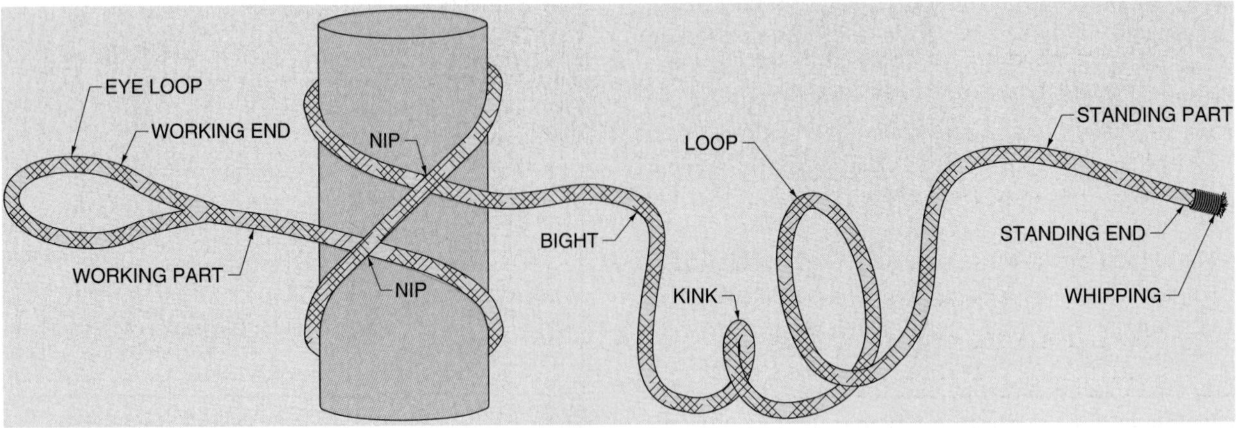

Figure 2-23. Most rope hitching and knotting terminology was derived from nautical (sailing) terms.

The working end of a rope is protected from untwisting or raveling by whipping (seizing), splicing, or crowning. Synthetic fiber rope is finished off by sealing fibers together with a match or soldering iron.

Whipping (seizing) is tightly binding the end of a rope with twine before it is cut. See Figure 2-24. Rope is whipped by applying the procedure:

1. Form a bight with the end of the twine and lay along the rope to be whipped.
2. Wrap twine tightly around the rope, gradually working toward the rope end. The turns are laid hard against each other without overlapping.
3. Tuck twine through the loop at the end of the rope.
4. Pull loop halfway through whipping by pulling other end of the twine.
5. Trim loose twine ends close to the turns.
6. Seal synthetic fibers with heat.

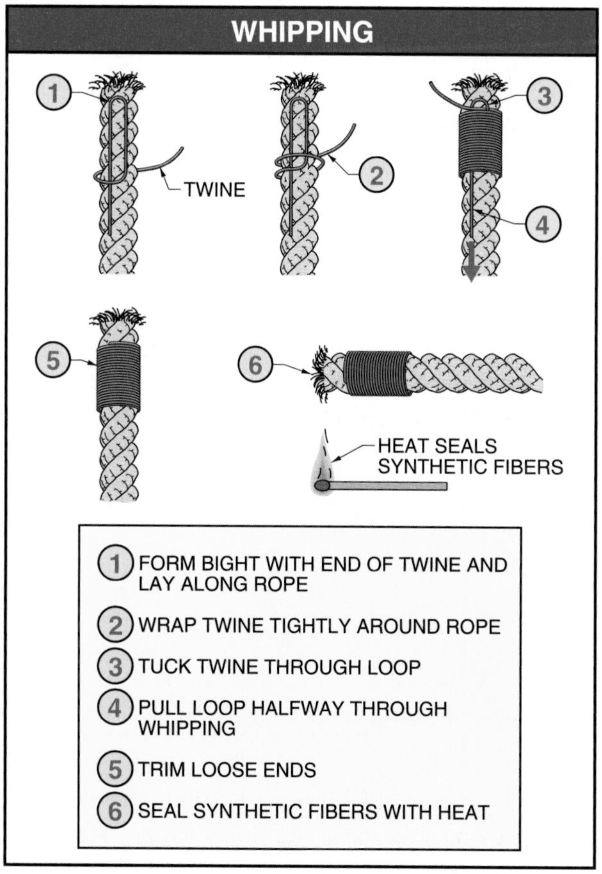

Figure 2-24. Whipping is tightly binding the end of a rope with twine before it is cut.

A *splice* is the joining of two rope ends to form a permanent connection. See Figure 2-25. Splices are used to join the ends of two ropes of similar strength and thickness. Splices include the short splice and the long splice. A short splice uses an unlay of six to eight rope strands on each rope. An *unlay* is the untwisting of the strands in a rope. A short splice, when braided together, increases the rope diameter. This makes the short splice unsuitable for pulley use. A long splice uses an unlay of 15 turns and does not increase the rope diameter. A long splice is formed by applying the procedure:

1. Unlay 15 turns and place a temporary whip on both rope standing parts.
2. Whip the strand ends.
3. Place the two rope ends (standing part terminations) together, alternating strands of one end with the strands of the other.
4. Remove the temporary whip from one rope and unlay one strand about 10 additional turns.
5. Fill the void in the grooves of the 10 turns with the matching strand of the other rope.
6. Remove the temporary whip from the other rope and unlay 10 additional turns.

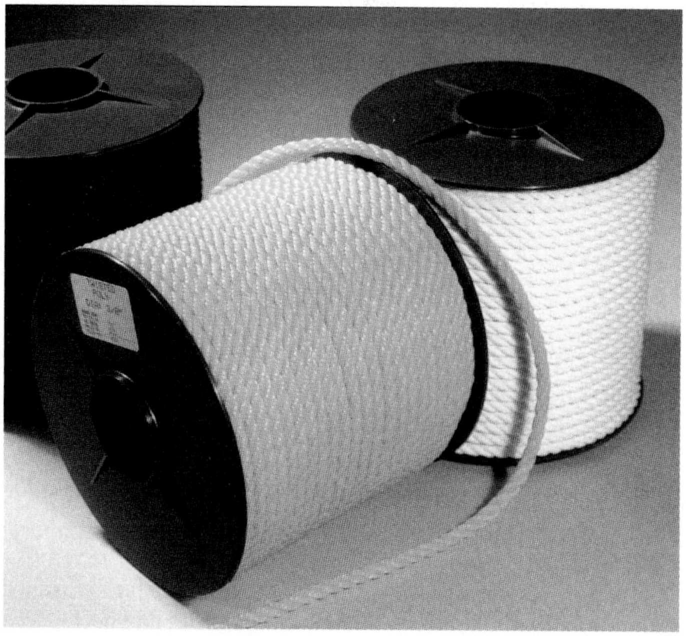

Crowe Rope Industries LLC

Polypropylene rope is a synthetic fiber rope that is available in a variety of colors, has a working temperature range of −20°F to 200°F, and excellent resistance against acids, alkalis, mildew, and rot.

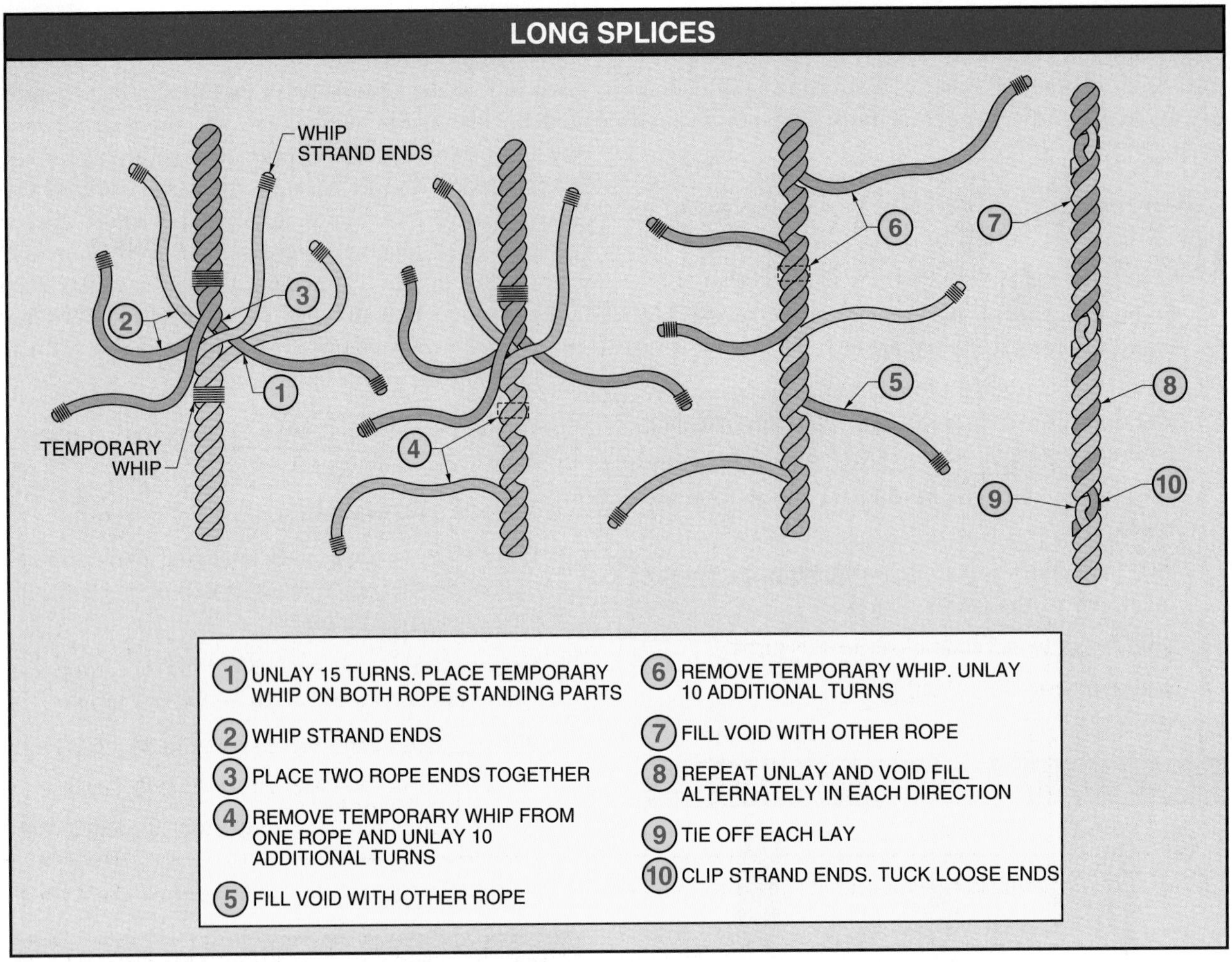
LONG SPLICES
WHIP STRAND ENDS
TEMPORARY WHIP
1 UNLAY 15 TURNS. PLACE TEMPORARY WHIP ON BOTH ROPE STANDING PARTS
2 WHIP STRAND ENDS
3 PLACE TWO ROPE ENDS TOGETHER
4 REMOVE TEMPORARY WHIP FROM ONE ROPE AND UNLAY 10 ADDITIONAL TURNS
5 FILL VOID WITH OTHER ROPE
6 REMOVE TEMPORARY WHIP. UNLAY 10 ADDITIONAL TURNS
7 FILL VOID WITH OTHER ROPE
8 REPEAT UNLAY AND VOID FILL ALTERNATELY IN EACH DIRECTION
9 TIE OFF EACH LAY
10 CLIP STRAND ENDS. TUCK LOOSE ENDS

Figure 2-25. A splice is the joining of two rope ends to form a permanent connection.

7. Fill the void in the grooves of the 10 turns with the matching strand of the other rope.
8. Repeat unlay and void fill alternately in each direction.
9. Tie off each lay of strands using an overhand knot and begin tucking the strand from one rope through the strands of the other rope. A minimum of two tuck sets is required. A *tuck set* is wedging a strand of rope into and between two other rope strands.
10. Clip the strand ends after rolling and pounding the splice.

Splicing the working end of a rope into a crown or eye loop produces a workable, neat, and permanent rope termination. *Crowning* is a reverse strand splice that is used when an enlarged rope end is desired or not objectionable. See Figure 2-26. A rope crown termination is formed by applying the procedure:

1. Unlay rope ends eight turns and whip the strand ends.
2. For a three-strand rope, loop strand 1 and lay strand 2 over strand 1 and down the side of the rope.
3. Lay strand 3 over strand 2 and through strand 1 loop.
4. Snug strands 1, 2, and 3.
5. Tuck strand 1 through strand 2 of the standing part of the rope.
6. Alternately tuck each strand. Trim ends. The crown of the rope becomes tighter with time and use.

Spliced fiber rope slings shall not be used unless they have been spliced in accordance with OSHA 1926.251.

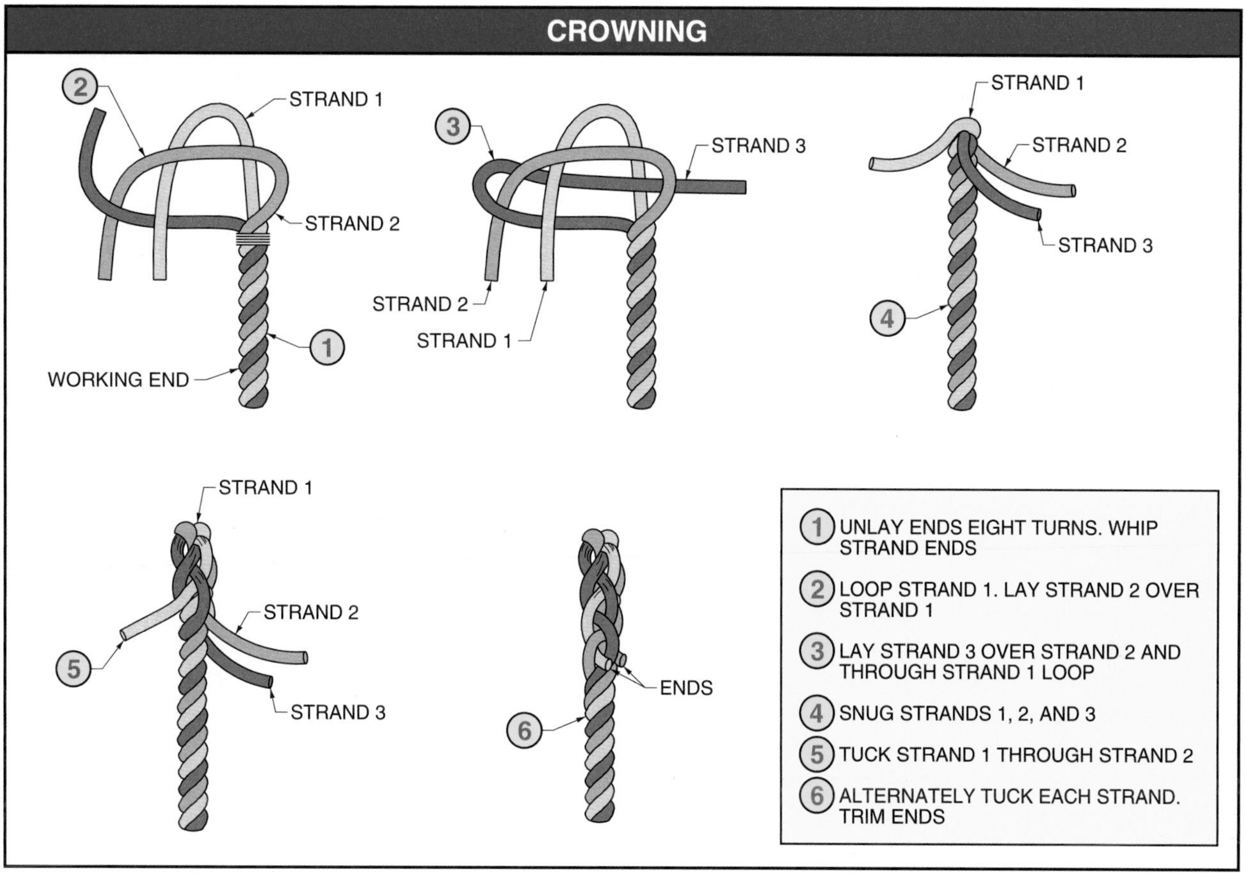

Figure 2-26. Crowning is a reverse strand splice that is used when an enlarged rope end is desired or not objectionable.

An *eye loop* is a rope splice containing a thimble. See Figure 2-27. Eye loops are spliced for a permanent termination. A thimble is inserted for strength and wear and is held in place by whipping. Wire rope thimbles must not be used with fiber rope. An eye loop is formed by applying the procedure:

1. Unlay four turns of strand. Place a temporary whip on the standing part and whip the strand ends.
2. Form the eye of thimble size.
3. Tuck strand 1 through the standing part at 90° to the lay of the rope.
4. Tuck strand 2 through the standing part in the same direction.
5. Turn the assembly over and tuck strand 3 through standing part.
6. Alternately tuck each strand through the standing part. Trim ends.
7. Remove the temporary whipping. Insert thimble and add whipping.

Crowe Rope Industries LLC

Eye loops may be formed to attach hooks to the ends of a length of rope.

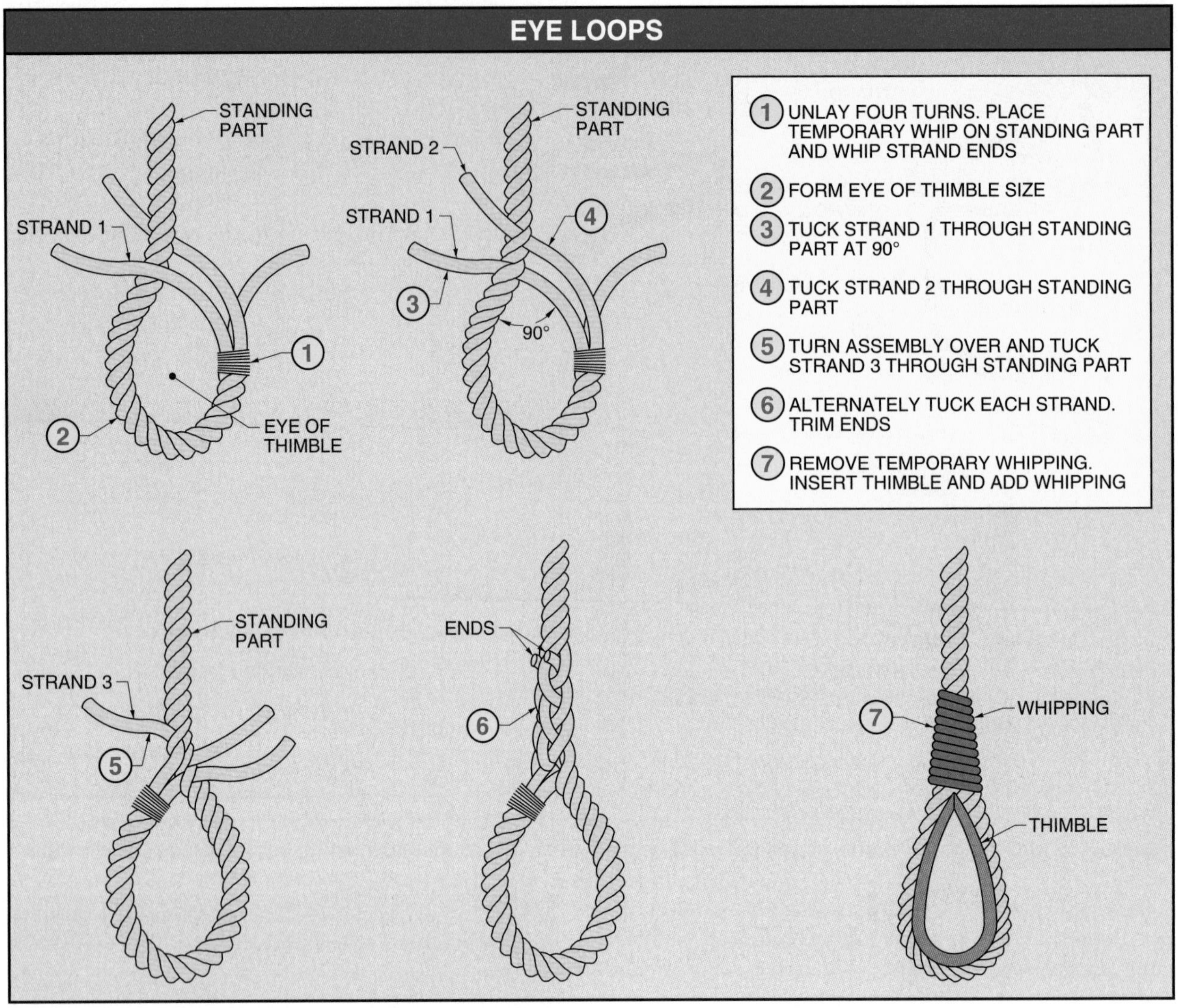
EYE LOOPS
STANDING PART
STRAND 1
2
1
EYE OF THIMBLE
STANDING PART
STRAND 2
STRAND 1
4
3
90°
1 UNLAY FOUR TURNS. PLACE TEMPORARY WHIP ON STANDING PART AND WHIP STRAND ENDS
2 FORM EYE OF THIMBLE SIZE
3 TUCK STRAND 1 THROUGH STANDING PART AT 90°
4 TUCK STRAND 2 THROUGH STANDING PART
5 TURN ASSEMBLY OVER AND TUCK STRAND 3 THROUGH STANDING PART
6 ALTERNATELY TUCK EACH STRAND. TRIM ENDS
7 REMOVE TEMPORARY WHIPPING. INSERT THIMBLE AND ADD WHIPPING
STANDING PART
STRAND 3
5
ENDS
6
7
WHIPPING
THIMBLE

Figure 2-27. An eye loop is a rope splice containing a thimble.

Knots and Hitches. A *knot* is the interlacing of rope to form a permanent connection. A *hitch* is the interlacing of rope to temporarily secure it without knotting the rope. Knots are designed to form a permanent connection that may be untied. Hitches are designed for quick release. Knots lose from 10% to 80% of the strength of a rope, depending on the knot used. A rope fails at the short bend in the knot if a rope fails under stress due to the presence of a knot.

Common rigging knots include the double hitch, slip, bowline, and wagoneer's hitch knots. Many knots are variations using the basic half hitch knot. A *double hitch knot* is a knot with two half hitch knots. A *half hitch knot* is a binding knot where the working end is laid over the standing part and stuck through the turn from the opposite side. The double turn allows for two gripping nips. A *nip* is a pressure and friction point created when a rope crosses over itself after a turn around an object. A nip is an essential ingredient of any knot because of the pressure and friction. See Figure 2-28. A half hitch knot is formed by applying the procedure:

1. Form a loop with the working end crossed over the standing part.
2. Tuck working end under and through loop.

The end of a rope that has been welded during torch cutting should be cut off prior to inserting into a socket so the individual wires and strands may slide and adjust when the rope is bent in the socket.

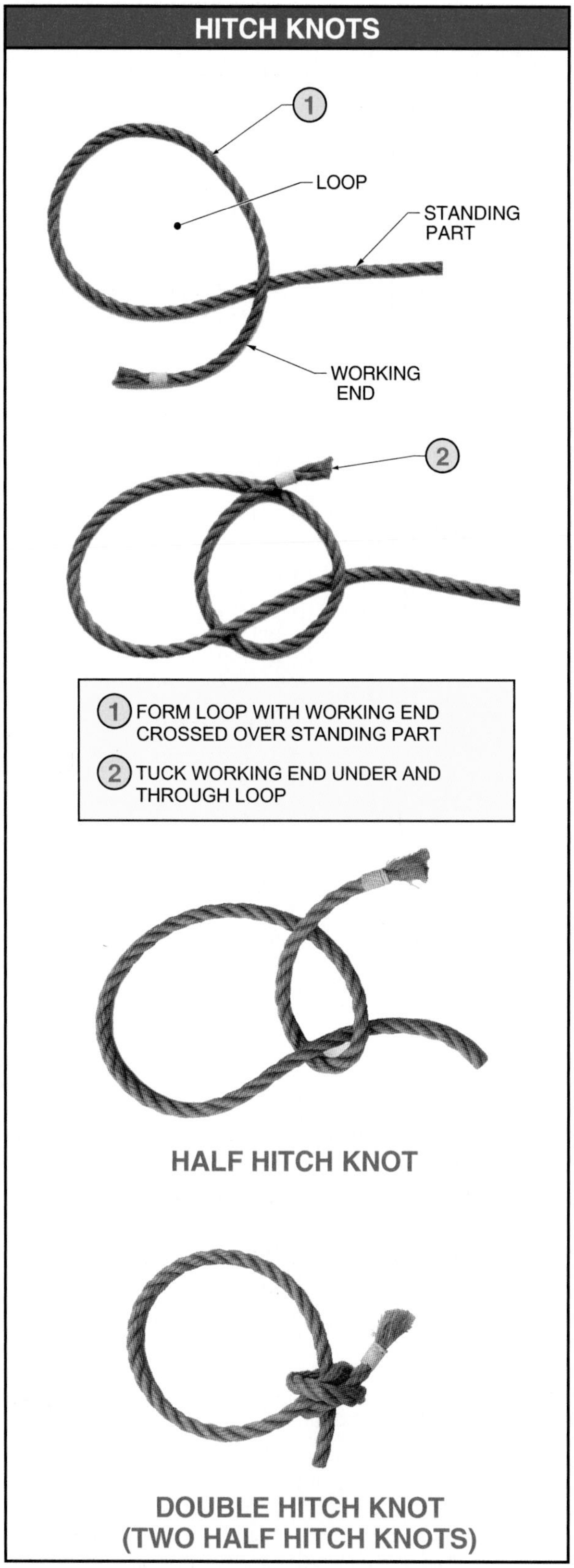
HITCH KNOTS
1
LOOP
STANDING PART
WORKING END
2
1 FORM LOOP WITH WORKING END CROSSED OVER STANDING PART
2 TUCK WORKING END UNDER AND THROUGH LOOP
HALF HITCH KNOT
DOUBLE HITCH KNOT (TWO HALF HITCH KNOTS)

Figure 2-28. Half hitch and double hitch knots are not secure knots, but are the base formation of other knots.

A *slip knot* is a knot that slips along the rope from which it is made. A slip knot forms a noose which, when placed around an object, is progressively tightened by strain on the standing part. See Figure 2-29. A slip knot is formed by applying the procedure:

1. Form a loop by placing the working end over the standing part.
2. Tuck working end under and through loop.
3. Pass standing part through loop.

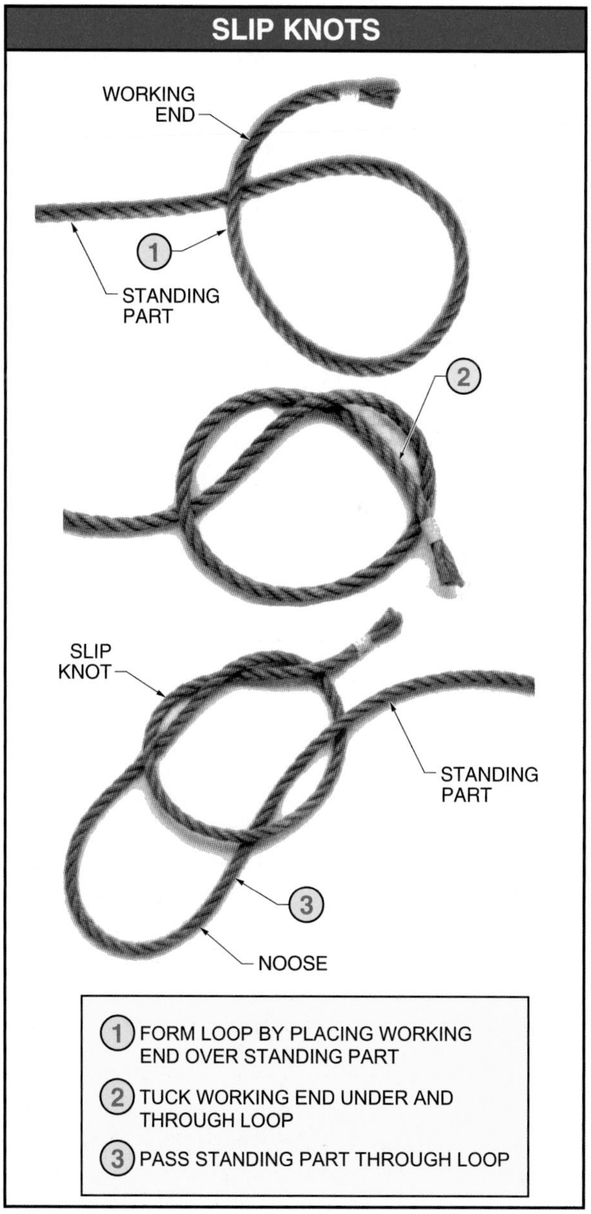
SLIP KNOTS
WORKING END
1
STANDING PART
2
SLIP KNOT
STANDING PART
3
NOOSE
1 FORM LOOP BY PLACING WORKING END OVER STANDING PART
2 TUCK WORKING END UNDER AND THROUGH LOOP
3 PASS STANDING PART THROUGH LOOP

Figure 2-29. A slip knot is a knot that slips along the rope from which it is made.

A *bowline knot* is a knot that forms a loop that is absolutely secure. See Figure 2-30. The more strain placed on the rope, the stronger the knot becomes. The knot is easily released when needed. A bowline knot is formed by applying the procedure:

1. Loop working part of rope over standing part. Allow enough rope to give the size loop required.
2. Thread the working end beneath and through the loop.
3. Pass the working end around the back side of the standing part.
4. Pass the working end back through the loop. Tighten by pulling the standing part and working end.

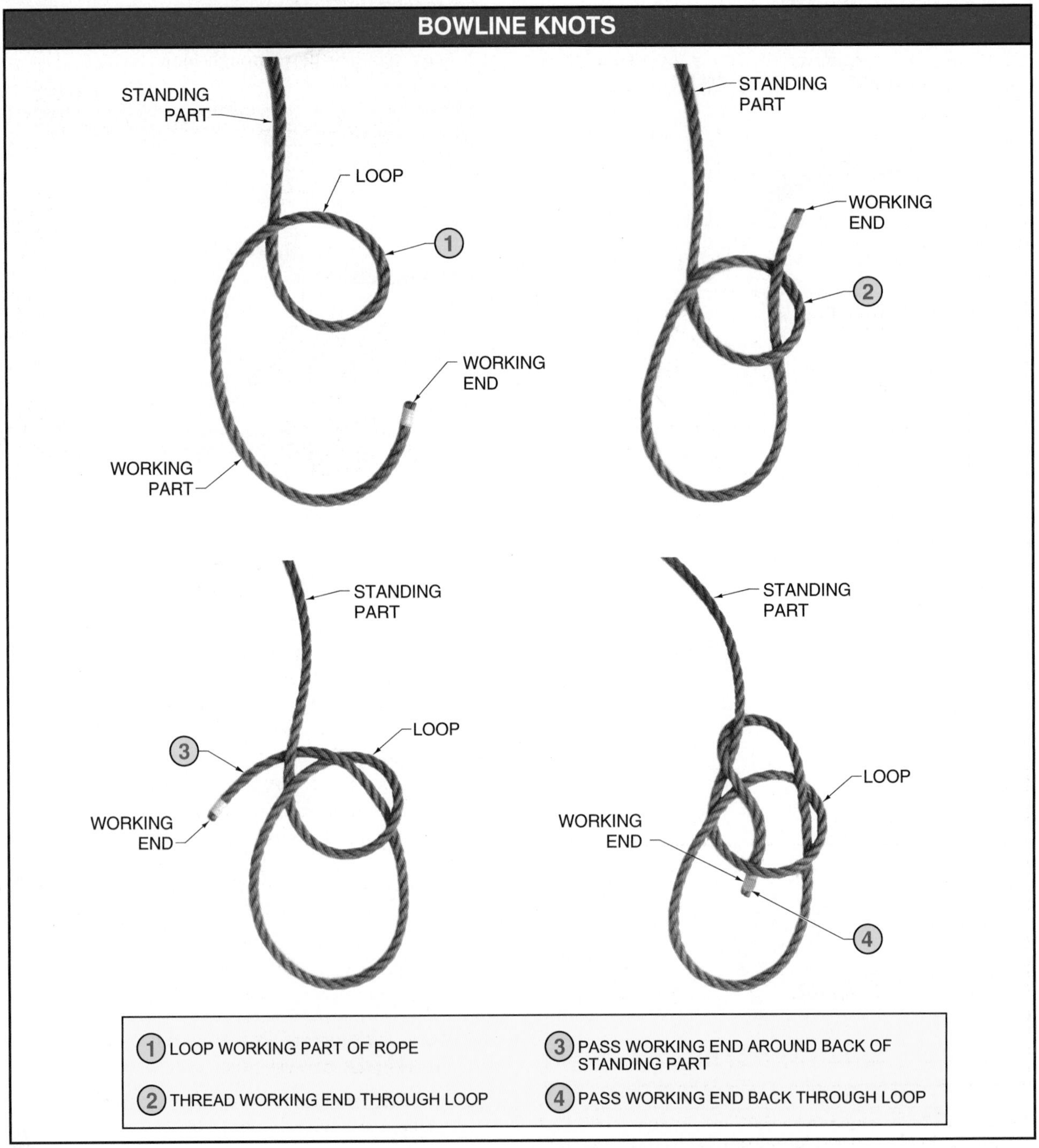

Figure 2-30. A bowline knot is a knot that forms a loop that is secure but is easy to release.

A *wagoneer's hitch knot* is a knot that creates a load-securing loop from the standing part of the rope. See Figure 2-31. A wagoneer's hitch knot is formed by applying the procedure:

1. Form a loop where the drawing loop is required by placing the working part on top of the standing part.
2. Form a second loop from the working part and insert into the first loop. Snug the knot assembly by pulling on the second loop and the standing part.
3. Bring the working end through the second loop after passing the working end through load-securing hooks or loops.
4. Pull the working end tight and into a half knot to secure the load.

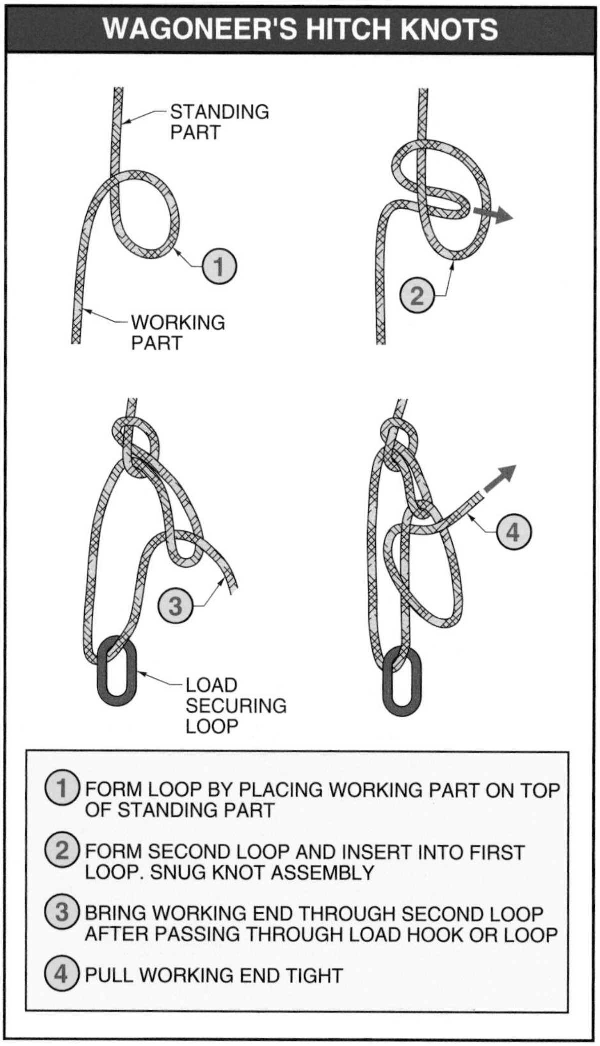
WAGONEER'S HITCH KNOTS
STANDING PART
WORKING PART
1
2
3
4
LOAD SECURING LOOP
1 FORM LOOP BY PLACING WORKING PART ON TOP OF STANDING PART
2 FORM SECOND LOOP AND INSERT INTO FIRST LOOP. SNUG KNOT ASSEMBLY
3 BRING WORKING END THROUGH SECOND LOOP AFTER PASSING THROUGH LOAD HOOK OR LOOP
4 PULL WORKING END TIGHT

Figure 2-31. A wagoneer's hitch knot is a knot that creates a load-securing loop from the standing part of the rope.

A clove hitch is a quick, simple method of fastening a rope around a post, pole, or stake.

Hitches work by the pressure of rope being pressed together. The standing part of the rope is nipped (jammed) over the working part. Because of the friction created by nipping, the greater the pull on the rope, the more tightly the standing part nips the working part and prevents it from slipping through. Hitches should never be formed with slippery rope or wire. Hitches created from rope are the timber hitch, clove hitch, cat's-paw hitch, cow hitch, scaffold hitch, and blackwall hitch.

A *timber hitch* is a binding knot and hitch combination used to wrap and drag lengthy material. A timber hitch may be used to wrap and drag logs, pipes, beams, etc. See Figure 2-32. A timber hitch is formed by applying the procedure:

1. Loop the working end around the standing part to form the binding knot.
2. Twist the working end in the direction of the lay of the rope three or four times.
3. Pull the working end tight to maintain form.

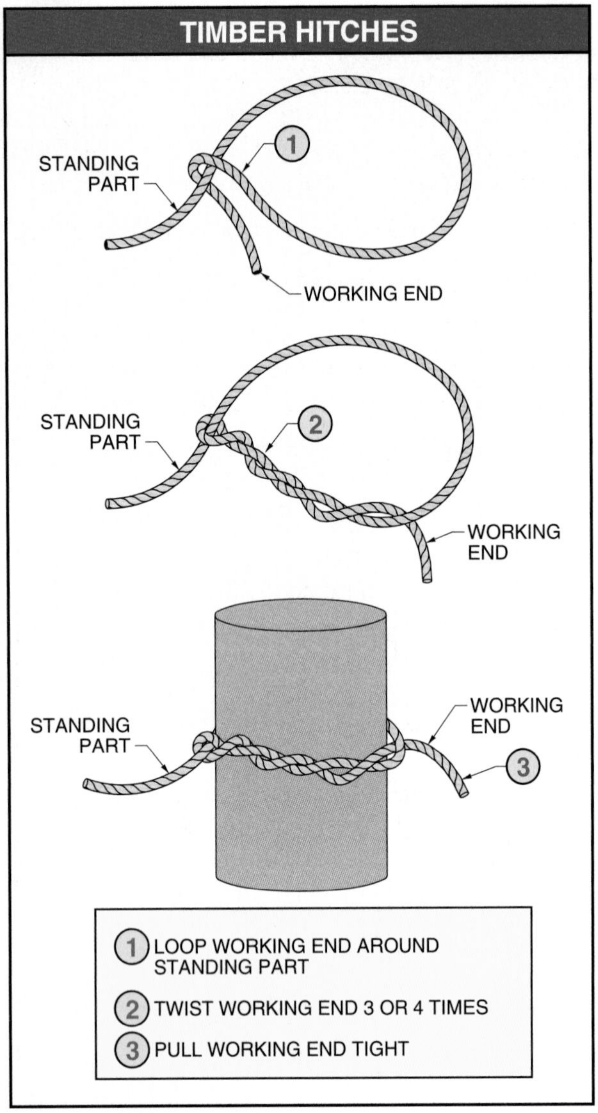
TIMBER HITCHES
STANDING PART
1
WORKING END
STANDING PART
2
WORKING END
STANDING PART
WORKING END
3
1 LOOP WORKING END AROUND STANDING PART
2 TWIST WORKING END 3 OR 4 TIMES
3 PULL WORKING END TIGHT

Figure 2-32. A timber hitch is a binding knot and hitch combination used to wrap and drag lengthy material.

A timber hitch, which does not jam and comes undone readily when the pull ceases, is used to tow or hoist cylindrical objects, such as logs, poles, etc.

A *clove hitch* is a quick hitch used to secure a rope temporarily to an object. A clove hitch is used because it is attached quickly, holds firmly, and has a rapid release. See Figure 2-33. A clove hitch is formed by applying the procedure:

1. Cross one hand over the other and grasp the rope with both hands.
2. Uncross the hands.
3. Bring the rope together to form two loops. Place over the object to be secured.

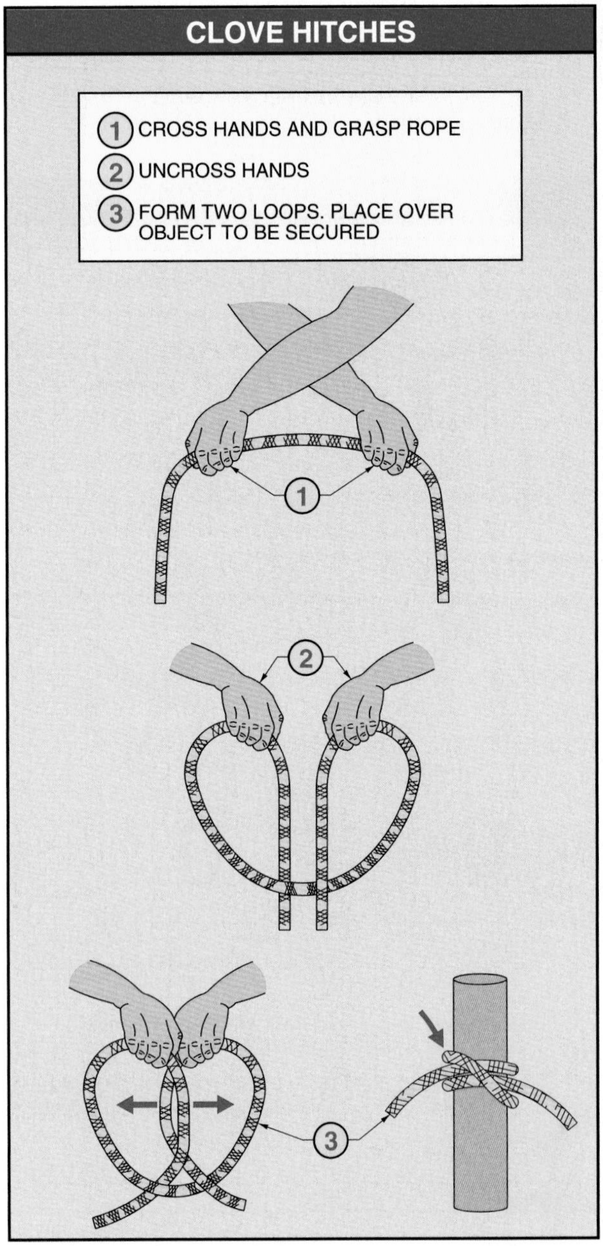
CLOVE HITCHES
1 CROSS HANDS AND GRASP ROPE
2 UNCROSS HANDS
3 FORM TWO LOOPS. PLACE OVER OBJECT TO BE SECURED
1
2
3

Figure 2-33. A clove hitch is a quickly-formed hitch used to secure a rope temporarily to an object.

A *cat's-paw hitch* is a hitch used as a light-duty, quickly-formed eye for a hoisting hook. See Figure 2-34. A cat's-paw hitch is formed by applying the procedure:

1. Grasp the rope with both hands, leaving plenty of bight.
2. Rotate both hands in the opposite direction and continue to rotate the two loops for two complete turns.
3. Place the eye over the end of a hook.

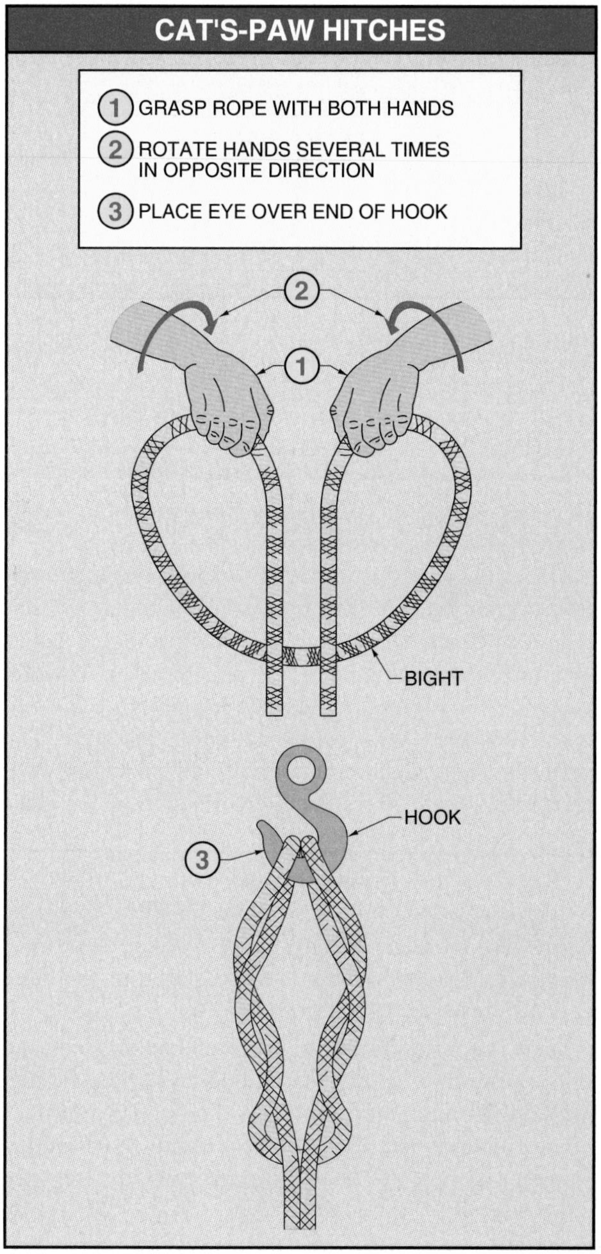

Figure 2-34. A cat's-paw hitch is a quickly-formed eye for light-duty lifting.

A *cow hitch* is a hitch used to secure a tag line to a load. A cow hitch is made and released easily but is firm enough to steady loads. See Figure 2-35. A *tag line* is a rope, handled by an individual, to control rotational movement of a load. A cow hitch is formed by applying the procedure:

1. Loop the line and pass the loop around the object.
2. Draw the rope through the loop.
3. Pull snug.

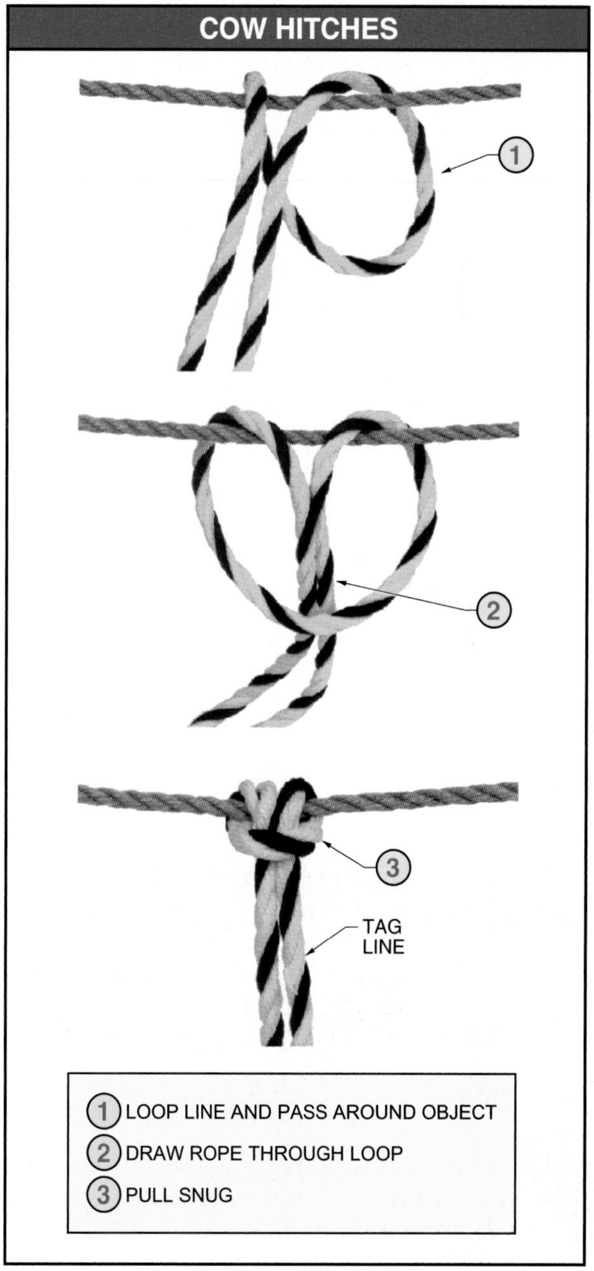

Figure 2-35. A cow hitch is a hitch used to secure a tag line to a load.

A *scaffold hitch* is a hitch used to hold or support planks or beams. See Figure 2-36. A scaffold hitch is made from a clove hitch and a bowline knot. A scaffold hitch is formed by applying the procedure:

1. Attach a clove hitch to the object.
2. Tie the working end to the standing part using a bowline knot.

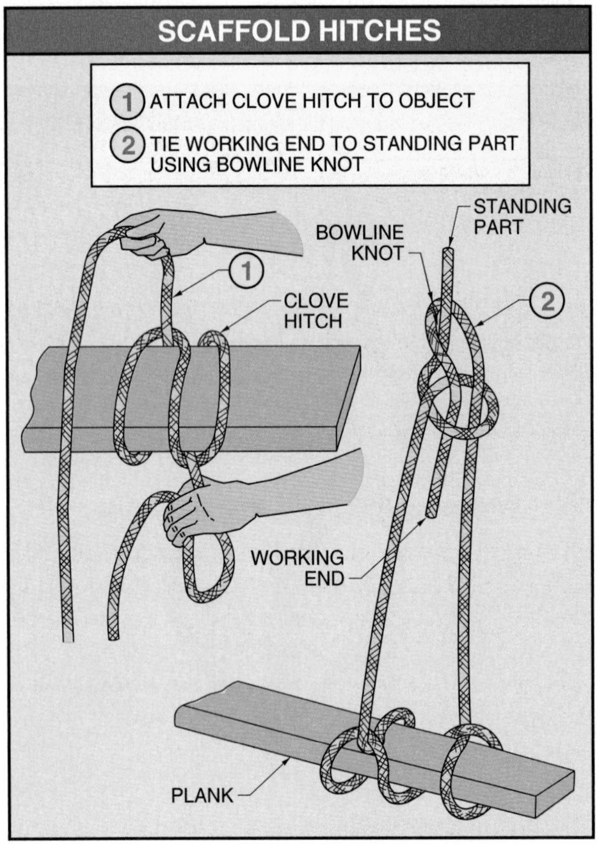

Figure 2-36. A scaffold hitch is used to support planks or beams.

A *blackwall hitch* is a hitch made for securing a rigging rope to a hoisting hook. See Figure 2-37. A blackwall hitch should be made from natural fiber ropes only because synthetic ropes may slip. A blackwall hitch is formed by applying the procedure:

1. Pass the working end twice around the shank of a hook.
2. Cross it under the standing part in the mouth of the hook.

Hands or fingers shall not be placed between the sling and its load while the sling is being tightened around the load.

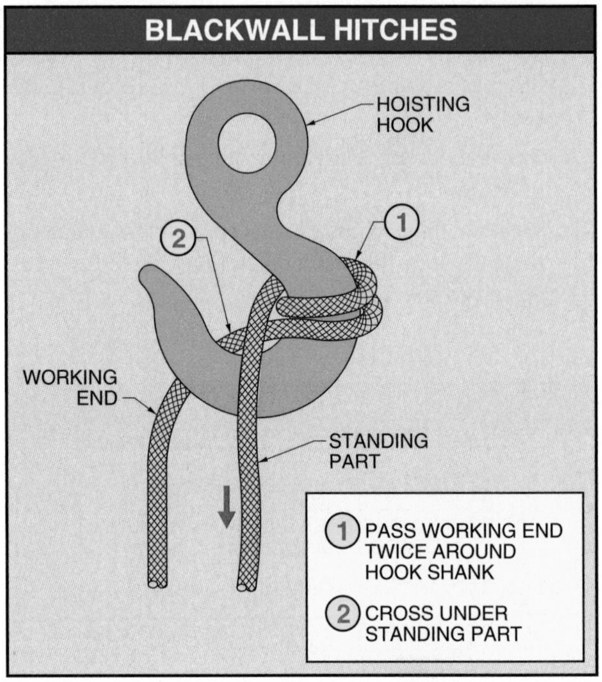

Figure 2-37. A blackwall hitch is a hitch made for securing a rigging rope to a hoisting hook.

WEBBING

Webbing is a fabric of high-tenacity synthetic yarns woven into flat narrow straps. See Figure 2-38. A *synthetic yarn* is yarn made of twisted, manufactured fibers such as nylon or polyester. Webbing is constructed in one to four plies, with protected edges and red warning cores to indicate wear or damage. Webbing for rigging purposes is normally used when maximum load damage protection is required. The softness of the webbing material, along with its wide and flat design, offers excellent protection for glass and polished or painted loads.

Webbing for rigging purposes is made of woven nylon or polyester with selvedges. *Selvedge* is a knitted or woven edge of a webbing formed to prevent raveling. Most web sling damage starts on the edge and progresses across the web face.

The thickness (number of plies) determines the duty rating of webbing. A *ply* is a layer of a formed material. The number of plies is the number of thicknesses of load-bearing webbing used in the sling assembly. Slings are available in one- to four-ply construction, with widths ranging from 1″ to 12″. Generally, web slings are constructed of one or two plies with three- or four-ply slings reserved for special conditions.

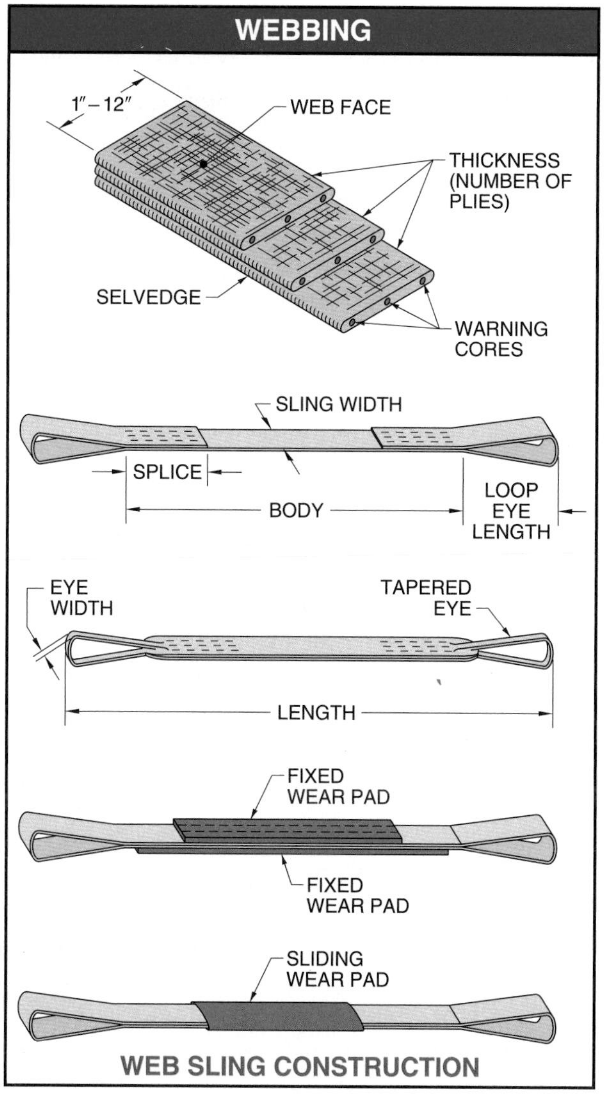

Figure 2-38. Webbing is a fabric of high-tenacity synthetic yarns woven into flat narrow straps.

In some cases, special colored yarns are woven into the webbing core to indicate excessive wear or cuts. The webbing must be removed from service when these warning cores of colored yarn are exposed.

Web sling construction consists of the length (reach), body, splice, and loop eye. The *web sling length* is the distance between the extreme points of a web sling, including any fittings. The *web sling body* is the part of the sling which is between the loop eyes or end fittings (if any). The *splice* of a web sling is the lapped and secured load-bearing part of a loop eye. The *loop eye* of a web sling is a length of webbing folded back and spliced to the sling body, forming an opening.

Lift-All Company, Inc.

Braided Tuflex roundslings from Lift-All are made from three (6-part) or four (8-part) individual Tuflex endless synthetic slings made from continuous loops of polyester yarn covered by a double wall tubular jacket.

Sling loop eyes may be too wide to properly fit into the bowl of a hoist hook. A tapered loop eye is formed by folding the webbing to a narrower width at its bearing point to accommodate the lifting device. Another common component of web slings is wear pads. A *wear pad* is a leather or webbed pad used to protect the web sling from damage. Wear pads are either sewn (fixed) or sliding for adjustable protection.

Basic web slings are assembled in various ways to be used as vertical, basket, or choker hitches. Web slings are fabricated in six configurations (Type I through Type VI). See Figure 2-39.

Type I is a web sling made with a triangle fitting on one end and a slotted triangle choker fitting on the other end. Type I web slings are used for vertical, basket, or choker hitches. Type I web slings are also used in cargo hold down situations where the webbing consists of a tightener and end fittings at each end. The strapping, being attached at two different hold down points, is then tightened over the cargo. Type II is a web sling made with a triangle fitting on both ends. Type II web slings are used for vertical or basket hitches. Type III is a web sling made with flat loops on each end with the loop eye openings in the same plane as the body sling. It is also known as an eye-to-eye sling. Type III web slings are used for basket hitch applications or as a choker hitch by passing one eye around the load and through the opposite eye. The eye of a Type III web sling is generally flat but is available tapered to permit use on crane hooks.

Type IV is a web sling made with both eyes twisted to form loop eyes which are at right angles to the plane of the body sling. It is also known as a twisted eye sling. Type IV web slings are used for choker hitch applications.

Type V is an endless web sling made by joining the ends with a load-bearing splice. It is also known as a grommet sling. Type V web slings are used for numerous applications and are the most widely used. Because they have an endless design, they may be used in a basket, vertical, or choker hitch application.

Type VI is a web sling sometimes made from a Type V sling by adding a wear pad the length of the sling body. The wear pad may be on one side or both sides of the sling body and is only long enough to form eye loops at each end, which are at right angles to the plane of the web body. It is also known as a reverse eye sling. Type VI web slings are used for rugged service, such as lifting irregularly-shaped objects such as stone, etc.

SLINGS

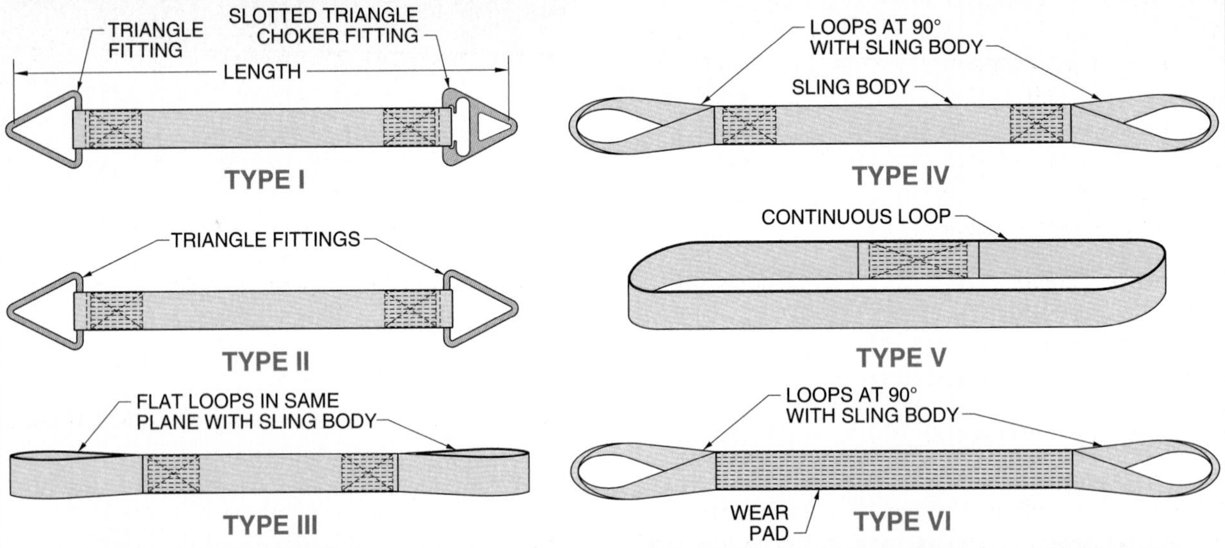

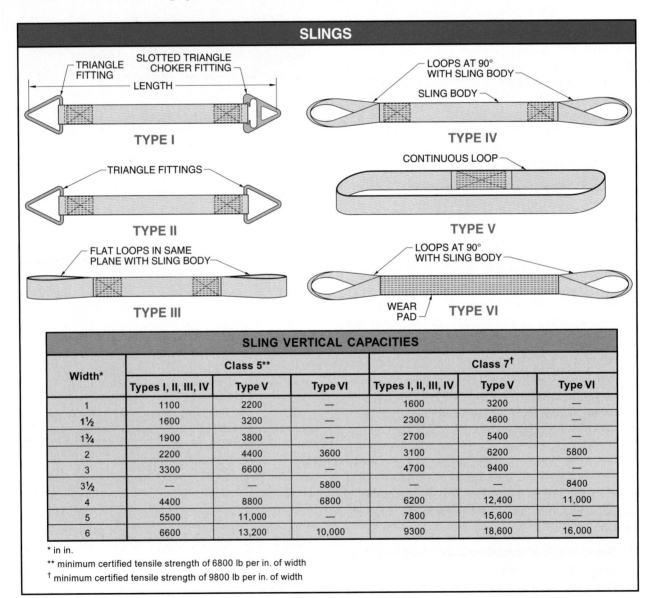

SLING VERTICAL CAPACITIES						
Width*	Class 5**			Class 7†		
	Types I, II, III, IV	Type V	Type VI	Types I, II, III, IV	Type V	Type VI
1	1100	2200	—	1600	3200	—
1½	1600	3200	—	2300	4600	—
1¾	1900	3800	—	2700	5400	—
2	2200	4400	3600	3100	6200	5800
3	3300	6600	—	4700	9400	—
3½	—	—	5800	—	—	8400
4	4400	8800	6800	6200	12,400	11,000
5	5500	11,000	—	7800	15,600	—
6	6600	13,200	10,000	9300	18,600	16,000

* in in.
** minimum certified tensile strength of 6800 lb per in. of width
† minimum certified tensile strength of 9800 lb per in. of width

Figure 2-39. Basic sling types are classified as Type I through Type VI.

Per OSHA 1926.251, employers shall have each synthetic web sling marked or coded to show the name or trademark of the manufacturer, rated capacities for the type of hitch, and type of material.

Webbing Strength

Sling vertical capacity tables give the minimum certified tensile strength for various classes, types, and number of plies in webbing. Webbing sling strength capacity is rated for one-ply or two-ply in Class 5 or Class 7. The specification of Class 5 or Class 7 is a specification for the manufacturer, with Class 7 being approximately 45% stronger than Class 5.

Webbing material strength is based on new webbing material. Factors that affect webbing strength include mishandling, environmental conditions, and ultraviolet light. Dragging webbing on the floor, tying in knots, pulling from under a load when the load is resting on the webbing, or dropping with metal fittings are all types of mishandling. Bunching of material between the ears of a clevis, shackle, or hook also weakens webbing strength.

Chemicals create environmental conditions affecting webbing strength. These conditions cause varying degrees of degradation. The proper webbing material must be used in chemically active areas. Polyester is resistant to many acids, but is still subject to some degradation. Nylon is resistant to many alkalis, but may be subject to moderate degradation. Webbing slings are not to be used if the end fittings are aluminum and alkalis or acid are present. Exposing synthetic webbing to ultraviolet light, such as sunlight or arc welding, affects the strength in varying degrees. The effects of ultraviolet degradation can occur without any visible indication.

The strength of webbing for vertical slings is used to determine strength capacities of choker, basket, and bridle sling hitches. Strength capacities are affected by the sling angle. Sling angle is measured from the horizontal to the point of attachment of multi-legged slings. The actual lifting capacity of a sling at a specific sling angle is found by applying the procedure:

1. Determine sling vertical capacity. (from Sling Vertical Capacities table)
2. Determine sling angle loss factor. (from Sling Angle Loss Factors table)
3. Calculate lifting capacity. Lifting capacity is found by applying the formula:

   $LC = vl \times l \times s$

   where

   LC = lifting capacity (in lb)

   vl = sling vertical capacity (in lb)

   l = sling leg(s)

   s = sling angle loss factor

For example, what is the lifting capacity of a basket hitch using a 1″ wide, Class 7, Type V endless sling without fittings having a 45° sling angle? See Figure 2-40.

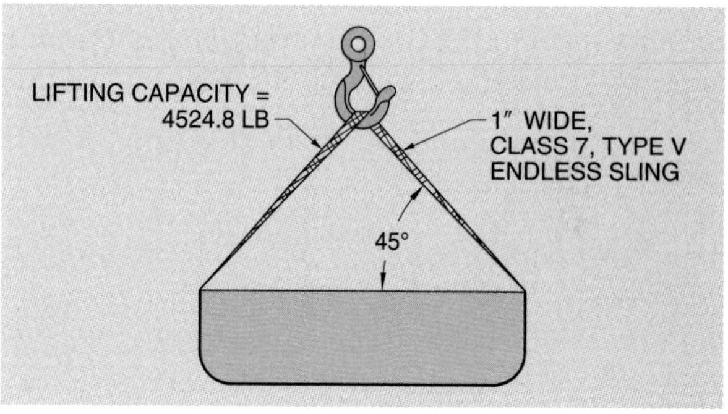

Figure 2-40. Sling load capacities vary based on the sling angles.

1. Determine sling vertical capacity.

   vl = 3200 lb (from Sling Vertical Capacities table)

2. Determine sling angle loss factor.

   s = .707 (from Sling Angle Loss Factors table)

3. Calculate lifting capacity.

   $LC = vl \times l \times s$

   $LC = 3200 \times 2 \times .707$

   LC = **4524.8 lb**

A choker hitch uses sling angle loss factors based on the angle measured from the vertical as the webbing passes through the choke eye. Choker hitch load capacity is based on sling vertical capacity, angle of choke, and the sling rated load factor. See Figure 2-41.

For example, what is the lifting capacity of a 2″, Class 5, Type V web sling choker hitch having a 110° angle at the point of choke?

1. Determine sling vertical capacity.

   vl = 4400 lb (from Sling Vertical Capacities table)

2. Determine sling angle loss factor.

   $s = .65$ (from Sling Angle Loss Factors table)

3. Calculate lifting capacity.

   $LC = vl \times l \times s$

   $LC = 4400 \times 1 \times .65$

   $LC =$ **2860 lb**

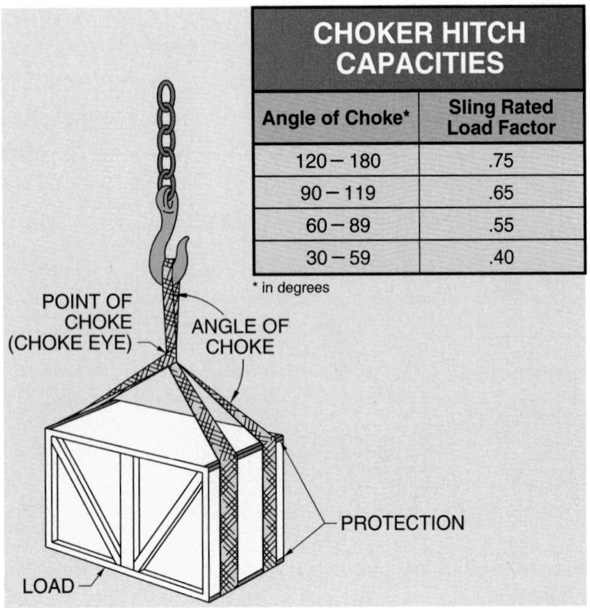

CHOKER HITCH CAPACITIES

Angle of Choke*	Sling Rated Load Factor
120 – 180	.75
90 – 119	.65
60 – 89	.55
30 – 59	.40

* in degrees

Figure 2-41. Choker hitch load capacity is based on the angle of choke as the sling body passes through the choke eye.

Basic Rigging with Web Slings

Consideration must be given to the kind of load, its weight, and its center of gravity when selecting a web sling. The type of web sling selected and its use must be made with safety as the main consideration. For example, basket and choker hitches are used with web slings. See Figure 2-42.

ROUND (TUBULAR) SLINGS

Round (tubular) slings make excellent choker slings because they are extremely flexible and conform to the shape of the load. Also, due to their construction, choker hitches do not bind or lock up, making sling release simple. The strength capacity of round slings is identified by color coding, which is an added safety feature.

Round Sling Construction

A *round sling* is a sling consisting of one or more continuous polyester fiber yarns wound together to make a core. Core yarns are wound uniformly to ensure even load-bearing distribution. A polyester round (tubular) sling is made of a core of continuous yarn, not woven, enclosed in a protective cover. See Figure 2-43.

The use of a round sling is similar to that of a Type I endless web sling when used as a choker or basket hitch. Round slings are also manufactured with fittings or coupling components. Additional sleeves are placed over the round sling to offer extra abrasion protection or to create a loop eye at each end (eye-and-eye design).

Bridle slings are assembled during manufacture by including more than one round sling leg to a master link. The tubular cover is color-coded to correspond with the rated strength capacity of the round sling. Each polyester round sling has rated capacities for use with vertical, choker, vertical basket, and 45° basket slings. The colors of round slings from smallest to largest are purple, green, yellow, tan, red, white, blue, and orange.

Round Sling Strength

A round sling is not to be used with a load greater than that marked on its identification tag. Identification tags generally carry the rated capacities for vertical, choker, and vertical basket slings. Remove any round sling from service that has a missing or unreadable identification tag. Capacities are rated similar to web slings.

For example, what is the lifting capacity of a round sling basket hitch with a green cover and 55° sling angle?

1. Determine sling vertical capacity.

   $vl = 10{,}600$ lb (from the Round Sling Color and Capacity Rating table)

2. Determine sling angle loss factor.

   $s = .819$ (from Sling Angle Loss Factors table)

3. Calculate lifting capacity.

   $LC = vl \times l \times s$

   $LC = 10{,}600 \times 2 \times .819$

   $LC =$ **17,362.8 lb**

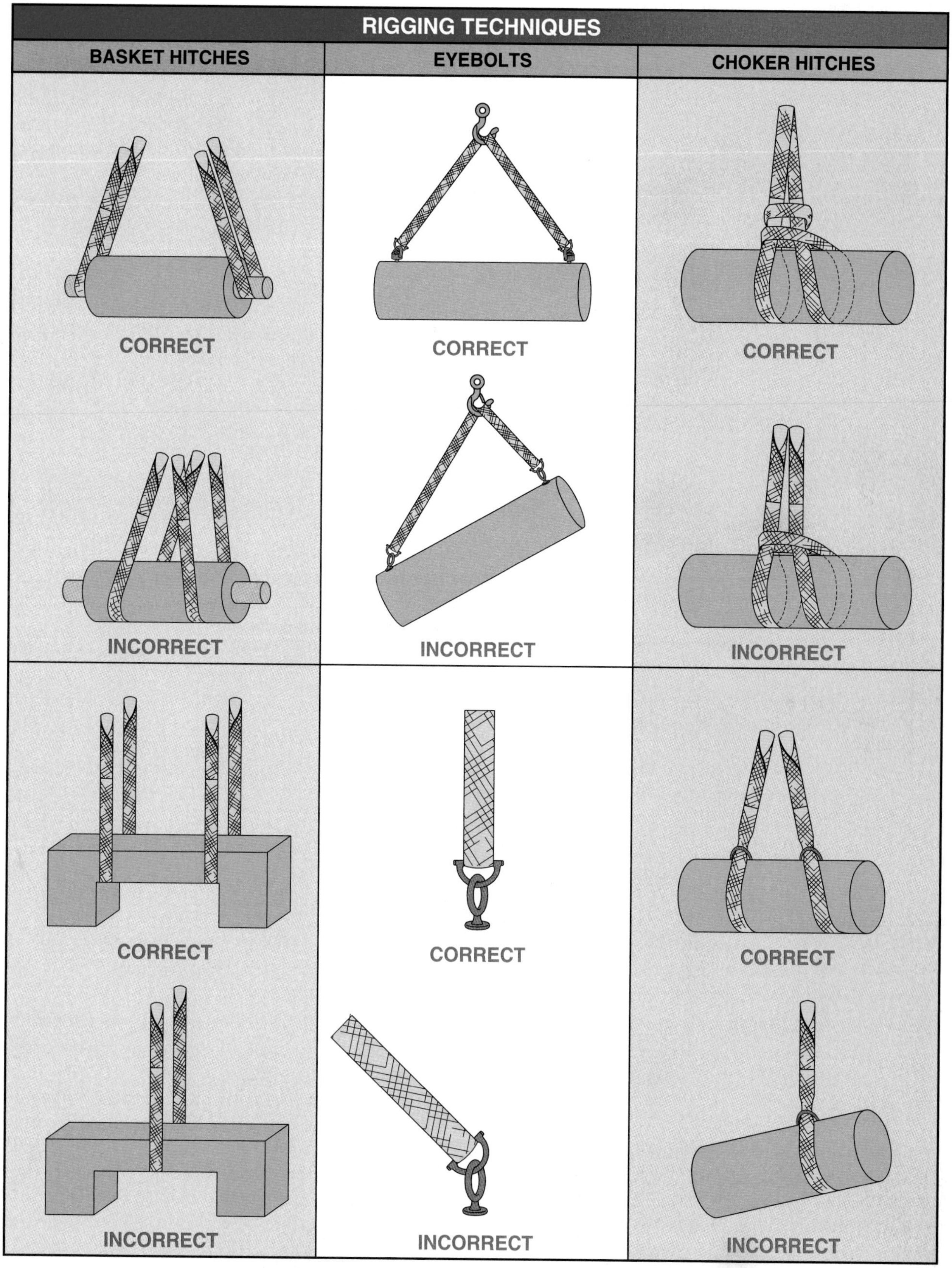

Figure 2-42. The type of web sling selected and its use must be made with safety as the main consideration.

ROUND SLINGS

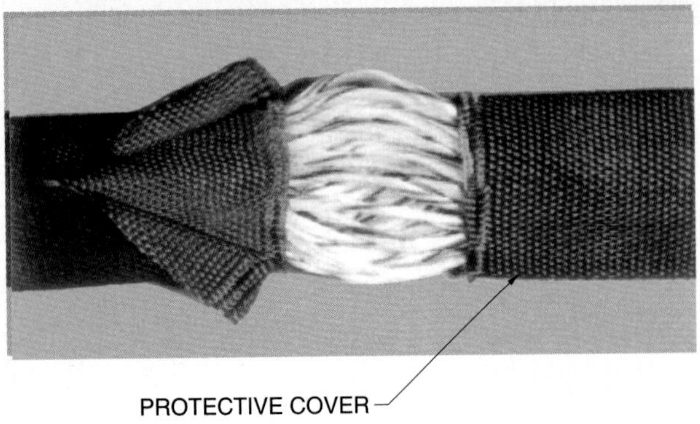

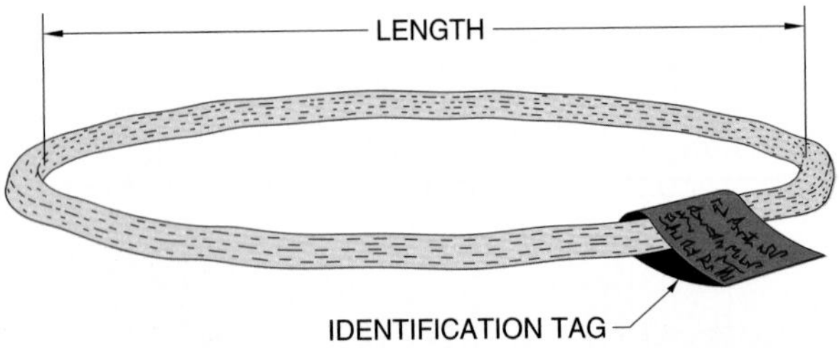

Lift-All Company, Inc.

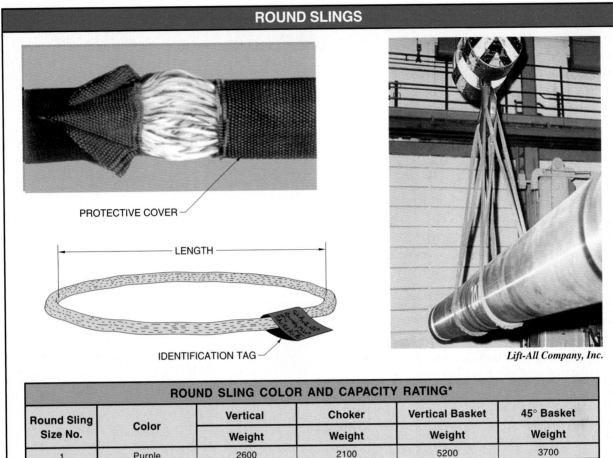

ROUND SLING COLOR AND CAPACITY RATING*					
Round Sling Size No.	Color	Vertical	Choker	Vertical Basket	45° Basket
		Weight	Weight	Weight	Weight
1	Purple	2600	2100	5200	3700
2	Green	5300	4200	10,600	7500
3	Yellow	8400	6700	16,800	11,900
4	Tan	10,600	8500	21,200	15,000
5	Red	13,200	10,600	26,400	18,700
6	White	16,800	13,400	33,600	23,800
7	Blue	21,200	17,000	42,400	30,000
8	Orange	25,000	20,000	50,000	35,400
9	Orange	31,000	24,800	62,000	43,800
10	Orange	40,000	32,000	80,000	56,600
11	Orange	53,000	42,400	106,000	74,900
12	Orange	66,000	52,800	132,000	93,000

* in lb

Figure 2-43. Round slings are slings consisting of one or more continuous polyester fiber yarns wound together to make a core.

CHAIN

A *chain* is a series of metal rings connected to one another and used for support, restraint, or transmission of mechanical power. Chain is used in situations in which other materials would be damaged by the load or environment, such as rough or raw castings or high-temperature processes.

The use of chain for rigging is normally favored over wire rope because chain has approximately three times the impact-absorption capability of wire rope and is more flexible. Also, wire rope costs more than chain of similar strength and the life of wire rope is only 5% of chain life. The chain used for rigging or hoisting is considered a special chain because specific chain steel must be used in its manufacture.

Chain Construction

Chain and chain attachment strength depends on the composition of the steel from which they are made and the heat treatment process. The material and the manufacturing process determines the tensile, shear, and bending strength of chain.

Tensile strength is a measure of the greatest amount of straight-pull stress metal can bear without tearing apart. *Shear strength* is a metal's resistance to a force applied parallel to its contacted plane. *Bending strength* is a metal's resistance to bending or deflection in the direction in which the load is applied. The temperature used in the chain manufacturing process determines the metal's hardness. Steel becomes very hard and brittle when heated to a red color and quenched (dipped) in water. As the metal is heated, the grain (carbon structure) within the metal becomes unstructured (without a pattern). The structure is stilled (frozen) in an unstructured position when the metal is quenched.

Tempering allows the carbon structure to be held in more structured patterns. *Tempering* is the process in which metal is brought to a temperature below its critical temperature and allowed to cool slowly. Slow cooling methods include no quenching (air), quenching in oil, or quenching in salt for a very slow cool. Tempering of chain metal determines whether a chain gives or flexes under pressure or breaks with a snap of the chain.

Steel hardness is measured with a Brinell hardness tester and given a Brinell number ranging from approximately 150 for soft metal to 750 for hardened metal. A typical sling chain varies from 250 to 450 Brinell depending on the manufacturer. This is equivalent to a material tensile strength of 125,000 lb/sq in. to 230,000 lb/sq in.

Steel used for rigging or hoisting chain is composed of premium quality, heat-treated, high-strength materials. This steel alloy contains .35% carbon maximum, .035% phosphorus maximum, and .040% sulfur maximum. A *steel alloy* is metallic material formulated from the fusing or combining of two or more metals.

Alloy material is chosen for sling chain to reduce the occurrence of fracturing of the metal. A *fracture* is a small crack in metal caused by the stress or fatigue of repeated pulling or bending forces. Through the combination of alloy metals and tempering, sling chain is capable of 15% to 30% elongation before breaking. This is a safety specification requirement for sling chain and is not to be used as a determination for replacement. Chain that has exceeded 1½% elongation from new should be removed from service. Any number of links may be used for measuring elongation, but the amount of used links should equal the number of new links. The chain should be taut during measurement. See Figure 2-44.

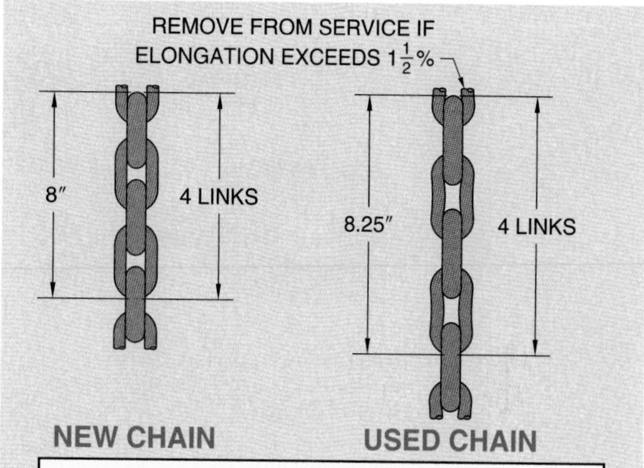

Four links of a new chain measure 8″. After use, four links measure 8.25″. Should chain be removed from service?

8″ x .015 = .12″

8″ + .12 = 8.12″

8.25″ > 8.12″

Yes, remove used chain from service.

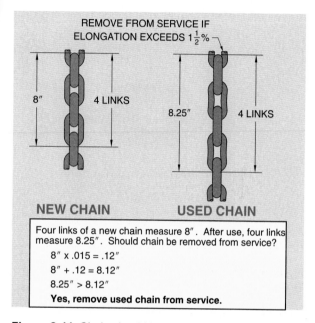

Figure 2-44. Chain should be removed from service if the measurement of used chain exceeds 1½% elongation from that of new chain.

The National Association of Chain Manufacturers (NACM), in conjunction with the International Organization for Standardization (ISO), develops programs to standardize materials and processes for chain. The four main types of chain are binding chain, rigging chain, lifting chain, and hoist apparatus chain. Each chain, except for hoist apparatus chain, has a periodic embossing of a grade number or letter, indicating its capability. NACM specifies that identification must appear at least once every 36 links. See Figure 2-45. Each chain is classified as follows:

- Grade 43 – High-test steel chain having a carbon content of approximately .15% to .22%. A ½″ grade 43 chain is rated as having a working load limit (WLL) of 9200 lb. The *working load limit (WLL)* is the maximum pull that should be applied to a vertical load. This chain is generally used for binding loads or tie downs and is embossed with an HT, $, $3, or M. Carbon steel chain is not to be used in overhead lifting.

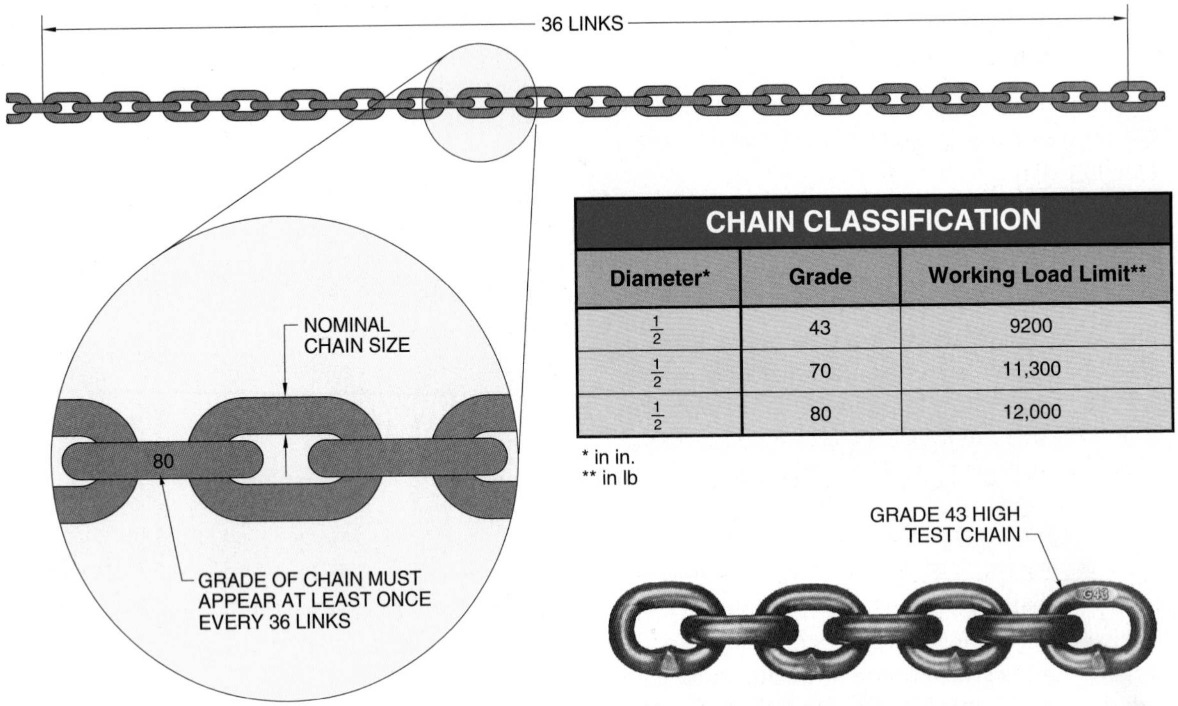

CHAIN CLASSIFICATION		
Diameter*	Grade	Working Load Limit**
$\frac{1}{2}$	43	9200
$\frac{1}{2}$	70	11,300
$\frac{1}{2}$	80	12,000

* in in.
** in lb

Columbus McKinnon Corporation Industrial Products Division

Figure 2-45. Each chain, except for hoist apparatus chain, has a periodic embossing of a grade number or letter, indicating its capability.

- Grade 70 – High-strength transport binding chain having a carbon content of about .25% to .30%. A ½″ grade 70 chain is rated as having a working load limit of 11,300 lb. Grade 70 chain is generally used for critical load-securing applications such as binding and tie down on log and steel transport trucks and heavy equipment towing. This chain is embossed with a 7 or 70. Carbon steel chain is not to be used in overhead lifting.
- Grade 80 – Alloy-steel chain is manufactured as a special-analysis alloy steel to provide superior strength, wear resistance, and durability. A ½″ grade 80 chain is rated as having a working load limit of 12,000 lb. See Figure 2-46. Grade 80 chain is used for all types of lifting slings, overhead lifting, or wherever maximum safety and strength is required.

Hoisting apparatus chain does not have a grade number because its physical measurements are more important than its strength. *Hoisting apparatus chain* is a precisely-measured chain calibrated to function in pocket wheels used in manual or powered chain hoists. Hoisting apparatus chain is not to be used in sling or lifting applications. Hoisting apparatus chain is designated as hoist load chain and is not used in sling applications. Hoist load chain is a precisely-measured chain designed to function in hand, air, or electric-powered hoist pockets.

In addition to the steel composition and heat treatment process, chain strength and working load limits are based on chain size. Nominal chain size is designated by the diameter of the material used to form the links.

Rigging Chain Strength

The working load capacity of chain is somewhat different from other slings because calculations are made using 1, 2, or 3 legs as a factor. A 4-leg factor is not used because a 4-leg (quad-branch chain sling), does not normally sustain the load evenly on each of its four legs. The maximum working load limits of a 4-leg chain are calculated as a 3-leg chain.

For example, what is the lifting capacity of a ¾″, 2-leg chain sling with a 30° sling angle?

1. Determine vertical capacity.
   vl = 28,300 lb (from Grade 80 Chain Load Limits table)

2. Determine sling angle loss factor.

   $s = .500$ (from Sling Angle Loss Factors table)

3. Calculate lifting capacity.

   $LC = vl \times l \times s$

   $LC = 28{,}300 \times 2 \times .500$

   $LC = \mathbf{28{,}300\ lb}$

Rigging Chain Attachments

Typical connecting attachments between rigging and the hoisting device include shackles, master links, and hooks. A *shackle* is a U-shaped metal link with the ends drilled to receive a pin or bolt. The removal of the pin or bolt allows an opening for one or more loop eyes that can be attached to complete a sling. Shackles are made in a straight-U design (chain shackle) or a curved-U design (anchor shackle). A shackle may be used to make the connection between the rigging assembly and the hoisting hook. Shackle strength varies to conform to load weight. See Figure 2-47.

Always inspect the point where a load contacts a hook to ensure the load is properly seated within the throat opening. Never force or hammer hooks or chain into position.

CooperTools

Care should be taken to select the type, grade, and size recommended by the manufacturer where attachments such as rings or hooks are designed for use with chain in sustaining loads.

GRADE 80 CHAIN LOAD LIMITS*

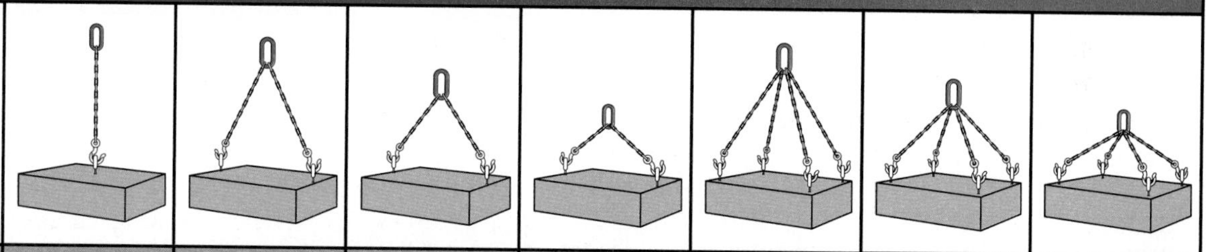

Chain Size**	90° Vertical Load	60° Vertical Load	45° Vertical Load	30° Vertical Load	60° Quad Leg Load	45° Quad Leg Load	30° Quad Leg Load
7/32	2100	3600	3000	2100	5450	4450	3150
9/32	3500	6100	4900	3500	9100	7400	5200
3/8	7100	12,300	10,000	7100	18,400	15,100	10,600
1/2	12,000	20,800	17,000	12,000	31,200	25,500	18,000
5/8	18,100	31,300	25,600	18,100	47,000	38,400	27,100
3/4	28,300	49,000	40,000	28,300	73,500	60,000	42,400
7/8	34,200	59,200	48,400	34,200	88,900	72,500	51,300
1	47,700	82,600	67,400	47,700	123,900	101,200	71,500
1 1/4	72,300	125,200	102,200	72,300	187,800	153,400	108,400

* in lb
** in in.

Figure 2-46. Working load limits for slings using Grade 80 chain can be determined for a 90° vertical load or quad leg load up to a 30° pull angle.

SHACKLES

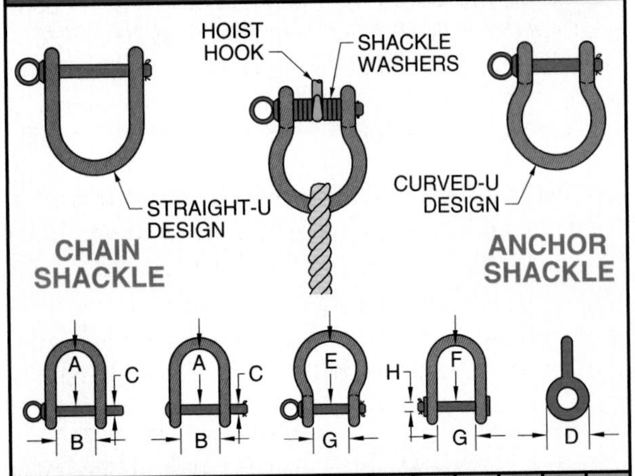

Working Load Limit*	Nominal Shackle Size**	A	B	C	D	E	F	G	H
$\frac{1}{2}$	$\frac{1}{4}$	$\frac{7}{8}$	$\frac{15}{32}$	$\frac{5}{16}$	$\frac{11}{16}$	—	—	—	—
$\frac{3}{4}$	$\frac{5}{16}$	$1\frac{1}{32}$	$\frac{17}{32}$	$\frac{3}{8}$	$\frac{13}{16}$	—	—	—	—
1	$\frac{3}{8}$	$1\frac{1}{4}$	$\frac{21}{32}$	$\frac{7}{16}$	$\frac{31}{32}$	—	—	—	—
$1\frac{1}{2}$	$\frac{7}{16}$	$1\frac{7}{16}$	$\frac{23}{32}$	$\frac{1}{2}$	$1\frac{1}{16}$	—	—	—	—
2	$\frac{1}{2}$	$1\frac{5}{8}$	$\frac{13}{16}$	$\frac{5}{8}$	$1\frac{3}{16}$	$1\frac{7}{8}$	$1\frac{5}{8}$	$\frac{13}{16}$	$\frac{5}{8}$
$3\frac{1}{4}$	$\frac{5}{8}$	2	$1\frac{1}{16}$	$\frac{3}{4}$	$1\frac{9}{16}$	$2\frac{3}{8}$	2	$1\frac{1}{16}$	$\frac{3}{4}$
$4\frac{3}{4}$	$\frac{3}{4}$	$2\frac{3}{8}$	$1\frac{1}{4}$	$\frac{7}{8}$	$1\frac{7}{8}$	$2\frac{13}{16}$	$2\frac{3}{8}$	$1\frac{1}{4}$	$\frac{7}{8}$
$6\frac{1}{2}$	$\frac{7}{8}$	$2\frac{13}{16}$	$1\frac{7}{16}$	1	$2\frac{1}{8}$	$3\frac{5}{16}$	$2\frac{13}{16}$	$1\frac{7}{16}$	1
$8\frac{1}{2}$	1	$3\frac{3}{16}$	$1\frac{11}{16}$	$1\frac{1}{8}$	$2\frac{3}{8}$	$3\frac{3}{4}$	$3\frac{3}{16}$	$1\frac{11}{16}$	$1\frac{1}{8}$
$9\frac{1}{2}$	$1\frac{1}{8}$	$3\frac{9}{16}$	$1\frac{13}{16}$	$1\frac{1}{4}$	$2\frac{5}{8}$	$4\frac{1}{4}$	$3\frac{9}{16}$	$1\frac{13}{16}$	$1\frac{1}{4}$
12	$1\frac{1}{4}$	$3\frac{15}{16}$	$2\frac{1}{32}$	$1\frac{3}{8}$	3	$4\frac{11}{16}$	$3\frac{15}{16}$	$2\frac{1}{32}$	$1\frac{3}{8}$
$13\frac{1}{2}$	$1\frac{3}{8}$	$4\frac{3}{8}$	$2\frac{1}{4}$	$1\frac{1}{2}$	$3\frac{5}{16}$	$5\frac{3}{16}$	$4\frac{3}{8}$	$2\frac{1}{4}$	$1\frac{1}{2}$
17	$1\frac{1}{2}$	$4\frac{13}{16}$	$2\frac{3}{8}$	$1\frac{5}{8}$	$3\frac{5}{8}$	$5\frac{3}{4}$	$4\frac{13}{16}$	$2\frac{3}{8}$	$1\frac{5}{8}$
25	$1\frac{3}{4}$	$5\frac{3}{4}$	$2\frac{7}{8}$	2	$4\frac{1}{8}$	7	$5\frac{3}{4}$	$2\frac{7}{8}$	2
35	2	$6\frac{3}{4}$	$3\frac{1}{4}$	$2\frac{1}{4}$	5	$7\frac{3}{4}$	$6\frac{3}{4}$	$3\frac{1}{4}$	$2\frac{1}{4}$

* in tons
** in in.

Figure 2-47. A shackle is a U-shaped metal link with the ends drilled to receive a pin or bolt.

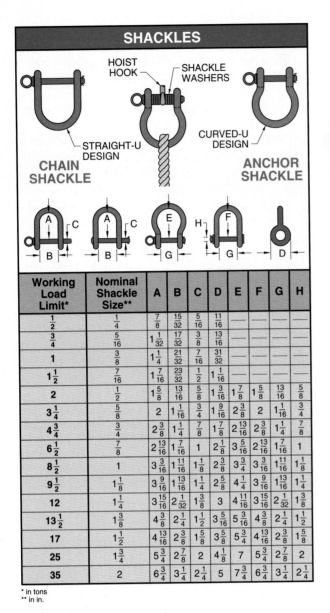

A chain shackle is used as a connector for a single lifting device such as a one loop eye for a vertical hitch. A chain shackle normally uses shackle washers as spacers to prevent side shifting. The anchor shackle uses a rounded eye to allow for more than one lifting device.

Like chain links, shackle strength is determined by its steel composition, heat treatment process, and rod diameter. Rigging shackles are made of alloy steel and are available in various strength capacities and rod sizes.

In most cases, chain slings are assembled by the manufacturer, but when a chain sling is required to be assembled, various attachments, such as a master link, connecting link, chain, and hook may be purchased separately and individually attached. A *master link* is a chain attachment with a ring considerably larger than that of the chain to allow for the insertion of a hook. See Figure 2-48. Master links are also large enough to incorporate more than one connecting link for creating extra-legged slings. Master links may be the connection between the hoisting hook and the rigged load. Master link capacity must be equal to or greater than that of any rigging components.

MASTER LINKS

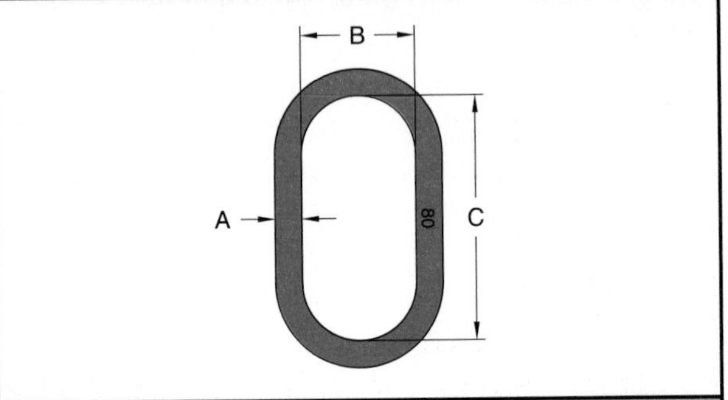

Link Size*			Size of Chain Sling on Which Used			
Diameter Material A	Inside Width B	Inside Length C	Single	Double	Triple	Quad
$\frac{13}{32}$	$1\frac{1}{2}$	3	$\frac{7}{32}$	$\frac{7}{32}$	—	—
$\frac{1}{2}$	$2\frac{1}{2}$	5	$\frac{9}{32}$	$\frac{9}{32}$	$\frac{7}{32}$	$\frac{7}{32}$
$\frac{3}{4}$	3	6	$\frac{3}{8}$	$\frac{3}{8}$	$\frac{9}{32}$	$\frac{9}{32}$
1	4	8	$\frac{1}{2}$ OR $\frac{5}{8}$	$\frac{1}{2}$	$\frac{3}{8}$	$\frac{3}{8}$
$1\frac{1}{4}$	$4\frac{3}{8}$	$8\frac{3}{4}$	$\frac{3}{4}$	$\frac{5}{8}$	$\frac{1}{2}$	$\frac{1}{2}$
$1\frac{1}{2}$	$5\frac{1}{4}$	$10\frac{1}{2}$	$\frac{7}{8}$	$\frac{3}{4}$	$\frac{5}{8}$	$\frac{5}{8}$
$1\frac{3}{4}$	6	12	1	$\frac{7}{8}$	$\frac{3}{4}$	$\frac{3}{4}$
2	7	14	$1\frac{1}{4}$	1	$\frac{7}{8}$	$\frac{7}{8}$
$2\frac{1}{4}$	8	16	—	$1\frac{1}{4}$	1	1
$2\frac{3}{4}$	9	16	—	—	$1\frac{1}{4}$	$1\frac{1}{4}$

* in in.

Figure 2-48. A master link is a chain attachment with a ring considerably larger than that of the chain to allow for the insertion of a hook.

A *connecting link* is a three-part chain attachment used to assemble and connect the master link to the chain. In either case, sling attachments should have a working load limit equal to or greater than that of the alloy steel chain. For example, a ½″ Grade 80 alloy steel chain is rated as having a 12,000 lb WLL. However, a ½″ master link is rated as having a 4920 lb WLL and a ¾″ master link is rated at 10,320 lb WLL. Therefore, a 1″ master link with a 24,360 lb WLL must be used in a ½″ chain sling assembly if it is to be of equal or greater strength.

The WLL of the weakest component should be the working load limit of the entire sling assembly if a sling is required to be assembled using attachments with different working load limits. Do not use makeshift hooks, links, or fasteners fashioned from bolts, rods, or other materials. The hook is a major link between the rigged load and the hoisting equipment. A *hook* is a curved or bent implement for holding, pulling, or connecting another implement. See Figure 2-49.

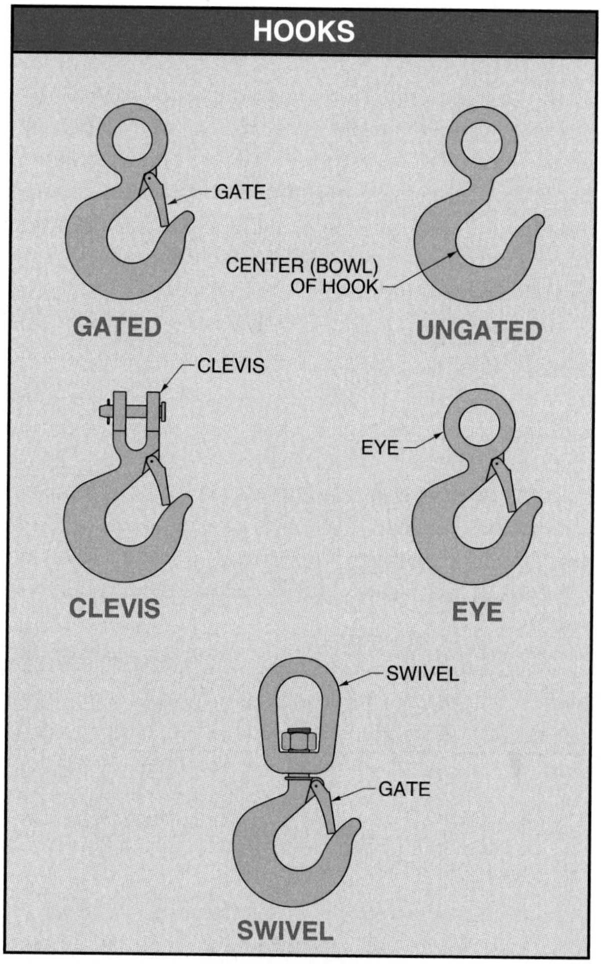

Figure 2-49. A hook is a curved or bent implement for holding, pulling, or connecting another implement.

Hooks are made in various shapes, designs, and sizes and are normally forged of alloy steel. Connecting (load) hooks are quenched and tempered so they are the first to bend or give in an overload situation. The hook should be the weakest member of the hoisting equipment.

Bend (deformation) is a permanent movement of the hook attempting to straighten. Hook deformation becomes the measurable part of rigging overload. For this reason, hooks should never be heated above 800°F or become part of a welding situation. Heat applied to a hook can temper the metal, reducing its strength.

Hooks may be gated or ungated, or contain a clevis, eye, or swivel. Some hooks are equipped with latches or gates to prevent the sudden release of shifted sling legs.

Hoisting hooks used for rigging purposes include choker, grab, foundry, swivel, and sorting hooks. See Figure 2-50. A *hoisting hook* is a steel alloy hook used for overhead lifting and is connected directly to the piece being lifted. See Figure 2-51.

The hoisting hook attached to rigging may be designed to swivel. The swivel on a hoisting hook allows the load to turn without damaging or twisting the chain. A swivel hook is used only as a hook-positioning device and is not intended for load rotation. Special load rotation swivel hooks are available for such applications.

A *choker hook* is a sliding hook used in a choker sling and is hooked to the sling eye. A *grab hook* is a hook used to adjust or shorten a sling leg through the use of two chains. The grab hook is connected to a short length of chain that is attached to the master link. The longer (rigging) sling leg is shortened by engaging one of the links in the longer chain into the hook. A *foundry hook* is a hook with a wide, deep throat that fits the handles of molds or casting. A *sorting hook* is a hook with a tapered throat and a point designed to fit into holes.

Special custom design clamps, hooks, or other lifting accessories shall be marked to indicate their safe working loads. "Mousing" the hook or latch by tying wire around the opening is an unsafe practice and is not permitted by OSHA regulations.

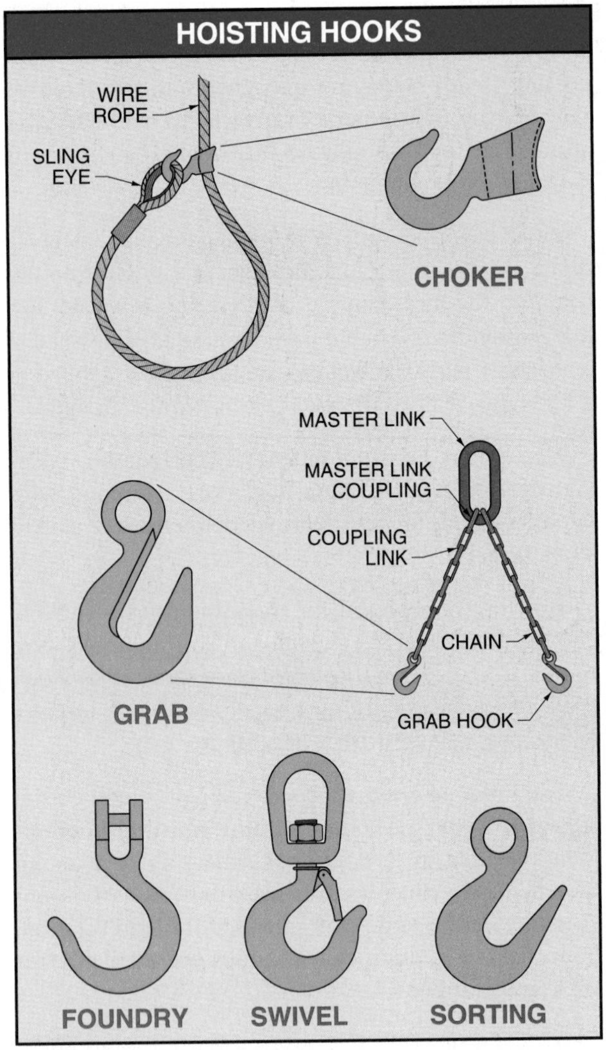

Figure 2-50. Hoisting hooks used for rigging purposes include choker, grab, foundry, swivel, and sorting hooks.

The load supported by a hoisting hook should be supported from the center or bowl of the hook. Hook failure is likely to occur if the load shifts or is applied to the tip or from an area between the bowl and the tip. Tip loading of hoisting hooks greatly decreases their lifting capacity. Hooks should be chosen for their proper strength rating. Specific size dimensions allow for specific load capacity.

Wire rope, shackles, rings, master links, and other rigging hardware must be capable of supporting, without failure, at least five times the maximum intended load. Where rotation resistant rope is used, the slings shall be capable of supporting, without failure, at least ten times the maximum intended load.

RIGGING COMPONENT INSPECTION

Inspection of rigging equipment covers inspection, recordkeeping, and storage. Maintenance of rigging components does not include making temporary repairs. Temporary repairs of rope, webbing, or chain should never be attempted. The component should be tagged "Defective – Do Not Use" and removed from service when there is an indication that repairs are necessary. Repair of rigging components can be completed and tested by the manufacturer of the component if needed.

An examination of all rigging equipment should be done initially, frequently, and periodically. Before any rigging component is placed in service, it shall be initially inspected to ensure that the specifications and condition requirements are correct. Frequent inspection is completed by the person using the rigging component each time it is used. Periodic inspections are conducted by designated, knowledgeable individuals.

The frequency of periodic inspection is determined by the service conditions and frequency of use. Experience, gained on the service life of the components, can also determine frequency of periodic inspection. However, periodic inspection of rigging components shall be conducted at least annually, with the exception of round slings, which shall be inspected at least monthly.

Wire Rope Inspection

Kinking, core protrusion, and bird caging may be encountered when inspecting wire rope. See Figure 2-52. *Kinking* is a sharp permanent bending. Kinking is normally caused by improper removal of wire rope from a spool or improper storage. Kinking weakens a wire rope and in many cases makes it useless.

Core protrusion is a damage condition of wire rope where compressive forces from within the rope force the strands apart. This happens when core material is squeezed out of the rope due to corrosion or degradation of the core. Core protrusion removes the support from the outer strands, reducing the efficiency of the rope.

Bird caging occurs from overloading, twisting, or squeezing when the rope is under load and is suddenly released. *Bird caging* is a damage condition of wire rope where the strands separate and open forming a shape similar to a bird cage.

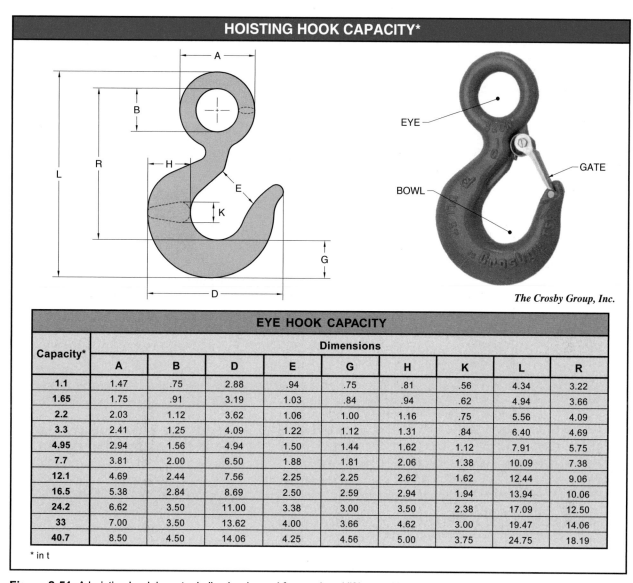

HOISTING HOOK CAPACITY*

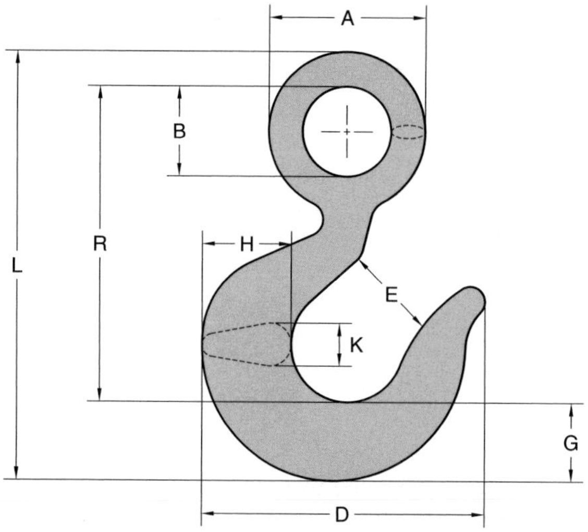

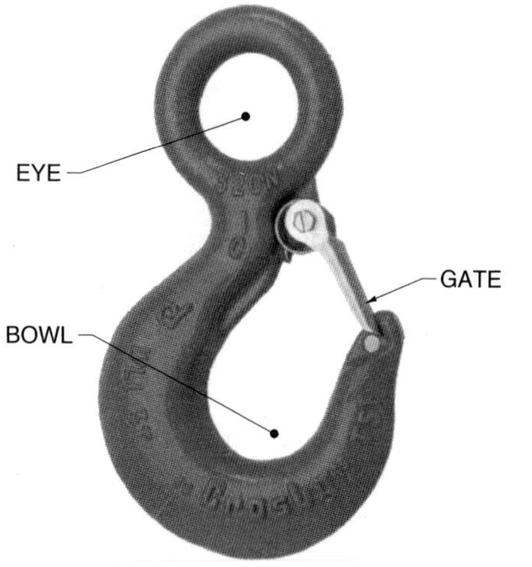

The Crosby Group, Inc.

EYE HOOK CAPACITY

Capacity*	Dimensions								
	A	B	D	E	G	H	K	L	R
1.1	1.47	.75	2.88	.94	.75	.81	.56	4.34	3.22
1.65	1.75	.91	3.19	1.03	.84	.94	.62	4.94	3.66
2.2	2.03	1.12	3.62	1.06	1.00	1.16	.75	5.56	4.09
3.3	2.41	1.25	4.09	1.22	1.12	1.31	.84	6.40	4.69
4.95	2.94	1.56	4.94	1.50	1.44	1.62	1.12	7.91	5.75
7.7	3.81	2.00	6.50	1.88	1.81	2.06	1.38	10.09	7.38
12.1	4.69	2.44	7.56	2.25	2.25	2.62	1.62	12.44	9.06
16.5	5.38	2.84	8.69	2.50	2.59	2.94	1.94	13.94	10.06
24.2	6.62	3.50	11.00	3.38	3.00	3.50	2.38	17.09	12.50
33	7.00	3.50	13.62	4.00	3.66	4.62	3.00	19.47	14.06
40.7	8.50	4.50	14.06	4.25	4.56	5.00	3.75	24.75	18.19

* in t

Figure 2-51. A hoisting hook is a steel alloy hook used for overhead lifting and is connected directly to the piece being lifted.

Conditions of wire rope that are considered reason for removing the wire rope from service include:

- One broken or cut strand
- Ten broken wires in one rope lay length or five broken wires within one strand in one rope lay
- Pitting in wires due to corrosion
- Corrosive failure of one wire adjacent to an end fitting
- Welding or weld splatter damage
- Flat spots worn on the outer strands of the rope, reducing the rope's diameter
- Distortion in the form of crushing, kinking, core protrusion, or bird caging

Fiber Rope Inspection

Fiber rope inspection is used to remove a rope from service before the rope's condition poses a hazard with continued operation. Fiber rope should be inspected monthly. See Figure 2-53. Conditions of fiber rope that are considered reason for removing from service include:

- Cuts or broken fibers
- Distortion in the form of kinking
- Excessive abrasion or wear
- Distortion in the form of heat damage, such as melting or charring on synthetic fiber rope

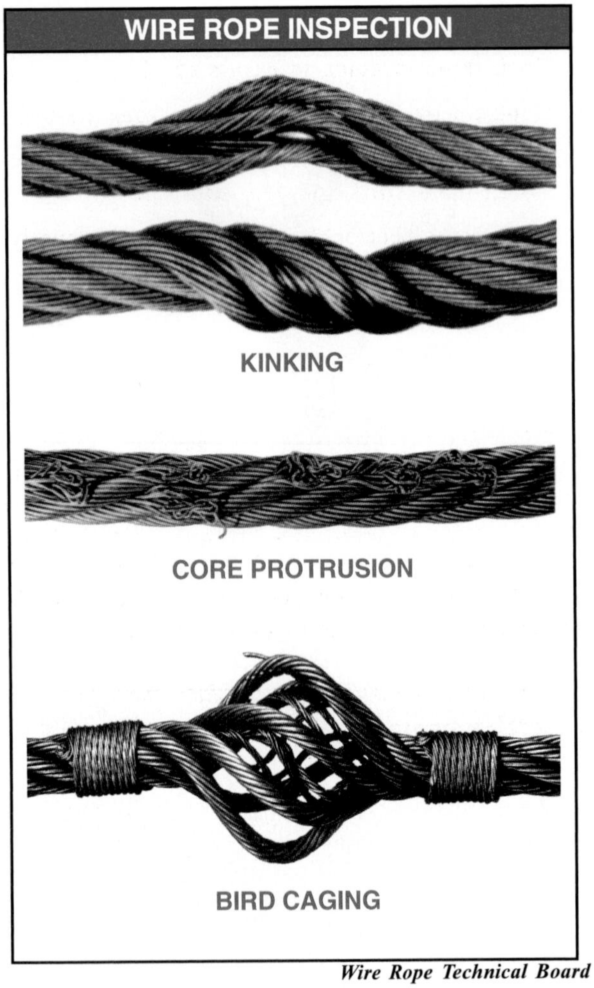

Wire Rope Technical Board

Figure 2-52. Kinking, core protrusion, and bird caging may be encountered when inspecting wire rope.

Lift-All Company, Inc.

Per OSHA 1926.251, each wire rope used in hoisting or lowering or in pulling loads shall consist of one continuous piece without knots or splices except for eye splices in the ends of wires and for endless rope slings.

Figure 2-53. Fiber rope inspection is used to remove a rope from service before the rope's condition poses a hazard with continued operation.

Webbing and Round Sling Inspection

Webbing should be inspected at least annually and round slings should be inspected monthly. See Figure 2-54. Conditions of webbing or round slings that are considered reason for removing from service include:

- Acid or alkali damage
- Melting, charring, or weld spatter on any part
- Holes, tears, cuts, snags, or embedded particles
- Broken or torn stitching in the load-bearing splices
- Excessive abrasive wear
- Knots in any part of the sling

- Discoloration or stiffness
- Edge tears
- Punctures
- Any conditions which cause doubt as to the strength of the sling

Chain Inspection

Chain should be inspected annually. Repairs to rigging and hoisting chain should only be made and tested by the chain manufacturer. Never use mechanical coupling links or repair links to repair any substandard rigging or hoisting chain. See Figure 2-55. Conditions of chain that are considered reason for removing from service include:

- Nicks, gouges, or wear having a depth in excess of the values given in maximum allowable link wear tables
- Stretching
- Hook throat opening in excess of 15%
- Hook tip twisted more than 10% from the plane of the unbent hook
- Hook shows cracks or signs of abuse
- Chain links bent or do not seat or flex properly

RIGGING EQUIPMENT STORAGE

Proper care, use, and storage of rigging equipment prevents damage and ensures safety. Rope, webbing, and chain slings must be kept in an assigned area that is kept clean, neat, dry, and away from harmful fumes or heat. Synthetic webbing, round slings, and natural fiber rope should be stored out of sunlight and away from areas used for arc welding.

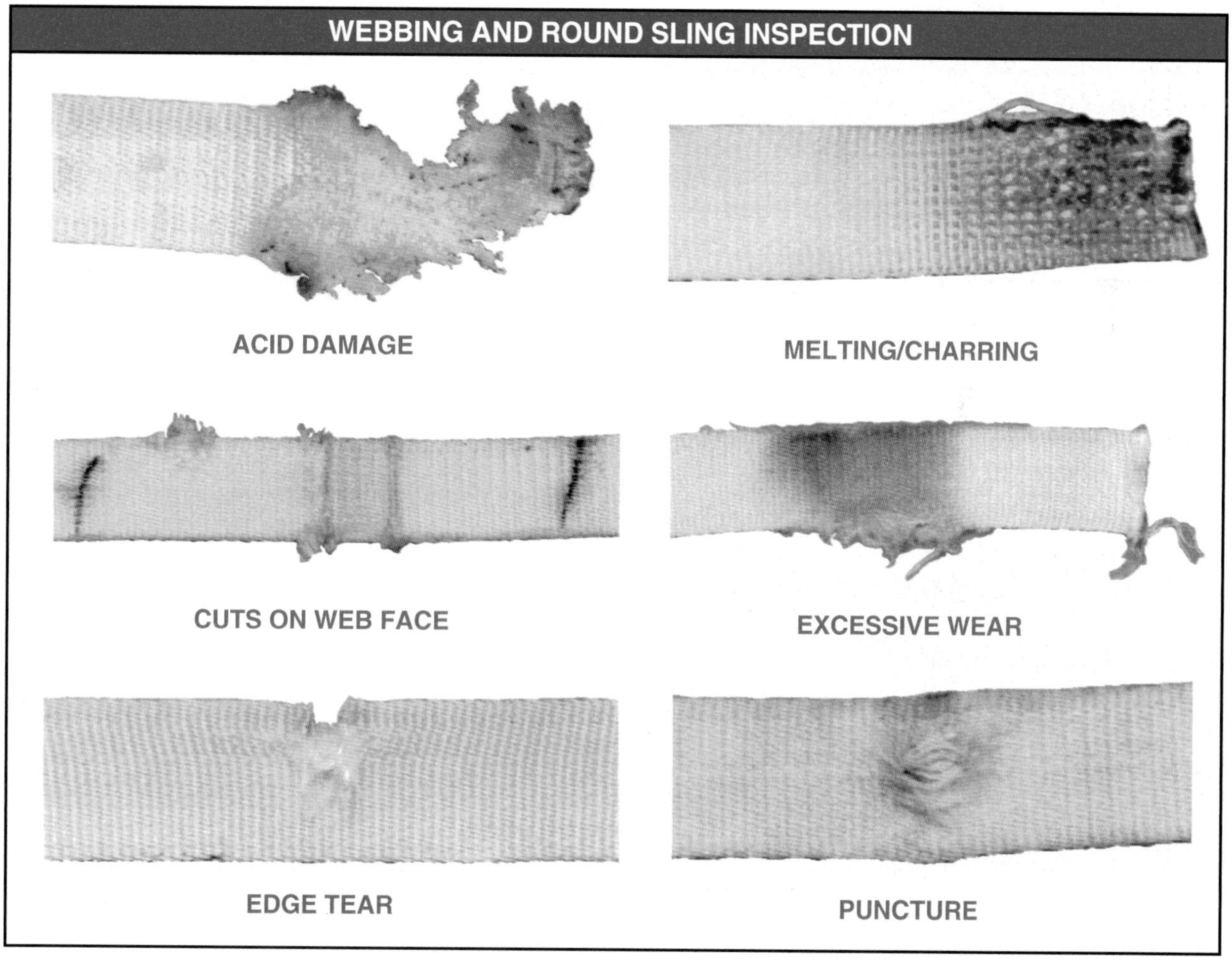

Lift-All Company, Inc.

Figure 2-54. Webbing should be inspected at least annually and round slings should be inspected monthly.

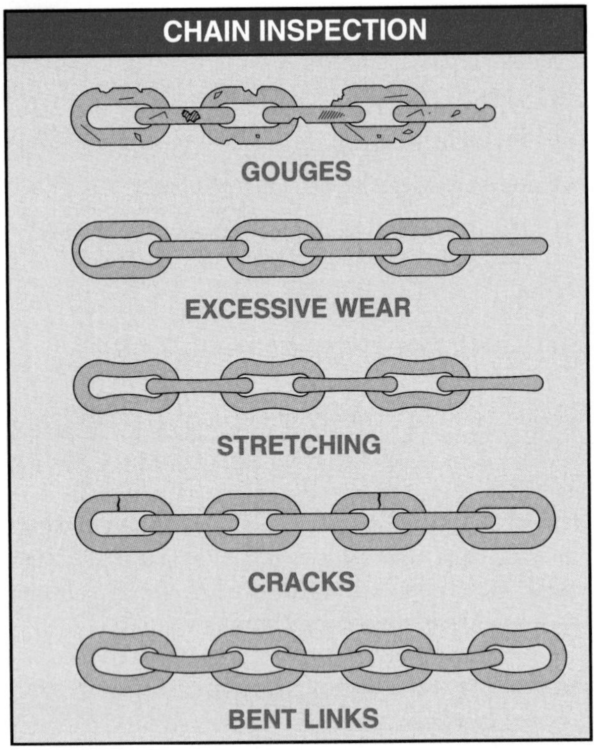

Figure 2-55. Chain should be inspected annually.

Kinking is prevented when slings are neatly hung from racks or lengthy rope is rolled onto spools. Slings or sling components must not be left laying where vehicles or forklifts may run over them or where heavy loads may be set on them. Avoid dragging slings over abrasive surfaces or sharp objects.

To prevent accidental use, damaged slings slated for manufacturer's repair must be stored in an area specifically posted, "Warning – Do Not Use." Immediately dispose of damaged or worn sling or sling attachments that are not to be repaired.

RIGGING COMPONENT RECORDKEEPING

Written inspection records and use of slings and sling components should be created for each new sling. Inspection records may cover the basic requirements of an annual inspection or be more comprehensive and frequent. The amount of rigging done by the user dictates the frequency and depth of recordkeeping.

A basic inspection document (log book) begins by issuing a serial number to the sling. The basic form includes the date purchased, condition at purchase, type, rated strength, list of included attachments or components, date of periodic inspections, inspecting personnel, and comments for each inspection.

Log book inspection forms may also include dates and hours of each use, department where used, initial sizes and diameters, and sizes and diameters at each inspection, with part numbers given to each sling attachment. Although frequent inspections made before and after each use are visual, unusual or abnormal conditions should be reported and included in the log book. All log book entries, if comprehensive, offer enough historical information to chart sling degradation and life expectancy.

Lifting

3 Chapter

Harrington Hoists Inc.

Lifting is the hoisting of equipment using mechanical means. Hoists may be manually-operated or power-operated hoists. Overhead hoists and cranes are regulated by a large number of standards. Hoist safety requires inspection at frequent and periodic intervals.

LIFTING DEVICES

Many lifting devices in use today use the centuries-old principle of the block and tackle (rope and pulley). Blocks and tackle were used to move sails, spars, and other components on sailing ships. Much of the block and tackle terminology used today is based on nautical applications. Today, blocks and tackle are primarily used for industrial lifting.

Lifting is the hoisting of equipment or machinery by mechanical means. Lifting is accomplished by using hand-operated or power-operated equipment. Industrial lifting equipment generally consists of the rigging assembly, hoist, and hoist support. Each component relies on the integrity of the other components. Lifting is attempted only after all components are determined to be safe.

Block and Tackle

A block and tackle is a combination of ropes and sheaves (pulleys). See Figure 3-1. A *block* is an assembly of hook(s), pulley(s), and frame suspended by hoisting ropes. *Tackle* is the combination of ropes and block assemblies arranged to gain mechanical advantage for lifting.

Block and tackle assemblies begin with the rope being reeved over a pulley. *Reeving* is passing a rope through a hole or opening or around a series of pulleys. The capacity, lifting speed, and lifting distance of a block and tackle is determined by the number of parts and the number of pulleys. A *part* is a rope length between the lower (hook) block and the upper block or drum. The greater the number of parts, the greater the lifting capacity.

Mechanical Advantage. *Mechanical advantage* is the ratio of the output force of a device to the input force. Applied force on the lead line of a block and tackle assembly is useful only if the force is either static (held) or dynamic (moving). A *lead line* is the part of the rope to which force is applied to hold or move a load. A static force on a block and tackle is strong enough to hold a load, but not strong enough to move a load. Block and tackle assemblies may have one-part, two-part (double), or three-part reeving.

The nominal breaking strength of the most heavily loaded rope in a system shall be no less than 3½ times the load applied to that rope.

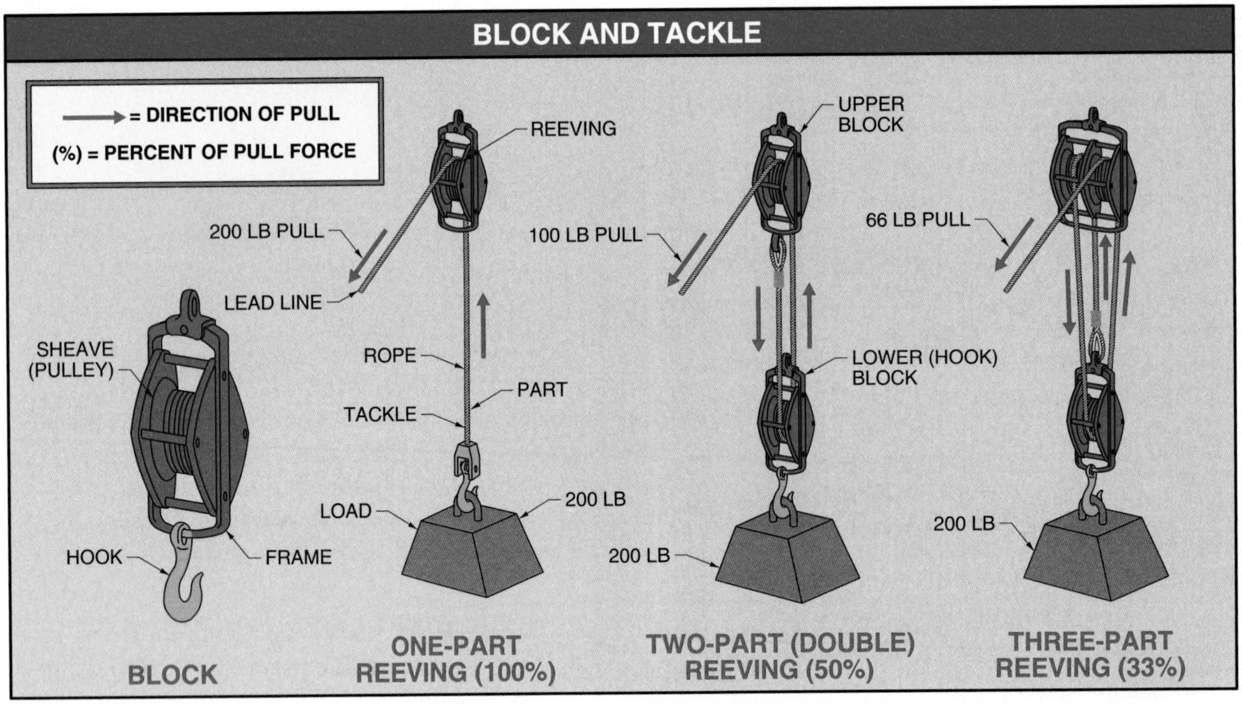

Figure 3-1. A block and tackle is a combination of ropes and sheaves (pulleys).

One-part reeving has one line between the load and the block. There is little or no mechanical advantage to one-part reeving. Two-part reeving has two lines between the load and upper block. Two-part reeving reduces the holding or lifting effort by approximately 50%. Three-part reeving has three lines between the load and upper block. Three-part reeving reduces the load by three, giving ⅓ of the support to each line.

Under ideal conditions (no friction), the mechanical advantage of a block and tackle equals the number of parts of rope that support the load. Therefore, a two-part block (reeving) system has an ideal mechanical advantage of 2.

A three-part block system (three-part reeving) consists of an upper block with two pulleys and a lower block with one pulley. Because three ropes support the load, each rope supports ⅓ of the load. The amount of static force required to hold a load is calculated by applying the formula:

$$L = \frac{w}{p}$$

where

L = lead line force (in lb)

w = total load weight including weight of slings, containers, etc. (in lb)

p = number of parts

For example, what is the force required to hold a 500 lb load using a four-part reeving system? *Note:* The rope and hook components total 30 lb.

$$L = \frac{w}{p}$$

$$L = \frac{530}{4}$$

$$L = \mathbf{132.5\ lb}$$

The characteristics of tackle are similar to the characteristics of a lever and fulcrum in that forces may be multiplied by the device, but the lesser the force required to produce a known pull, the greater the distance required for the force to be moved to do the required work.

As a force becomes greater than the static force on a load, the force becomes dynamic and strong enough to overcome any opposing forces, such as friction and weight, and the load moves. All dynamic mechanical power is subject to the basic principle of force and time. This principle states that whatever is lost to time is gained as force, and whatever is lost to force is gained as time. This principle is observed in a block and tackle assembly where no two pulleys travel at the same speed. In a block and tackle assembly, the lead line pulley rotates faster than the load pulley. For example, a two-part, double-reeved load moves one-half the speed of the lead line. As reeves are added, the lifting force of the load is increased, the lifting speed is decreased, and the lifting distance is decreased. All of these are equal to the number of parts added. See Figure 3-2.

A load is raised 1′ if the lead line on a four-part reeve is pulled 4′. Also, the force required by the lead line is 20 lb (¼ of the load) if the load is 80 lb. The mechanical advantage of a four-part reeve under ideal conditions is 4. See Figure 3-3. The distance the load moves is decreased by additional pulleys to the same extent that the ideal mechanical advantage is increased by additional pulleys.

Lifting hazards include overloading, dropping or slipping of the load caused by improper hitching or slinging, obstruction of the free passage of the load, and using equipment for a purpose for which it was not intended or designed.

For example, a 20 lb lead line force lifts 40 lb with two-part reeving. The 20 lb force must be moved 3′ to move the 40 lb load 1½′. With three-part reeving, the 20 lb lead line force lifts 60 lb, but the 3′ lead line movement moves the load only 1′.

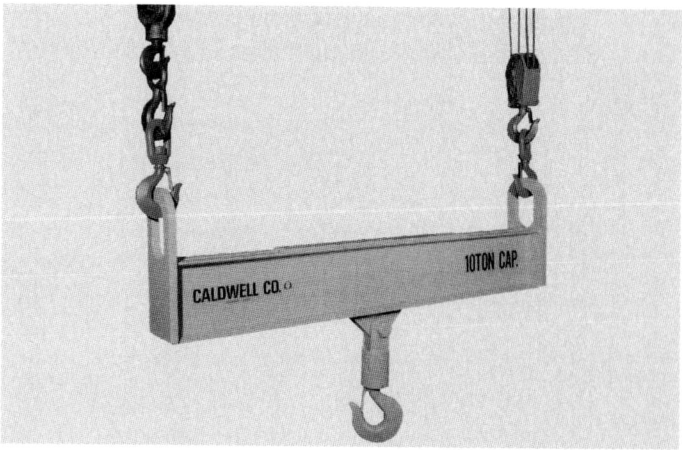

The Caldwell Group, Inc.

Mechanical advantage gained from using a block and tackle during lifting allows a small hoist motor to lift a heavy load.

Force. Force on the lead line is increased once the load starts to move. This force differs according to the type of pulley used. Pulleys may be plain bearing pulleys or rolling-contact bearing pulleys. A 109 lb (100 × 1.09 = 109 lb) force is required to move a 100 lb load using one-part reeving and plain bearing pulleys. A 104 lb force is required if rolling-contact bearing pulleys are used (100 × 1.04 = 104 lb). See Figure 3-4.

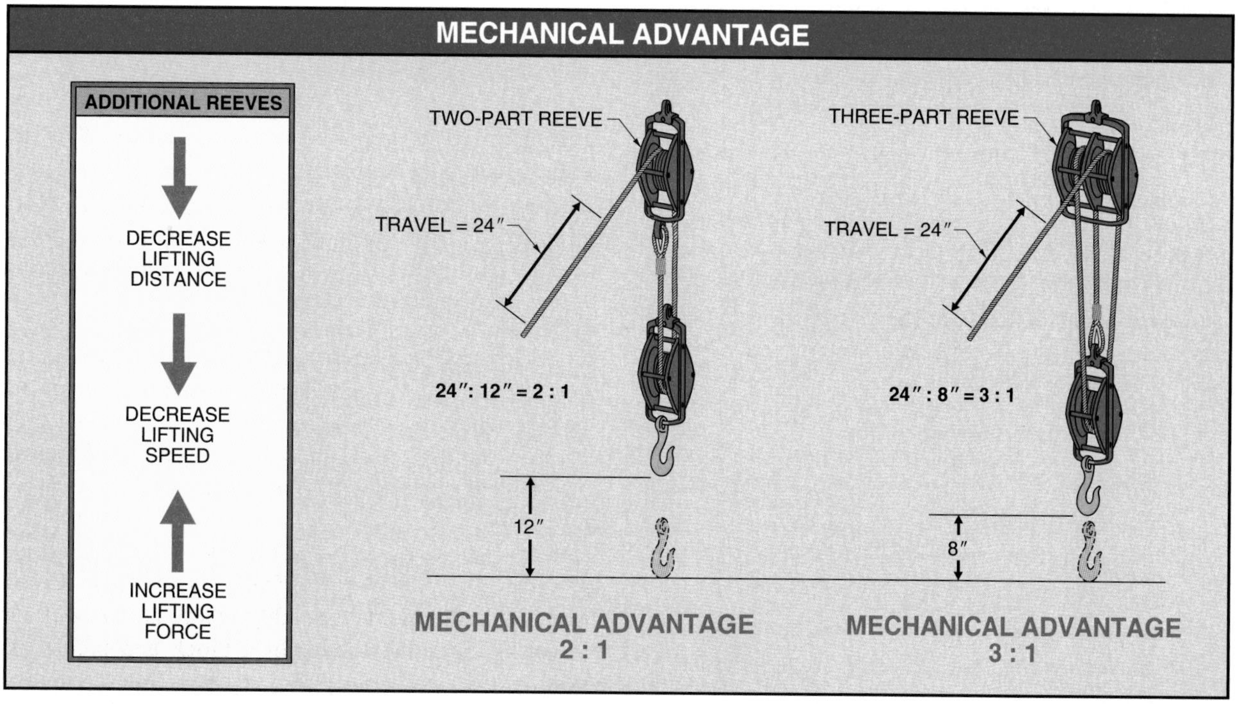

Figure 3-2. As reeves are added, the lifting force of the load is increased, the lifting speed is decreased, and the lifting distance is decreased.

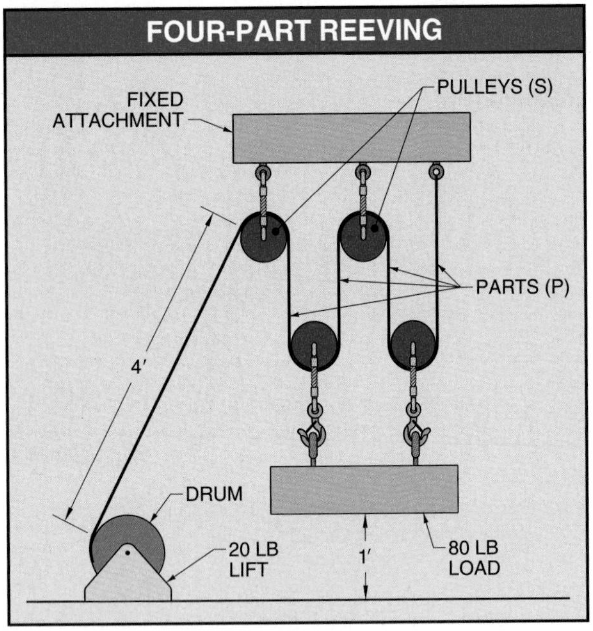

Figure 3-3. The mechanical advantage of a four-part reeve under ideal conditions is 4.

LEAD LINE FACTORS*

Parts of Line	Plain Bearing Pulleys	Rolling-Contact Bearing Pulleys
1	1.09	1.04
2	.568	.530
3	.395	.360
4	.309	.275
5	.257	.225
6	.223	.191
7	.199	.167
8	.181	.148
9	.167	.135
10	.156	.123
11	.147	.114
12	.140	.106
13	.133	.100
14	.128	.095
15	.124	.090

* based on equal number of pulleys

Figure 3-4. Force on the lead line differs according to the type of pulley used.

The weight capable of being lifted is equal to the lead line force multiplied by the number of parts supporting the load. The formula used to find the dynamic force required to move a load assumes that the number of parts of line is equal to the number of pulleys. Multiple reeving also permits the use of smaller diameter rope, pulleys, and drum. Lead line force is calculated by applying the formula:

$L = f \times w$

where

L = lead line force (in lb)

f = lead line factor (from Lead Line Factors table)

w = weight of load (in lb)

For example, what is the force required to move a 6000 lb load using an eight-part reeving system equipped with plain bearing pulleys?

$L = f \times w$

$L = 0.181 \times 6000$

$L =$ **1086 lb**

> ***Each hoist shall have its rated load marked on it or its front load block. This marking shall be clearly legible from the ground or floor.***

Manually-Operated Hoists

Lifting force is increased when the lead pull is accomplished mechanically. Two mechanical devices used to aid in lifting are the worm gear and bevel gear. Worm and bevel gear drives offer extra force to the lead pull, thereby multiplying the reeving force. See Figure 3-5. A *worm gear* is a set of gears consisting of a worm (drive gear) and a wheel (driven gear) that are used extensively as a speed reducer. The helix (spiral angle) of the worm gear is generally 15° to 25° to that of the worm wheel, thus creating extensive sliding and high torque. *Torque* is the twisting (rotational) force of a shaft. A 15° to 25° spiral angle prevents reverse rotation of the gears so when the hand chain on the hoist is not being pulled, the gears cannot slip in reverse.

The principle of lost time equaling extra force is shown in the worm gear drive. In the worm gear drive, the worm gear turns many revolutions to the worm wheel. The applied force is multiplied many times.

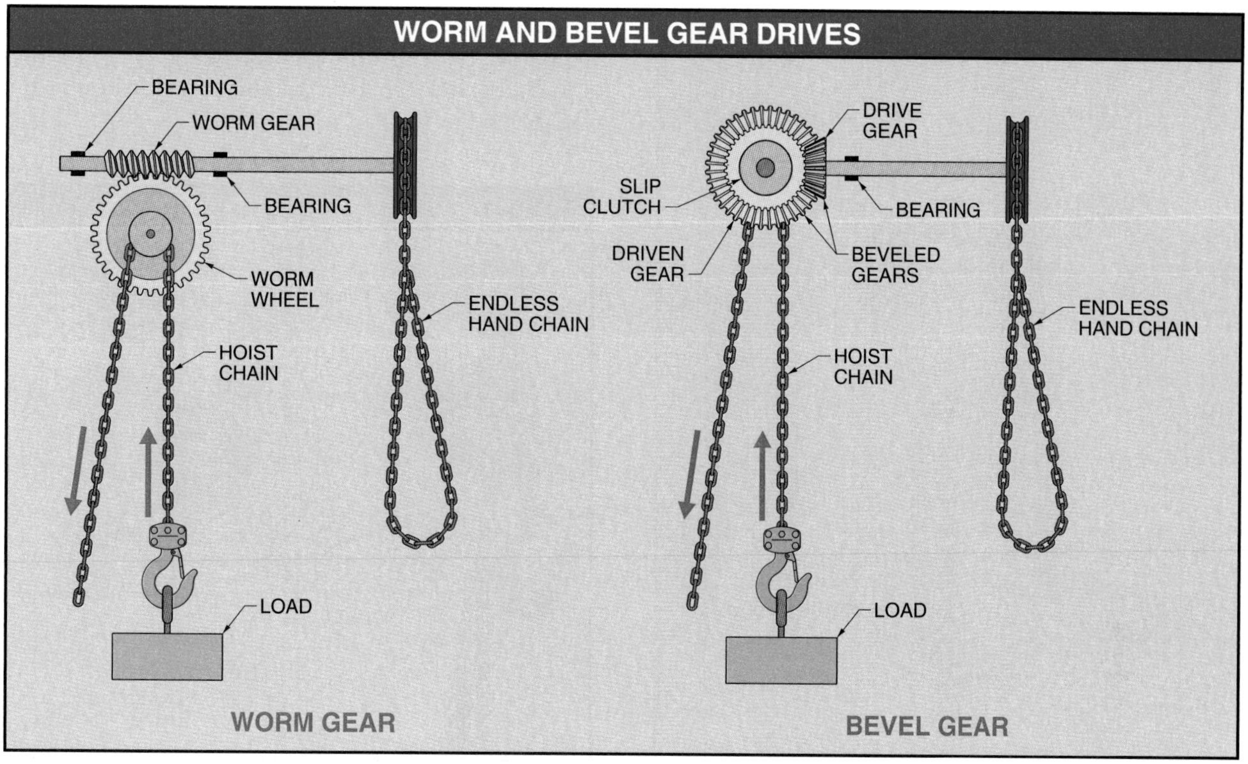

Figure 3-5. Worm and bevel gear drives offer extra force to the lead pull, thereby multiplying the reeving force.

A *bevel gear* is a gear that connects shafts at an angle in the same place. Bevel gears are generally at right angles with each other, but may be positioned at angles other than 90°. Unlike the worm gear hoist, bevel gear hoists normally require a braking or latching mechanism to prevent reverse rotation. The worm and bevel gear drives use gear reduction principles with the smaller gear being the drive gear.

Hand-Chain Hoists. A *hand-chain hoist* is a manually-operated chain hoist used for moving a load. Hand-chain hoists are suspended overhead from a top hook attached to a supporting structure. Supporting structures may be tripods, trolleys, cranes, or other fixed points. Hand-chain hoists are normally rated for ¼ t to 50 t.

A hand-chain hoist uses two chains, the hand chain and the hoist chain. See Figure 3-6. A *hand chain* is a continuous chain grasped by the operator to operate the pocket wheel. A *pocket wheel* is a pulley-like wheel with chain link pockets that are connected to the hoist mechanism. A *hoist chain* is the chain that raises the load. The rotation of the pocket wheel causes the hoist chain to raise or lower the load.

The *hand-chain drop* is the distance between the lower portion of the hand chain to the upper limit of the hoist hook travel. *Lift* is the distance between the hoist's upper and lower limits of travel. *Headroom* is the distance from the cup of the top hook to the cup of the hoist hook when the hoist hook is at its upper limit of travel. The *top hook* is the hook assembled to the top of a hoisting mechanism to allow for overhead suspension. Top hooks, like hoist hooks, should be drop-forged from alloy steel and should be heat-treated to open slowly when overloaded. *Reach* is the distance between the cup of the top hook and the cup of the hoist hook when the hoist hook is at its lower limit of travel. Reach is the sum of the lift and headroom.

Lever-Operated Hoists. A *lever-operated hoist* is a lifting device that is operated manually by the movement of a lever. See Figure 3-7. Lever-operated hoists are used in confined and awkward areas and many are small enough to be kept in a tool box. Lever-operated hoists are generally used to lift light loads. A light load is a load that weighs from 200 lb to 500 lb. An automobile engine is an example of a light load.

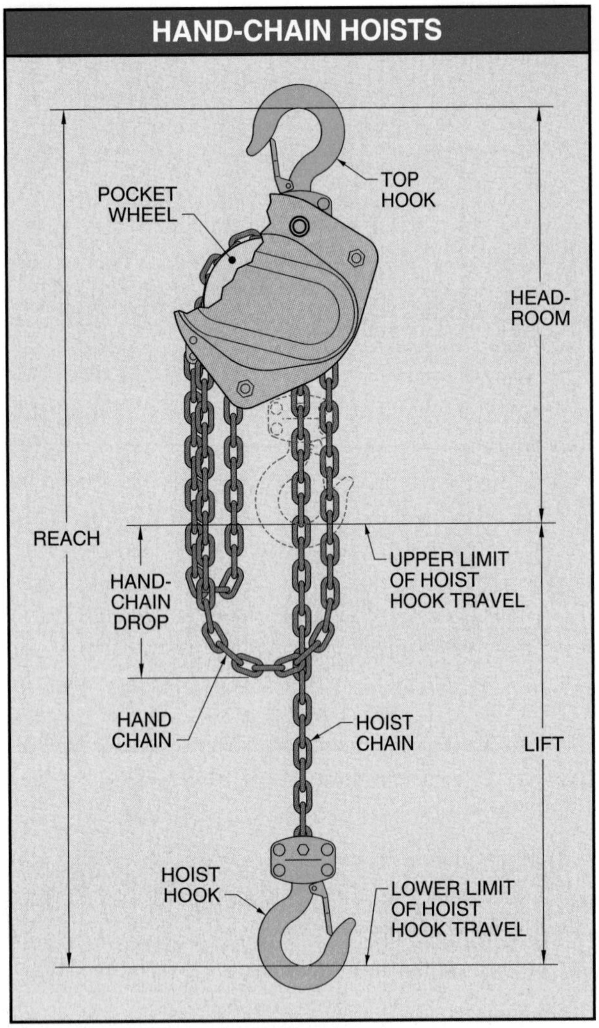

Figure 3-6. A hand-chain hoist is a manually-operated chain hoist used for moving a load.

Two major types of lever-operated hoists are ratchet and slip-clutch models. Lever-operated hoists have ratchet mechanisms to prevent reversal. A *ratchet* is a mechanism that consists of a toothed wheel and a spring-loaded pawl. A *pawl* is a mechanism used to prevent the ratchet wheel from turning backwards. Some manufacturers design safety-slip clutches into lever-operated hoists to prevent unsafe overloading.

A *slip clutch* is a spring-loaded, friction-held fiber disc that is adjusted to slip at 125% to 150% of the hoist-rated load. Many slip clutches are preset by the manufacturer, while others are adjustable. Adjustable slip clutches must never be adjusted to hold over 175% of the hoist-rated load.

Never lift more than the hoist-rated load, except for clutch adjustment and testing. Before attempting to adjust and test a slip clutch, determine that the supporting structure is capable of safely supporting a load equal to 175% of the hoist-rated load plus the weight of the hoist.

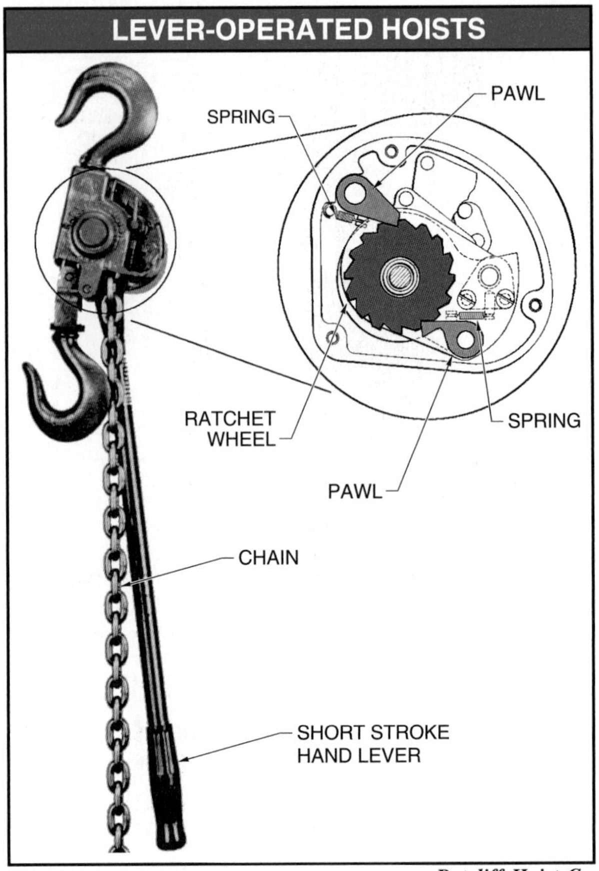

Ratcliff Hoist Co.

Figure 3-7. A lever-operated hoist is a lifting device that is operated manually by the movement of a lever.

Lever-operated hoists are subject to shock loads and severe conditions. Although the load may not be overweight, the shock can cause the friction brakes to slip or freeze. For this reason, some lever-operated hoist manufacturers believe a slip clutch may be unreliable or unsafe and do not manufacture slip-clutch models.

Power-Operated Hoists

A *power-operated hoist* is a hoist operated by pneumatic or electric power and use either chain or wire rope as the lifting component. Chain or wire rope power-operated hoists are selected based on the heaviest load to be lifted. See Figure 3-8.

POWER-OPERATED HOISTS

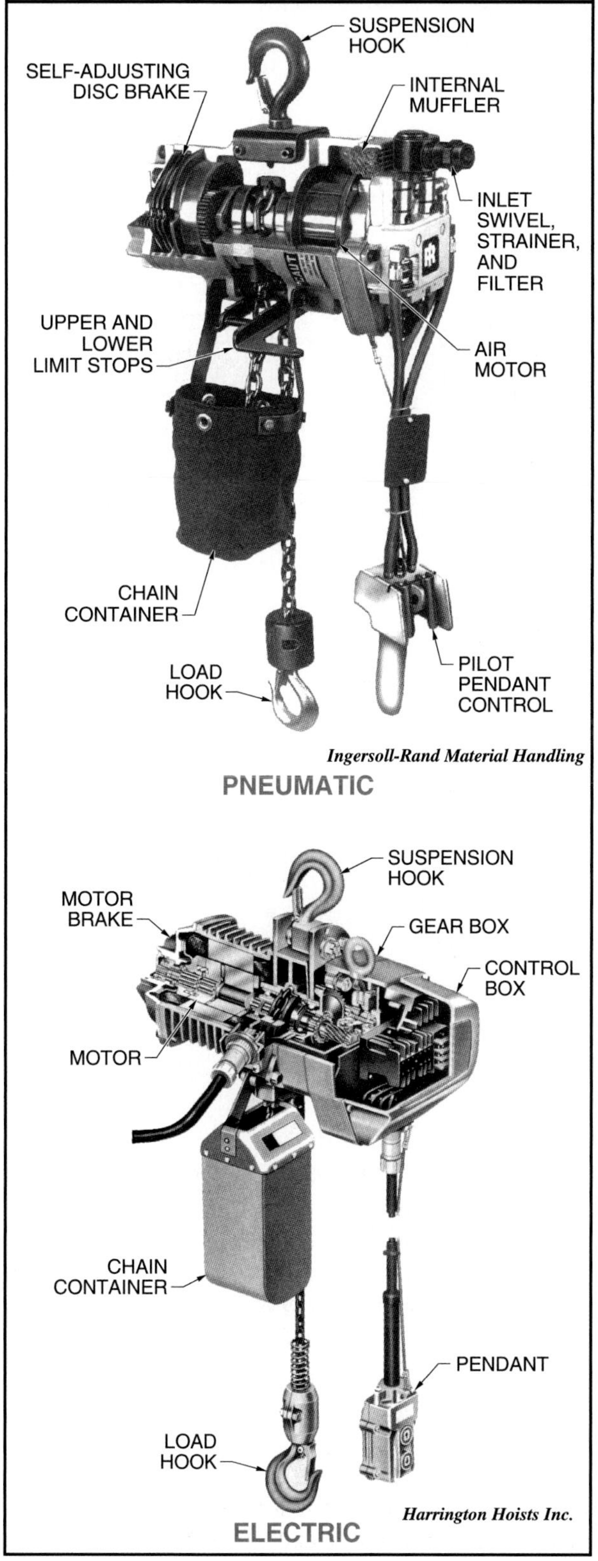

Figure 3-8. Power-operated hoists are operated by pneumatic or electric power.

Pneumatic Hoists. A *pneumatic hoist* is a power-operated hoist operated by a geared reduction air motor. Pneumatic hoists have distinct advantages over electric hoists. An advantage is that there is no explosion hazard in areas such as paint spraying or petrochemical facilities because compressed air does not produce electric arcing.

A requirement for heavy-duty applications is the ability for unlimited start/stop situations. Air motors are inherently self-cooling and able to start, stop, and reverse without overheating and are ideal for high ambient temperature applications. *Ambient temperature* is the temperature of the air surrounding a piece of equipment.

Pneumatic controls have the advantage of variable speed to lift a load from slow to full speed while allowing smooth movements. A disadvantage of pneumatic operation is the chance that the operation will consume a greater amount of air than the available air supply. Because air compressor output is rated in horsepower, compressor horsepower must be converted to hoist standard cubic feet per minute (scfm) required by the hoist when selecting the required capacity compressor for a given hoist. See Figure 3-9.

AIR FLOW REQUIREMENTS

Pipe Size*	SCFM	Min Comp HP	Lift Speed**	Capacity†
3/8	48	12	50	1/2
3/4	75	19	30	2
1	260	65	20	8
1 3/4	420	105	8	15

* in in.
** in fpm based on 90 psi line pressure with full speed at throttle
† 2000 lb ton

Figure 3-9. Hoist standard cubic feet per minute (scfm) must be determined to select the required capacity compressor for a given hoist.

The minimum compressor size for a hoist is found by applying the formula:

$$HP = \frac{scfm}{4}$$

where
HP = horsepower
$scfm$ = standard cubic feet per minute
4 = constant

For example, what is the minimum compressor size required for a pneumatic hoist that requires 75 scfm?

$$HP = \frac{scfm}{4}$$

$$HP = \frac{75}{4}$$

$$HP = \mathbf{18.75\ HP}$$

When equipment such as a large hoist, pneumatic motor, or air drill is to be purchased to replace electrical units, the pneumatic units require a certain scfm for operation. To maintain a sufficient supply of compressed air, the scfm of new and existing equipment must not be higher than the present compressor system output. The scfm output of a compressor is found by applying the formula:

$$scfm = 4 \times HP$$

For example, what is the scfm of a compressor rated at 15 HP?

$$scfm = 4 \times HP$$

$$scfm = 4 \times 15$$

$$scfm = \mathbf{60\ scfm}$$

All outdoor cranes shall be provided with secure fastenings adequate to hold the crane against 30 psf wind pressure. Special anchorage shall be provided by the user in areas where wind forces are anticipated to be in excess of 30 psf.

Electric Hoists. Common industrial electric hoist capacities range from ¼ t to 20 t. Depending on the capacity of the hoists, the power supply range is 110 VAC, 208 VAC, 230 VAC, 460 VAC, or 575 VAC (volts alternating current). The operator's controls are low-voltage (normally 24 V) to provide electrical shock protection. Many electric hoists are equipped with mechanical and electrical safety overload protectors. Mechanical overloads (slip clutches) refuse to lift (by means of a slipping action) when a load is applied beyond their capacity. Electrical overloads are brake mechanisms activated when power is removed or lost. Hoist operating controls require pushbutton or lever action and an upper limit switch.

A pendant is used by the hoist operator to control load movement from the floor or other level beneath the crane. A *pendant* is a pushbutton or lever control suspended from a crane or hoisting apparatus.

A *limit switch* is a device that cuts off the power automatically at or near the upper limit of hoist travel. All pneumatic and electric hoists are equipped with an upper limit switch to prevent damage from overwrap. Limit switches are safety switches and are not to be used as stop switches. See Figure 3-10.

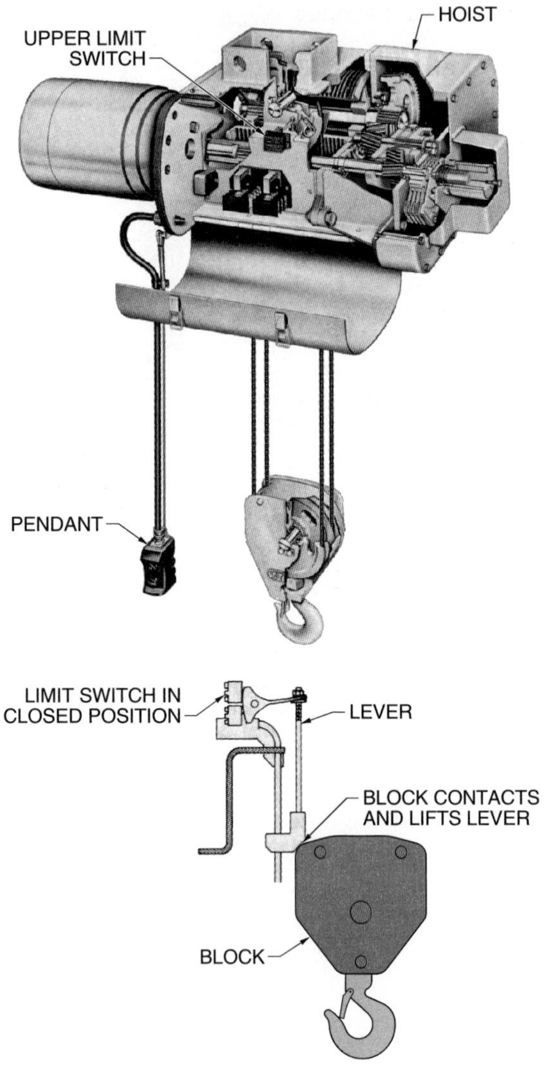

Lift-Tech International, Division of Columbus McKinnon

Figure 3-10. Limit switches are devices which cut off the power automatically at or near the upper limit of hoist travel.

Lifting must be done under the control of an operator at all times. Sudden or jerky movements, rapid dropping, or high speeds in short distances can damage the load or lifting equipment. The movement of a load must always be in view of the hoist operator.

Drum Wrap. *Drum wrap* is the rope length required to make one complete turn around the drum of a hoist or crane. An overwrap condition occurs when the drum wraps enough rope or wire so that the load block comes in contact with the hoist or crane. If the load block comes in contact with the hoist or crane, ropes can be severely stretched or broken, causing the load to drop.

The proper direction for winding the first layer on a drum is determined by the lay of the rope. The rope lay determines if the rope is wound over or under on the spool. The rope should be overwound from left to right if the rope is anchored on the left and the rope is a right lay rope. See Figure 3-11.

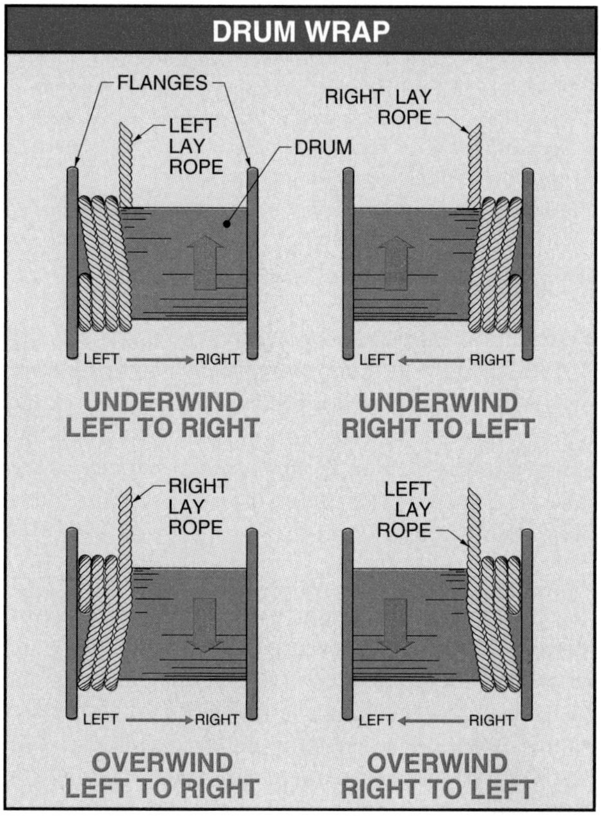

Figure 3-11. The proper direction for winding the first layer on a drum is determined by the lay of the rope.

The drum must be observed to ensure a two-wrap minimum if the system is not equipped with a low-limit switch. The two-wrap minimum ensures the strength of a two-reeve pull against the rope and drum attachment. The attachment would not hold a load if all the rope is unwound from the drum. A drum that is rewound must be rewound in the same direction and the rope seated properly in the drum grooves if the drum is equipped with grooves. Incorrect winding damages the rope.

Care must be exercised when transferring rope from a reel to a hoist drum. A *reel* is a wooden assembly on which wire rope is wound for shipping and storage. During rope transfer, the unreeling process should be straight and under tension. A light squeezing pressure well away from the drum is used when guiding the wire rope onto the drum. Always wear gloves when handling wire rope. Never handle wire rope with bare hands.

While grooved drums offer few winding problems, smooth-face drums can be more difficult to wind and the proper procedure must be followed. The first layer of rope should be wound with sufficient tension to ensure a close helix with each wrap being wound as close as possible to the preceding wrap. The first layer acts as a helical groove, which guides the successive layers. For this reason, the first layer should not be unwound on a smooth drum. As the rope is forced up to the second layer at the flange, a reverse helix is created, causing the rope to crossover. A *crossover* is one wrap winding on top of the preceding wrap. The crossover is the point at which the rope winds back over two rope grooves to advance. See Figure 3-12.

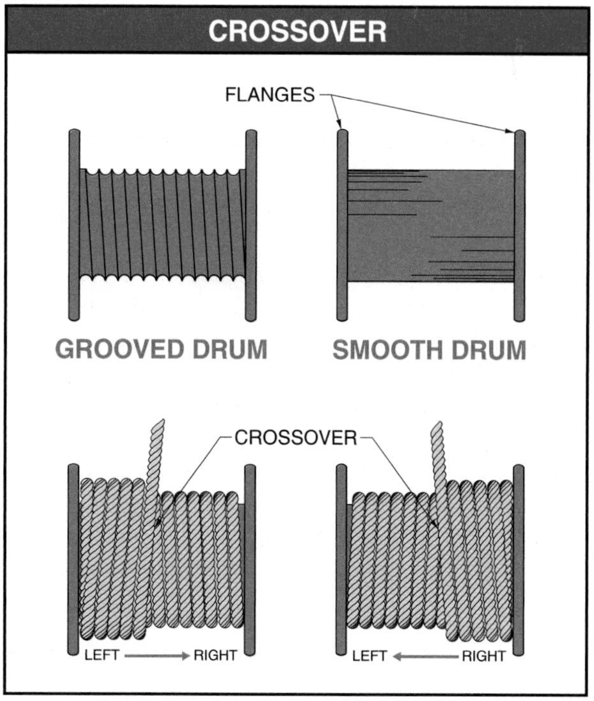

Figure 3-12. A crossover is one wrap winding on top of the preceding wrap.

HOIST SAFETY

Overhead hoists and cranes are regulated by a large number of standards. A *standard* is a guideline adopted by regulating authorities. Regulating authorities include OSHA, ANSI, ISO, CMAA, ASME, and NFPA. Standards organizations pertinent to lifting loads include:

- **OSHA** – The Occupational Safety and Health Administration specifies safety standards through the U.S. Department of Labor and the Occupational Safety and Health Act. OSHA is concerned with the development and enforcement of safety standards for industrial workers.
- **ANSI** – The American National Standards Institute is a standards-developing organization that adopts and co-publishes standards that are written and approved by member organizations. ANSI branches out and connects its member organizations by unifying their adopted standards. ANSI manages United States participation in ISO standards activities. See ANSI A10.22-1990, *Rope-Guided and Nonguided Worker's Hoists – Safety Requirements.*
- **ISO** – The International Organization for Standardization is a nongovernmental international organization that is comprised of national standards institutions of over 90 countries (one per country). The ISO provides a worldwide forum for the standards developing process.
- **CMAA** – The Crane Manufacturers Association of America, Inc. is an organization of the leading crane manufacturers in the United States developed for the purpose of promoting standardization and providing a basis for proper equipment selection and use. CMAA is instrumental in establishing many crane-operating practice standards.
- **ASME** – The American Society of Mechanical Engineers helps establish safe structural design of hoists and cranes. In conjunction with ANSI, ASME develops safety standards for hoists and cranes. See ANSI/ASME B30.2-1990, *Overhead and Gantry Cranes (Top Running Bridge, Single or Multiple Girder, and Top Running Trolley Hoist).*
- **NFPA** – The National Fire Protection Association publishes the National Electrical Code® (NFPA 70), which contains standards for the practical safeguarding of persons and property from hazards arising from the use of electricity. Article 610 covers the installation of electrical equipment and wiring used in connection with cranes, monorail hoists, hoists, and all runways.

The two most prevalent standards in hoist safety and operating rules are that industrial hoists and cranes are not designed for and should not be used for lifting, supporting, or transporting humans, and loads or empty hook blocks should not be used over any individual, especially if the load is held magnetically or by vacuum.

All running ropes in continuous service should be visually inspected once each working day. A thorough inspection of all ropes shall be made at least once each month and a full written, dated, and signed report of rope condition kept on file where readily available to appointed personnel.

Inspection Programs and Procedures

Verification that hoisting equipment meets current code requirements must be made if an inspection program is being developed and records of initial installation or subsequent modifications are not available. Verification includes a step-by-step procedure similar to that of inspecting and checking a new hoist. See Appendix.

Regularly Scheduled Inspection. Existing equipment must be inspected at frequent and periodic intervals. Frequent inspections are conducted daily and monthly. Daily inspections are accomplished with the assistance of a checklist and may not require the signature of the person making the inspection. See Figure 3-13. These checks are mostly visual and include identifying unusual sounds or temperatures that may indicate problems.

The exception to this procedure is the brake mechanism. The brake mechanism should be checked frequently by operating the hoist with and without a load and by testing its holding power at various levels. Any repairs or major adjustments performed as a result of the inspection must be recorded on a written report. The report should identify the hoist serviced and indicate work performed, the date, reason, individual performing the inspection, and the parts replaced.

ELECTRIC HOIST CHECKLIST				
Item	**Daily**	**Monthly**	**Semi-Annual**	**Deficiencies**
All functional operating mechanisms	•	•	•	Maladjustment interfering with proper operation, excessive component wear
Controls	•		•	Improper operation
Safety Devices	•		•	Malfunction
Hooks	•	•	•	Deformation, chemical damage, 15% in excess of normal throat opening, 10% twist from plane of unbent hook, cracks
Load-bearing components (except rope or chain)	•	•	•	Damage (especially if hook is twisted or pulling open)
Load-bearing rope	•	•	•	Wear, twist, distortion, improper dead-ending, deposits of foreign material
Load-bearing chain	•	•	•	Wear, twist, distortion, improper dead-ending, deposits of foreign material, stretch
Fasteners	•	•	•	Not tight
Drums, sheaves, sprockets			•	Cracks, excessive wear
Pins, bearings, shafts, gears, rollers, locking and clamping devices			•	Cracks, excessive wear, distortion, corrosion
Brakes	•		•	Excessive wear, drift
Electrical			•	Pitting, loose wires
Contactors, limit switches, pushbutton stations			•	Deterioration, contact wear, loose wires
Hook retaining members (collars, nuts) and pins, welds, or rivets securing them			•	Not tight or secure
Supporting structure or trolley			•	Continued ability to support imposed loads
Warning label	•		•	Removed or illegible
Pushbutton markings	•		•	Removed or illegible
Capacity marking	•		•	Removed or illegible

Figure 3-13. Existing equipment must be inspected at frequent and periodic intervals.

Monthly checks are generally concerned with load-bearing components, such as the condition of hooks, wire rope, chain, and nut and bolt tightness. Periodic or semiannual inspections are conducted by specifically-designated personnel. Written reports of monthly and periodic inspections must be signed, dated, and placed in the equipment identification file. In many cases, the manufacturer of the equipment provides a checklist. This checklist can be used or it can serve as a guide in preparing a checklist better suited to a particular situation. In addition to inspection of hooks, wire rope, and chain, basic periodic or semiannual inspection checklists must include braking systems and limit switches.

Hook, wire rope, and chain inspection for hoists is identical to inspection for rigging components. Hook inspection is concerned with the throat opening, load-bearing thickness, tip twisting, chemical damage, and latch damage. Wire rope inspection covers reduction in rope diameter, broken wires, worn outside wires, distortion damage, weld spatter, and corroded rope. Examination for elongation, gouges, bending, and worn chain is included in chain inspection. Chain assemblies should not be stored where they may be subject to damage or where exposed to corrosive action.

WESTON BRAKE

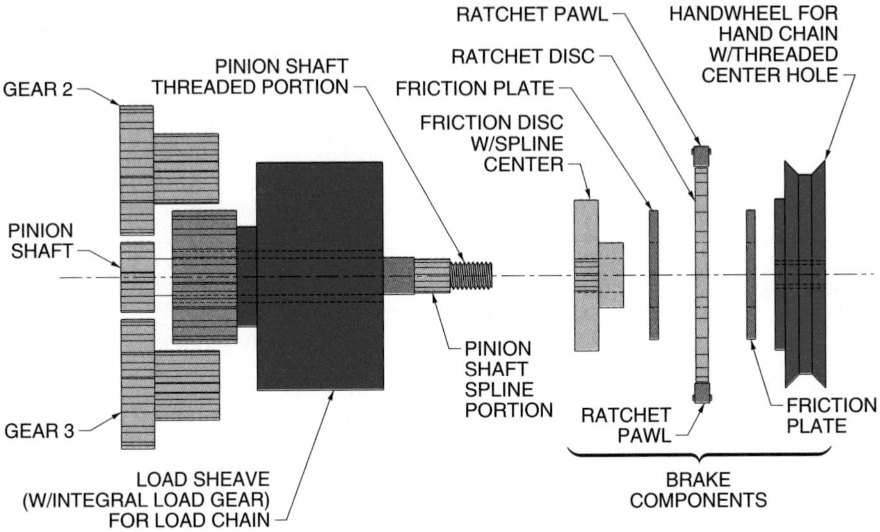

The mechanical brake employed in Harrington hoists is a Weston Brake. The Weston Brake was invented by Mr. Weston in the beginning of the 1900s and was introduced into the hoist industry around 1910. In one form or another, it is used in practically all hand hoists worldwide.

The Weston Brake is comprised of brake components, such as the friction disc with splined center hole, friction plates, ratchet disc, ratchet pawls, and handwheel with threaded center hole.

When the hand chain is used to rotate the handwheel in the forward (hoisting) direction, the threaded center portion of the handwheel screws the handwheel tighter onto the threaded portion of the pinion shaft. This squeezes the handwheel, the two friction plates, and the ratchet disc up against the friction disc. The friction disc, which is mated to the splined portion of the pinion shaft, cannot move along the pinion shaft because of the shoulder at the end of the splined portion of the pinion shaft. Therefore, the squeezing cinches the brake components together and the rotational motion imparted by the hand chain is transmitted to the pinion shaft.

The pinion shaft, which runs through a hole in the center of the load sheave, engages gear 2 and gear 3. When the pinion shaft rotates, it transmits its rotation to these two gears, which in turn are engaged to the geared portion of the load sheave. The rotation of gear 2 and gear 3 is transmitted to the load sheave and the load is lifted.

When the pulling on the hand chain ceases, as would happen between pulling strokes or when the load has reached its intended position, gravity acting on the load tends to cause the load sheave to rotate in the backward (lowering) direction. This rotational torque is transmitted through the gears to the pinion shaft and keeps the brake components cinched tightly together. With the pinion shaft and brake components cinched tightly together as a single body, the entire assembly attempts to rotate in the backward (lowering) direction. The ratchet pawls engage the ratchet disk and prevent this. Thus, the brake stops the load from lowering.

When the hand chain is used to rotate the handwheel in the backward (lowering) direction, the threaded center portion of the handwheel begins to back the handwheel off the threaded portion of the pinion shaft, which decompresses the brake components. This allows the pinion shaft to rotate in the backward (lowering) direction, which it begins to do by virtue of the load itself. As the load begins to fall and causes the pinion shaft to rotate in the backward direction, the threaded portion of the pinion shaft causes the handwheel and the other brake components to cinch up tight again, and the lowering of the load stops. The lowering of the load ceases until the handwheel is again rotated in the lowering direction. In this way, the lowering of the load is actually accomplished by a series of very small controlled falls that are perceived as a smooth motion.

Harrington Hoists Inc.

The Peerless LB leverhoist from Harrington contains a Weston load brake using two brake pads with four braking surfaces for positive brake action.

Hoist Motor Brake Inspection

Braking systems must be inspected before each shift change or prior to use after periods of nonuse. Daily hoist brake inspections are concerned with braking integrity (hook drift). *Hook drift* is the slippage of a hook caused by insufficient braking.

To inspect for hook drift, the hoist is operated in the lifting and lowering direction without load on the hook. The hook is stopped to check operation of the hoist braking system. The drift of the hook should not exceed 1″ in either direction. The motor brake normally requires adjustment or lining replacement if hook drift exceeds 1″.

Typically, motor brakes are direct-acting (ON or OFF) disc-type mechanisms controlled by a rectified DC current. The air gap is set at a specific spacing (typically .03″). The air gap increases as the brake lining wears. Significant brake lining wear causes hook drift. In many cases, a limit switch is added to the assembly at the air gap location. The limit switch does not close and the hoist motor does not start when the gap reaches a worn limit (.090″ in many cases). See Figure 3-14.

Disconnect and lock out power prior to adjusting the air gap. Use a brush or compressed air to remove accumulated brake lining dust. Specific attention should be directed to removing dust from the air gap between the coil and plates. Solenoid burning may occur if the air gap is not equal at all three adjusting points. Use proper eye and breathing equipment when using compressed air to clean the air gap. Hook drift occurs when adjustment springs are weak or broken, or when overheating has warped the brake lining or compression plates.

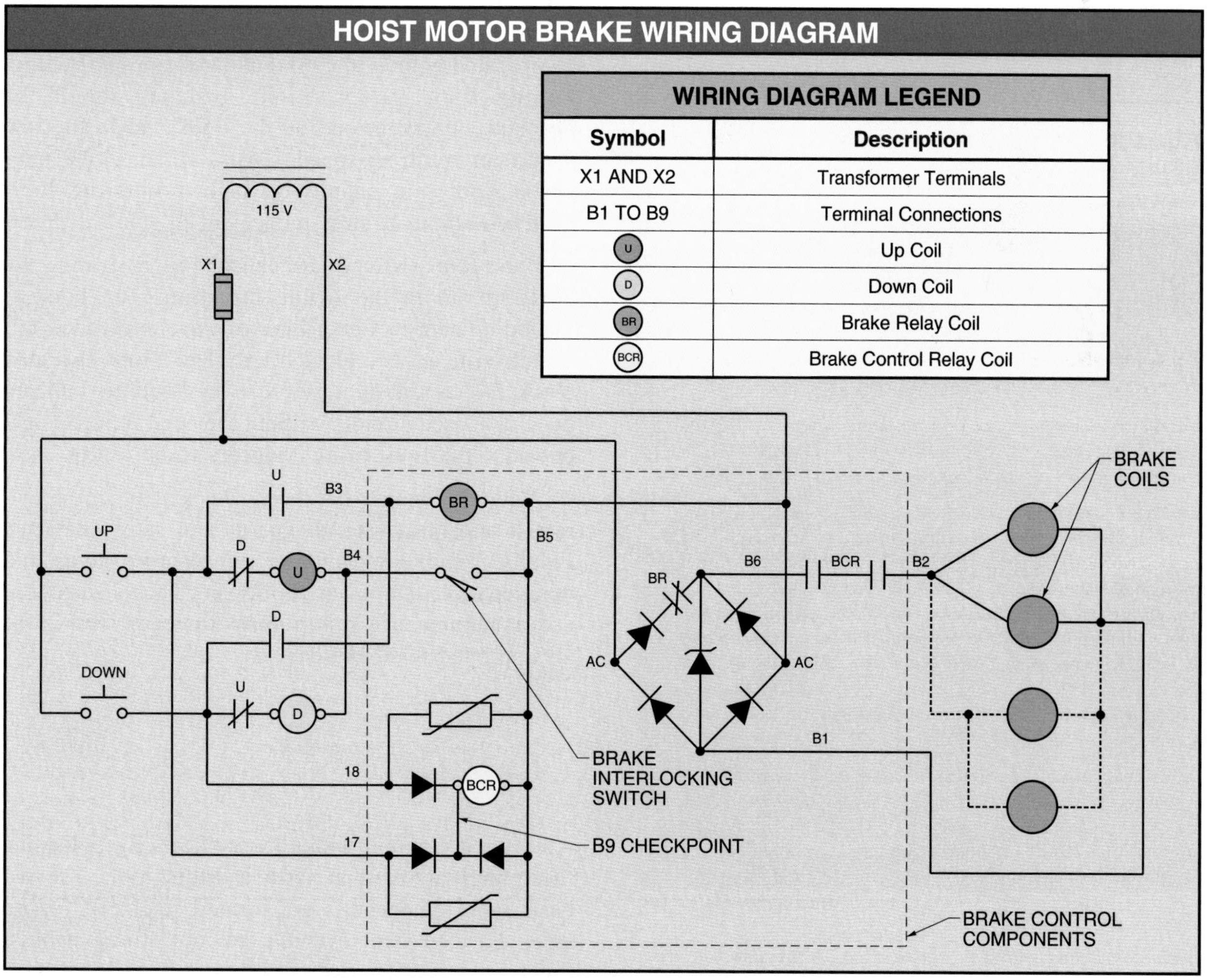

Figure 3-14. The brake interlocking switch must be closed for the hoist motor to start.

Inspect brake linings for warpage by laying them on a clean, flat, level surface. Lay a straightedge across the center and check for gaps as the straightedge is rotated. Typically, gaps of 1⁄32″ or more require that the brake linings be replaced. See Figure 3-15.

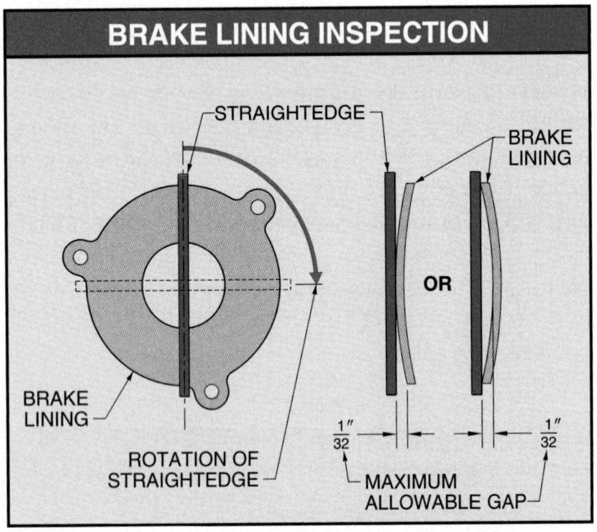

Figure 3-15. A straightedge is placed across the brake linings to check for warpage.

Harrington Hoists Inc.

Harrington electric hoists with an overhead crane are used to move bundles of aluminum and steel tubing through the warehouse of a greenhouse manufacturer.

Conducting Load Tests

Warning signs and barriers must be used on the floor beneath a hoist, crane, or lifting system. A mechanical load brake test checks the hoist braking system for proper operation. All personnel should be alerted that a free-fall condition could exist during a mechanical load brake test. Attach a rated capacity load to the hoist hook. Before lifting the load, slowly remove any slack from the line. Raise the load a few inches and stop. If the load stops and the brakes hold, continue raising and lowering the load several feet, stopping the hoist several times in each direction to check the brakes. Next, check the hoist with a load equal to 125% of its rated load capacity. This check tests any load-limiting devices. Load-limiting devices for power-operated hoists are normally rated at 110% of the hoist's rated load capacity.

Hoist limit switches and other safety devices should be checked daily or at the start of each shift when the hoist is in operation. Hoists should never be operated without the protection of a properly functioning limit switch. Limit switches should be checked without a load on the hook and in the low speed on multiple-speed hoists. Hoists with one-speed operation should be inched into the limit switch. *Inching* is slow movement in small degrees.

Hoist limit switches are checked by activating the hoist to run in the lifting direction. The hoist is inched up near its upper limit of travel until the limit switch arm or weight is lifted. This stops the load block. A continuity tester can be used to indicate open or closed circuits without the need for electrical power if the limit switch appears to be faulty.

After the switch has been checked or corrected, install and reactivate all guards and safety devices. Finally, remove warning or out-of-order signs and maintenance equipment. Following each inspection, use a maintenance report form to report results of all approved tests and checks.

EYEBOLTS

An *eyebolt* is a bolt with a looped head. Eyebolts, like lifting lugs, are used primarily as lifting tools. The two basic types of eyebolts are formed steel and forged steel. Formed steel eyebolts are not strong enough for heavy weights and should not be used in lifting applications. Only forged steel eyebolts should be used for lifting applications. Three common types of forged eyebolts are machinery, regular nut, and shoulder nut eyebolts. See Figure 3-16. See Appendix.

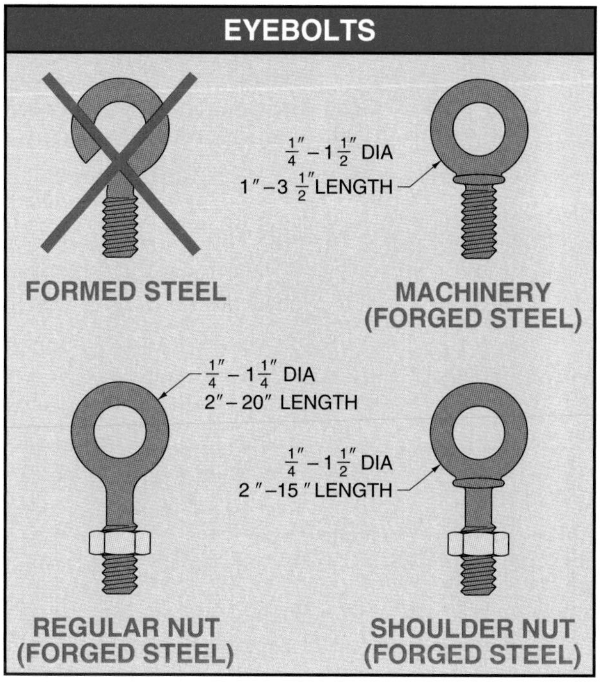

Figure 3-16. An eyebolt is a bolt with a looped head.

The regular nut eyebolt and the shoulder nut eyebolt (where the shoulder is not used) should never be used for angular lifting. Angular lifting force should only be applied to an eyebolt that is firmly supported by a shoulder. See Figure 3-17.

Attaching eyebolts is accomplished by screwing the eyebolt into a threaded hole or inserting an eyebolt shank through a hole and securing it with a nut. The machinery eyebolt is the most commonly used eyebolt. Machinery eyebolts must be screwed in until the shoulder is tight against the load.

Always inspect an eyebolt before use and never use one that shows signs of wear, fatigue, or damage, or one that is bent or elongated. Inspect for clean threads and sharp shoulder corners. Washers must be used to create a tight shoulder-to-load fit if the shoulder is not sharp and the load has not been countersunk. Never undercut an eyebolt shank.

Always use shoulder nut eyebolts for angular lifting and never use an eyebolt that shows signs of wear or damage.

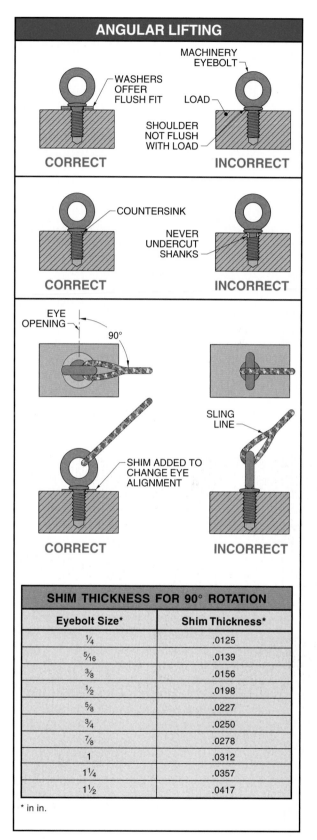

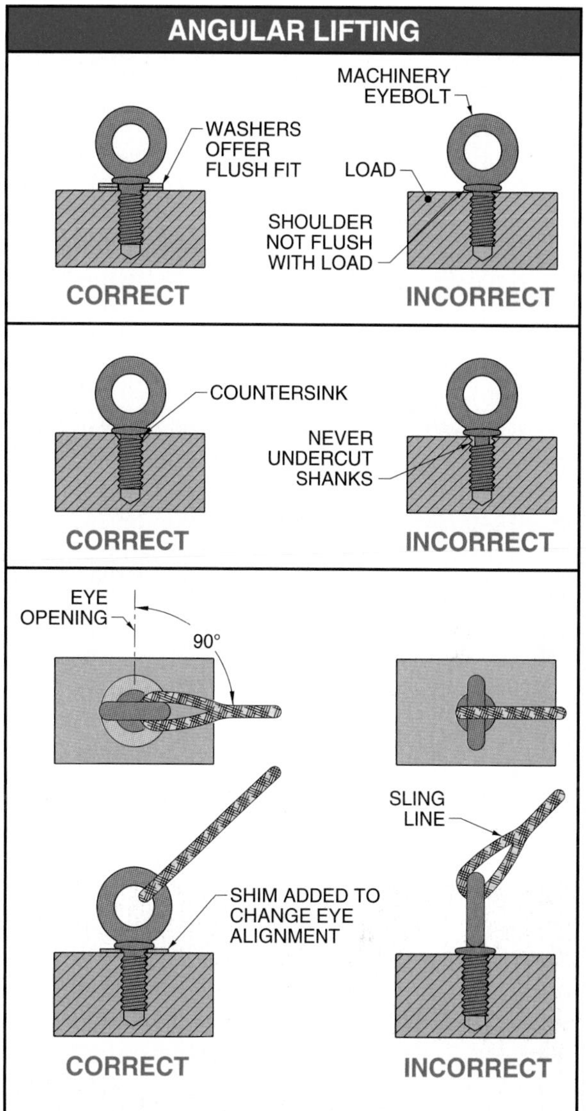

SHIM THICKNESS FOR 90° ROTATION

Eyebolt Size*	Shim Thickness*
1/4	.0125
5/16	.0139
3/8	.0156
1/2	.0198
5/8	.0227
3/4	.0250
7/8	.0278
1	.0312
1 1/4	.0357
1 1/2	.0417

* in in.

Figure 3-17. Angular lifting force should only be applied to an eyebolt that is firmly supported by a shoulder.

Acco Chain & Lifting Products Division

Stacker cranes from Louden have continuous 360° rotation to provide flexibility in removing and positioning steel bundles.

After a machinery eyebolt has been installed and tightened, the eye of the eyebolt must be perpendicular (at 90°) to the sling line for any degree of side pull. Adjusting the pivot of the eyebolt for proper alignment is accomplished by placing a shim under the shoulder. The shim thickness determines the angle of unthread rotation and varies according to the eyebolt size. *Unthread rotation* is counterclockwise rotation of an eyebolt having right-handed threads, or the clockwise rotation of an eyebolt having left-handed threads. Unthread rotation can be anywhere from 5° to 90° to ensure a right angle to the sling line.

Lifting equipment should not be installed, operated, or maintained by any person who does not understand the contents of the installation and operation manual. Serious bodily injury, death, and/or property damage may result from failure to read and comply with the manufacturer's instructions.

As a sling moves from a vertical to an angular position (decreased sling angle), the capacity of the eyebolt is reduced. Each degree of change requires a reevaluation of the eyebolt's safe limits. The safe limit is increased if the sling angle is increased. The safe limit is reduced if the sling angle is reduced. See Figure 3-18.

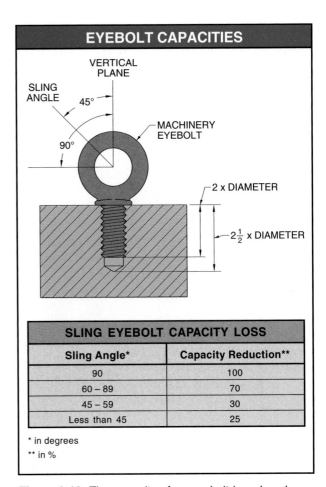

EYEBOLT CAPACITIES

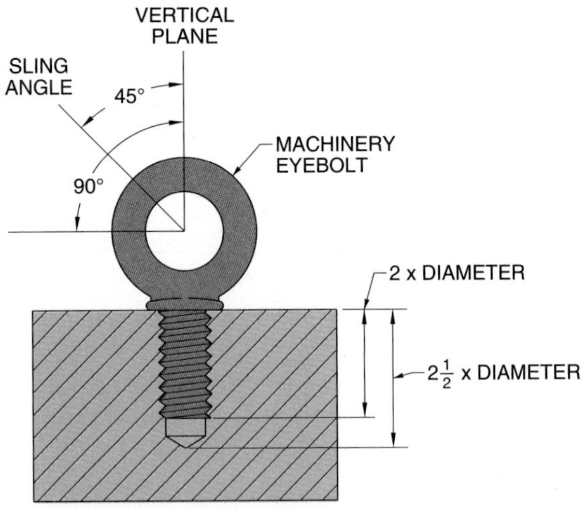

SLING EYEBOLT CAPACITY LOSS

Sling Angle*	Capacity Reduction**
90	100
60 – 89	70
45 – 59	30
Less than 45	25

* in degrees
** in %

Figure 3-18. The capacity of an eyebolt is reduced as a sling moves from a vertical to an angular position.

For example, what is the percent change in eyebolt capacity of a sling that has been relocated from 55° to 65°?

A 55° angle equals a 30% reduction and a 65° angle equals a 70% reduction in eyebolt capacity (from Sling Eyebolt Capacity Loss table). The percent change is 40% (70% – 30% = 40%). The eyebolt capacity is reduced an additional 40% with a total capacity loss of 70%.

Nut eyebolts may be regular or shoulder. A regular nut eyebolt is normally screwed into the load and held firm using a secured hex nut. See Figure 3-19. A tapped hole in the load should be at a minimum depth of 2½ times the eyebolt diameter. The eyebolt should be screwed into the load at a minimum depth of 2 times the eyebolt diameter. Two nuts (one on either side of the load) must be used if the load thickness is less than one eyebolt diameter of threads. One nut is required when the load thickness is greater than one eyebolt diameter of threads. A regular nut eyebolt may also be slid through an unthreaded hole in a load and held in place by two nuts firmly secured against each other on the bottom of the load with a third nut secured on top of the load.

When a shoulder nut eyebolt is used, the threaded portion of the shank must protrude through the load sufficiently to allow full engagement of the nut. Washers must be used to take up the excess between the nut and the load if the unthreaded portion of the shoulder nut eyebolt protrudes so far that the nut cannot be tightened securely against the load. The washers must exceed the distance between the bottom of the load and the last thread of the eyebolt. The shoulder of the eyebolt must be tightly secured against the load surface. See Figure 3-20.

Eyebolt Loads

Eyebolt load ratings are affected by the angle of pull on the eyebolt. A straight vertical pull of a shoulder nut eyebolt offers 100% of the eyebolt load rating. An angular pull of 45° reduces the eyebolt load rating by as much as 65%.

Until international standards are specifically defined and adhered to by all manufacturers, actual eyebolt load ratings will vary significantly between manufacturers. Specific load rating information must be obtained from distributors' or manufacturers' catalogs when a particular eyebolt is to be used. General lifting capacities of eyebolts may be used as a primary estimation when designing lifting assemblies.

Typical eyebolt angular lift capacity is calculated using a constant of .21 for sling angles of less than 45° and .25 for slings angles greater than 45°. The constant is multiplied by the eyebolt manufacturer's working load limit for vertical lifts. See Figure 3-21. This figure is divided by the sling angle loss factor to determine the working load capacity. See Appendix. Working load capacity is found by applying the formula:

$$L = \frac{c \times wl}{s}$$

where

L = working load capacity (in lb)

c = constant (.21 for sling angles less than 45°; .25 for sling angles greater than 45°)

wl = eyebolt working load limit (in lb)

s = sling angle loss factor (from Sling Angle Loss Factors Table)

Always inspect eyebolts before use and never use an eyebolt if its eye is bent or elongated.

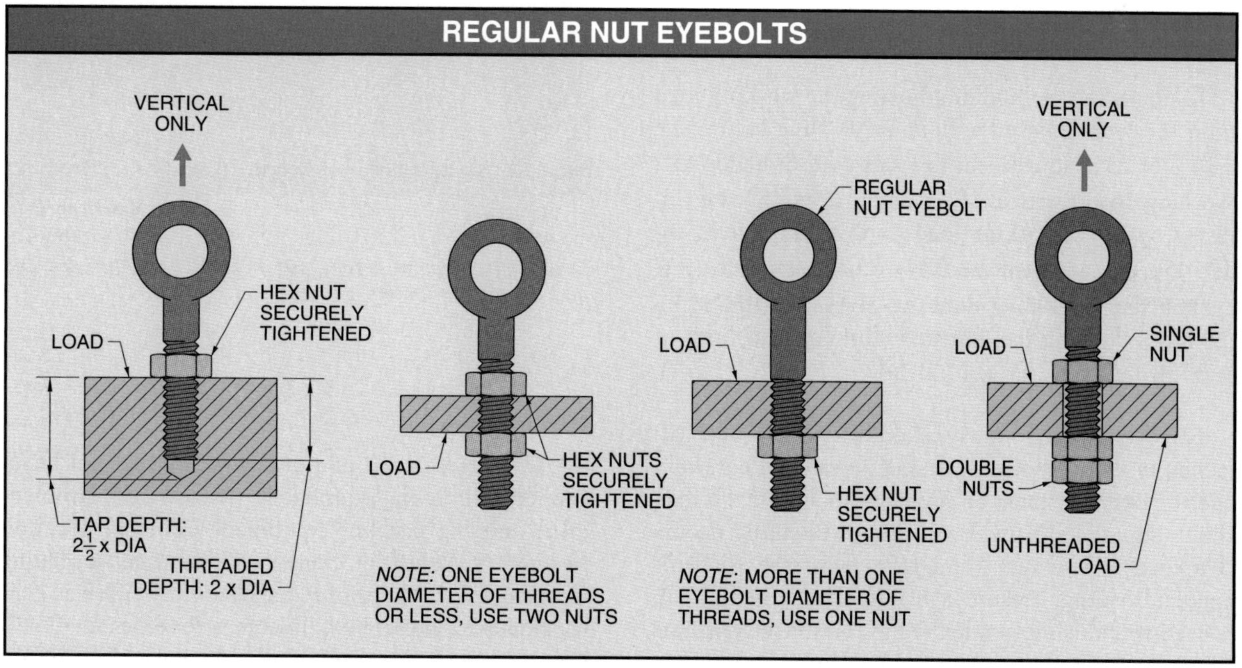

Figure 3-19. A regular nut eyebolt is normally screwed into the load and held firm using a secured hex nut.

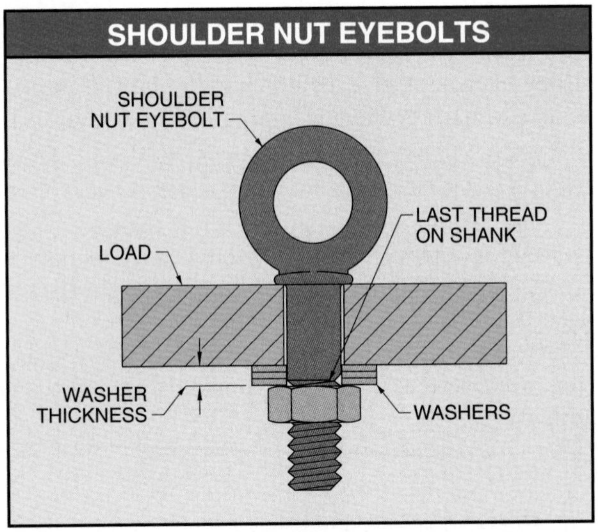

Figure 3-20. Washers must be used to take up the excess between the nut and the load if the unthreaded portion of the shoulder nut eyebolt protrudes so far that the nut cannot be tightened securely against the load.

For example, what is the working load capacity of a 50° bridle sling using a ¾″ shoulder nut eyebolt?

$$L = \frac{c \times wl}{s}$$

$$L = \frac{.25 \times 5200}{.766}$$

$$L = \frac{1300}{.766}$$

$$L = \mathbf{1697.128\ lb}$$

A simplified method of adjusting the working load limit for eyebolts used with an angle sling is to make a 30% or 25% adjustment to loads over or under 45°. Working loads with a sling angle over 45° are adjusted by 25%. Working loads with a sling angle under 45° are adjusted by 30%. For example, a ¾″ eyebolt having a sling angle of 50° is multiplied by .25 (25%), giving the adjusted working load limit of 1300 lb (5200 × .25 = 1300 lb).

Force is applied slowly after slings have been attached to the eyebolts. The load is watched carefully and the operator must be prepared to stop if the load shows signs of shifting or buckling. Buckling occurs if a load is not stiff enough to resist the force of angular loading. Ensure a bridle sling combination is used when using two-leg slings. Never reeve slings from one eyebolt to another. Reeving slings alters the load strength of sling materials. See Figure 3-22.

SHOULDER NUT EYEBOLT WORKING LOAD LIMITS – VERTICAL LIFTS

Size*	Working Load Limit**
¼	500
5⁄16	800
⅜	1200
½	2200
⅝	3500
¾	5200
⅞	7200
1	10,000
1¼	15,200
1½	21,400

* in in.
** in lb

Figure 3-21. Eyebolt working load limits are given by the manufacturer and are based on the size of the eyebolt.

North American Industries, Inc.

Overhead industrial cranes use hoist trolleys that are supported by and operate on bridge girders.

INDUSTRIAL CRANES

An *industrial crane* is a crane with structural beam supports for lifting equipment. Cranes are instrumental in moving parts or equipment from one location to another. Industrial cranes are classified depending on their service. Industrial cranes are classified as gantry cranes for portability, jib cranes for extending loads out and over areas, and overhead cranes for moving loads from one area to another in confined spaces.

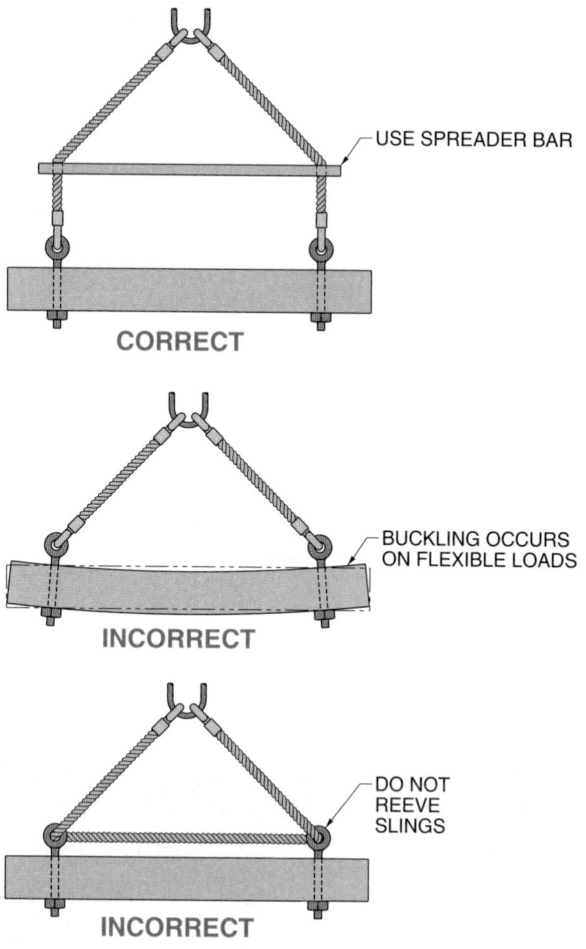

Figure 3-22. Buckling occurs if a load is not stiff enough to resist the force of angular loading.

Gantry Cranes

A *gantry crane* is a crane with bridge beams supported on legs. See Figure 3-23. The legs are supported by end trucks that normally travel on floor rails. Floor rails are small-gauge railroad rails, which are recessed into the floor or set directly on top of the floor surface.

An *end truck* is a roller assembly consisting of a frame, wheels, and bearings generally installed or removed as complete units. Some gantry crane legs are supported by wheels for movement across flat floor surfaces.

Gantry cranes may be single-leg or double-leg gantry cranes. A single-leg gantry crane is normally a supplemental crane for a large capacity overhead crane. In this application, one side of the bridge beam is attached to a leg which rolls on a floor rail and the other side is attached to an overhead crane rail.

Double-leg gantry cranes are used on two parallel floor rails. In some cases, the double legs are on wheels, which allows the gantry crane to operate across any flat floor surface.

Baldor Electric Co.

Gantry cranes are used outdoors for the lifting of cut stone.

Jib Cranes

A *jib crane* is a crane that is mounted on a single structural leg. The three basic types of jib cranes are wall-mounted, base-mounted, and mast. See Figure 3-24. Wall-mounted jib cranes are top-braced or cantilevered and may have a stationary or a partial rotating boom. A *cantilever* is a projecting beam (boom) or member supported at only one end.

Base-mounted jib cranes are free-standing cranes on a heavily-supported base mounting. The boom of a base-mounted jib crane may be stationary or be capable of 360° rotation.

Mast jib cranes have one structural leg (mast) mounted to the floor and/or ceiling. Mast jib cranes are cantilevered, underbraced, or top-braced. Mast jib cranes may be stationary or installed with upper and lower bearings for partial or 360° rotation.

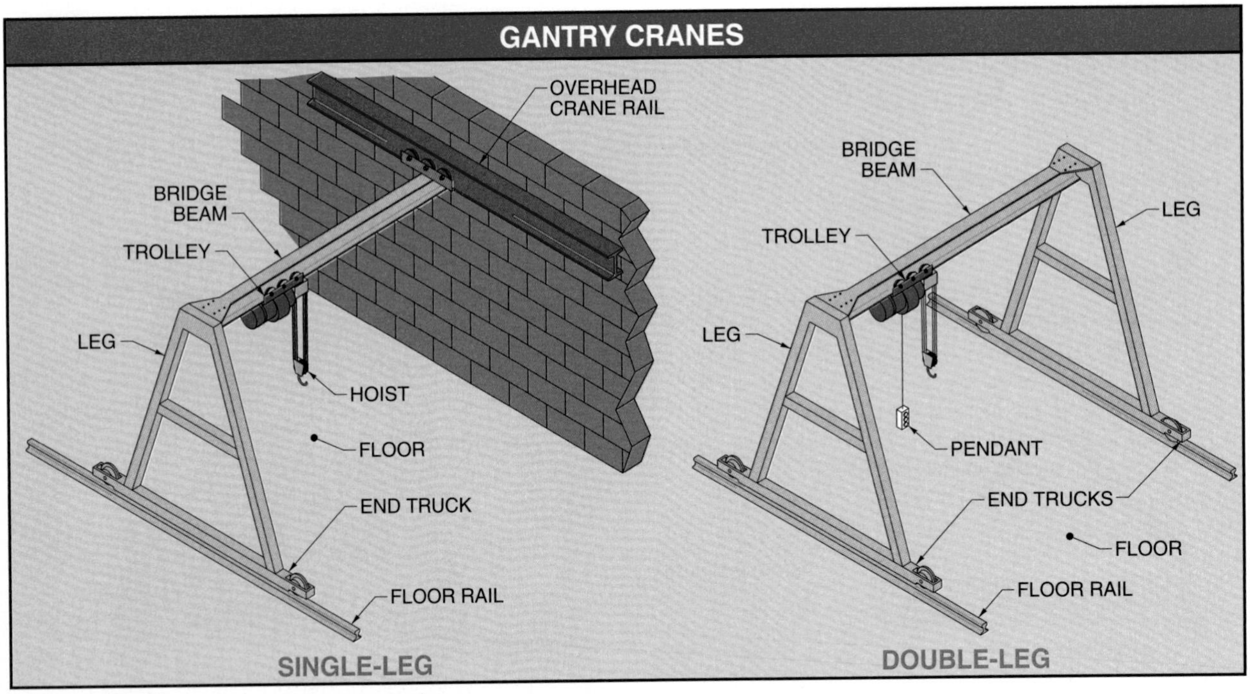

Figure 3-23. A gantry crane is a crane with bridge beams supported on legs.

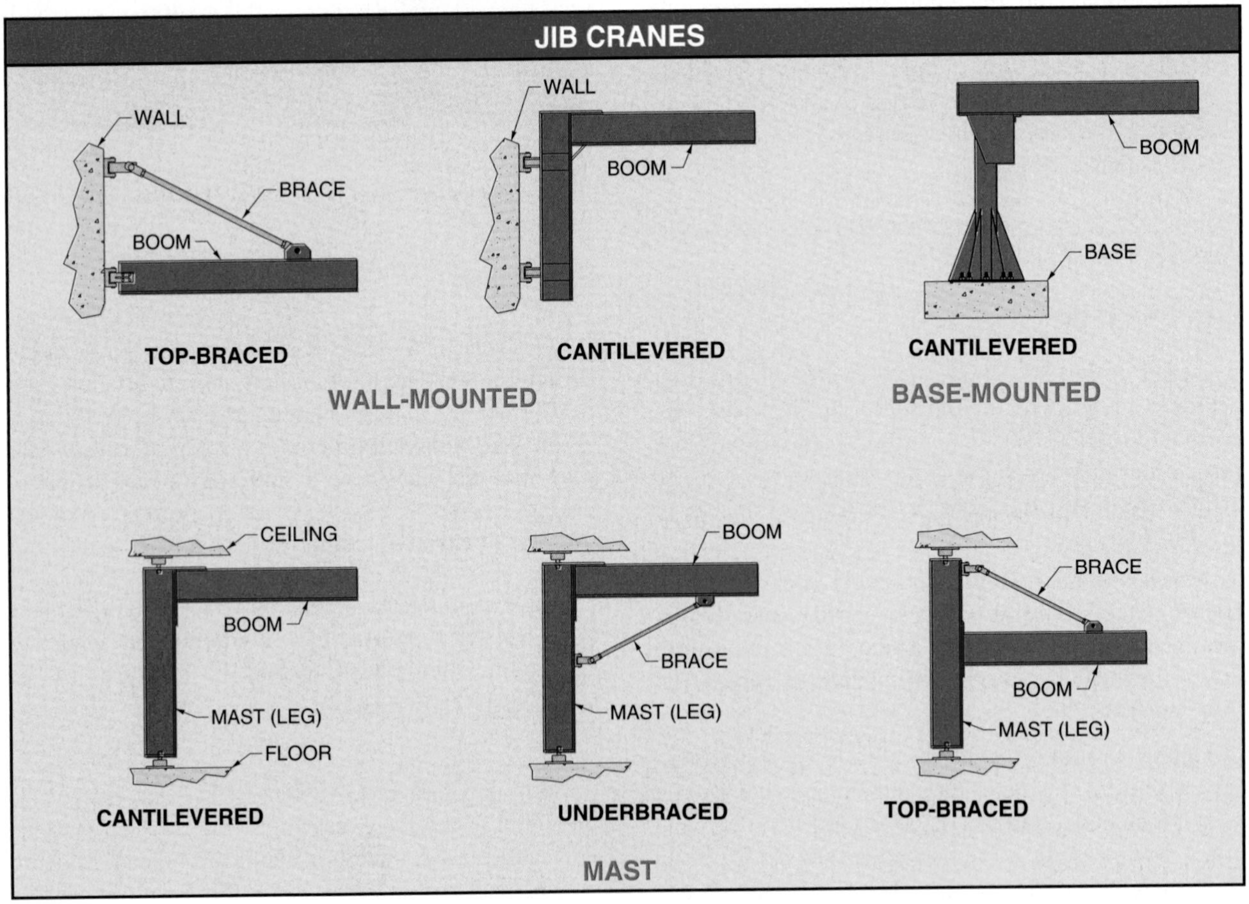

Figure 3-24. A jib crane is a crane that is mounted on a single structural leg.

The booms on all three types of cranes serve as tracks for hoisting mechanisms. Prior to lifting a jib crane load, slack must be taken up slowly. Removing slack minimizes shock to the crane boom. Rotating the boom on a jib crane should be done slowly and easily to prevent damage to the load, surroundings, or individuals.

> ***All exposed noncurrent-carrying metal parts of cranes, hoists, and accessories shall be joined together into a continuous conductor so that the entire crane or hoist is grounded per NEC® 250. Branch circuits for cranes and hoists shall be protected by fuses or ITCBs having ratings in accordance with NEC® Table 430-152.***

Overhead Cranes

An *overhead crane* is a crane that is mounted between overhead runways. Three common overhead crane configurations are the top-running crane with top-running hoist, the top-running crane with underhung hoist, and the underhung crane with underhung hoist. See Figure 3-25. Overhead cranes are normally operated from the cab or pendant pushbutton station. The *cab* is a compartment or platform attached to the bridge girder from which an operator may ride while controlling the crane.

Top-running cranes with top-running hoists are the most common overhead crane configuration. A *bridge girder* is the principal horizontal beam that supports the hoist trolley and is supported by the end trucks. A *hoist trolley* is the unit carrying the hoisting mechanism that travels on the bridge girder. End trucks are units consisting of truck frame, wheels, bearings, axles, etc., which support the bridge girder(s).

End trucks on top-running cranes travel on small gauge railroad rails mounted on top of the overhead runways. A *runway* is the rail and beam on which the crane operates. Either single- or double-bridge girders support the hoist trolley. The hoist trolley travels on rails mounted on top of the bridge girders in top-running cranes with top-running hoists.

Top-running cranes with underhung hoists are top-running single-bridge girder cranes with underhung hoist trolleys. The end trucks travel on rails mounted on top of the runways. A single-bridge girder supports the hoist trolley, which travels on the upper surface of the lower flange of the bridge girder.

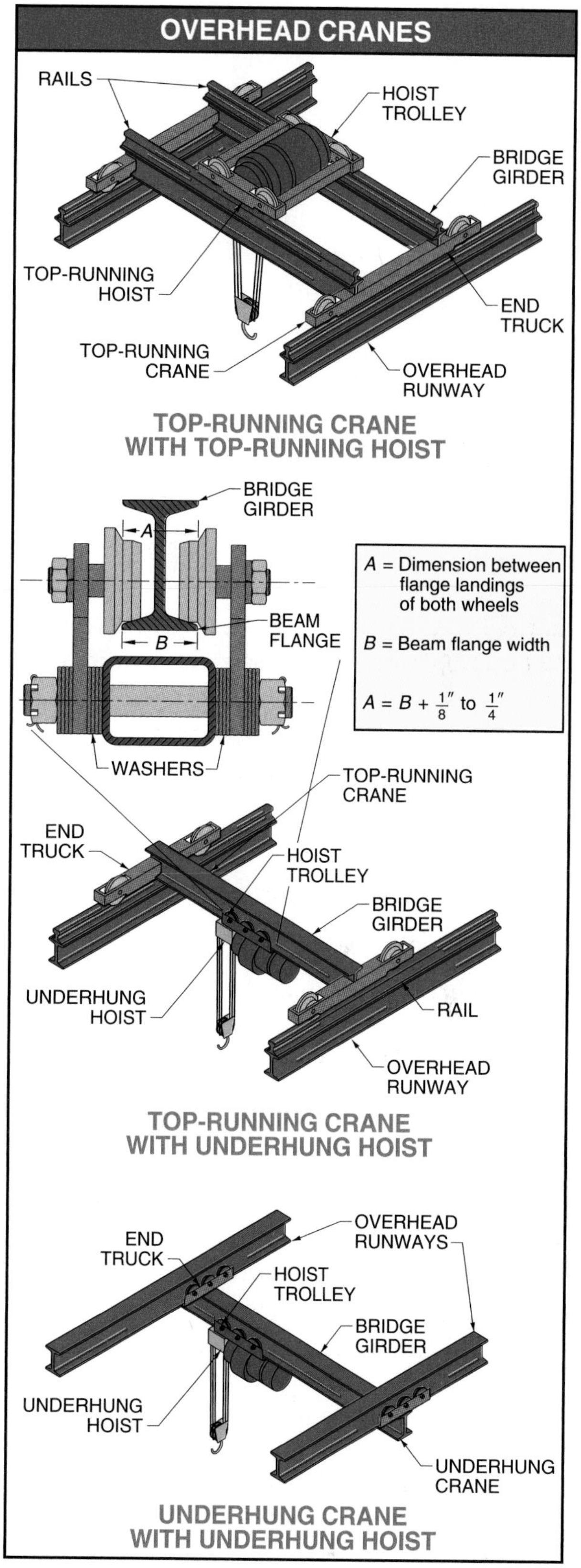

Figure 3-25. An overhead crane is a crane that is mounted between overhead runways.

Columbus McKinnon Corporation Industrial Products Division

Motorized sheet lifters from Columbus McKinnon Corporation are designed to lift bundles 48′ long by 12′ wide.

Underhung cranes with underhung hoists are another common crane configuration. This design has a single- or double-bridge girder supporting an underhung hoist trolley. Single- or double-bridge girders may support the hoist trolley, but the hoist trolley only travels on one bridge girder. The hoist trolley travels on the upper surface of the lower flange of the bridge girder.

Hoist trolleys should be inspected frequently and periodically. The inspection should include observation of the condition of the wheel bearings and checking for flat spots, cracks, or breaks on the trolley wheels and the trolley bumpers. Periodic inspections should include checking the dimension between wheel flanges.

On most hoist trolleys, the dimension between the flange landings of both wheels (A) must be at least ⅛″ greater than the beam flange width (B) and not more than ¼″. The addition or removal of washers may be necessary to obtain the proper dimension. Distribute washers equally between flanges to ensure that the hoist is centered under the bridge girder. Some hoist trolleys are not adjustable. Check the manufacturer's manual for proper use.

Crane Operation

Each crane operator is held directly responsible for the safe operation of the crane. Lifting may be performed after ensuring that the condition of all rigging, hoisting, and crane components are within specifications.

Once the hoist is brought directly over the load's center of gravity, check that no lines or chains are twisted, overwrapped, or unseated. Connect the rigging to the block hook, ensuring that the hook latch is closed and the lifting devices are fully seated in the hook bowl. The hook should be started upward slowly until all slack has been removed from the slings when slings are used to lift a load. The load is lifted slowly until it is clearly off the floor and properly balanced. At this time, the operator may increase the lifting speed. Do not use a crane for a side pull. See Figure 3-26. Relocate the crane if the load does not appear to be directly under the crane after the slack is removed. A swinging load is dangerous.

Cranes should be controlled to move smoothly and gradually to avoid abrupt, jerky movements. When loads are lowered, the lowering speed should be decreased gradually. Do not attempt to lift a load that is beyond the load capacity of any of the rigging, lifting, or crane components. Hoist brakes should be tested with the load a few inches off the floor. This is especially necessary if the load is near or at capacity. Check the load for drift in the raise and lower positions. *Drift* is the slippage of a load caused by insufficient braking. Return the load to the floor and notify the supervisor if a drift of 1″ or more is noticed. A suspended load must not be left unattended.

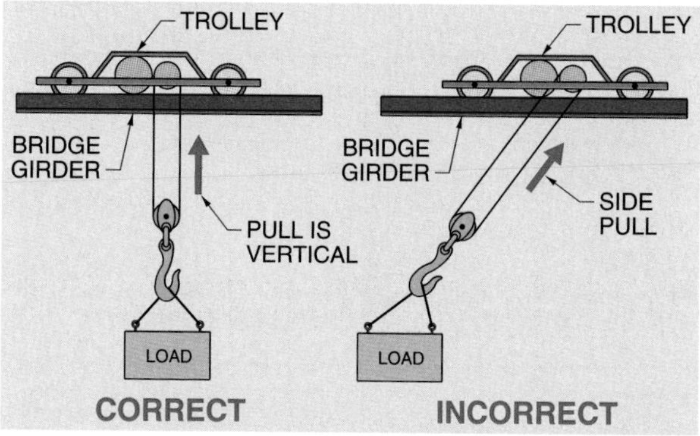

Figure 3-26. A crane should not be used for a side pull.

Place all crane controls in the OFF position if power is interrupted during operation. After power is restored, check all controls for satisfactory operation and correct direction before use.

To be safe and efficient, crane operation requires skill, the exercise of extreme care and good judgment, alertness, and concentration. Crane operators must adhere to proven safety rules and practices as outlined in applicable and current ANSI and OSHA standards.

Hand Signals. Many cab-operated cranes require the assistance of an additional person. The assistant, working in conjunction with the cab operator, gives crane and hoist communication in the form of standard hand signals. See Figure 3-27.

Loads should not be moved unless standard crane signals are clearly given, seen, and understood. The operator must pay particular attention to the required moves signaled by the assistant. The operator takes signals only from the assistant. The only exception to this rule is that the operator must obey a stop signal, at all times, no matter who gives it. A stop signal may be given by anyone.

General Practice Conditions. Generally, no person should be permitted to operate a crane who cannot speak the appropriate language, read and understand the printed instructions, or who is not of legal age to operate the equipment. Anyone who is hearing or eyesight impaired or may be suffering from heart or other ailments which might interfere with safe performance should not operate a crane. The operator must have carefully read and studied the operation manual and have been properly instructed.

Crane designs should conform to applicable specifications of the American Institute of Steel Construction (AISC), Uniform Building Code (UBC), Crane Manufacturers Association of America (CMAA), and American National Standards Institute (ANSI).

With the mainline switch open (power OFF), the crane operator should operate each master switch or pushbutton in both directions to get the "feel" of each device and to determine that they do not bind or stick in any position. The operator should report the condition to the proper supervisor immediately if any switch or pushbutton binds or sticks in any position.

North American Industries, Inc.

Overhead bridge cranes from North American Industries are available with variable speed controls for precise increases and decreases in travel speed to prevent swinging loads caused by ON/OFF controls.

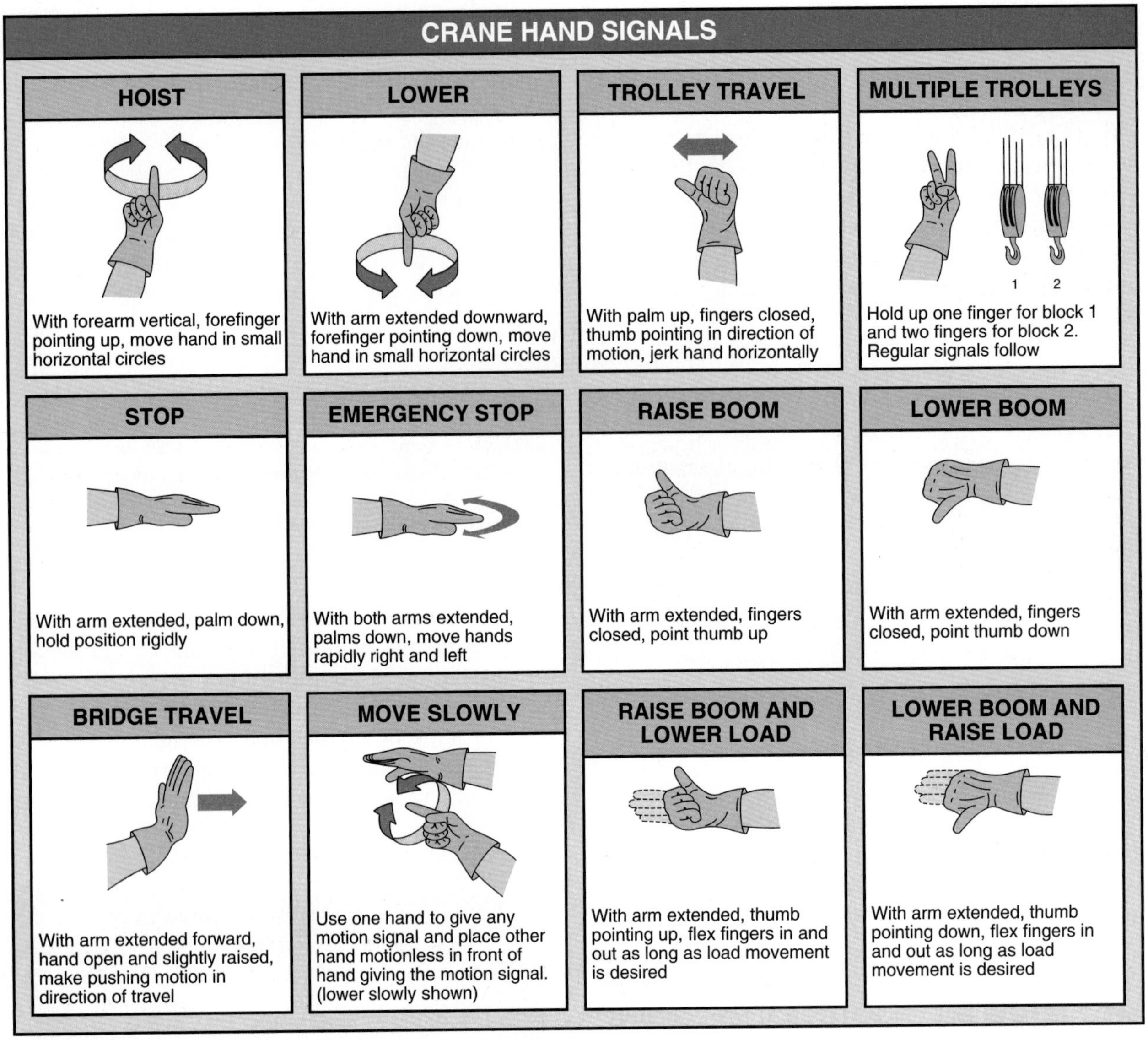

Figure 3-27. Standard hand signals are given as communication between the crane operator and the assistant.

Ladders and Scaffolds

Chapter 4

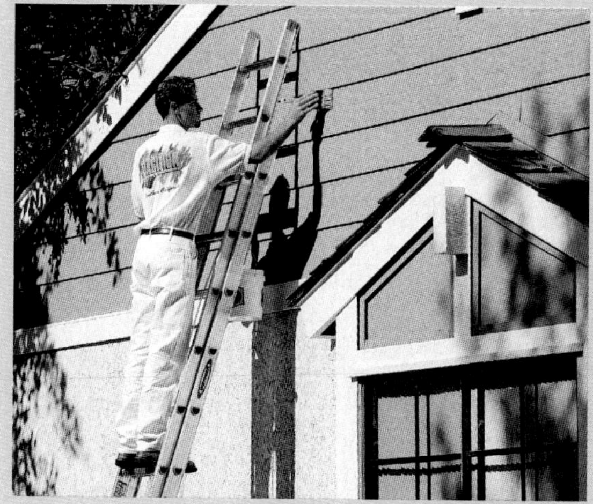

Werner Ladder Co.

Ladders and scaffolds are used in industry to raise workers to correct working heights. Ladders are constructed of wood, metal, or fiberglass and scaffolds are constructed of wood or metal. Proper safety precautions must be observed when working from ladders and scaffolds.

LADDERS

A *ladder* is a structure consisting of two siderails joined at intervals by steps or rungs for climbing up and down. Ladders are manufactured in lengths of 3′ to 50′. Ladders are constructed of wood, metal, or fiberglass. All ladders, regardless of the construction material, are manufactured to meet the same standards. Ladder purchase is generally decided by cost, use, and in some cases, portability. Industrial ladders include fixed, single, extension, and stepladders. See Figure 4-1.

All ladders must be used only for the purpose for which they are designed. Ladders must not be used as pry bars or horizontal platforms. All ladders must be equipped with nonslip safety feet such as butt spurs or foot pads. See Figure 4-2. A *butt spur* is a notched, pointed, or spiked end of a ladder which helps prevent the ladder butt from slipping. Butt spurs are generally attached to long ladders such as extension ladders. A *foot pad* is a metal swivel attachment with rubber or rubber-like tread which helps prevent the ladder butt from slipping. Foot pads are normally attached to short ladders such as stepladders.

Wood Ladders

A *wood ladder* is a ladder constructed of wood. The advantages of wood ladders include relatively low cost, ability to take abuse, nonconductivity of electricity, and good temperature insulating qualities. Nonconductivity of electricity makes wood ladders relatively safe to use when working around power lines and service-entrance conductors. Wood ladder temperature insulating qualities prevent the transmission of excessive heat or cold to a worker.

The disadvantages of wood ladders include deterioration with age, shrinkage causing loose rungs when the wood becomes dry and warm, and the need to maintain their integrity by regularly coating them with clear shellac or linseed oil. Wood ladders should never be painted. Paint covers any defects that may otherwise be seen.

Wood ladders must be stored on their edge away from excessive dampness, dryness, and heat to reduce the possibility of warping. All wood ladders should be stored by hanging horizontally on hooks spaced 4′ to 6′ apart. This minimum storage hook spacing prevents sagging and offers easy access to a ladder for inspection.

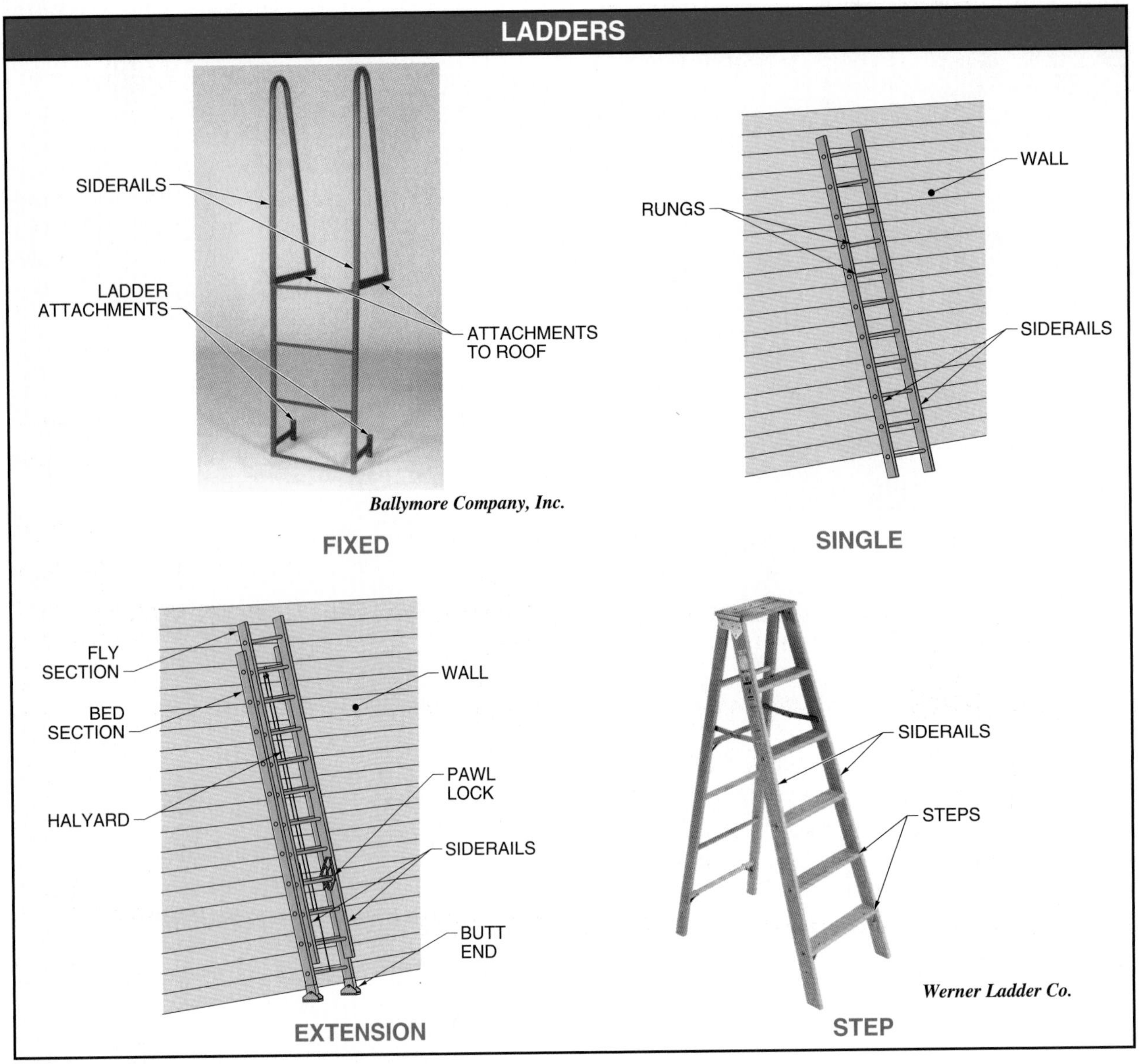

Figure 4-1. A ladder is a structure consisting of two siderails joined at intervals by steps or rungs for climbing up and down.

Metal Ladders

A *metal ladder* is a ladder constructed of metal (normally aluminum). The advantages of metal ladders include relatively light weight, extreme toughness, resistance to splintering or cracking when subjected to impact, and resistance to deterioration with age. These qualities reduce maintenance of metal ladders to inspection and lubrication, without the need for sanding and refinishing.

A disadvantage of metal ladders is their tendency to become very cold in winter and very hot in summer. In addition, metal ladders should not be used within 4′ of electrical circuits or equipment. Extreme caution is necessary if metal ladders are used near electrical power lines, service-entrance conductors, and electrical equipment because of their good electrical conductivity.

Fiberglass Ladders

A *fiberglass ladder* is a ladder constructed of fiberglass. Fiberglass ladders are rapidly becoming the most popular type of ladder, particularly in stepladders. Advantages of fiberglass ladders are that they do not conduct electricity when dry, can withstand considerable abuse, do not require surface finishing, and are more comfortable to use than metal ladders in cold or hot environments.

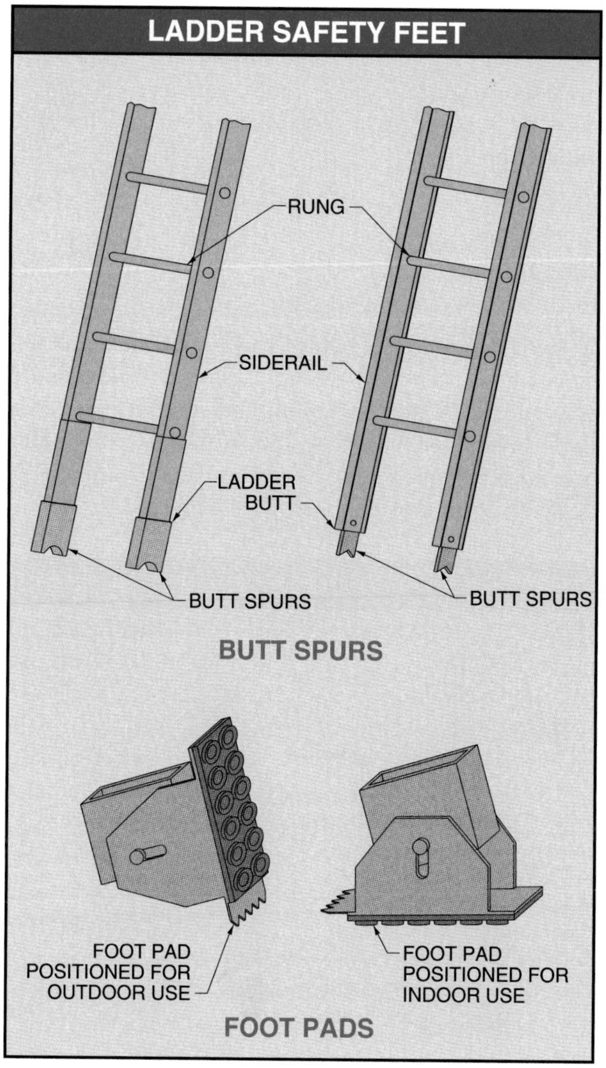

Figure 4-2. Ladders must be equipped with nonslip safety feet such as butt spurs or foot pads.

The disadvantages of fiberglass ladders are that they may crack and fail when overloaded, and may crack and chip when severely impacted.

Regulations and Standards

Regulations and standards for the use, design, and testing of ladders are published by various federal, state, and standards organizations. OSHA publishes federal industry standards for ladders in publications CFR Title 29 Part 1910.25, *Portable Wood Ladders*, Part 1910.26, *Portable Metal Ladders*, and Part 1910.27, *Fixed Ladders*.

ANSI publishes standards for ladders in ANSI A14.1-1994, *Ladders – Portable Wood, Safety Requirements for*, ANSI A14.3-1992, *Ladders – Fixed – Safety Requirements*, and ANSI A14.5-1992, *Ladders – Portable Reinforced Plastic – Safety Requirements.*

Duty Rating. *Ladder duty rating* is the weight (in lb) a ladder is designed to support under normal use. See Figure 4-3. The four ladder duty ratings are:

- Type IA – Extra heavy-duty, industrial, 300 lb capacity
- Type I – Heavy-duty, industrial, 250 lb capacity
- Type II – Medium-duty, commercial, 225 lb capacity
- Type III – Light-duty, household, 200 lb capacity

Werner Ladder Co.

Figure 4-3. Ladder duty rating is the weight a ladder is designed to support under normal use.

Over 11% of injuries that require time away from work are related to falling from ladders.

Climbing Techniques

Climbing may begin only after a ladder is properly secured. Climbing movements should be smooth and rhythmical to prevent ladder bounce and sway. Safe climbing employs the three-point contact method. In the three-point contact method, the body is kept erect, the arms straight, and the hands and feet make the three points of contact. Two feet and one hand, or two hands and one foot are in contact with the ladder rungs at all times.

Avoid reaching above shoulder level to grasp a rung to maintain balance and unobstructed knee movements. Each hand should grasp the rungs with the palms down and the thumb on the underside of the rung. Upward progress should be caused by the push of the leg muscles and not the pull of the arm muscles. When climbing, tools, parts, or equipment must be secured in a pouch or raised and lowered with a rope.

Fixed Ladders

A *fixed ladder* is a ladder that is permanently attached to a structure. Fixed ladders are commonly constructed of steel or aluminum. Fabrication of a fixed ladder, including design, materials, and welding, must be done under the supervision of a qualified licensed structural engineer. See Figure 4-4.

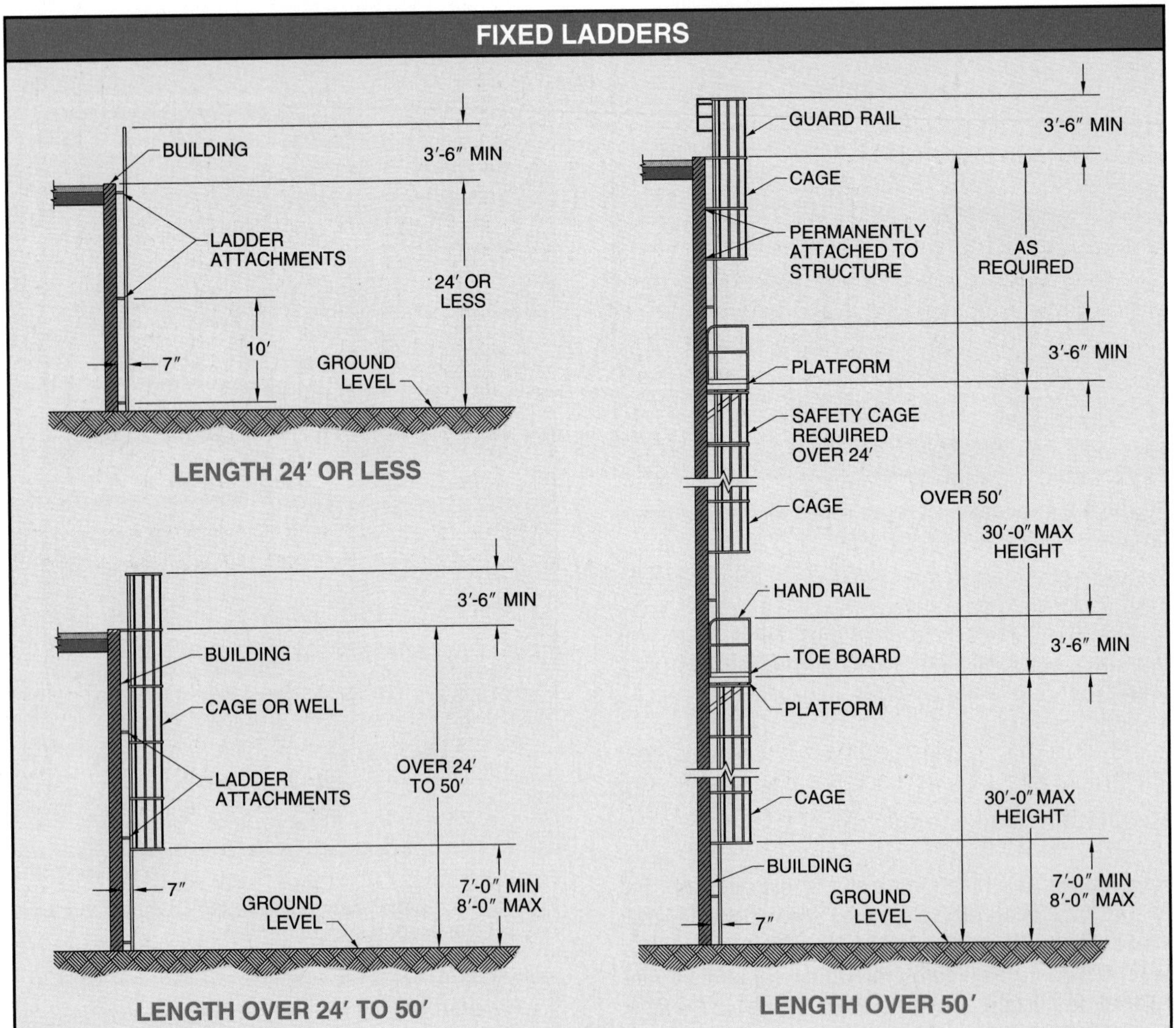

Figure 4-4. A fixed ladder is a ladder that is permanently attached to a structure.

For fixed ladders, the width between siderails is normally 16″ and the spacing between rungs is 12″. Each rung cross-section must be a minimum of 3⁄4″ in diameter. The rungs must be a minimum of 1″ in diameter if a steel fixed ladder is located in an unusually corrosive environment. Fixed ladder siderails for use in normal conditions must be made of material sized for gripping and are typically 2½″ wide and 3⁄8″ thick. Unusually corrosive environments require siderail material ½″ thick.

Fixed ladders are normally attached to a building or structure at spaces of 10′ or less with a minimum distance of 7″ from the center of the rung to the building. Siderails of a fixed ladder must extend a minimum of 3′-6″ above the landing for a walk-through ladder and 4′-0″ above the landing for a side access ladder. Each rung of a fixed ladder must support 250 lb.

Fixed ladders over 24′ in length must have a cage, well, or ladder safety system. A *cage* is a barrier or enclosure mounted on the siderails of a fixed ladder or fastened to the structure. Cages are also referred to as cage guards or basket guards. A *well* (shaft) is a walled enclosure around a fixed ladder. A well provides a climber the same protection as a cage. Cages must start no less than 7′-0″ and no more than 8′-0″ from the ground or platform.

A *ladder safety system* is an assembly of components whose function is to arrest the fall of a worker. Ladder safety systems consist of a carrier and its associated attachments. A *carrier* is the track of a ladder safety system consisting of a flexible cable or rigid rail secured to the ladder or structure. The carrier is attached to a safety sleeve. A *safety sleeve* is a moving element with a locking mechanism that is connected between a carrier and the worker's body belt.

A cage, well, or ladder safety system must be provided where a single length of climb is greater than 24′ but less than 50′. The ladder must consist of multiple sections (50′ maximum each) if a cage or well is used. A landing platform must be provided at least every 50′ within the length of climb. A *platform* is a landing surface which provides access/egress or rest from a fixed ladder. The length of climb may be continuous if a ladder safety system is used. Rest platforms must be provided at maximum intervals of 150′ on fixed ladders using a ladder safety system.

Platforms should be a minimum of 24″ wide by 30″ long with railings and toe board. Railings around platforms are commonly 3′-6″ high. Adjacent ladder sections are offset at each landing by no more than 18″.

Fixed ladders are installed in a range between 60° and 90° from horizontal. The range between 60° and 75° is considered to be the fixed ladder substandard pitch range and is to be used only for special conditions. Fixed ladders are not to be installed over 90° from horizontal. See Figure 4-5.

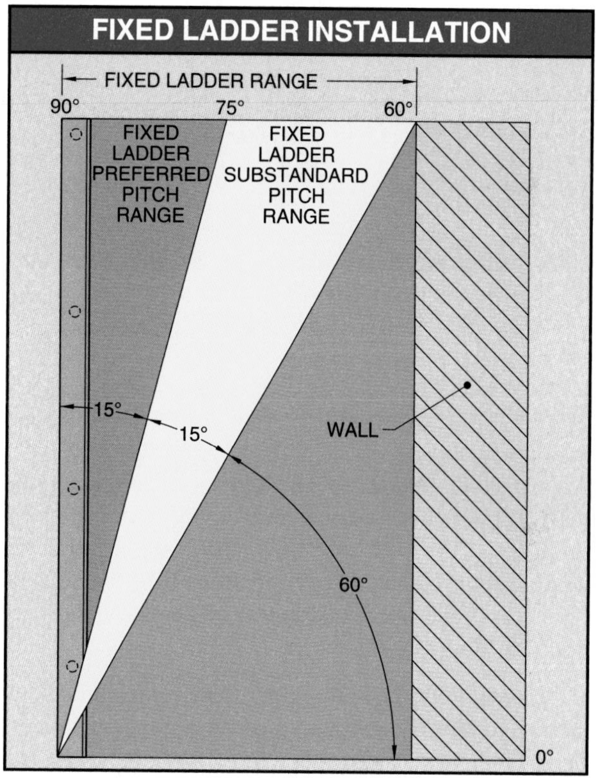

Figure 4-5. Fixed ladders are installed in a range between 60° and 90° from horizontal.

Single Ladders

A *single ladder* is a ladder of fixed length having only one section. Typical lengths of single ladders vary from 6′ to 24′. Single ladders offer the convenience of use by one person. However, they are limited in their versatility because a given length ladder may be safely used only within a small height range.

To ascend or descend a ladder safely, use the 3-point climbing method, in which one hand and two feet or two hands and one foot are in contact with the ladder rungs at all times.

Extension Ladders

An *extension ladder* is an adjustable-height ladder with a fixed bed section and sliding, lockable fly section(s). The *bed section* is the lower section of an extension ladder. The *fly section* (first fly, second fly, etc.) is the upper section(s) of an extension ladder. See Figure 4-6.

A *pawl lock* is a pivoting hook mechanism attached to the fly section(s) of an extension ladder. Pawl locks are used to hold the fly section(s) at the desired height. The three main parts of a pawl lock are the hook, finger, and spring. See Figure 4-7.

The rungs nest into the hook of each pawl, preventing downward movement of the fly section. The fly section is lowered by first raising the fly section just enough for a rung to pass below the finger of the pawl lock. The fly section is then lowered. When lowered, the rung forces the finger up, preventing the rungs from nesting into the hook. To hold the fly section in place, the pawl lock must be lowered slightly below a rung to allow the finger to drop. The fly is then raised slightly to allow the spring to force the pawl hook over the rungs.

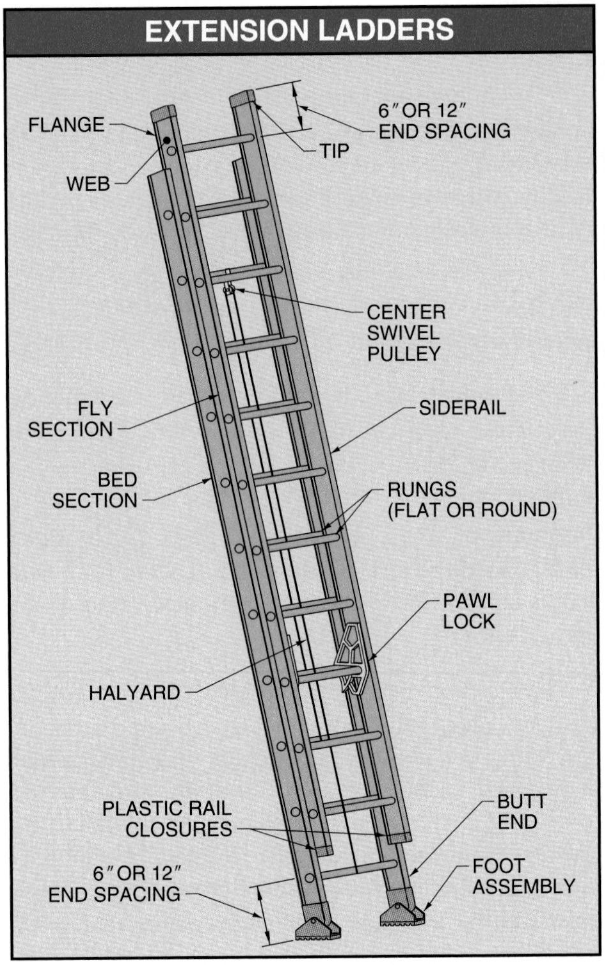

Figure 4-6. An extension ladder is an adjustable-height ladder with a fixed bed section and a sliding, lockable fly section(s).

Many fly sections are raised and lowered with the use of a halyard, halyard anchor, and pulley. A *halyard* is a rope used for hoisting or lowering objects. A halyard must be a minimum of $\frac{3}{8}''$ in diameter with a minimum breaking strength of 825 lb. The halyard is threaded through the pulley attached to the top rung of the bed section. One end of the rope is attached to the bottom rung of the fly section and the other end is usually tied off at the bottom. See Figure 4-8.

> ***All hardware on wood ladders shall be made of aluminum, wrought iron, steel, malleable iron, or another material that is adequate in strength for the purpose intended and free from sharp edges and from sharp projections in excess of $\frac{1}{64}''$.***

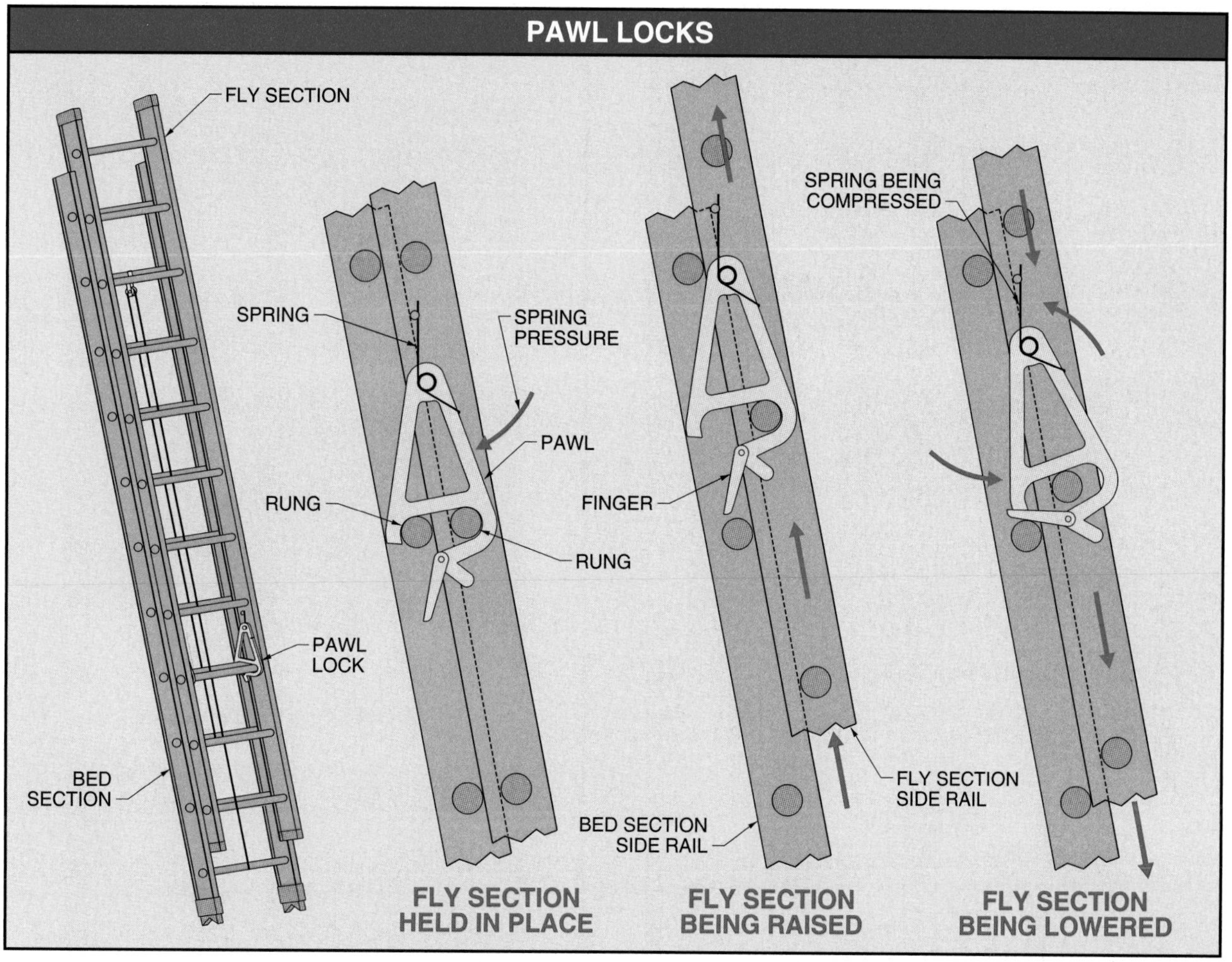

Figure 4-7. A pawl lock is a pivoting hook mechanism attached to the fly section(s) of an extension ladder.

Raising Ladders. Raising a ladder involves a smooth, proper, and safe operation. Care must be taken before beginning a raise to ensure that electrical conductors or equipment are not present.

Single or extension ladders may be raised with the ladder tip away from the building or with the ladder tip against the building. The method used is determined by whether the ladder is raised indoors or outdoors, the presence of overhead obstructions, the setup area, and the ladder size. See Figure 4-9.

To raise a ladder with the ladder tip away from the building, place the butt end of the ladder against the building with the fly section retracted and to the down side. Grasp the rung at the ladder tip with both hands. Raise the tip and walk under the ladder, grasping succeedingly lower rungs while walking toward the building.

When the ladder is erect, hold the ladder against the wall by placing force with one hand at about eye level. Place one foot approximately 10″ to 12″ in front of one of the ladder legs. Use the free hand to apply an upward pressure and an outward pull to slide the butt of the ladder to the foot. This procedure is continued until the ladder is in the approximately correct position and angle. Adjust the fly section for the proper height and readjust the ladder as necessary for the proper angle.

To raise a ladder with the ladder tip against the building, place the ladder tip against the building with the fly section fully retracted and to the upside. Standing with back to the wall, lift the ladder tip while pulling the butt end toward the wall. The ladder tip remains against the wall while repositioning for another lift and pull. This procedure is continued until the ladder is in the approximately correct position and angle. Adjust the fly section for the proper height and readjust the ladder as necessary for the proper angle.

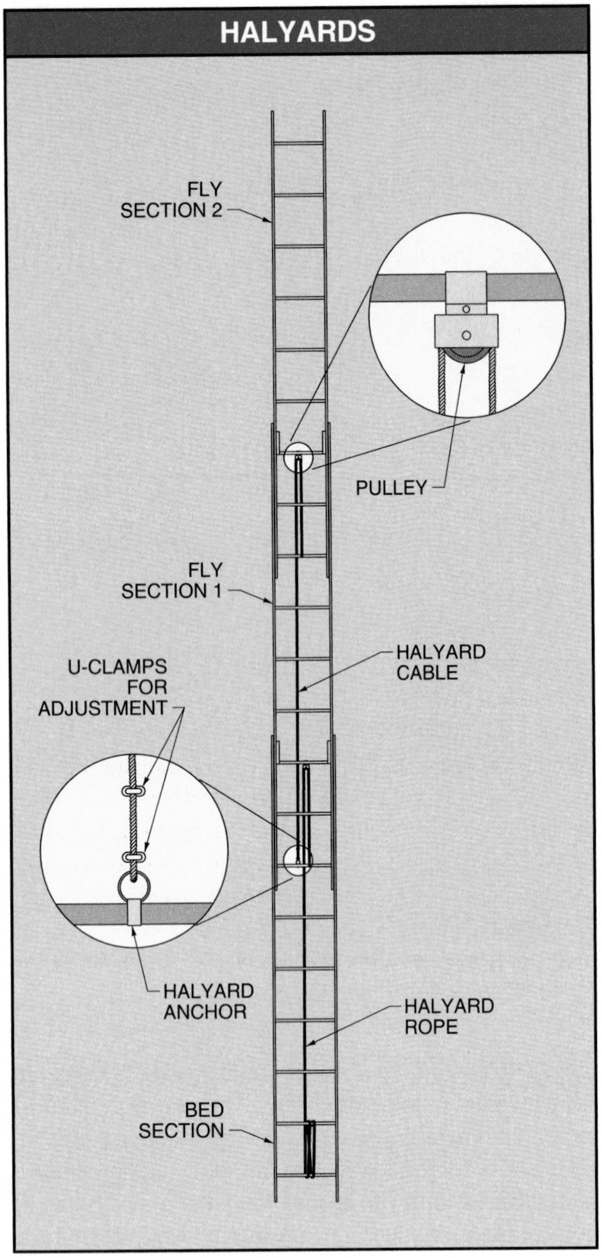

Figure 4-8. Extension ladder fly sections are raised and lowered with the use of halyards.

Figure 4-9. Single or extension ladders are raised with the tip away from or against the building.

Never attempt to raise long extension ladders alone. At least two people are required to raise long extension ladders into position, with one person on each side of the ladder. Lifting and pulling are done together until the ladder reaches its final position. Three or more people may be required to raise extension ladders to overhead beams or pipes. Two people raise the ladder and one or two people firmly secure the butt end.

Extension Overlap and Height. Extension ladders must have positive stops to prevent overextension of the fly section(s). The overlap of the fly section must be at least 3′ for extension ladders up to 36′, 4′ for extension ladders over 36′ and up to 48′, and 5′ for extension ladders over 48′ and up to 60′.

Extension ladders are positioned on a 4 : 1 ratio (75° angle). For every 4′ of working height, 1′ of space is required at the base. *Working height* is the distance from the ground to the top support. The *top support* is the area of a ladder that makes contact with a structure. For example, a 12′ ladder should be placed at an angle that places the butt end 3′ from the wall. See Figure 4-10.

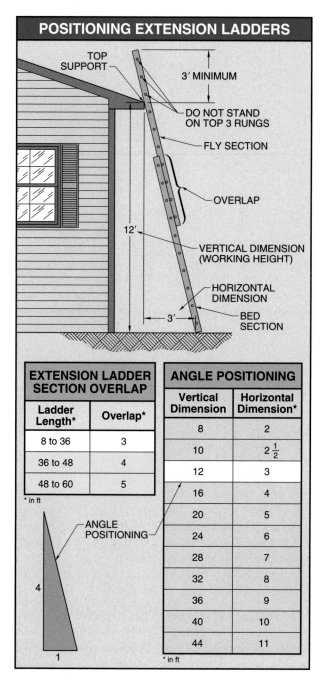

POSITIONING EXTENSION LADDERS

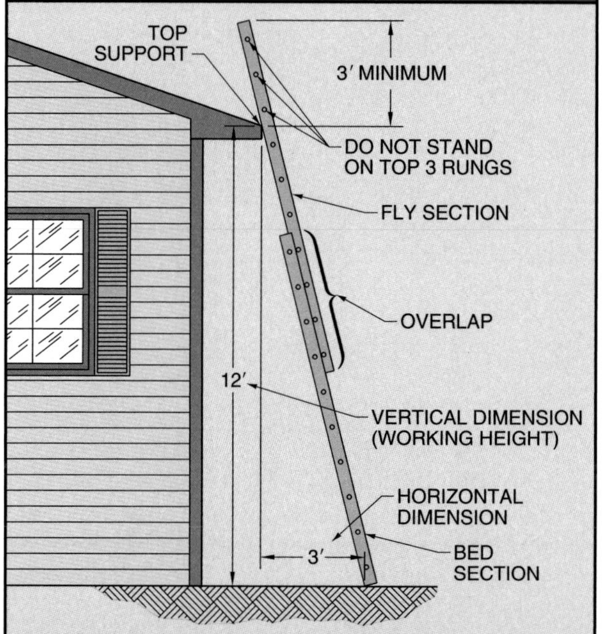

EXTENSION LADDER SECTION OVERLAP

Ladder Length*	Overlap*
8 to 36	3
36 to 48	4
48 to 60	5

* in ft

ANGLE POSITIONING

Vertical Dimension	Horizontal Dimension*
8	2
10	$2\frac{1}{2}$
12	3
16	4
20	5
24	6
28	7
32	8
36	9
40	10
44	11

* in ft

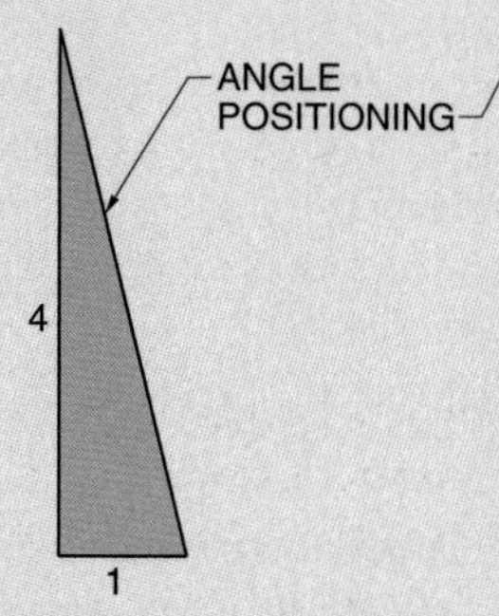

Figure 4-10. Extension ladders are positioned on a 4 : 1 ratio.

Selection of a properly-sized extension ladder requires knowledge of the vertical height of the top support. In applications where the top support is a roof eave, an additional working length is needed to obtain the required 3′ to 5′ extension beyond the top support.

The tip of a single or extension ladder should be secured at the top to prevent slipping and must be at least 3′ above the roof line or top support. Never stand on the top three rungs of a single or extension ladder. Ladders over 15′ should also be secured at the bottom.

The spacing between rungs shall be on 12″ centers ±1⁄8″ except for step stools where the spacing shall be uniform but not less than 8″ ±1⁄8″ nor more than 12″ ±1⁄8″ measured along the side rail.

Ladder Jacks. A *ladder jack* is a ladder accessory that supports a plank to be used for scaffolding. A *plank* is a board 2″ to 4″ thick and at least 8″ wide. A ladder jack is positioned on the inside or outside of the rungs. A ladder jack has hooks at the top and the bottom which attach to the ladder rungs. The hooks on the jacks are placed close to the siderails to allow suitable load force. A plank is placed on the horizontal projection, providing a work platform. See Figure 4-11. To prevent cracking and failing of a ladder jack assembly, use only IA, heavy-duty extension ladders with ladder jacks.

LADDER JACKS

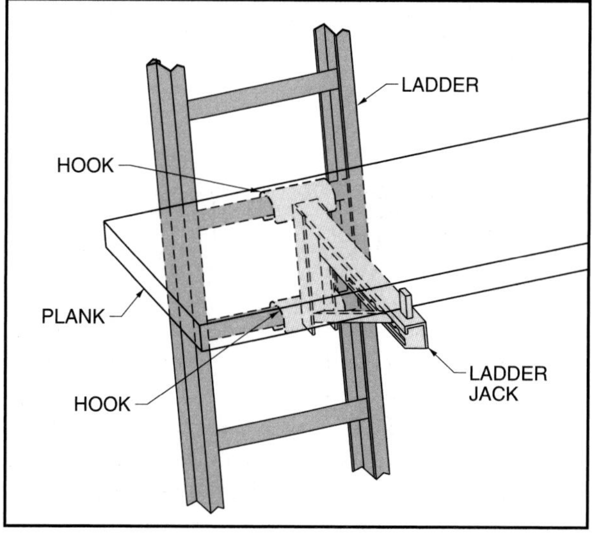

Figure 4-11. A ladder jack is a ladder accessory that supports a plank to be used for scaffolding.

Standoffs. A *standoff* is a ladder accessory that holds a single or an extension ladder a fixed distance from a wall. A standoff is attached to the ladder with adjustable U-bolts. Adjustable, non-slip tips on each end of the standoff help protect the wall surface from marring or scratching. Standoffs provide a comfortable working distance between a wall and the worker. Standoffs are particularly useful for painting around windows, etc. See Figure 4-12.

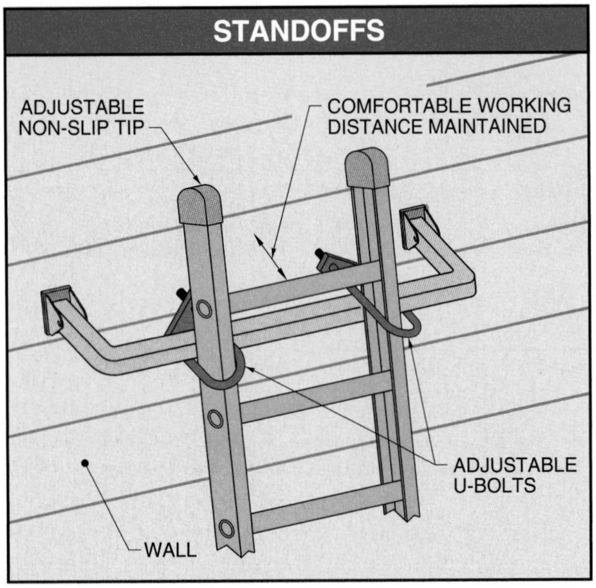

Figure 4-12. A standoff is a ladder accessory that holds a single or an extension ladder a fixed distance from a wall.

Werner Ladder Co.

Ladder jacks are used on extension ladders to create a support for a platform.

Stepladders

A *stepladder* is a folding ladder that stands independently of support. Stepladders are commonly 2′-0″ to 8′-0″ in length. Because stepladders are easily portable and provide adequate working height for many applications, they are used more often than any other ladder. See Figure 4-13.

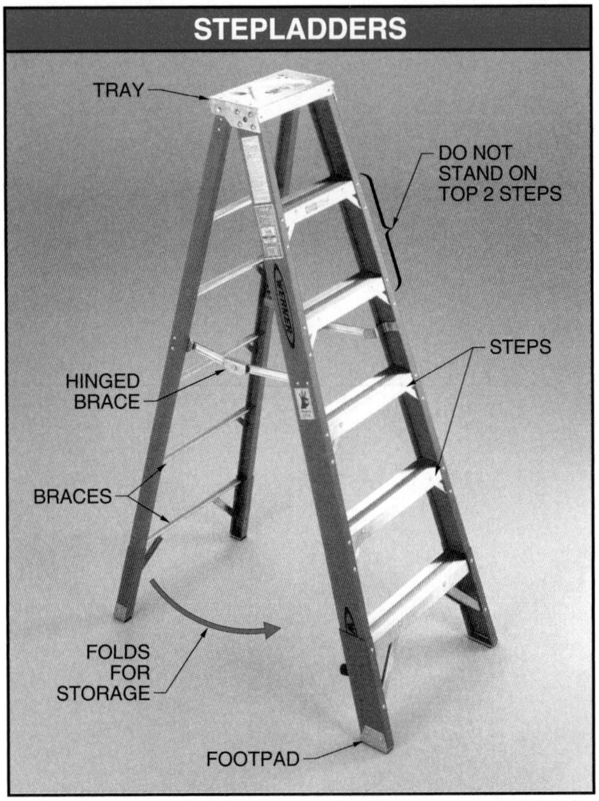

Werner Ladder Co.

Figure 4-13. A stepladder is a folding ladder that stands independently of support.

Stepladders must be self-supporting, stable, and free from racking. *Racking* is the ability to be forced out of shape or form. Stepladders have steps instead of rungs. The steps must be horizontal when the ladder is open. Top-quality stepladders may have braces at the sides, folding shelves (trays) for utensils, platforms at the top with three-sided guard rails, or steps on both front and back of the ladder.

Never lean out or reach to one side of a stepladder. Reposition the stepladder so the work area can be easily and conveniently reached. Never stand on the top two steps of a stepladder. There is no support and it is easy to lose balance. Do not use metal stepladders near electrical conductors or equipment.

SCAFFOLDS

A *scaffold* is a temporary or movable platform and structure for workers to stand on when working at a height above the floor. Three basic types of scaffolds are pole, sectional metal-framed, and suspension scaffolds. See Figure 4-14.

Scaffold design shall be such as to produce a platform unit that will safely support the specified loads. The material selected shall be of sufficient strength to meet the performance requirements and shall be protected against corrosion unless inherently corrosion-resistant to normal atmospheric conditions.

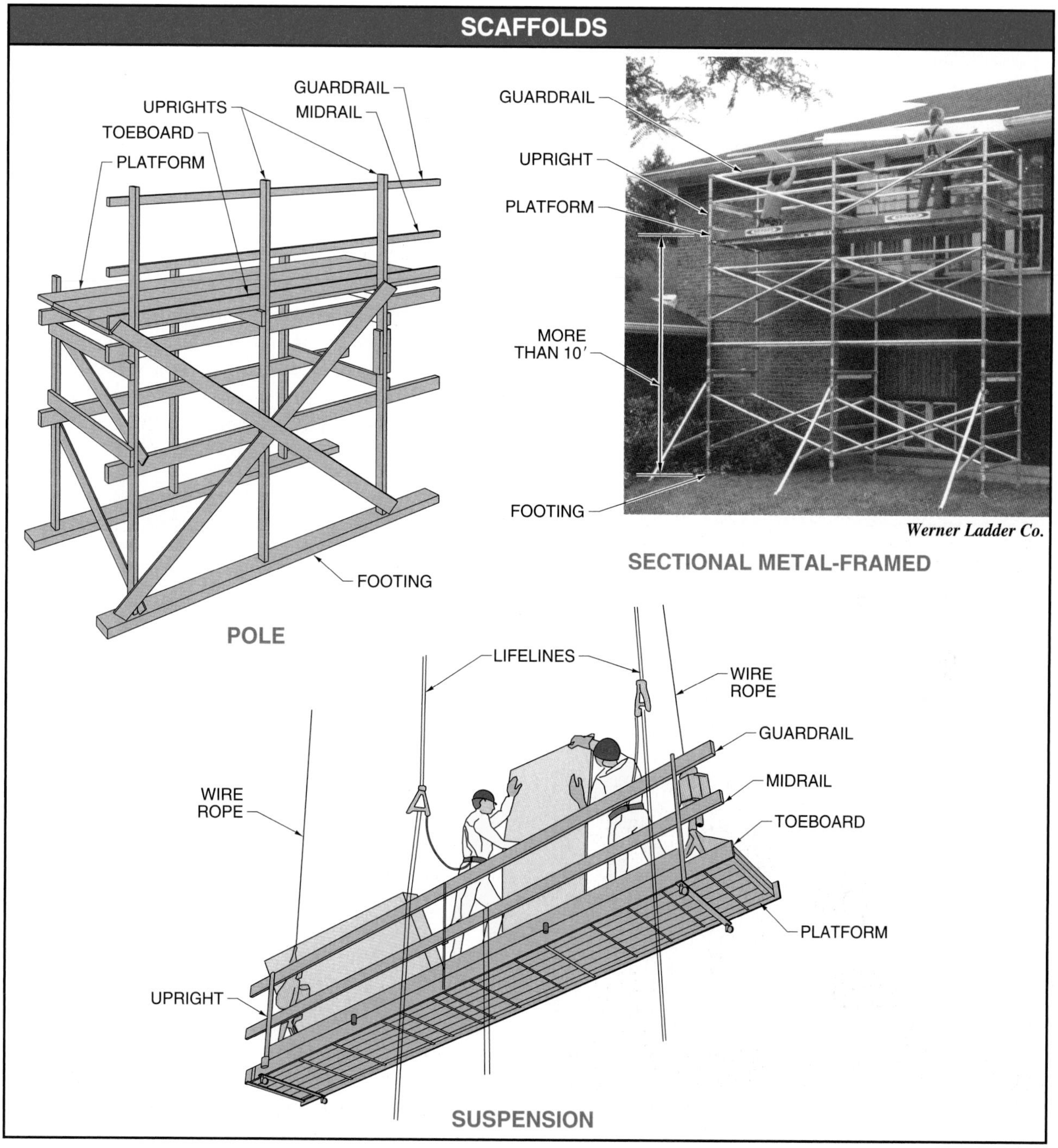

Figure 4-14. Three basic types of scaffolds are pole, sectional metal-framed, and suspension scaffolds.

A scaffold generally consists of wood planks or metal platforms to support workers and their materials. Scaffolds must not be supported by any unstable object, such as boxes, concrete blocks, or loose bricks. Scaffold footing must be sound and stable and must not settle or displace while carrying the maximum intended load. A *maximum intended load* is the total of all loads, including the working load, the weight of the scaffold, and any other loads that may be anticipated. Scaffolds and their components must be capable of supporting at least four times their maximum intended load.

All scaffolds 10′ or more above ground must have guardrails, midrails, and toeboards. A *guardrail* is a rail secured to uprights and erected along the exposed sides and ends of a platform. A *midrail* is a rail secured to uprights approximately midway between the guardrail and the platform. A *toeboard* is a barrier to guard against the falling of tools or other objects. Toeboards are secured along the sides and ends of a platform. Guardrails must be installed no less than 38″ or more than 45″ high, with a midrail. Guardrail and midrail support is to be at intervals of no more than 10′.

Regulations and Standards

OSHA regulations governing the use of scaffolds require that a scaffold may be erected, moved, or dismantled only under the supervision of a competent person. A *competent person* is a person capable of recognizing and evaluating employee exposure to hazardous substances or to other unsafe conditions and of specifying the necessary protection and precautions to be taken to ensure the safety of all employees. Refer to OSHA and ANSI for guidelines. OSHA industry standards are found in CFR Title 29 Part 1910.28 and 29, *Safety Requirements for Scaffolding* or Part 1926.451, *Scaffolding*. ANSI standards are found in ANSI A10.8-2001, *Safety Requirements for Scaffolding.*

Pole Scaffolds

A *pole scaffold* is a wood scaffold with one or two sides firmly resting on the floor or ground. See Figure 4-15. A *single-pole scaffold* is a wood scaffold with one side resting on the floor or ground and the other side structurally anchored to the building. A *double-pole scaffold* is a wood scaffold with both

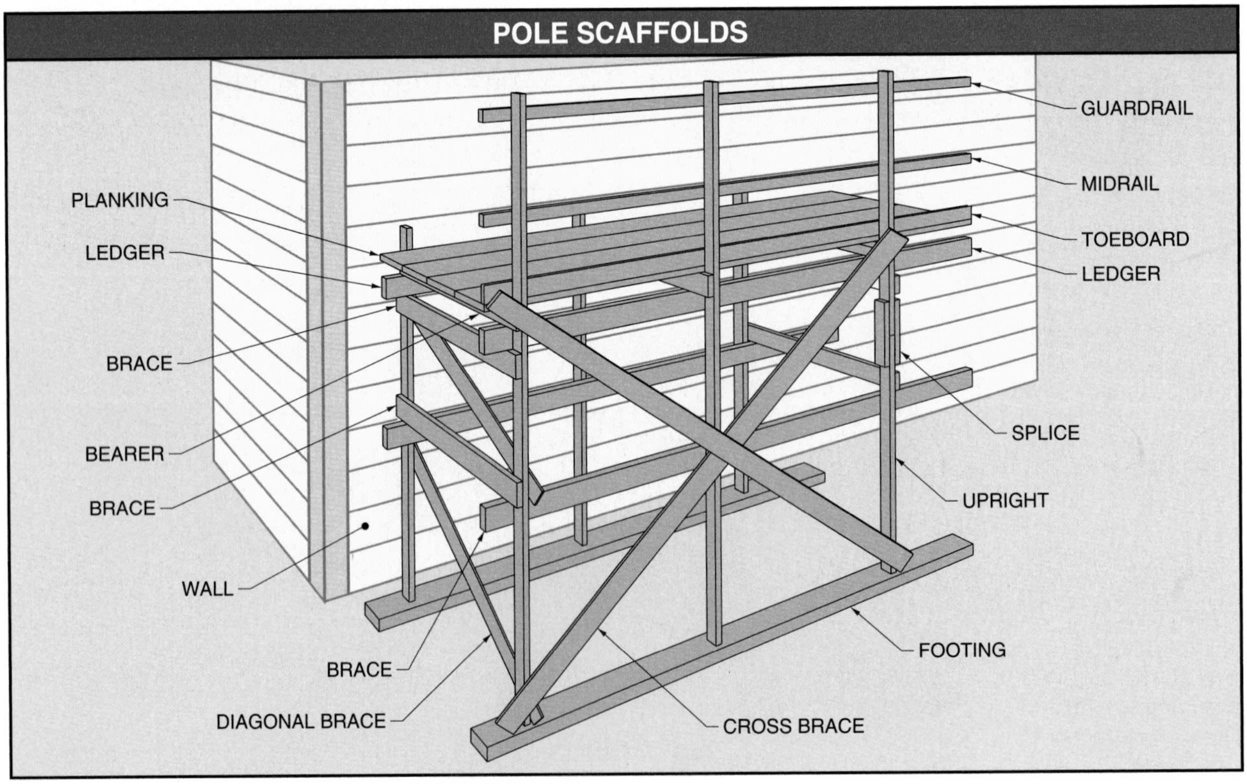

Figure 4-15. A pole scaffold is a wood scaffold with one or two sides firmly resting on the ground.

sides resting on the floor or ground and is not structurally anchored to a building or other structure.

The uprights of pole scaffolds are assembled from wood or metal legs (poles). Uprights must be plumb and securely braced to prevent displacement or swaying. The poles are to be erected on suitable bases or footings, which must be strong enough and large enough to support the maximum scaffold load without settling or displacement. Unstable objects such as barrels, boxes, loose brick, or concrete blocks must not be used to support scaffolds. Steel plate supports are used under steel poles and a minimum of 2″ planking support is used under wood poles. Each base or footing must be of sufficient size and thickness to support at least four times the maximum intended load.

Pole scaffolds must be constructed with guardrails, midrails, toeboards, planking, bearers, ledgers, cross braces, diagonal braces, and footings. The complete assembly must be plumb, level, square, and rigid.

Toeboards prevent tools or materials from being knocked or kicked off of the platform. The bottom of toeboards should make contact with the platform and the top of the toeboard should be more than 3½″ from the platform. Bearers (putlogs) must be long enough to project at least 3″ over the ledgers. Ledgers must be long enough to extend over two pole spaces and must not be spliced between poles. Cross braces are assembled between the left and right uprights. Cross braces must be long enough to extend over two pole spaces. Both diagonal and cross braces are used to prevent buckling and lateral movement. Diagonal braces are included in the scaffold assembly between the inner and outer uprights (pole sets). All pole scaffolds must be constructed using minimum- and maximum-sized components according to their duty rating. See Figure 4-16.

Wood pole scaffolds are constructed from select clear lumber for maximum strength. Duplex-head nails are used to make dismantling easier. All nails must be driven in their full length and in directions where the pull is across their length, not with their length. Nails smaller than 8d common must not be used to construct scaffolds. See Appendix.

Scaffold platform planks consist of 2″ nominal structural planks. Maximum permissible planking spans vary according to wood thickness and width. For example, the maximum permissible span for a 2 × 10 (nominal) plank on a light-duty scaffold is 10′. See Figure 4-17.

POLE SCAFFOLD COMPONENTS*

Type	Poles	Bearers	Ledgers (Stringers)	Braces	Planking	Rails
Light-duty** single-pole	20′ or less – 2 × 4 60′ or less – 4 × 4	3′ width – 2 × 4 5′ width – 4 × 4	20′ or less – 1 × 4 60′ or less – 1¼ × 9	1 × 4	2 × 10	2 × 4
Medium-duty† single-pole	60′ or less – 4 × 4	2 × 10	2 × 10	1 × 6	2 × 10	2 × 4
Heavy-duty‡ single-pole	60′ or less – 4 × 4	2 × 10	2 × 10	2 × 4	2 × 10	2 × 4
Light-duty* double-pole	20′ or less – 2 × 4 60′ or less – 4 × 4	3′ width – 2 × 4 5′ width – 4 × 4	20′ or less – 1¼ × 4 60′ or less – 1¼ × 9	1 × 4	2 × 10	2 × 4
Medium-duty† double-pole	60′ or less – 4 × 4	2 × 10	2 × 10	1 × 6	2 × 10	2 × 4
Heavy-duty‡ double-pole	60′ or less – 4 × 4	2 × 10	2 × 10	2 × 4	2 × 10	2 × 4

* all members except planking are used on edge
** not to exceed 25 lb/sq ft
† not to exceed 50 lb/sq ft
‡ not to exceed 75 lb/sq ft

Figure 4-16. Components of pole scaffolds are sized based on their duty rating.

PLANKING SPANS		
Duty Rating	Working Load*	Permissible Span**
Light	25	10
Medium	50	8
Heavy	75	7

* in lb/sq ft
** in ft

Figure 4-17. Maximum permissible planking spans vary according to wood thickness and width.

Planking must extend between 6″ and 18″ over the end support. These dimensions ensure the prevention of tipping if a weight is placed on the end. The underside of the planks is cleated at both ends to the support to prevent the planks from sliding. A *cleat* is a narrow wood piece, nailed across another board or boards, to provide support or to prevent movement. Platform planks are laid side-by-side with a maximum opening between planks of 1″. If plank overlapping is necessary due to a continuous run, the planks must be secured from movement, overlapped a minimum of 12″, and supported in the center by a brace.

When platform planks are added as upward work progresses, the previous bearers and ledgers must remain for pole bracing. Bearers must be face nailed to the poles as well as toenailed to the ledgers. For heavy-duty scaffolds, the ledgers must be nailed to the poles and supported by cleats that are also attached to the poles.

In single-pole scaffolds, the bearer that butts against the wall must be supported by a 2 × 6 bearing block, a minimum of 12″ long, notched out the width of the bearer. The bearing block must be securely fastened to the wall. The bearer must be nailed to the bearing block and toenailed to the wall. See Figure 4-18.

Ladders must be provided and attached to the ends of a scaffold so their use will not subject the scaffold to tipping. Cross braces must not be used as a means of access. Scaffolds must not be used during storms or high winds or when covered with ice or snow. All tools and materials should be secured or removed from the platform before a scaffold is moved.

All solid sawn scaffold planks shall be of a "scaffold plank grade" and shall be certified by, or bear the grade stamp of, a grading agency approved by the American Lumber Standards Com-

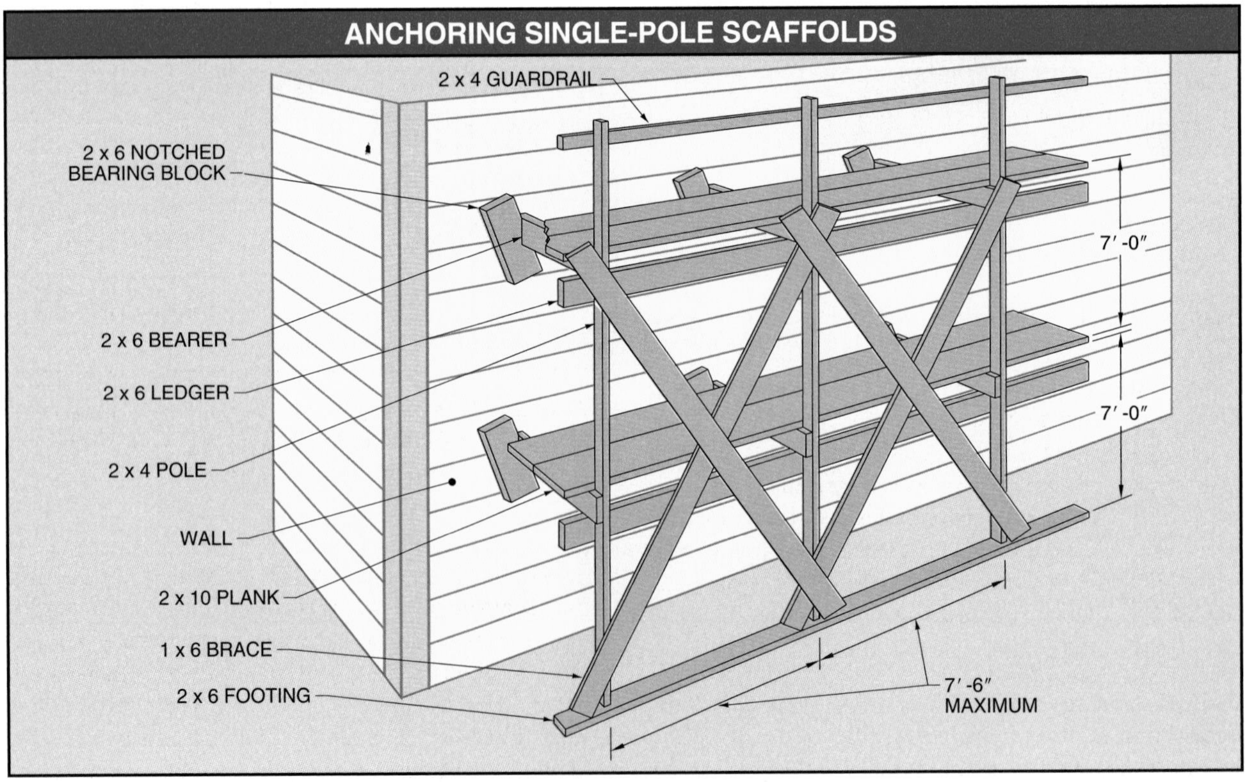

Figure 4-18. Single-pole scaffolds are anchored to the wall using notched bearing blocks.

Scaffolds over 25′ in height must be securely guyed or tied to the structure or building with No. 12 double-wrapped wire. A *guyline* is a rope, chain, rod, or wire attached to equipment as a brace or guide. See Figure 4-19. Guylines for scaffolds are wire ropes 1/4″ in diameter or larger. They are commonly positioned at a 45° angle to the vertical. Guylines may be anchored temporarily in the ground by screw ground anchors or may be tied off securely to a structure. Where the height of the scaffold exceeds 25′, the scaffold must be secured at intervals no greater than 25′, vertically or horizontally.

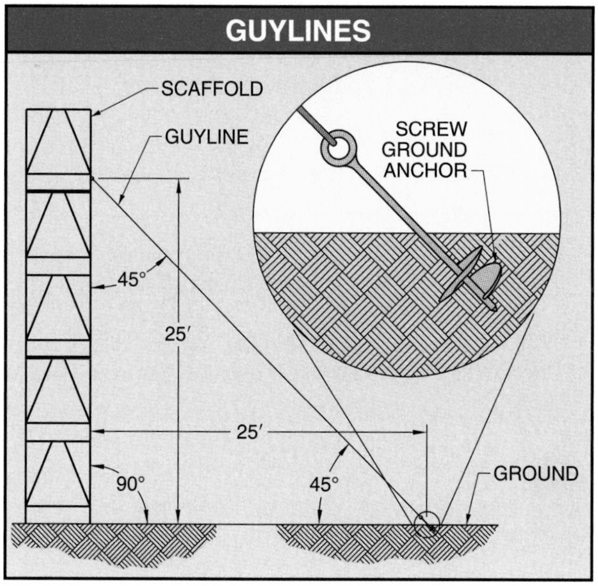

Figure 4-19. Scaffolds over 25′ in height must be secured with guylines.

Sectional Metal-Framed Scaffolds

A *sectional metal-framed scaffold* is a metal scaffold consisting of preformed tubes and components. See Figure 4-20. Sectional metal-framed scaffolds are also known as tube and coupler scaffolds. Sectional metal-framed scaffolds may be free-standing or mobile. They are easy to use, easily assembled with bolts, pins, or brackets, and may be fitted with locking casters for ease in moving.

Always check the manufacturer's product specifications sheet for scaffold load limits. Cross and diagonal braces of sectional metal frames allow several levels of planking installation. Planking may be wood, metal-reinforced wood, or metal. Each frame member is fastened together using nonslip clamps or pipe fittings. Prior to each use, the clamps or fittings must be checked and tightened if necessary.

When used as free-standing units, the height of a metal-framed scaffold must not exceed four times its minimum base dimension. For example, if the base of a scaffold measures 4′ × 8′, the maximum height is 16′ (4′ × 4 = 16′). Outriggers are sometimes used to increase the working height of a scaffold. Outrigger beams must rest on a sound foundation or on wood bearing blocks.

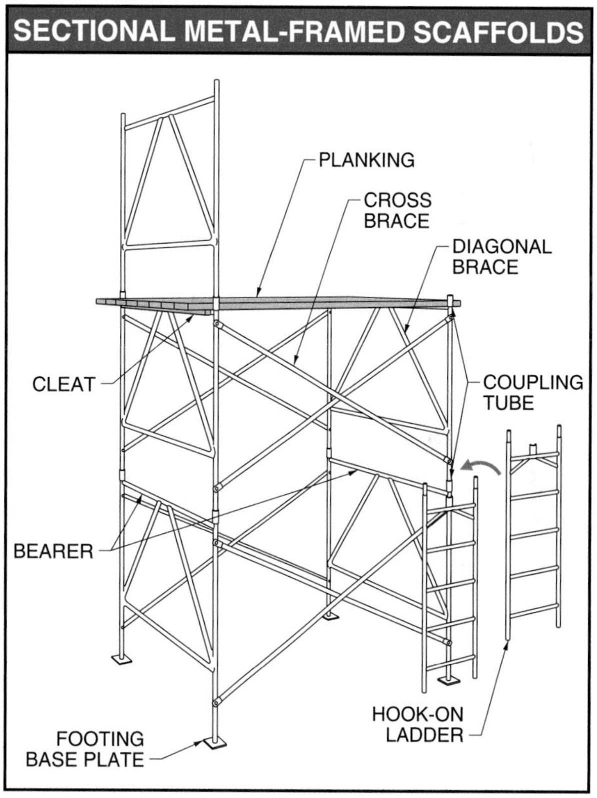

Figure 4-20. A sectional metal-framed scaffold is a metal scaffold consisting of preformed tubes and components.

Mobile scaffolds are equipped with casters. A mobile scaffold may be moved with a worker on the platform. However, the worker on the mobile scaffold must be advised and aware of each movement in advance. The minimum dimension of the base, when ready for rolling, must be at least one-half of the height. Outriggers may be included as part of the base dimension. See Figure 4-21.

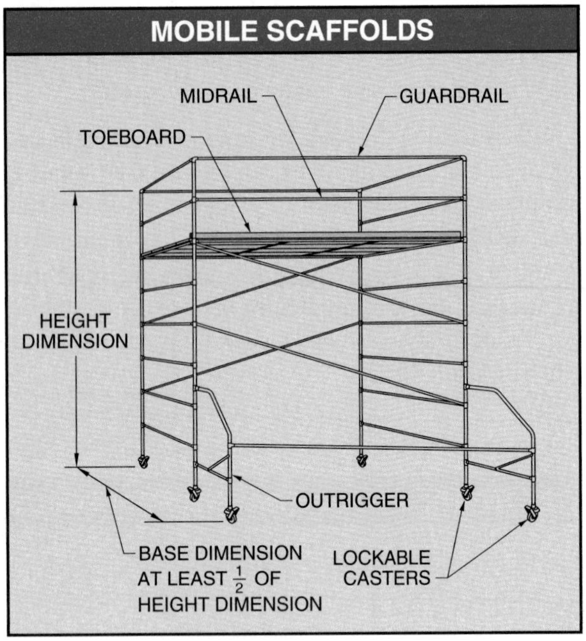

Figure 4-21. The minimum base dimension of a mobile scaffold must equal half the scaffold height before moving.

All tools and materials must be removed or secured before a mobile scaffold is moved. When mobile scaffolds are used on concrete flooring, the floor surface must be free from pits, holes, or obstructions that may create an unsafe condition. The surface must also be within 3% of level. After the scaffold has been moved, the casters must be locked to prevent movement while the scaffold is being used.

A hydraulic scissor lift is another type of mobile scaffold. A *hydraulic scissor lift* is a mobile hydraulically-operated platform controlled by remote switches attached at the platform. The platform is generally 4′ in its lowered position, with a maximum working height at about 20′. To offer a firm base, hydraulic scissor lifts may be equipped with screw-type outriggers. See Figure 4-22.

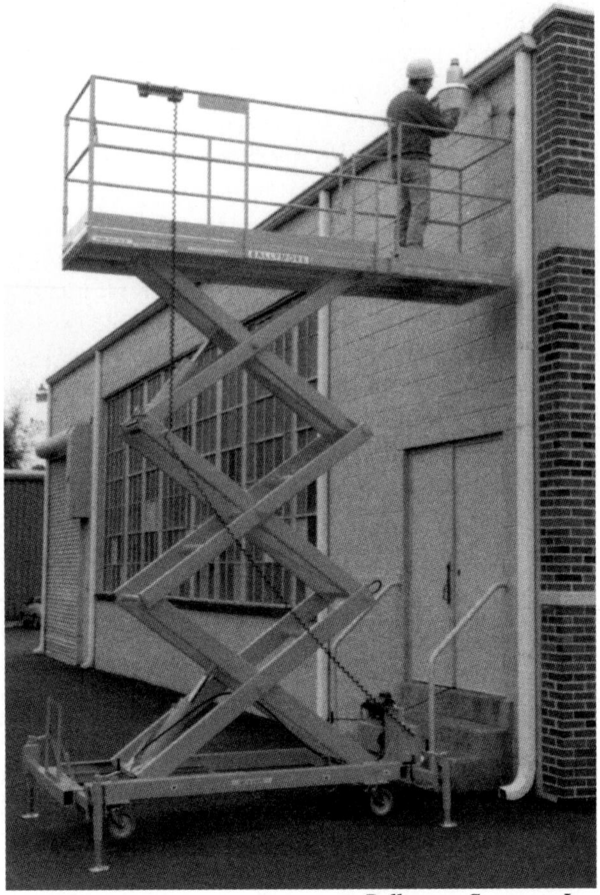

Ballymore Company, Inc.

Figure 4-22. A hydraulic scissor lift is a mobile hydraulically-operated platform controlled by remote switches attached at the platform.

Suspension Scaffolds

A *suspension scaffold* is a scaffold supported by overhead wire ropes. Suspension scaffolds are also referred to as swinging scaffolds. Suspension scaffolds use either the two-point or multiple-point suspension design.

A *two-point suspension scaffold* is a suspension scaffold supported by two overhead wire ropes. The overall width of two-point suspension scaffolds must be greater than 20″ but not more than 36″. See Figure 4-23.

A *multiple-point suspension scaffold* is a suspension scaffold supported by four or more ropes. Multiple-point suspension scaffolds must be capable of sustaining a working load of 50 lb/sq ft and are used mainly for repair and maintenance projects. They can be raised or lowered by permanently-installed, electrically-operated hoisting equipment. Multiple-point suspension scaffolds must not be overloaded.

Only qualified and trained operators are to be authorized to use and operate mobile work platforms. Platforms must be elevated only on a firm, level surface.

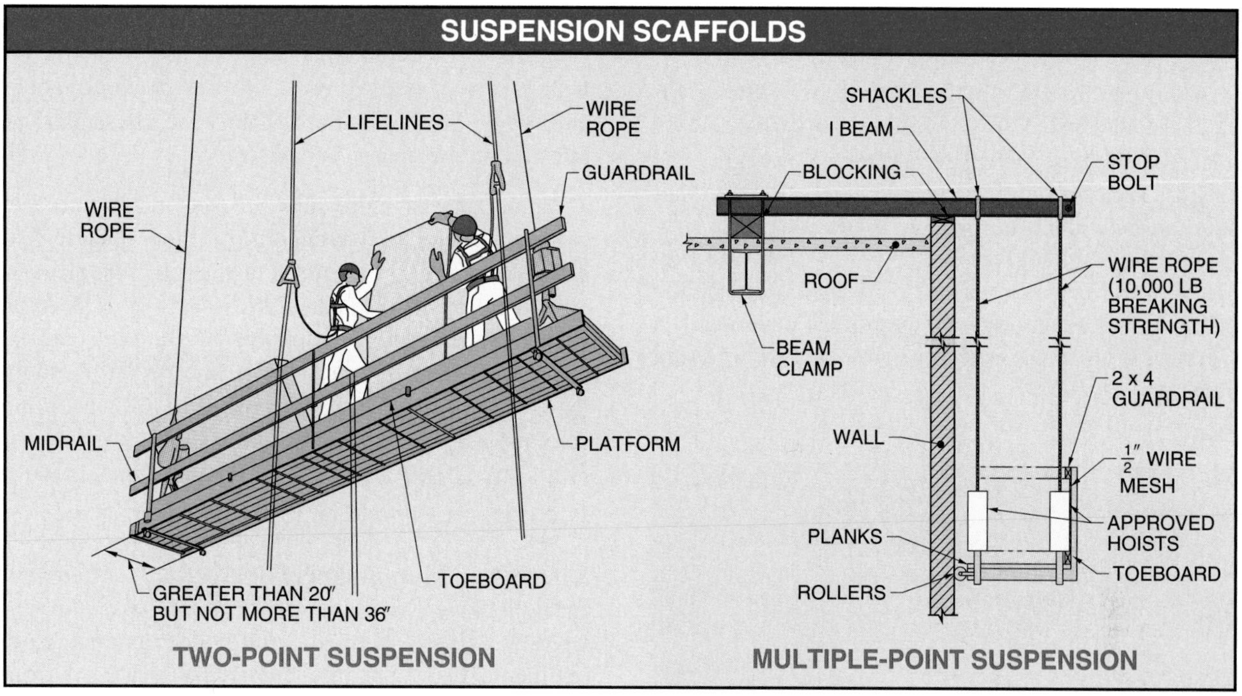

Figure 4-23. A suspension scaffold is a scaffold supported by overhead wire ropes.

Suspension scaffolds have a platform supported near the ends by overhead wire ropes. The ropes are attached to the platform by hangers or metal stirrups. Wire, fiber, or synthetic rope used for suspension scaffolds must be capable of supporting at least six times the maximum intended load. The hangers, U-bolts, brackets, and other hardware used for constructing a two-point or multiple-point suspension scaffold must be capable of sustaining four times the maximum intended load.

When power-driven hoisting equipment is used on suspension scaffolds, the power-driven equipment must have an emergency brake in addition to the normal operating brake. The emergency brake must operate automatically when the normal speed of descent is exceeded. The running end of the hoisting rope is attached to the hoist drum. At least four turns of rope must remain on the hoist drum at all times.

Suspension scaffold load testing is accomplished by raising the scaffold about 1′ off the ground and placing a load that is at least four times the normal workload on the platform. After approximately five minutes, check for cracked or splitting rope or sagging platforms. Do not use the scaffold if any of these conditions exist.

Lifelines and harnesses that can safely support a worker's weight must be provided for each worker. The lifeline is suspended from a substantial overhead structural member other than the scaffold, and should extend to the ground. Each worker's harness is tied to a lifeline by a lanyard, and to a fall prevention device that will limit the free fall to no more than 6′.

Each person on a two-point suspension scaffold shall use an approved harness with a lanyard and shall be attached by means of a fall-arresting device to an independent lifeline.

Tractel Inc., Griphoist® Division

Suspended scaffolding systems by Tractel Inc., contain tirak® powered traction hoists that are efficient and have a simple and constant traction principle that allows the hoist to be used in any orientation.

Safety Nets

A *safety net* is a net made of rope or webbing for catching and protecting a falling worker. A safety net must be used anywhere a person is working 25′ or more above ground, water, machinery, or any other solid surface when the worker is not otherwise protected by a lifeline, harness, or scaffolding. See Figure 4-24. Safety nets must also be used when public traffic or other workers are permitted underneath a work area that is not otherwise protected from falling objects.

The Sinco Group, Inc.

Figure 4-24. A safety net is a net made of rope or webbing for catching and protecting a falling worker.

Safety Net Requirements. Safety net size is generally 17′ × 24′, and may be coupled with another net to form a larger net. Netting mesh size for bodily fall protection is normally 6″ × 6″. Netting is constructed of 3⁄8″ No. 1 grade manila, 1⁄4″ nylon, or 5⁄16″ polypropylene rope. *Mesh* is the size of the openings between the rope or twine of a net.

In applications where workers or others are to be protected from falling tools or other objects, a lining of smaller mesh must be added to the fall protection net. The size and strength of the net lining mesh must restrict tools and materials capable of causing injury. Net lining mesh must normally be less than 1″ and constructed of twine equal to or greater than No. 18. Installation of netting must have level border ropes and, when hung, no more than 3′ of sag should be allowed at the center of the net.

When two or more nets are secured together to form a larger net, lacing, drop-forged shackles, or safety hooks may be used, but must be less than 6″ apart. A drop-forged shackle or safety hook is to be used to attach nets to supporting structures, cables, or beams and must be spaced at intervals of no more than 4″. Border rope is to have a 5000 lb breaking strength when new. The minimum diameter for manila border rope is 3⁄4″. The minimum diameter for synthetic border rope is 1⁄2″.

Safety Net Maintenance. Factors that affect net safety include environmental contaminants, sunlight, welding, mildew, abrasion, and impact loading. Contaminants from airborne chemicals create environmental conditions that affect net strength. Even though polypropylene and nylon are resistant to many acids and alkalis, moderate and unknown degradation can occur to rope used in these environmental conditions. Manila rope, being organic, degrades rapidly in a chemically-active environment.

Synthetic and natural fibers degrade in the presence of ultraviolet rays from sunlight or from arc welding. When safety nets are used regularly outdoors, an ultraviolet absorbing dye may be used for outer-layer protection. Welding slag or sparks may also harm safety nets because each is sufficient to burn the net.

Mildew and abrasion damage is caused by improper storage and rough handling. Storing safety nets in a warm, moist location causes mildew growth and a weakening of rope fibers. Dragging nets over rough or sharp surfaces abrades and degrades rope fibers. Also, impact loading is a form of damage created by the continuous shock of loads being dropped into the net. Even the impact from net testing may degrade the net's integrity.

Safety Net Testing. Nets must be impact load tested to assure that there is sufficient strength or that there has been no loss of strength. Impact load tests are first done on a sample by the manufacturer. Each safety net is certified by the manufacturer to withstand a 50′ drop of a 350 lb bag of sand, 24″ in diameter. On-the-job testing is also required by the user immediately following installation or after a major repair. Impact load testing must be done at six month intervals if the net is in regular use. Testing consists of dropping a 400 lb bag of sand, not more than 30″ (±2″) in diameter, from a height of 25′ above the net into the center of the net.

SAFETY

Safe use of ladders and scaffolds includes the use of fall-protection equipment. The three categories of fall-protection equipment include ladder climber fall protection, position protection, and scaffold worker fall protection.

Ladder Climber Fall Protection

Ladder climbers should use a carrier for fall protection. A *carrier* is the track of a ladder safety system consisting of a flexible cable or rigid rail secured to the ladder or structure. See Figure 4-25. The carrier is the track for the safety sleeve. A *safety sleeve* is a moving element with a locking mechanism that is connected between a carrier and the worker's harness. The connecting line between the carrier and the safety belt must be less than 9″. Fall-arrest devices utilize the worker's weight for activation. The closer a worker is connected to the fall-arrest device, the less distance traveled in a fall.

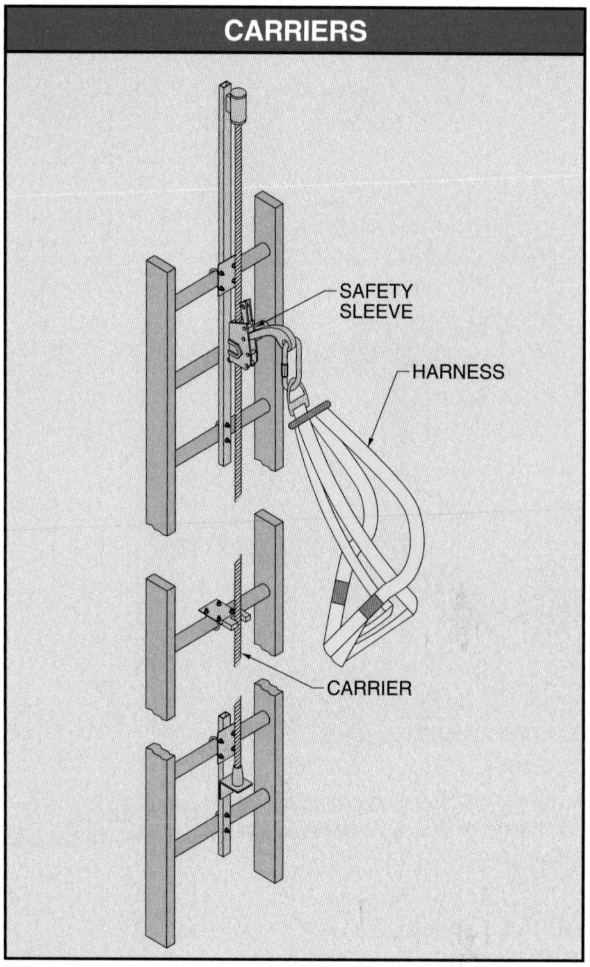

Figure 4-25. A carrier is the track of a ladder safety system consisting of a flexible cable or rigid rail secured to the ladder or structure.

Position Protection

Position protection supports a person in a working position by wrapping a body strap around a post, tree, or attachment to a structure. The weight and angle of the individual secures the worker. A position protection device holds an individual in position, but does not prevent falling. Position protection is commonly used by electrical lineworkers, telephone lineworkers, tree climbers, and window washers. See Figure 4-26.

Suspension scaffolds shall be operated only by persons who have been instructed in the operation, use, and inspection of the particular suspended scaffold to be operated. Employers shall instruct and supervise their employees in the safe use of all equipment provided.

The Sinco Group, Inc.

Sinco personnel nets consist of strong, high tenacity nylon with a 3½″ diagonal net design made in accordance with ANSI A10.11-1989, Safety Nets Used During Construction, Repair, and Demolition Operations.

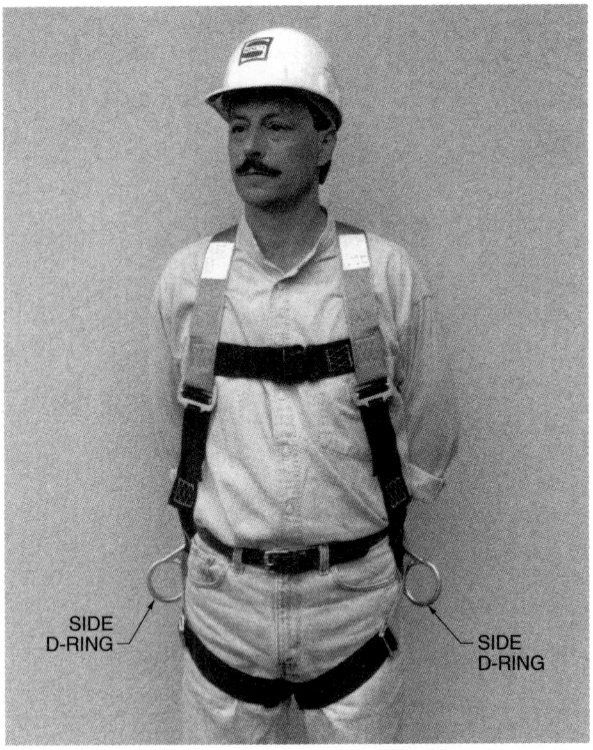

The Sinco Group, Inc.

Figure 4-26. Position protection devices include harnesses containing side D-rings that are used to attach a body strap around a post, tree, or attachment to a building.

Scaffold Worker Fall Protection

Scaffold worker fall protection may include items such as lifelines, harnesses, lanyards, and rope grabs. The proper fall-protection equipment should be worn when working at heights greater than the minimum safe distance from the ground. See Figure 4-27.

Lifelines are anchored above the work area, offering a free-fall path, and must be strong enough to support the force of a fall arrest. Vertical lifelines must never have more than one person attached per line and must be long enough to reach the ground or landing below the work area. The lifeline must then be terminated (tied up) to prevent the safety sleeve from sliding off of its end.

The path of a fall must be visualized when anchoring a lifeline. Use an anchored system without any obstructions to the fall. Obstacles below and in the fall path can be deadly.

Harnesses, when used properly, protect internal body organs, the spine, and other bones in a fall. Chest harnesses must not be worn for free-fall protection. Harnesses must fit snugly and be attached to the lanyard at the person's back. A *lanyard* is a rope or webbing device used to attach a worker's harness to a lifeline. A *rope grab* is a device that clamps securely to a rope. Rope grabs contain a ring to which a lifeline can be attached. Rope grabs protect workers from falls while allowing freedom of movement.

Fall-Arrest Sequence

When a fall-arrest device is used, the breaking of a fall is preceded by a free fall and the taking up of slack between the harness and the safety device. This is followed by the distance of deceleration. Deceleration distance is generally $3\frac{1}{2}'$ to $4'$. Always limit the total fall to $6'$ or less. For this reason, a lanyard must be kept high enough or short enough to limit the free fall.

A person may also be protected against a fall by tying off. *Tying off* is securely connecting a harness directly or indirectly to an overhead anchor point. Certain precautions must be made that lines and lanyards are not weakened by knotting or tying off to sharp or rough surfaces.

Ladder Safety

Proper maintenance of a ladder is critical due to its direct relationship to life safety. The following list of precautions should be observed for proper safety when using ladders.

- Use ladders only for the purpose for which they were designed.
- Inspect ladders carefully when new and before each use.
- Use leg muscles, not back muscles, for lifting and lowering ladders.
- Stand ladders on a firm, level surface.
- Face the ladder when ascending or descending.
- Exercise extreme caution when using ladders near electrical conductors or equipment. All ladders conduct electricity when wet.
- Ladders are intended for use by only one person unless specifically designated otherwise.
- Never use a ladder as a substitute for scaffold planks or for horizontal work.
- Always check for the proper angle of inclination before climbing a ladder.

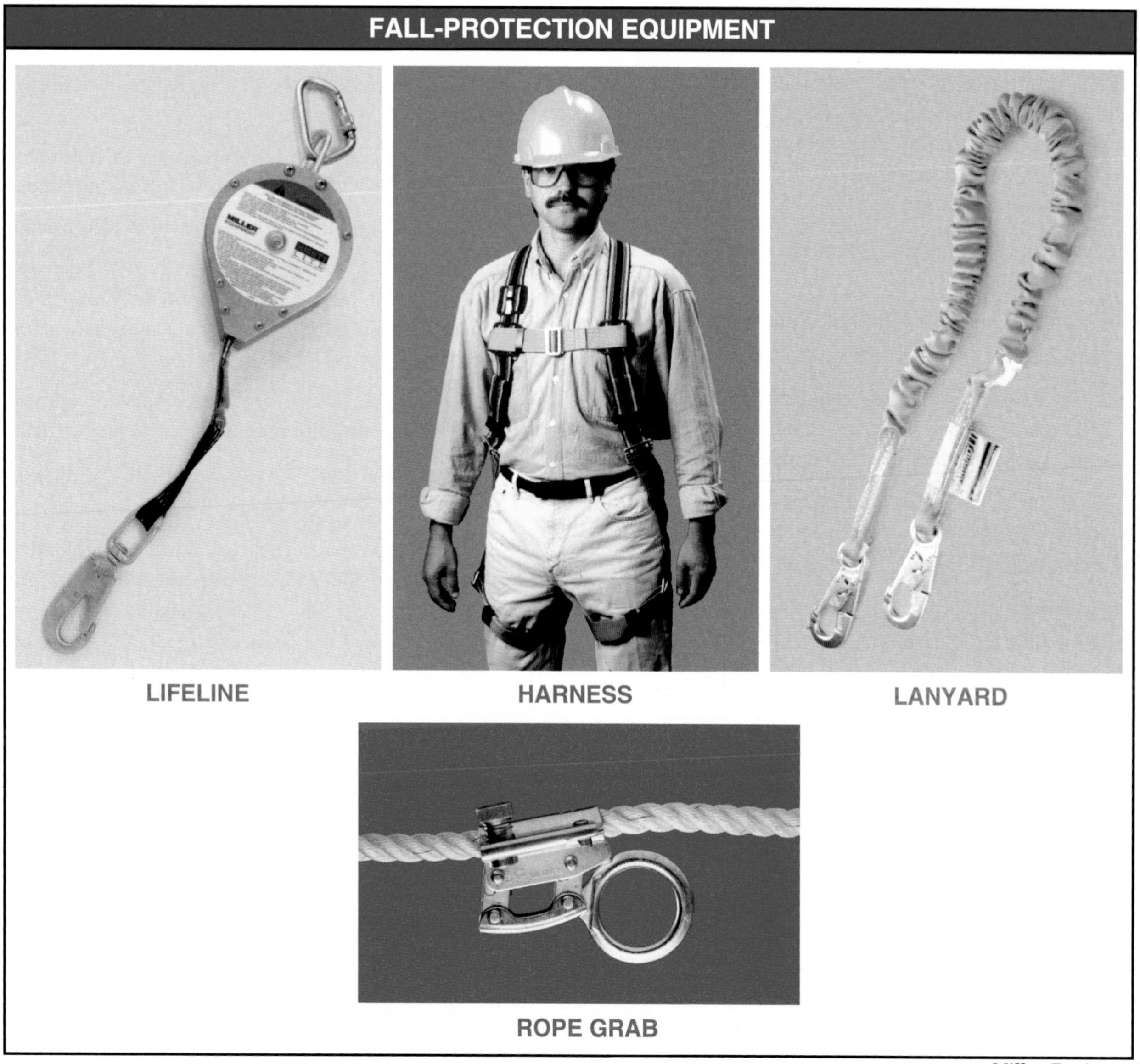

Miller Equipment

Figure 4-27. The appropriate fall-protection equipment should be worn when working at heights greater than the minimum safe distance from the ground.

- Verify that all pawl locks on extension ladders are securely hooked over rungs before climbing.
- Always check for proper overlap of extension ladder sections before climbing.
- Keep all nuts, bolts, and fasteners tight. Lubricate all moving metal parts as required.
- Ensure that stepladders are fully open with spreaders locked before climbing.
- Do not stand on the top two rails of a stepladder or on the top three rungs of an extension ladder.
- Use the three-point climbing method when ascending or descending a ladder.
- Never place a ladder in front of a door unless appropriate precautions have been taken.

Only extra-heavy-duty (type IA) and heavy-duty (type I) ladders shall be used with ladder jacks. Medium-duty (type II) and light-duty (type III) ladders shall never be used with ladder jacks.

Scaffold Safety

OSHA regulations state that scaffolding may be erected, moved, altered, or dismantled only under the supervision of a competent person. The following precautions should be observed because the lives of workers depend on the construction of scaffolding.

- Use only 2″ nominal structural planking that is free of knots for scaffold platforms.
- Platform end extensions must be cleated with a minimum of 6″ extension and a maximum of 18″.
- Always observe working load limits. Scaffolds and their components must be capable of supporting four times the maximum intended load.
- Guardrails, midrails, and toeboards must be installed on all open sides and ends of platforms more than 10′ above the ground.
- Platform planks are to be laid with no openings more than 1″ between adjacent planks.
- Overhead protection must be provided for persons on a scaffold exposed to overhead hazards.
- Work must not be done on a scaffold during high winds or storms.
- Work must not be done on ice-covered or slippery scaffolds.
- Scaffolds with a height-to-base ratio of more than 4 : 1 must be restrained by the use of guylines.
- Mobile scaffolds must be locked in position when in use.
- All tools and materials must be secured or removed from the platform before a mobile scaffold is moved.
- All personnel in close proximity must be advised and aware of the movement of a mobile scaffold.
- Fall protection must be used on working heights of more than 10′.
- Safety nets for workers at any level over 25′ must be used when the workers are not otherwise protected.
- Safety nets restricting falling objects must be used when persons are permitted to be underneath a work area.

The Sinco Group, Inc.

Sinco personnel nets have been designed to withstand 350 lb dropped from a height of 50 ft. However, it is essential that nets be created so a worker cannot fall more than 25′.

Hydraulic Principles

5

Chapter

Hydraulics is the branch of science that deals with the practical application of water or other liquids at rest or in motion. Hydrostatics is the study of liquids at rest and the forces exerted on them or by them. Hydrodynamics is the study of the forces exerted on a solid body by the motion or pressure of a fluid. A liquid is a fluid that can flow readily and assume the shape of its container. Fluid flow is the movement of fluid caused by a difference in pressure between two points. In a hydraulic system, fluid flow is produced by the action of a pump and is expressed as a measurement of gallons per minute or liters per minute.

HYDRAULICS

Hydraulics is the branch of science that deals with the practical application of water or other liquids at rest or in motion. The two major divisions of hydraulics are hydrostatics and hydrodynamics.

Hydrostatics

Hydrostatics is the study of liquids at rest and the forces exerted on them or by them. *Equilibrium* is the condition when all forces and torques are balanced by equal and opposite forces and torques. Most hydraulic systems apply hydrostatic principles. For example, the fluid in an automobile's hydraulic braking system is at rest and the pressure throughout the system is in equilibrium. See Figure 5-1. The brake system is activated by applying pressure to the foot pedal. The fluid in the system transmits the applied force from the foot pedal to the slave cylinder piston. The slave cylinder piston transmits the force to the brake pad, which applies pressure to the brake drum. The pressure of the fluid is equal in all parts of the system, but higher than the pressure of the fluid when the system is at rest.

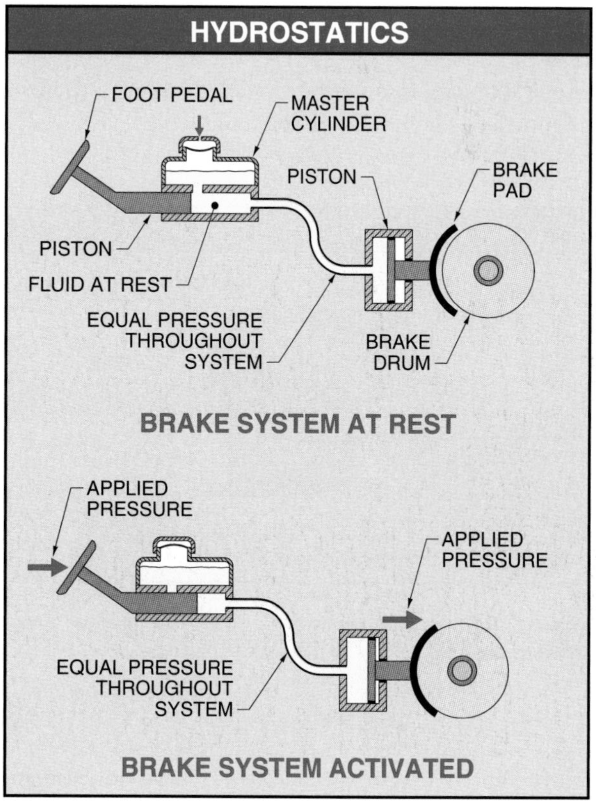

Figure 5-1. Hydrostatics is the study of liquids at rest and the forces exerted on them or by them.

Atlas Technologies, Inc.

Hydraulics provides the force required for the stamping and fabrication of today's products.

Hydrodynamics

Hydrodynamics is the study of the forces exerted on a solid body by the motion or pressure of a fluid. For example, fluids are transferred through a non-positive displacement pump by centrifugal force. A *nonpositive displacement pump* is a pump that is not sealed between its inlet and outlet. *Centrifugal force* is the outward force produced by a rotating object. See Figure 5-2. The fluid is forced to the discharge (outlet) port by rotating impeller vanes. The output of the pump may be reduced or completely blocked if the pressure in the discharge circuit is increased because there is no positive displacement of fluid.

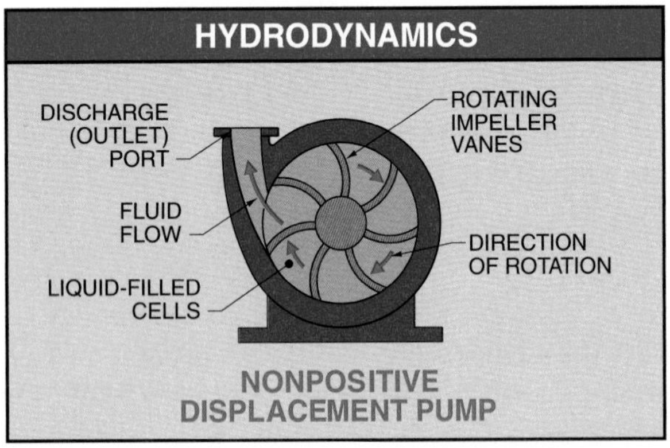

Figure 5-2. Hydrodynamics is the study of the forces exerted on a solid body by the motion or pressure of a fluid.

Hydraulic System History

Early hydraulic systems consisted of diverting streams for village irrigation and water supply and digging wells. In prehistoric Europe, plumbing was formed by hollowing tree trunks and connecting them together. The ancient Romans, Greeks, and Egyptians used lead, copper, or bronze plumbing for conveying water. About 500 B.C., Persian aqueducts using metal plumbing were tunneled through mountains. Around 100 B.C., hydraulic machines, such as the water-lifting machine and the Archimedes water-screw, were developed.

The water-lifting machine (water wheel) was a device equipped with paddles which raised water by the force of current from a stream. The Archimedes water-screw was another device used to raise water from a stream or lake up to an irrigation ditch. The Archimedes water-screw consisted of a wood core with layers of pitch-covered wood strips attached to form a spiral. See Figure 5-3. This assembly was covered to create a spiral tube. With one end of the tube lowered into the water, the complete assembly was rotated, allowing water to hydrostatically work its way up the screw.

Devices applying hydrodynamic principles appeared around 1500 A.D., when the piston concept was used to pump water to the top of a 40′ Roman aqueduct. The Ramelli quadruple suction pump used a water wheel to drive a wooden peg gear mechanism. See Figure 5-4. The peg gear mechanism drove a worm gear that was connected to a rotating crank. The rotating crank was attached to a reciprocating piston (suction pump) that would raise water with each rotation of the crank.

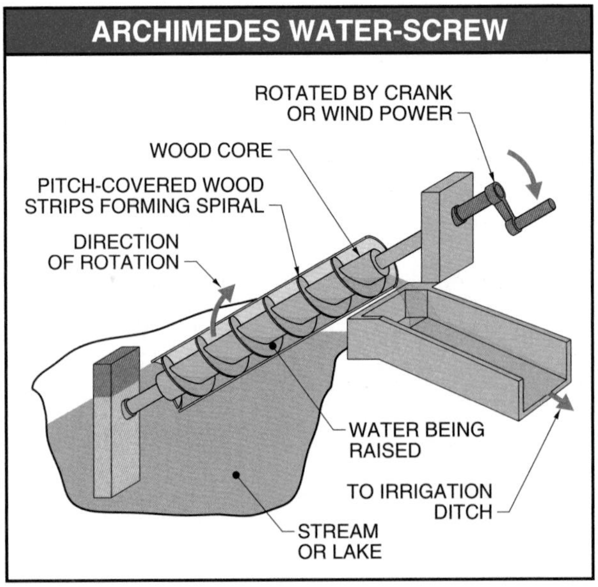

Figure 5-3. The Archimedes water-screw hydrostatically raised water as the device was rotated.

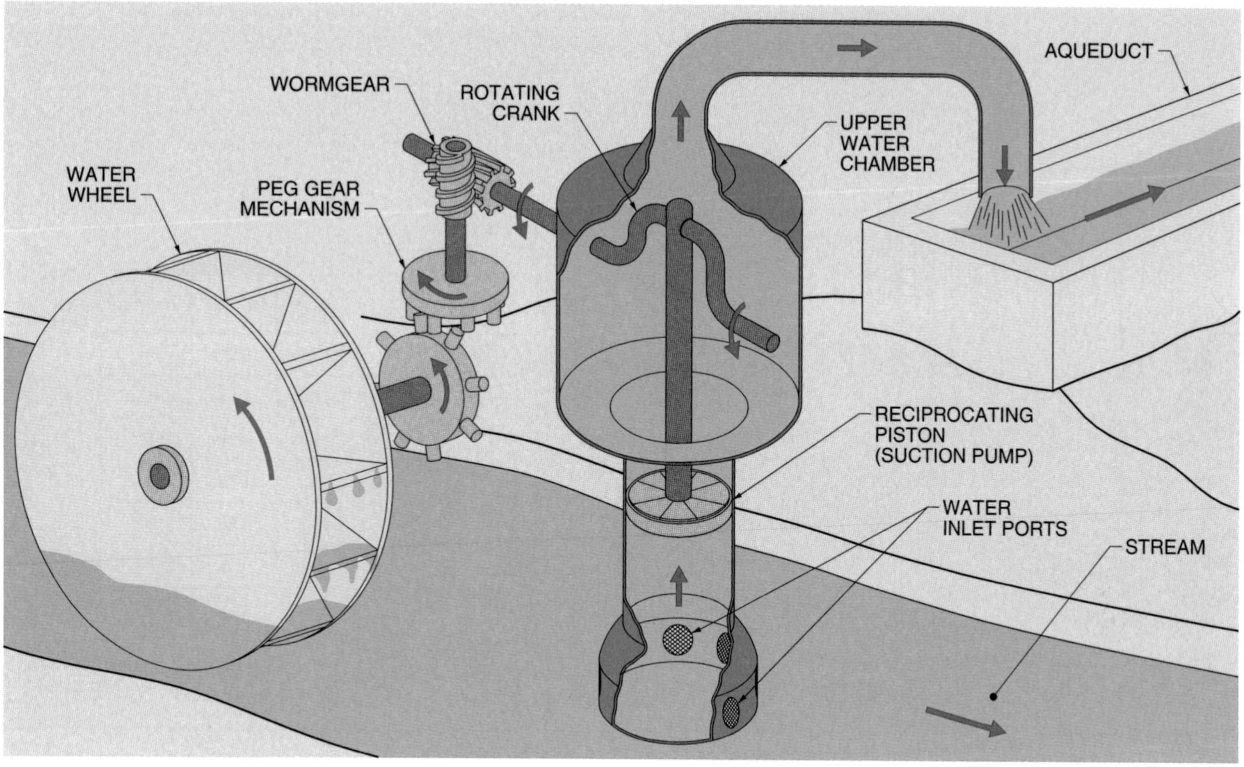

Figure 5-4. Devices applying hydrodynamic principles appeared around 1500 A.D., when the piston concept was used to pump water to the top of a 40′ Roman aqueduct.

LIQUID CHARACTERISTICS

In hydraulics, the term "fluid" refers to gases as well as liquids. A *fluid* is a substance that tends to flow or conform to the outline of its container (such as a liquid or a gas). Fluids yield easily to pressure. A *liquid* is a fluid that can flow readily and assume the shape of its container. Fluids have no independent shape but do have a definite volume. Liquids do not expand indefinitely and are only slightly compressible. A *gas* is a fluid that has neither independent shape nor volume and tends to expand indefinitely. Oxygen, hydrogen, etc. are gases.

Liquids make convenient fluids for transmitting force because they are not highly compressible like gases. The term "fluid" is used in reference to a liquid because liquids are specifically used in hydraulic systems. Work produced in a hydraulic system is dependent on the pressure and flow of the fluid in the system.

Pressure

Pressure is the force per unit area. In 1653, French scientist Blaise Pascal realized that enclosed fluids under pressure follow a definite law. Pascal's law states that pressure at any one point in a static liquid is the same in every direction and acts with equal force on equal areas. *Force* is the energy that produces movement. Although this law and its potential for technology were realized in the 17th century, it was not until the 20th century that fluid power became a means of energy transmission.

Pressure is expressed as atmospheric, gauge, and absolute. *Atmospheric pressure* is the force exerted by the weight of the atmosphere on the Earth's surface. The weight of the atmosphere, acting over a height of several hundred thousand feet above the Earth's surface, varies slightly with weather conditions. The weight of the atmosphere at sea level is 14.7 pounds per square inch absolute (psia). Atmospheric pressure is also expressed in inches of mercury absolute (in. Hg abs) and is measured with a mercury barometer. A *mercury barometer* is an instrument that measures atmospheric pressure using a column of mercury. See Figure 5-5.

Mineral-base oil is the most widely used hydraulic fluid. It has excellent lubricating properties, does not cause rusting, dissipates heat readily, and can be cleaned easily by mechanical filtration and gravity separation.

PRESSURE

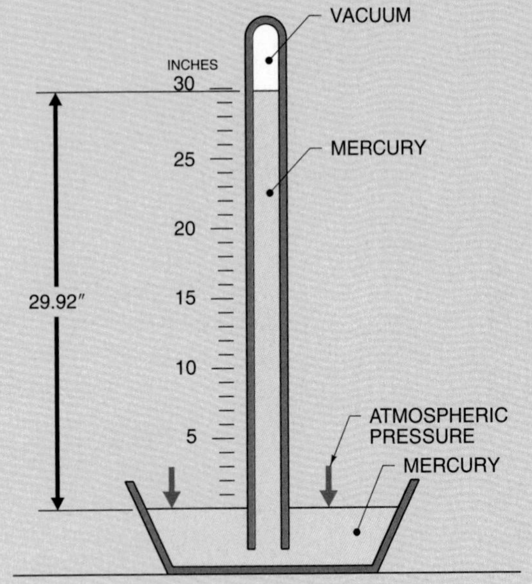

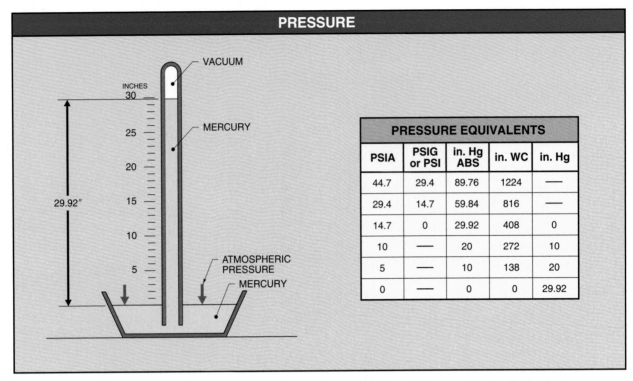

PRESSURE EQUIVALENTS				
PSIA	**PSIG or PSI**	**in. Hg ABS**	**in. WC**	**in. Hg**
44.7	29.4	89.76	1224	—
29.4	14.7	59.84	816	—
14.7	0	29.92	408	0
10	—	20	272	10
5	—	10	138	20
0	—	0	0	29.92

Figure 5-5. A mercury barometer is an instrument that measures atmospheric pressure using a column of mercury.

A mercury barometer consists of a glass tube that is closed on one end and completely filled with mercury. The tube is inverted and the open end is submerged in a dish of mercury. A vacuum is created at the top of the tube as the mercury tries to run out of the tube. *Vacuum* is a pressure lower than atmospheric pressure. The pressure of the atmosphere on the mercury in the open dish prevents the mercury in the tube from running out of the tube. The height of the mercury in the tube corresponds to the pressure of the atmosphere on the mercury in the open dish.

A mercury barometer is commonly calibrated in inches of mercury (in. Hg). At sea level, the atmosphere can support 29.92″ Hg in the tube. A barometric pressure of 29.92″ Hg equals one atmosphere or 14.7 psi. Pressures above one atmosphere are generally expressed in psi and pressures below one atmosphere are generally expressed in in. Hg. Minute pressure changes are expressed in inches of water column (in. WC). Atmospheric pressure at sea level should be able to hold water in a column 407.37″ (33.9′) high because atmospheric pressure is able to hold mercury in a column 29.92″ high and water is 13.6 times lighter than mercury.

Gauge pressure is pressure above atmospheric pressure that is used to express pressures inside a closed system. Gauge pressure assumes that atmospheric pressure is zero (0 psi). Most pressures are measured as gauge pressure unless otherwise specified. Gauge pressure is expressed in pounds per square inch gauge (psig) or psi.

Absolute pressure is pressure above a perfect vacuum. Absolute pressure is the sum of gauge pressure plus atmospheric pressure. Absolute pressure is expressed in pounds per square inch absolute (psia).

Pressure outside a closed system (such as normal air pressure) is expressed in pounds per square inch absolute. The difference between gauge pressure and absolute pressure is the pressure of the atmosphere at sea level at standard conditions (14.7 psia). A pressure gauge reads 0 psig at normal atmospheric pressure. To find absolute pressure when gauge pressure is known, the atmospheric pressure of 14.7 psia is added to the gauge pressure. Absolute pressure is found by applying the formula:

$$psia = psig + 14.7$$

where

$psia$ = pounds per square inch absolute

$psig$ = pounds per square inch gauge

14.7 = constant (atmospheric pressure at standard conditions)

For example, what is the absolute pressure in a system when a pressure gauge reads 100 psig?

$psia = psig + 14.7$

$psia = 100 + 14.7$

$psia =$ **114.7 psia**

Pressure other than atmospheric pressure is considered to be artificial and is produced to transfer or amplify force in hydraulic systems. This transferred or amplified force is used to do work such as lifting a car with a hydraulic jack, running a conveyor with a hydraulic motor, or shaping steel into car or truck components.

Area, force, and pressure are the basis of all hydraulic systems. The force exerted by a liquid is based on the size of the area on which the liquid pressure is applied. In hydraulic systems, this area usually refers to the face of a piston, which is circular in shape. Area is always expressed in square units, such as sq in. or sq mm.

A circle with a diameter the same as a square has less area. The area of a circle is exactly 78.54% of the area of a square with the same measurement. See Figure 5-6. The area of a circle is found by applying the formula:

$A = .7854 \times D^2$

where

A = area (in sq in.)

.7854 = constant

D^2 = diameter squared

For example, what is the area of a circle with a diameter of 3″?

$A = .7854 \times D^2$

$A = .7854 \times (3 \times 3)$

$A = .7854 \times 9$

$A =$ **7.069 sq in.**

The area of a piston can be found if the force and pressure applied to a cylinder are known. The applied pressure on a piston can be found if the amount of force and the piston area are known. Also, the force produced by a piston can be found if the area and pressure applied to a piston are known. Two of the values must be known to find the unknown value.

The relationship between force, pressure, and area can be recalled using a force, pressure, and area formula pyramid. By covering the letter of the unknown value, the formula for finding the solution is shown. See Figure 5-7. For example, to find area, covering *A* indicates that *F* is divided by *P*. To find pressure, covering *P* indicates *F* is divided by *A*. To find force, covering *F* indicates *P* is multiplied by *A*.

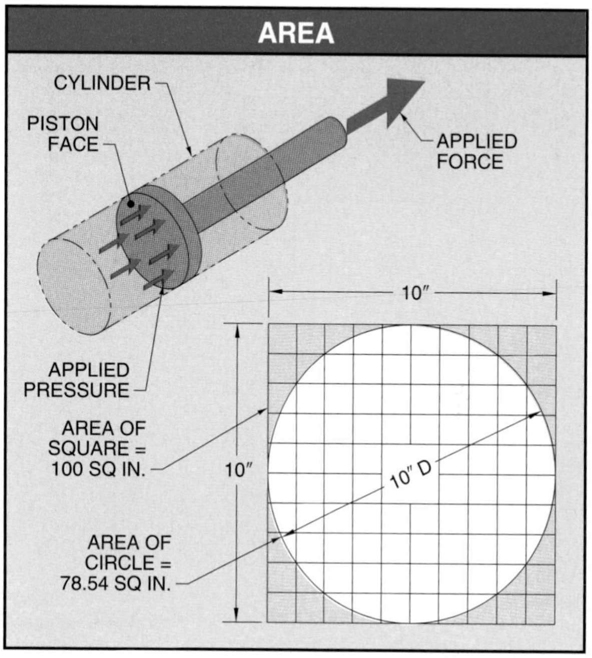

Figure 5-6. In hydraulic systems, area usually refers to the face of a piston, which is circular in shape.

For example, what is the area of a piston face with 5000 lb of force exerting 250 psi?

$A = \frac{F}{P}$

$A = \frac{5000}{250}$

$A =$ **20 sq in.**

Force and diameter must be known to calculate the required pressure of a cylinder. The area of a piston is calculated from the diameter and then used to find the pressure in a cylinder.

For example, how much pressure is required to move a 5000 lb force with a 4″ D piston?

1. Find area of piston face.

   $A = .7854 \times D^2$

   $A = .7854 \times (4 \times 4)$

   $A = .7854 \times 16$

   $A = 12.566$ sq in.

2. Find required pressure.

$$P = \frac{F}{A}$$

$$P = \frac{5000}{12.566}$$

P = **397.899 psi**

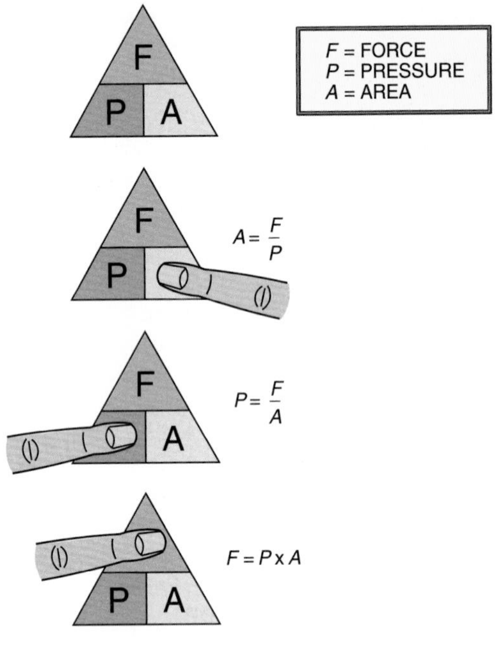

Figure 5-7. The force, pressure, or area of a piston can be calculated if any two values are known.

Pressure and diameter must be known when calculating the required force of a cylinder. The area of a piston is calculated using the diameter of the piston and then the area is used to find the required force.

For example, how much force does 450 psi produce in a 4″ D piston?

1. Find area of piston face.

$$A = .7854 \times D^2$$

$$A = .7854 \times (4 \times 4)$$

$$A = .7854 \times 16$$

A = 12.566 sq in.

2. Find required force.

$$F = P \times A$$

$$F = 450 \times 12.566$$

F = **5654.7 lb**

Head Pressure. *Head* is the difference in the level of a liquid (fluid) between two points. Head is expressed in feet. *Head pressure* is the pressure created by fluid stacked on top of itself. See Figure 5-8.

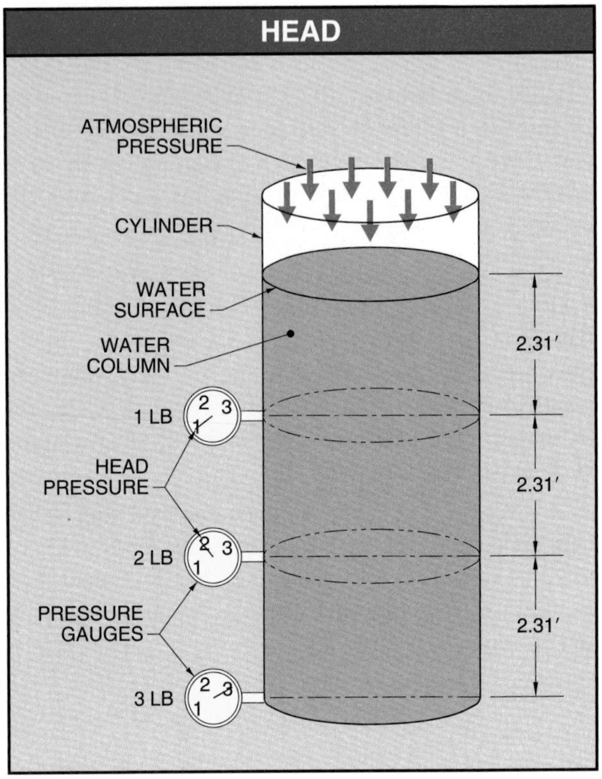

Figure 5-8. Head pressure is the pressure created by fluid stacked on top of itself.

In an open cylinder, the pressure of the fluid at any depth in the cylinder is proportional to the height of the column of fluid. The pressure in a column of fluid is determined using the column's height and the fluid's weight, not the shape of the vessel. The pressure at the same level in each vessel is identical if the pressure surrounding the different-shaped vessels is the same and the fluid in each vessel is the same. The pressure of the fluid at any level in a vessel is based on the height of the fluid above that level and is the same at that level regardless of the shape of the vessel. See Figure 5-9.

The head pressure at any point in a container is directly proportional to the density of the fluid and the depth below the surface of the fluid.

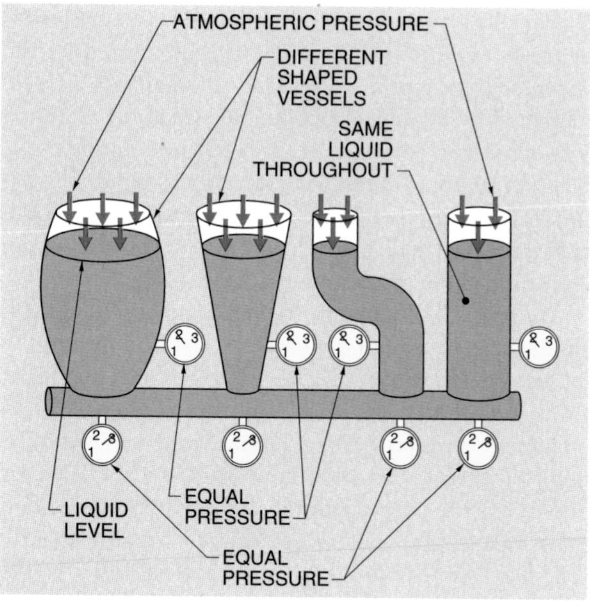

Figure 5-9. The pressure of the fluid at any level in a vessel is based on the height of the fluid above that level and is the same at that level regardless of the shape of the vessel.

The pressure at the base of a column of fluid is calculated by multiplying the weight of the fluid by its height. The weight of a fluid is obtained from a Fluid Weights/Temperature Standards table. See Appendix. The pressure of a fluid in a cylinder is found by applying the formula:

$P = w \times h$

where

P = pressure at base (in psi)

w = weight of fluid (in lb/cu in. from Fluid Weights/Temperature Standards table)

h = height (in in.)

For example, what is the pressure at the base of a 72″ D cylindrical vessel that contains 96″ of water? *Note:* The weight of water (in lb/cu in.) equals .0361 at 39°F (from Fluid Weights/Temperature Standards table).

$P = w \times h$

$P = .0361 \times 96$

$P = \mathbf{3.466\ psi}$

In a hydraulic system, head pressure is the energy or pressure that supplies a hydraulic pump. Atmospheric pressure and head pressure combine to feed the suction (intake) line connecting a hydraulic pump to a reservoir.

Head is classified as static or dynamic. *Static head* is the height of a fluid above a given point in a column at rest. *Static head pressure* is a force over an area created by the weight of the fluid itself. Static head pressure is potential energy. The pressure of water per ft of static head is calculated by using .0361 lb/cu in. or 2.31′ head of water for each psi. See Figure 5-10.

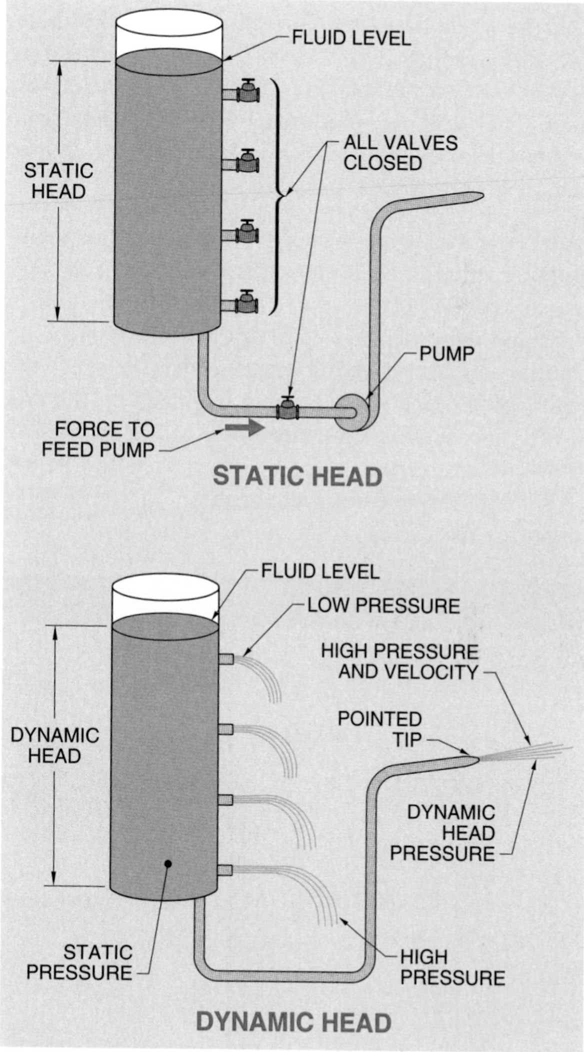

Figure 5-10. Static head is the height of a fluid above a given point in a column at rest. Dynamic head is the head of fluid in motion.

For example, a storage tank located on the second floor of a building contains 10′ of water. The tank feeds a pump 14′ below on the first floor. The total static head from the surface of the water to the pump below is 24′. Therefore, the theoretical head pressure is 10.397 psi (.0361 × 12″ × 24′ = 10.397 psi). Static head pressure provides pressure to move a fluid when a port or valve is opened.

Dynamic head is the head of fluid in motion. Dynamic head represents the pressure necessary to force a fluid from a given point to a given height. *Dynamic head pressure* is the pressure and velocity of a fluid produced by a liquid in motion. Dynamic head pressure results when a valve is opened and fluid is allowed an open flow. Dynamic head pressure may be used to direct an open flow of fluid. For example, dynamic head pressure was used in early prospecting days to wash away the sides of mountains to retrieve gold. This was accomplished by piping water from higher lakes and using dynamic head pressure to produce a high pressure and high velocity.

Lift Pressure. *Lift* is the height at which atmospheric pressure forces a fluid above the elevation of its supply source. See Figure 5-11. A pipe with one end in fluid and the other end open to the atmosphere is in equilibrium. Atmospheric pressure lifts (pushes) the liquid in the pipe when a pump is placed on the end of the pipe open to the atmosphere and a vacuum is drawn. With respect to pump operation, lift is the height measured from the elevation of the supply source to the center of the pump's inlet port.

The maximum height a fluid at a standard temperature of 62°F can be lifted is determined by the barometric pressure. Temperature standards are established by agencies such as the American National Standards Institute (ANSI) and the International Organization for Standardization (ISO) to create national and international uniformity. Temperature standards for fluids are required due to the fluctuation of fluid volume at different temperatures.

Static lift is the height to which atmospheric pressure causes a column of fluid to rise above the supply to restore equilibrium. The weight of a column of fluid required to create equilibrium is equal to atmospheric pressure. For example, when an elevated pump is turned ON, the pump removes air from its plumbing, creating a partial vacuum. The fluid then rises to a height that is determined by atmospheric pressure. Atmospheric pressure essentially lifts the fluid to a height of equilibrium, or the balance between the atmosphere's pressure and the water's weight. See Figure 5-12.

The energy applied to a fluid by a pump goes either into the production of usable pressure or velocity in the fluid or into friction losses.

LIFT

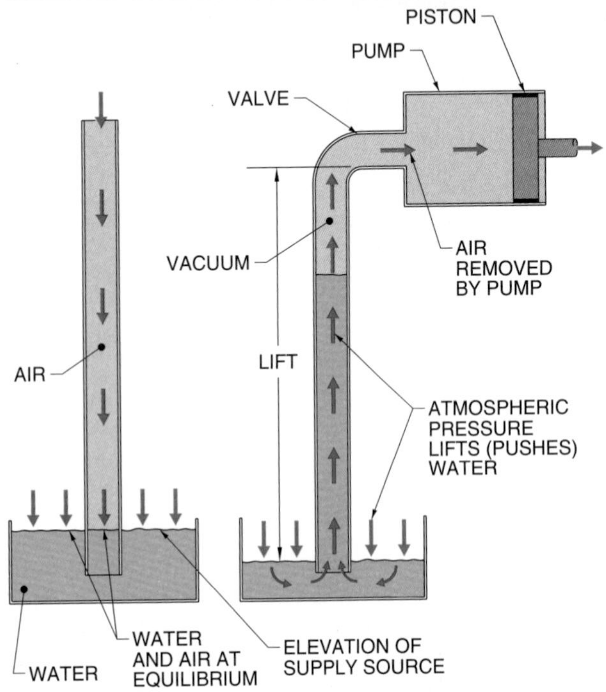

ATMOSPHERIC PRESSURE VS LIFT

Altitude above Sea Level*	Barometer Reading**	Atmospheric Pressure†	Theoretical Lift at Standard Temperature of 62°F*
0	29.92	14.7	34
1000	28.8	14.2	33
2000	27.7	13.6	31.5
3000	26.7	13.1	30.2
4000	25.7	12.6	29.1
5000	24.7	12.1	28
6000	23.8	11.7	27
7000	22.9	11.2	26
8000	22.1	10.8	25
9000	21.2	10.4	24
10,000	20.4	10.0	23

* in ft
** in in. Hg
† in psi

Figure 5-11. Lift is the height at which atmospheric pressure forces a liquid above the elevation of its supply source.

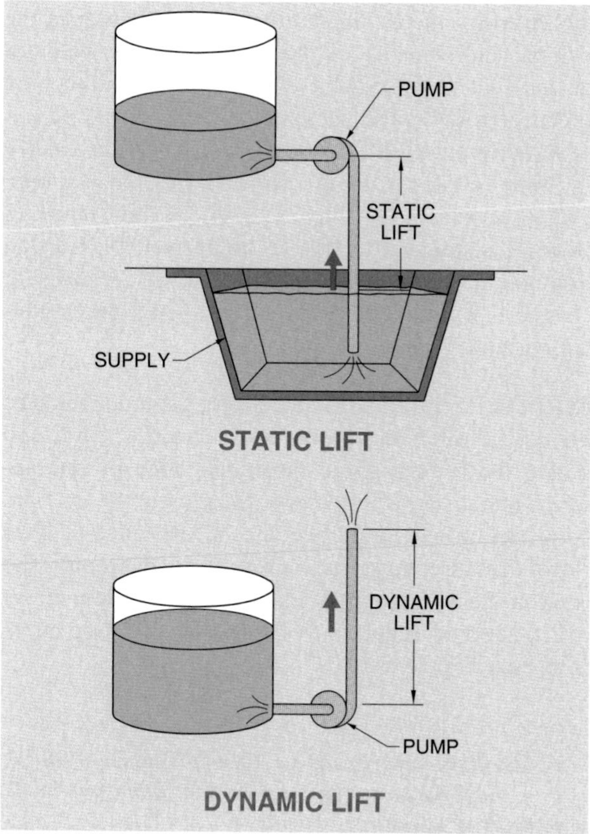

Figure 5-12. Static lift is the height to which atmospheric pressure causes a column of fluid to rise above the supply to restore equilibrium. Dynamic lift is the lift of fluid in motion.

Dynamic lift is the lift of fluid in motion. Dynamic lift represents the pressure necessary to lift a fluid from a given point to a given height. The dynamic lift and distance a fluid can be raised vary due to pump imperfections and pipe friction. Therefore, the limit of actual pump lift ranges to approximately 25′. A pump can force a fluid to greater heights depending on the force exerted on the fluid.

Practical dynamic lift is considerably less than practical static lift because of the friction within piping lengths, piping sizes, number of fittings such as elbows or valves, and because of the fact that most installations are higher than sea level. The practical lift in a pump moving water generally falls in the 20′ to 25′ range.

Total column is the fluid head plus lift. Total column may be static or dynamic. See Figure 5-13. *Static total column* is static head plus static lift. Static columns are determined from pressures created by a fluid at rest. A pump removing water from a retention pond to supply an overhead cooling tower is an example of a static total column. The pump, which is mounted above the pond, assists in lifting the water up to its center (static lift), and at the same time supports the column of water up to the cooling tower (static head). The sum of a pump's static lift and static head is the static total column.

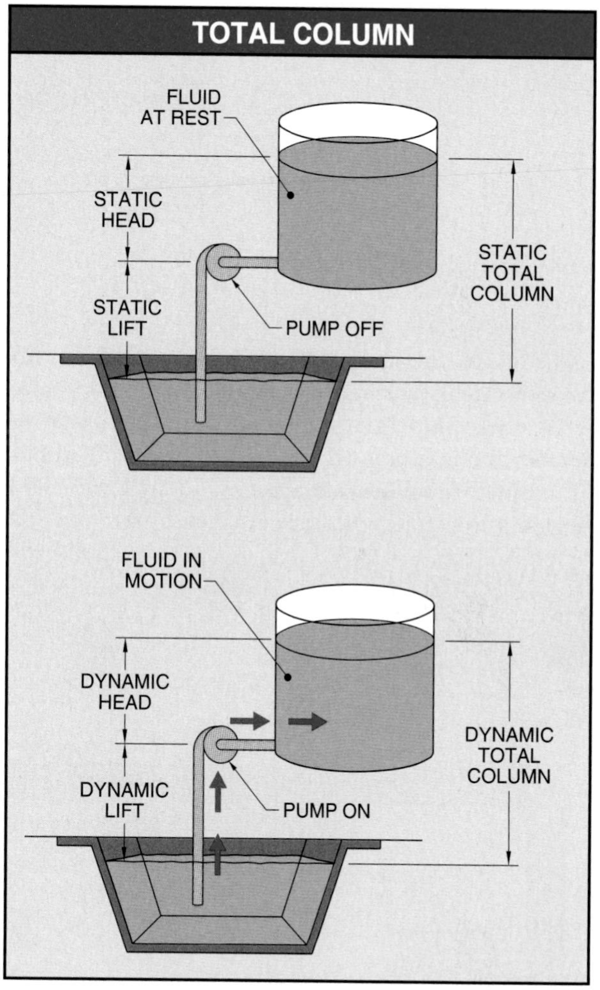

Figure 5-13. Static or dynamic columns are determined from pressures created by a fluid at rest or in motion.

Dynamic total column is dynamic head plus dynamic lift. Dynamic columns are determined from pressures created by a fluid in motion. Dynamic total column is the total column of fluid in motion and represents the pressure from the total column plus frictional resistance.

Eaton Corporation

Series 26 fixed displacement gear pumps from Eaton® produce fluid flow in a hydraulic system from 6.4 gpm to 24.1 gpm and operate at a pressure of 3000 psi continuously.

Flow

Fluid flow is the movement of fluid caused by a difference in pressure between two points. In a hydraulic system, fluid flow is produced by the action of a pump and is expressed as a measurement of gallons per minute (gpm) or liters per minute (lpm). Fluid flow in a hydraulic system is affected by friction and the viscosity of the fluid. Fluid flow is based on the volume and capacity of the system and the velocity of the fluid in the system. Fluid flow also affects the speed of a hydraulic system.

In a system with flowing fluid, pressure is caused by total resistance to the fluid flow from a pump. Pressure results only where there is resistance to flow. Resistance to flow is comprised of friction throughout the system and actuator loads. A pressure change that occurs to a fluid due to its flow is generally expressed in psi.

Friction. Friction is generated throughout a hydraulic system between the piping wall and the fluid, and within the fluid as fluid molecules slide by one another. The faster a fluid flows, the greater the friction. Any friction generated becomes a resistance to fluid flow. Pressure must be increased to overcome the friction. Each component in a hydraulic system offers resistance and reduction of available working pressure. See Figure 5-14.

A fluid vaporizes if an attempt is made to lift the fluid more than the distance atmospheric pressure is capable of raising it.

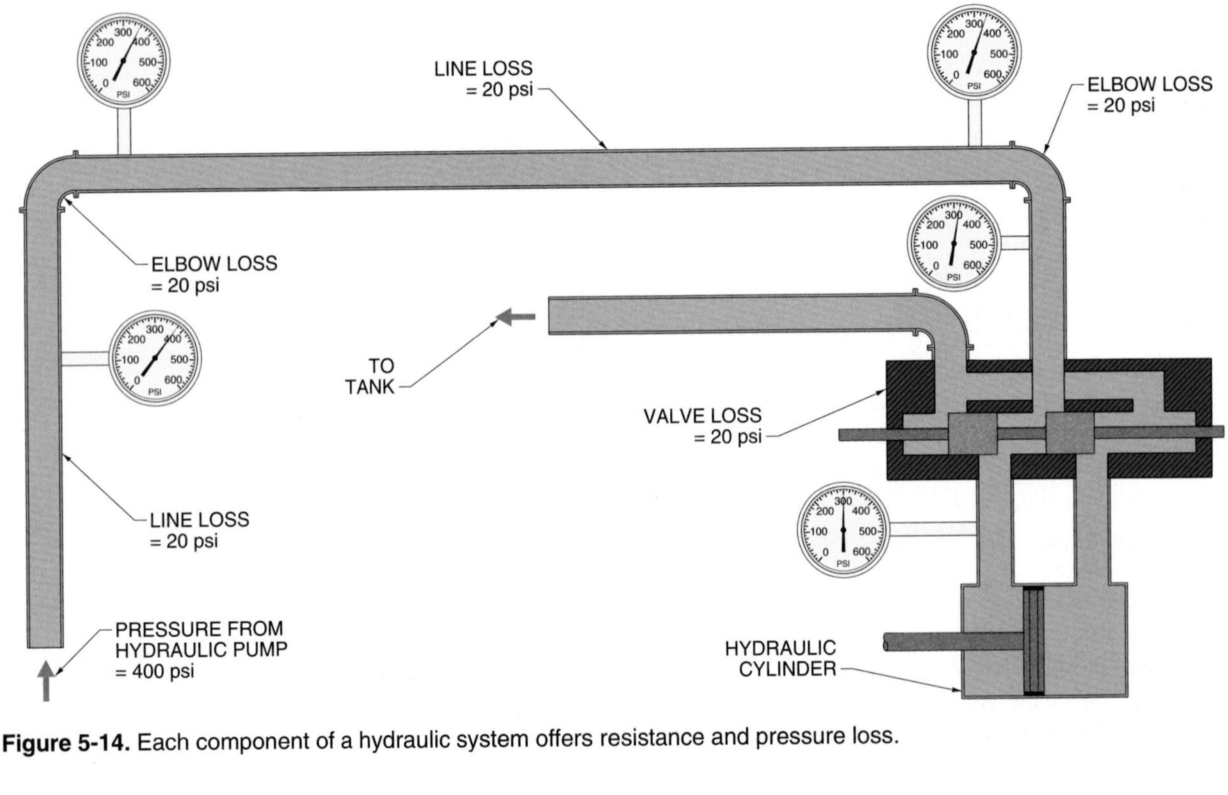

Figure 5-14. Each component of a hydraulic system offers resistance and pressure loss.

A fluid flows because of a difference in pressure. The pressure of a moving fluid is always higher upstream. *Pressure drop* is the pressure differential between upstream and downstream fluid flow caused by resistance. The pressure developed in a hydraulic system is designed to be used as hydraulic leverage. Pressure and fluid flow rate are independent of each other, but both assist in the output. Pressure provides the force and flow rate is used to provide speed. Flow rate is expressed in gpm and is typically determined by the capacity of the pump.

Fluids follow the path of least resistance. For example, a hydraulic system consists of two equal-diameter cylinders with a load of 500 lb on Cylinder A and 200 lb on Cylinder B. Cylinder B moves to the end of its travel before Cylinder A begins to move because of the reduced resistance produced by the lighter weight. See Figure 5-15.

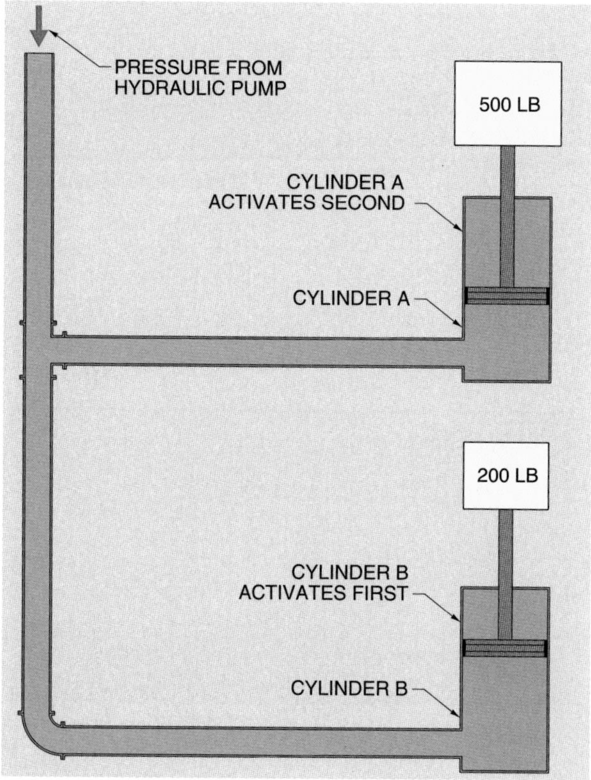

Figure 5-15. Fluids follow the path of least resistance.

Viscosity. *Viscosity* is the measurement of the resistance of a fluid's molecules to move past each other. Fluids that flow with difficulty have a high viscosity. Fluids that are thin and flow easily have a low viscosity. For example, cold honey has a high viscosity and water has a low viscosity.

Viscosity is determined under laboratory conditions by measuring the time required for a specific amount of a fluid at a specific temperature to flow through a specific size orifice. Viscosity is measured in Saybolt Seconds Universal (SSU) using a Saybolt viscometer. See Figure 5-16.

A *Saybolt viscometer* is an instrument used to measure the viscosity of a fluid. Upon reaching the proper temperature, a cork is pulled, allowing 60 mL of test fluid to flow out of the cylinder while being timed with a stopwatch. The measured time is the SSU. The Society of Automotive Engineers (SAE) has established standard numbers for oil viscosity readings.

For example, an SAE 10 oil at 130°F placed in a Saybolt viscometer takes between 90 sec and 120 sec to empty. An oil that is thicker (more viscous) and at the same temperature takes longer to empty the viscometer. This oil has a higher SAE number such as SAE 30, which takes between 185 sec and 255 sec to empty the viscometer.

Hydraulic oil operating viscosity should be between 100 SSU and 300 SSU. Hydraulic oil temperatures should not exceed 150°F for optimum life of the oil.

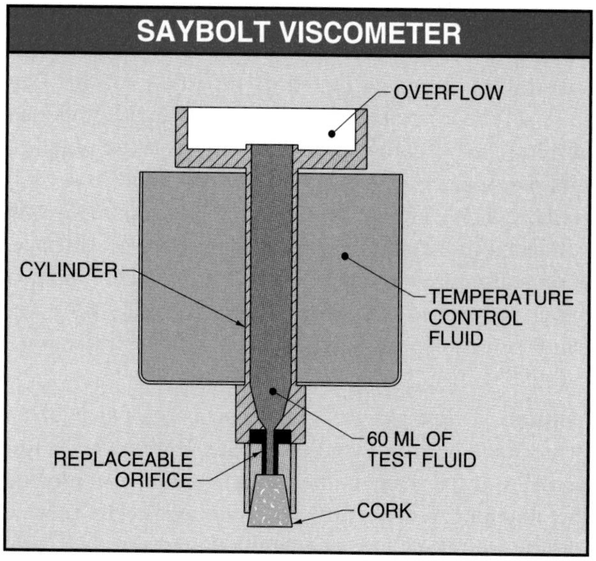

Figure 5-16. A Saybolt viscometer is an instrument used to measure the viscosity of a fluid.

The *viscosity index* is a scale used to show the magnitude of viscosity changes in lubrication oils with changes in temperature. The viscosity index indicates the relative change in SSU readings. Desirable oils are those that have a high viscosity index (relatively low SSU reading change). Oils with a low viscosity index register a large change in SSU readings as temperatures change. Oils with a high viscosity index change slightly as the temperature of the fluid changes.

High-viscosity fluids lead to high internal resistance (friction). This results in high resistance to flow through hydraulic components, creating slow component movement. High resistance increases power consumption, creating a considerable pressure drop throughout the system. Also, fluid temperatures rise when friction is high. Increased fluid temperatures can cause fluid breakdown and damage to pumps and seals.

Fluid viscosity that is too low can be equally harmful. Slippage and leaking may occur, producing an increase in wear because of a reduced lubricating fluid film between mechanical parts. *Slippage* is the internal leaking of hydraulic fluid from a pump's outlet to a pump's inlet. Slippage occurs between the gear teeth and the housing and along the sides of the gears in a gear pump. A *gear pump* is a positive-displacement pump containing intermeshing gears that force the fluid from the pump. *Displacement* is the volume of oil moved during each cycle of a pump. A *positive-displacement pump* is a pump that delivers a definite volume of fluid for each cycle of the pump at any resistance encountered. Slippage is desirable in limited amounts to lubricate moving parts. However, as pressures increase beyond the pump's pressure rating, slippage increases rapidly.

A difference exists between actual pump output and absolute pump output (zero slippage) because slippage indicates a decrease in pump output and fluid flow. *Volumetric efficiency* is the percentage of actual pump output compared to the pump output if there were no slippage. A typical positive-displacement pump in good condition has a volumetric efficiency of approximately 85% because some slippage is required for pump lubrication.

Volume. *Volume* is the three-dimensional size of an object measured in cubic units. See Figure 5-17. Regardless of the shape of the figure, volume is expressed in cubic units (cu in., cu ft, mm^3, m^3, etc.). The volume of a figure is found by calculating the area of the figure and multiplying the area by the length. The volume of a cylinder is found by applying the procedure:

1. Find area of cylinder.

   $A = .7854 \times D^2$

   where

   A = area (in sq units)

   .7854 = constant

   D^2 = diameter squared

2. Find volume of cylinder.

   $V = A \times l$

   where

   V = volume (in cu units)

   A = area (in sq units)

   l = length (in units)

VOLUME

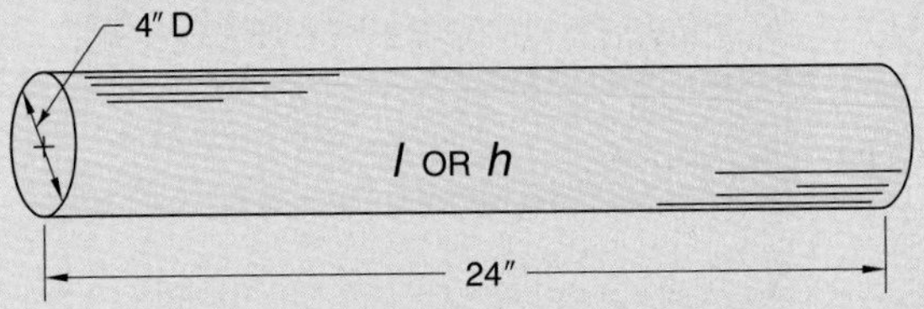

DETERMINING AREA AND VOLUME OF A CYLINDER

1. Find area of cylinder.	2. Find volume of cylinder.
$A = .7854 \times D^2$	$V = A \times l$
$A = .7854 \times (4 \times 4)$	$V = 12.566 \times 24$
$A = .7854 \times 16$	$V =$ **301.584 cu in.**
$A =$ 12.566 sq in.	

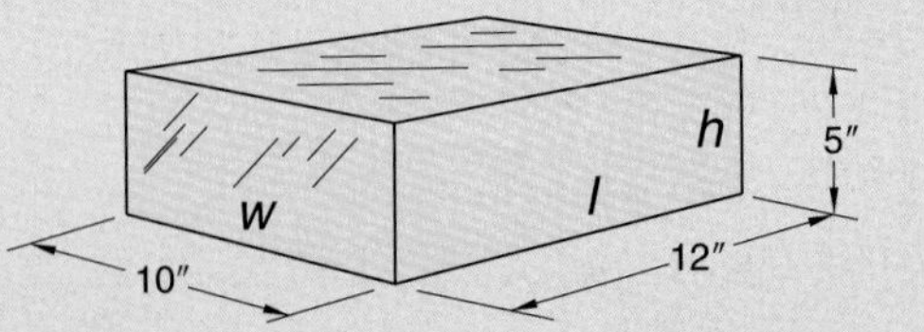

DETERMINING AREA AND VOLUME OF A RECTANGULAR SOLID

1. Find area of rectangular solid.	2. Find volume of rectangular solid.
$A = l \times w$	$V = A \times h$
$A = 12 \times 10$	$V = 120 \times 5$
$A =$ 120 sq in.	$V =$ **600 cu in.**

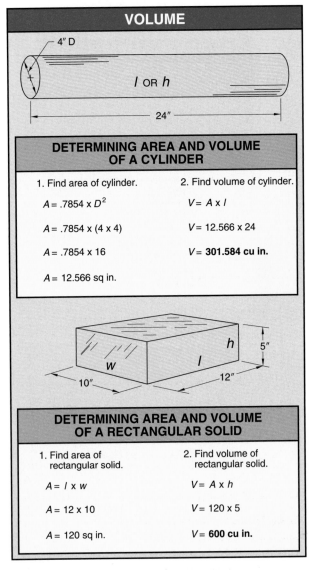

Figure 5-17. Volume is the three-dimensional size of an object measured in cubic units.

For example, what is the volume of a 4″ D cylinder that is 24″ long?

1. Find area of cylinder.

   $A = .7854 \times D^2$

   $A = .7854 \times (4 \times 4)$

   $A = .7854 \times 16$

   $A = 12.566$ sq in.

2. Find volume of cylinder.

   $V = A \times l$

   $V = 12.566 \times 24$

   $V =$ **301.584 cu in.**

Note: The formula for finding the volume of a cylinder may also be expressed as $V = .7854 \times D^2 \times 1$.

The volume of a rectangular solid is found by calculating the area and multiplying by the height. The volume of a rectangular solid is found by applying the procedure:

1. Find area of rectangular solid.

   $A = l \times w$

   where

   A = area (in sq units)

   l = length (in units)

   w = width (in units)

2. Find volume of rectangular solid.

   $V = A \times h$

   where

   V = volume (in cu units)

   A = area (in sq units)

   h = height (in units)

For example, what is the volume of a 12″ long, 10″ wide, and 5″ high rectangular solid?

1. Find area of rectangular solid.

   $A = l \times w$

   $A = 12 \times 10$

   $A = 120$ sq in.

2. Find volume of rectangular solid.

   $V = A \times h$

   $V = 120 \times 5$

   $V =$ **600 cu in.**

Capacity. *Capacity* is the ability to hold or contain something. Capacity is expressed in cubic units and is calculated from a container's volume. Fluids are measured in ounces, pints, quarts, gallons, liters, etc. based on the size of their containers. See Appendix. Fluid measurements can also be expressed in cubic units (cu in., cu ft, etc.) because fluids occupy three dimensions. For example, one gallon of fluid equals 231 cu in.

The quantity of fluid required to fill a specific volume is determined by calculating the volume and dividing by 231. Capacity of a cylinder is found by applying the procedure:

1. Find area of cylinder.

   $A = .7854 \times D^2$

2. Find volume of cylinder.

   $V = A \times l$

3. Find capacity of cylinder.

   $$C = \frac{V}{231}$$

   where

   C = capacity (in gal.)

   V = volume (in cu in.)

   231 = constant (cu in. of fluid per gallon)

For example, what is the capacity of a 4″ D cylinder that has a 24″ stroke?

1. Find area of cylinder.

   $A = .7854 \times D^2$

   $A = .7854 \times (4 \times 4)$

   $A = .7854 \times 16$

   $A = 12.566$ sq in.

2. Find volume of cylinder.

   $V = A \times l$

   $V = 12.566 \times 24$

   $V = 301.584$ cu in.

3. Find capacity of cylinder.

   $$C = \frac{V}{231}$$

   $$C = \frac{301.584}{231}$$

   $C =$ **1.306 gal.**

Less hydraulic fluid is required to retract a piston than is required to extend a piston. This is due to the piston rod taking up part of the cylinder volume (reduced capacity). The volume that the piston rod occupies must be subtracted from the total volume of the cylinder when determining the volume of fluid that a cylinder displaces when retracting. See Figure 5-18. The capacity of a cylinder when retracting is found by applying the procedure:

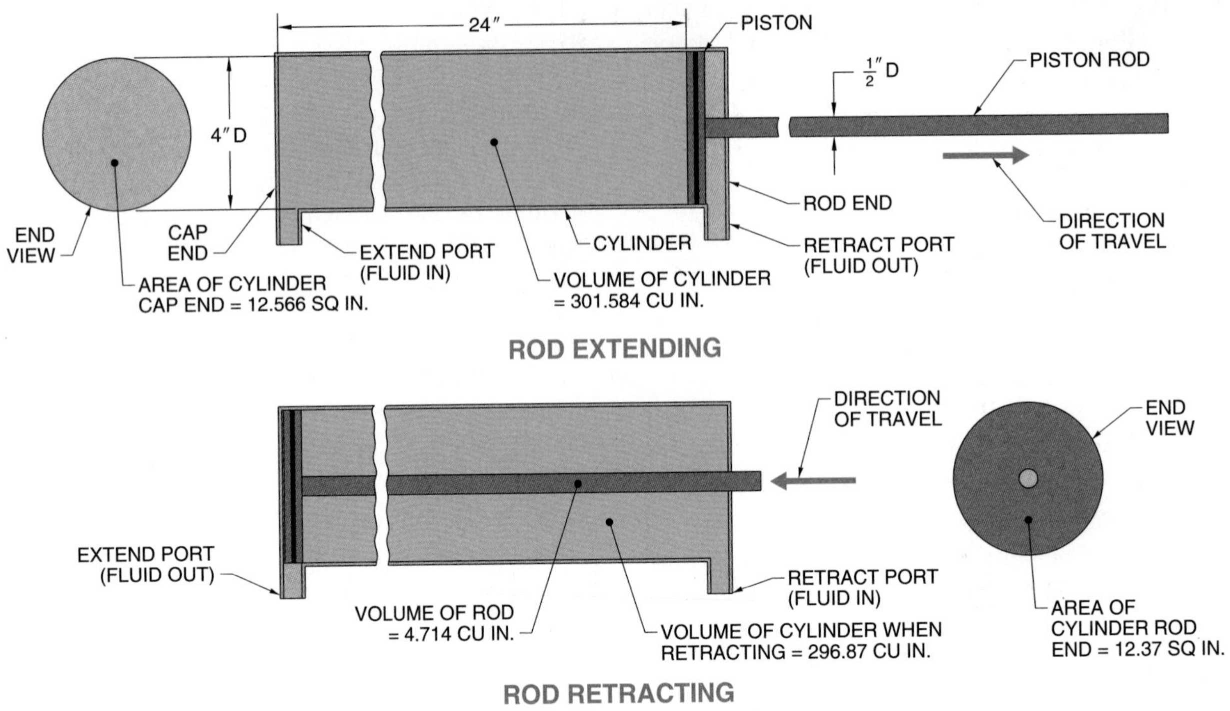

Figure 5-18. Less fluid is required to retract a piston than to extend a piston due to the piston rod occupying a part of the cylinder volume.

1. Find area of piston.

   $A_p = .7854 \times D^2$

2. Find volume of cylinder.

   $V_c = A_p \times l_c$

   where

   V_c = volume of cylinder (in cu units)

   A_p = area of piston (in sq units)

   l_c = length of cylinder (in units)

3. Find area of rod.

   $A_r = .7854 \times D^2$

4. Find volume of rod.

   $V_r = A_r \times l_r$

   where

   V_r = volume of rod (in cu units)

   A_r = area of rod (in sq units)

   l_r = length of rod (in units)

5. Find volume of cylinder when retracting.

   $V_{cr} = V_c - V_r$

   where

   V_{cr} = volume of cylinder when retracting (in cu units)

   V_c = volume of cylinder (in cu units)

   V_r = volume of rod (in cu units)

6. Find capacity of cylinder when retracting.

   $$C = \frac{V_{cr}}{231}$$

For example, what is the capacity of a 4″ D hydraulic cylinder when retracting with a 24″ stroke and a ½″ piston rod?

1. Find area of piston.

   $A_p = .7854 \times D^2$

   $A_p = .7854 \times (4 \times 4)$

   $A_p = .7854 \times 16$

   $A_p = 12.566$ sq in.

2. Find volume of cylinder.

   $V_c = A_p \times l_c$

   $V_c = 12.566 \times 24$

   $V_c = 301.584$ cu in.

3. Find area of rod.

   $A_r = .7854 \times D^2$

   $A_r = .7854 \times (.5 \times .5)$

   $A_r = .7854 \times .25$

   $A_r = .1964$ sq in.

4. Find volume of rod.

$V_r = A_r \times l_r$

$V_r = .1964 \times 24$

$V_r = 4.714$ cu in.

5. Find volume of cylinder when retracting.

$V_{cr} = V_c - V_r$

$V_{cr} = 301.584 - 4.714$

$V_{cr} = 296.87$ cu in.

6. Find capacity of cylinder when retracting.

$$C = \frac{V_{cr}}{231}$$

$$C = \frac{296.87}{231}$$

$C =$ **1.285 gal.**

Velocity. *Velocity* is the distance a fluid travels in a specified time. See Figure 5-19. Velocity generally means the change of position of a fluid particle during a certain time interval. This may be represented as distance in feet per second (ft/sec).

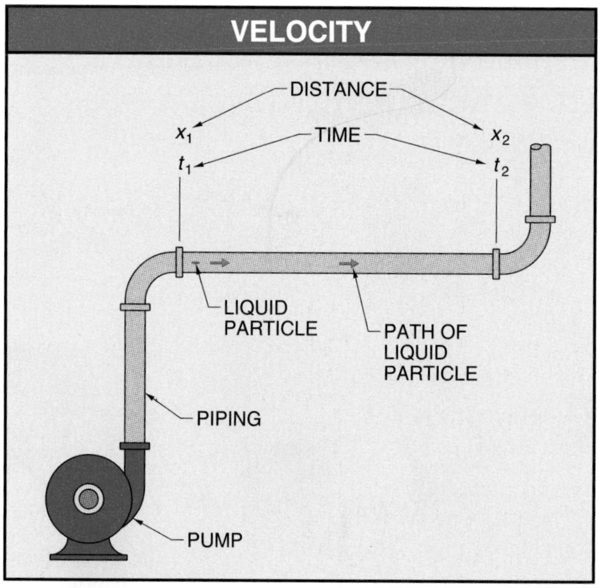

Figure 5-19. Velocity is the distance a fluid travels in a specified time.

Velocity is measured as a vector. A *vector* is a quantity that has a magnitude and direction. A vector is commonly represented by a line segment whose length represents its magnitude and whose orientation represents its direction.

The velocity of a fluid particle is determined by subtracting its initial position from its final position and dividing by the value of the initial time subtracted from the final time. Velocity is found by applying the formula:

$$v = \frac{x_2 - x_1}{t_2 - t_1}$$

where

v = velocity (in ft/sec)

x_2 = final position (in ft)

x_1 = initial position (in ft)

t_2 = final time (in sec)

t_1 = initial time (in sec)

For example, what is the velocity of a fluid particle in a hydraulic system that is at point x_1 at 9:33:54 AM, and after traveling 50′ reaches point x_2 at 9:34:30 AM?

$$v = \frac{x_2 - x_1}{t_2 - t_1}$$

$$v = \frac{50 - 0}{9{:}34{:}30 - 9{:}33{:}54}$$

$$v = \frac{50}{36}$$

$v =$ **1.389 ft/sec**

The velocity of the hydraulic fluid in a system should not exceed recommended values because turbulent conditions result with loss of pressure and excessive heating.

Exercise caution around swinging arms and booms because anything that is supported by fluid pressure may fall if a hose breaks.

The velocity of a fluid varies from one moment to another as its speed or direction of flow changes. *Acceleration* is an increase in speed. Acceleration of a fluid is determined as its change in velocity per unit of time.

The symbol delta (Δ) is generally used to indicate a change. Acceleration is given in units of ft/sec^2 because velocity is measured in ft/sec and time is measured in sec. Acceleration, like velocity, is constantly changing within a hydraulic system. Pipes of various diameters, elbows, valves, and other components all affect the velocity and acceleration of the fluid within a hydraulic system. Acceleration of a fluid is found by applying the formula:

$$a = \frac{\Delta v}{\Delta t}$$

where

a = acceleration (in ft/sec^2)

Δv = average velocity during Δt (in ft/sec)

Δt = time interval elapsed in traveled distance (in sec)

For example, what is the acceleration between measuring points when the fluid flow within a hydraulic system has an initial velocity of 15 ft/sec and changes to 30 ft/sec in 8 sec?

$$a = \frac{\Delta v}{\Delta t}$$

$$a = \frac{30 - 15}{8}$$

$$a = \frac{15}{8}$$

a = **1.875 ft/sec^2**

Flow is the movement of a fluid. *Flow rate* is the volume of fluid flow. A fluid in motion is always flowing, but its rate of flow may change. Fluid velocity depends on the rate of flow in gallons per minute (gpm) and the cross-sectional area of a pipe or component.

The velocity of a fluid increases at any restriction in a pipe or component if the flow rate remains the same in the system. Common restrictions include valves, elbows, pipes, reducers, etc. Also, the velocity of a fluid decreases as the cross-sectional area of a pipe or component increases. See Figure 5-20.

The law of conservation of matter states that the mass or volumetric flow rate of an incompressible fluid through a pipe is constant at every point in the pipe. The velocity of a fluid must increase at any restriction if there are no leaks in the system and the flow rate remains constant. The velocity increases four times to maintain a constant rate of flow if a pipe diameter is changed to one-half of its original size. Velocity of a fluid in a pipe is found by applying the formula:

$$v = \frac{l_2}{\frac{A \times l_1}{231} \times \frac{60}{Q}}$$

where

v = velocity (in ft/sec)

l_2 = length of pipe (in ft)

A = cross-sectional area of pipe (in sq in.)

l_1 = length of pipe (in in.)

231 = constant (cu in. of fluid per gallon)

Q = flow rate (in gpm)

60 = constant (sec in 1 min)

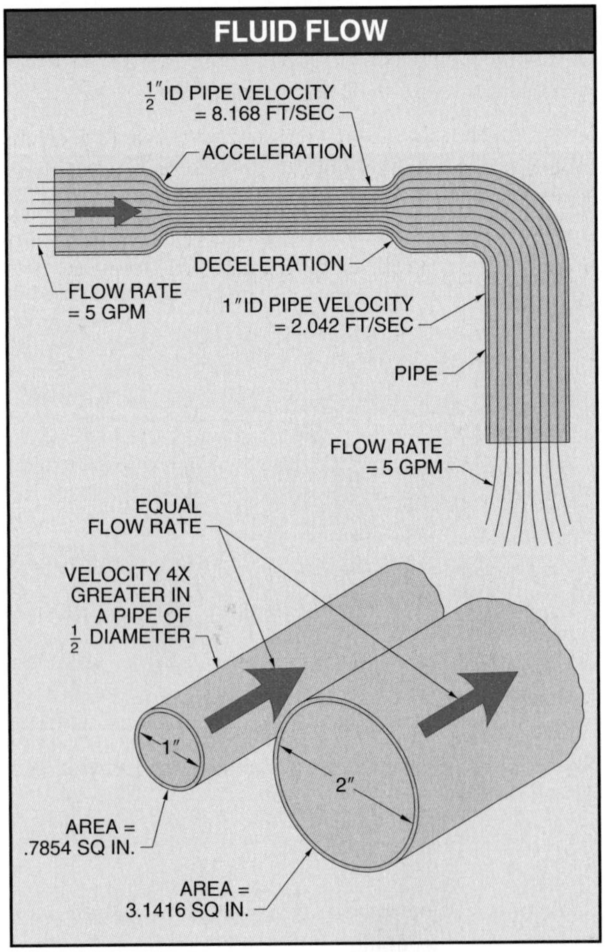

Figure 5-20. The velocity of a fluid increases at any restriction in a pipe or component if the flow rate remains the same in the system.

For example, what is the velocity of a fluid having a flow rate of 5 gpm through a 12″ section of 1″ D pipe?

$$v = \frac{l_2}{\frac{A \times l_1}{231} \times \frac{60}{Q}}$$

$$v = \frac{1}{\frac{.7854 \times 12}{231} \times \frac{60}{5}}$$

$$v = \frac{1}{\frac{9.4248}{231} \times 12}$$

$$v = \frac{1}{.0408 \times 12}$$

$$v = \frac{1}{.4896}$$

v = **2.042 ft/sec**

Speed. The speed of a cylinder rod is determined by the volume of the cylinder, and the fluid flow rate (gpm). To determine the speed at which a cylinder rod moves, the flow rate at which hydraulic fluid is directed into the cylinder must be known.

The speed of a cylinder rod is independent of pressure. The speed of rod extension is usually expressed in inches per minute (in./min). The speed of rod extension is directly proportional to the flow rate. Cylinder rod extension speed is calculated by applying the formula:

$$s = 231 \times \frac{Q}{.7854 \times D^2}$$

where

s = speed of extension (in in./min)

231 = constant (cu in. of fluid per gallon)

Q = flow rate (in gpm)

.7854 = constant

D^2 = diameter of cylinder squared

For example, what is the rod speed of a 5″ D cylinder supplied by a 5 gpm pump?

$$s = 231 \times \frac{Q}{.7854 \times D^2}$$

$$s = 231 \times \frac{5}{.7854 \times 5 \times 5}$$

$$s = 231 \times \frac{5}{19.64}$$

$s = 231 \times .255$

s = **58.9 in./min**.

Two methods of increasing the speed at which a load (or cylinder rod) in a hydraulic system moves are by using a smaller diameter cylinder or by increasing the rate of fluid flow to the cylinder. A small diameter cylinder produces an increase in speed and a decrease in the applied force as compared to a larger cylinder. Two cylinders of different diameters having the same length have different fluid capacities and, if both receive the same rate of fluid flow, the rate of travel and pressure output are different.

MECHANICAL ADVANTAGE

Mechanical advantage is the ratio of the output force of a device to the input force. Mechanical advantage is achieved when an applied force is multiplied, resulting in a larger output force. See Figure 5-21. Devices that produce mechanical advantage include levers, block and tackles, gears, etc.

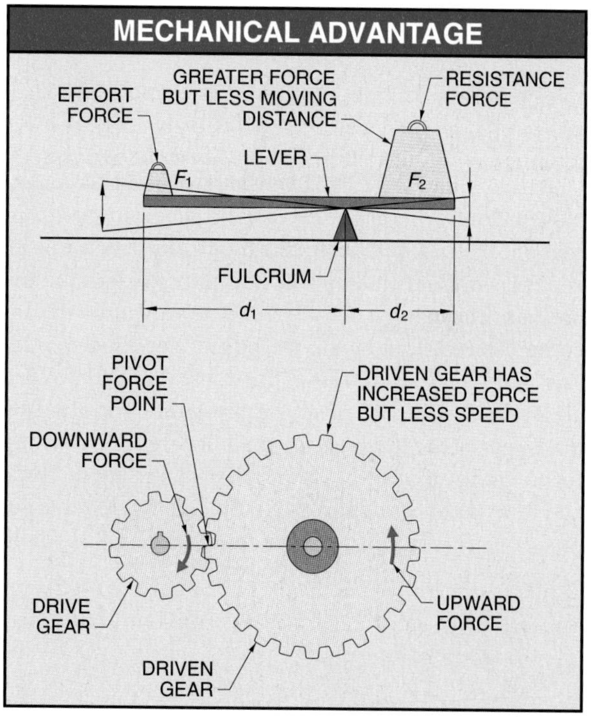

Figure 5-21. Mechanical advantage is the advantage gained by the use of a mechanism in transmitting force.

Mechanical advantage results from a force applied a certain distance from a fulcrum. A *fulcrum* is a support on which a lever turns or pivots and is located somewhere between the effort force and the resistance force. In determining the force needed to balance a lever/fulcrum mechanism, the effort force must be farther from the fulcrum than the resistance force or must have an effort force equal to or greater than the resistance force. The force required to overcome a resistance force is calculated by applying the formula:

$$F_1 = \frac{F_2 \times d_2}{d_1}$$

where

F_1 = effort force (in lb)

F_2 = resistance force (in lb)

d_1 = distance between effort force and fulcrum (in ft)

d_2 = distance between resistance force and fulcrum (in ft)

For example, what is the effort force, placed 15′ from the fulcrum, required to lift a resistance force of 800 lb placed 1½′ from the fulcrum?

$$F_1 = \frac{F_2 \times d_2}{d_1}$$

$$F_1 = \frac{800 \times 1.5}{15}$$

$$F_1 = \frac{1200}{15}$$

$$F_1 = \mathbf{80\ lb}$$

Pascal's law states that pressure exerted on an enclosed fluid is transmitted undiminished in every direction. This is demonstrated by a fluid-filled bottle. As the cork is pressed further into the bottle, the pressure throughout the bottle increases until the incompressible fluid bursts the bottle. See Figure 5-22. The bottle bursts because the force applied to one area (the cork) is equal to the pressure multiplied by the larger area (the body of the bottle). The resulting force within a vessel is a product of the input force and the input pressure area divided by the output pressure area. Resulting force within a vessel is found by applying the formula:

$$F_2 = F_1 \times \frac{A_2}{A_1}$$

where

F_2 = resulting force (in lb)

F_1 = input force (in lb)

A_2 = area of output pressure (in sq in.)

A_1 = area of input pressure (in sq in.)

For example, what is the force placed on a bottle with a cork area of .4418 sq in. (.7854 × .75 × .75 = .4418) and a bottle surface area of 70.686 sq in. (3″ D × 3.1416 × 7.5″ h = 70.686 sq in.) when 5 lb is applied to the cork?

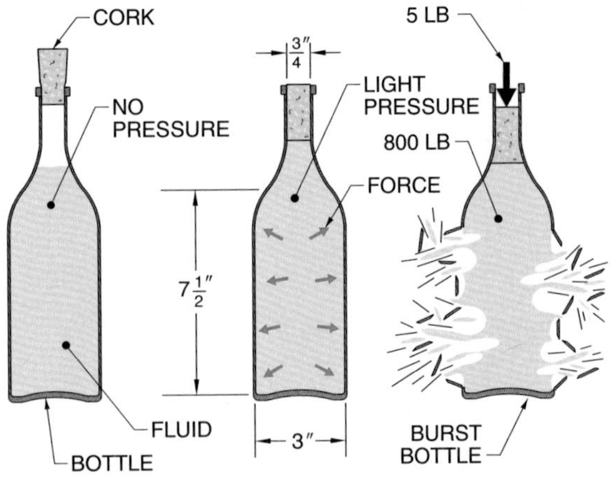

Figure 5-22. Pressure exerted on an enclosed fluid is transmitted undiminished in every direction.

$$F_2 = F_1 \times \frac{A_2}{A_1}$$

$$F_2 = 5 \times \frac{70.686}{.4418}$$

$$F_2 = 5 \times 159.99$$

$$F_2 = \mathbf{800\ lb}$$

Fluids are well suited for being transmitted through pipes, hoses, and passages because of these force characteristics. This force is energy, which can produce movement, work, or leverage when applied to a hydraulic application. For example, interconnected hydraulic cylinders of different diameters produce hydraulic leverage in a typical car jack. See Figure 5-23.

The output pressure of two interconnected cylinders is found by calculating the area of both cylinders, dividing the area of the output cylinder by the area of the input cylinder, and multiplying the result by the input force. Output pressure of two interconnected cylinders is found by applying the procedure:

1. Find area of input piston.

   $A_1 = .7854 \times D_1^2$

2. Find area of output piston.

   $A_2 = .7854 \times D_2^2$

3. Find output piston force.

   $$F_2 = F_1 \times \frac{A_2}{A_1}$$

For example, what is the force of a 3″ D output piston interconnected to a ½″ D input piston if a 50 lb force is applied by the input piston?

1. Find area of input piston.

   $A_1 = .7854 \times D_1^2$

   $A_1 = .7854 \times (.5 \times .5)$

   $A_1 = .7854 \times .25$

   $A_1 = .196$ sq in.

2. Find area of output piston.

   $A_2 = .7854 \times D_2^2$

   $A_2 = .7854 \times (3 \times 3)$

   $A_2 = .7854 \times 9$

   $A_2 = 7.069$ sq in.

3. Find output piston force.

   $$F_2 = F_1 \times \frac{A_2}{A_1}$$

   $$F_2 = 50 \times \frac{7.069}{.196}$$

   $F_2 = 50 \times 36.066$

   $F_2 = \mathbf{1803.3\ lb}$

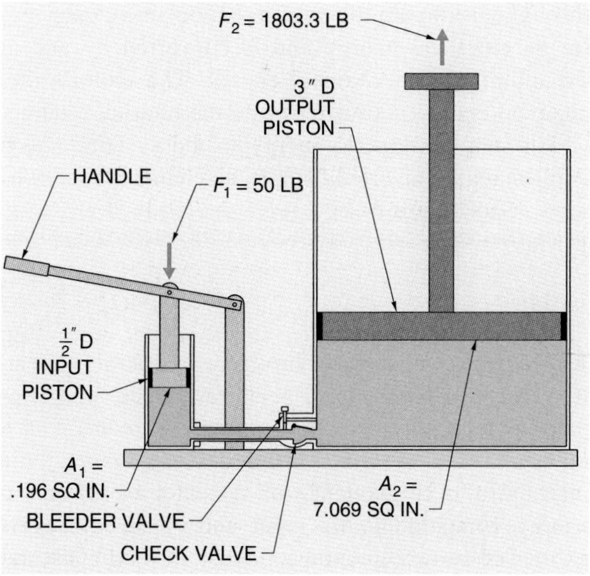

Figure 5-23. Interconnected hydraulic cylinders of different diameters produce hydraulic leverage in a typical car jack.

At times, system pressure must be determined before calculating either the input force or the output force. This may be required when determining the input force required to produce a given output force with given size cylinders. The required input force is determined by calculating the area of the output cylinder, calculating the pressure in the system, and determining the input force based on the system pressure and area of the input cylinder. Input force is found by applying the procedure:

1. Find area of output piston.

   $A_2 = .7854 \times D_2^2$

2. Find pressure in system.

   $$P = \frac{F_2}{A_2}$$

3. Find area of input piston.

   $A_1 = .7854 \times D_1^2$

4. Find input force required.

   $F_1 = P \times A_1$

For example, what is the necessary input force on a 3″ D piston if a static load of 5000 lb being lifted by a 10″ piston stalls due to loss of input force?

1. Find area of output piston.

   $A_2 = .7854 \times D_2^2$

   $A_2 = .7854 \times (10 \times 10)$

   $A_2 = .7854 \times 100$

   $A_2 = 78.54$ sq in.

2. Find pressure in system.

   $$P = \frac{F_2}{A_2}$$

   $$P = \frac{5000}{78.54}$$

   $P = 63.662$ psi

3. Find area of input piston.

   $A_1 = .7854 \times D_1^2$

   $A_1 = .7854 \times (3 \times 3)$

   $A_1 = .7854 \times 9$

   $A_1 = 7.069$ sq in.

4. Find input force required.

   $F_1 = P \times A_1$

   $F_1 = 63.662 \times 7.069$

   $F_1 = \mathbf{450.027\ lb}$

ENERGY AND WORK

The mechanics of hydrostatics, where the flow of fluid within an enclosed system is used to do work, is based on the theory of the conservation of energy. *Energy* is a measure of the ability to do work. The theory of the conservation of energy states that the total energy of a fluid at any point in a system is equal to the total energy of the fluid at another point, unless work has been done by the fluid on some external component.

Total energy is a measure of a fluid's ability to do work. In hydrostatics, total energy is the sum of static energy, kinetic energy, heat energy, and pressure energy. See Figure 5-24.

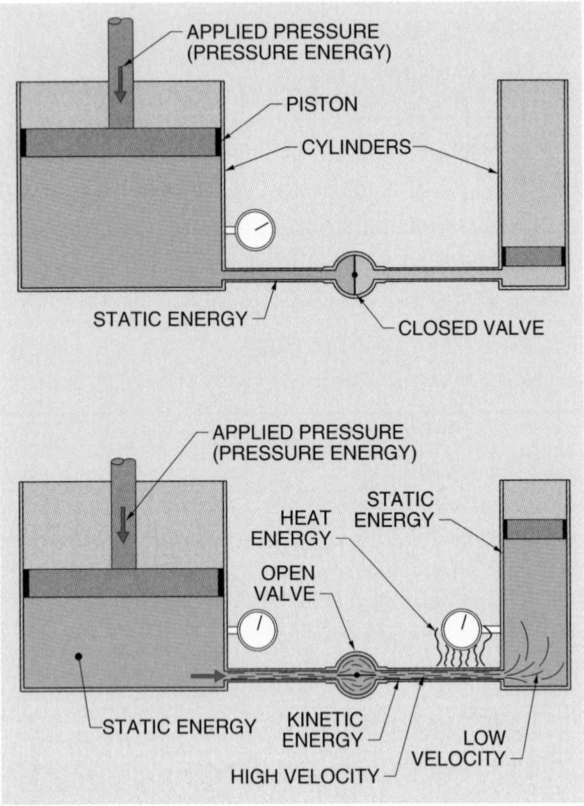

Figure 5-24. In hydrostatics, total energy is the sum of static energy, kinetic energy, heat energy, and pressure energy.

Static energy (potential energy) is the ability of a fluid to do work using the height and weight of the fluid above some reference point. Static energy is stored energy ready to be used. In a hydraulic system, static energy is transformed into kinetic energy when a valve is opened, allowing fluid to flow. This flow causes velocity, acceleration, and the ability to do work.

Kinetic energy is the energy of motion. Any moving object, such as a fluid in a hydraulic system, has kinetic energy. As a fluid flows through a system and through a hydraulic motor, it is kinetic energy. However, when the fluid enters a cylinder to do work, flow and velocity decrease and the system's kinetic energy is changed to static energy. Kinetic energy can be changed into heat energy because of its movement (friction and pressure).

Heat energy is the ability to do work (usually destructive) using the heat stored or built up in a fluid. Heat energy cannot be harnessed or used in a hydraulic system. Once a portion of kinetic energy is converted to heat energy, it is lost energy.

Pressure energy is the ability to do work by applying pressure to a fluid. Energy exists in various forms and has the ability to change from one form to another. Pressure energy begins the moment pressure is applied at the beginning of a system. Pressure energy can be produced by manual force, such as a foot brake or car jack, or by the use of a pump. A pump, however, only creates fluid flow and does not add to the pressure until there is a resistance to the flow.

Pressure energy is introduced when resistance is met by a hydraulic pump. The transmission of energy throughout a hydraulic system begins at a pump motor as electrical energy and is converted by and at the motor into mechanical energy. The motor's mechanical energy is transferred to the hydraulic pump, which supplies kinetic energy to the system, which is ultimately converted back to mechanical energy as work. See Figure 5-25.

Efficiency

As energy is transmitted through a hydraulic system, it is reduced by friction, heat, resistance, and slippage. The degree to which energy is reduced is a measure of a system's efficiency. *Efficiency* is a measure of a component's or system's useful output energy compared to its input energy. Efficiency is expressed as a percentage. When new, the natural slippage within a hydraulic pump reduces its efficiency by as much as 15%. Electric motors are typically 85% efficient. No electrical, hydraulic, pneumatic, or mechanical system is 100% efficient. Total efficiency of more than one energy component in a system is found by applying the formula:

$$Eff_T = Eff_1 \times Eff_2 \times \ldots \times 100$$

where

Eff_T = total efficiency (in %)

Eff_1 = efficiency of component 1

Eff_2 = efficiency of component 2

100 = constant (to convert to percent)

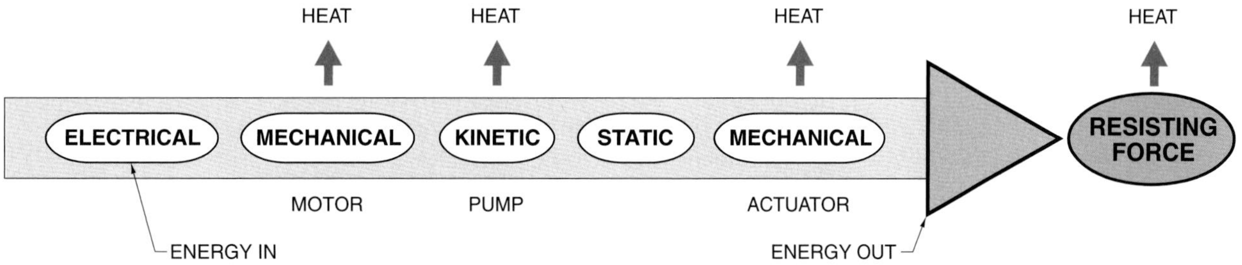

Figure 5-25. Energy changing within a hydrostatic system greatly reduces its overall efficiency.

For example, what is the total efficiency of a system power unit containing a motor listed as 85% efficient and a hydraulic pump listed as 90% efficient?

$Eff_T = Eff_1 \times Eff_2 \times 100$

$Eff_T = .85 \times .90 \times 100$

$Eff_T = \mathbf{76.5\%}$

Energy that changes form from hydraulic to mechanical represents work. *Work* is the energy used when a force is exerted over a distance. Work is expressed in pound-feet (lb-ft). Work is found by applying the formula:

$W = F \times d$

where

W = work (in lb-ft)

F = force (in lb)

d = distance (in ft)

For example, how much work is performed by a forklift exerting a 3000 lb force over a vertical lift distance of 9′?

$W = F \times d$

$W = 3000 \times 9$

$W = \mathbf{27{,}000\ lb\text{-}ft}$

Work may also be expressed by the amount of power required. *Power* is the rate or speed of doing work. Power is found by applying the formula:

$$P = \frac{F \times d}{t}$$

where

P = power (in lb-ft/time)

F = force (in lb)

d = distance (in ft or in.)

t = time (in sec, min, or hr)

For example, how much power is required to move a 3000 lb force 9″ in 8 sec?

$$P = \frac{F \times d}{t}$$

$$P = \frac{3000 \times 9}{8}$$

$$P = \frac{27{,}000}{8}$$

$P = \mathbf{3375\ lb\text{-}ft/sec}$

Select a 25% larger cylinder and a 25% higher system pressure than is mathematically required to move the load when determining cylinder size and system pressure.

Snorkel

Wildcat rough terrain hydraulic scissor lifts from Snorkel feature articulating rear axles for better traction and are available with 4-wheel drive that provides gradeability up to 40%.

Horsepower

Mechanical energy is often expressed in horsepower (HP). One horsepower is the amount of energy required to lift 33,000 lb 1′ in 1 min. One horsepower equals 550 ft lb/sec. See Figure 5-26. Mechanical horsepower is found by applying the formula:

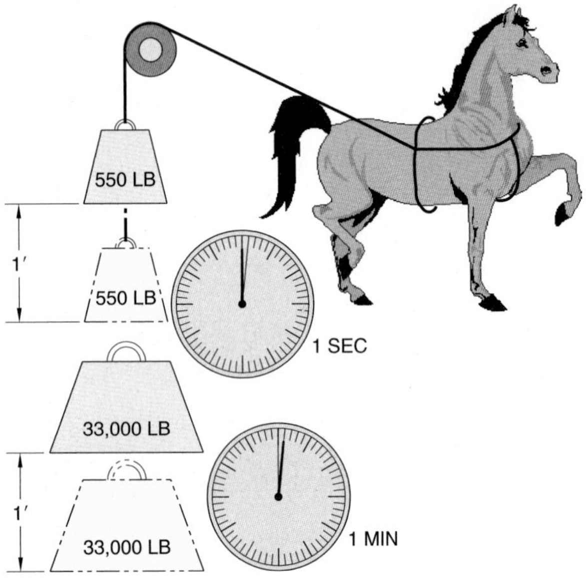

Figure 5-26. One horsepower is the amount of energy required to lift 33,000 lb 1′ in 1 min.

$$HP = \frac{F \times d}{550 \times t}$$

where

HP = horsepower

F = force (in lb)

d = distance (in ft)

550 = constant

t = time (in sec)

For example, what is the horsepower required to lift 3000 lb 9′ in 8 sec?

$$HP = \frac{F \times d}{550 \times t}$$

$$HP = \frac{3000 \times 9}{550 \times 8}$$

$$HP = \frac{27{,}000}{4400}$$

HP = **6.136 HP**

Horsepower in a hydraulic system is used to calculate the rate at which a system is doing work. To calculate hydraulic horsepower, pressure (in psi) and flow rate (in gpm) are used instead of ft, lb, and sec to determine mechanical HP. Also, hydraulic horsepower formulas use a conversion factor of .000583, which indicates the relationship between ft, lb, psi, and gpm. Hydraulic horsepower is found by applying the formula:

$$HP = P \times Q \times .000583$$

where

HP = horsepower

P = pressure (in psi)

Q = flow rate (in gpm)

.000583 = constant

For example, what horsepower is needed in a hydraulic system to deliver 10 gpm at 800 psi?

$$HP = P \times Q \times .000583$$

$$HP = 800 \times 10 \times .000583$$

HP = **4.664 HP**

Torque

Torque is the twisting (rotational) force of a shaft. See Figure 5-27. The twisting effort at the shaft causes the shaft to rotate. The presence of torque indicates that there is a force present, even without rotation.

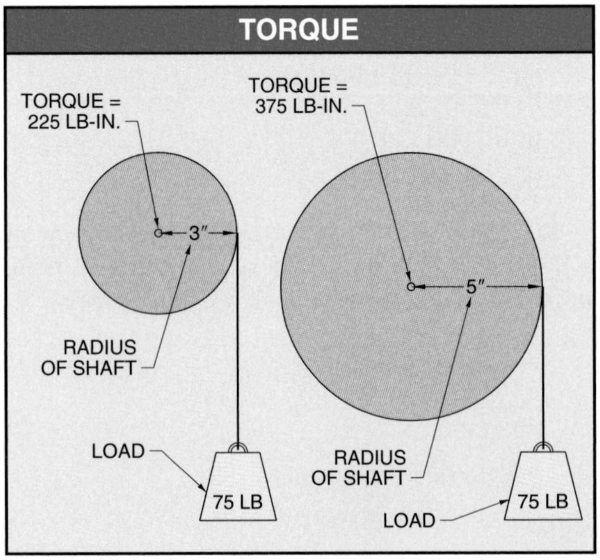

Figure 5-27. Torque is the twisting (rotational) force of a shaft.

Torque is measured at a distance from the motor shaft center. The farther a force is from the shaft's center, the greater its torque. Torque is normally measured in lb-in. and is equal to the product of its force (lb) times the distance from the shaft's center. Torque is found by applying the formula:

$$T = F \times d$$

where

T = torque (in lb-in.)

F = force (in lb)

d = distance from the shaft center (in in.)

For example, what is the torque required to overcome a 75 lb force connected 3″ from the motor shaft's center?

$$T = F \times d$$

$$T = 75 \times 3$$

$$T = \mathbf{225\ lb\text{-}in.}$$

A motor with a large shaft or pulley would have to apply a greater torque. For example, if the distance between the shaft center and point of force is 5″, the torque required is 375 lb-in. (75 × 5 = 375 lb-in.).

Torque applied by a hydraulic motor can also be calculated by replacing force and distance with pressure (in psi) and hydraulic motor displacement per revolution, divided by 2π. The torque developed by a hydraulic motor is found by applying the formula:

$$T = \frac{P \times d}{2\pi}$$

where

T = torque (in lb-in.)

P = pressure (in psi)

d = motor displacement (in cu in.)

π = constant (3.1416)

For example, what is the available torque delivered by a hydraulic motor with a displacement of 2.146 cu in. per revolution and an applied pressure of 500 psi?

$$T = \frac{P \times d}{2\pi}$$

$$T = \frac{500 \times 2.146}{2 \times 3.1416}$$

$$T = \frac{1073}{6.283}$$

$$T = \mathbf{170.778\ lb\text{-}in.}$$

Cincinnati Milacron

Torque requirements on many machine processes vary with the type of material being machined.

The torque a hydraulic motor develops depends on its applied pressure and displacement. In most cases, if the available delivered torque is not enough, the pressure is increased and in some cases, depending on the motor type, the displacement can be increased. Either of these may be necessary to staart a hydraulic motor and overcome breakaway, starting, and running torque.

Breakaway torque is the initial energy required to get a nonmoving load to turn. *Starting torque* is the energy required to start a load turning after it has been broken away from a standstill. *Running torque* is the energy that a motor develops to keep a load turning.

Changing displacement may allow for greater breakaway or starting torque, but it also has an adverse effect on the system's speed and operating pressure. An increase in displacement decreases the motor's speed and decreases the operating pressure. Decreasing displacement increases the speed of the motor and also increases the effect on operating pressure. See Figure 5-28.

EFFECT ON MOTOR OUTPUT FROM SYSTEM CHANGES*			
Change	**Speed**	**Effect on Operating Pressure**	**Available Torque**
Increase Displacement	Decreases	Decreases	Increases
Decrease Displacement	Increases	Increases	Decreases
Increase Pressure Setting	No effect	No effect	Increases
Decrease Pressure Setting	No effect	No effect	Decreases
Increase gpm	Increases	No effect	No effect
Decrease gpm	Decreases	No effect	No effect

* effects on changes assume working loads remain the same

Figure 5-28. Hydraulic systems may be adjusted to allow for higher torque requirements, but operating conditions may also change.

Increasing pressure or flow rate may be used to increase a system's working force or speed. Increasing pressure increases cylinder output force. Increasing flow rate increases cylinder speed. The output force of a hydraulic motor is determined by the amount of pressure acting on the area of its rotating parts. The output force of a hydraulic cylinder is determined by the amount of pressure acting on the area of its piston. The output speed of a hydraulic motor is determined by the rate at which the fluid flows through the motor. The output speed of a hydraulic cylinder is determined by how quickly the flow rate fills the volume ahead of the cylinder's piston.

Practical Hydraulics

Chapter 6

Atlas Technologies, Inc.

Hydraulic circuits consist of controlling the movement of a contained liquid. Hydraulic diagrams explain, demonstrate, or clarify the relationship or functions between hydraulic components. Any hydraulic circuit must contain hydraulic fluid, a reservoir, piping, a pump, valves, and actuators. The hydraulic circuit application, complexity, and power requirements dictate the type and number of components used. Every energy source must be identified, understood, and disabled prior to working on a hydraulic system. Always follow the equipment manufacturer's recommendations when servicing hydraulic equipment and circuits.

HYDRAULIC CIRCUITRY

Hydraulics is the branch of science that deals with the practical application of water or other liquids at rest or in motion. A *liquid* is a fluid that can flow readily and assume the shape of its container. Hydraulic circuits consist of controlling the movement of a contained fluid (liquid). A *circuit* is a closed path through which hydraulic fluid flows or may flow. Basic hydraulic circuits include the storing of hydraulic fluid, a method of controlling its flow, and devices that transfer force. Practical hydraulics deals with the operation and repair of hydraulic circuits.

Hydraulic Diagrams

A *hydraulic diagram* is the layout, plan, or sketch of a hydraulic circuit and is designed to explain, demonstrate, or clarify the relationship or functions between hydraulic components. The three basic hydraulic diagrams are pictorial, cutaway, and graphic.

Pictorial. A *pictorial diagram* is a diagram that uses drawings or pictures to show the relationship of each component in a circuit. See Figure 6-1. Pictorial diagrams generally use single lines to show the elements of a circuit. The components are shown using simple outlines to indicate their relative position and appearance in a circuit.

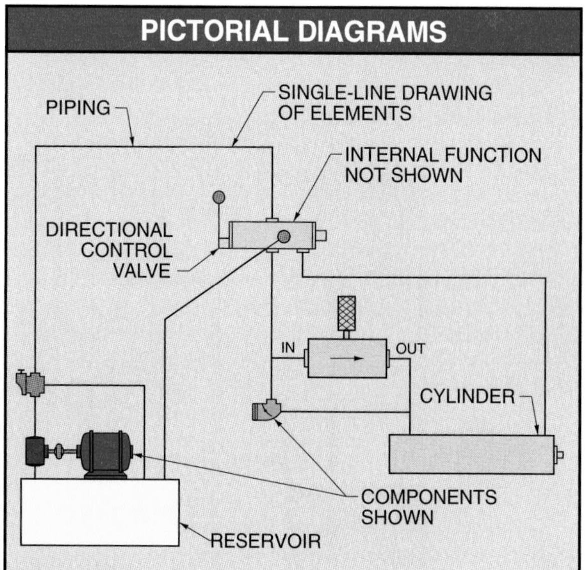

Figure 6-1. A pictorial diagram is a diagram that uses drawings or pictures to show the relationship of each component in a circuit.

Pictorial diagrams show the component's purpose within a circuit but do not provide the internal function or specific information about the components. For example, an outline of a directional control valve may be illustrated in a pictorial diagram, but its type is not defined.

Cutaway. A *cutaway diagram* is a diagram showing the internal details of components and the path of fluid flow. See Figure 6-2. Cutaway diagrams provide more detail than pictorial diagrams. Cutaway diagrams consist of double-line drawings of the circuit components showing their operation and internal positions. A cutaway diagram provides an excellent understanding of simple circuits. Cutaway diagrams may be color-coded to show direction of flow or pressure of the fluid in the piping. Cutaway diagram color coding is:

- Red – Fluid flowing at system operating pressure or highest working pressure. *System operating pressure* is the pressure of a fluid after the pump until the flow is reduced, metered, or returned to the reservoir.
- Yellow – Controlled flow by a metering device or lowest working pressure. *Controlled flow* is the fluid flow after a flow control device has reduced the flow rate of the fluid.
- Orange – Intermediate pressure that is lower than system operating pressure.
- Green – Intake flow to pump or drain line flow. *Intake flow* is the fluid flow from the reservoir, through the filters, to the pump.
- Blue – Exhaust or return flow to the reservoir. *Exhaust flow* is the fluid flow from the actuator, back through the valve, to the reservoir.
- White – Inactive fluid (reservoir fluid).

A disadvantage of a cutaway diagram is that a considerable amount of space is required to show a system consisting of more than the minimum basic components. Also, cutaway diagrams do not indicate some elements such as type or direction of rotation of a pump or motor.

Graphic. A *graphic diagram* is a diagram that uses simple line shapes (symbols) with interconnecting lines to represent the function of each component in a circuit. See Figure 6-3. Graphic diagrams are used when designing and troubleshooting fluid power circuits because the connecting lines and symbols are used to explain how a circuit works. A *symbol* is a graphic element which indicates a particular device, etc. Graphic symbols simplify the explanation of a circuit and can be used by individuals that speak different languages because a person does not have to speak a particular language to understand them. This promotes a universal understanding of fluid power systems.

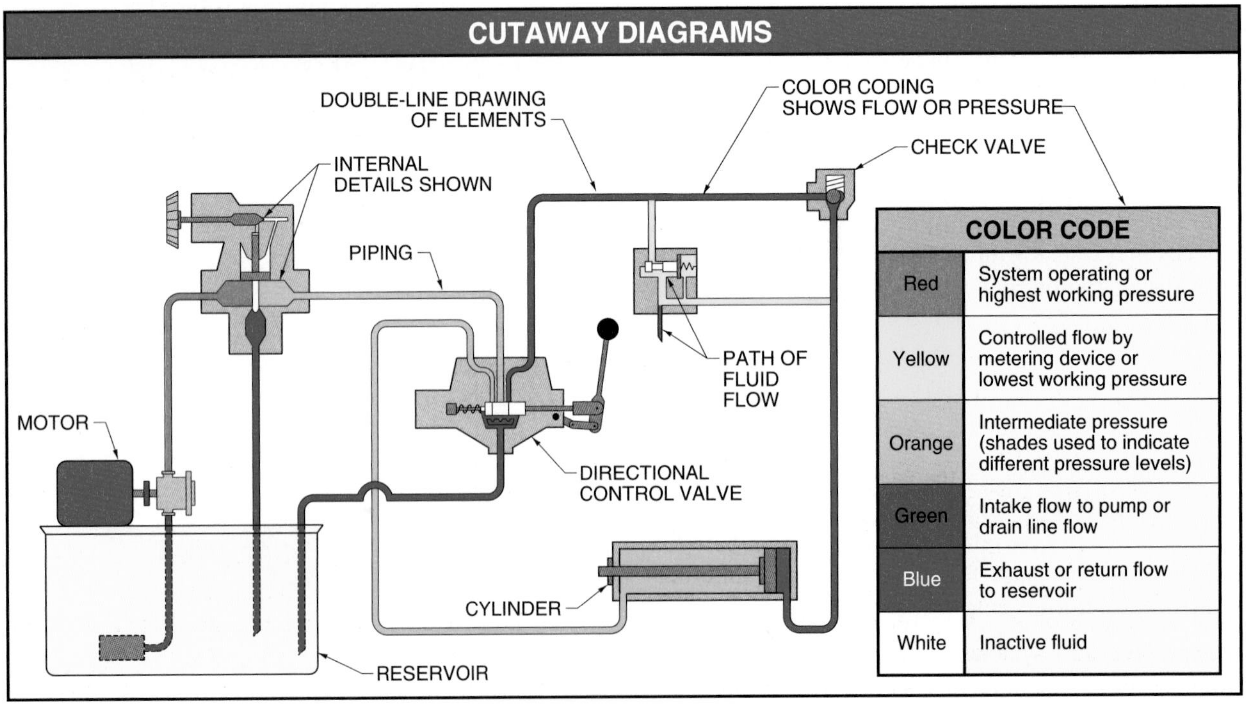

Figure 6-2. A cutaway diagram is a diagram showing the internal details of components and the path of fluid flow.

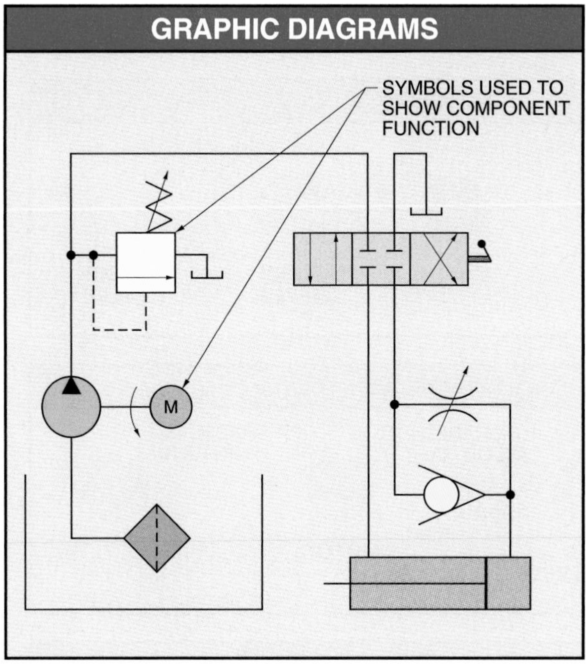

Figure 6-3. A graphic diagram is a diagram that uses symbols with interconnecting lines to represent the function of each component in a circuit.

Hydraulic circuits are used in the automobile repair industry to safely raise an automobile to any convenient working height for repair.

The American National Standards Institute (ANSI), in collaboration with the American Society of Mechanical Engineers (ASME), has adopted a standard regarding symbols for fluid power diagrams, ANSI/ASME Y32.10-1967, *Graphic Symbols for Fluid Power Diagrams*. See Appendix. This standard illustrates the basic fluid power symbols and describes the principles on which the symbols are based.

Graphic symbols show flow paths, connections, and functions of components. They are not used to indicate the rate of flow or to offer pressure settings. These must be added to the graphic diagram. Also, graphic diagrams do not give the actual position of a component in the system. The components are generally positioned to show the flow of the system and how each component is related to the others. Little written explanation is required because standard symbols and lines are used. For example, hydraulic circuit graphic diagrams use four different lines, with each representing a working pipe. The four lines are solid, dashed, dotted, and center lines. A solid line represents a main pipe, outline, shaft, or conductor. This pipe is essentially the working pipe. A dashed line represents pilot piping for controlling a component's function. A dotted line represents exhaust or drain piping. A center line shows the outline of an enclosure.

Graphic symbols present considerable information in a small space and may contain information about the flow of fluid. See Figure 6-4. For example, the simplified symbol for a check valve shows a ball being held against a seat, indicating the direction of flow through that part of the line. Also, some symbols closely resemble the actual component. For example, the symbol for a cylinder is similar to a cutaway view of a cylinder, and a pressure gauge looks much like the face of a gauge with its indicating needle.

Components and their parts are represented by shapes such as circles, triangles, squares, or rectangles. Circles generally represent a component that is round, such as a gauge. Circles also represent rotary devices, such as pumps or motors. Triangles generally represent direction of fluid flow. Triangles that are completely shaded represent liquid flow within a hydraulic system. Triangles that are unshaded represent gas flow within a pneumatic system. Triangles are used to distinguish between pumps and motors. For example, a pump (circle) having an unshaded triangle pointing out is an air compressor. A circle with a solid triangle pointing in the direction of system fluid flow is a hydraulic motor. Squares or rectangles generally represent valves and may be grouped together to show multiple internal functions of a valve such as that of a directional control valve.

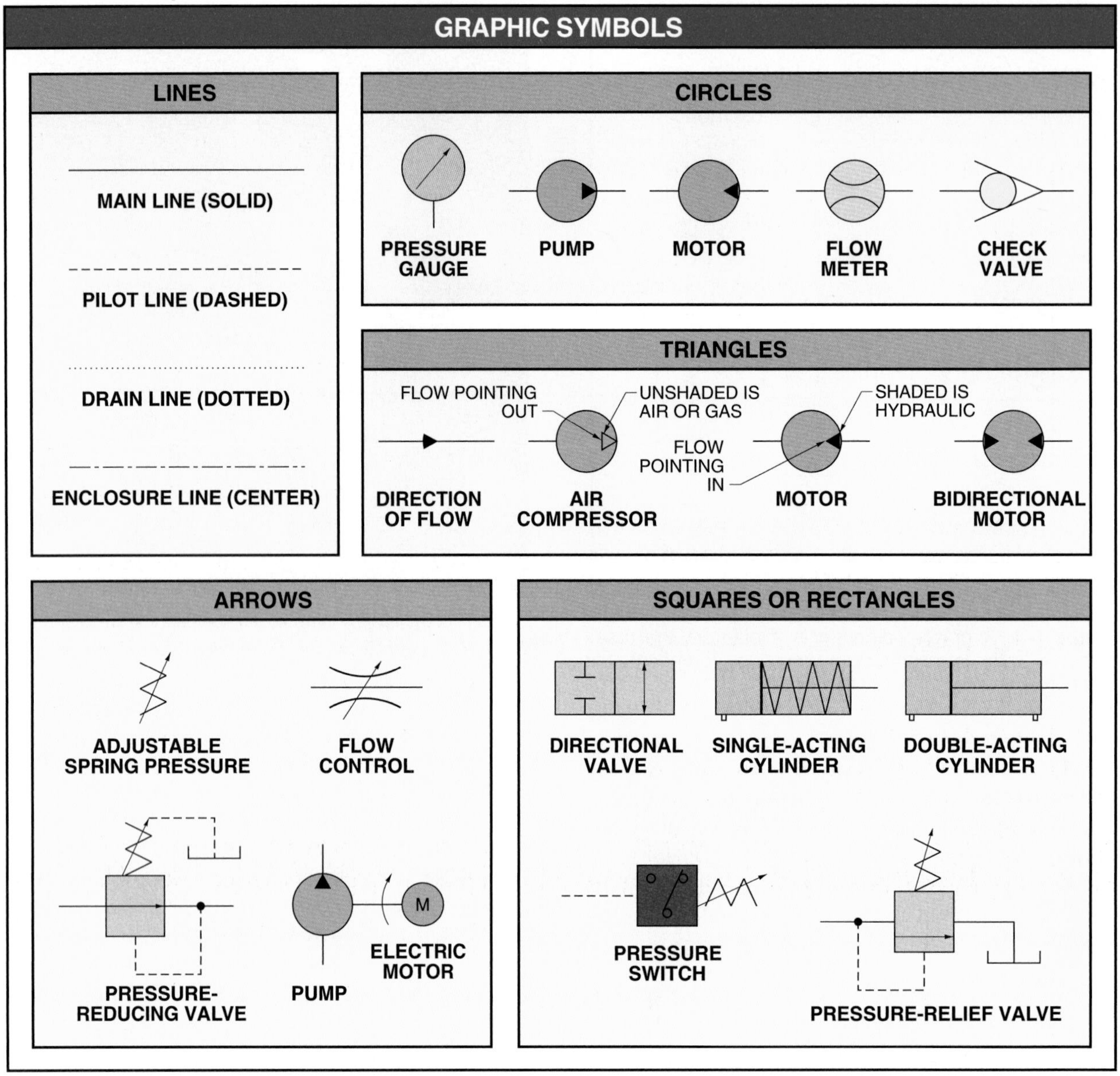

Figure 6-4. Graphic symbols present considerable information in a small space and may contain information about the flow of fluid.

Arrows are used in graphic diagrams to indicate an adjustable or variable component or to show shaft rotation on the near side of the shaft. A component that may be adjusted or varied is represented by an arrow passing through the symbol at approximately a 45° angle. For example, an angled arrow passing through four zig-zag lines indicates adjustable spring pressure. An angled arrow passing through a circle represents an adjustable pump or motor. An arrow parallel to the short side of a symbol, within the symbol, indicates that the component is pressure-compensated. The arrow within the symbol is actually a piston that is activated by line or pilot pressure. An example of pressure compensation is that of a pressure-relief valve where the arrow (piston) within the symbol does not line up with the circuit lines until enough pressure from a pilot line allows the internal arrow to overcome spring pressure and align the internal passage with the circuit lines.

Oil-soaked clothes must be removed immediately because they are toxic and can also catch on fire from a match, cigarette, sparks, or any open flame.

HYDRAULIC CIRCUIT COMPONENTS

Hydraulic components are used in a wide range of combinations for different applications. The circuit application, complexity, and power requirements dictate the type and number of components used. Any hydraulic circuit must contain six essential elements: hydraulic fluid to transmit force and motion, a reservoir (tank) to store the fluid, piping to transport the fluid through the circuit, a pump to move the fluid, valves to control the pressure and direction of the fluid, and actuators to convert hydraulic force into mechanical force. See Figure 6-5.

Fluid

Hydraulic fluid, the major component of a hydraulic circuit, transmits force that is used to do work, conducts heat away from metal surfaces, and lubricates moving parts. Hydraulic fluid (normally petroleum oil) by itself is not sufficient to effectively perform these functions. Also, the quality and content of petroleum oil before and after refining is not consistent. Hydraulic fluids must have the following characteristics to be effective:

- Lubricate by offering a substantial film of fluid even when subjected to high heat.
- Sufficient viscosity to resist leakage. Leakage results in loss of pump efficiency, loss of pressure, and generated heat.
- Resist oxidation, thereby preventing the reactions of oil products to form gum, varnish, and sludge. *Oxidation* is the combining of oxygen with elements in oil which break down the basic oil composition.
- Resist or depress foaming caused by turbulence, agitation, or splashing. *Foaming* is excessive air in hydraulic fluid.
- Resist rust, corrosion, and pitting caused by the chemical (usually acid) union of iron or steel with oxygen. *Pitting* is localized corrosion that has the appearance of cavities (pits).
- Remain relatively stable over a broad temperature range.

These characteristics of hydraulic fluids are developed by special compounding of refined oil and various additives. A large number of compounded fluids are available due to the wide variety of materials used in hydraulic systems, such as seals, rings, or flexible hoses. Care must be taken to ensure that proper fluids are used with compatible components. Use only the compounded fluids specified by the manufacturer.

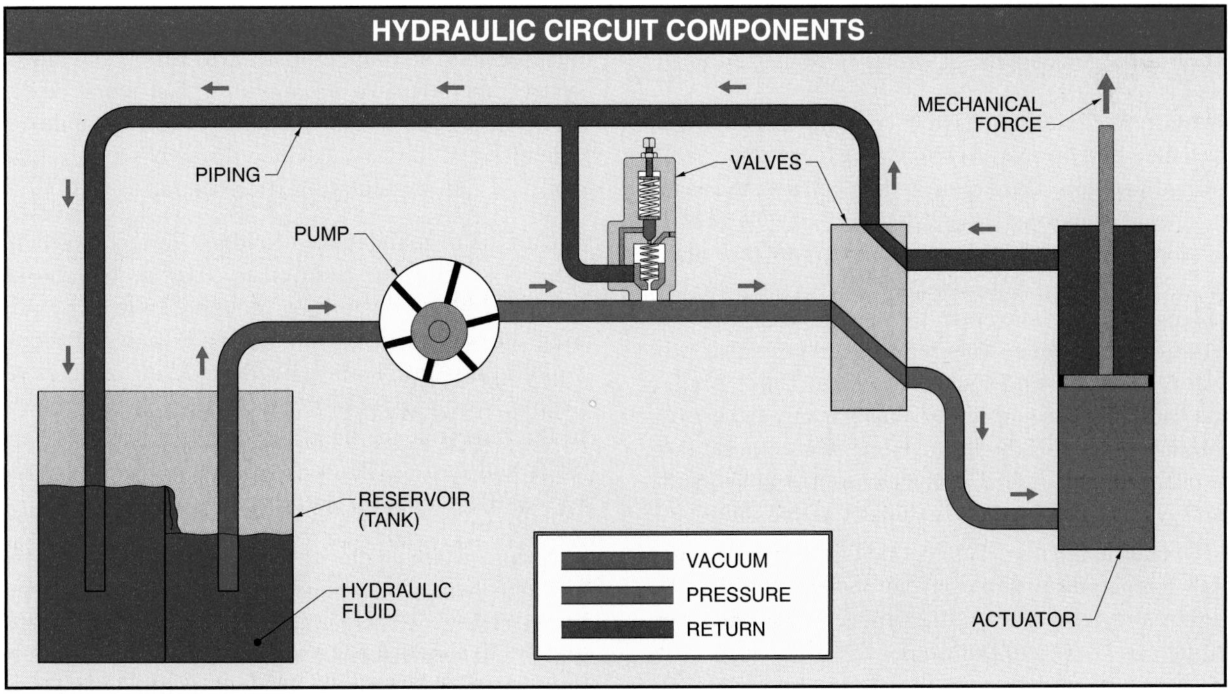

Figure 6-5. Hydraulic circuits require six elements to draw fluid into the system (vacuum), build operating pressures to do work, and return the fluid back to storage.

Additives. An *additive* is a chemical compound added to a fluid to change its properties. Additives protect the fluid in addition to protecting components. Fluids are protected so they are free of destructive chemicals, resistant to foaming, and, in some cases, fire-resistant. Additives are also included to speed demulsification and stabilize viscosity. *Demulsification* is the act of separating water and oil quickly. *Emulsification* is the act of mixing oil and water. In most hydraulic systems, water is damaging and must be rapidly separated from the fluid.

Viscosity is the measure of the resistance of a fluid's molecules to move past each other. The viscosity of the fluid must match the specific application. Fluids without the proper viscosity cannot transmit power satisfactorily. Heavy fluids do not flow properly and cause a slow, sluggish operation. Light fluids create leakage around components and do not lubricate properly.

A fluid's resistance to foaming is increased by adding anti-foaming additives. Fluids allowed to foam from air entering the system cause spongy component movements and higher-than-normal temperatures.

Fire-resistant fluids are sometimes required in environments where flash point, fire point, or auto-ignition exist. *Flash point* is the temperature at which oil gives off enough gas vapor to ignite briefly when touched with a flame. *Fire point* is the temperature at which oil ignites when touched with a flame. *Auto-ignition* is the temperature at which oil ignites by itself.

Oxidation. Oxidation is the combining of oxygen with elements in the oil which break down the basic oil composition. Oxidation greatly reduces the service life of a hydraulic fluid because oxygen readily combines with the hydrogen and carbon that make up the oil. The oxidation processes create resins, which are then converted to varnish and gum and settle out as sludge. This harmful process also produces an acidic and highly corrosive fluid. Sludge and varnishes are considered contaminants and are byproducts of an already-destroyed oil. The oil must then be changed and flushed. A major additive to hydraulic fluids is one that prevents oxidation of the fluid.

Oxidation begins when the hydraulic fluid reaches high temperatures while in the presence of air. The higher the temperature, the greater the oxidation. Oxidation rates double for approximately each 20°F increase in temperature. Boiling burnt hydraulic oils separates certain resins from the fluid. These resins create varnish and acid, which are two corrosive contaminants of a hydraulic system. Scale-like varnish results when these resins touch hot metal components. Filter-clogging sludge is formed as these particles drop off. Formed and baked resins in a system create an acid condition in the fluid. This acid attacks and dissolves the metal it contacts.

Corrosion is an acidic condition which dissolves and washes away metal, leaving metal marked with pits. The acid condition of hydraulic fluid may be checked with the use of litmus paper. *Litmus paper* is a color-changing, acid-sensitive paper that is impregnated with lichens. *Lichens* are fungi normally seen as a growth on tree trunks or rocks. Lichens turn from blue to a reddish color when submerged in acidic oil.

The presence of moisture and oxygen causes iron to rust. *Rust* is a form of oxidation in which metal oxides are chemically combined with water to form a reddish-brown scale on metal. Ferrous metals increase in size and weight when rust occurs. Ferrous metals decrease in size and weight when oxidation occurs. *Ferrous metals* are metals containing iron. The prevention of rust depends on an oil's ability to form a film on metal surfaces, which prevents the metal from coming in contact with air or water.

Strainers and Filters. Generally, contaminants enter a hydraulic system from the air surrounding the system and must be filtered or strained before and during system operation. A *strainer* is a fine metal screen that blocks contaminant particles. A *filter* is a device containing a porous substance through which a fluid can pass but particulate matter cannot.

The major maintenance function of any hydraulic fluid is keeping the fluid clean. Particle buildup interferes with lubrication by blocking flow or by rubbing or scraping against moving parts. Particle buildup also interferes with the cooling process by making heat transfer difficult. As heated fluid returns to the reservoir, the heat is normally given up to the walls and baffles. Particle buildup on the walls and baffles acts as an insulator, preventing cooling.

Strainers are made of fine mesh wire screening elements wrapped around a metal frame. Strainers are used because screening is not as fine as a filter and offers less resistance to fluid flow. Most strainers can be cleaned periodically, while filters, made of porous materials, absorb particles from flowing fluids and must be replaced.

Strainer screens are rated in mesh and filters are rated in microns. *Mesh* is the number of horizontal and vertical threads per square inch. A *micron* (μ) is a unit of length equal to one millionth of a meter (.000039″). Most strainers remove particles above 100 mesh, while filters remove particles above 3μ. See Figure 6-6. Filter performance is based on the amount of particulate matter that can be removed from the fluid.

Strainers remove particles with a straight flow path through one layer of material. Materials used for strainers are generally cloth thread, metallic thread, or perforated metal. The threads are laid in equal amounts vertically and horizontally with the amount of threads being counted per square inch. For example, a 200 mesh strainer has 200 vertical threads and 200 horizontal threads per square inch.

RELATIVE MICRON PARTICLE SIZES

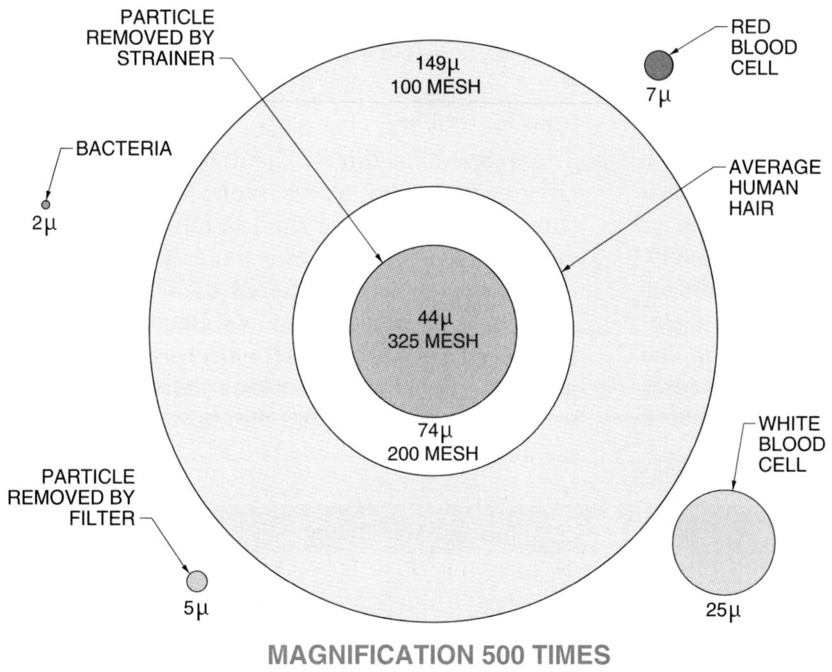

Flow Ezy Filters, Inc.

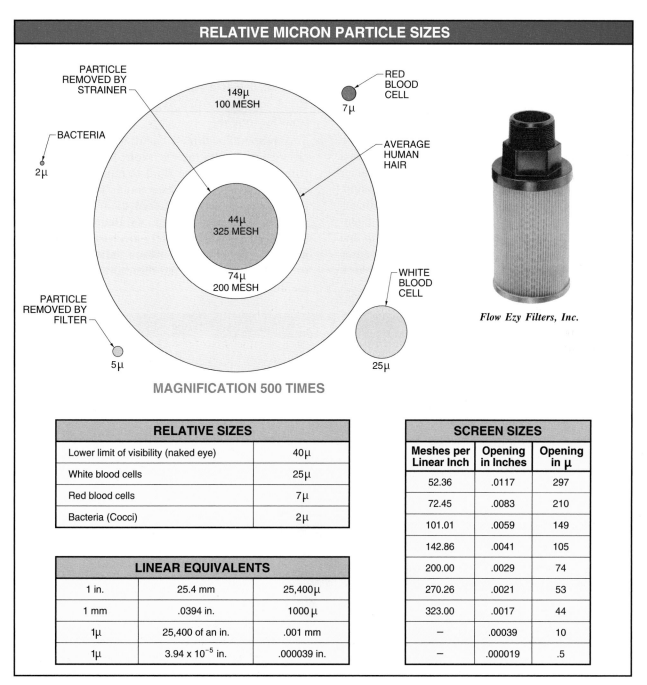

RELATIVE SIZES	
Lower limit of visibility (naked eye)	40μ
White blood cells	25μ
Red blood cells	7μ
Bacteria (Cocci)	2μ

LINEAR EQUIVALENTS		
1 in.	25.4 mm	25,400μ
1 mm	.0394 in.	1000 μ
1μ	25,400 of an in.	.001 mm
1μ	3.94×10^{-5} in.	.000039 in.

SCREEN SIZES

Meshes per Linear Inch	Opening in Inches	Opening in μ
52.36	.0117	297
72.45	.0083	210
101.01	.0059	149
142.86	.0041	105
200.00	.0029	74
270.26	.0021	53
323.00	.0017	44
–	.00039	10
–	.000019	.5

Figure 6-6. Very small particles are removed from hydraulic fluids by strainers and filters.

Certain mesh sizes produce certain pore (opening) sizes because of the accurate control of the strainer manufacturing process. A 200 mesh strainer has a pore size of 74μ. Strainer mesh is given an absolute rating because the pore size is accurately controlled. An *absolute rating* is an indication of the largest opening in a strainer element. Therefore, 74μ is the largest particle that can pass through a 200 mesh strainer. The higher the mesh number, the smaller the opening.

Ideally, a filter should be placed in the hydraulic line before every component. This is not economical, so filters of various types and ratings are placed strategically in a system to offer the best and most economical results. Three basic locations for filters used in a hydraulic circuit are the suction strainer, pressure filter, and return-line filter. See Figure 6-7.

A *suction strainer* is a coarse filter attached to a pump inlet. A suction strainer is used to protect a pump from particle contamination. A suction strainer is placed inside of the reservoir, before the pump. This protects the pump from reservoir contaminants and makes the pump easily serviceable. Poor maintenance of the suction strainer starves the pump and prevents the lubrication of the pump, which causes it to generate high temperatures and contaminates. Suction strainers range from 25μ to 235μ.

A *pressure filter* is a very fine filter placed after a pump for protection of the system components. Pressure filters are placed between the pump and components or between individual components in a circuit. Pressure filters placed between components have the advantage of filtering out particles introduced upstream by a deteriorating component.

Some pressure filters are capable of handling bidirectional flow. Bidirectional flow filters may be used between a directional control valve and the actuator. Pressure filters can filter out very fine particles because system pressure is used to push the fluid through the minute openings. Particles can be pushed through the element or the element may collapse or tear if a filter becomes contaminated and is not equipped with a bypass. Pressure filters range from 5μ to 40μ.

A *return-line filter* is a filter positioned in the circuit just before the reservoir. Return-line filters filter fluids before the fluid is returned to the reservoir and do not operate under pressure. These fluids may be finely filtered using normal operating back pressure. Poor maintenance increases back pressure and adversely affects circuit components. Also, return-line filters deliver clean fluid to a dirty reservoir if the reservoir is not properly maintained.

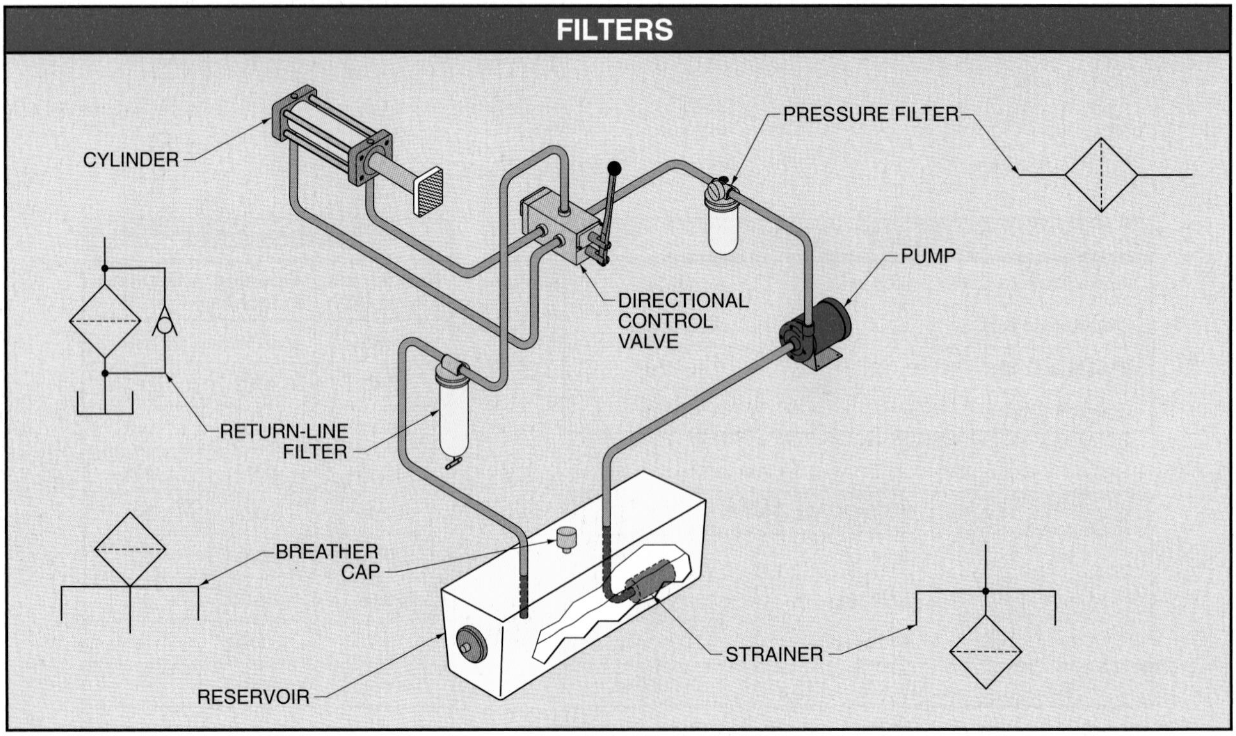

Figure 6-7. Suction, pressure, and return-line filters are the three basic locations for filters found in a hydraulic circuit.

Some filters are designed to allow fluids to bypass the filter when a difference in specific pressure is sensed. In this case, a relief or bypass valve allows full fluid flow across the filter if proper filter maintenance is not performed and pressures increase above normal. In some units, indicators are designed into the filters to show their condition. Filter-clogging conditions may be indicated by dial indicator movement, a light or buzzer, or an equipment OFF button. Regardless of the hydraulic filter or fluid used, lack of a proper maintenance program rapidly destroys hydraulic equipment.

Always cap or plug open lines or connectors when installing or removing components to reduce the possibility of contaminants entering a system.

Reservoirs

A *reservoir* is a container for storing fluid in a hydraulic system. The primary purpose of a reservoir is to provide a storage space for the fluid required by the system. The reservoir capacity should normally be two to three times the volume of fluid pumped through the system in one minute. In addition to fluid storage, a reservoir also prevents fluid contamination, helps with fluid/air separation, and maintains safe fluid temperatures.

Reservoirs are constructed with a dished bottom to allow for drainage, a fluid level gauge, a breather cap, baffle plate(s), and return, drain, and suction lines. Reservoirs may also be equipped with a strainer to prevent fluid contamination, clean-out covers or removable tops to allow cleaning the reservoir, and a drain plug at the lowest point to allow changing fluid or draining accumulated moisture. See Figure 6-8.

A reservoir is normally equipped with a breather cap to allow atmospheric pressure to push the fluid up to the pump. Also, the fluid level is constantly rising and falling when the circuit is operating. A breather cap roughly filters the dirt-laden air, which enters and exits the reservoir as the circuit actuator is filled and exhausted. Breather caps become plugged and cause improper pump operation if they are not cleaned regularly. Breather caps are not used and the reservoirs are pressurized when reservoirs are used in applications in unclean or corrosive environments.

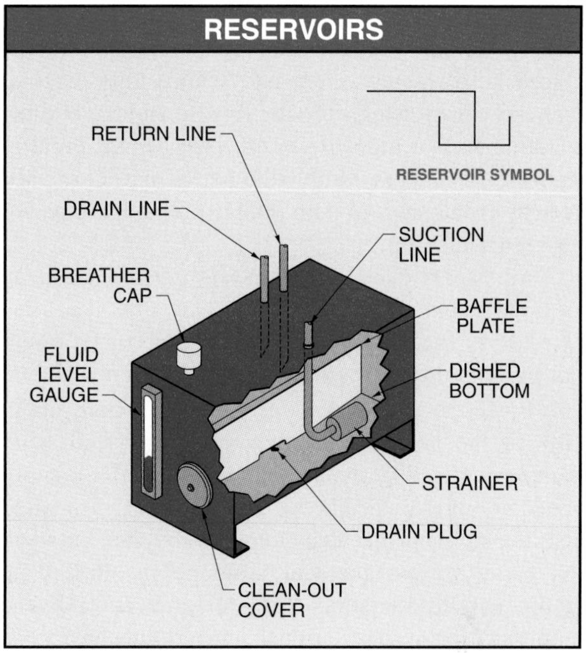

Figure 6-8. A reservoir is a container for storing fluid in a hydraulic system.

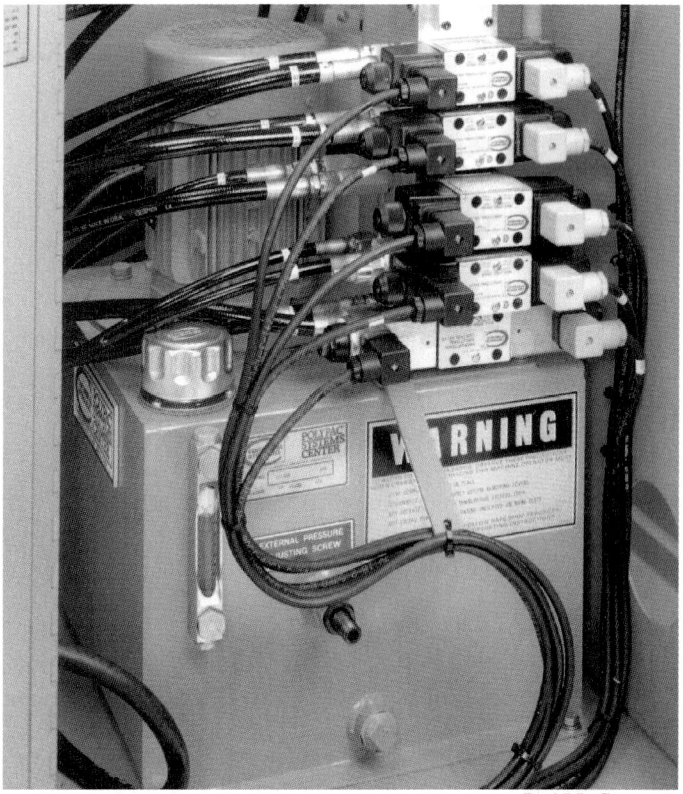

DoALL Company

The C-305 Series power saws from DoALL Company contain 2 HP hydraulic pumps and 10 gal. fluid reservoirs to operate the hydraulic vise cylinders.

Baffle plates block the returning fluid from going directly to the suction line. The baffle plates are placed between the return and suction line ports in the reservoir to help settle the flow of fluid and allow contaminants or moisture to drop out and air bubbles to rise to the top. Additionally, baffle plates are used as heat exchangers to help reduce the temperature of the returning fluid.

To dissipate heat, reservoirs are narrow and deep as opposed to short and wide. This offers a large exterior surface to contact ambient air for cooling. For best operation, reservoir fluids should not exceed 160°F–180°F. Generally, reservoirs dissipate about 70% of the heat generated within the system with the remaining 30% being radiated from the components or plumbing. The amount of heat dissipated also depends on the temperature difference between the reservoir surface and ambient air. Hydraulic equipment installed too close to furnaces or ovens may reach or surpass critical temperature levels.

Heat Exchangers. A *heat exchanger* is a device which transfers heat through a conducting wall from one fluid to another. Heat produced in an operating circuit is radiated away by the reservoir baffle plate and into the air surrounding the circuit components. This heat transfer is generally sufficient for a hydraulic circuit. However, a reservoir may not be capable of dissipating enough of the heat produced. In such cases, heat exchangers are placed in the hydraulic lines to remove excess and damaging heat. Heat exchangers are either air-cooled or water-cooled devices. See Figure 6-9.

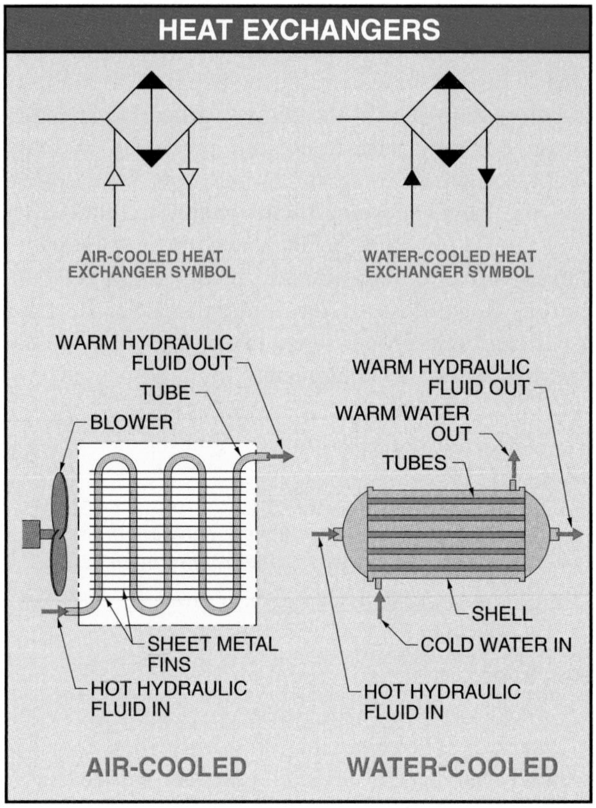

Figure 6-9. Heat exchangers help to lower the temperature of hydraulic fluids when other methods are not sufficient.

Air-cooled heat exchangers (coolers) operate by pumping the hot circuit fluid through tubes attached to sheet metal fins. Cooling of the hot fluid is accomplished through the use of a blower, which blows air over the tube and fins. Air-cooled heat exchangers are similar to automobile radiators. Water-cooled heat exchangers operate by pumping hot circuit fluid through a shell and over tubes containing circulated cool water. The circulating water carries away unwanted heat from the circuit fluid. This type of heat exchanger may be reversed to warm circuit fluids. By circulating warm water through the tubes, circuit fluid within the shell can be heated.

An estimated 300 million gallons of used oil are disposed of improperly each year. Used oil is a toxic substance, and must be disposed of properly.

Piping

Hoses, pipes, and tubing are the basic piping devices used to connect components and to conduct fluid in a hydraulic system. Hydraulic circuit piping must be leakproof and strong enough to withstand required temperatures, vibrations, and pressures. Proper materials and procedures must be used to prevent excess restriction, turbulence, leakage, or dangerous situations.

Hoses. A *hose* is a flexible tube for carrying fluids under pressure. Hoses are fabricated in layers for use in high-pressure or extra-high-pressure hydraulic circuits. High-pressure hoses are capable of withstanding pressures up to 5000 psi. Hoses generally consist of an inner layer of soft synthetic rubber that is compatible with hydraulic fluids, two or more layers of multiple wire braid reinforcement, a layer of cotton braid, and a rubber cover. Extra-high-pressure hoses contain four or more wire braid layers. See Figure 6-10.

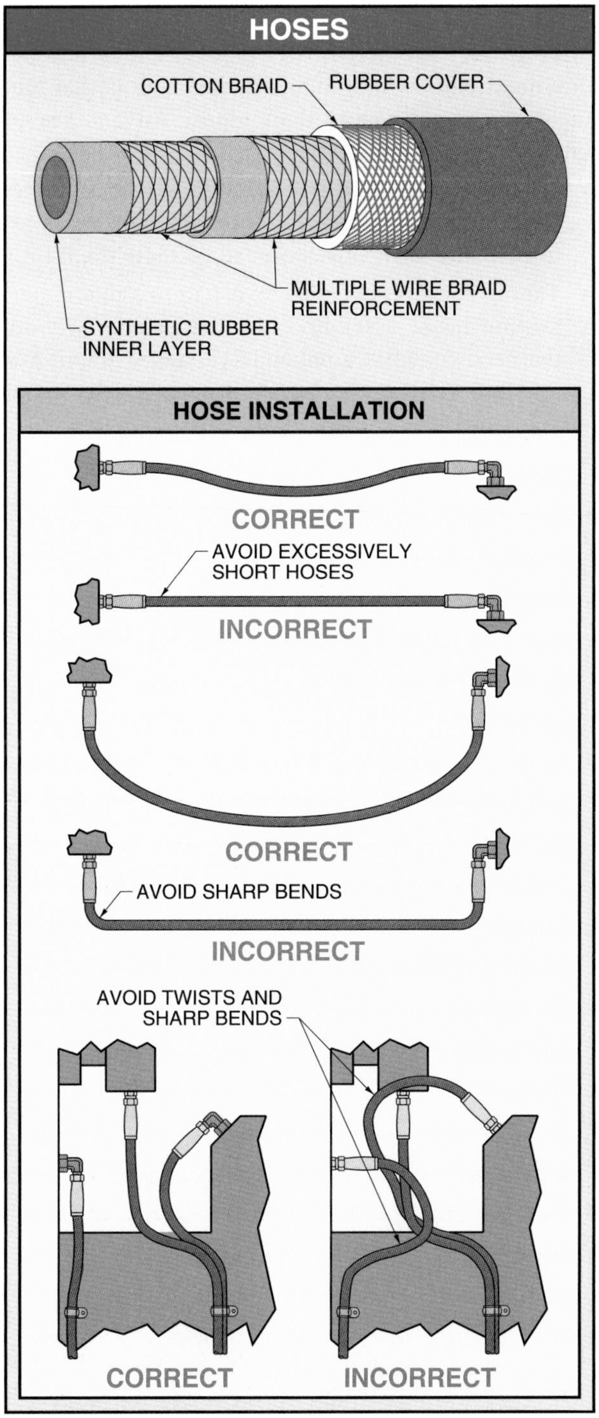

Figure 6-10. Hoses are fabricated in layers for use in high-pressure hydraulic circuits.

Proper tools and skill are required when installing reusable hose fittings. Always follow fitting manufacturer's instructions when installing specific fittings.

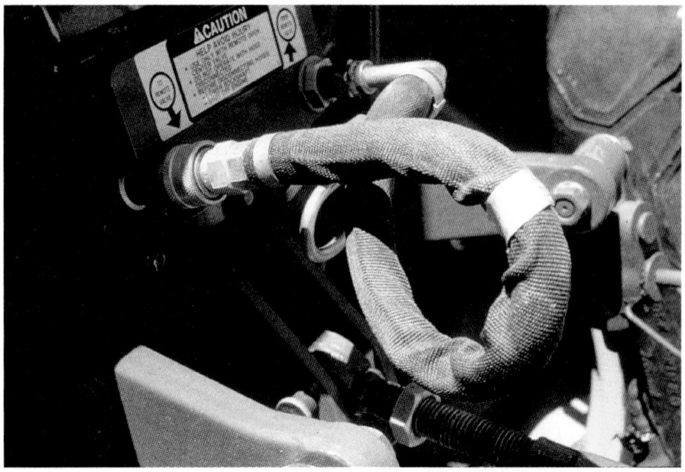

The Gates Rubber Company

Hydraulic hoses should be enclosed in protective sleeves when subject to rubbing and should be installed with a bending radius of greater than six times the inside diameter.

Hoses are installed to avoid twists and sharp bends. The bending radius of flexible hose must be greater than six times the inside diameter. Protective sleeves must encase any hose that is subject to rubbing. Hoses must not be excessively long or excessively short. Hoses that are excessively long have more internal resistance. Hoses tend to decrease in length when pressurized, so a hose which is excessively short, without any bend or flex, will fail prematurely.

Pipes. A *pipe* is a hollow cylinder of metal or other material of substantial wall thickness. Pipe wall thickness is normally thick enough that the pipe may be threaded. Originally, pipe was manufactured with one wall thickness and its size was the actual inside diameter. This changed due to an increase in the strength requirements of pipe. Wall thicknesses were increased, which reduced the inside diameter (ID) of the pipe, leaving the outside diameter (OD) unchanged.

Pipe is designated according to its nominal size and wall thickness. Presently, the nominal pipe size indicates the thread size for connections. A standard ½″ pipe has an ID of .622″ and is classified as Schedule 40 pipe. An extra-heavy ½″ pipe is classified as Schedule 80 pipe with an ID of .546″. A double-extra-heavy pipe has an ID of .252″ (approximately Schedule 160). The actual inside diameter varies with piping, but the actual outside diameter remains constant for any given size pipe. See Figure 6-11. See Appendix.

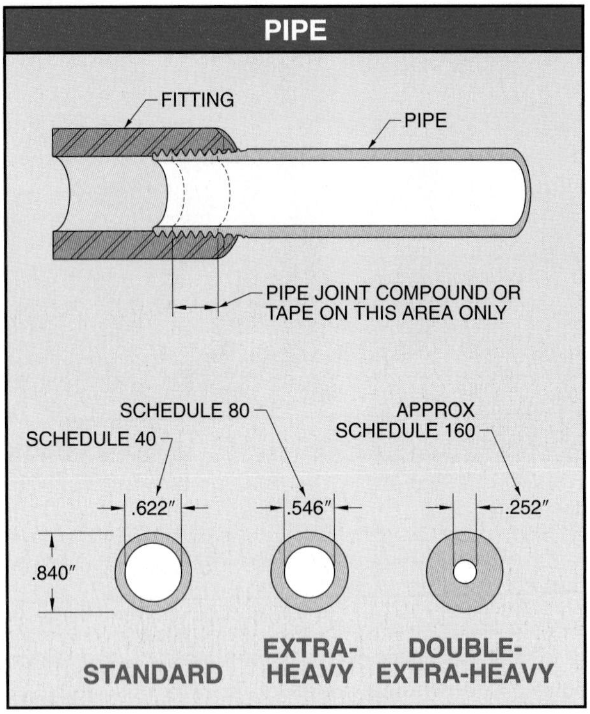

Figure 6-11. Pipe is designated according to its nominal size and wall thickness.

Pipe is generally used for permanent installations. This is because pipe threads are tapered, and once a connection is broken, it must be tightened further to reseal. Each time a pipe is retightened, its length is changed. In some cases, replacing a pipe with a slightly longer pipe may be required.

Proper installation procedures ensure the permanency of pipe fittings. Two-thirds of the threaded area should be covered with a pipe joint compound, such as Teflon® tape or paste. This is applied to the middle portion of the male thread to prevent any compound from entering and contaminating the system. Overtightening threads may cause premature leaking or undue stress.

Hydraulic lines and fittings should be made of steel, with the exception of flexible hoses. Galvanized pipe must not be used due to the possibility of metal flaking. Also, the zinc used for galvanizing reacts adversely with certain hydraulic fluid additives. Copper tubing should not be used because it reacts to hydraulic fluids. In addition to its reaction to the fluids, copper tends to harden and crack under the heat and vibrations of hydraulic circuits.

Tubing. A *tube* is a thin-walled, seamless or seamed, hollow cylinder. Tubing is soldered, welded, or formed for compression because their wall thickness is usually too thin for threading. Carbon steel tubing offer distinct advantages over pipes when used in hydraulic circuits. One minor disadvantage is that tubing is more expensive than pipes. Advantages of tubing include:

- Tubing requires fewer connections than pipe because they can be bent. Tubing also absorb vibrations better than pipe because of their flexibility.
- Tubing connections make every joint a union, permitting faster assembly and disassembly without the need for joint compounds or tape. A *union* is a fitting used to connect or disconnect two tubes that cannot be turned.
- Tubing is lighter in weight than pipe and has a smoother inner surface, which produces less friction and less pressure loss.

Hydraulic tubing is connected using fittings. Tubing fittings may be flared or flareless fittings.

Flared Fittings. A *flared fitting* is a fitting that is connected to tubing whose end is spread outward. The body of the fitting is screwed tightly against the flared end of the tubing. Proper flaring provides a firm, leakproof connection. See Figure 6-12.

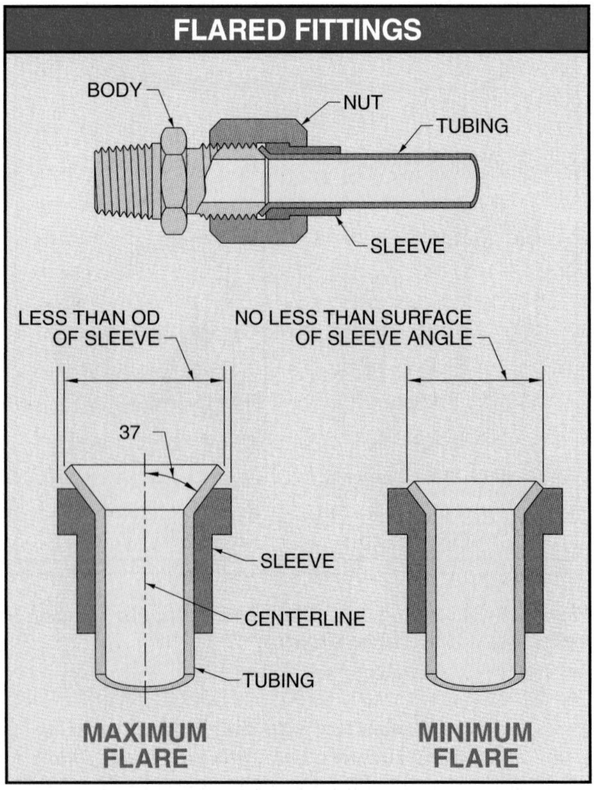

Figure 6-12. A flared fitting is a fitting that is connected to tubing whose end is spread outward.

Flared fittings generally consist of a body, sleeve, and nut. A seal is made when the flared tubing is pressed against the angular seat of the body by the sleeve. The angles of the body, sleeve, and flared tubing ensure a good seal when the tubing, which is the softest of the three pieces, is pressed into the body. The standard flare angle for hydraulic tubing fittings is 37° from the centerline. The flare extends to cover the total angular surface of the sleeve, but not beyond the sleeve's outside diameter. The flare seats firmly and positively between the sleeve and the body when tubing is flared properly and tubing nuts are tightened securely. Flares that are too short do not provide enough mating area to prevent leaks, and flares that are too long hang up during assembly. Clean, square tubing cuts are achieved with a tube cutter. Hacksaws produce rough cuts which generally are not square.

Flared tubing connections may also have a flare angle of 45°. A 45° flare angle is used for low-pressure applications such as pneumatic, refrigeration, or automotive applications, and is not to be used for high-pressure hydraulic circuits.

Incorrect flares may appear to assemble satisfactorily and may even pass initial pressure tests. They are not, however, reliable for continuous service. All tubing flares must conform to the sleeve and body used to join tubing sections.

Flared Joint Tightening. A positive seal is vital to prevent fluid loss, keep out contamination, and maintain hydraulic circuit pressure. A positive seal does not allow the slightest amount of fluid to pass and is normally compressed between two rigid parts. A nonpositive seal allows a certain amount of leakage, which provides a lubricating film between surfaces. An example of a nonpositive seal is a piston and an O-ring moving within a cylinder.

Tightening a flared fitting is a positive seal and is accomplished by using the proper torque. Undertightening or overtightening a flared fitting nut is avoided by using a torque wrench with proper torque settings or by manually turning the fitting nut while observing witness marks applied to the sealing nut and body. See Figure 6-13.

Witness marks may be used when a torque wrench is not available. The joint is assembled with the nut bottomed out and tightened to fingertight. A line is marked, using a felt marker, lengthwise on a flat on the body and onto the corresponding flat of the nut. The nut is rotated with a wrench until a determined number of flats on the body have been passed. The number of flats passed is based on the size of the fitting. For example, a size 8, ½″ flared fitting should be rotated 2 flats after fingertight. This fitting may also be tightened to 200 lb-in. – 300 lb-in. using a torque wrench.

The Gates Rubber Company

After hand-tightening, a wrench is used to properly torque hose fittings.

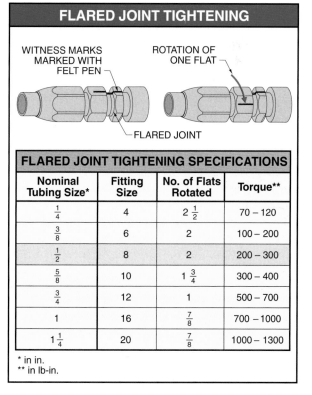

FLARED JOINT TIGHTENING

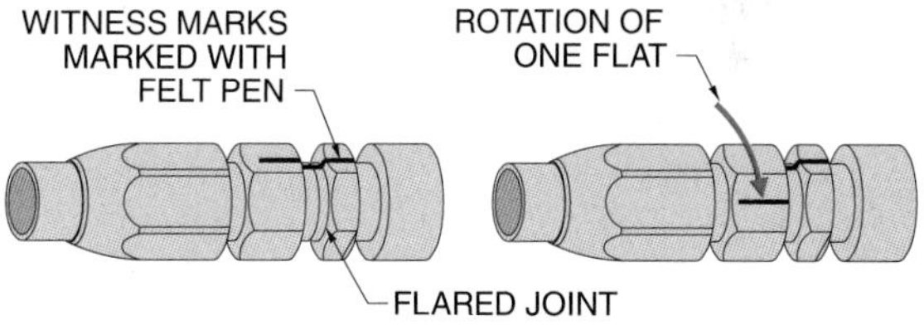

FLARED JOINT TIGHTENING SPECIFICATIONS

Nominal Tubing Size*	Fitting Size	No. of Flats Rotated	Torque**
1/4	4	2 1/2	70 – 120
3/8	6	2	100 – 200
1/2	8	2	200 – 300
5/8	10	1 3/4	300 – 400
3/4	12	1	500 – 700
1	16	7/8	700 – 1000
1 1/4	20	7/8	1000 – 1300

* in in.
** in lb-in.

Figure 6-13. A flared fitting is tightened using a torque wrench or by turning the fitting nut while observing witness marks.

Tubing should never be assembled in a straight line. Bending tubing for assembly reduces vibration strains and compensates for thermal expansion. A gradual bend is preferred over elbow fittings because elbow fittings have sharp turns with high resistance to flow. Tubing must be bent with the correct radius and without kinks, wrinkles, or flattened bends. The bending radius should be greater than four times the tubing ID. Tubing must also be properly supported to minimize the stresses of vibration. See Figure 6-14.

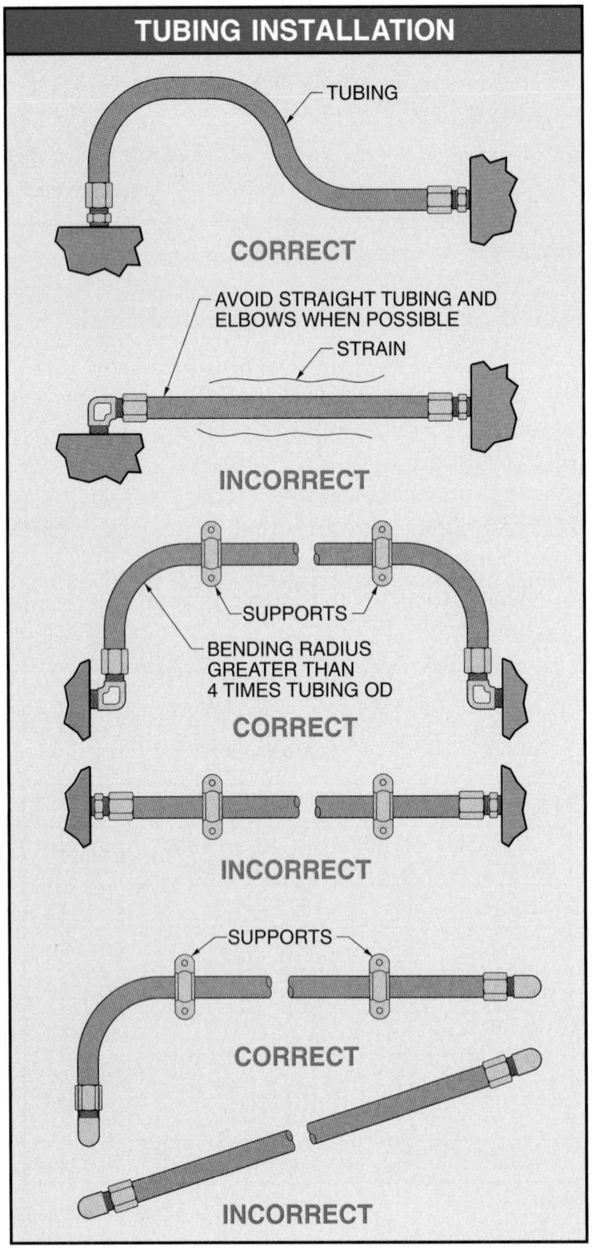

Figure 6-14. Bending tubing for assembly reduces vibration strains and compensates for thermal expansion.

Impact Flaring Method. The *impact flaring method* is a basic flaring method in which a flaring tool is inserted into the tubing end and hammered into the tubing until the tubing end is spread (flared) as required. Flaring tool kits consist of a split female die, a tubing clamp, and a variety of different-sized flaring tools. See Figure 6-15.

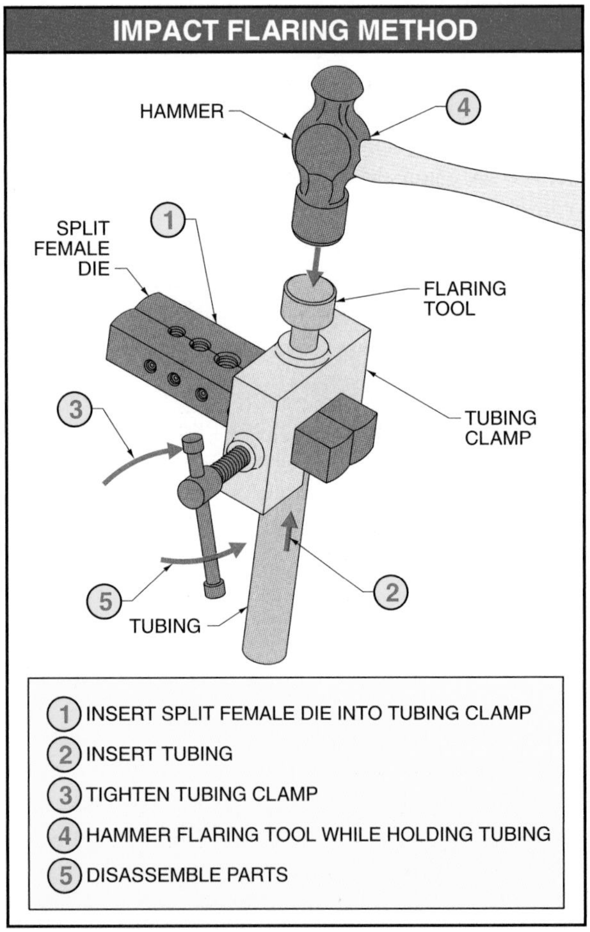

Figure 6-15. The impact flaring method uses a flaring tool which is inserted into the tubing end and hammered into the tubing until the tubing end is spread (flared) as required.

A tubing is flared using the impact method by applying the procedure:

1. Insert the split female die into the tubing clamp and slide it to the end, enabling the die to be spread open.
2. Insert the tubing into the appropriate die hole with the tubing end approximately 1⁄32″ above the top surface of the die.
3. Place the tubing clamp directly over the tubing end and tighten.

4. Hold the tubing with one hand while hammering the flaring tool. This hand takes up the concussion from the blows and is used to feel the thud of the flaring tool when bottomed out.
5. Disassemble the parts and check the flare when the bottomed out thud of the flare tool is felt. Reinsert the tube 1/32″ above the die surface and repeat Steps 3 and 4 if a wider flare is required.

Flareless Fittings. A *flareless (compression) fitting* is a fitting that seals and grips by manual adjustable deformation. Flareless fittings are designed for thicker wall tubes that are not suitable for flaring. Flareless fittings create a seal with a ferrule. A *ferrule* is a metal sleeve used for joining one piece of tube to another. The ferrule cuts into and compresses the tube when the nut is tightened onto the body. The nut is tightened a full turn after the completed assembly is fingertight. See Figure 6-16.

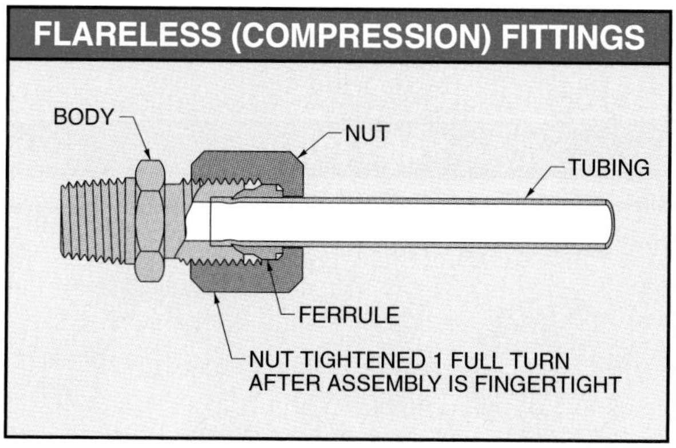

Figure 6-16. A flareless (compression) fitting uses a ferrule which cuts into and compresses the tube when the nut is tightened onto the body.

Pumps

A *pump* is a mechanical device that causes fluid to flow. Hydraulic pumps do not create energy, they convert the energy of a prime mover into hydraulic energy. A *prime mover* is an electric motor or engine that supplies rotational force at a constant speed. Generally, pumps used in hydraulic circuits are positive-displacement pumps. *Positive displacement* is the moving of a fixed amount of a substance with each cycle. Displacement of hydraulic pumps is expressed in cu in. of oil discharged per revolution.

Pumps are rated by the manufacturer with a pressure and fluid output rating in revolutions per minute (rpm). Pressure is produced by resistance against fluid flow. Flow is the output of a positive-displacement pump and is expressed in gpm. Flow is proportional to the prime mover's rpm. For example, a system designed for a pump output of 50 gpm at 1700 rpm should remain at 1700 rpm. A change of rpm greatly affects the pump output. At 1600 rpm the pump's output may be as low as 40 gpm, while at 1800 rpm the pump's output may be as high as 60 gpm. The flow rate (gpm) dictates the actuator speed. Changing speed could cause cavitation and may be damaging. Most pumps should be sized for a greater output than what is actually required for the circuit output devices.

Positive displacement pumps used in hydraulic circuits are generally gear, vane, or piston pumps. Hydraulic pumps must be constructed to high-quality standards, used properly, and maintained using a scheduled maintenance program because pump efficiency is directly related to pump cleanliness.

Personal injury may result if hose assemblies are not periodically checked for damage or replaced with properly rated assemblies.

Boeing Commercial Airplane Group

Many aircraft contain hydraulic systems for the control of flaps, stabilizers, landing gear, and brakes.

Gear Pumps. A *gear pump* is a pump consisting of two meshing gears enclosed in a close-fitting housing. One gear is driven by the prime mover. Gear pumps are the most widely used hydraulic pumps because of their simple design and ease of repair. See Figure 6-17.

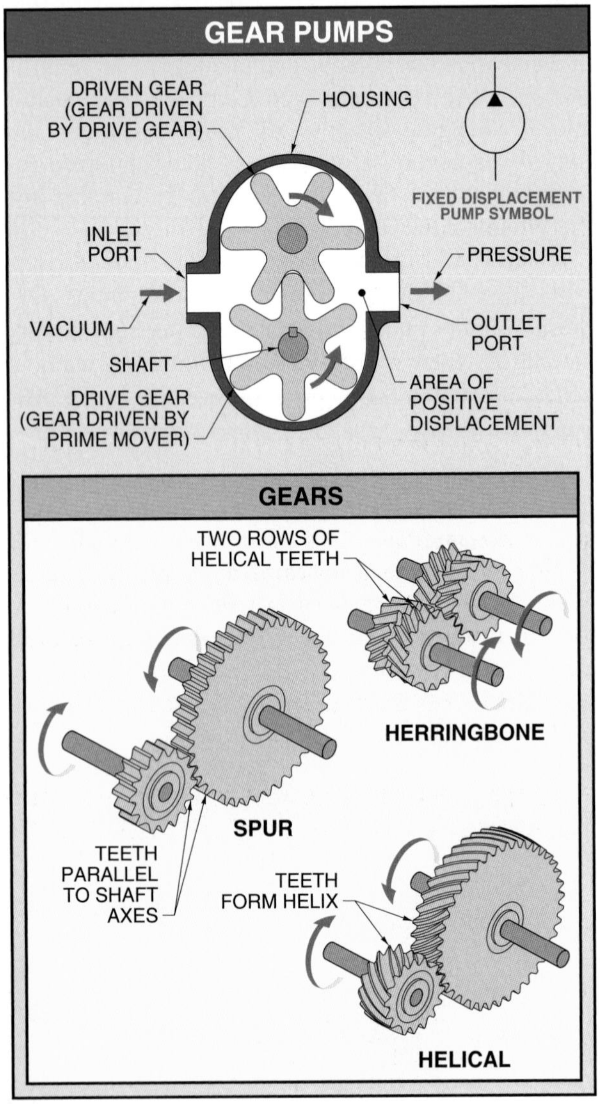

Figure 6-17. A gear pump contains two gears within a housing that are rotated to produce fluid flow.

Gear pumps develop fluid flow by carrying oil between the teeth of the two closely meshed gears and their housing. The housing includes an inlet port and an outlet port. The fluid in the reservoir is constantly pushed by atmospheric pressure. A vacuum is created at the inlet port as the gears begin rotating. The vacuum allows the fluid to be pushed into the spaces between the gear teeth by atmospheric pressure. As the teeth rotate, the fluid is moved between each tooth from the inlet side of the pump to the outlet side. The oil between the teeth is displaced and forced through the outlet port as the teeth of the rotating gears mesh on the outlet side of the pump.

The gears used in a gear pump are generally spur gears. A *spur gear* is a gear that has straight teeth that are parallel to the shaft axes. For quieter operation and increased performance, helical or herringbone gears are used at greater expense. A *helical gear* is a gear with teeth that are cut at an angle to its axis of rotation. A *herringbone gear* is a double helical gear that contains a right- and left-handed helix.

Vane Pumps. A *vane pump* is a pump that contains vanes in an offset rotor. Vane pumps produce fluid flow as the rotor and vanes rotate. See Figure 6-18. As the rotor rotates, a vacuum is produced at the inlet port allowing atmospheric pressure to push the fluid and fill the voids between the vanes. Centrifugal force, springs, and/or pressure under the vanes hold them firmly against a cam ring to form a positive seal between the tip of the vanes and the cam ring.

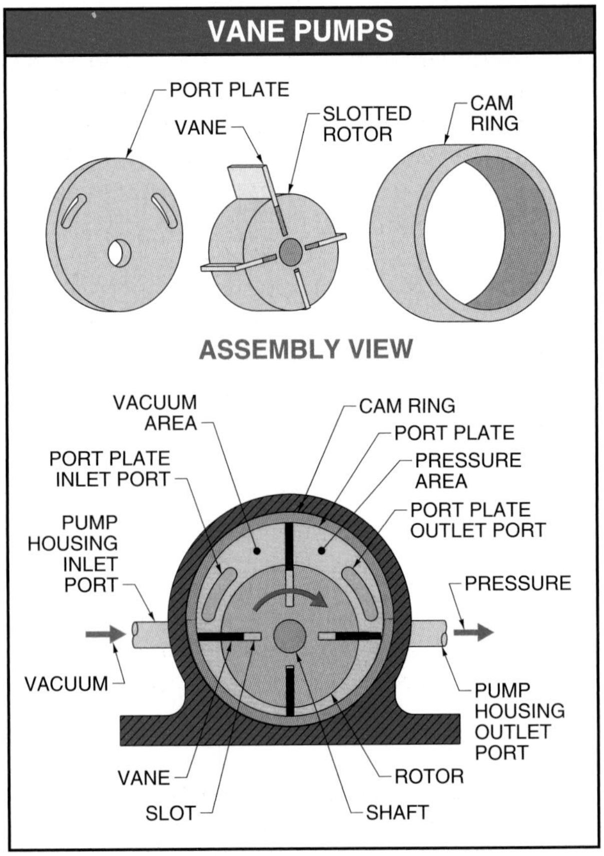

Figure 6-18. A vane pump contains vanes in the slots of a rotor that produce fluid flow as the rotor rotates.

A *cam ring* is a metal ring that provides an area for fluid flow and a surface against which the vanes ride. The fluid enters and is discharged from the pump through a port plate. A *port plate* is a device that contains ports that connect the pump's internal inlet and discharge areas to the pump housing inlet and outlet ports. The inlet port of the port plate is connected to the inlet port of the pump housing. The outlet port of the port plate is connected to the outlet port of the pump housing.

As the rotor continues to rotate and fluid is carried forward, the vanes begin to retract as the volume of the void produced between the off-centered rotor and the cam ring is reduced. The fluid is forced out of the port plate outlet port at increased pressure. Vane pumps remain efficient throughout their use due to the movement of the vanes, which compensates for wear.

Vane pumps are classified as unbalanced or balanced. An unbalanced vane pump has one set of internal ports. The rotor is off-center from the cam ring and each vane creates a pumping action once in each revolution. Unbalance is developed by the low inlet pressure on one side of the rotor and the high discharge pressure on the other side of the rotor.

A *balanced-vane pump* is a vane pump that has two sets of internal ports and contains an elliptical cam ring. See Figure 6-19. In a balanced-vane pump, the rotor is centered inside the cam ring. The vanes create a pumping action twice in one revolution because the two inlet and two outlet ports on the port plate are 180° apart. Both inlet ports are connected together and both outlet ports are connected together so each leads to one respective port in the pump housing. The pump remains balanced because the high discharge pressures are applied to both sides of the rotor.

The volume of a vane pump is determined by how far the rotor and cam ring are offset. Changing the offset between the rotor and cam ring changes the volume of fluid supplied by the pump. The offset is changed by moving the cam ring to reduce or increase the size of the area between the cam ring and the rotor. Displacement control is performed by the turn of an external hand wheel or by a pressure compensator. A *pressure compensator* is a displacement control that alters displacement in response to pressure changes in a system.

Relieve all hydraulic pressure before working on pressurized hydraulic lines or components by cycling the control levers before disconnecting the parts of the system.

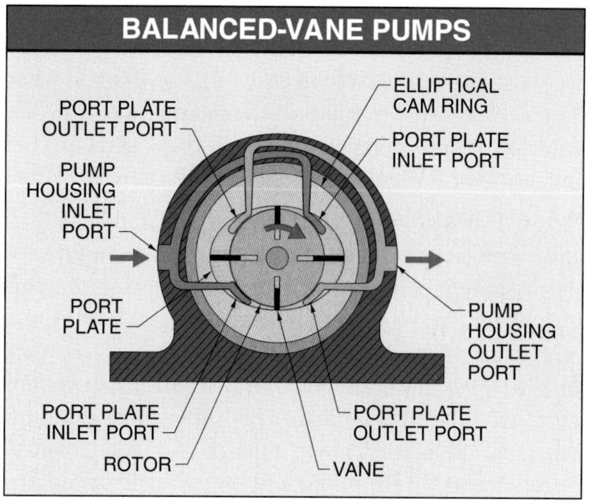

Figure 6-19. A balanced-vane pump has two sets of internal ports and contains an elliptical cam ring.

A *pressure-compensated vane pump* is a vane pump equipped with a spring on the low-displacement side of the cam ring. The spring pressure is adjustable and determines the amount of fluid pressure required to create cam movement. The cam ring centers and pumping ceases when the pressure acting on the inner wall of the cam ring is high enough to overcome the spring force. In many cases, a pressure-compensated pump is used to limit system pressure. As a result, there is little fluid heating or wasted horsepower. See Figure 6-20.

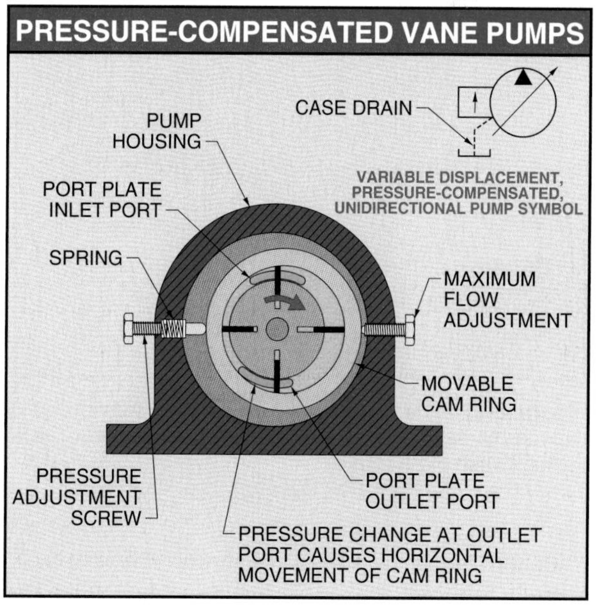

Figure 6-20. A pressure-compensated vane pump is equipped with a spring on the low-displacement side of the cam ring.

At full compensation, the displacement of a pump is zero plus minor internal leakage. All pressure-compensated pumps require the case oil to be drained when fully compensated because of the internal leakage. The fluid left in the pump during zero displacement continues to rise in temperature if there is no fluid flow.

Piston Pumps. A *piston pump* is a pump in which fluid flow is produced by reciprocating pistons. A piston pump consists of a cylinder barrel, pistons with shoes, valve plate, and swash plate. See Figure 6-21. The pistons are connected to the swash plate by shoes. As the cylinder barrel rotates, the pistons slide over the valve plate. The valve plate contains two crescent-shaped ports. One port is connected to the inlet port of the pump and the other is connected to the outlet port. The pistons reciprocate, drawing oil from the crescent-shaped inlet port due to the angle of the swash plate. At mid-rotation, the pistons are completely extended and filled with fluid. As the cylinder barrel continues to rotate, the cylinders begin to retract and force the fluid out of the crescent-shaped outlet port.

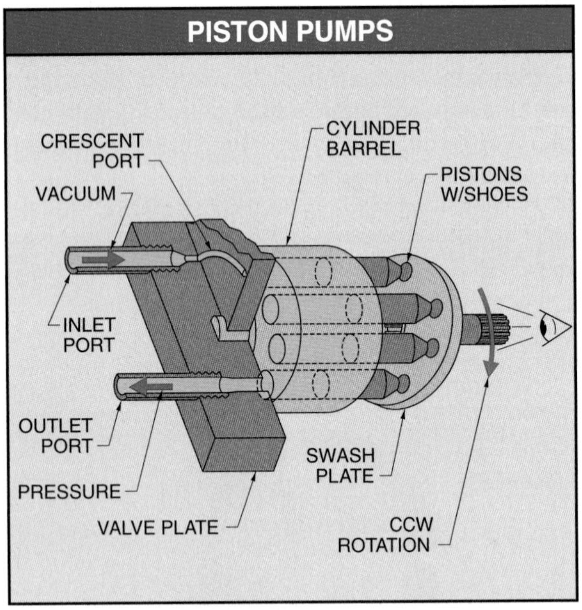

Figure 6-21. The pumping action of a piston pump is caused by reciprocating pistons within the cylinder barrel as the swash plate rotates.

Piston pumps are used when variable displacement is required. Variable displacement allows the control of fluid from a no-flow condition up to full flow. Regulating flow is accomplished by adjusting the swash plate from a no-angle position to a full-angle position. The pistons do not reciprocate if the swash plate is not at an angle. The pistons reciprocate, drawing oil from the inlet port and discharging it at the outlet port, if the swash plate is at an angle. The displacement varies according to the angle of the swash plate.

Cavitation. *Cavitation* is the process in which microscopic gas bubbles expand in a vacuum and suddenly implode when entering a pressurized area. *Implosion* is an inward bursting. Cavitation occurs when the inlet port of a pump is restricted. An indication of pump cavitation is a high shrieking sound or a sound similar to loose marbles or ball bearings in the pump. Cavitation is normally created when the suction line is damaged, plugged, or collapsed. Cavitation may also be caused by an increase in pump rpm, requiring more fluid than the system piping allows, fluids with an increased viscosity due to lower ambient temperatures, and an increase in the viscosity of a fluid in a system when the system has a long suction line.

As the pump pulls against a fluid that does not flow, a greater vacuum is created. Any microscopic air or gas within the fluid expands. Expanded bubbles on the inlet side collapse rapidly on the outlet side of the pump. The small but tremendous implosions can cause great damage to pump parts. Theoretically, an air bubble exposed to 5000 psi system cavitation may create an implosion pressure of 75,000 psi and travel at a speed of 600 fps to 4000 fps. See Figure 6-22.

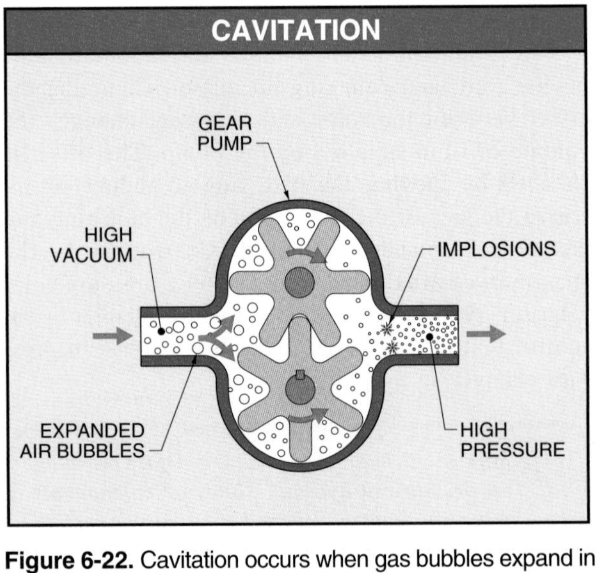

Figure 6-22. Cavitation occurs when gas bubbles expand in a vacuum and implode when entering a pressurized area.

Pseudocavitation is artificial cavitation caused by air being allowed into the pump suction line. Pseudocavitation is caused by low reservoir fluid, contaminated fluid, or leaking pump suction lines. Pseudocavitation is indicated by an unchanged or lower pump intake vacuum.

Valves

A *valve* is a device that controls the pressure, direction, or rate of fluid flow. Hydraulic circuit components, including valves, are equipped with one or more external openings (ports), which allow the flow of fluid to and from the device. Each external port is a primary or secondary port. A *primary port* is the source or inlet port. A primary port may be labeled with a P for primary or pressure. A *secondary port* is an external passage that allows fluid flow to other components. Secondary ports may be labeled A, B, or T. A- and B-labeled ports are ports that lead to other pressure components. A T- (tank) labeled port is a port that leads to the reservoir. Basic hydraulic valves operate by moving elements that open or block fluid passages that are connected to other components. Hydraulic circuit valves are grouped by their function. Hydraulic circuit valves may be pressure, directional, or flow control valves.

Pressure Control. Pressure control in a hydraulic circuit is concerned with maintaining or reducing (regulating) system pressure to operate a circuit. Pressure control valves include pressure-relief, sequence, and pressure-reducing valves.

Hydraulic circuit pressure is maintained by the use of a pressure-relief valve. A *pressure-relief valve* is a valve that sets a maximum operating pressure level for a circuit to protect the circuit from overpressure. Pressure-relief valves are normally closed valves that require higher-than-spring pressure to open. In a pressure-relief valve, pressure on a ball or poppet overcomes spring pressure, allowing fluid to flow. The inlet (primary) port is connected to circuit pressure and the discharge (secondary) port is connected to the reservoir. Ball or poppet movement is controlled by a predetermined pressure level. The pressure level, or spring pressure, is usually varied by screw adjustment. See Figure 6-23.

Hydraulic valves may be direct-acting or pilot-operated. A *direct-acting valve* is a valve that is activated or directly moved by fluid pressure from the primary port. For example, in a pressure-relief valve, the spool or poppet is directly activated by an increase in circuit or upstream pressure. See Figure 6-24.

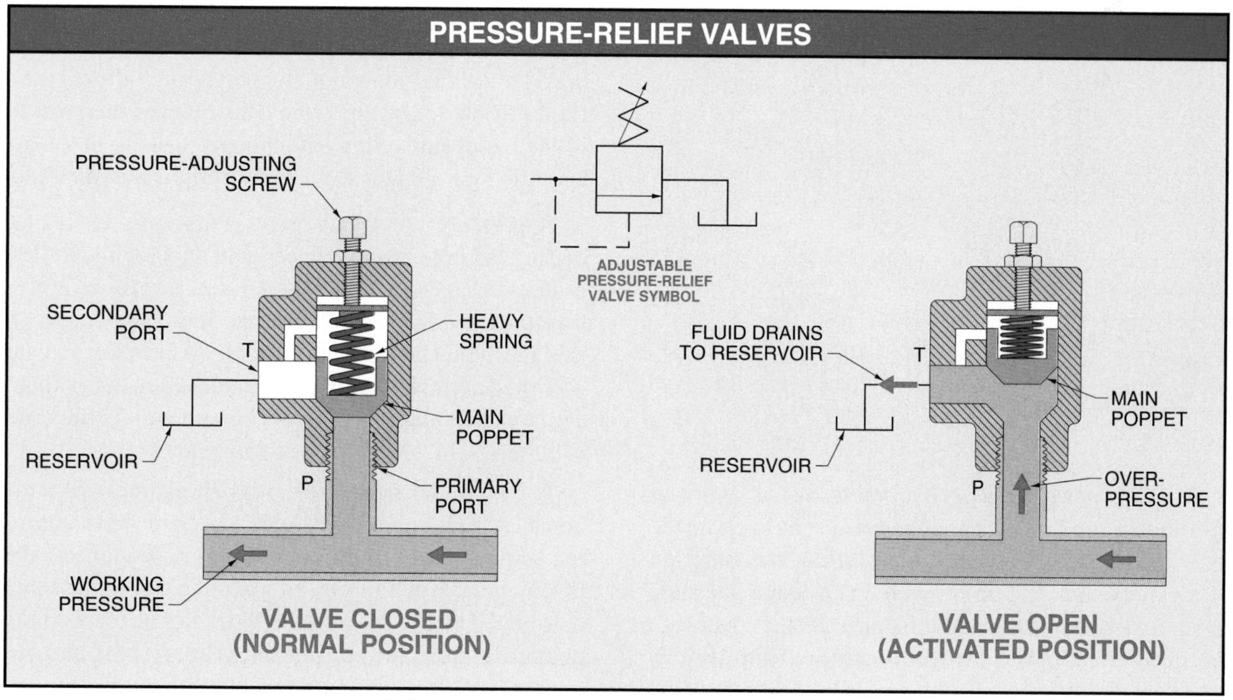

Figure 6-23. A pressure-relief valve limits the maximum pressure in a hydraulic circuit.

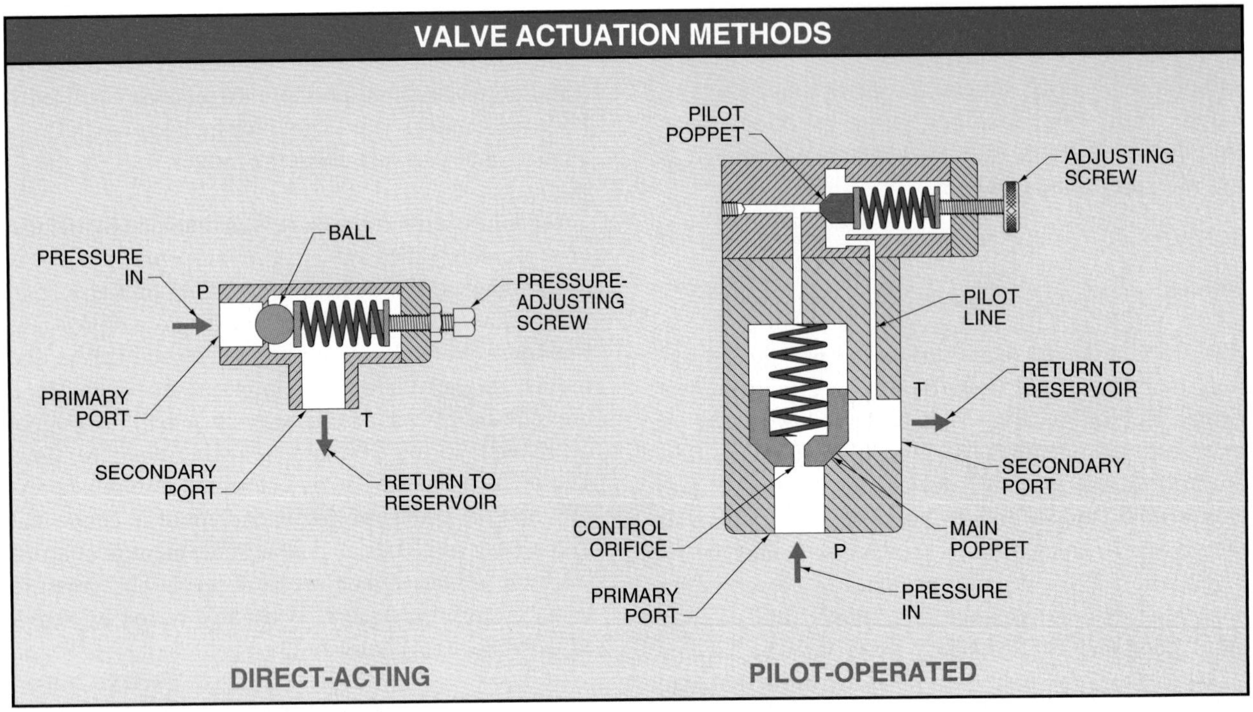

Figure 6-24. Hydraulic valves may be direct-acting (actuated by fluid pressure from the primary port) or pilot-operated (actuated by fluid in the line that is otherwise sent back to the reservoir).

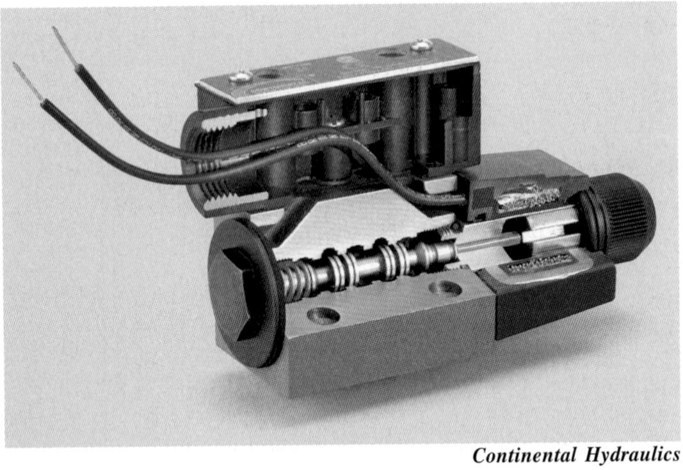

Continental Hydraulics

Directional control valves manufactured by Continental Hydraulics are available with up to 5 actuators (solenoid, air, cam, oil, and lever) and 12 spool options for use with pressures up to 4600 psi.

A *pilot-operated valve* is a valve that is actuated by fluid in the line that is otherwise sent back to the reservoir. *Pilot operation* is controlling the function of a valve using system pressure or pressure supplied by an external (pilot) source. A *pilot line* is a passage used to carry fluid to control a valve. A pilot line is not used to power an actuator. Pilot lines may be externally plumbed to transfer the flow of fluid from another component or may be passages that are machined within a component.

Pilot-operated pressure-relief valves are used as circuit pressure overload protection and are also used for circuit operating pressure regulation. In a pilot-operated pressure-relief valve, pilot pressure is sensed through a control orifice by the pilot poppet. As circuit pressure builds, the pilot poppet opens, allowing the fluid to flow to the reservoir. This reduces the pressure of the outlet side of the main poppet, causing it to open and allowing greater fluid flow to the reservoir.

A pressure-relief valve that is normally closed can be situated between two linear actuators (cylinders) and used as a sequence valve. A *sequence valve* is a pressure-operated valve that diverts flow to a secondary actuator while holding pressure on the primary actuator at a predetermined minimum value after the primary actuator completes its travel. A *sequence* is the order of a series of operations or movements.

A clamp and stamp circuit is an example of a circuit in which a sequence valve can be used to control the sequence of circuit operations. A sequence valve is positioned in the circuit just ahead of the stamp cylinder. The pressure setting does not allow the main poppet to shift, preventing the primary port and secondary port from being connected until the set pressure has been reached. See Figure 6-25.

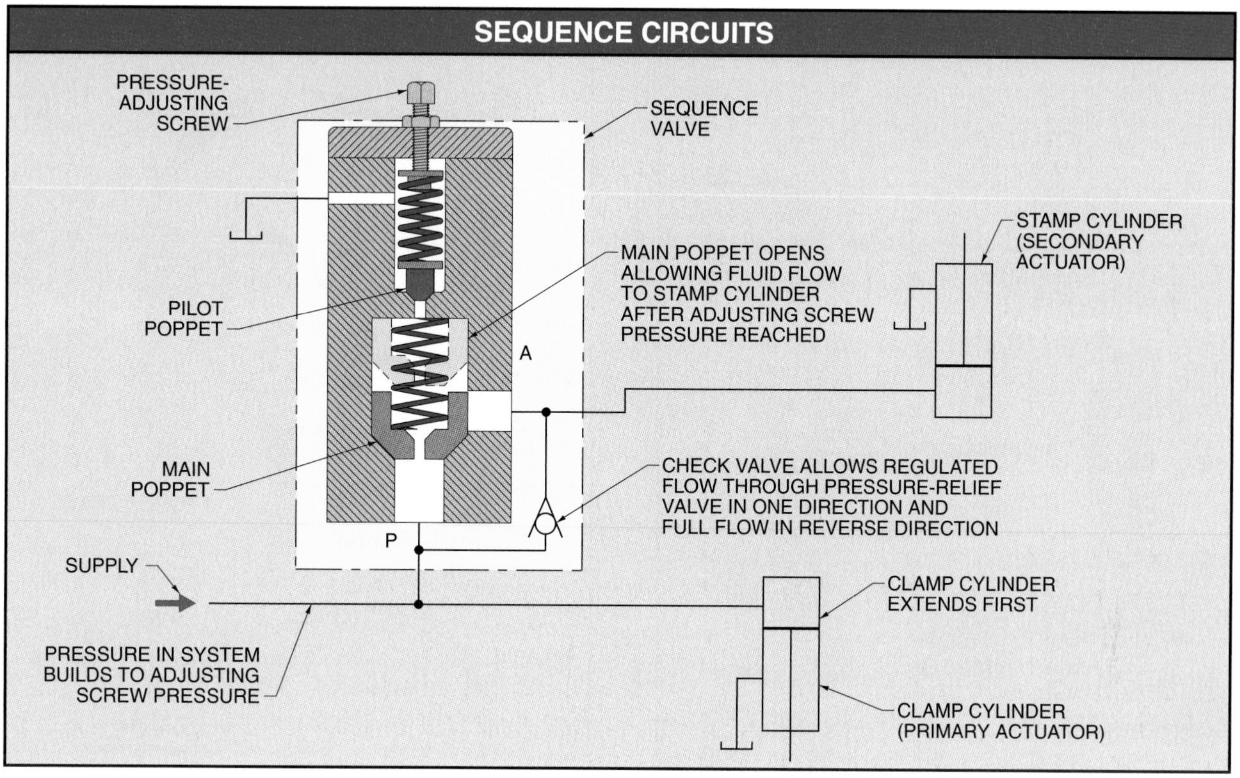

Figure 6-25. A sequence circuit diverts flow to a secondary actuator while holding pressure on the primary actuator at a predetermined minimum value after the primary actuator completes its travel.

Fluid flow to the clamp cylinder extends the clamp cylinder while the stamp cylinder does not move. Pressure in the circuit continues to build after the clamp cylinder is fully extended. The sequence valve main poppet opens, allowing fluid flow to the stamp cylinder when the adjusting screw pressure is reached. This allows a part to be clamped and stamped in the correct sequence.

System pressure set by pressure-relief valves may not always be sufficient to operate multiple actuators. A pressure-reducing valve is used where each actuator or circuit may require a lower pressure than the circuit's operating pressure. A *pressure-reducing valve* is a valve that limits the maximum pressure at its outlet, regardless of the inlet pressure. Pressure-reducing valves are normally open and may be direct-acting or pilot-operated. Pressure-reducing valves, which sense pressure from their secondary port, are normally direct-acting. However, pressure in another part of a circuit can be sensed and pilot pressure used to operate a pressure-reducing valve by means of an external pilot line. See Figure 6-26.

Pressure-reducing valves operate by pressure being sensed at their secondary port. The spool is moved off its normal position, reducing or blocking working pressure when higher-than-system pressure is reached. Excess fluid flow is diverted to the reservoir through the drain port. Only enough flow is passed to the outlet to maintain the preset pressure. A light flow is sent to the reservoir through the drain port if the valve closes completely. This prevents pressure from building up in the circuit. Pressure levels are maintained by a pressure adjusting screw.

Pressure intensity is measured by a pressure gauge when adjustments are made to pressure-control components. A *pressure gauge* is a device that measures the intensity of a force applied to a fluid. Pressure gauges are required for adjusting control valves to within proper or required values, determining the forces exerted by a cylinder, or determining the torque produced by a hydraulic motor.

In the U.S., the Comprehensive Environmental Response, Compensation, and Liability Act of 1980 holds any party creating or contributing to a hazardous waste site financially responsible for clean-up costs.

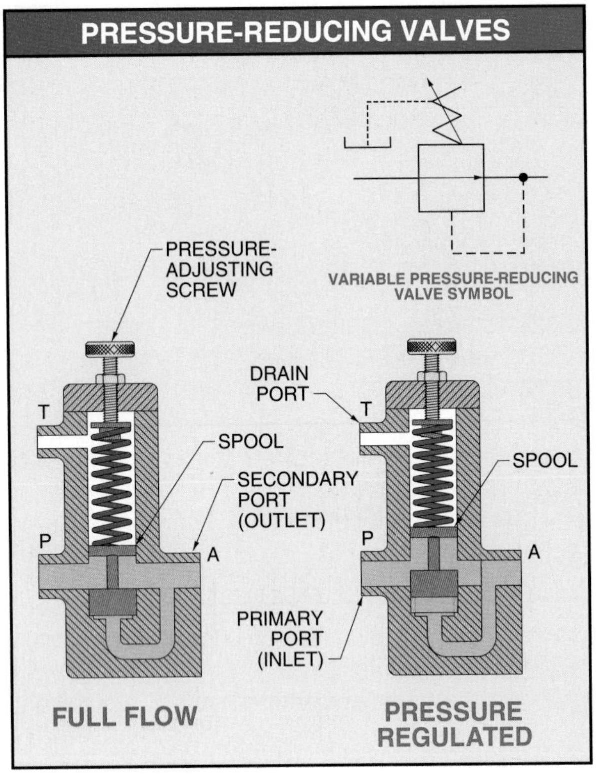

Figure 6-26. Pressure-reducing valves are used to regulate pressure in one leg of a circuit, or individual component pressure.

Many gauges used for measuring high pressure use a Bourdon tube. A *Bourdon tube* is a hollow metal tube made of brass or similar material. A Bourdon tube is oval or elliptical in its cross-sectional area and is bent in the shape of the letter C. One end of the Bourdon tube is fixed to a frame where the fluid enters. The other end is closed and free to move. As fluid pressure inside the tube changes, the elliptical cross section changes, and the free end of the Bourdon tube tends to straighten. This actuates a linkage to a pointer gear, which moves the pointer to indicate the pressure on a scale. See Figure 6-27.

Directional Control. A *directional control valve* is a valve whose primary function is to direct or prevent flow through selected passages. A directional control valve allows fluid to be directed to actuators and other system components at the appropriate time and valve port. Directional control valves include check, 2-way, 3-way, and 4-way valves.

> ***Bourdon tube pressure gauges can be damaged by pressure surges in the hydraulic system. To limit the damage, bourdon tube gauges may be oil filled.***

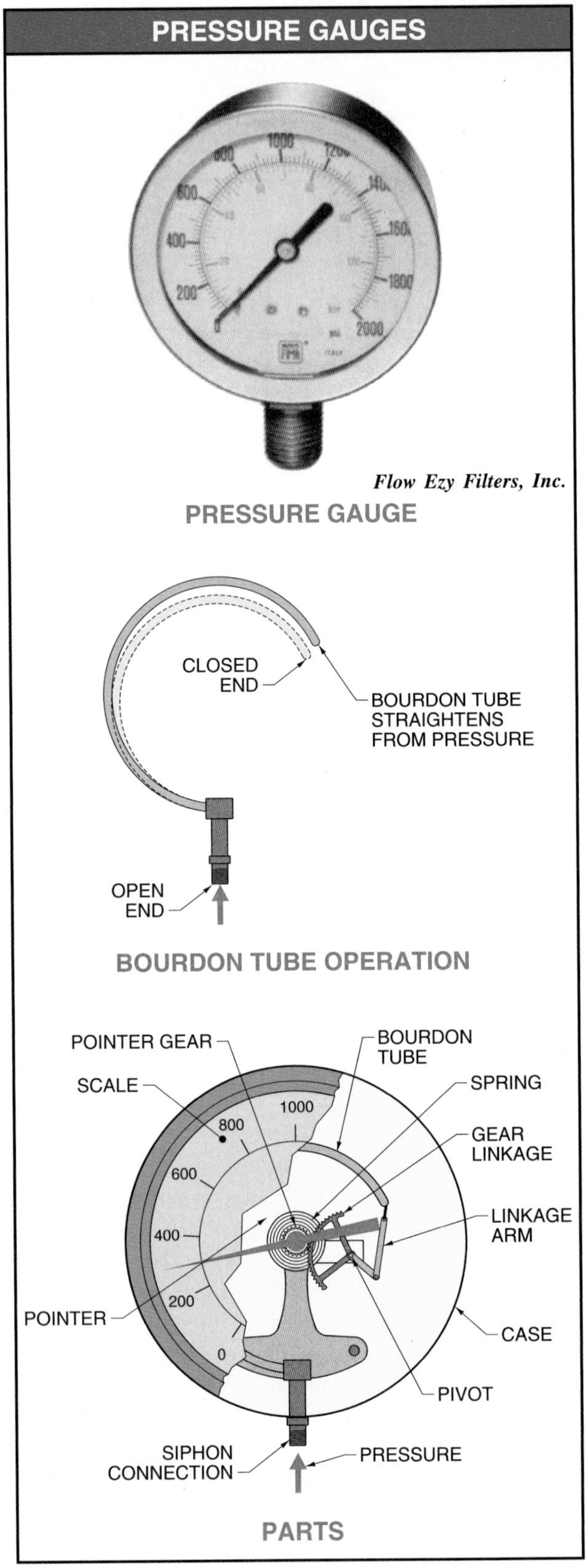

Figure 6-27. Pressure gauges use a Bourdon tube to measure pressures.

A *check valve* is a valve that allows flow in only one direction. Check valves are normally closed and may be direct-acting or pilot-operated. A direct-acting check valve consists of a valve body, spring, and ball or poppet. The valve body contains an inlet (primary) and outlet (secondary) port. The spring holds the ball or poppet in one position. The ball or poppet blocks fluid flow when held against the seat or allows fluid flow when pushed off its seat as the inlet pressure rises high enough to overcome the spring pressure. The spring and fluid pressure forces the ball or poppet to seat, preventing fluid flow if fluid attempts to flow in the reverse direction. See Figure 6-28.

The flow through a check valve in both directions may be accomplished with the use of pilot operation. The external pilot supply may be hydraulic or pneumatic. Pilot-operated check valves operate normally as a check valve. Pilot pressure is needed at the pilot poppet when reverse fluid flow is required. The check valve poppet is unseated, allowing reverse flow when sufficient pilot pressure is produced at the pilot port. The pilot poppet and the main poppet control the flow of fluid.

Directional control valves are described by their number of ways and spool positions. The movement of the spool determines which way the fluid flows.

A *way* is a route that fluid can take through a valve. For example, a check valve is referred to as a one-way valve because fluid flow is routed in only one direction. Ways may connect more than one port.

A *two-way directional control valve* is a valve that has two main ports that allow or stop the flow of fluid. Two-way valves are used as shutoff, check, and quick-exhaust valves. A *three-way directional control valve* is a valve that has three main ports that allow or stop fluid flow or exhaust. Three-way valves are used to control single-acting cylinders, fill-and-drain tanks, and nonreversible fluid power motors. A *four-way directional control valve* is a valve that has four main ports that change fluid flow from one port to another. Four-way valves are used to control the direction of double-acting cylinders and reversible fluid motors.

Two-way, 3-way, and 4-way directional control valves are constructed similarly with some minor changes. The basic parts of a directional control valve are the body and the internal spool. The body has a series of holes (ports) that are usually labeled A, B, P, and T. Ports labeled A and B are passage ports to and from an actuator. Ports labeled P are pump or pressure ports. Ports labeled T are tank ports. There are many different flow paths used to connect these ports. See Figure 6-29.

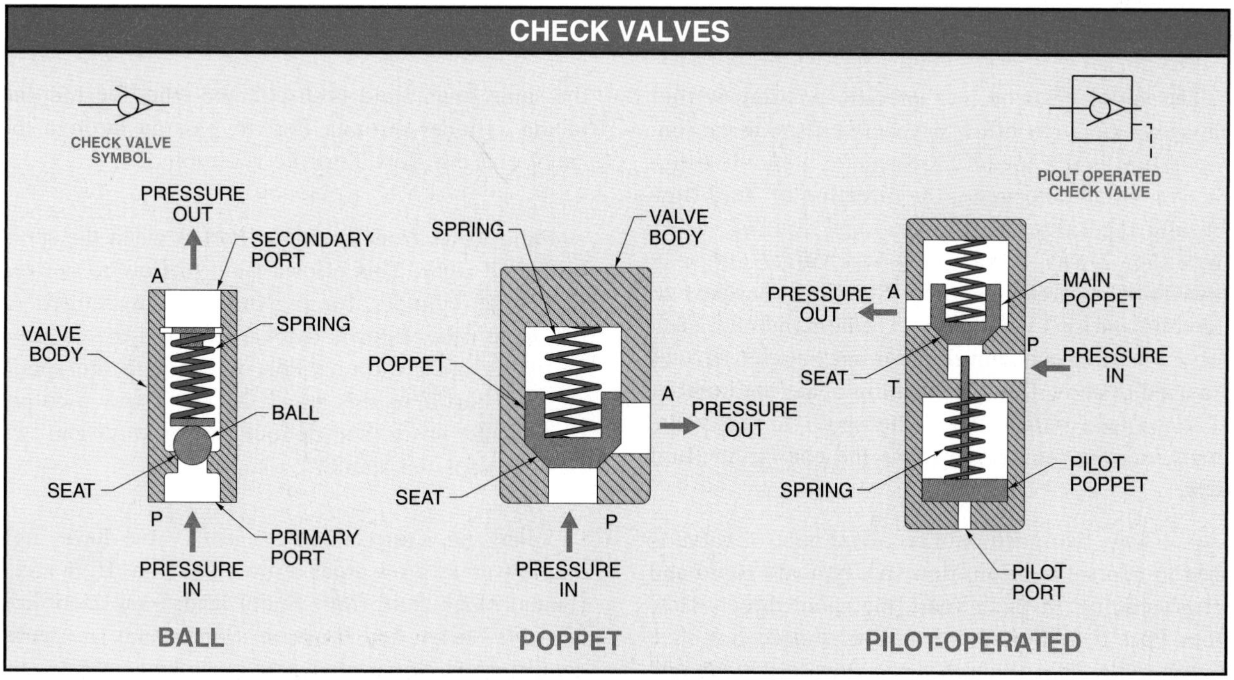

Figure 6-28. Check valves are normally closed and allow fluid flow in one direction.

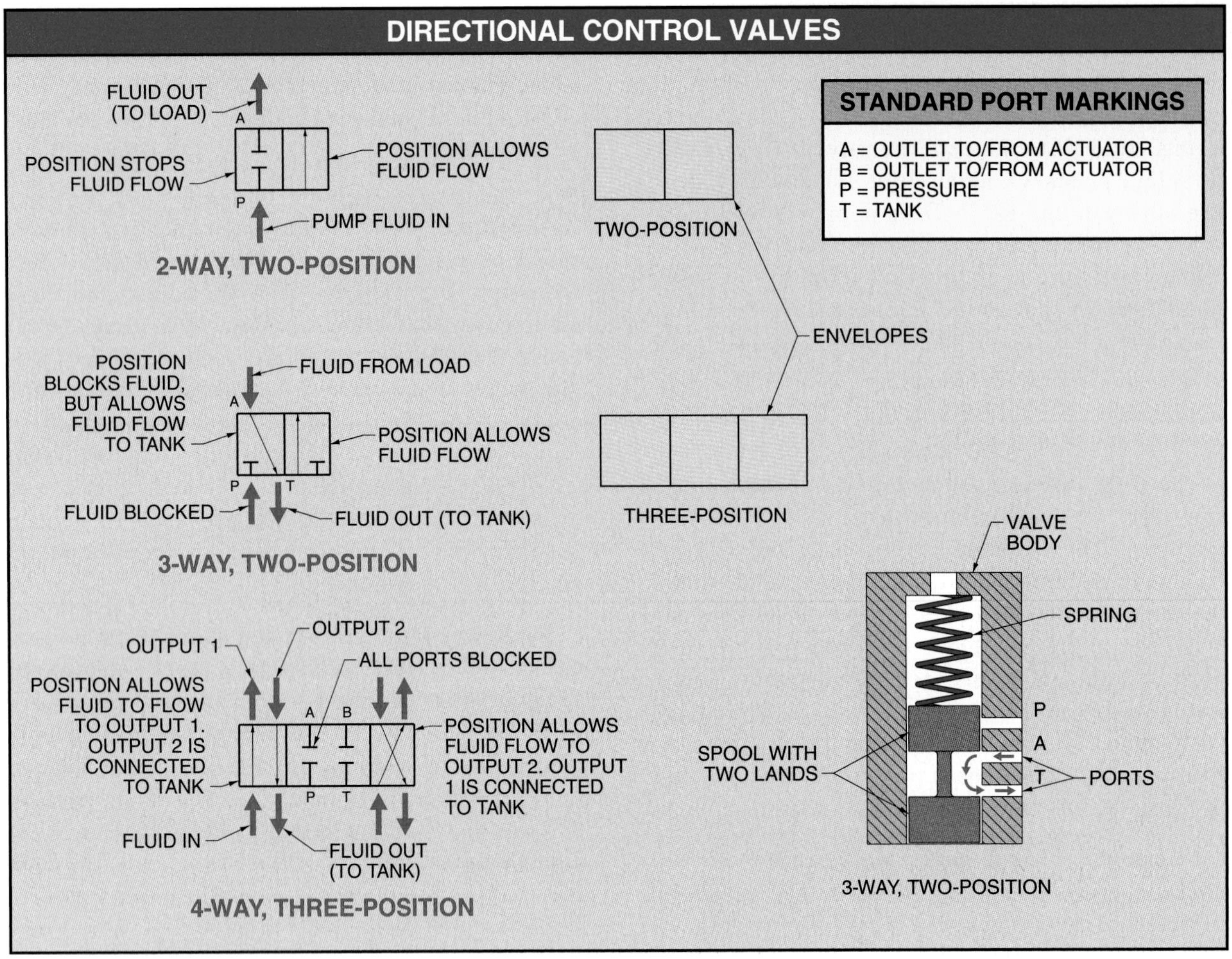

Figure 6-29. Directional control valve symbols describe fluid ways and spool position.

The spool must be in a specific position within the valve body to allow a specific flow direction. A *position* is the specific location of a spool within a valve which determines the direction of fluid flow through the valve. Most directional control valves are either 2-way, 3-way, or 4-way valves and may have two or three positions. Symbols are used to show the various valves. The symbols indicate each valve position as an envelope or one square. Arrows are used to show fluid flow and indicate the number of ways in a valve. When the spool in the valve shifts, another envelope shows the change in fluid flow.

A 4-way, two-position directional control valve is used to control the fluid flow in a typical extend and retract circuit. In this circuit, the spool directs flow from Port P to Port A when it is shifted left. See Figure 6-30. This allows fluid to flow to the cap end of the cylinder, forcing the piston rod to extend. At the same time, fluid is discharged from the rod end of the cylinder through Port B, passing through the spool and out Port T to the reservoir.

Fluid flows from Port P to Port B when the spool is shifted right. This allows fluid to flow to the rod end of the cylinder, forcing the piston to retract. At the same time, fluid is discharged from the cap end of the cylinder through Port A, through the spool, and out Port T to the reservoir. The 4-way, two-position valve has a total of four passageways and two positions for the spool.

Spools in a directional control valve have high and low areas. Low areas allow fluid flow. High areas (lands) block fluid flow. Spool lands vary according to valve design and function. Some lands are wide, some narrow. Some spools have two lands, while others have four.

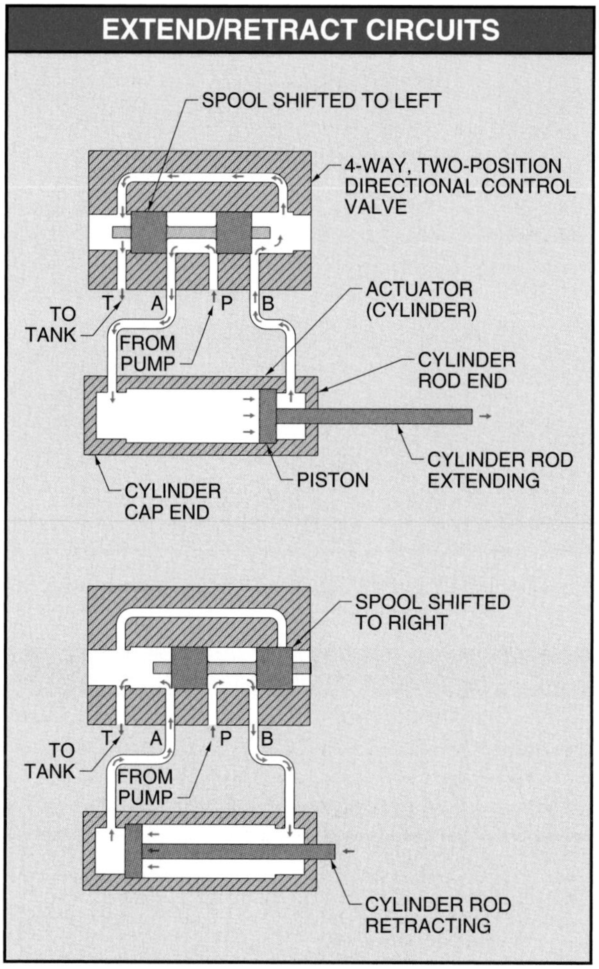

Figure 6-30. A 4-way, two-position directional control valve is used to control the fluid flow in a typical extend and retract circuit.

Spool fit is very precise. The clearance between a spool and valve body ranges from 8μ – 10μ. Although this clearance allows for oil leakage (used for sealing and lubrication), the spool land would prohibit even a red blood cell from passing between ports. For this reason, most spools are not interchangeable and must be handled with care.

Three-position directional control valves have a center position. The actuator motion is controlled by the left or right positions of the valve spool. The center position satisfies the circuit's requirements. Many different valve spool center configurations are available. The most common is the tandem center configuration. The tandem center configuration blocks ports A and B and connects ports P and T. This allows the actuator to remain in its last placed position with the pump flow returning to the reservoir without going through the pressure-relief valve. In one tandem design, the center of the spool is drilled out, allowing a connection between ports P and T through the core of the spool.

An example of a tandem center directional control valve is that of a forklift operation. When the forks are lifted off the floor and lifting stops, the forks remain in position until the valve is directed to raise or lower the forks. Fluid flows through the spool's core from port P to port T while the forks are immobile and the hand lever is in the center position. This allows fluid to return to the reservoir without loading the pump. See Figure 6-31.

The spool can be held in the center position by hydraulic pressure or springs. Spring centering is the most common. Three position valves are normally centered by equal spring pressure located at both ends.

Directional control valves must have a means to change the position of the valve spool. A *valve actuator* is a device that changes the position of a valve spool. Valve actuators may be manual, mechanical, pneumatic, hydraulic, or electrical. Manual valve actuators include levers, pushbuttons, or foot pedals. Mechanical valve actuators include cams depressing a plunger connected to the spool. Pneumatic or hydraulic valve actuators include air or oil pilot operation. A common means of positioning spools is electrically with the use of a solenoid. Each method of valve actuation is shown by a symbol for the easy identification of the valve actuation method. See Figure 6-32.

The Gates Rubber Company

A backhoe uses extend and retract circuits to operate the boom, bucket, crowd, and stabilizing cylinders. The cylinders are double-acting to give full force in both directions.

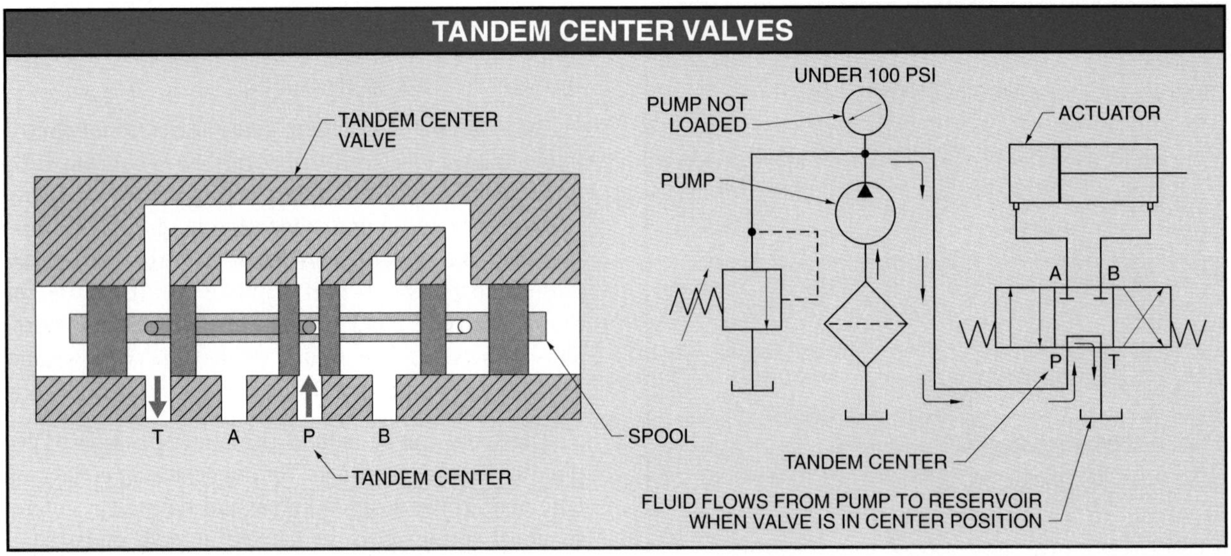

Figure 6-31. The spool in a 4-way, three-position, tandem center directional control valve allows fluid to return to the reservoir when the valve is in the center position.

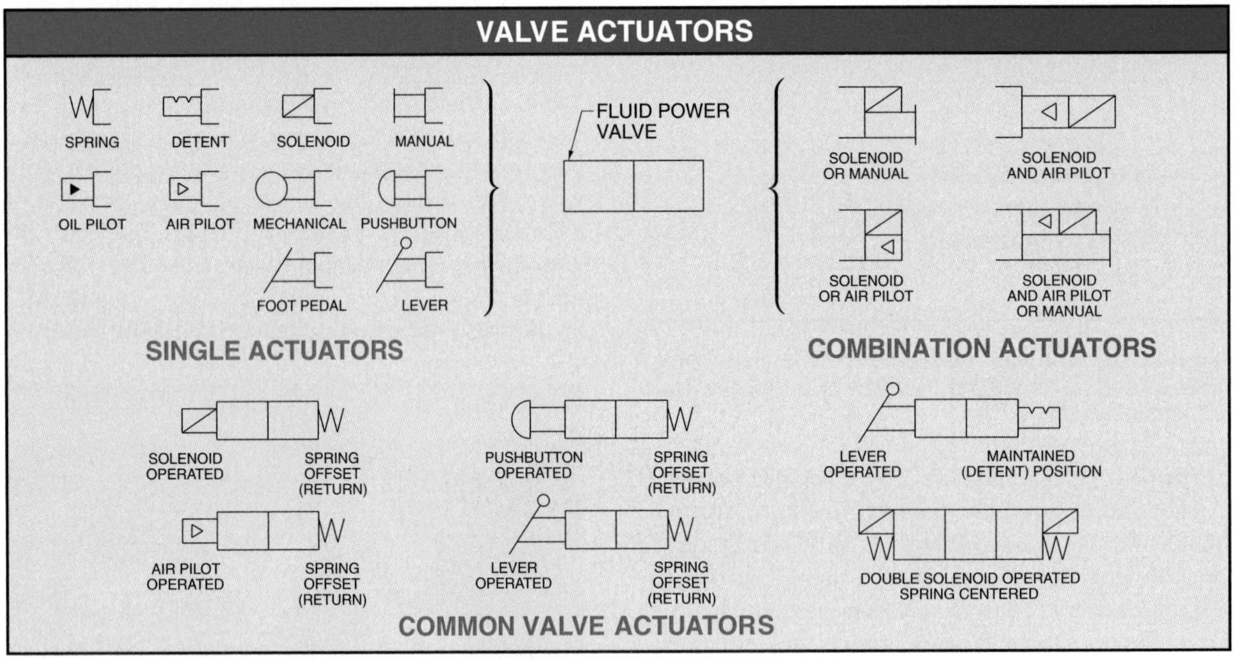

Figure 6-32. Valve spool shifting is accomplished through the use of valve actuators.

System or pilot pressure is often used in combination with solenoids because many directional control valves require more pressure to operate the spool than a solenoid can produce. Typical pilot operation consists of a piggyback design where a smaller, solenoid-operated, directional control valve is positioned on top of the main valve. Flow from the small valve is directed as pilot pressure to either side of the main directional control valve spool when shifting is required. See Figure 6-33.

Flow Control. Flow control in a hydraulic circuit is primarily used to regulate speed. Flow control may be accomplished by using a variable displacement pump or a flow control valve. A *variable displacement pump* is a pump in which the displacement per cycle can be varied.

A *flow control valve* is a valve whose primary function is to regulate the rate of fluid flow. By regulating the rate of fluid flow, the flow control valve becomes a resistor and increases fluid pressure upstream from the valve. The increase in pressure opens the pressure-relief valves to allow excess fluid to return to the reservoir and reduced flow continues to a branch circuit or actuator. In some cases, the excess fluid is used in another circuit rather than being sent to the reservoir. Controlling the flow of fluid in a circuit is accomplished by using an orifice or needle valves. See Figure 6-34.

A *restrictive check valve* is a check valve with a specific size hole drilled through its center. Factory preset orifices are sized to control a flow rate at a specific inlet pressure. Fluid flow increases if the pressure increases at any factory preset orifice.

A *gate valve* is a two-position valve that has an internal gate that slides over the opening through which fluid flows. Gate valves are generally used for full flow or no flow operation and are not designed for restricting fluid flow. Fluid flows in a straight path through the valve, which offers very little pressure drop in the circuit when fully open. Vibration and wear occur when a gate valve is used in a partially open position. Any restricting of flow by a gate valve should be of very coarse metering.

A *globe valve* is an infinite-position valve that has a disk that is raised or lowered over a port through which fluid flows. Globe valves do not have a straight through path for fluid flow. Fluid flow makes two 90° turns when flowing through the valve. The opening between the seat and disk is controlled to meter or throttle the fluid flow from zero to full flow. *Metering* is regulating the amount or rate of fluid flow. *Throttling* is permitting the passing of a regulated flow. The flow is regulated in one direction only because of the two 90° turns in the globe valve's flow path. The flow direction through a globe valve is generally indicated by an arrow on the side of the valve's housing.

A *needle valve* is an infinite-position valve that has a narrow tapered stem (needle) positioned in line with a tapered hole or orifice. An *orifice* is a precisely-sized hole through which fluid flows. The size of the orifice controls the flow rate by creating a pressure drop. The remaining pump flow, which is not passed through the orifice, is either dumped to the reservoir or used in another circuit. Needle valves offer precise flow control because of their cone-shaped needle and seat and the fine threaded adjusting stem. The fine threaded adjusting stem offers a very gradual change in orifice size.

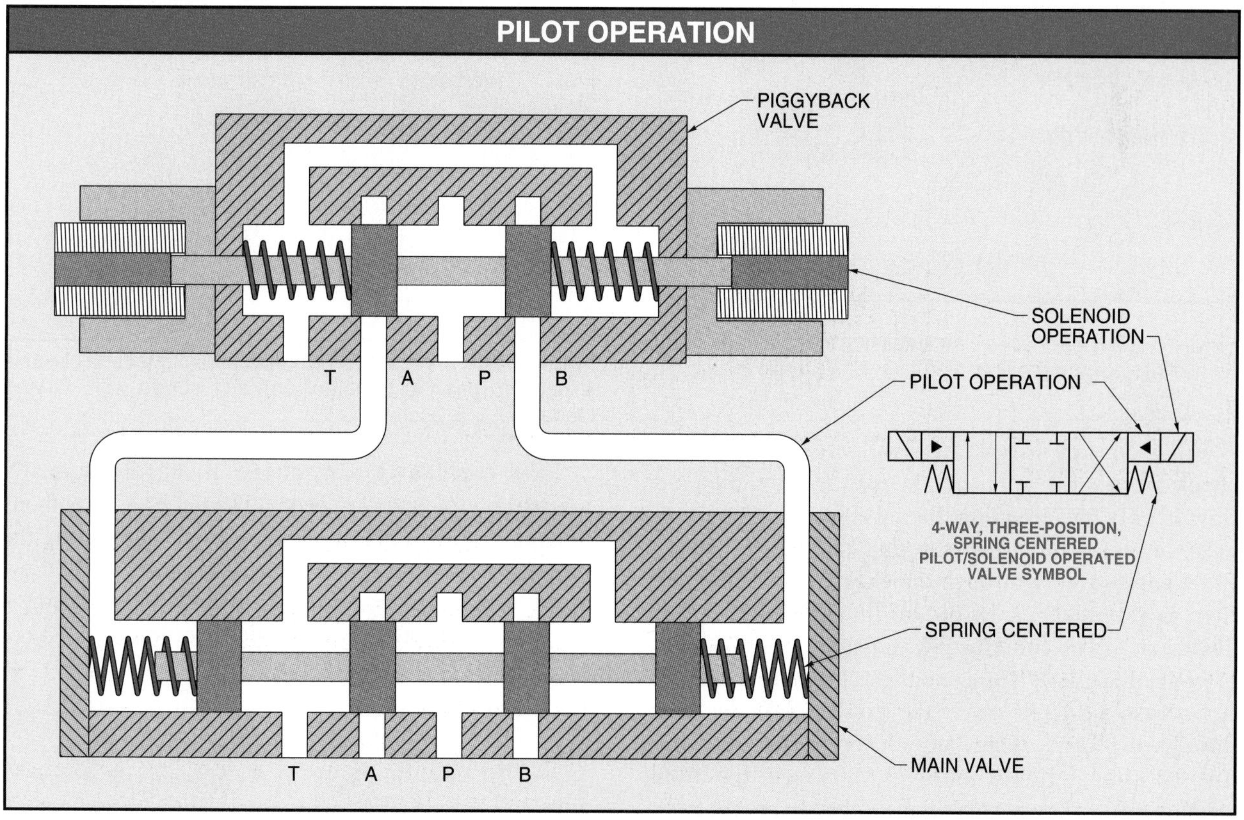

Figure 6-33. Many directional control valves are actuated through the use of a piggyback pilot operation.

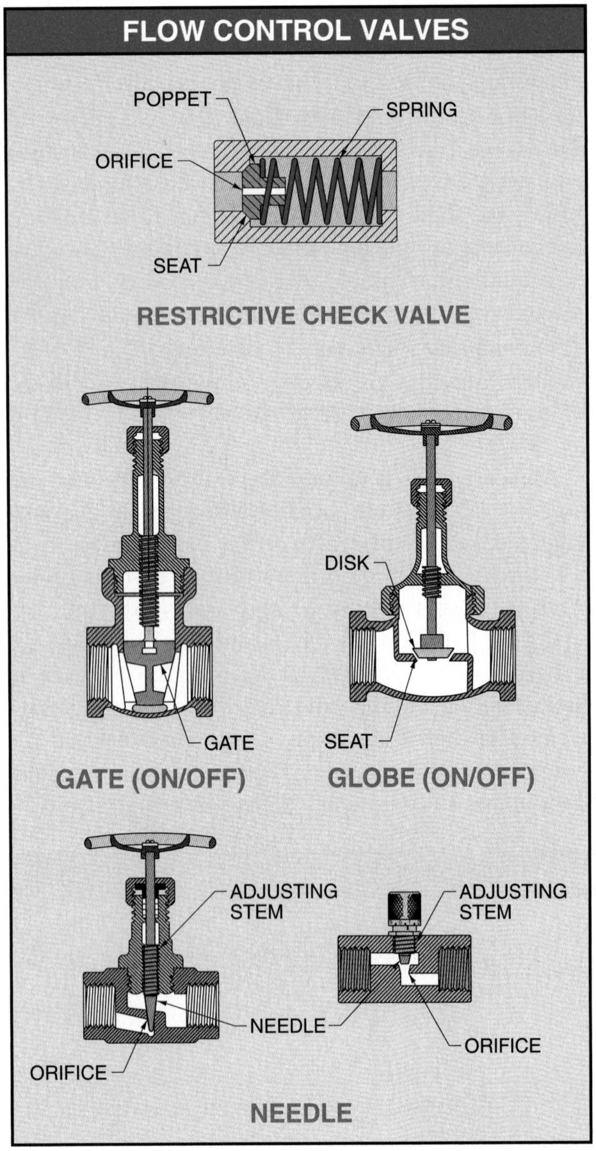

Figure 6-34. A restrictive check valve or needle valve may be used to control the flow of fluid in a hydraulic circuit.

Flow control valves are available with built-in check valves for applications that require metered flow in only one direction. In one direction, the check valve directs fluid flow through the needle valve. Fluid passes freely through the check valve when the flow is reversed. An arrow on the side of the needle/check valve combination indicates the direction of controlled flow. The greater the pressure differential across a needle valve, the greater the flow rate through the valve. Thus, any change in pressure before or after a flow control valve affects the flow through the valve, resulting in a change in actuator speed. For example, an increase in the load on an actuator lowers the pressure differential and reduces the flow through the valve. This reduces the speed of the actuator. To increase the actuator speed, the flow control valve is partially opened. The speed of the actuator can become too great if the load on the actuator is reduced. This scenario can be corrected by using a pressure compensated flow control valve.

A *pressure compensated flow control valve* is a needle valve that makes allowances for pressure changes before or after the orifice through the use of a spring and spool. The needle valve's adjustment knob provides a controlled orifice, but any change in pressure is compensated for by the spool. Stable flow occurs when the spool has created a restriction equivalent to the balance of inlet and outlet pressure forces assisted by the spring and controlled orifice. See Figure 6-35.

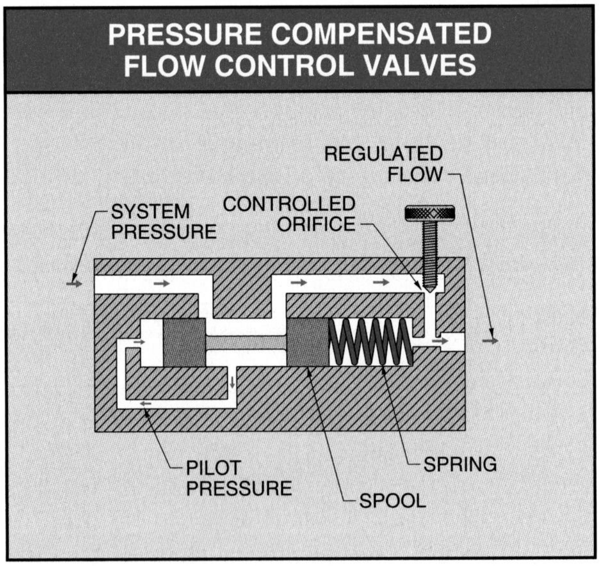

Figure 6-35. A pressure compensated flow control valve is used to provide constant actuator speed with varying loads.

Flow controls in a hydraulic circuit are normally used to control actuator speeds. The three basic methods of actuator flow control include meter-in (metering the flow at the actuator inlet), meter-out (metering the flow at the actuator discharge), and bleed-off (metering a portion of the inlet flow to the reservoir.

Hydraulic oil becomes hot during use. Be careful not to burn hands on hot hydraulic oil. Always let a system cool before beginning service procedures.

Metering the flow at the actuator inlet controls the amount of fluid going into the actuator. This provides a very accurate and constant movement of the actuator when the load resistance is continually present, for example, when a load is required to move at a controlled speed or a vertical load is being lifted. Metering the flow at the actuator's discharge prevents a jerky movement when the load is not constant. The use of either inlet or discharge flow controls requires a check valve to allow full return flow.

Metering a portion of the inlet flow is sometimes called bleed-off flow control and is not as accurate as the discharge flow control method. Bleed-off flow control is accomplished by placing a tee in the actuator's inlet plumbing. The flow control valve is installed at the tee between the supply line to the actuator and the reservoir. The advantage of this method is that less work is being done by the pump and pressure-relief valve due to the excess fluids being metered back to the reservoir. The disadvantage is that the metered or set orifice flow is to the reservoir, which offers the same problem as inlet metering when the load is not constant.

Actuators

A *hydraulic actuator* is a device that converts hydraulic energy into mechanical energy. Hydraulic actuators include cylinders and motors.

Cylinders. A *hydraulic cylinder* is a device that converts hydraulic energy into straight-line (linear) mechanical energy. A basic hydraulic cylinder consists of a cylinder body, piston, piston rod, and seals. The end through which the rod protrudes is the rod end and the opposite end is the cap end. Fluid ports are located in the rod and cap ends. As the cylinder rod reciprocates, the rod is supported by a steady bearing which also holds the rod seal and wiper seal in place. See Figure 6-36.

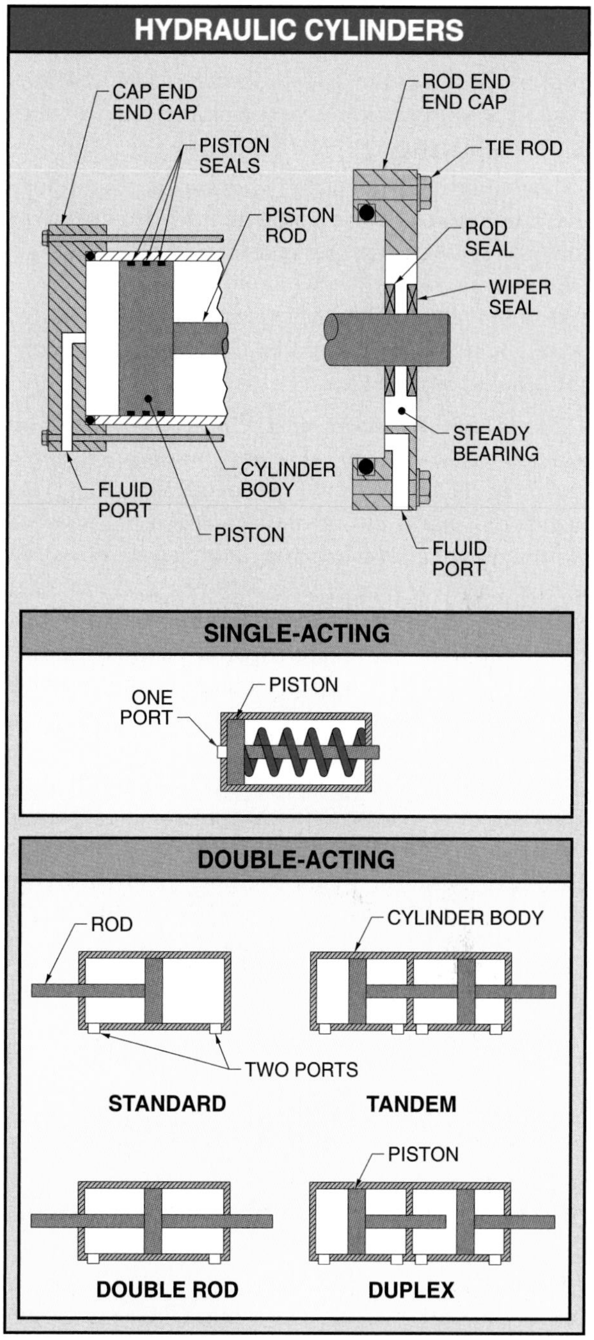

Figure 6-36. A hydraulic cylinder is an actuator that converts hydraulic energy into linear mechanical energy.

Cylinders are classified as single- or double-acting. A *single-acting cylinder* is a cylinder in which fluid pressure moves the piston in only one direction. A small rod end port allows atmospheric air in and out of the rod end of the cylinder. The piston and piston rod extend as hydraulic fluid is pumped into the cap end port. The piston rod retracts when fluid in the cap end of the cylinder is released to the reservoir. The rod in a single-acting cylinder retracts either by gravity, spring, or some other mechanical force. An example of a single-acting cylinder is that of most lifting cylinders on a forklift.

A *double-acting cylinder* is a cylinder that requires fluid flow for extending and retracting. A double-acting cylinder has ports and fluid at each end of the cylinder (both sides of the piston). The piston rod extends when fluid is pumped into the cap end port. The piston rod retracts with the fluid in the cap end of the cylinder returning to the reservoir when fluid is pumped into the rod end port. Double-acting cylinders include tandem, duplex, and double rod cylinders. The cylinder used in an application is based on the requirements of the application.

Seals, an integral part of a cylinder, create positive contact across the piston and the rod, allowing for maximum pressure and preventing leakage. Seals must be well-lubricated, contaminate-free, and without any nicks or damage to provide maximum contact. They should be used on smooth, true, and unmarred surfaces.

Seals may be static or dynamic. A *static seal* is a seal used as a gasket to seal nonmoving parts. Static seals are used where there is contact between two parts but no motion, such as between stationary items taken apart and reassembled. A *dynamic seal* is a seal used between moving parts that prevents leakage or contamination. For example, dynamic seals are used on pistons and piston rods to allow the piston and rod to slide inside the cylinder. Seals may be positive or nonpositive. A *positive seal* is a seal that does not allow the slightest amount of fluid to pass. A *nonpositive seal* is a seal that allows a minute amount of fluid through to provide lubrication between surfaces. Seals include quad-ring, lip, compression, and packing seals. See Figure 6-37.

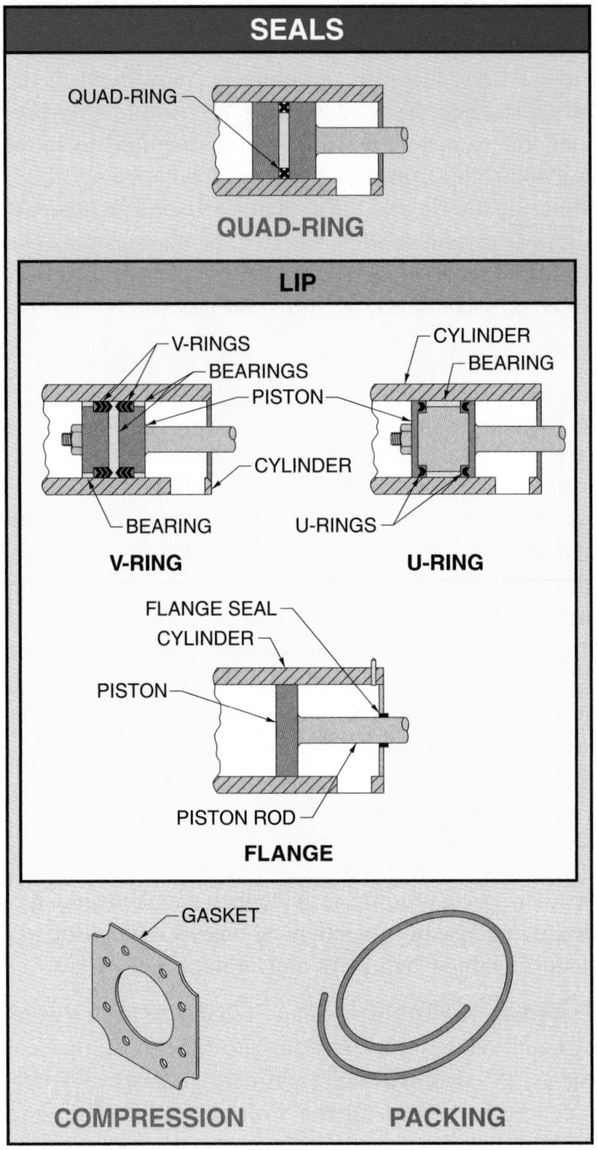

Figure 6-37. Seals are used to create a positive contact between moving parts to provide pressure and prevent leakage.

O-rings are the most commonly used seal in mechanical assemblies. In hydraulics, however, the pressures reached in a system cause an O-ring to extrude. O-rings are only used in hydraulics for special static applications. Quad-rings are similar to O-rings. A *quad-ring* is a molded synthetic rubber seal having a basically square cross-sectional shape. Quad-rings look like they are made of four O-rings. During use and under pressure, quad-rings offer a dynamic seal by being pressed (forced) against one side of their groove. This pressure forces the seal outward and against the sealing surface. Increased pressure on quad-rings results in greater sealing forces.

Quad-rings require a back-up ring for pressures over 1500 psi because of the disfiguration of the seal under pressure. The back-up ring prevents the resilient rubber of the quad-ring from being extruded out of the groove. Quad-rings have twice the sealing power with greater resistance to rolling, twisting, or extruding. *Resilience* is the capability of a material to regain its original shape after being bent, stretched, or compressed.

A *lip seal* is a seal that is made of a resilient material that has a sealing edge formed into a lip. Lip seals are made of leather, rubber, or synthetic material. Lip seals use the fluid's pressure on its lip to form a seal. As pressures increase, the seal tries to expand at the lip, creating a tighter seal. Lip seals are made in various shapes for various pressure duties. Lip seals are dynamic seals used for sealing rotating or reciprocating shafts, pistons, cylinder rod ends, and pump shafts. Typical lip seals are the V-ring, U-ring, and flange seal.

A *V-ring seal* is a lip seal shaped like the letter V. V-ring seals are dynamic seals used in severe operating condition applications. V-ring seal material includes impregnated and treated leather, rubber, and asbestos. The advantages of V-ring seals are that dissimilar seals of different materials may be used in combination to provide the best pressure, wear, and friction characteristics for an application.

A *U-ring seal* is a lip seal shaped like the letter U. U-ring seals are dynamic seals used in reciprocating and rotating applications. System pressure increases the sealing force on the lips of the seal. U-ring seal material includes impregnated and treated leather, rubber, and other elastic compounds.

Flange seals are generally used as cylinder rod seals instead of piston seals. Flange seal materials include impregnated and treated leather, rubber, and other elastic compounds.

Compression seals are static seals commonly referred to as gaskets. A *gasket* is a seal used between machined parts or around pipe joints to prevent the escape of fluids. A gasketed joint is sealed by the molding of the gasket material into the imperfections of the mating surfaces of the joint. Gasket material may be rubber, leather, synthetic, or metal. Metal is used by itself or in combination with other softer materials. For example, in some cases, a sheet of brass or copper may be soft enough to form a proper seal when sandwiched between two harder metals.

Packing is a bulk deformable material or one or more mating deformable elements reshaped by manually adjustable compression. Packing is used where some form of motion occurs between rigid members of an assembly. Packing must only be deformed enough to allow or throttle leakage between the moving and stationary parts. This leakage becomes a lubricant and coolant for the packing. On large applications, the leakage rate may be as high as 10 drops per minute. On small or light applications, one drop per minute is sufficient.

All seals should be handled and installed in a clean environment. Always lubricate seals before installation with a compatible lubricant for the seal material.

Packing must be pliable enough to provide a radial seal when axially compressed. Packing material is made of braided, woven, or twisted cotton or flax. Solid lubricants, such as graphite or mica, are added to the packing to protect the moving parts. Packing must be adjusted frequently to compensate for wear.

Never try to locate a leak by running a hand over the suspected area. Always use a piece of cardboard. Escaping fluid under pressure can penetrate the skin and cause serious injury. If any fluid is injected into the skin, it must be surgically removed within a few hours or gangrene may result.

Motors. A *hydraulic motor* is a device that converts hydraulic energy into rotary mechanical energy. Hydraulic motors are often referred to as rotary actuators. Hydraulic motor construction is similar to the construction of hydraulic pumps. Hydraulic motors include gear, vane, and piston models. The transformation of hydraulic energy to rotary mechanical energy is a reversal of the pumping function. See Figure 6-38.

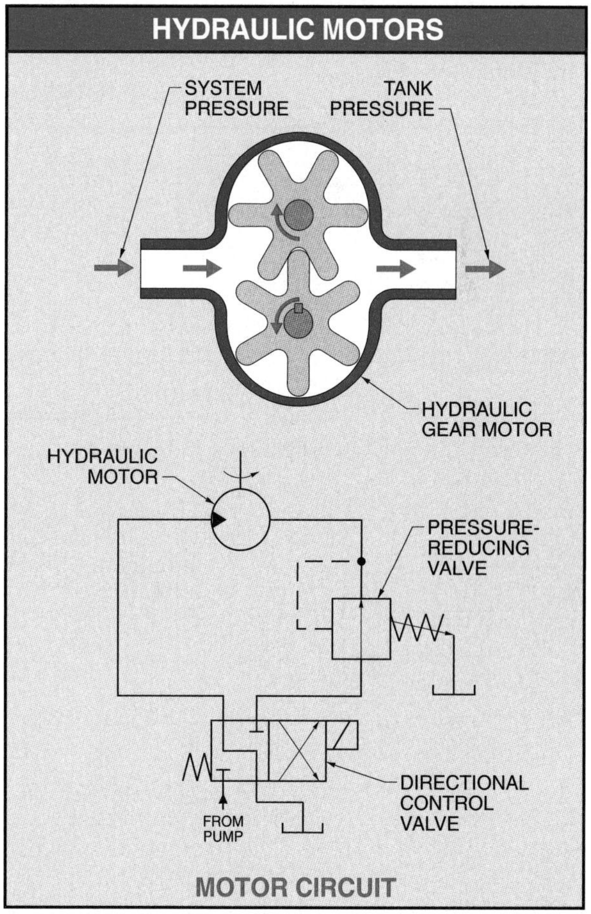

Figure 6-38. A hydraulic motor is a device that converts hydraulic energy into rotary mechanical energy.

Rotary mechanical energy is produced in a gear motor when fluid at high pressure is forced against the teeth of the upper and lower gears, causing them to rotate. Similarly, fluid at high pressure forced against the vanes of a hydraulic vane motor causes the shaft to rotate. The difference between a vane pump and a motor is that the vanes in a motor need to be extended by spring action. The vanes are generally extended by fluid pressure after motor torque is developed.

Accumulators

An *accumulator* is a container in which fluid is stored under pressure. Accumulators store hydraulic fluid under pressure (potential energy) until it is needed. Accumulators also maintain circuit pressure, develop circuit flow, and absorb circuit shock. Accumulators may also maintain circuit pressure in an emergency if a pump fails. This allows a circuit to complete cycling before shutdown occurs. Accumulators may be spring-loaded, weight-loaded, or hydro-pneumatic. See Figure 6-39.

Warning: All accumulators are extremely dangerous when their energy is uncontrolled. For this reason, all accumulator energy must be released or blocked before performing repairs.

A *spring-loaded accumulator* is an accumulator that applies force to a fluid by means of a spring. Spring-loaded accumulators consist of a cylinder body, piston, and spring. The piston rides between the spring and the hydraulic fluid inside the cylinder body. Circuit pressure and the compression rate of the spring determine the amount of stored energy. In many cases, a mechanical stop at the spring area is used to prevent excessive pressure from overcompressing and damaging the spring. Most spring-loaded accumulators are adjustable, allowing for varied amounts of stored fluid pressure.

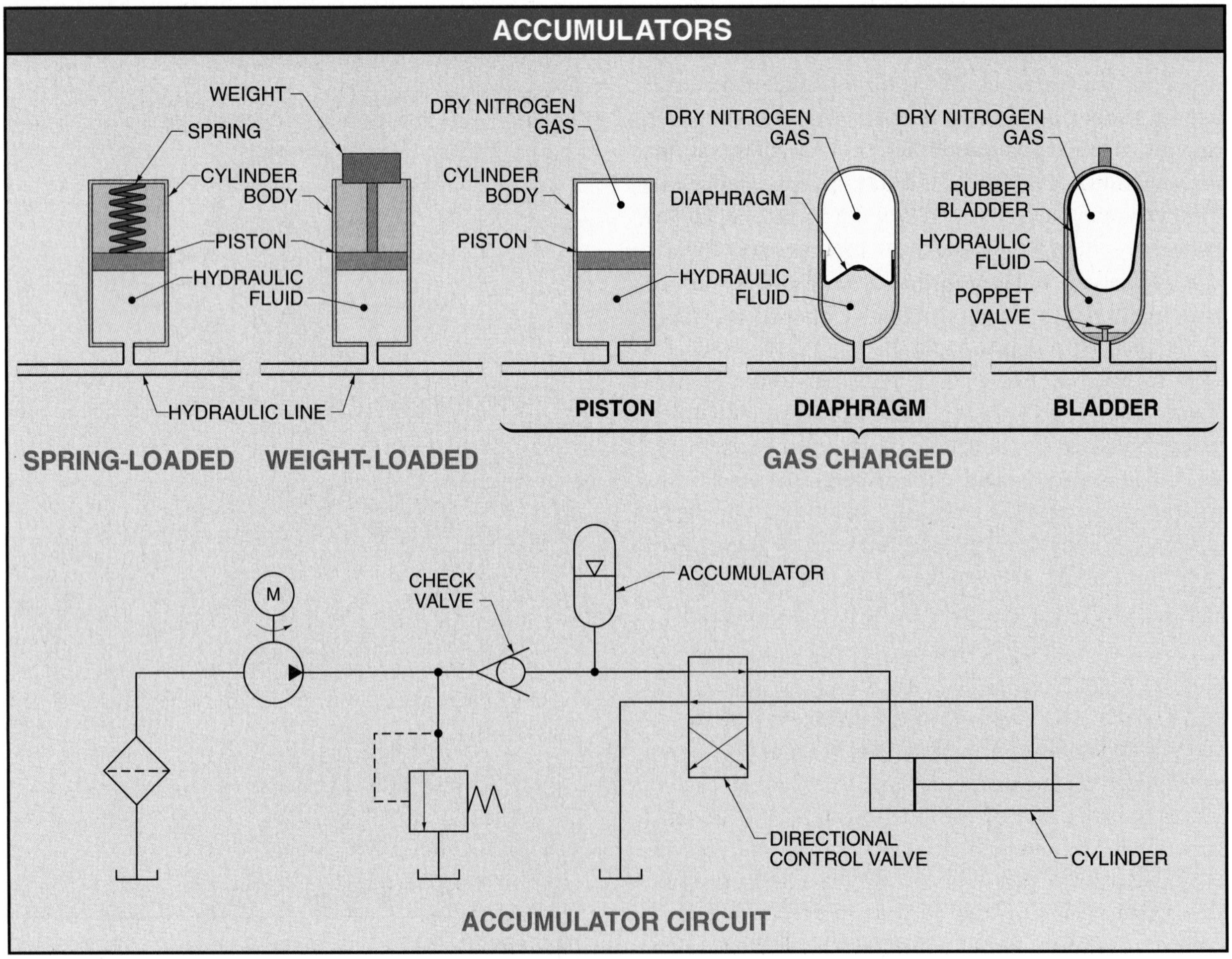

Figure 6-39. Accumulators store hydraulic fluids under pressure for use at a later time.

A *weight-loaded accumulator* is an accumulator that applies force to a fluid by means of heavy weights. The weights are generally iron or concrete and offer a constant pressure throughout the piston stroke. A weight-loaded accumulator may cause excessive pressure surges in a system when the accumulator is quickly discharged and suddenly stopped because of the constant pressure of the heavy weights.

A *gas charged accumulator* is an accumulator that uses compressed gas over hydraulic fluid to store energy. Dry nitrogen is used as the gas in hydro-pneumatic accumulators because air-oil vapors are explosive. Gas charged accumulators are divided into categories according to the method used to separate the gas from the fluid. The methods include piston, diaphragm, and bladder.

A *piston gas charged accumulator* is an accumulator with a floating piston acting as a barrier between the gas and fluid. Gas occupies the volume of space above the piston and is pressurized as system fluid enters and occupies the space below the piston. The gas pressure equals the system pressure when the accumulator is pressurized.

A *diaphragm gas charged accumulator* is an accumulator with a flexible diaphragm separating the gas and fluid. Diaphragm gas charged accumulators are generally small and lightweight with capacities up to 1 gal. They are constructed of two steel hemispheres bolted together with a flexible, dish-shaped, rubber diaphragm clamped between them. The top half of the sphere is pressurized (precharged) with gas. Fluid supplied by system pressure is applied to the bottom hemisphere, compressing the gas. The gas acts as a spring against the diaphragm as fluid pressures equalize the pressure in each hemisphere.

A *bladder gas charged accumulator* is an accumulator consisting of a seamless steel shell, a rubber bladder (bag) with a gas valve, and a poppet valve. The steel shell is cylindrical in shape and rounded at both ends. A large opening at the bottom of the shell is used to insert the bladder. A small opening at the top of the shell is used for the bladder's gas valve. The one-piece, pear-shaped bladder is molded of synthetic rubber. It includes a molded gas valve, which is fastened to the inside upper end of the steel shell by a locknut. The bottom end of the shell is equipped with a poppet valve at the discharge port. This valve closes off the port when the accumulator is fully discharged, preventing the bladder from being squeezed into the discharge port opening.

Accumulators are precharged while empty of hydraulic fluid. *Precharge pressure* is the pressure of the compressed gas in an accumulator prior to the admission of hydraulic fluid. The higher the precharge pressure, the less fluid the accumulator can hold. Precharge pressures vary with each application. Typically, an application consists of the circuit pressure range and the volume of fluid required in that range. Precharge pressures should never be less than ⅓ of the maximum circuit pressure.

Atlas Technologies, Inc.

Hydraulic presses used in manufacturing stamping plants contain accumulators for the storage of pressurized hydraulic fluid and to absorb circuit shock.

HYDRAULIC CIRCUIT MAINTENANCE

Special precautions must be followed when servicing powered equipment. Every energy source must be identified, understood, and disabled prior to working on a machine. All energy is categorized as kinetic or potential. *Kinetic energy* is the energy of motion. For example, a saw blade or grinding wheel in operation is kinetic energy. *Potential energy* is stored energy a body has due to its position, chemical state, or condition. For example, hydraulic accumulators, raised loads, or equipment counterweights are potential energy.

Controlling an energy source that is potentially dangerous is accomplished by using an energy-isolating device. An *energy-isolating device* is a device that prevents the transmission or release of energy. An energy source may be electrical, mechanical, hydraulic, pneumatic, chemical, nuclear, thermal, or other energy source that could cause injury to personnel.

Warning: The heat and pressures of hydraulic circuits can cause severe burns and injury.

Energy-isolating devices include manually-operated electrical circuit breakers, disconnect switches, slip blinds, line valves, blocks, and similar devices that indicate the position of a device. Pushbuttons, selector switches, and other circuit control devices are not energy-isolating devices. The four basic steps to controlling hazardous energy in a hydraulic system are:

1. Prepare to completely disable the system through identification of all energy sources.
2. Isolate the equipment by turning OFF all switches (including main disconnects), closing all necessary valves, and disconnecting, capping, or blocking auxiliary energy sources such as fluid product lines, steam lines, or pneumatic lines.
3. Apply lockout/tagout/blockout devices to energy-isolating devices.
4. Verify that all necessary energy sources are isolated and locked. This is accomplished by attempting machine startup. All area personnel must be warned of the startup attempt and cleared to safety. Return switches and controls to their OFF position after verification has been satisfied.

In many hydraulic circuit repair operations, the machinery requires testing and must be energized before other work is performed. This can be safely accomplished by using five basic steps:

1. Remove personnel to safety.
2. Remove tools and materials from the machinery.
3. Remove necessary energy-isolation devices and re-energize machine functions following complete safety procedures.
4. Perform test or tryout.
5. Return all energy sources to isolation, including the purging of necessary systems. Replace locks and tags as required.

Cleanliness is the key to successful troubleshooting and repair of hydraulic circuits. Steps that must be taken to maintain a clean system include:

- Adhere to a preventive maintenance program for fluid, filters, and strainers.
- Use only clean equipment to prevent contamination when replacing or changing hydraulic fluid.
- Clean and cover fluid containers and store in a clean, dry area when a maintenance project is complete.
- Replace packing or seals before obvious replacement is necessary.
- Maintain clean hydraulic equipment. Keep dust and dirt buildup down by removing fluids from equipment surfaces.

Fluid Maintenance

Always follow the equipment manufacturer's recommendations when servicing hydraulic equipment and circuits. Contaminants such as dirt, sand, scale, and other foreign materials are often found in new circuits when installed. These contaminants must be flushed from the circuit after an initial break-in period. Also, check the manufacturer's recommendations as to whether cylinders should be extended during a fluid change.

Remove the fluid according to the equipment manufacturer's recommendations. Disconnect any hoses that need draining. Remove as much fluid from the circuit as possible. Any leftover contaminants could cause early equipment failure. Shift levers to release trapped fluids. Clean all fitting surfaces with a dry, lint-free cloth before reassembling. Clean all exterior surfaces with a cleaning solvent after plugging any breather or dipstick holes to prevent the introduction of contaminants to the interior of the reservoir. Replace or clean any filters or breathers before refilling the reservoir. Run the equipment at idle for a few minutes to warm the system upon completion of fluid change. Operate the actuators until foaming stops, then add fluid to the full level. Finally, check for leaks.

Pneumatic Principles

Chapter 7

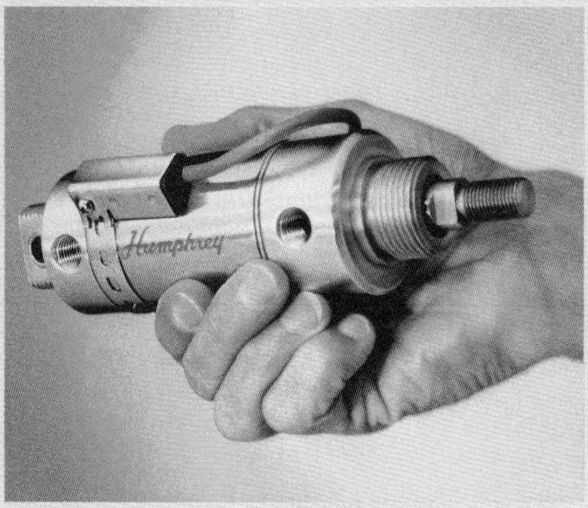

Humphrey Products Company

Pneumatics is the branch of science that deals with the transmission of energy using a gas. Pneumatic systems are based on fluid (gas) theory, which states that a fluid can flow, has no definite shape, and is susceptible to an increase in volume with an increase in temperature. The physical characteristics of gases are affected by pressure, volume, and temperature. Useful pneumatic pressure is produced by compressing atmospheric air and pushing it into a tank for later use. Contaminants of a compressed air system includes particulates, oil, and water. Contaminants must be removed to prevent damage to a compressed air system.

PNEUMATICS

Pneumatics is the branch of science that deals with the transmission of energy using a gas. A *pneumatic system* is a combination of components that controls energy through the use of a pressurized gas within an enclosed circuit. A pneumatic system uses gas, such as air or dry nitrogen, for power. Pneumatic systems use compressed (pressurized) air or other gas to transmit or control power.

Pneumatic System History

One of the earliest forms of pneumatics was the bellows. A *bellows* is a device that draws air in through a flapper valve when expanded and expels the air through a nozzle when contracted. A bellows consists of two wooden handles and a flexible leather cover. It was used to start or increase the heat of a fire through the use of compressed air. See Figure 7-1.

Today, pneumatic systems are used in the entertainment industry to operate air motors for ski lift drives and air cylinders for moving concert backdrops. In the medical field, doctors and dentists use non-sparking pneumatic drills and saws because of the flammability of oxygen, a widely used gas. Also, car washes use air motors instead of electric motors for rotating scrub brushes to prevent electrical shorts due to the high water content within the working area.

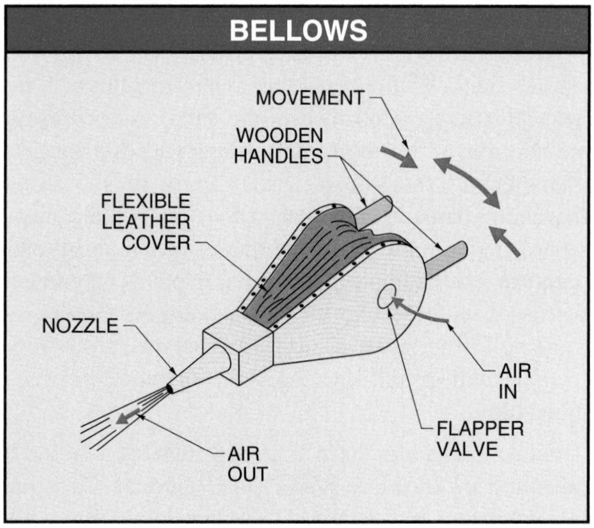

Figure 7-1. An early use of pneumatics was the bellows, which was used to start and increase the heat of a fire.

The manufacturing industry, however, is the largest user of pneumatic systems. Pneumatic systems in manufacturing include the use of air motors to move conveyors, rotate assembly fixtures, act as vibrators for parts feeding, rotate fluid mixing or agitating shafts, etc. Air cylinders are used to clamp and hold parts for machining or assembly, close or open safety guards, operate cutting blades, inject ink in a printer, etc.

Pneumatic systems have replaced many hydraulic systems because pneumatic systems are cleaner. Pneumatic logic systems have been used for electronic circuit replacement because overheating is seldom a problem in pneumatic circuits. In addition, pneumatic circuits are not likely to cause an explosion in hazardous environments. Pneumatic systems are based on fluid (gas) theory, which states that a fluid can flow, has no definite shape, and is susceptible to an increase in volume with an increase in temperature. Unlike molecules of a solid or liquid, air molecules are not freely attracted to each other and are easily compressible.

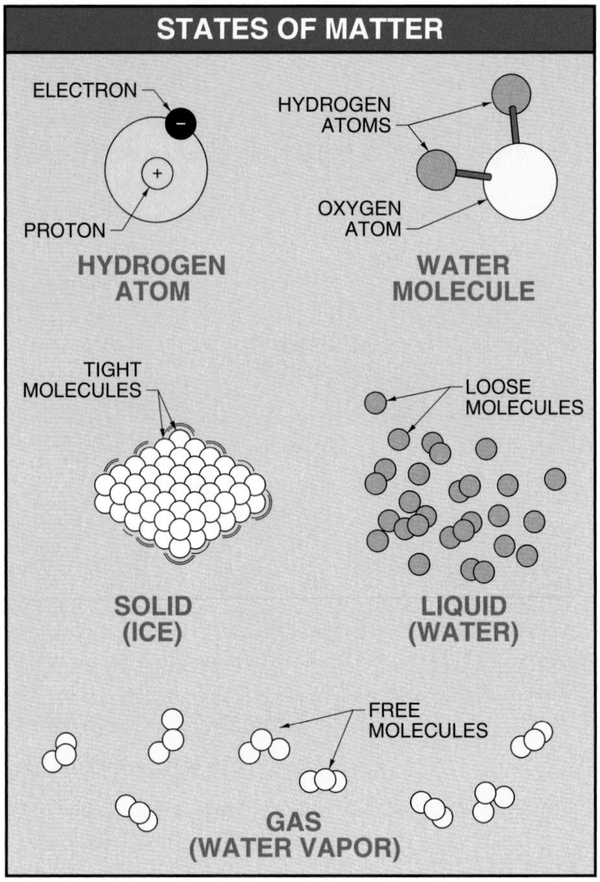

Figure 7-2. The closeness of the molecules that make up a substance determines whether the substance is a solid, liquid, or gas.

GAS CHARACTERISTICS

All substances are made up of atoms. An *atom* is the smallest building block of matter that cannot be divided into smaller units without changing its basic character. Atoms combine to form molecules. For example, an oxygen molecule (O_2) is made up of two atoms of oxygen. Water (H_2O) is made up of two atoms of hydrogen and one atom of oxygen.

Molecules are always in motion, and this motion creates heat. Without molecular motion, there is no heat. Heating a substance is the same as increasing the motion of its molecules. Increasing the motion of molecules sends them farther apart, thus creating thermal expansion. *Thermal expansion* is the dimensional change of a substance due to a change in temperature. For example, a metal bar expands when heat increases its molecular motion, causing its molecules to move farther apart. Cooling a substance slows molecular motion and decreases the distance between molecules.

Some molecules have a strong attraction to each other and are known as solids. Others have an attraction to each other but require some freedom. These are known as liquids. Molecules that move quickly and freely are known as gases. See Figure 7-2.

Although gas molecules are far apart, they can be pushed closer together, allowing gas to be compressed. Also, because gas molecules are always moving, they are constantly bumping each other and the walls of the container they occupy. This motion allows gas to expand to fill the volume and shape of its container. A balloon remains inflated because of the constant impact of gas molecules striking the balloon wall. When a gas is contained, any force applied to that gas is transmitted equally throughout its container.

Pressure

Pressure is the force per unit area. Pressure increases as gas molecules are forced closer together through compression. Air is a mixture of gases that has weight and creates a pressure on Earth's surface through compression. The pressure exerted on Earth's surface varies with altitude, temperature, and humidity.

The force acting on a unit area is generally a unit of weight. For example, the weight of the air molecules that make up the atmosphere is measured from a 1 sq in. column and is referred to as atmospheric pressure. *Atmospheric pressure* is the force exerted by the weight of the atmosphere on Earth's surface. Atmospheric pressure at sea level is equal to about 14.7 pounds per square inch (psi). At higher altitudes, there is less air, so the atmosphere exerts less weight on each square inch of Earth's surface. Atmospheric pressure decreases at higher altitudes. The pressure (weight) of air in Denver, Colorado, (5280′ above sea level) is 12.2 psi and the pressure of air on top of Mt. Everest (29,002′ above sea level) is approximately 5 psi. See Figure 7-3.

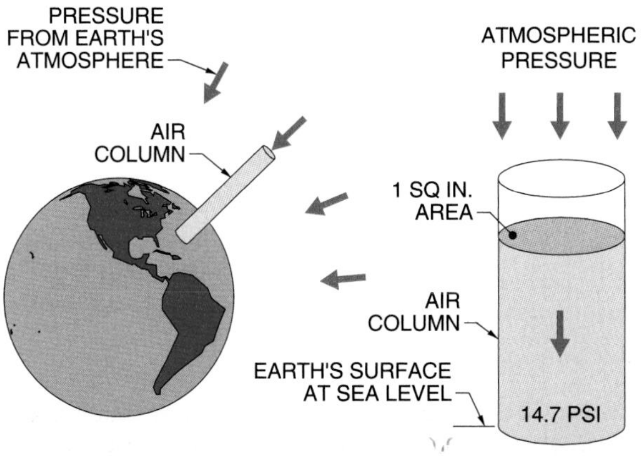

ATMOSPHERIC PRESSURE		
Pressure*	Location	Elevation**
5.0	Mt. Everest	29,002
11.1	Mexico City	7556
12.2	Denver, CO	5280
14.7	Sea Level	0

* in psi
** in ft

Figure 7-3. Atmospheric pressure at sea level is equal to about 14.7 pounds per square inch and decreases at higher altitudes.

The total pressure of air that is compressed is equal to the pounds per square inch times the number of square inches being acted on by the pressure. For example, air applying 1 psi on a 5 sq in. surface exerts a total force of 5 lb. Another example of an exerted pressure on a unit of area is that of the weight of a solid. For example, a 1 lb block sitting on 1 sq in. applies a force of 1 lb per square inch (1 psi).

The total force of a compressed gas is the pounds per square inch times the number of square inches being acted on by the pressure. For example, if a 100 lb force is applied to an area of 5 sq in., the resulting pressure is 20 psi.

Pascal's law states that pressure applied to a confined static fluid (liquid or gas) is transmitted with equal intensity throughout the fluid. This means that because a fluid takes the shape of its container and is compressed (under pressure), pressure is the same at any point, regardless of the size or shape of the container.

A slight change in pressure at any point of the container is instantly transmitted throughout the container. For example, a 50′ long pressurized hose becomes a gas station announcing bell when it is plugged at one end, connected to a bell at the other, and laid across a driveway. As the weight of a car drives over the hose, pressure is increased and is instantly transmitted to activate the bell. Pneumatic forces can also be transmitted over considerable distances with little loss. For example, a mile-long freight train uses one common pneumatic system to operate the air brakes of each car.

Measuring Pressure. Pneumatic systems use gases to transmit and control energy to produce mechanical work. The power of a pneumatic system is pressure (psi). Normally, pressure (gauge pressure) is any pressure greater than atmospheric pressure. A *vacuum* is any pressure lower than atmospheric pressure.

Atmospheric pressure (pressure at sea level) is only a reference level and is used when determining more specific types of pressure measurements. The two most commonly used pressure measurements are gauge and absolute pressure.

Gauge pressure is the pressure above atmospheric pressure that is used to express pressures inside a closed system. A typical pressure gauge measures pressures above the surrounding atmosphere. Any pressure below 0 psig is a vacuum. Gauge pressure is expressed in pounds per square inch gauge (psig) or psi.

Liquids and gases expand when heated. Any excessive pressure within a system creates undue strain on seals and packing. The amount of gas expansion can be calculated by applying Charles' law.

A pressure gauge reads 0 psig at normal atmospheric pressure. For example, a simple plunger pressure gauge used to check tire pressures indicates 0 when not being used even though its plunger has 14.7 psi on both sides (at sea level). When a pressure of 30 lb is indicated when checking an inflated tire with a pressure gauge, the actual pressure within the tire is equal to 44.7 psia (gauge pressure reading plus atmospheric pressure). Gauge pressure shows the numerical value of the difference between atmospheric pressure and absolute pressure. See Figure 7-4.

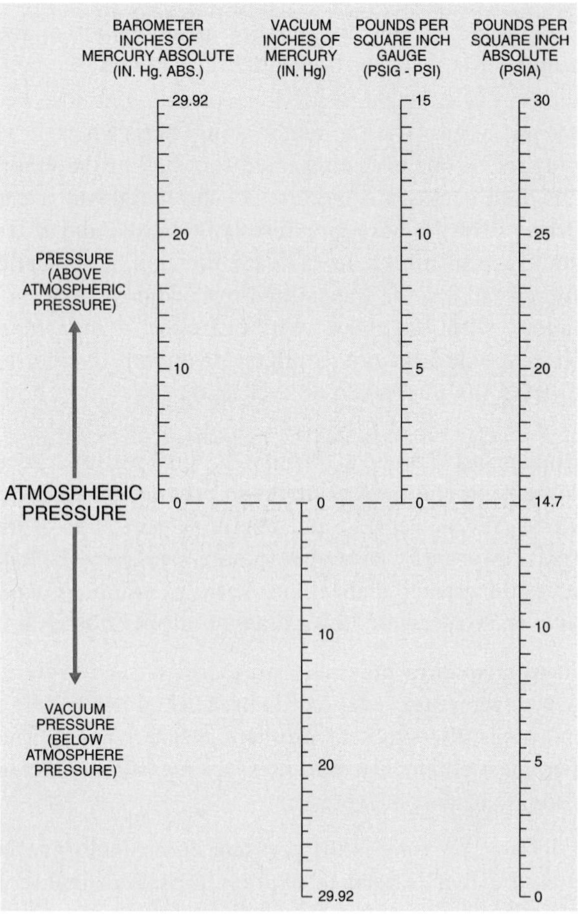

Figure 7-4. Gauge pressure shows the numerical value of the difference between atmospheric pressure and absolute pressure.

Absolute pressure is pressure above a perfect vacuum. Absolute pressure is the sum of gauge pressure plus atmospheric pressure. Absolute pressure is expressed in pounds per square inch absolute (psia).

Pressure outside a closed system (such as normal air pressure) is expressed in pounds per square inch absolute. The difference between gauge pressure and absolute pressure is the pressure of the atmosphere at sea level at standard conditions (14.7 psia).

Industrial pneumatic systems are designed to operate with minimal frictional resistance and working pressures of around 90 psi.

Volume

Any substance consisting of molecules, such as solids, liquids, or gases, is considered to have volume. *Volume* is the three-dimensional size of an object measured in cubic units. Liquids and solids have definite volumes that do not vary significantly when compressed. Gases have indefinite volumes that vary significantly when compressed.

Volume values in other units of measure must be converted to cubic units for calculations, etc. A conversion table is used to determine the conversion constant required to convert one unit of measure to another. See Appendix. For example, the volume of a 2000 gal. (capacity) vessel is divided by the conversion constant of 7.48 to convert gallons to cubic feet. The 2000 gal. vessel has a volume of 267.4 cu ft (2000 ÷ 7.48 = 267.4). To convert cubic feet to gallons, multiply cubic feet by .1337. For example, a 100 cu ft vessel can hold 747.9 gal. (100 × .1337 = 747.9 gal.). The volume of a cube measuring 1′ × 1′ × 1′ is 1 cu ft. A cubic foot of air may be contained in a cube, cylinder, or any other shaped vessel. The actual volume of air in a container is the same as the volume of the space within the container. See Figure 7-5.

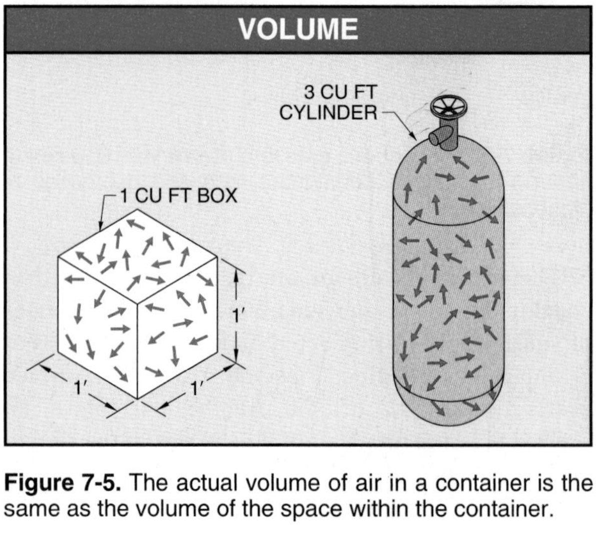

Figure 7-5. The actual volume of air in a container is the same as the volume of the space within the container.

Gas Laws

The physical characteristics of gases are affected by pressure, volume, and temperature. The relationships between each property are established as gas law equations. *Gas laws* are the relationships between the volume, pressure, and temperature of a gas. Gas laws are used to determine the change in volume, pressure, or temperature of a gas. The behavior of a gas is affected by the characteristics of the gas and the interaction between these characteristics. Gas law equations assume the gas to be under perfect (ideal) conditions.

Boyle's Law. Boyle's law is concerned with the compression of gas. Boyle's law states that the volume of a given quantity of gas varies inversely with the pressure as long as the temperature remains constant. Two vessels or cylinders of the same size may contain air of different pressures because air is compressible. When the volume of a gas is reduced by one half, the pressure of the gas doubles. When air is forced to occupy a smaller volume its pressure increases. See Figure 7-6.

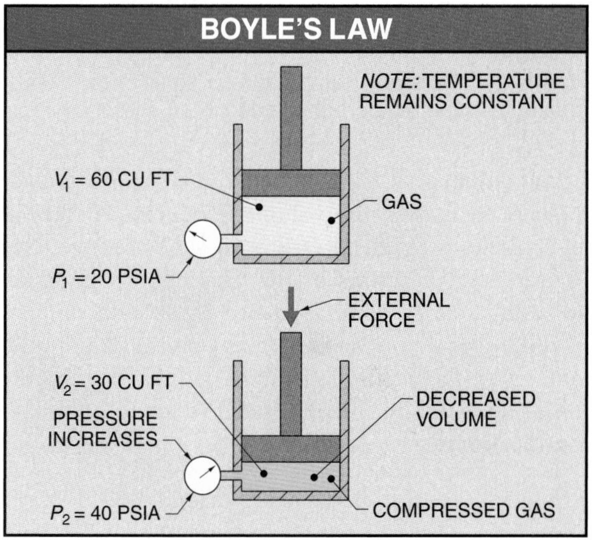

Figure 7-6. The volume of a given quantity of gas varies inversely with the pressure as long as the temperature remains constant.

Boyle's law is used when temperature changes are not a factor, such as in determining the consumption of air by an air cylinder or determining the volume/pressure capacity of a vessel. The final pressure is determined by multiplying the initial pressure by the initial volume of the gas and dividing by the final volume. For calculating purposes, pressures must be absolute values or converted from gauge pressure to absolute pressure at sea level. Final pressure is determined by applying the formula:

$$P_2 = \frac{P_1 \times V_1}{V_2}$$

where

P_2 = final pressure (in psia)
P_1 = initial pressure (in psia)
V_1 = initial volume (in cubic units)
V_2 = final volume (in cubic units)

For example, what is the final pressure of 60 cu ft of air at 20 psia when compressed to 30 cu ft?

$$P_2 = \frac{P_1 \times V_1}{V_2}$$

$$P_2 = \frac{20 \times 60}{30}$$

$$P_2 = \frac{1200}{30}$$

P_2 = **40 psia = 25.3 psi**

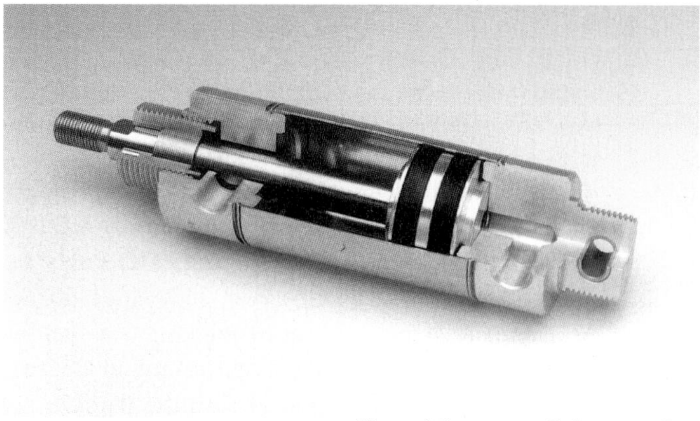

Clippard Instrument Laboratory, Inc.

A cylinder rod retracts when a gas (air) at an increased pressure fills the volume in the cylinder around the piston rod.

In addition, when the initial and final pressures are known along with the initial volume, the final volume may be determined by multiplying the initial pressure by the initial volume and dividing by the final pressure. Final volume is determined by applying the formula:

$$V_2 = \frac{P_1 \times V_1}{P_2}$$

where

V_2 = final volume (in cubic units)
P_1 = initial pressure (in psia)
V_1 = initial volume (in cubic units)
P_2 = final pressure (in psia)

For example, what is the final volume of 60 cu ft of air at 20 psia when compressed to 40 psia?

$$V_2 = \frac{P_1 \times V_1}{P_2}$$

$$V_2 = \frac{20 \times 60}{40}$$

$$V_2 = \frac{1200}{40}$$

V_2 = **30 cu ft**

In a pneumatic system, energy is stored and distributed in a potential state (compressed air). Useful work results from a pneumatic system when the compressed air is allowed to convert its potential energy into kinetic energy.

Charles' Law. Boyle's law remains practical as long as temperatures do not change. However, temperatures increase when air is compressed. Charles' law states that the volume of a given mass of gas is directly proportional to its absolute temperature provided the pressure remains constant. *Absolute temperature* is the temperature on a scale that begins with absolute zero. *Absolute zero* is the temperature at which substances possess no heat.

The absolute temperature scale was first determined in 1872 by William Rankine, and is referred to as the Rankine scale (°R or °abs). Molecules are still in motion at 0°F. However, molecules do not move at 0°R. A comparison of Rankine and Fahrenheit scales shows that the temperature in degrees Rankine is always 460° greater than the temperature in degrees Fahrenheit. See Figure 7-7.

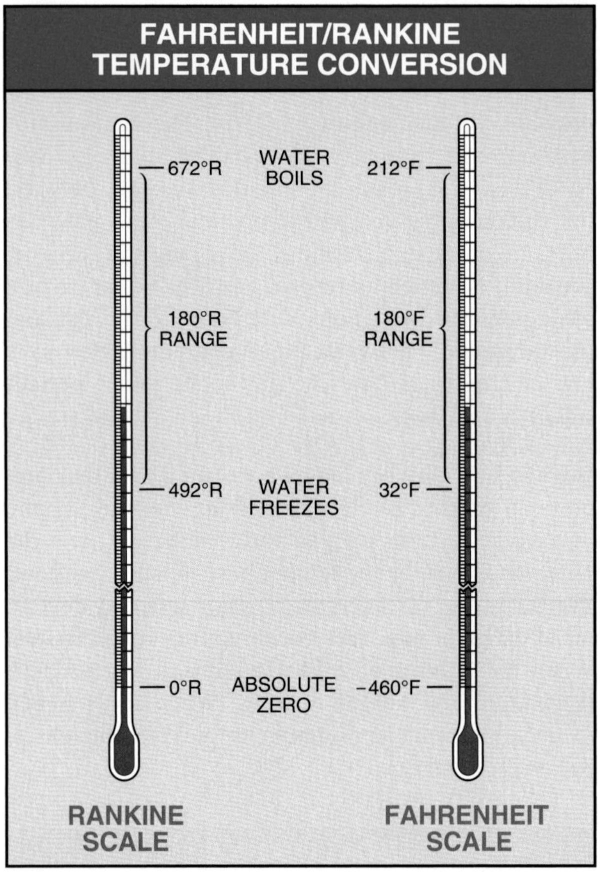

Figure 7-7. Absolute temperature (°R) is always 460° greater than the temperature in degrees Fahrenheit (°F).

Degrees Fahrenheit is converted to degrees Rankine by adding 460° to the Fahrenheit temperature. Degrees Fahrenheit is converted to degrees Rankine by applying the formula:

$$°R = 460 + °F$$

where

$°R$ = degrees Rankine

460 = constant

$°F$ = degrees Fahrenheit

For example, what is the Rankine equivalent of 96°F?

$$°R = 460 + °F$$

$$°R = 460 + 96$$

$°R$ = **556°R**

According to Charles' law, if the temperature of a gas is increased, the volume increases proportionately as long as the pressure does not change. See Figure 7-8. Variations of the Charles' law equation are used to determine a change in volume or temperature. Final volume is found by multiplying initial volume by the final temperature and dividing by the initial temperature. Final volume is found by applying the formula:

$$V_2 = \frac{V_1 \times T_2}{T_1}$$

where

V_2 = final volume (in cubic units)

V_1 = initial volume (in cubic units)

T_2 = final temperature (in °R)

T_1 = initial temperature (in °R)

For example, what is the final volume of a gas that occupies 40 cu ft at 60°F when the temperature is increased to 90°F?

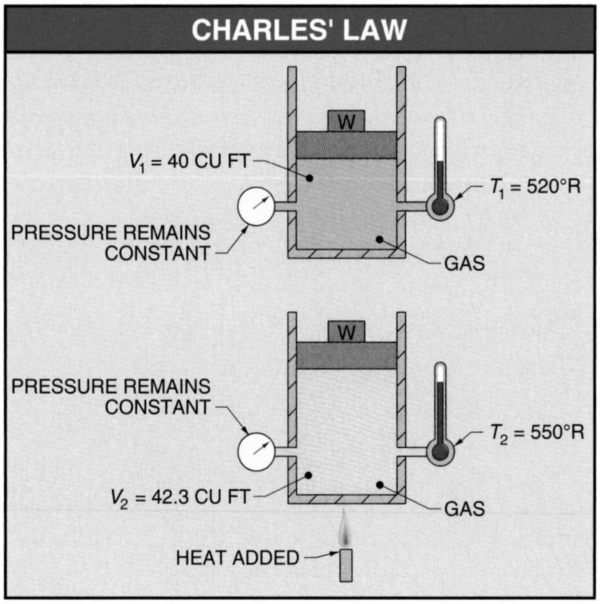

Figure 7-8. The volume of a given mass of gas is directly proportional to its absolute temperature provided the pressure remains constant.

1. Convert initial temperature to °R.

   $°R = 460 + °F$

   $°R = 460 + 60$

   $°R = 520°R$

2. Convert final temperature to °R.

   $°R = 460 + °F$

   $°R = 460 + 90$

   $°R = 550°R$

3. Calculate final volume.

   $$V_2 = \frac{V_1 \times T_2}{T_1}$$

   $$V_2 = \frac{40 \times 550}{520}$$

   $$V_2 = \frac{22{,}000}{520}$$

   $V_2 =$ **42.3 cu ft**

In addition, final temperature is found by multiplying the final volume by the initial temperature and dividing by the initial volume. Final temperature is found by applying the formula:

$$T_2 = \frac{V_2 \times T_1}{V_1}$$

where

T_2 = final temperature (in °R)

V_2 = final volume (in cubic units)

T_1 = initial temperature (in °R)

V_1 = initial volume (in cubic units)

For example, what is the final temperature of 10 cu ft of gas at 492°R that is increased to 14 cu ft?

$$T_2 = \frac{V_2 \times T_1}{V_1}$$

$$T_2 = \frac{14 \times 492}{10}$$

$$T_2 = \frac{6888}{10}$$

$T_2 =$ **688.8°R = 228.8°F**

Gay-Lussac's Law. Gay-Lussac's law states that if the volume of a given gas is held constant, the pressure exerted by the gas is directly proportional to its absolute temperature. See Figure 7-9. Gay-Lussac's law is used to determine the pressure based on an increase in temperature. Final pressure is determined by multiplying the initial pressure by the final temperature and dividing by the initial temperature. Final pressure is found by applying the formula:

$$P_2 = \frac{P_1 \times T_2}{T_1}$$

where

P_2 = final pressure (in psia)

P_1 = initial pressure (in psia)

T_2 = final temperature (in °R)

T_1 = initial temperature (in °R)

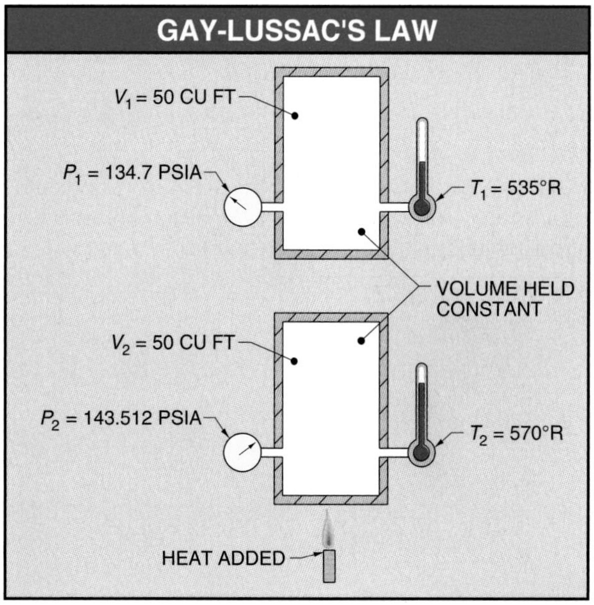

Figure 7-9. The pressure exerted by a gas is directly proportional to its absolute temperature if the volume of the gas is held constant.

Advanced Assembly Automation Inc.

Automated industrial systems use pneumatics for applications such as pneumatic wrenches, air clamps, ejection molding machines, and automatic gates and doors.

For example, what is the final pressure in a 50 cu ft vessel holding air at 120 psig at 75°F if the temperature is increased to 110°F?

1. Convert initial pressure to absolute pressure.

   $psia = 14.7 + psig$

   $psia = 14.7 + 120$

   $psia = 134.7$ psia

2. Convert initial temperature to °R.

   $°R = 460 + °F$

   $°R = 460 + 75$

   $°R = 535°R$

3. Convert final temperature to °R.

   $°R = 460 + °F$

   $°R = 460 + 110$

   $°R = 570°R$

4. Calculate final pressure.

   $$P_2 = \frac{P_1 \times T_2}{T_1}$$

   $$P_2 = \frac{134.7 \times 570}{535}$$

   $$P_2 = \frac{76{,}779}{535}$$

   $P_2 =$ **143.512 psia = 128.812 psi**

Combined Gas Law. The relationship between pressure, volume, and temperature is best determined when Boyle's, Charles', and Gay-Lussac's laws are combined. The combined gas law covers all the variables regarding the relationships between the pressure, volume, and temperature of a gas. Final pressure is found by applying the formula:

$$P_2 = \frac{P_1 \times V_1}{T_1} \times \frac{T_2}{V_2}$$

where

P_2 = final pressure (in psia)

P_1 = initial pressure (in psia)

V_1 = initial volume (in cubic units)

T_1 = initial temperature (in °R)

T_2 = final temperature (in °R)

V_2 = final volume (in cubic units)

For example, what is the final pressure in a 40 cu in. cylinder containing air at 100 psig and 100°F when compressed to 20 cu in. while heated to 150°F?

1. Convert initial gauge pressure to absolute pressure.

   $psia = 14.7 + psig$

   $psia = 14.7 + 100$

   $psia = 114.7$ psia

2. Convert initial temperature to °R.

   $°R = 460 + °F$

   $°R = 460 + 150$

   $°R = 610°R$

3. Convert final temperature to °R.

   $°R = 460 + °F$

   $°R = 460 + 150$

   $°R = 610°R$

4. Calculate final pressure.

   $$P_2 = \frac{P_1 \times V_1}{T_1} \times \frac{T_2}{V_2}$$

   $$P_2 = \frac{114.7 \times 40}{560} \times \frac{610}{20}$$

   $$P_2 = \frac{4588}{560} \times 30.5$$

   $P_2 = 8.193 \times 30.5$

   $P_2 =$ **249.886 psia = 235.186 psi**

The combined gas law equation may be rearranged to determine any value when the others are known. Final volume is found by applying the formula:

$$V_2 = \frac{P_1 \times V_1}{T_1} \times \frac{T_2}{P_2}$$

where

V_2 = final volume (in cubic units)

P_1 = initial pressure (in psia)

V_1 = initial volume (in cubic units)

T_1 = initial temperature (in °R)

T_2 = final temperature (in °R)

P_2 = final pressure (in psia)

For example, what is the final volume of air if 40 cu ft of air at 12 psia is compressed to 42 psia and the compressor suction temperature is 60°F and the discharge temperature is 180°F?

1. Convert initial temperature to °R.

   $°R = 460 + °F$

   $°R = 460 + 60$

   $°R = 520°R$

2. Convert final temperature to °R.

   $°R = 460 + °F$

   $°R = 460 + 180$

   $°R = 640°R$

3. Calculate final volume.

   $$V_2 = \frac{P_1 \times V_1}{T_1} \times \frac{T_2}{P_2}$$

   $$V_2 = \frac{12 \times 40}{520} \times \frac{640}{42}$$

   $$V_2 = \frac{480}{520} \times 15.238$$

   $V_2 = .923 \times 15.238$

   $V_2 =$ **14.065 cu ft**

Final temperature may be found when initial and final pressure and volume and initial temperature are known. Final temperature is found by applying the formula:

$$T_2 = \frac{P_2 \times V_2}{P_1 \times V_1} \times T_1$$

where

T_2 = final temperature (in °R)

P_2 = final pressure (in psia)

V_2 = final volume (in cubic units)

P_1 = initial pressure (in psia)

V_1 = initial volume (in cubic units)

T_1 = initial temperature (in °R)

For example, what is the final temperature of 80 cu ft of air at 65°F and 14.7 psia when compressed to 40 cu ft at 40 psia?

1. Convert initial temperature to °R.

   $°R = 460 + °F$

   $°R = 460 + 65$

   $°R = 525°R$

2. Calculate final temperature.

   $$T_2 = \frac{P_2 \times V_2}{P_1 \times V_1} \times T_1$$

   $$T_2 = \frac{40 \times 40}{14.7 \times 80} \times 525$$

   $$T_2 = \frac{1600}{1176} \times 525$$

   $T_2 = 1.36 \times 525$

   $T_2 =$ **714°R**

Saylor-Beall Manufacturing Company

Air compressors may be powered by electric motors or gasoline engines for use in a variety of locations.

COMPRESSION

Useful pneumatic pressure is produced by first compressing quantities of atmospheric air and then pushing it into a tank (receiver) for later use. As a piston retracts, it draws air (suction) through the inlet port at one pressure. As the piston extends, it decreases (compresses) the volume of air, which increases the pressure of the discharged air.

The difference between the suction (inlet) pressure and the discharge pressure is the ratio of compression (R_c). The ratio of compression is 3 when a compressor triples the absolute inlet pressure. Ratio of compression is calculated by dividing the absolute discharge pressure by the absolute inlet pressure. Ratio of compression is found by applying the formula:

$$R_c = \frac{P_2}{P_1}$$

where

R_c = ratio of compression

P_2 = final pressure (in psia)

P_1 = initial pressure (in psia)

For example, what is the ratio of compression if a compressor inlet pressure is 1.5 psi vacuum and the discharge pressure is 40 psig?

1. Convert final pressure to absolute.

$$psia = 14.7 + psig$$
$$psia = 14.7 + 40$$
$$psia = 54.7 \text{ psia}$$

2. Convert initial pressure to absolute.

$$psia = 14.7 + psig$$
$$psia = 14.7 - 1.5$$
$$psia = 13.2 \text{ psia}$$

3. Calculate ratio of compression.

$$R_c = \frac{P_2}{P_1}$$
$$R_c = \frac{54.7}{13.2}$$
$$R_c = \mathbf{4.14}$$

Air temperature increases as a piston extends and the air molecules are forced closer together (compressed). This happens because as the compressor forces gas molecules closer together, it also increases the collisions of the molecules, thus increasing their temperature. See Figure 7-10. The temperature increase is based on the inlet temperature and the ratio of compression. Increasing the ratio of compression increases the temperature of the discharged air.

The amount of energy required to compress a quantity of air to a given pressure depends on the rate that heat is dissipated. Higher compression temperatures require greater amounts of energy (horsepower) to accomplish the same task as those having lower temperatures. Greater horsepower is required to meet output demands if discharge temperatures are excessive. The inlet air temperature or ratio of compression may be reduced to reduce the need for greater horsepower and to reduce the discharge temperature. The ratio of compression can be decreased by reducing the discharge pressure, increasing the inlet pressure, or both.

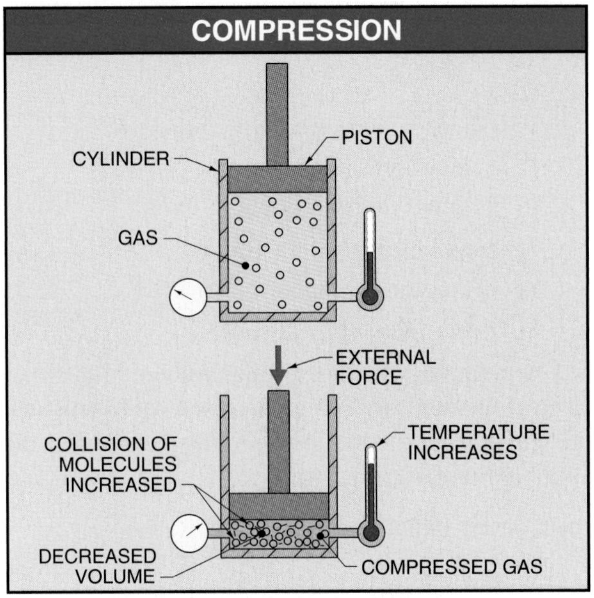

Figure 7-10. Compressing gas molecules increases the collisions of the molecules, thus increasing their temperature.

Heat generated during compression is controlled by limiting the pressure of the compressed air or by cooling. Lowering the suction temperature increases the standard volume of air processed and decreases the discharge temperature if the ratio of compression value and inlet pressure remain the same.

Multistage Compression

Air must be compressed in two or more steps (stages) for a reciprocating compressor to obtain pressures over 100 psi. A *reciprocating compressor* is a device that compresses gas by means of a piston(s) that moves back and forth in a cylinder. In multistage compression, the air moves from one pumping chamber to another with each chamber receiving and discharging a higher pressure. See Figure 7-11.

Multistage compression is required when the ratio of compression is greater than 6. When a compressing unit is required to have a total ratio of compression of 18, compression is usually accomplished in three stages. For example, the upper limit of a two-stage compressor at sea level with each stage at a ratio of compression of 4 is 220.5 psi.

Compressed air shall not be used for cleaning purposes except where reduced to less than 30 psi and then only with effective chip guarding and personal protective equipment.

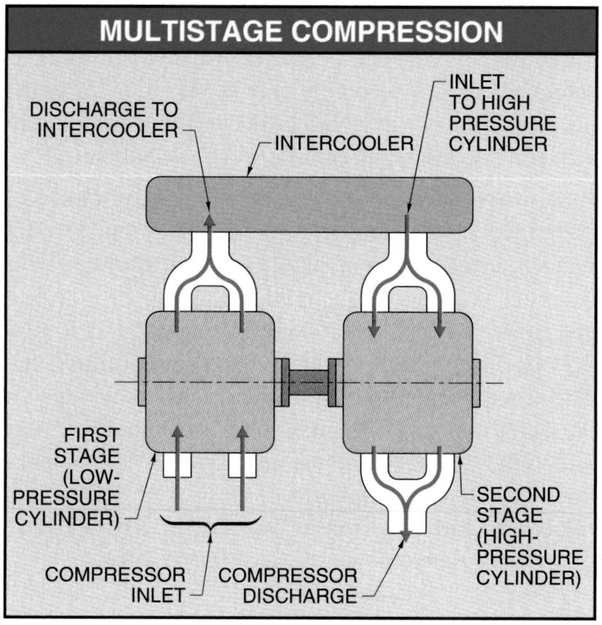

Figure 7-11. In multistage compression, the air moves from one pumping chamber to another with each chamber receiving and discharging a higher pressure.

Compression Intercooling

Heat generated in a cylinder due to compression of a reciprocating compressor is removed by lubrication, conduction through the cylinder walls, and intercooling. *Intercooling* is the process of removing a portion of the heat of compression as the air is fed from one compression stage to another. Intercooling is generally performed by a tube and shell heat exchanger (intercooler). An *intercooler* is a piped connection between the discharge of one compression stage and the inlet of the next compression stage. A tube and shell intercooler uses cool water to dissipate heat. See Figure 7-12.

Intercooling is used to reduce the temperature of the compressed air and reduce the volume of the air before it reaches the next stage. Also, for every 5°F absorbed at the intercooler, approximately 1% of horsepower is saved. Without intercooling, multistage compression does not allow an overall reduction in discharge temperature.

In a gas, molecules are continuously moving. A gas with increased temperature has faster moving molecules than a gas that is cool. The amount of molecule collisions determine the temperature of the gas.

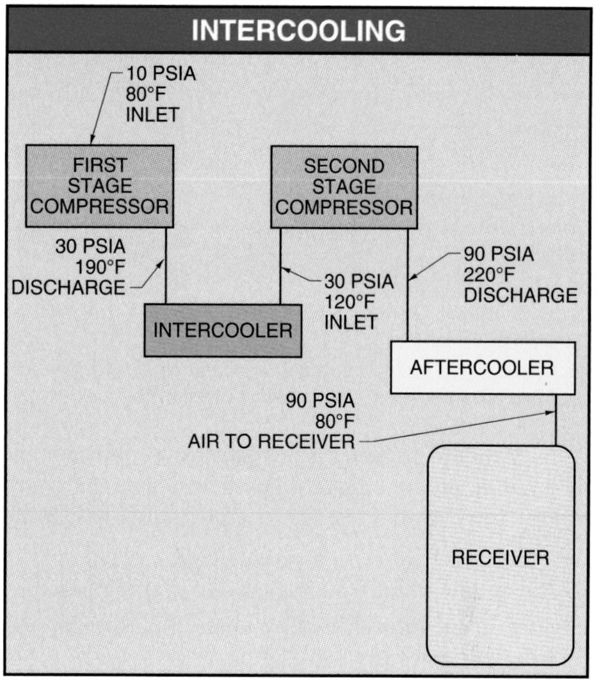

Figure 7-12. Intercooling removes a portion of the heat of compression as air is fed from one compression stage to another.

AIR TEMPERATURE AND MOISTURE CONTENT

Atmospheric air contains water in some form or another. The water may be solid (snow or ice crystals), visible (clouds or fog), or invisible (water vapor). Invisible vapor makes up most of the moisture in atmospheric air. *Humidity* is the amount of moisture in the air. Humidity comes from water that has evaporated into the air.

Water in a pneumatic system is a natural occurrence because the air that a compressor takes in contains significant amounts of water from the atmosphere. It takes 7.8 cu ft of atmospheric air to produce 1.0 cu ft of compressed air at 100 psig. This means that after compression there is 7.8 times more moisture in the compressed air as compared to the atmospheric air.

Air generally contains less moisture than it is capable of holding. However, as the moisture level in air rises to its maximum, the air becomes saturated. *Saturated air* is air that holds as much moisture as it is capable of holding. The amount of moisture air is capable of holding is also greatly affected by the temperature of the air. The higher the air temperature, the greater the amount of moisture it is able to hold.

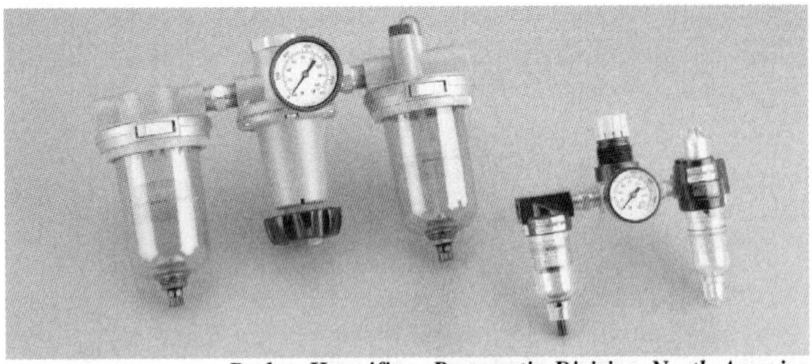

Parker Hannifin - Pneumatic Division North America

An FRL conditions air before it enters components to remove contaminants, control pressure changes, and provide constant oil-air mixture.

At standard pressure and temperature, the weight of 1 cu ft of air equals .076 lb. An average room (12′ × 13′ × 8′ = 1248 cu ft) contains 94.848 lb of air (1248 cu ft × .076 lb = 94.848 lb).

Air at dew point (saturation) and at a temperature of 80°F is capable of holding about .022 lb of moisture per pound of dry air. The air in an average room (1248 cu ft) at saturation is capable of holding 2.086 lb of water vapor (94.848 lb × .022 lb = 2.086 lb). Moisture in the air is more commonly expressed in grains. See Figure 7-13. There are 150 grains (gr) of moisture per pound of saturated air at 80°F. Therefore, an average room (1248 cu ft) contains 14,227.2 gr of moisture (150 gr × 94.848 lb = 14,227.2 gr).

SATURATED AIR MOISTURE HOLDING PROPERTIES			
Dew Point*	Pounds of Moisture**	Grains of Moisture**	Grains of Moisture†
100	.044	300	19.766
90	.03	210	14.790
80	.022	150	10.934
70	.015	110	7.980
60	.011	76	5.745
50	.0075	53	4.076
40	.005	36	2.749
30	.0033	24	1.935
20	.002	15	1.235
10	.0014	9	.776
0	.0008	5.5	.481

* in °F
** per lb of dry air
† per cu ft of dry air

Figure 7-13. The total amount of moisture that air is capable of holding varies based on the temperature of the air.

Air that is at 100°F is saturated when it contains 19.766 gr of moisture per cu ft. Air that is at 60°F is saturated when it contains only 5.745 gr of moisture per cu ft. This is why cooling the air at the compressor reduces the moisture content in the receiver.

Relative humidity is the percentage of moisture contained in air compared to the maximum amount of moisture (saturation) it is capable of holding. For example, if air at 80°F contains 4.374 gr of moisture per cu ft, the air is at 40% relative humidity, because 4.374 gr is only 40% of the 10.934 gr saturation level. The relative humidity of this air increases when it is cooled because the air is capable of holding less moisture, even though the total grains of moisture does not change. Similarly, if this air is heated, the relative humidity decreases because the air is capable of holding more moisture. See Figure 7-14.

AIR CONTAMINANTS

Contaminants of a compressed air system include particulates, oil, and water. Each contaminant damages a pneumatic system in a different way. A *particulate* is a fine solid particle which remains individually dispersed in a gas. Particulates in a system grind away metal surfaces, lodge between moving parts, or plug orifices. Oil is a contaminant when it is carried over from the compressor. Oil carried over from the compressor is too thick for air line conditioning. Thick compressor oil gets thicker as it travels through a system, picking up dust, dirt, and other solids. The result is plugged orifices or gummy, sticky moving parts. Water is a pneumatic system's most obvious contaminant. Water in a system blocks passageways, slows actuators, rusts metal components, and mixes with other contaminants.

Contaminant Removal

Removing particulates begins with a pneumatic system's intake filter. An *intake filter* is a filter that removes solids from the free air at the compressor inlet port. *Free air* is air at atmospheric pressure and ambient temperature. The condition of the free air at the compressor normally determines the type or method of intake filter used. Excessive moisture or corrosive gases may require either a purifying filter or extra piping to reach cleaner air from a remote area. Intake filters must be cleaned or replaced periodically.

MOISTURE HELD BY AIR*										
Temperature**	**Percent Saturation**									
	10	**20**	**30**	**40**	**50**	**60**	**70**	**80**	**90**	**100**
−10	.028	.057	.086	.114	.142	.171	.200	.228	.256	.285
0	.048	.096	.144	.192	.240	.289	.337	.385	.433	.481
10	.078	.155	.233	.210	.388	.466	.543	.621	.698	.776
20	.124	.247	.370	.494	.618	.741	.864	.988	1.112	1.235
30	.194	.387	.580	.774	.968	1.161	1.354	1.548	1.742	1.935
40	.285	.570	.855	1.140	1.424	1.709	1.994	2.279	2.564	2.749
50	.408	.815	1.223	1.630	2.038	2.446	2.853	3.261	3.668	4.076
60	.574	1.149	1.724	2.298	2.872	3.447	4.022	4.596	5.170	5.745
70	.798	1.596	2.394	2.192	3.990	5.788	5.586	6.384	7.182	7.980
80	1.093	2.187	3.280	4.374	5.467	6.560	7.654	8.747	9.841	10.934
90	1.479	2.958	4.437	5.916	7.395	8.874	10.353	11.832	13.311	14.790
100	1.977	3.953	5.930	7.906	9.883	11.860	13.836	15.813	17.789	19.766

* in grains
** in °F

Figure 7-14. The percentage of moisture contained in air compared to the maximum amount of moisture (saturation) it is capable of holding varies based on the temperature of the air.

Conditioning Compressed Air. Contaminant removal in a compressed air system is accomplished by conditioning the air. This requires the ambient air to be filtered, compressed, and cooled for moisture separation before it enters the receiver. Cooling the compressed air is accomplished by use of an aftercooler. An aftercooler is a heat exchanger that cools air that has been compressed. Aftercoolers generally use water as the cooling medium. Aftercoolers remove up to 80% of the moisture by means of condensation. See Figure 7-15.

Condensation is the change in state from a gas or vapor to a liquid. The ability of air to hold water vapor decreases when the air is cooled. The point at which water vapor begins to change to a liquid is referred to as dew point (dp). *Dew point* is the temperature to which air must be cooled in order for the moisture in the air to begin condensing. Dew point is also known as saturation temperature. An example of condensation is the moisture that condenses on a glass of water with ice in it. The outside surface of the glass lowers the temperature of the surrounding atmospheric air enough to cause the water vapor in the air to condense (turn to liquid) on the surface of the glass.

The greatest volume of water and oil is removed at the discharge of the aftercooler by means of a moisture separator. A *moisture separator* is a device that separates a large percentage of water from the cooled air through a series of plates or baffles. A moisture separator is located at the discharge port of an aftercooler. The plates or baffles are shaped and placed in such a way that air entering the separator is directed into a swirling motion. Centrifugal force causes large contaminants (mostly water) to strike the wall of the separator and flow to the bottom, where they are drained.

Conditioning Circuit Air. The remaining contaminants, such as oil carry-over, water vapor, and particulates, are removed in sequence to maximize the remaining removal equipment. Filters are required in each circuit to remove particulates or liquids that remain in the system or are introduced by the system. For example, particulates may be introduced from wear of moving metal components. Oil is introduced as oil carry-over. *Oil carry-over* is the released lubricating oil from the walls of the compressor cylinder and piston. Oil carry-over travels with the compressed air into the system as a vapor. As compressor cylinder walls or piston seals (rings) wear, more oil is carried over into the system.

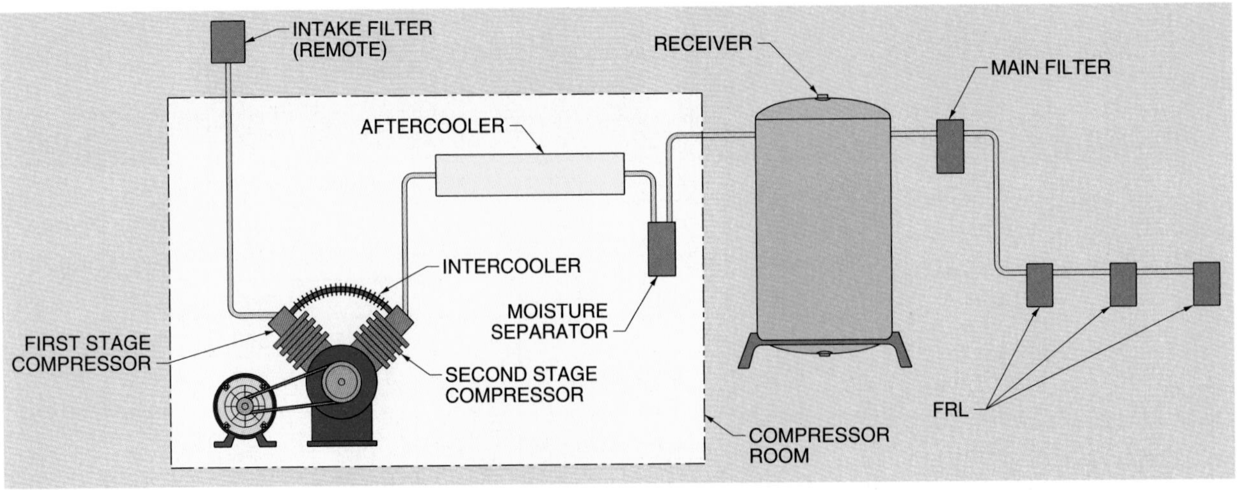

Figure 7-15. Compressor intake air is conditioned by removing particulates from the free air with an intake filter and removing 80% of its moisture with an aftercooler.

A basic pneumatic circuit contains a general purpose filter, regulator, and lubricator. The filter removes contaminants before the air enters the regulator. The regulator adjusts system pressure to the required circuit pressure. The lubricator releases or injects the proper protective lubricant into the circuit. These three components, when used in combination, are commonly referred to as an FRL (filter/regulator/lubricator).

Air in a pneumatic system must always be conditioned through filtering. The filtering components must be of the correct type and placed properly in the system to maximize their filtering functions. Component placement for conditioning compressed air requires removing contaminants in sequence. See Figure 7-16.

Dust and dirt are removed using a general purpose or particulate filter. These filters also remove some liquids, but their major function is to remove solids. General purpose filters remove condensed (liquid) water, oil, and solid particles as small as 5 microns (μ), or .0002″. General purpose filters allow most water and oil vapors, as well as submicron solid particles, to pass. Due to the significant oil, water, and particle collection during filtering, general purpose filter elements must be changed often. A general purpose filter is always used in a pneumatic system because it removes the solids that would otherwise clog a more expensive filter. Solid particles create an increase in flow resistance. For this reason, the life of a general purpose filter is determined by the quantity of solids retained, not by the amount of liquid visible.

The coarsest grade of filter element that satisfactorily protects the system should be used. This grade provides the longest element life without a sizable pressure drop. *Pressure drop* is the pressure differential between upstream and downstream fluid flow caused by resistance. Resistance creating a pressure drop may be caused by friction between the air and its plumbing, intentional creation of resistance by placing an orifice (restriction) in a system, or blockage of flow created by contaminants.

Pressure drops on new filters should not exceed 2 psi. In some cases, multiple filters can be used in series. The beginning filter having the coarsest element is followed by a filter with a finer element. The success of filter elements to remove solids at this stage determines the life (and expense) of any downstream liquid removing elements.

All air receivers shall be constructed in accordance with the* ASME Boiler and Pressure Vessel Code, Section VIII, Edition 1968 *and installed so that all drains are easily accessible. Under no circumstances shall an air receiver be buried underground or located in an inaccessible place.

Removing Oil And Water. Liquid-removing filters, such as coalescing filters, become plugged and ineffective prematurely if solids are allowed to continue downstream. A *coalescing filter* is a device that removes submicron solids and vapors of oil or water by uniting very small droplets into larger droplets.

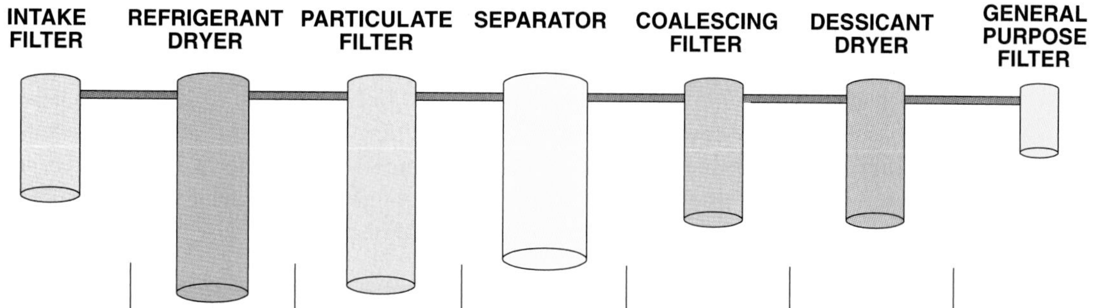

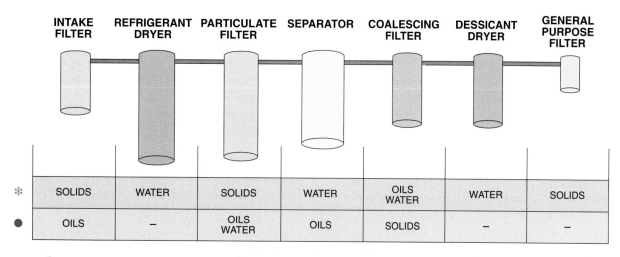

	INTAKE FILTER	REFRIGERANT DRYER	PARTICULATE FILTER	SEPARATOR	COALESCING FILTER	DESSICANT DRYER	GENERAL PURPOSE FILTER
*	SOLIDS	WATER	SOLIDS	WATER	OILS WATER	WATER	SOLIDS
●	OILS	–	OILS WATER	OILS	SOLIDS	–	–

* PRIMARY REMOVAL ● SECONDARY REMOVAL

Figure 7-16. Filtering components must be of the right type and placed properly in the system to maximize filtering functions.

Coalescing filters are ideal filters for removing oil and water from a system. Fine liquid droplets are continuously trapped in the element. The droplets grow in size and emerge on the outside surface of the element to flow to the filter drain. Coalescing elements function at their original efficiency even when saturated and continue functioning well until restricted by particulates. When coalescing filters become plugged, they must be discarded and replaced. See Figure 7-17.

Even with excellent pre-filtering, some solids make their way to the coalescing element, shortening its life. With proper maintenance, coalescing filter element life should be approximately 2000 operating hours. Oil not removed by a coalescing filter and allowed to continue downstream enters actuators to become a gummy, sticky resistance producing early actuator failure.

The temperature of ambient air must be considered when placing equipment that conditions compressed air. Coalescing filters are most effective when placed in the coolest location of the system.

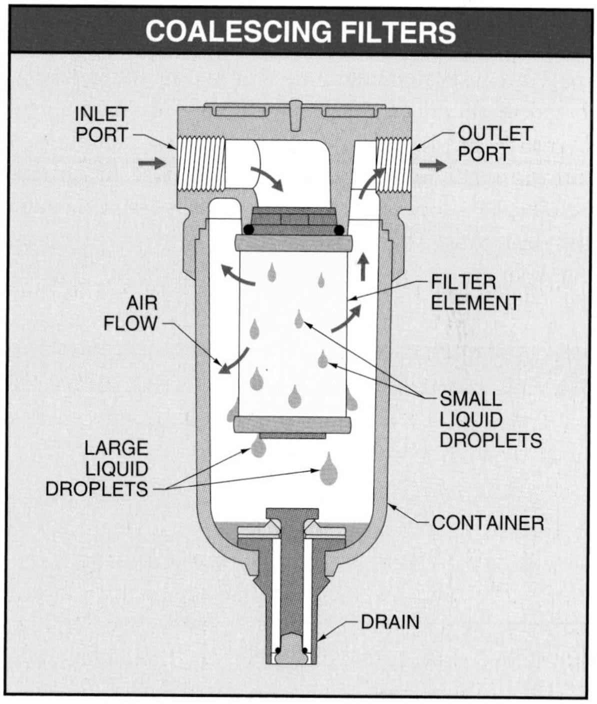

Figure 7-17. A coalescing filter is a device that removes submicron solids and vapors of oil or water by uniting very small droplets into larger droplets.

Dryers. Pneumatic technology is used extensively in automatic processing equipment. Automatic processing equipment such as instrumentation, measuring devices, controllers, etc. are some of the devices used in automatic production operations. *Instrumentation* is the area of industry that deals with the measurement, evaluation, and control of process variables. A *process variable* is any characteristic that changes its value during any operation within the process. Process variables include temperature, pressure, flow, force, etc.

To be dependable and effective, most automatic processing equipment depends on clean, dry air. General purpose and coalescing filters are not completely effective in removing all water vapor because of the ability of water to change form at different temperatures. In many cases, proper installation and placement of coalescing filters offers sufficient protection. However, a dryer is used when water vapor may contaminate a sensitive device.

A *dryer* is a device that dries air through cooling and condensing. Dryers leave air dry enough for applications such as instrumentation, air logic, etc. *Dry air* is air free of water vapor or oil droplets. Water vapor is not always visible. Most water vapor droplets are in the .5μ – 2μ range, and the smallest size droplet visible is about 15μ. Dryers remove water vapor from the air using desiccants or refrigeration.

A desiccant dryer is used to remove invisible water vapor when maximum drying is required. A *desiccant dryer* is a device that removes water vapor by adsorption. *Adsorption* is the adhesion of a gas or liquid to the surface of a porous material. A desiccant dryer removes water vapor using material such as silica gel or alumina. Adsorption offers the capability of removing 99.9% of the water vapor in the air. The desiccant material adsorbs water and becomes saturated and ineffective. For this reason, two cylinders may be interconnected to allow for the heating, evaporation, and regeneration of used desiccant. See Figure 7-18. While one side is drying compressed air, the other is being reactivated by use of an embedded heating coil or dry air being passed through the desiccant. The desiccant used as drying material can also adsorb any oil present in the system, leaving the dryer contaminated and ineffective.

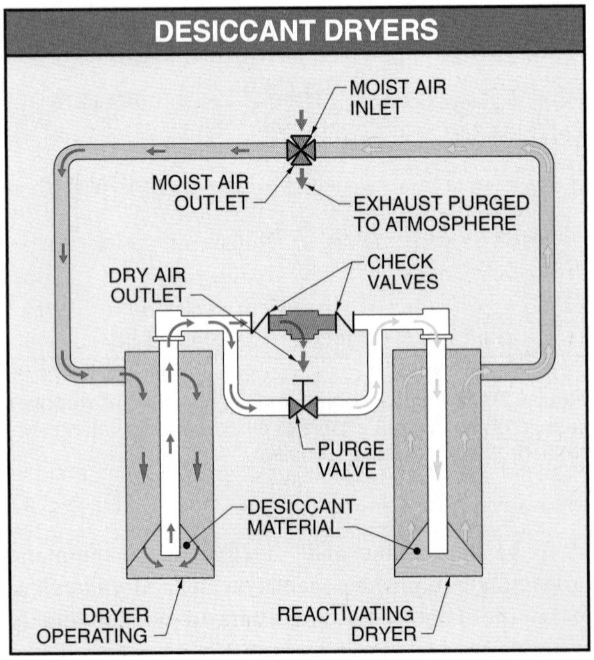

Figure 7-18. A desiccant dryer is a device that removes water vapor by adsorption.

A *refrigerant dryer* is a device designed to lower the temperature of the compressed air to 35°F. The cooling provided by a refrigeration system causes the water in the air to condense by lowering the relative humidity and the dew point of the air. The condensed liquids are drained automatically. Generally, the cold dry air flows through a heat exchanger to precool the incoming air. See Figure 7-19.

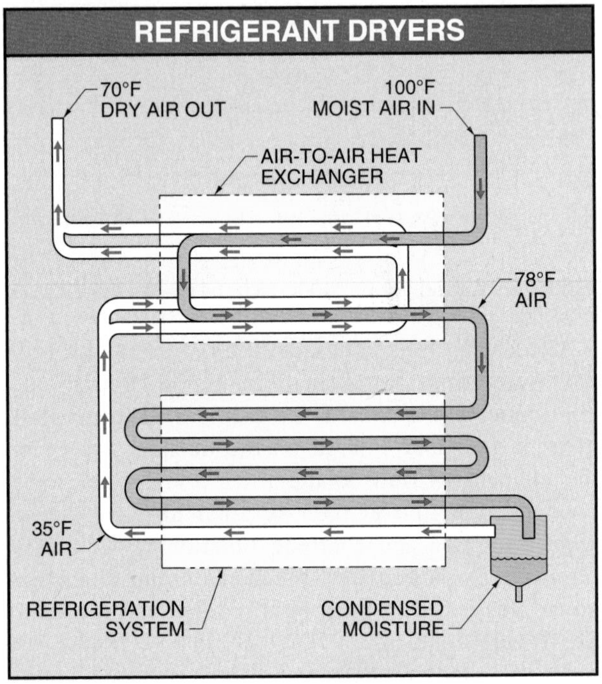

Figure 7-19. A refrigerant dryer is a device designed to lower the temperature of the compressed air to 35°F using a refrigeration system.

Dryers are normally placed in the coolest downstream location to lessen the energy required to lower the air temperature and condense the moisture in the air. To prevent contamination, dryers must be used for water removal only. Oil that is not removed upstream from the dryer builds up on the dryer tube walls. The oil buildup acts as an insulator, which reduces the dryer capacity. Dessicant dryers are most effective when placed after coalescing filters and closest to the point of air use. The air entering a dessicant dryer should have a maximum temperature of 100°F.

Always consult the manufacturer before using any product for nonindustrial applications, with fluids other than those specified, or for life-support systems not within published specifications.

Practical Pneumatics

8
Chapter

ARO Fluid Products Div., Ingersoll-Rand

A pneumatic system transmits and controls energy through the use of a pressurized gas within an enclosed circuit. A pneumatic system consists of a compressor, receiver, pressure switch, piping, check valve, receiver safety valve, pressure gauge, and pneumatic circuit. A pneumatic circuit is a combination of air-operated components that are connected to perform work. Pneumatic circuit components include check valves, filters, lubricators, pressure valves, directional control valves, flow control valves, and actuators. Pneumatic logic elements are miniature air valves used as switching devices to provide decision making signals in a pneumatic circuit.

PNEUMATIC CIRCUITRY

A *pneumatic system* is a system that transmits and controls energy through the use of a pressurized gas within an enclosed circuit. Pneumatic systems compress, store, and provide clean and safe air for a pneumatic circuit. A *pneumatic circuit* is a combination of air-operated components that are connected to perform work. The use of pneumatic circuits on industrial assembly lines has increased greatly since the development of mass production. Pneumatic circuits are being used today where hydraulic circuits were used in the past because pneumatic circuits offer benefits over hydraulic circuits. Pneumatic circuit benefits include:

- Easy air storage and use in remote locations because air is compressible
- Provides potential energy without the use of electricity
- Cleaner than hydraulic circuits
- Economical because initial costs are relatively low for equipment and spare parts
- Overheating is generally not a problem
- High reliability due to fewer moving parts
- Electrical shock and spark-free controls, enabling use in wet or explosive locations
- Pneumatic logic components can be used in place of electrical switches, relays, resistors, and timers

Graphic Diagrams

Pneumatic system and circuit understanding is clarified by using graphic diagrams. A *graphic diagram* is a drawing that uses simple line shapes (symbols) with interconnecting lines to represent the function of each component in a system or circuit. A *symbol* is a graphic element which indicates a particular device, etc. To simplify a graphic diagram, each device or component is given a symbol in place of a full drawing or picture. The path along the connecting lines from any one of the symbols (components) to any other component can be traced to determine the system or circuit operation. See Figure 8-1.

Pneumatic system hose fittings must be tightened securely because a whipping hose can damage equipment and cause injury to personnel.

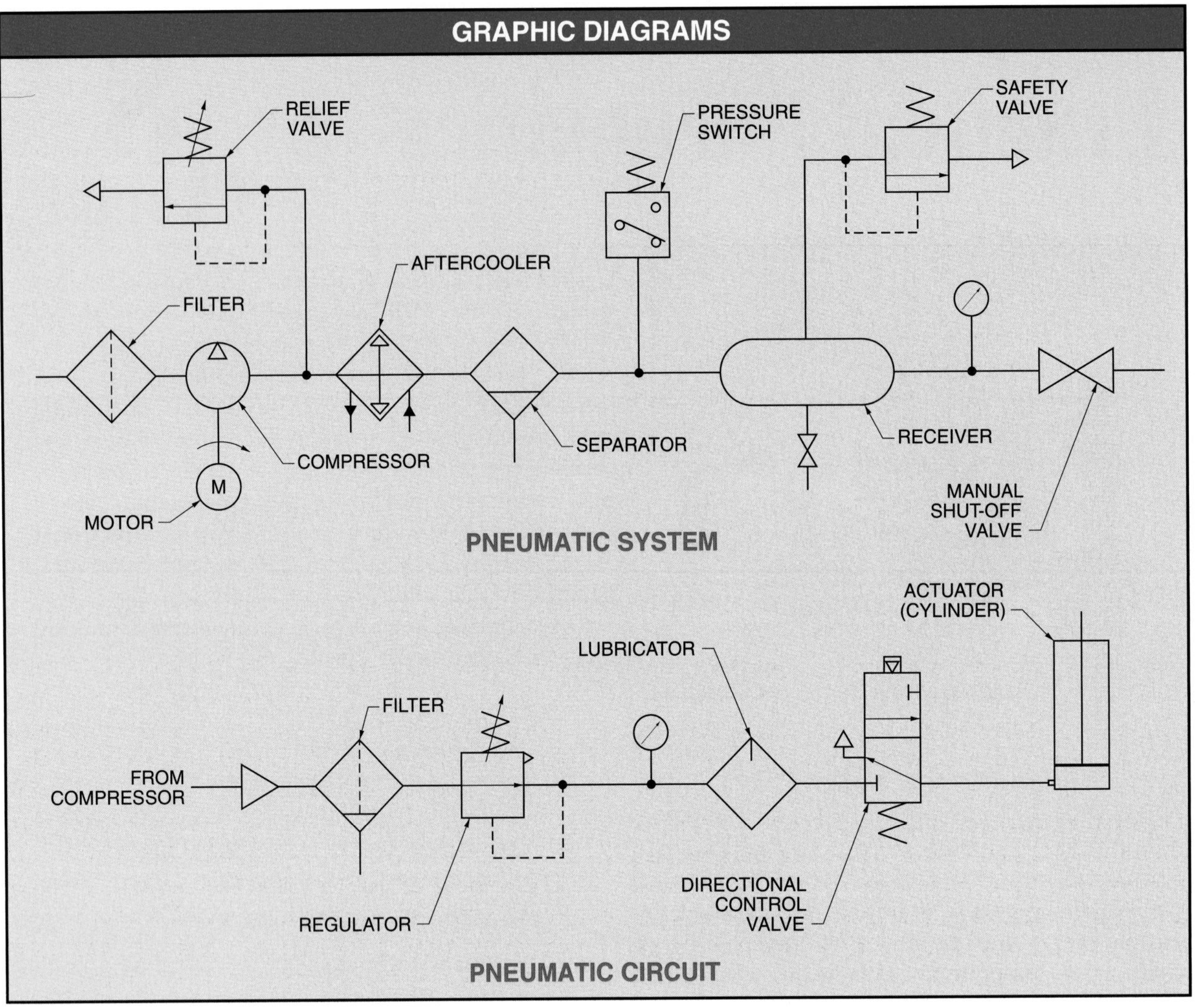

Figure 8-1. A graphic diagram uses symbols and interconnecting lines to represent the function of each component in a system or circuit.

The graphic diagram of a pneumatic system begins at the atmospheric air input into the system. This is normally at a filter or breather symbol that is located upstream from the compressor. The diagram shows all components used to compress and store air for use in a pneumatic circuit. The components include the breather, compressor, safety relief valve, aftercooler, separator, pressure switch, receiver, and manual shut-off valve. A pneumatic system ends at the shut-off valve downstream from the receiver.

The graphic diagram of a pneumatic circuit shows the circuit components beginning with the compressor, which indicates the circuit's beginning and direction of air flow. The compressor symbol is followed by a filter, a regulator, a lubricator, a directional control valve, and an actuator (cylinder).

Complex circuit graphic diagrams may show origination and direction of air flow by the use of an arrow or an "S" (supply). The piping connecting the components in a pneumatic graphic diagram is traced to determine the circuit operation. For example, the circuit piping is traced beginning at the supply, through the pilot-operated 4-way valve, and on to the actuator. This reveals that pressure is being applied to the rod end port of the actuator. See Figure 8-2.

The solenoid must be activated to shift the 3-way valve to allow pilot air to shift the 4-way valve to allow pressure to flow to the cap end of the actuator. The actuator piston extends when the 4-way valve is shifted by pilot pressure. Both valves return to their original position by spring pressure when power is removed from the solenoid.

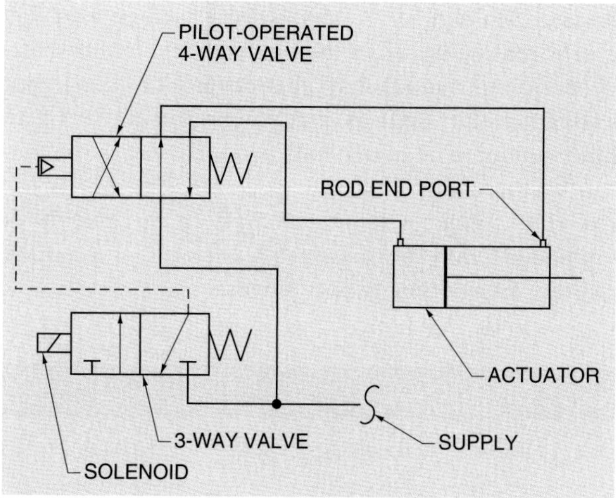

Figure 8-2. The piping connecting the components in a pneumatic graphic diagram is traced to determine the circuit operation.

PNEUMATIC SYSTEMS

A pneumatic system consists of a compressor to compress air, a receiver (tank) to store the compressed air, a pressure switch to shut down the compressor when the preset pressure has been met or to activate unloaders, piping to transfer air within the system, a check valve to prevent compressed air from backing into the compressor, a receiver safety valve to prevent dangerous overpressure, a pressure gauge to observe system pressure, and a circuit to perform the work required. See Figure 8-3.

Air Compressors

An *air compressor* is a device that takes air from the atmosphere and compresses it to increase its pressure. Compressing air for storage is accomplished using a positive displacement compressor, which increases the air's pressure by reducing its volume in a confined space. A *positive displacement compressor* is a compressor that compresses a fixed quantity of air with each cycle. Compressors generally have fixed operating speeds and constant pumping rates. Positive displacement compressors include piston, helical screw, and vane compressors.

Compressor air intake filters must be kept clean because a dirty intake filter decreases a compressor's efficiency and performance.

Piston Compressors. A *piston compressor* is a compressor in which air is compressed by reciprocating pistons. To *reciprocate* is to move forward and backward alternately. Piston compressors, also referred to as reciprocating compressors, may be single-stage or multistage. A *single-stage compressor* is a compressor that uses one piston to compress air in a single stroke before it is discharged. A *multistage compressor* is a compressor that uses two or three cylinders, each with a progressively smaller diameter, to produce progressively higher pressures.

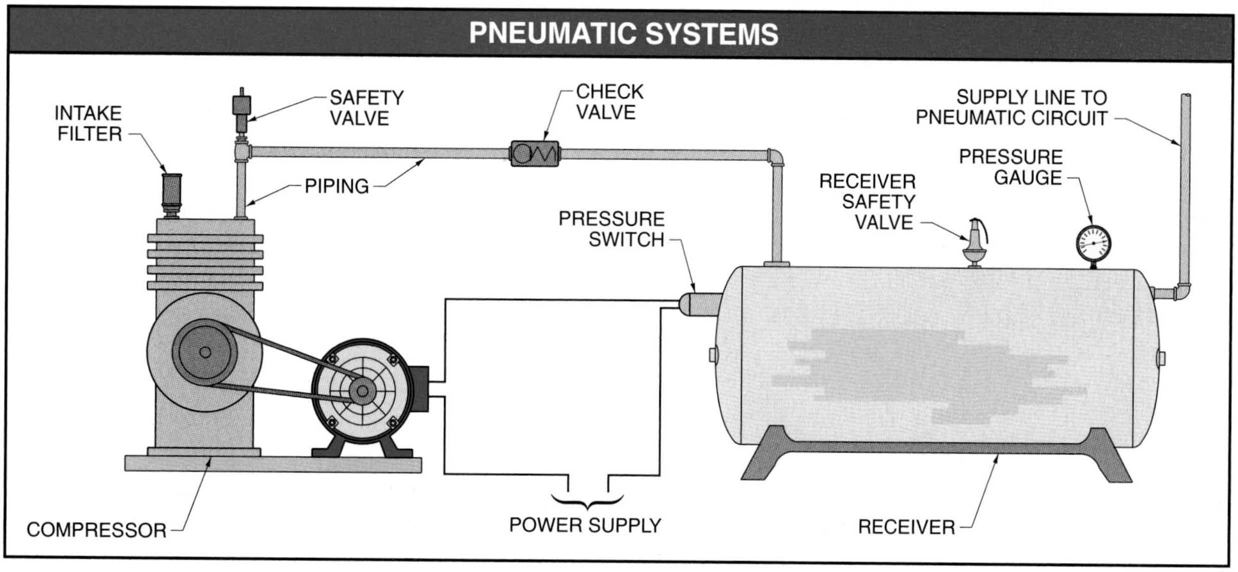

Figure 8-3. A basic pneumatic system consists of a compressor, receiver, pressure switch, piping, check valve, receiver safety valve, pressure gauge, and a circuit to perform the work required.

Piston compressors consist of a crankcase, cylinder(s), crankshaft, connecting rod(s), piston(s), piston rings, and an inlet and outlet valve. See Figure 8-4. The reciprocating motion of the piston fills the cylinder and compresses the air with each alternation. Pistons are generally driven in a reciprocating motion by a crankshaft. A *crankshaft* is a shaft that has one or more eccentric surfaces that produce a reciprocating motion when the shaft is rotated. An *eccentric surface* is a surface that has a different center than the center of the crankshaft. A section of a crankshaft that centers on a different axis than the shaft is said to be eccentric or to have runout. The crankshaft is connected to the piston by the connecting rod. A *connecting rod* is the rod that connects the crankshaft to the piston. A crankshaft may be driven by a motor, a gasoline engine, or another prime mover.

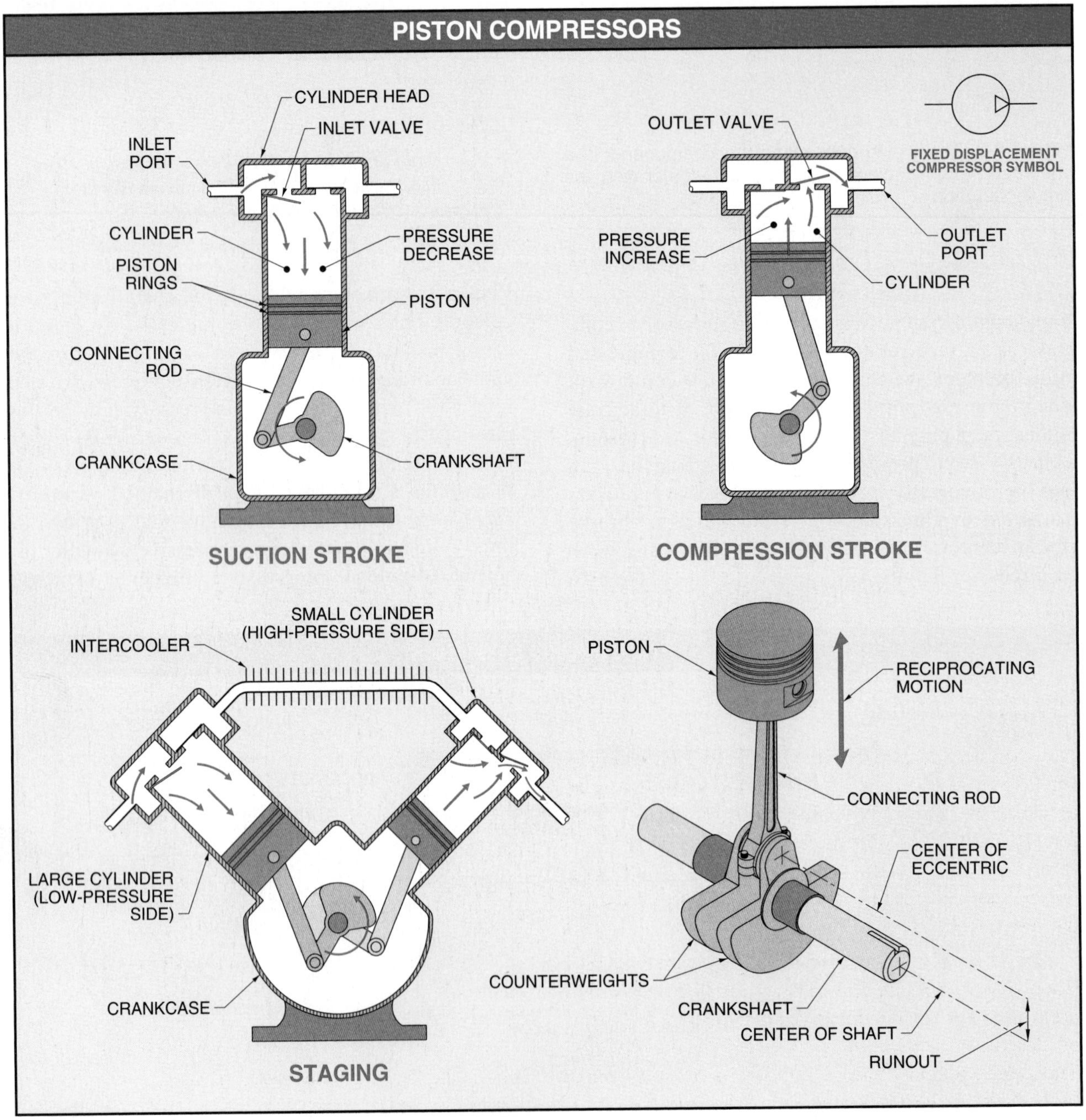

Figure 8-4. Air is compressed as a reciprocating piston draws air in on one stroke and pushes it out under pressure on its alternating stroke.

As the crankshaft rotates, the piston reciprocates within the cylinder. An increasing volume is produced in the cylinder as the crankshaft pulls the piston downward. At the end of the stroke, the cylinder is filled with air and the intake valve closes. The piston compresses the air as it moves upward. The discharge (outlet) valve opens to force the air to the next stage or to a receiver when the air pressure reaches a high pressure.

Staging is required to move compressed air to higher pressures. *Staging* is the process of dividing the total pressure among two cylinders by feeding the outlet from the first large (low-pressure) cylinder into the inlet of a second small (high-pressure) cylinder. The air compressed by the first cylinder is boosted to a higher pressure. A connecting intercooler is required between each stage to increase compressor efficiency and to lower the air temperature at the outlet of the first stage.

Helical Screw Compressors. A *helical screw compressor* is a compressor that contains meshing screw-like helical rotors that compress air as they turn. The meshing rotors (screws) draw air in at one end of a close-fitting chamber. The air is forced along the rotors as they rotate. The air trapped between the rotors is compressed as the size of the cavities between the rotors is progressively reduced. Air flow is positive and continuous because the air is constantly being drawn in and forced axially along the rotors. See Figure 8-5.

Helical screw compressors contain one, two, or three rotors. Two-rotor helical screw compressors are the most common in industry. A two-rotor helical screw compressor contains one male and one female rotor, which generally have four and six lobes, respectively. A *lobe* is the screw helix of a rotor. In the operation of a two-rotor helical screw compressor, air entering the chamber is compressed as the lobes of the male rotor mesh with the lobes of the female rotor. This pushes the trapped air along the rotors and compresses it. Air is progressively compressed until the lobes pass the outlet port, discharging the compressed air.

Helical screw compressors may have dry or oil-flooded compressing mechanisms. Dry screw compressing mechanisms compress air without the lobes of the rotors contacting each other as they rotate. The constant and close separation is accomplished by both rotors being rotated by a set of timing gears that are driven by the prime mover. Prime movers normally rotate the rotors at speeds between 3000 rpm and 12,000 rpm. At these relatively high speeds, the rotors turn freely with a carefully controlled clearance between the rotors and the housing. The clearance is protected by a light film of oil. Lubrication is not a major factor and screw life is long because there is no contact between the meshed rotors. Dry screw mechanisms are used in applications where oil-free air is required, such as instrumentation, paint spraying, clean rooms, etc.

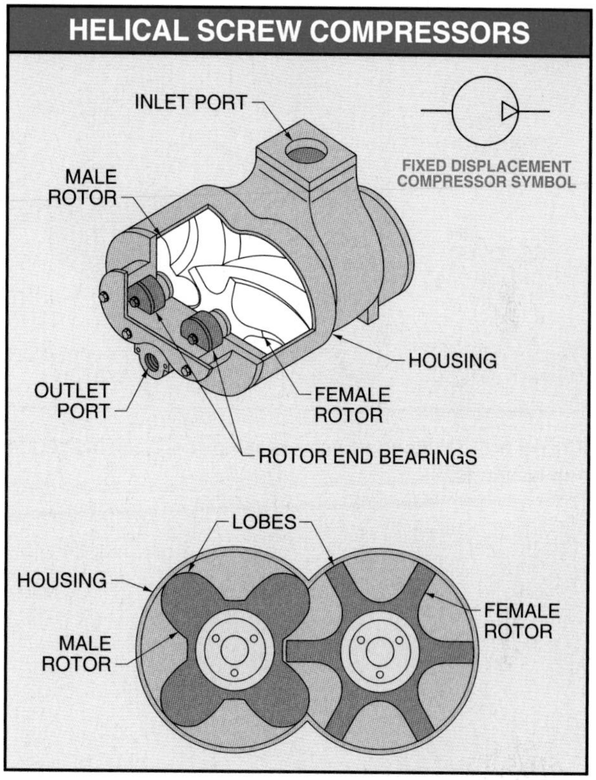

Figure 8-5. Air is compressed in a helical screw compressor as the size of the cavities between the rotors is progressively reduced.

Oil-flooded compressing mechanisms compress air with the lobes of the rotors contacting each other. The male rotor is driven by the prime mover and meshes with and drives the female rotor. This driving motion causes surface contact between the rotors, which must be well-lubricated. Lubrication is accomplished by injecting oil through internal passages to the rotors and rotor end bearings. See Figure 8-6. The oil bath lubricates the rotors, seals the rotor clearances for high-compression efficiency, and absorbs the heat of compression. This results in low outlet air temperature. The oil is separated from the compressed air, filtered, cooled, and returned to the compressor for reuse.

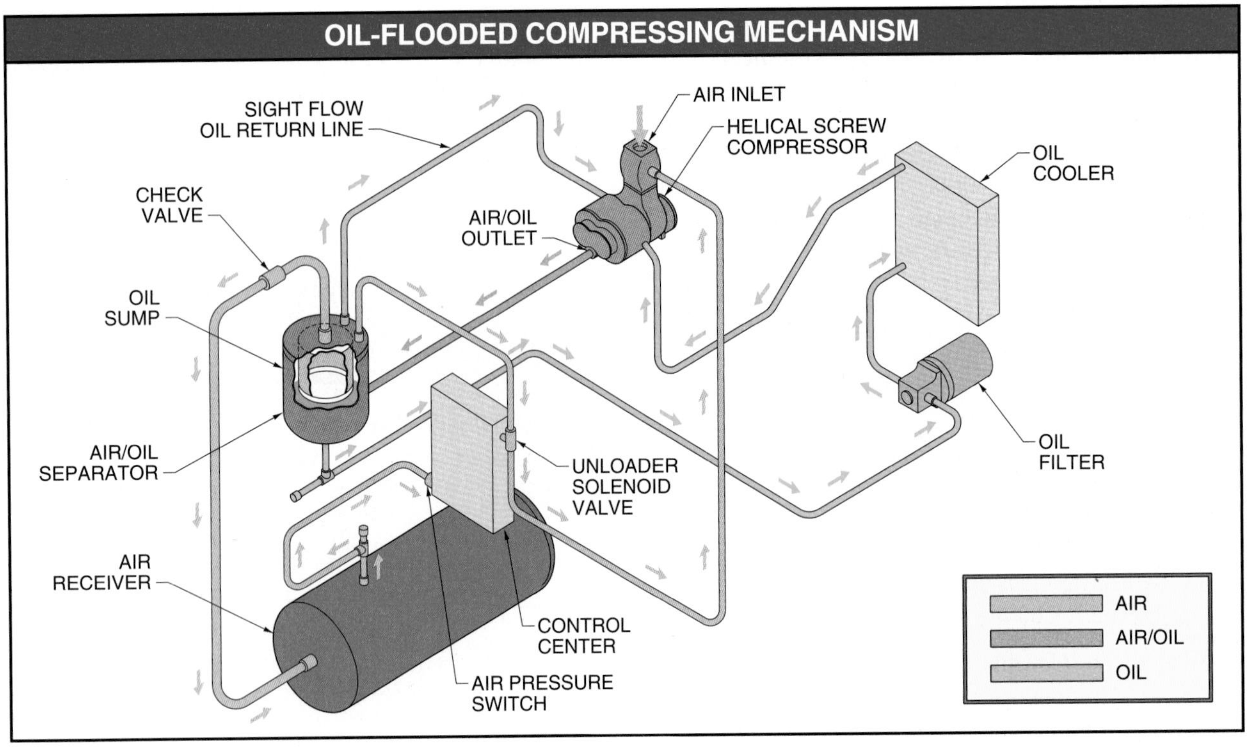

Figure 8-6. Oil-flooded compressing mechanisms require oil injected through internal passages to the rotors and rotor end bearings.

Gast Manufacturing Company

Rotary vane compressors from Gast Manufacturing Company are available in oilless or lubricated models and are used in packaging, food processing equipment, and air bearings.

Vane Compressors. A *vane compressor* is a positive-displacement compressor that has multiple vanes located in an offset rotor. The vanes form a seal as they are forced against the cam ring. The offset of the rotor in the cam ring produces different distances between the rotor and cam ring at different points inside the compressor. As the rotor rotates, its offset position allows the vanes to slide out and draw air from the inlet port. As the rotor continues to rotate, the volume between the vanes and the cam ring decreases, pushing the vanes into their slots in the rotor. The decreasing volume compresses the air and forces it out of the outlet port. Compressor vanes are normally made of high-temperature metal and are spring-loaded to ensure outward sliding and contact with the housing. See Figure 8-7. Vane compressors may be either single-stage (up to 50 psi) or two-stage (50 psi to 125 psi).

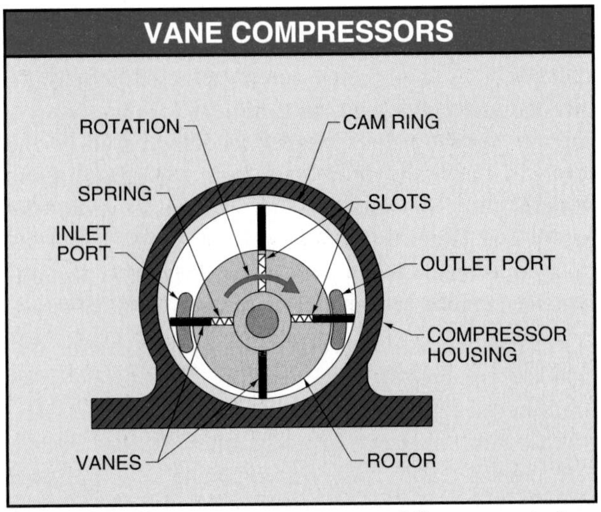

Figure 8-7. Vane compressors use sliding vanes to trap inlet air and compress it as the volume between the vanes decreases.

Pressure Control

The compressor in a pneumatic system must be able to maintain the required (set) level of pressure at all times. Extensive damage occurs to the receiver, motor, or compressor if the compressor is allowed to continue compressing air beyond the system's high-pressure limit.

A pressure compensator is used to avoid damage to a system using a vane compressor. A *pressure compensator* is a displacement control that alters displacement in response to pressure changes in a system. A pressure-compensated vane compressor consists of a pressure ring, a pressure-adjustment spring, and a thrust block. As the pressure builds within the compressor to the set level, the pressure ring is forced to a center position, where compression ceases. The pressure level is set by a pressure compensator adjustment screw, which adjusts the spring force required to center the pressure ring. A pressure switch or unloading valve is used to prevent damage to a piston or helical screw compressor. See Figure 8-8.

A *pressure switch* is a device that senses a high- or low-pressure condition and relays an electrical signal to turn the compressor motor ON or OFF. The motor is shut OFF when the receiver pressure reaches its preset maximum. The motor does not restart until the preset minimum pressure is reached. Pressure switch operation works well in applications involving intermittent air demand or low air consumption.

Excessive stopping and starting of a compressor motor overheats and burns up the motor. Generally, a compressor motor should not be cycled ON and OFF more than three times per hour. A system with an unloader must be used if the air consumption is high enough to create excessive stopping and starting.

An *unloading valve* is a device that senses a high-pressure condition and removes the compression energy. An unloading valve is placed in the compressor system to allow the motor to continue to run even after the high-pressure setting has been reached. Releasing the compression energy is accomplished by closing the compressor's inlet valve to prevent flow through the compression chamber, holding the inlet valve open. Each unloading valve is operated by pilot pressure from the receiver to a piston within the unloading valve and allows the prime mover to remain operational without building pressure. Unloading valves are used in applications where the air demand is high, constant, or both.

Safety Relief Valves

Pressure developed by a compressor is designed to be regulated by the system's pressure control system. However, in emergencies such as the failure of the pressure control system, pressure may build to a dangerous level. To relieve unsafe overpressure, safety relief valves are placed at the compressor and receiver. A *safety relief valve* is a device that prevents excessive pressure from building up by venting air to the atmosphere.

Safety relief valves operate as normally-closed valves with spring-loaded poppets. The poppet is moved off its seat when the force of the air on the poppet becomes greater than the spring force. The undesirably high air pressure is exhausted through the valve vent port. See Figure 8-9.

Safety relief valves are designed strictly for safety and are not intended for frequent operation. However, safety valves do require periodic maintenance. Maintenance of safety valves consists of verifying that the valve can move freely. On most valves, this is done by moving the test lever or by pulling a ring to unseat the poppet. The valve vent port normally remains clean from the air's exhausting velocity if regular testing begins when the valve is new. However, a dirty and untested valve may leak after being tested for the first time if it is allowed to sit and accumulate dust, dirt, and oil.

The oil used in a compressor is not the same oil used to lubricate pneumatic valves. If air is allowed to carry compressor oil vapor to a valve, the valve could varnish, causing it to stick. Oil supplied by a lubricator is transported by gravity and the working air to various points in a system.

Pneumatic System Piping

Piping in a pneumatic system begins at the compressor and runs to each component so energy can be transmitted by use of the compressed air. As the air is used throughout the system, it is returned (exhausted) to the atmosphere. For this reason, pneumatic system piping does not return to the beginning stage (compressor) like hydraulic system piping. Pneumatic system piping must be installed to minimize pressure loss and air leaks and to provide liquid drainage.

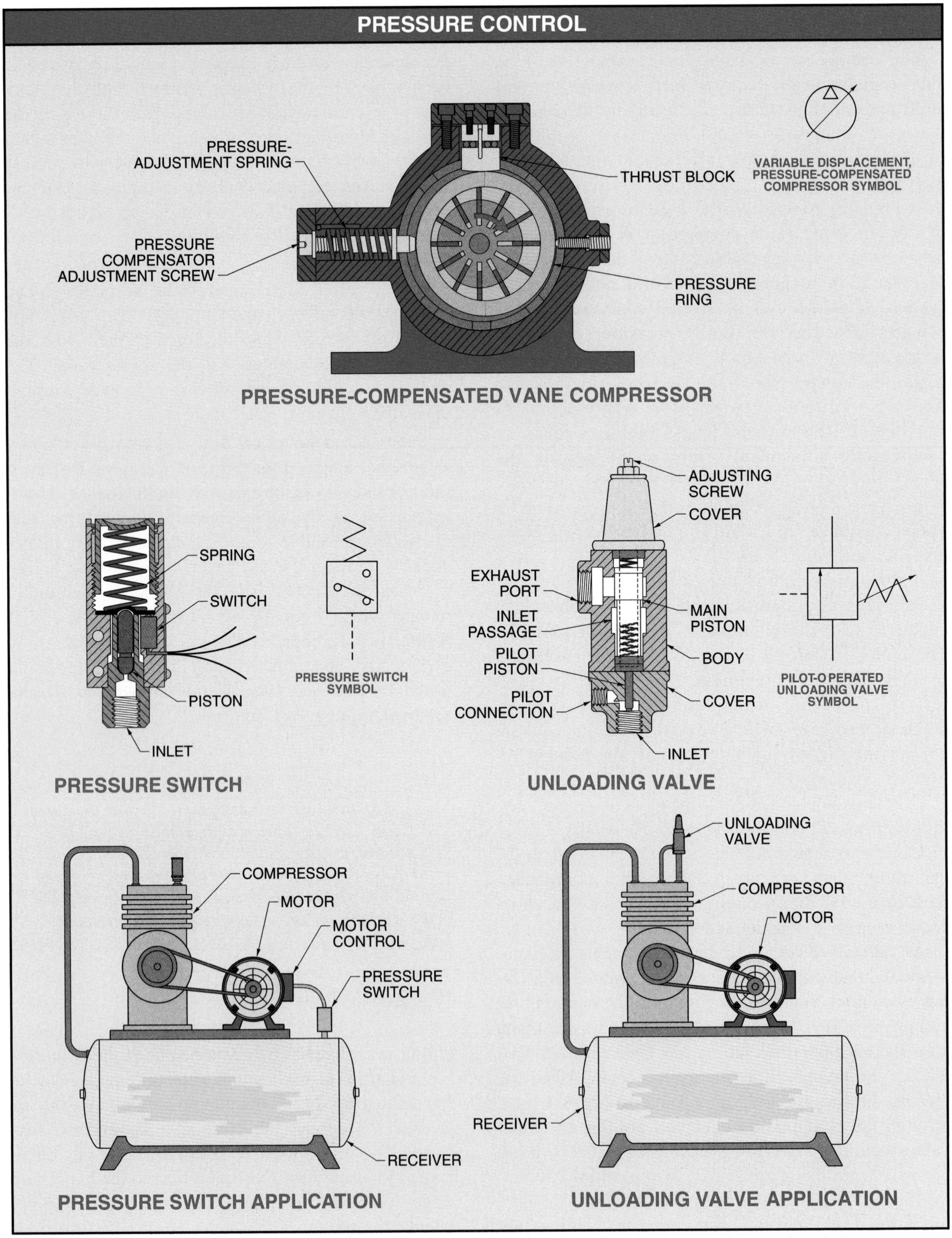

Figure 8-8. Continuous compressor pressure buildup is regulated by a pressure compensator, pressure switch, or unloading valve.

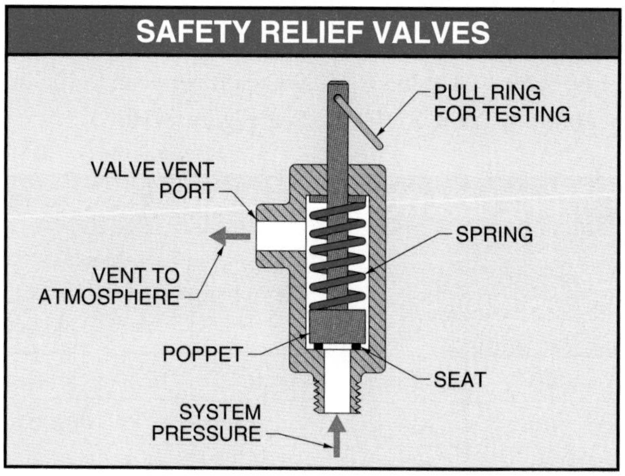

Figure 8-9. Safety relief valves operate when system pressure builds high enough to overcome spring pressure, which pushes a poppet off of its seat.

Pressure Loss. Pressure loss in a pneumatic system is a result of the resistance created within the system and circuit components, the work load demands, and the size and length of pipe and pipe fittings. Pressure loss is also created during system operation by moving controls such as valves. Pressure loss occurs only while air is moving through the piping. The maximum pressure drop in a system should be less than 10% over the system operating pressure. Pressure loss figures are taken from Pressure Loss Constants tables when determining total system pressure loss resulting from friction through lengths of pipe, pipe fittings, and other components such as valves. Pressure Loss Constants tables are also used to determine if the system has the required capacity. See Appendix.

Calculating the total pressure drop in a pneumatic system from piping, fittings, and other components is determined by calculating the individual pressure drops in each pipe, fitting, and component and adding these values together. The individual pressure drops in pipe, fittings, and components are calculated from the flow rate, pressure, and constant based on the component. Pressure loss in a component is found by applying the formula:

$$\Delta P = \frac{CQ^2}{1000} \times \frac{14.7}{14.7 + P}$$

where

ΔP = pressure drop (in psi)
C = constant (from Pressure Loss Constants table)
Q = air flow rate (scfm)
14.7 = constant (atmospheric pressure)
1000 = constant
P = working pressure (in psi)

For example, what is the pressure drop in a pneumatic system having a working pressure of 100 psi, an air flow rate of 50 scfm, and containing 100′ of 1″ Schedule 40 pipe, one 1″ gate valve, three 1″ 90° elbows, and one 1″ 50μ filter?

1. Calculate pressure drop for 1″ pipe.

Note: C value for 1″ pipe = 1.66 (from Pipe Pressure Loss Constants table)

$$\Delta P = \frac{CQ^2}{1000} \times \frac{14.7}{14.7 + P}$$

$$\Delta P = \frac{1.66 \times 50^2}{1000} \times \frac{14.7}{14.7 + 100}$$

$$\Delta P = \frac{1.66 \times 2500}{1000} \times \frac{14.7}{114.7}$$

$$\Delta P = \frac{4150}{1000} \times .128$$

$$\Delta P = 4.15 \times .128$$

$$\Delta P = .531 \text{ psi}$$

2. Calculate pressure drop for 1″ gate valve.

Note: C value for 1″ gate valve = .018 (from Pipe Fitting Pressure Loss Constants table)

$$\Delta P = \frac{CQ^2}{1000} \times \frac{14.7}{14.7 + P}$$

$$\Delta P = \frac{.018 \times 50^2}{1000} \times \frac{14.7}{14.7 + 100}$$

$$\Delta P = \frac{.018 \times 2500}{1000} \times \frac{14.7}{114.7}$$

$$\Delta P = \frac{45}{1000} \times .128$$

$$\Delta P = .045 \times .128$$

$$\Delta P = .006 \text{ psi}$$

3. Calculate pressure drop for three 90° 1″ elbows.

Note: C value for one 90° 1″ elbow = .043 (from Pipe Fitting Pressure Loss Constants table). Total pressure drop for three 90° elbows = .129 (.043 × 3 elbows = .129).

$$\Delta P = \frac{CQ^2}{1000} \times \frac{14.7}{14.7 + P}$$

$$\Delta P = \frac{.129 \times 50^2}{1000} \times \frac{14.7}{14.7 + 100}$$

$$\Delta P = \frac{.129 \times 2500}{1000} \times \frac{14.7}{114.7}$$

$$\Delta P = \frac{322.5}{1000} \times .128$$

$$\Delta P = .323 \times .128$$

$$\Delta P = .041 \text{ psi}$$

4. Calculate pressure drop for a 1″ 50μ filter.

Note: C value for a 1″ 50μ filter = .20 (from Filter Pressure Loss Constants table)

$$\Delta P = \frac{CQ^2}{1000} \times \frac{14.7}{14.7 + P}$$

$$\Delta P = \frac{.20 \times 50^2}{1000} \times \frac{14.7}{14.7 + 100}$$

$$\Delta P = \frac{.20 \times 2500}{1000} \times \frac{14.7}{114.7}$$

$$\Delta P = \frac{500}{1000} \times .128$$

$$\Delta P = .5 \times .128$$

$$\Delta P = .064 \text{ psi}$$

5. Find total system pressure drop by adding each pressure drop from piping and components.

$$\Delta P = .531 + .006 + .041 + .064$$

$$\mathbf{\Delta P = .642 \text{ psi}}$$

Selecting slightly oversized pipe adds a safety margin and allows for additional demands on a system. Also, to ensure that an adequate supply of air is available, pressure loss may be calculated when the purchase of additional equipment is anticipated.

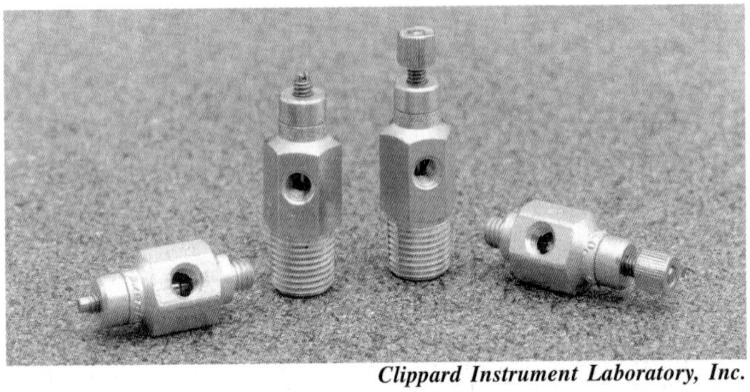

Clippard Instrument Laboratory, Inc.

Needle valves are used in pneumatic systems to control actuator speeds by the adjustment of a tapered needle.

Leaks. Leaks in a pneumatic system are very costly, and most leaks in a facility go unnoticed. In most cases, large savings are realized when a company takes the time to look, listen, and document major leaks within the facility. Leaks that cannot be heard may be located by brushing soapy water on each threaded or compression fitting. The presence of bubbles indicates a leak. Repairs are scheduled starting with the most major leak. Cost savings from fixing air leaks can be substantial. For example, a 1/32″ hole in a system having an initial pressure of 100 psig loses 1.6 standard cubic feet per minute (scfm). The annual cost savings from repairing this hole, based on \$.25/1000 scfm, is \$210.24 (1.60 × 60 × 24 × 365 = 840,960 cu ft per year. [840,960 ÷ 1000] × .25 = \$210.24). See Figure 8-10.

PNEUMATIC SYSTEM AIR LOSS*

Initial Pressure**	Hole Size†				
	1/64	1/32	1/16	1/8	1/4
40	.195	.775	3.10	12.4	50.0
60	.265	1.05	4.25	17.0	68.0
80	.335	1.35	5.35	21.5	85.5
100	.405	1.60	6.50	26.0	104.0
125	.495	2.0	7.90	31.5	126.0

* in scfm
**in psig
† in in.

AIR LEAK COSTS*

Hole Size**	Annual Cost†
1/64	53.22
1/32	210.24
1/16	854.10
1/8	3416.40
1/4	13,665.60

* @100 psig based on \$.25/1000 scfm, 24 hr/day, 52 weeks/year
** in in.
† in dollars

Figure 8-10. In most cases, large savings are realized by fixing major leaks within a facility.

Liquid Drainage. Draining liquid is generally accomplished by installing the main header at a downward pitch of 1″ for every 10′. A *main header* is the main air supply line that runs (generally overhead) between the receiver and the circuits in a pneumatic system. The main header should terminate with a moisture water leg and drain valve. To prevent moisture from draining into a circuit branch line, each branch line is connected to the top of the main header or branch header. Moisture water legs should also be placed at locations in the system where the header passes through areas of low temperatures because water drops out through condensation at these locations. See Figure 8-11.

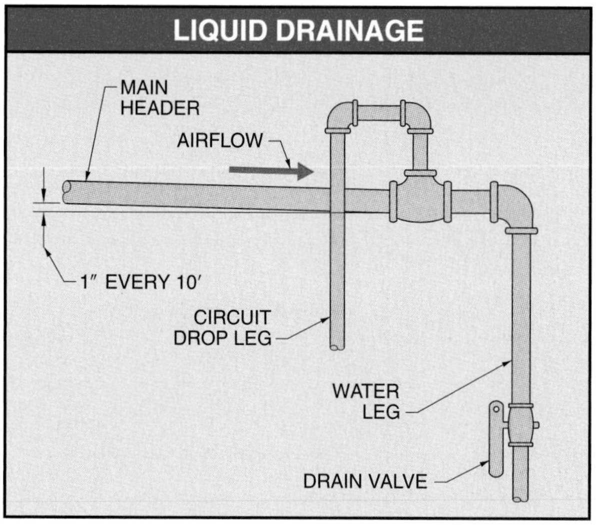

Figure 8-11. Headers must have the proper slope so that moisture that condenses within the header flows to a moisture water leg and drain valve.

Pipe Installation. A pneumatic system graphic diagram shows the layout of a pneumatic system piping system. The graphic diagram shows the main header from the receiver and the feeder lines that supply individual circuits. The graphic diagram also shows the location of the various components used to condition (filter, regulate, lubricate, etc.) the air in the individual circuits. Pneumatic system piping should be installed in a two-way loop around a plant whenever possible. The two-way loop provides two paths for air flow. This prevents significant pressure drop in the circuits at the end of the main header. See Figure 8-12.

Piping Materials. Standard piping is generally used for pneumatic system main headers and feeder lines. Pipe material and connections are normally determined by the working load. Generally, pipe over 6″ in diameter has welded joints, pipe between 3″ and 6″ has bolted flange connections, and pipe under 3″ is threaded. Rigid feeder lines should be connected to the main header or feeder lines as close to the point of use as possible.

A thread-lubricating material must be used to prevent leaks when threaded connections are used. Thread-lubricating material should be placed on the male fitting only. The first two lead-in threads must be left bare to allow for thread starting and to prevent contamination from pipe dope or Teflon tape.

Copper, nylon, or plastic tubing may be used to pipe individual circuits. Copper tubing is used because it resists abrasion and heat damage. Nylon or plastic tubing is normally used for circuits because it is easy to work with, can be cut to length with a sharp knife and installed quickly and easily, and, in most cases, can be used with operating pressures up to 200 psi. Also, the cost of nylon or plastic tubing is considerably less than the cost of copper tubing.

Check Valves

A *check valve* is a valve that allows flow in only one direction. A check valve is used in a pneumatic system to prevent the flow of stored compressed air in the receiver from flowing back into the compressor.

A check valve consists of a body with a primary (inlet) port and a secondary (outlet) port. A ball or poppet is held against the inlet port by a spring. Air pressure at the inlet port that is greater than the spring pressure moves the ball or poppet off of its seat, thus allowing air flow. The ball or poppet is held against the inlet port by spring and air pressure if air attempts to flow in the opposite direction. See Figure 8-13.

An example of a check valve application is a hand tire pump used to pump air into bicycle tires. Air is free to flow through the hose and into the tire when the handle is pushed down. The check valve prevents air from flowing out of the tire when the handle is lifted to draw air into the cylinder.

Saylor-Beall Manufacturing Company

Air compressors are used in water pumping station installations to supply compressed air for air tools, valve actuation, and air motors.

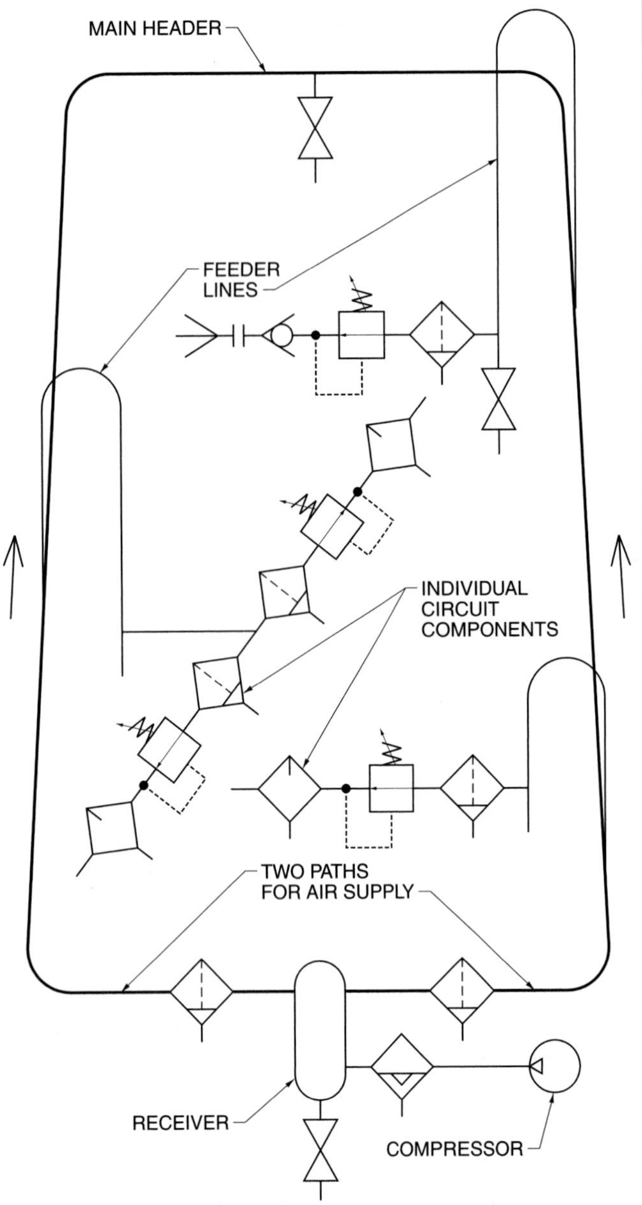

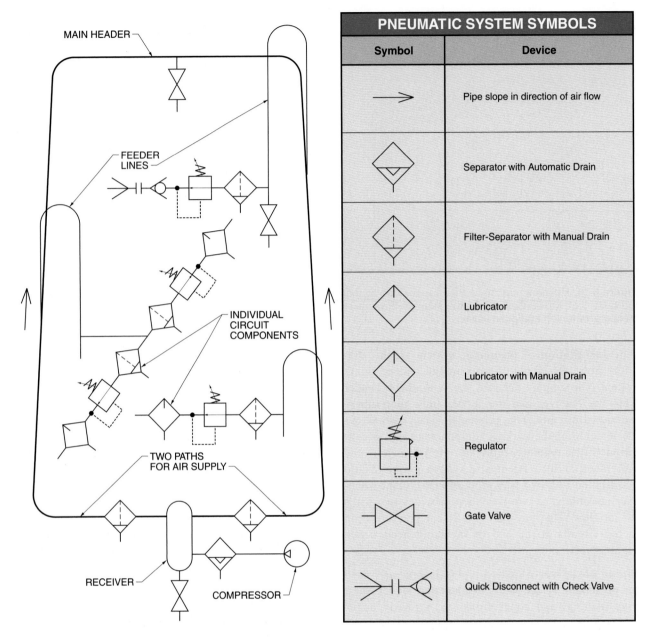

PNEUMATIC SYSTEM SYMBOLS

Symbol	Device
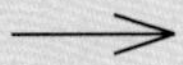	Pipe slope in direction of air flow
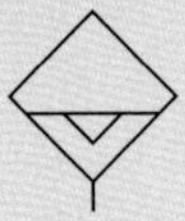	Separator with Automatic Drain
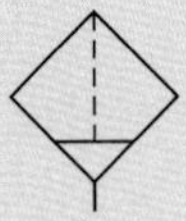	Filter-Separator with Manual Drain
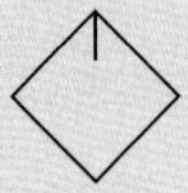	Lubricator
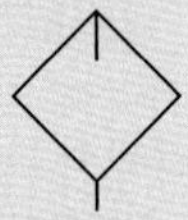	Lubricator with Manual Drain
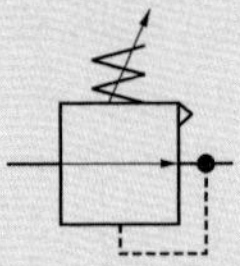	Regulator
	Gate Valve
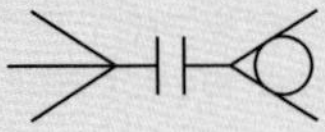	Quick Disconnect with Check Valve

Figure 8-12. A two-way looped pneumatic system piping installation prevents significant pressure drop in the individual circuits.

PNEUMATIC CIRCUIT COMPONENTS

Pneumatic circuit components use stored compressed air from a pneumatic system to support the work process or to perform the actual work. Work process supporting components condition, control, or direct the air. The symbols for most of the components used in a pneumatic circuit are similar to those used in hydraulic circuits. Actuators perform the actual work. An *actuator* is a device that transforms fluid energy into linear or rotary mechanical force.

Conditioning Compressed Air

Conditioning compressed air in a pneumatic system begins at the compressor, where the air is filtered and cooled and moisture is removed. Cooling and removing the moisture at this point in a pneumatic system is the minimum required for proper compressed air storage. Additional air conditioning is required at each circuit in the system. Circuit air conditioning consists of filtering and lubricating the air according to the requirements of the individual circuits. To be effective, air conditioning devices must be of the proper type and size and must be installed and maintained correctly.

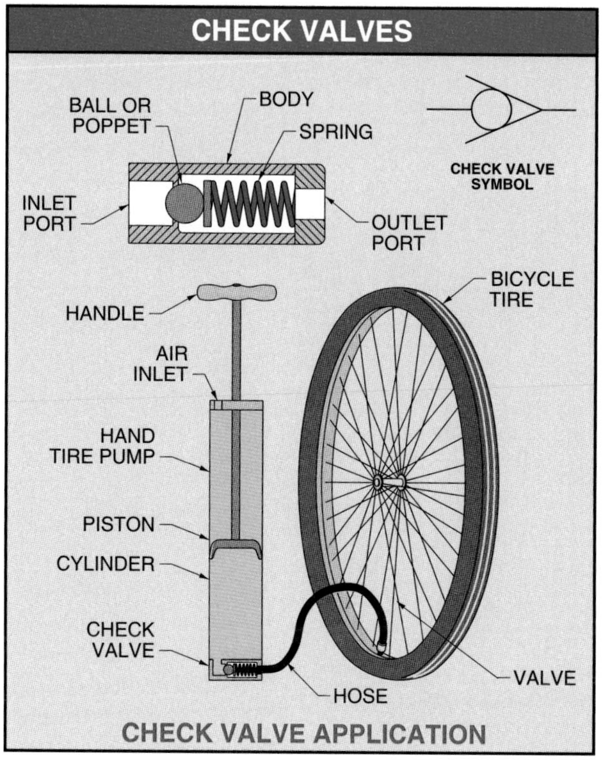

Figure 8-13. Check valves are used in a pneumatic system to allow air to flow in one direction.

Filters. A *filter* is a device containing a porous substance through which a fluid can pass but particulate matter cannot. See Figure 8-14. Ideally, filters, or a combination of filters, should remove all solids and liquids from the air. In many circuits, a particulate filter, which removes solid particles of 5μ and smaller, is placed just ahead of a coalescing filter. The coalescing filter removes oil and moisture. A screen filter should be placed before the particulate filter if the air is extra dirty and the particulate filter requires frequent changing. Filters must be changed or cleaned per manufacturer specifications.

An intake filter acts as a silencer for air as it rushes into a compressor. Filters are required in a pneumatic system because polluted air may also contain undesirable gases which, when mixed with the moisture found in a pneumatic system, could be corrosive.

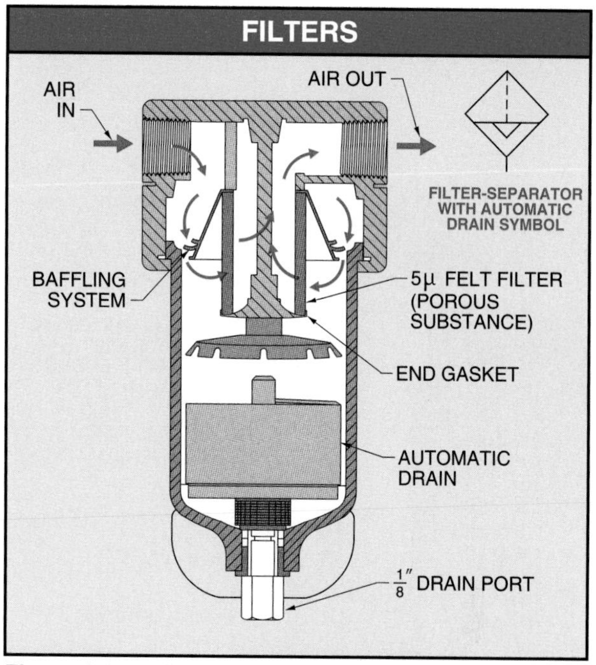

Figure 8-14. A filter contains a porous substance through which air can pass but particulate matter cannot.

Lubricators. A *lubricator* is a device that injects atomized oil into the air sent to pneumatic components. Proper lubrication of moving components is generally accomplished by atomizing light oil into the air stream. Lubricators use an orifice to create a pressure differential in the device. The pressure differential creates a siphon effect on the feeder tube. The siphon effect draws the oil up to the drip tube, where it is dripped into the air stream. See Figure 8-15. Oil flow is metered by an oil adjustment screw. The oil adjustment screw controls the number of oil drops that are released to be atomized into the system air.

Oil flow must be regulated because under- or over-lubrication may be a problem. Depending on the actuator requirements, begin by using 1 drop for every 10 cfm. This may be regulated up or down based on actuator requirements. Generally, a light-grade oil is satisfactory for lubricator atomizing. The grade of oil selected must atomize properly for the correct lubrication of the components.

Lubricators should be placed downstream from the filter and should be as close as possible to the components being lubricated. Lubricators should be placed no more than 10′ from the lubricated components because atomized oil begins to drop out of the air beyond 10′.

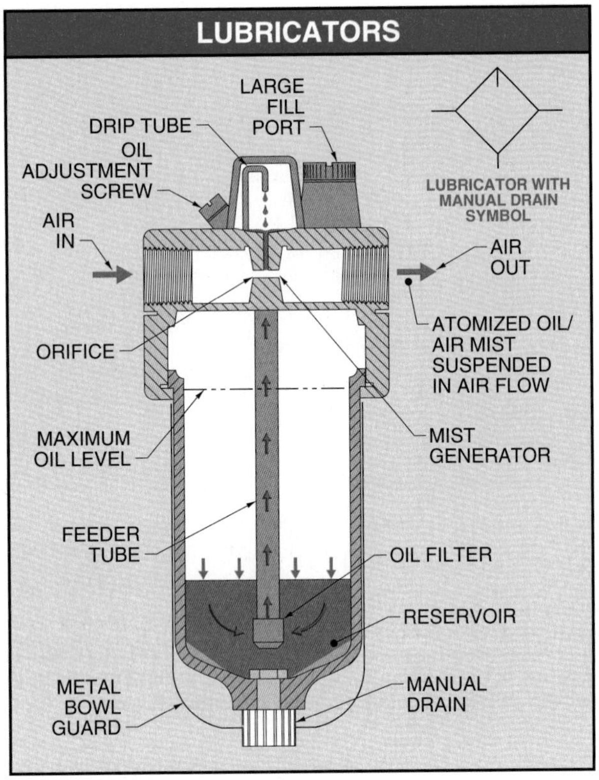

Figure 8-15. A lubricator is a device that injects atomized oil into the air sent to pneumatic components.

Controlling and Directing Air

Pneumatic circuit controls are valves designed to control the pressure, direction, and flow of air through a circuit. Pneumatic circuit valves include pressure regulators, directional control valves, and flow control valves. Also, special valves are used as switching devices to control time or sequence of events. Control valves are operated by system or circuit pressure or by the exhaust from another valve. The operation of one valve from the exhaust of another is a form of pilot operation.

Pressure Regulators. A *pressure regulator* is a valve that restricts and/or blocks downstream air flow. Pressure regulators (pressure-reducing valves) are used in a pneumatic circuit to provide a constant and proper air pressure to pneumatic components. Pressure regulators generally control circuit pressures from 0 psi to 150 psi, depending on the circuit's maximum pressure and application. Pressure regulators operate on the pressure differential between the downstream pressure, the regulating spring force, and the upstream pressure. The upstream pressure and the regulating spring force equal the downstream pressure. A pressure regulator always adjusts to a balanced pressure (equilibrium). Pressure regulators may be diaphragm or piston design. See Figure 8-16.

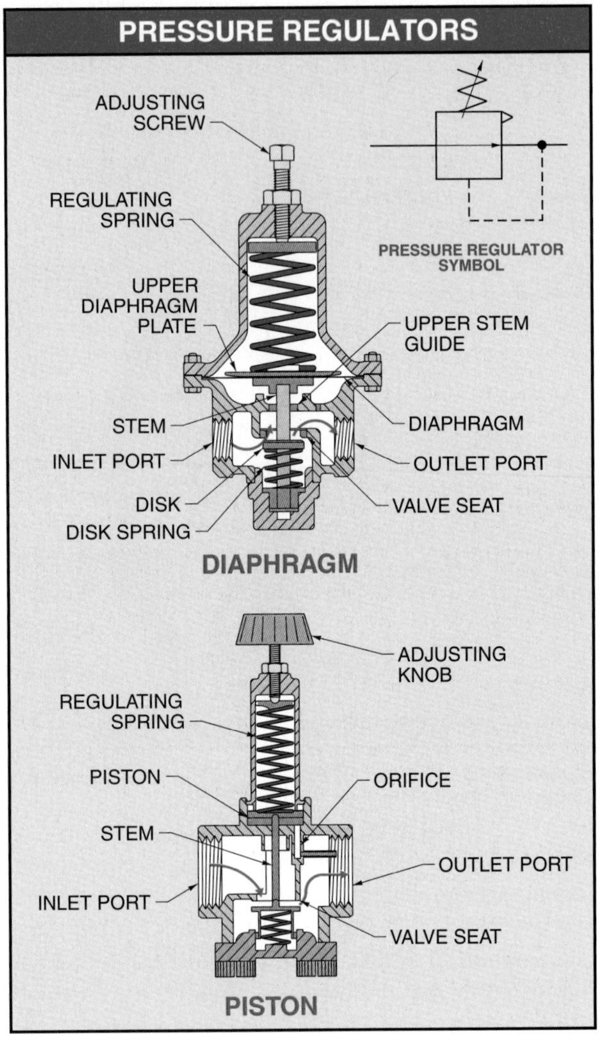

Figure 8-16. Pressure regulators provide constant and proper air pressure to pneumatic components.

A diaphragm pressure regulator uses a metallic (bronze) or nylon reinforced rubber diaphragm to sense a pressure differential between a regulating spring and a disc spring. The regulating spring exerts a force on the upper side of the diaphragm. The diaphragm is connected to the disc by a stem. An increase in outlet pressure increases the upward force on the diaphragm. This closes or reduces the flow through the regulator passageway. A decrease in outlet pressure reduces the upward pressure on the bottom of the diaphragm. This opens or increases the flow through the regulator passageway.

Piston regulators use a piston to sense the pressure differential between the valve's inlet and outlet pressures. The regulator body acts as a cylinder for the piston. Pressure from the regulating spring forces the stem to open the valve seat and allow air flow. Downstream pressure is sensed through the orifice and is applied to the bottom of the piston. This pressure offsets the regulating spring and inlet air pressure to maintain a constant outlet pressure. A decrease in outlet pressure reduces the pressure through the orifice and on the bottom of the piston. This increases the spring pressure, opening the valve seat, allowing increased air flow. An increase in outlet pressure increases the pressure through the orifice and on the bottom of the piston. This reduces the spring pressure, closing the valve seat, allowing reduced air flow.

Directional Control Valves. Directional control valves direct the flow of air to an actuator or another valve in a pneumatic circuit. Most directional control valves are 2-way, 3-way, or 4-way valves. A *way* is a flow path through a valve. Two-way valves have two main ports for air flow. Three-way valves have three main ports for air flow. Four-way valves have four main ports (possibly five) for air flow. See Figure 8-17.

Directional control valves are placed in different positions to start, stop, or change the direction of fluid flow. A *position* is the specific location of a spool within a valve which determines the direction of fluid flow through the valve. A 2-position valve can be placed in two positions and a 3-position valve can be placed in three positions.

A 3-way, 2-position, solenoid-operated, spring-return directional control valve may be used to activate a single-acting cylinder. The valve directs air to the cap end of the cylinder when the solenoid is energized. De-activating the solenoid causes the spring to shift the valve spool, exhausting the air in the cylinder to the atmosphere. See Figure 8-18.

A 3-way, 3-position, manually-operated, spring-centered directional control valve may be used to control a single-acting cylinder. The left position extends the cylinder and right position retracts the cylinder. The center position allows the cylinder to be stopped and held in any position between fully extended and fully retracted.

A constant air leak at the vent hole of a piston regulator is an indication of damaged or worn piston seals that require replacement.

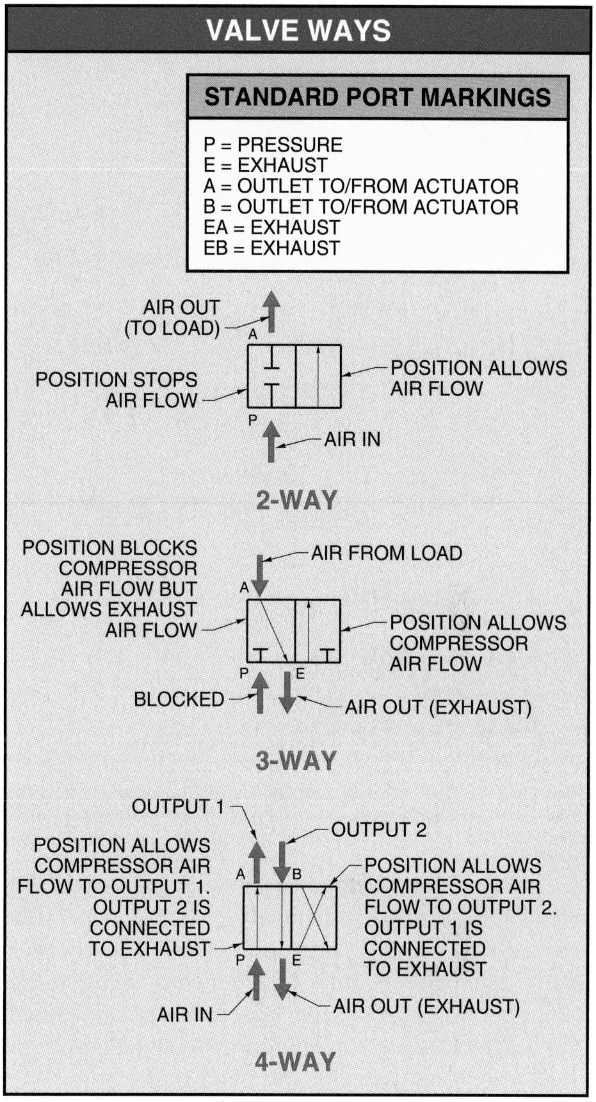

Figure 8-17. Directional control valves have 2, 3, or 4 ways to direct the flow of air to an actuator or another valve in a pneumatic circuit.

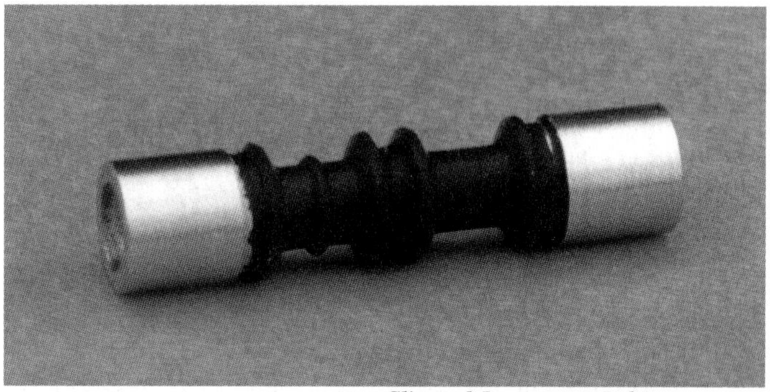

Clippard Instrument Laboratory, Inc.

The shifting of the spool in a directional control valve determines the ports to which fluid flows.

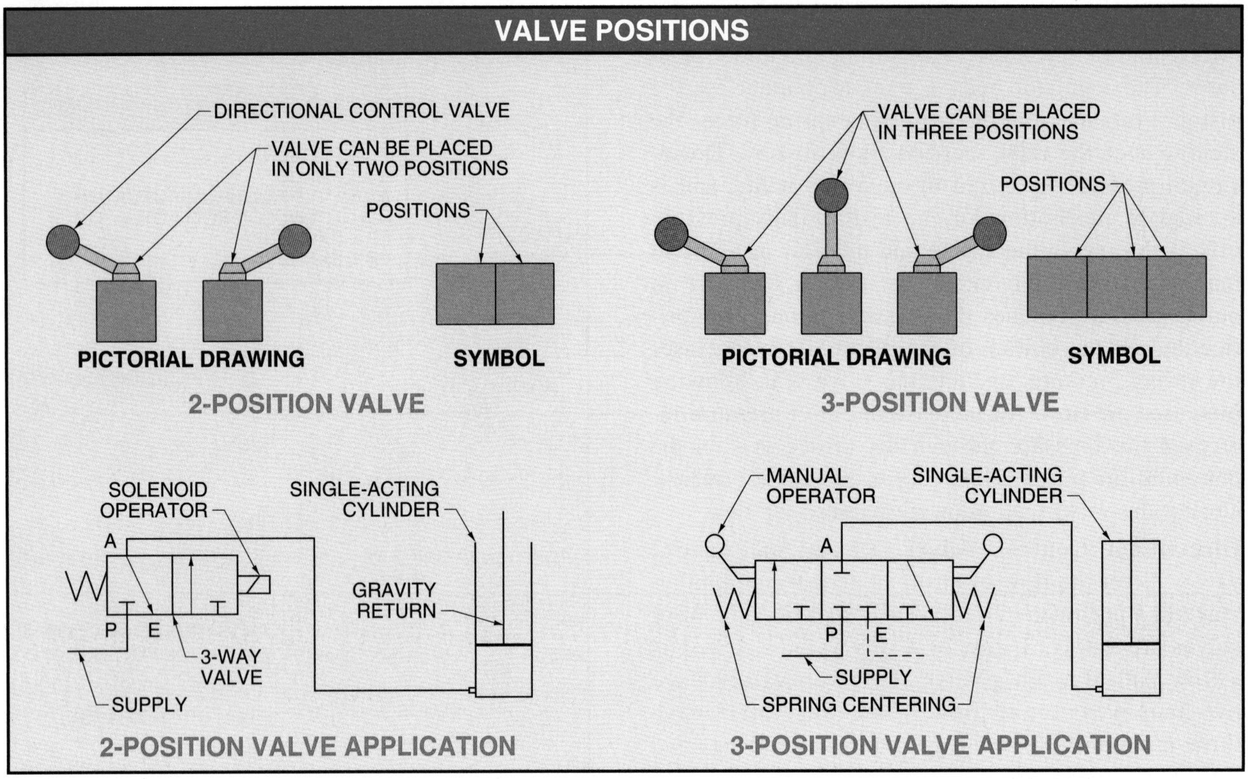

Figure 8-18. Directional control valves are placed in different positions to start, stop, or change the direction of fluid flow.

Directional control valves may be operated electrically, mechanically, manually, or by pilot operation and may be normally open or normally closed. Normally open valves allow flow between the inlet and outlet ports when the valve operator is not energized. Normally closed valves require the valve operator to open a path between the inlet and outlet ports.

Most directional control valves are 2- or 3-position valves. Two-position valves have two positions in which the spool can be placed. These positions are referred to as the extreme positions. Three-position valves have a center (neutral) position in addition to the two extreme positions. The neutral position is generally the deactivated position, where the internal spool is normally centered by spring action on both ends of the spool.

The neutral position produces various functions based on the design of the valve spool. For example, one neutral function has all ports blocked. This function, known as the closed center supply, allows for infinite positioning of a cylinder. In this case, the cylinder remains in its last actuated position when the operator is first activated and deactivated because the air is not allowed to exhaust. Another neutral function is the open center supply design. In the open center supply design, the neutral ports from the supply to the outputs are open. This allows the supply pressure to hold the outputs in their actuated positions. In the open center exhaust design, the neutral ports from the outputs are to the exhausts and are open, allowing the cylinder ports to exhaust while blocking the inlet port. See Figure 8-19.

Pneumatic circuits using 4-way directional control valves are the most common in industry. A 4-way, 2-position, hand-lever operated, spring-return directional control valve may be used to activate a double-acting cylinder in both directions. In this circuit, the hand-lever operator fully extends or retracts the piston rod depending on the position of the operator. See Figure 8-20.

A 4-way, 3-position, hand-lever operated, spring-centered directional control valve may be used with a double-acting cylinder for infinite positioning. As the hand lever is operated, the piston rod extends or retracts. The piston remains in its present location whenever the lever is released and the springs center the valve spool. This use is similar to positioning forks on a forklift.

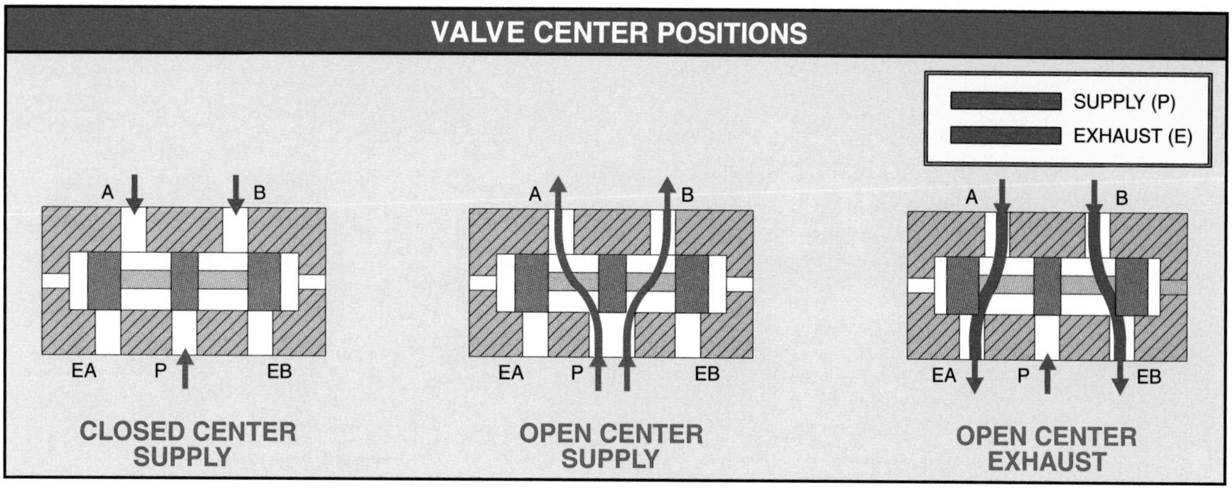

Figure 8-19. Center positions of directional control valves are designed to exhaust, block, or allow inlet or exhaust air to travel to another component.

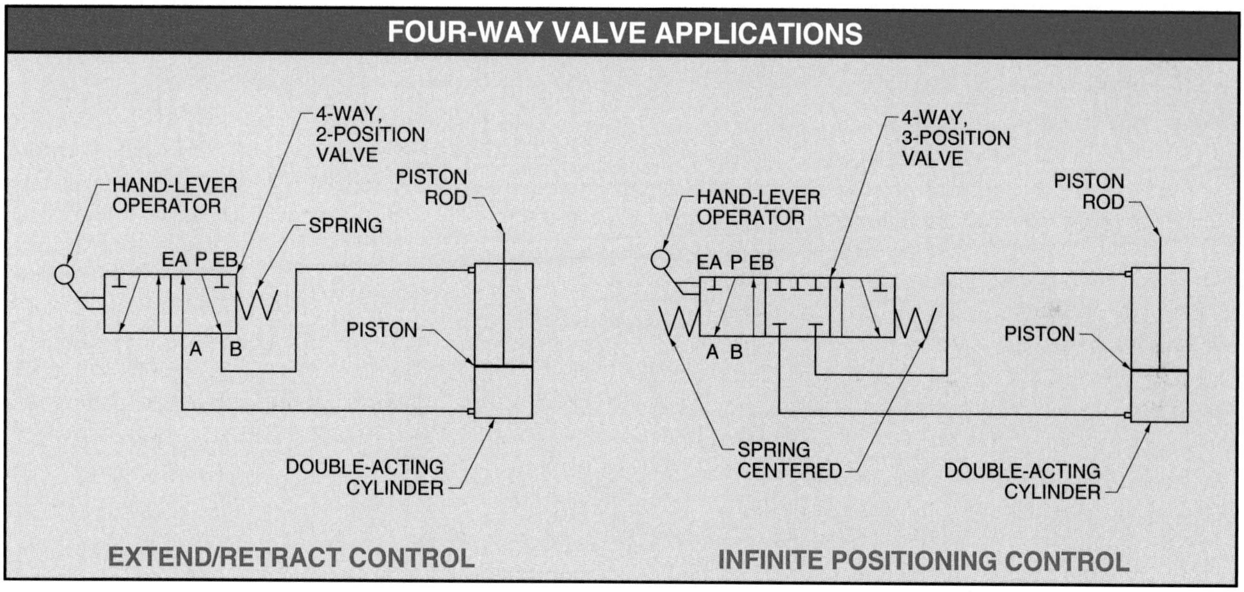

Figure 8-20. Pneumatic circuits using 4-way, 5 ported directional control valves are commonly used in industry to control the operation of double-acting cylinders.

Directional control valves may be controlled electrically by a solenoid. A *solenoid* is a device that converts electrical energy into a linear, mechanical force. The mechanical force in a solenoid is created by a magnetic field that is set up by the flow of electric current through a coil of wire. In pneumatic circuits, solenoids are used to allow or prevent (open/close) air flow in 2-way valves or control the position of the spool in 3-way valves. See Figure 8-21.

In 2-way valves, solenoids are used to control the operation of a plunger to open or close ports. This produces a flow or no flow condition. In 3-way valves, solenoids may directly move the spool of the main valve (solenoid-operated) or move the spool of a pilot valve (pilot-operated). In a 3-way solenoid-operated valve, an electrical signal to the solenoid pushes the solenoid rod, which shifts the main valve spool. This controls the flow of air to the outlet ports of the main valve. In a 3-way, solenoid-controlled, pilot-operated valve, an electrical signal to the solenoid pushes the solenoid rod, which shifts the spool of a pilot valve attached to the main valve. The movement of the pilot valve spool allows air flow to shift the spool of the main valve.

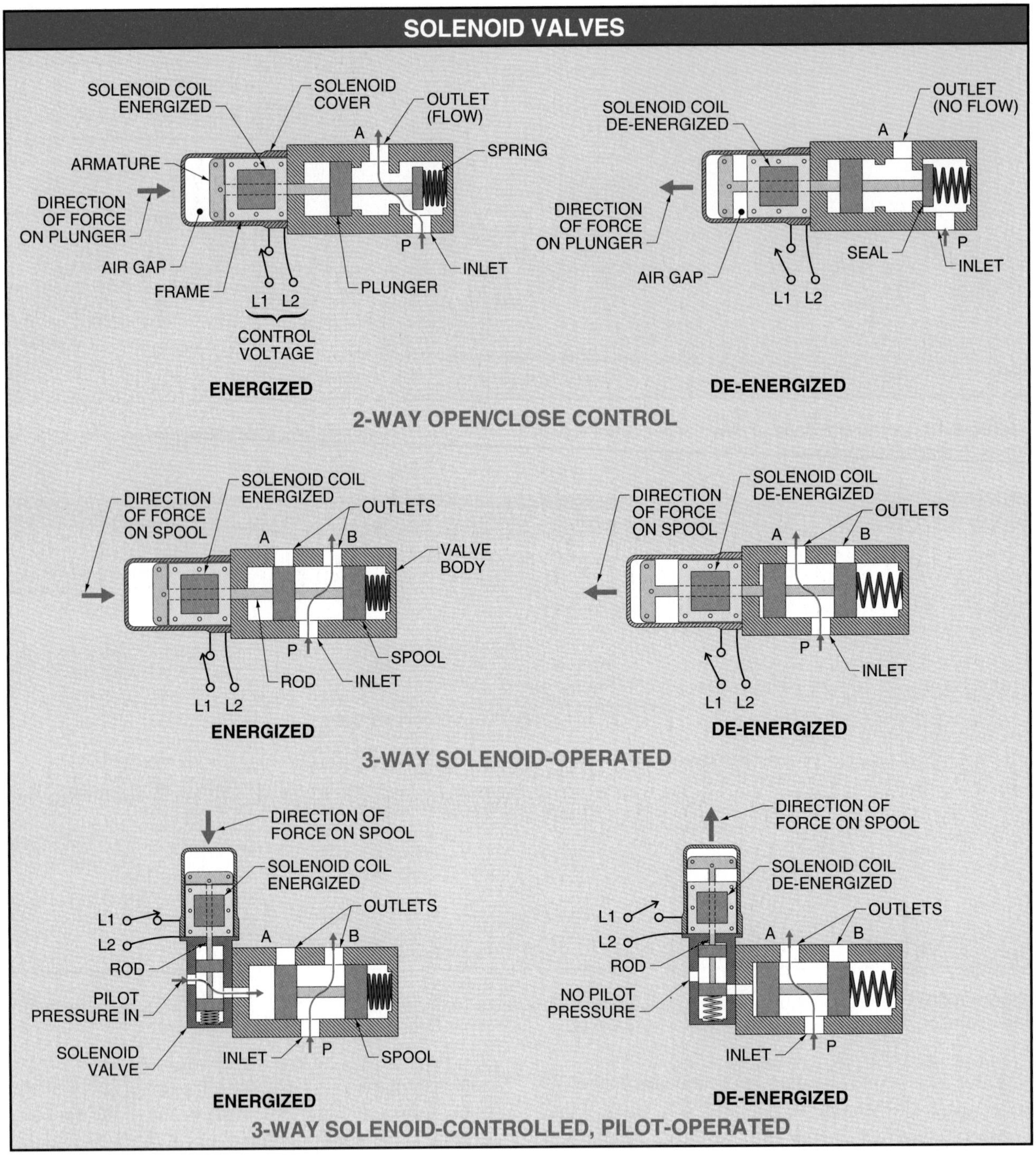

Figure 8-21. Solenoids are used to electrically open or close valves or to shift the spool within a valve to control air flow.

Flow Control Valves. A *flow control valve* is a valve whose primary function is to regulate the rate of fluid flow. Flow control valves, sometimes referred to as needle valves, are normally used for metering air flow to control motor speed, cylinder piston speed, or valve spool shifting speed (for timing). See Figure 8-22.

Cylinder piston movement or motor speed is controlled precisely and smoothly if the air flow is regulated at the exhaust rather than at the inlet. Regulating exhaust air is used on cylinders under light or no load where the volume of air supplied is less than the amount required to rapidly and smoothly move the piston. Heavy loads may be regulated at the inlet or outlet port as long as rapid speed is not required. Also, a flow control valve can be coupled with a check valve to give regulated flow in one direction and full flow in the reverse direction.

Actuators

An actuator transforms fluid energy into linear or rotary mechanical force. An air cylinder produces linear mechanical force. An air motor produces rotary mechanical force.

Air Cylinders. An *air cylinder* is a device that converts compressed air energy into linear mechanical energy. Pneumatic cylinders operate by air pressure and flow acting on a piston. Work performed is a product of the area of the cylinder bore and the air pressure.

Cylinders are classified as single-acting or double-acting and are manufactured in a variety of diameters, stroke lengths, and mounting arrangements. A *single-acting cylinder* is a cylinder in which fluid pressure moves the piston in only one direction. The piston is returned by spring or gravity force. A *double-acting cylinder* is a cylinder that requires fluid flow for extending and retracting. The major parts of a air cylinder are the cylinder body, ends, piston, piston rod, and seals. See Figure 8-23.

Bimba Manufacturing Company

Double-wall cylinders from Bimba use easy-to-assemble bolt-on mounting kits to convert one basic cylinder into various National Fluid Power Association (NFPA) mounting styles, such as front flange, side lugs, pivot, end lugs, and clevis.

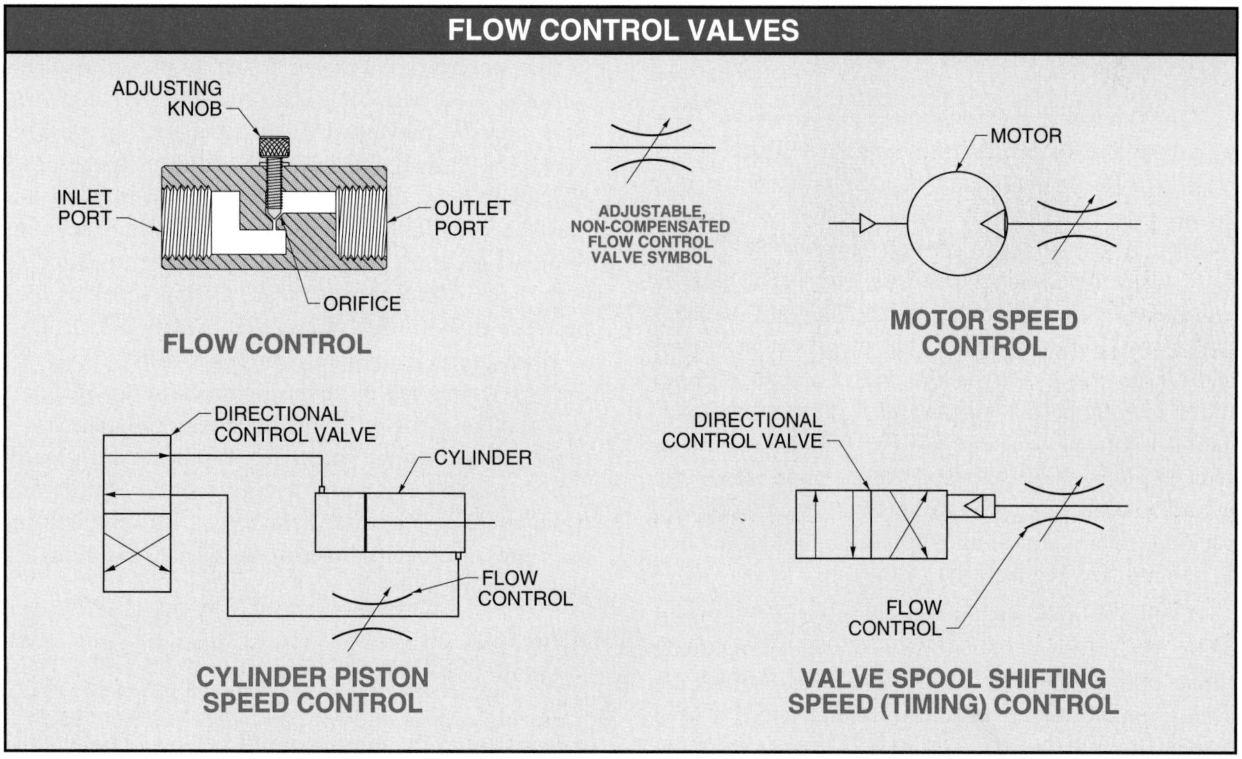

Figure 8-22. Flow control valves use a fine threaded adjusting screw to precisely meter the flow of air within a circuit.

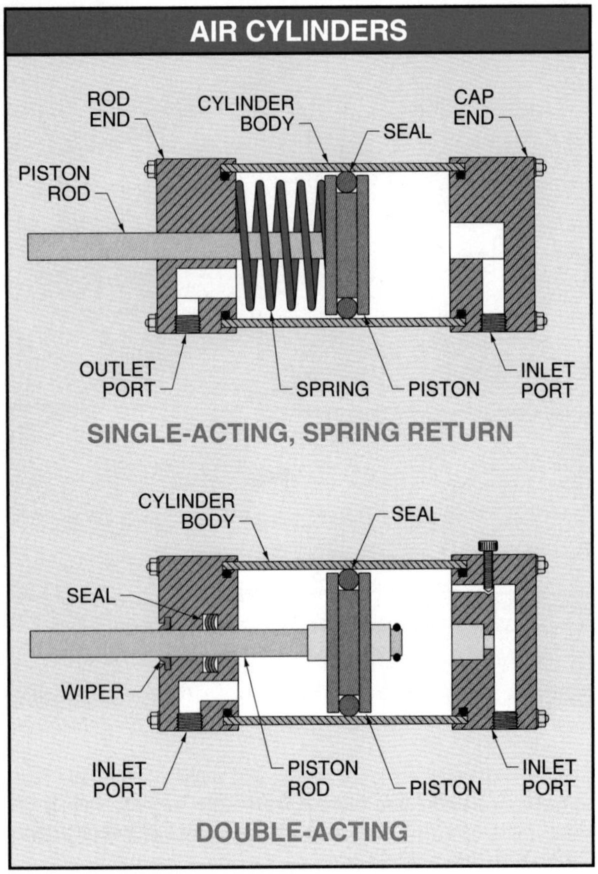

Figure 8-23. Air cylinders are actuators that are operated by the force of compressed air in one (single-acting) or two (double-acting) directions.

Air cylinders use various seals to prevent air pressure loss and to protect the internal cylinder parts from outside contaminants. A *seal* is a device that creates positive contact between cylinder components to contain pressure and prevent leakage. Seals may be static or dynamic. A *static seal* is a seal used as a gasket to seal nonmoving parts. They are used between two stationary parts that may be taken apart and reassembled. A *dynamic seal* is a seal used between moving parts that prevents leakage or contamination. Dynamic seals are used on pistons and piston rods to allow the piston and rod to slide inside the cylinder. Seals include O-rings, lip seals, wipers, and packing. Seal material may be Teflon, nylon, leather, or rubber. See Figure 8-24.

O-rings are the most commonly used seal in pneumatic applications. An *O-ring* is a molded synthetic rubber ring having a round cross section. O-rings are used as static and dynamic seals and may be used in some high-pressure operations. O-rings used in dynamic applications depend on the smoothness of the moving parts and the closeness of their fit for the best service life. O-rings should have a 10% compression between the cylinder and piston groove walls when installed. As pressure builds, the O-ring becomes distorted in an attempt to completely fill all voids at one end of its groove. This forced distortion becomes the seal under dynamic conditions.

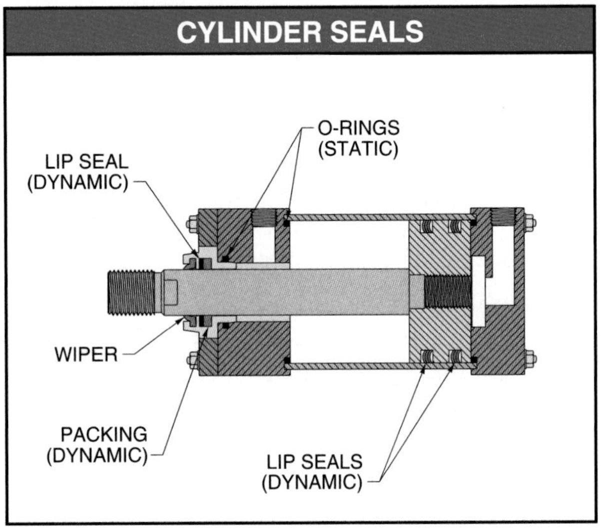

Figure 8-24. Seals are used in cylinders to prevent leakage between moving parts (dynamic seals) or prevent leakage between two immovable parts (static seal).

Pressure that becomes excessive may distort the O-ring enough to squeeze it out of its groove and into any void between the piston and cylinder wall. This may be prevented by using a backup ring between the O-ring and the wall of the piston groove. A backup ring supports the O-ring's distortion and must be installed on the side of the O-ring receiving the least pressure. Backup rings must be installed on both sides of the O-ring if the O-ring receives high pressure in both directions. See Figure 8-25.

High-pressure dynamic forces within a cylinder are best contained by using lip seals. A *lip seal* is a seal that is made of a resilient material that has a sealing edge formed into a lip. Lip seals are specifically designed for reciprocating motion and form a tighter seal as pressure increases. A lip seal uses the air's pressure on its lip to form a seal. Lip seals include V-ring and cup seals.

Long-stroke cylinders used with high air pressure create great force and high piston and rod bearing stresses and are susceptible to piston, bearing, rod, and body damage.

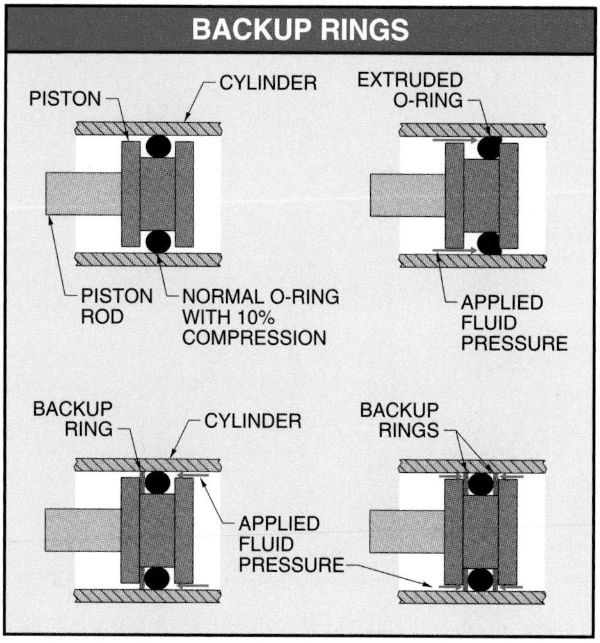

Figure 8-25. Backup rings support O-rings during compression.

A *V-ring seal* is a lip seal shaped like the letter V. V-ring seals are dynamic seals used in high-pressure and severe operating condition applications. A *cup seal* is a lip seal whose lip forms the shape of a cup. Cup seals are used specifically as a piston seal and may be used as a seal for single-acting cylinders. Two cup seals may be used back-to-back in double-acting cylinders. See Figure 8-26.

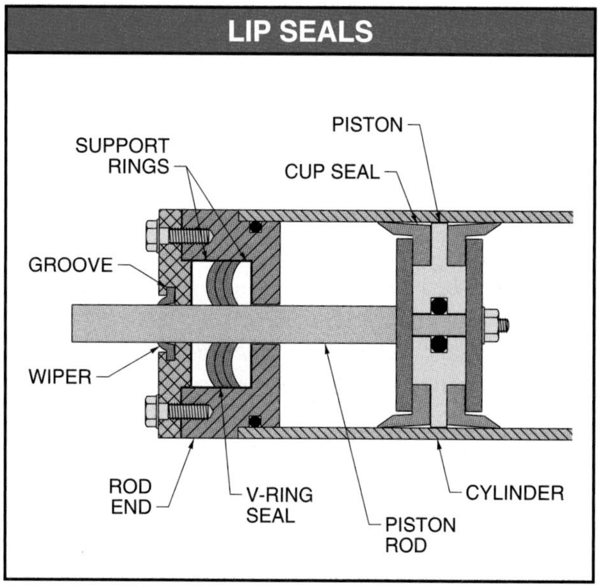

Figure 8-26. Lip seals are designed for reciprocating motion and form a tighter seal as pressure increases.

A *wiper* is a seal designed to prevent foreign abrasive or corrosive material from entering a cylinder. Wipers are designed with a lip to wipe the rod clean of foreign materials with each stroke of the piston rod. Wipers are normally installed with a slip fit into a machined groove on the outermost portion of the rod end. Wipers protect the end sealing material in addition to removing contamination from the rod. Wipers are made of metal or synthetic material and are not designed to seal against pressure.

Packing is a bulk deformable material or one or more mating deformable elements reshaped by manually adjustable compression. Packing seals the piston rod to prevent air from escaping around the rod. Packing uses various designs and materials to seal in the cylinder pressure.

A *piston cushioning device* is a device within a cylinder that provides a gradual deceleration of the piston as it nears the end of its stroke. Piston cushioning devices help to reduce the shock produced when a piston reaches the end of its stroke. See Figure 8-27.

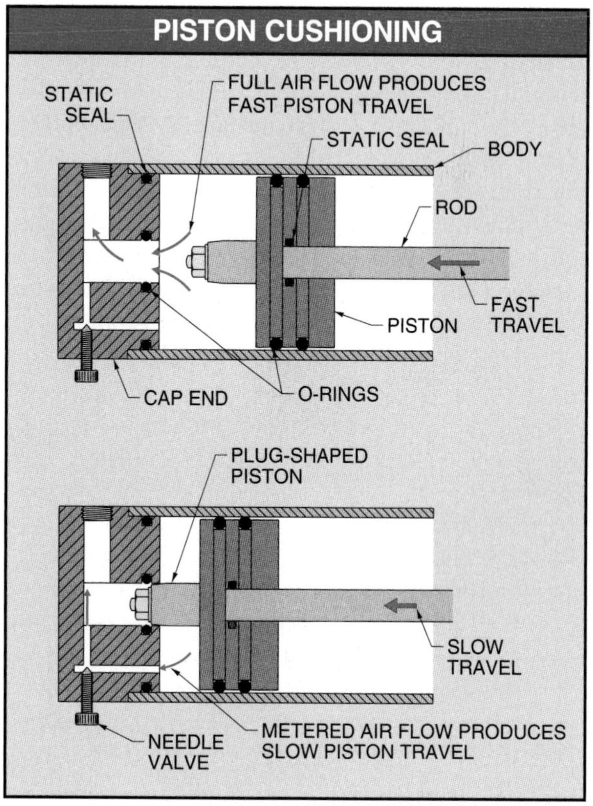

Figure 8-27. Piston cushioning devices are used at the end of a piston stroke to prevent possible damage of the load or cylinder component by slowing piston movement.

A piston cushioning device consists of a plug-shaped piston and a slightly larger bore on the inside of the cylinder end. The piston cushioning device may be sealless, have a V-ring or O-ring seal on the plug, or have an O-ring set in a machined groove in the rod end. The plug may be on one or both sides of the piston, depending on whether the cylinder is cushioned at one or both ends. The cushioning device operates by trapping a volume of air and compressing it as the piston approaches the end of its stroke. The air is trapped between the piston and the cylinder end as the plug enters the bore. Needle valves installed in the cylinder end(s) are adjusted to allow the trapped air to be metered through the port at the desired rate. This adjustment determines the intensity of the piston cushioning.

High-pressure punching machines or high-pressure forming tools require increased actuator output force. In most cases, increasing the supply pressure to the actuator is sufficient to increase the actuator output force. An intensifier may be added to a system if a pressure greater than the supply system pressure is required. An *intensifier (booster)* is a device that converts low-pressure fluid power into high-pressure fluid power. An intensifier uses the difference in the areas of the two cylinders to increase output pressure without an increase in input pressure. Intensifiers normally consist of a large surface area piston (operating piston) connected to a small surface area piston (ram). The pressure exerted by an intensifier is determined by dividing the area of the operating piston by the product of the area of the ram and the system operating pressure. See Figure 8-28. Intensifier pressure is found by applying the formula:

$$P_O = \frac{A_C}{A_R} \times P_I$$

where

P_O = outlet pressure (in psi)

A_C = area of operating piston (in sq in.)

A_R = area of ram (in sq in.)

P_I = inlet pressure (in psi)

Intensifiers enable high force levels to be produced in a low-pressure pneumatic system. Intensifiers are mounted close to the work cylinder so high pressure is confined to the small part of the circuit between the intensifier and the work cylinder. A pressure relief in the high pressure portion of the intensifier circuit helps prevent excessive pressures.

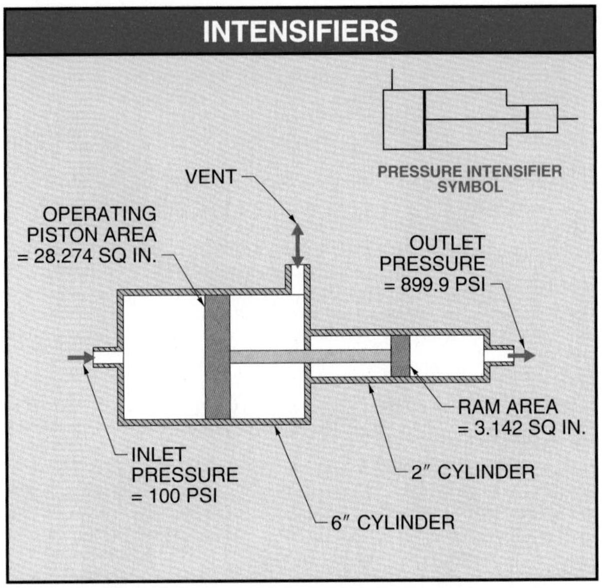

Figure 8-28. An intensifier converts low-pressure fluid power into high-pressure fluid power by using the difference in areas of two pistons to increase output pressure without an increase in input pressure.

For example, what is the outlet pressure produced by a 6″ D operating piston and a 2″ D ram operating at 100 psi? *Note:* the area of the 6″ operating piston = 28.274 sq in. and the area of the 2″ ram = 3.142 sq in.

$$P_O = \frac{A_C}{A_R} \times P_I$$

$$P_O = \frac{28.274}{3.142} \times 100$$

$$P_O = 8.999 \times 100$$

$$P_O = \mathbf{899.9\ psi}$$

Air Motors. An *air motor* is an air-driven device that converts fluid energy into rotary mechanical energy. In many cases, air motors are selected over electric motors because air motors are two to four times lighter than a direct replacement electric motor. Also, air motors can stall for an indefinite period of time without overheating or burning up. Air motors can be reversed without any strain or shock and rarely break down suddenly if maintained. Air motors normally wear slowly with a gradual reduction in power, allowing for scheduled maintenance. Disadvantages of air motors are that air motors are less efficient than electric motors, air motors slow down as the work load increases, and supplying sufficient air for operating an air motor may be a problem.

The most popular air motor is the vane air motor. A *vane air motor* is an air motor that contains a rotor with vanes that are rotated by compressed air. Vane air motors are simple in design, available in a wide range of sizes, and are easily maintained. Most vane air motors are connected to a gear train because the rotor and vanes rotate at a high speed. The gear train develops greater output torque by reducing the motor speed. Vane air motors are popular for pump motors, portable tools, mixing motors, and air-operated hoists because they are able to produce different speeds or torque from the same input pressure. Rotary vane motors are available in sizes up to 10 HP and capable of speeds of up to 15,000 rpm at operating pressures of 100 psi.

Vane air motor design is similar to a vane compressor. Vane motors develop torque by the air pressure acting on the exposed surfaces of the vanes. The vanes slide in and out of the rotor, which is connected to the drive shaft. See Figure 8-29. As the rotor rotates, the vanes follow the surface of the housing due to centrifugal force. Good lubrication is required because of the constant sliding of the vanes and housing. Lubrication may be provided by an in-line lubricator, which injects atomized oil into the air stream. Recommended oil usage for a vane air motor is one drop per minute for every 50 cfm to 75 cfm of air flow. In many cases, air motor manufacturers provide lubrication and flow rate charts.

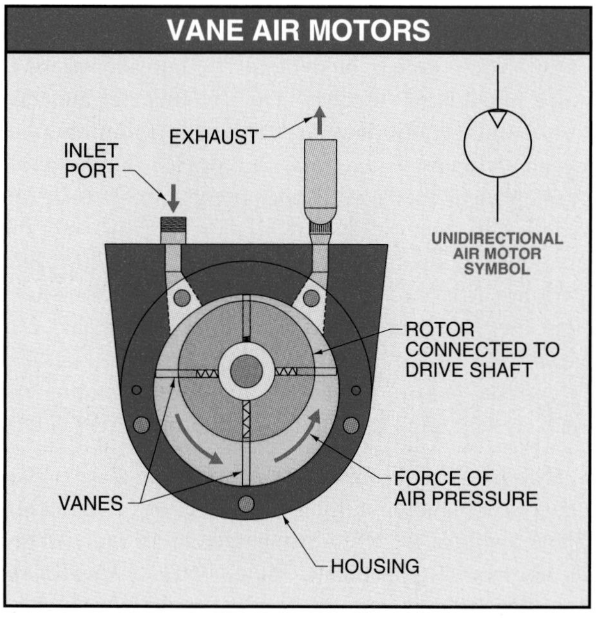

Figure 8-29. A vane air motor is an air motor that contains a rotor with vanes that are rotated by compressed air.

Gast Manufacturing Company

Air motors from Gast Manufacturing Company are available in lubricated and oilless models for use in mixing equipment, conveyor drives, pump drives, hoists, and winches.

PNEUMATIC LOGIC

Logic is the science of correct reasoning. Logic is used to describe electrical or pneumatic switching (switching logic) because of the varied number of functions available. Pneumatic logic is referred to as a binary system. A *binary system* is a system that has two values such as pressure or no pressure. In pneumatic logic, ones and zeros each have the coded meaning of pressure or no pressure. Binary functions are logical functions that may be considered ON or OFF, normally open or normally closed (NO or NC), or 1 or 0. In pneumatic logic terms, a 3-way directional control valve performs the same logical function as a light switch (ON or OFF).

Signal, Decision, and Action

All pneumatic circuits are designed based on how energy is used to perform work. The design should include a method of signaling the elements in the circuit to start their function followed by the elements making their decisions in causing an action. An *element* is a logic device that is capable of making a 0 or 1 output decision based on its input. Similar to electrical switching devices, pneumatic logic elements are the decision maker within the three basic control divisions of signal, decision, and action.

A *signal* in pneumatic logic is a condition that initiates a start or stop of fluid flow by opening or closing a valve. A signal component is a start/stop switch, relief valve, directional control valve, pressure switch, flow switch, etc. All signals rely on another condition to occur. This condition may be manual, mechanical, or automatic. For example, a manual condition occurs when a person operates a palm button or foot switch.

A mechanical condition occurs when a limit valve is mechanically operated. Limit valves are the most common pneumatic logic control. A *limit valve* is a mechanically-actuated 3-way valve that is used to either monitor motion or measure position of an object. Limit valves generally sense an object by the use of a lever with a ball or roller at its tip. For example, a container on a moving conveyor makes contact with the switching mechanism of a limit valve, thereby sending a signal. A limit valve signal is 1 or 0, where 1 is when the valve is actuated (pressure), and 0 is when the valve is released (no pressure). An example of an automatic signal is that of a flow control switch or pressure switch that automatically produces an input when flow or pressure is detected or met.

Honeywell's MICRO SWITCH Division

A limit switch, which may be a pneumatic or electric switch, is mechanically operated by an object that presses against the limit switch lever.

A *decision* is a judgment or conclusion reached or given. A decision is the selection of the action or work to be accomplished based on an input. The decision process selects, sorts, and redirects the input information to a directional control valve, which causes an action to take place. An *action* is the work of an actuator or a pilot operator, which becomes the input for another section of a control circuit. The action produced by a pneumatic logic control circuit generally results in the pilot operation of a directional control valve. In some circuits, pneumatic logic controls operate a pneumatic/electric switch. The valve or switch is used to activate an actuator or become a signal in another circuit.

Pneumatic logic controls make their decision based on input signals received and relay the order for action. In many cases, the action that ultimately occurs changes the original input signal. In other words, the completion of the final action can signal a reversal of circuit operation. For example, when a cylinder receives the order to extend, there must also be an order to retract.

Pneumatic Logic Elements

A *pneumatic logic element* is a miniature air valve used as a switching device to provide decision making signals in a pneumatic circuit. A pneumatic logic element accepts input signals, makes logical decisions based on the input signals, and provides an output signal. The output signal is used to power output devices. Pneumatic logic elements are similar to electrical relays in that they provide an output based on any input information. Pneumatic logic elements are static in nature and only require low air pressure for operation (generally between 75 psi and 90 psi) because they have no continuous air flow. The air supplied to logic elements should be filtered to remove particulates and moisture, regulated, and unlubricated. Particulate filtration should be 40μ or less.

Pneumatic logic elements are approximately one or two cubic inches in size and are considered miniature pneumatic switches. Each element is attached to a manifold, eliminating the need for plumbing between switches. A *manifold* is a device that contains passageways that enable one input signal to be divided into several output signals. Piping into the manifold comes from components such as door switches (for safety), palm switches (for activation), limit switches (for detecting product movement), etc. Piping out of the manifold is sent to pilot-operated directional control valves, which activate components such as clamping cylinders, air motors, drills, etc.

The National Fluid Power Association (NFPA) has designated symbols for pneumatic logic elements. These symbols are NFPA's standardization and are used in diagramming pneumatic logic controls. These symbols can be compared to similar electrical switching logic, electrical relay logic, or hydraulic controls. See Appendix.

Pneumatic logic elements were designed for operation sequencing, automated production, and controlling certain machine functions. Pneumatic logic elements make the decisions as to the work and order of work to be done. The output from a single element or combination of elements can provide a decision (pilot signal) required of a directional control valve that determines the machine's action. The basic logic elements used in pneumatic circuits include the AND, OR, and NOT elements.

AND. An *AND logic element* is a logic element that provides a logic level 1 only if all inputs are at logic level 1. An AND logic element has two or more inputs and one output. The output supplies pressure (ON or 1) only if all of its inputs have pressure. The output supplies no pressure (OFF or 0) if one or more input has no pressure. See Figure 8-30.

In an AND logic element, the output is 0 if both inputs are 0 and if only one input is 0. The output is 1 if both inputs are 1. Output decisions based on various inputs are shown using a truth table. A *truth table* is a table that lists the output condition of a logic element or combination of logic elements for every possible input condition. On logic elements, the A and B ports are the input ports, and the C port is the output port.

An example of AND logic use is a safety circuit on a punch press. In this circuit, the operator is required to press two palm buttons (switches) to activate the press. The operator's left hand presses one switch and the right hand is required to press the other switch before the press activates. This safety circuit keeps both hands out of a machine. An AND logic installation is similar to using two 3-way directional control valves that use manual palm buttons as actuators. In this application, valve 1 receives supply air at its inlet. The outlet of valve 1 is connected to the inlet of valve 2. The outlet of valve 2 is connected to the inlet of the actuator. As the palm button on valve 1 is pressed and held down, supply air is sent to valve 2. As the palm button on valve 2 is pressed and held down, supply air is sent to the actuator for circuit operation. Activating either valve by itself does not send air pressure to the actuator.

OR. An *OR logic element* is a logic element that provides a logic level 1 if one or more inputs are at logic level 1. An OR logic element has two or more inputs and one output. The output is 1 if any one or more input(s) are 1. The output is 0 if all inputs are 0. See Figure 8-31.

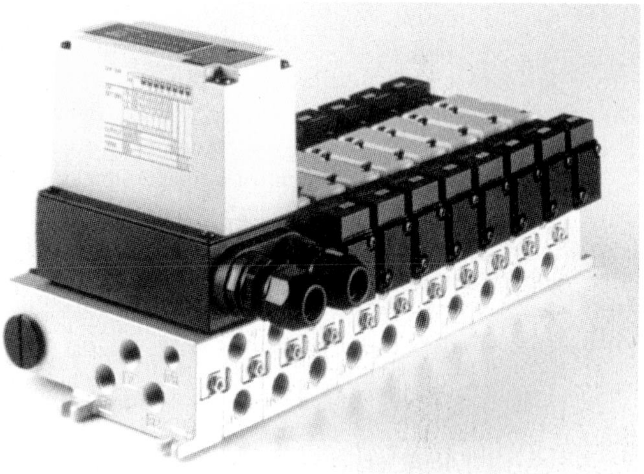

Humphrey Products Company

The Humphrey Serial Relay system features manifold-mounted plug-in solenoid valves that can be changed rapidly for circuit modification and use a single power supply.

AND LOGIC ELEMENTS

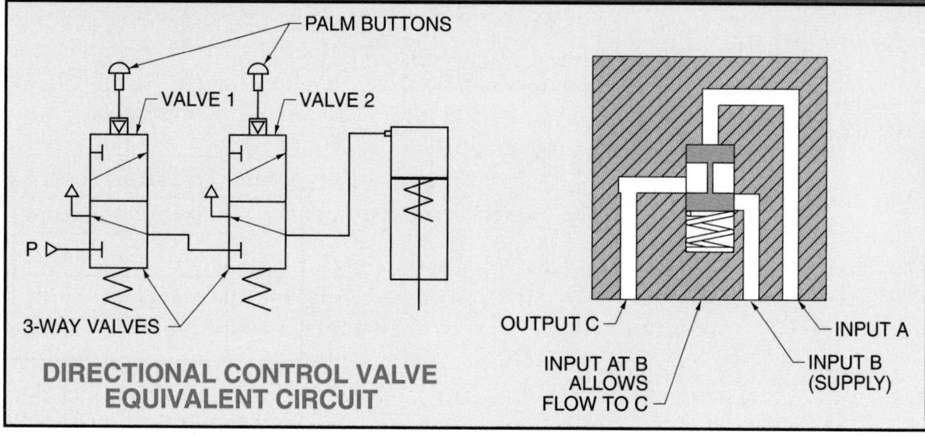

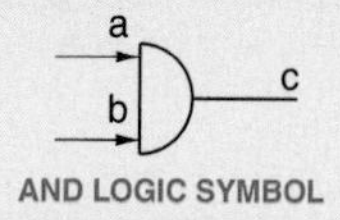

AND TRUTH TABLE

	Inputs		Output
	a	b	c
AND	0	0	0
	0	1	0
	1	0	0
	1	1	1

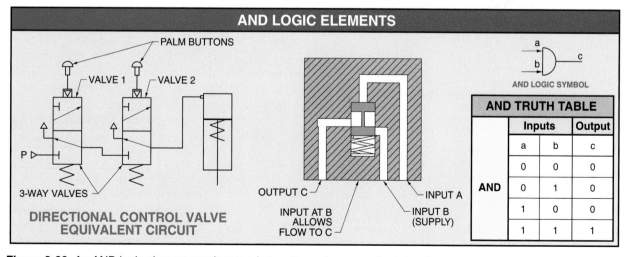

Figure 8-30. An AND logic element requires two inputs to produce an output signal.

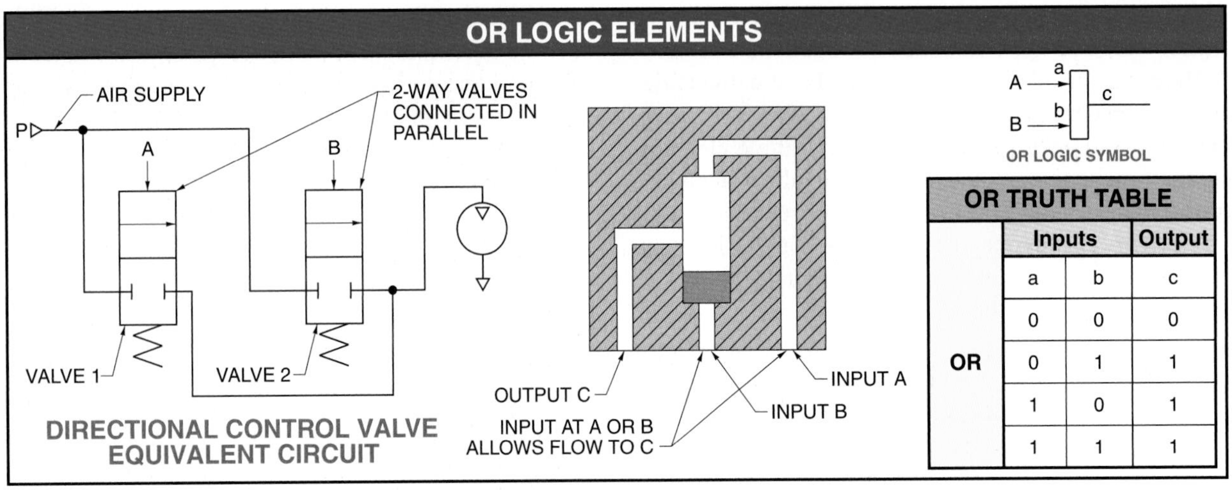

	Inputs		Output
	a	b	c
OR	0	0	0
	0	1	1
	1	0	1
	1	1	1

Figure 8-31. An OR logic element requires one or more inputs to produce an output signal.

An example of OR logic use is a circuit in which two switches at different locations operate the same cylinder or motor. Actuation of one switch or the other can start or stop the actuator. Two or more 2-way NC directional control valves connected in parallel may be used to show the operation of OR logic. The supply air is connected to the inlet of each valve and the output of each valve is connected to the actuator or device to be controlled. Operating either valve activates the actuator.

NOT. A *NOT logic element* is a logic element that provides an output that is the opposite of the input. A NOT logic element has one input, one output, and one supply. The supply normally flows through the element until an input signal stops the air flow. The output is 1 if the input is 0. The output is 0 if the input is 1. See Figure 8-32.

NOT logic applications include machine safety circuits. For example, a NOT logic element may be used for a safety block-out on a punch press. A valve is activated, which allows air flow to the input of the NOT element when a safety block is shifted under the ram on a press. This element does not allow air flow to the palm buttons, which prevents accidental ram operation. The NOT truth table shows that any signal (pressure) to port A removes any output at C. An equivalent circuit is a normally open valve. The supply is present as long as the valve is not actuated.

The life expectancy of pneumatic logic control systems is 50 to 100 times longer than corresponding electrical devices such as high-maintenance switches.

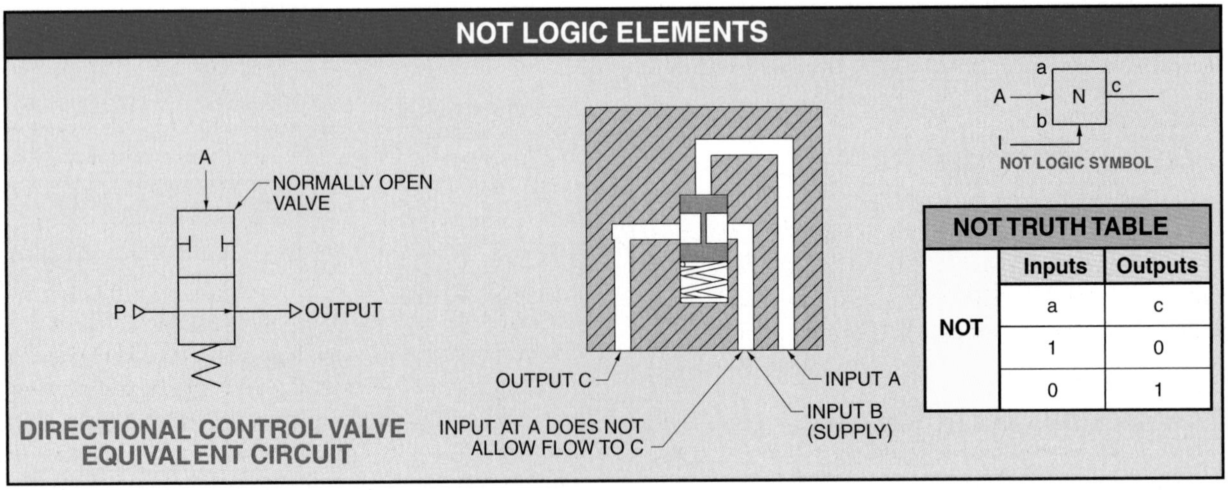

	Inputs	Outputs
	a	c
NOT	1	0
	0	1

Figure 8-32. A NOT logic element provides an output that is the opposite of the input.

Logic Element Combination

AND, OR, and NOT elements may be connected in combination to form additional logic combinations. An example of using an AND, OR, and NOT combination in a circuit is a safety anti-tie down, two-hand, palm button circuit. An anti-tie down circuit requires both hands of a machine operator to activate two palm buttons to operate a machine. See Figure 8-33.

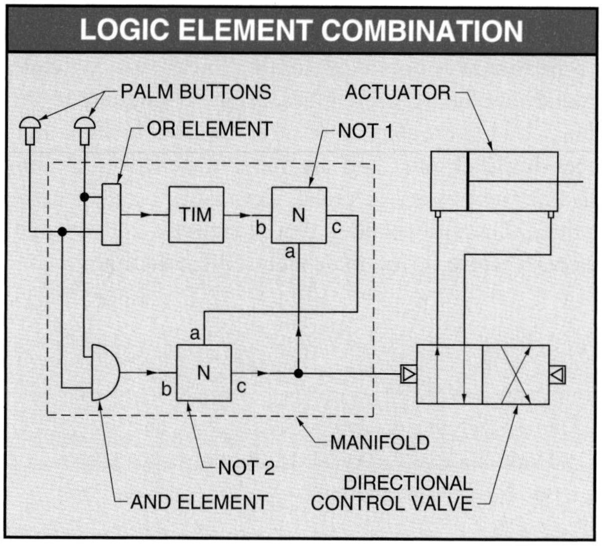

Figure 8-33. AND, OR, and NOT logic elements may be combined in a safety anti-tie down, two-hand, palm button circuit.

The anti-tie down circuit uses a timer to prevent either palm button from being tied down and defeating the safety feature. This circuit uses one OR, one AND, two NOT, and a timing element (TIM.). This combination consists of two inputs (palm buttons) and one output (pilot control to directional control valve). The machine activates if both palm buttons are pressed and held simultaneously. However, the timer prevents an output if one pushbutton is pressed before the other (normally more than $\frac{1}{10}$ of a second). The elements are arranged so that the output signal is 1 only when both palm buttons are 1 (pressed). The output is 0 as soon as either palm button is released. The output remains 0 until both palm buttons are released and pressed again.

In actual industrial applications, pneumatic logic control systems often respond twice as fast as electrical controls.

Atlas Technologies, Inc.

Two hand anti-tie down circuits are used for safety on industrial punch presses to prevent an operator's hands from being inside the press when it operates.

The output of the AND element becomes 1 and continues through NOT 2 to supply the output signal if both palm button inputs become 1 simultaneously. The pressure at the output of NOT 2 is also the input of NOT 1, preventing flow through NOT 1 after the timed period. If either palm button is held down without the other, the flow of air through the timer after its timed period continues through NOT 1 because there is no pressure at its A port. The flow from NOT 1 flows to the A port of NOT 2, holding NOT 2 closed and not allowing an output signal. To reactivate palm button operation, both palm buttons must be released, allowing the elements to exhaust. This also prevents double-tripping of a machine.

AND, OR, and NOT logic elements are the basic logic elements. Other logic elements include NAND, NOR, and flip-flop elements. As elements are added to a circuit, the complexity of the total circuit increases.

The NAND element is a combination of the NOT and AND elements. The NOR element is a combination of the NOT and OR elements. The flip-flop element is an element with two inputs and two outputs. When a signal is applied to one input of a flip-flop, a corresponding output is turned ON and the other output is turned OFF. In a flip-flop, one output is always ON and the other is OFF.

Honeywell's MICRO SWITCH Division

Electric and pneumatic controls are often used in machines designed for the manufacture of products.

Memory and timer elements are also used to produce pneumatic logic. Memory elements are capable of memorizing information for recall at a later time. Memory decisions are the 1 and 0 (ON and OFF, pressure and no pressure) produced by other elements, except that once a decision is set, the decision is retained until it is reset. Timer elements (differentiators) are created by restricting the flow of the input signal into a timing chamber (accumulator). The time it takes to build enough pressure for an output signal is the time delay.

Pneumatic logic elements may be combined to form simple or complex circuits. Pneumatic logic elements are being used more often as pneumatic circuits and are easier and safer to troubleshoot than electronic circuits and are more dependable as long as the supply air is clean, dry, and properly maintained. One drawback is that pneumatic logic circuits occupy more space than electronic circuits.

Lubrication

9

Chapter

LPS Laboratories, Inc.

Lubrication maintains a fluid film between solid surfaces to prevent their physical contact. Lubricants reduce friction, prevent wear, act as a coolant for moving parts, act as a barrier under load pressure, prevent adhesion or galling of materials, and prevent corrosion. Lubricants are classified as gas, liquid, semisolid, or solid. Lubricants must be distributed within a mechanical apparatus so all parts requiring lubrication receive the proper amount. Lubrication programs should be established within an organization to ensure that the criteria needed for dependable operation are met.

LUBRICATION

Lubrication is the process of maintaining a fluid film between solid surfaces to prevent their physical contact. A *lubricant* is a substance placed between two solid surfaces to reduce their friction. Friction occurs when an object in contact with another object tries to move. For example, walking requires friction between the feet and the floor in order to move. Stopping requires even more friction. Walking and stopping are more difficult if the friction is reduced by placing wheels, such as rollerblades, under the feet.

In addition to reducing friction, a lubricant is used to prevent wear, act as a coolant for moving parts, act as a barrier under load pressure, prevent adhesion or galling of materials, and prevent corrosion. Machines and tools depend on lubrication to ensure smooth and safe operation. Lubricants are classified as gas, liquid, semisolid, or solid. In many cases, a lubricant is a mixture of the different classes. Fluid lubricants, which include gas, liquid, and semisolid lubricants, must create a film between material surfaces to prevent contact with each other. See Figure 9-1.

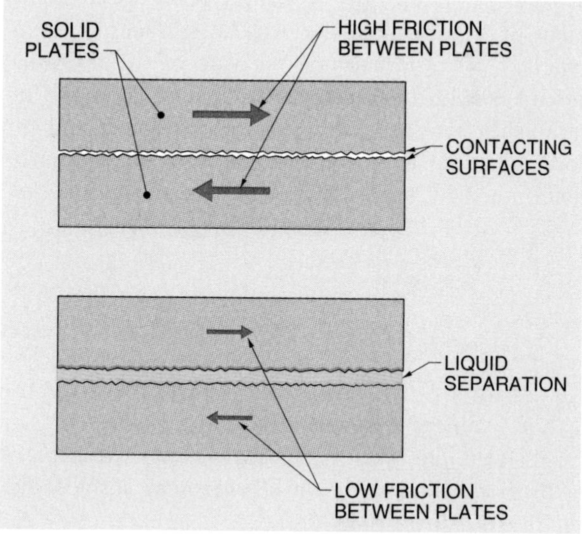

Figure 9-1. Fluid lubricants must create a film thick enough to completely separate moving surfaces.

High-pressure injection of grease through the skin, such as that from a grease gun, can cause serious delayed damage to soft tissue. A physician should be contacted immediately regardless of the size or appearance of the wound.

Dow Corning Corporation

Dow Corning® 1122 chain and open gear lube is a synthetic grease used for lubricating high-speed gears and chains and slow-moving bearings.

Coefficient of Friction

Lubrication generally involves coating surfaces with a material (grease, oil, etc.) that has a lower coefficient of friction than the original surfaces. The *coefficient of friction* is the measure of the frictional force between two surfaces in contact. It is the relationship between the weight of an object and the force required to move it. Coefficient of friction is determined by applying the formula:

$$f = \frac{F}{N}$$

where

f = coefficient of friction

F = force at which sliding occurs (in lb)

N = object weight (in lb)

For example, what is the coefficient of friction of a 25 lb object resting on a horizontal surface that requires 10 lb to move?

$$f = \frac{F}{N}$$

$$f = \frac{10}{25}$$

$$f = \mathbf{.40}$$

Greater force is required to move a body from rest (static condition) than the force required to keep it in motion (kinetic condition). Even greater force is required to move a body from rest that is lubricated to one that is not. However, once a lubricated body is in motion, less force is required to keep it in motion than if it were unlubricated. Generally, a static condition between two solid objects means neither object is moving. However, a static condition relating to coefficient of friction refers to the forces required to start a solid object in motion. See Figure 9-2.

COEFFICIENTS OF FRICTION				
Material	Unlubricated		Lubricated*	
	Static	Kinetic	Static	Kinetic
Steel-to-Steel	.8	.4	.16	.02
Copper-to-Copper	1.5	.3	.08	.02
Aluminum-to-Aluminum	1.3	—	.3	—
Nylon-to-Nylon	.3	.1	—	—
Teflon-to-Teflon	.04	.03	—	—
Graphite-to-Graphite	.1	.06	—	—

* values are approximations and vary according to lubricant type

Figure 9-2. The coefficient of friction is the measure of the frictional force between two surfaces in contact.

For example, two unlubricated solid steel objects have a static coefficient of .8. This drops to a kinetic coefficient of .4 once sliding has started. However, when the two steel objects are lubricated, the static coefficient is .16 and the kinetic coefficient drops to .02.

Boundary Lubrication

Boundary lubrication is the condition of lubrication in which the friction between two surfaces in motion is determined by the properties of the surfaces and the properties of the lubricant other than viscosity. These properties determine the behavior of the sliding system and its kinetic coefficient of friction. Effects of boundary lubrication occur mostly during stopping, starting, and periods of severe operation, where most sliding system failure is caused by inadequate lubrication during these moments.

Boundary lubrication occurs when molecules are adsorbed on metal surfaces through an exchange of electrons. This process occurs with metals that are either lubricated or unlubricated. Unlubricated metals are metals without the addition of a lubricant. However, even metals that have been cleaned of all foreign material contain a degree of lubrication. This lubrication is in the form of water vapor, adsorbed gases, or contaminants from handling. These lubricants affect the material's coefficient of friction. See Figure 9-3.

UNLUBRICATED METAL CHARACTERISTICS	
Steel-to-Steel	
Condition of Contact	**Coefficient of Friction**
Inert gas (no oxygen)	Seizure
Oxygen	.8
Moist air	.6
Water	.5
Copper-to-Copper	
Condition of Contact	**Coefficient of Friction**
Inert gas (no oxygen)	10.5
Oxygen	2.0
Moist air	1.2
Water	1.0

Figure 9-3. Cleaned, unlubricated metals contain a degree of lubrication from elements in the atmosphere, such as oxygen and moisture.

Chemisorption. In most cases, chemicals are added to a lubricant to enhance its qualities or to blend with certain chemicals in the lubricant to create chemisorption. *Chemisorption* is a chemical adsorption process in which weak chemical bonds are formed between liquid or gas molecules and solid surfaces. Certain additives applied to a lubricant increase its chemisorption rate. This produces a thicker boundary layer and forms a solid, low shear-strength film. For example, when stearic acid is mixed with iron oxides in the presence of water, a film of iron stearate is formed on a specific metal surface by chemisorption. Another example is when the chemisorption of fatty acids is enhanced to form a film of soap on metal surfaces when water and various oxides are combined in the presence of oxygen. This film is the second lubrication barrier that must be broken before permanent metal destruction begins. The first barrier is the viscosity of the liquid lubricant. See Figure 9-4.

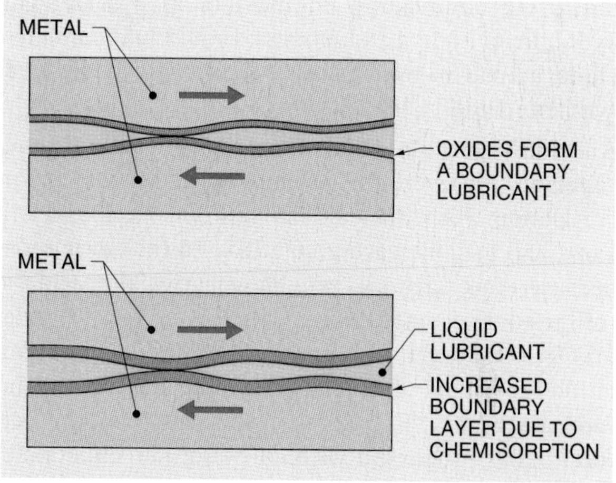

Figure 9-4. Oxides that form on sliding metals are considered boundary lubricants and are thickened and strengthened by the addition of certain chemicals.

Gas Lubricants

A *gas lubricant* is a lubricant that uses pressurized air to separate two surfaces. Other gases such as nitrogen or helium are used where inert properties are required. *Inert gases* are gases that lack active properties, such as argon or helium, that do not readily combine with other elements. For example, nitrogen is an inert gas that does not contain oxygen.

Gas lubricants are commonly used in low-friction, high-speed, high-technology applications such as grinding spindles, air-turbine dentist's drills, and computers. In addition to the low-friction and high-speed advantages, gas lubricants offer less wear and are able to operate in temperatures from –400°F to over 3000°F. Disadvantages of gas lubricants include the high-precision, high-cost relationship, and the requirement of a very clean gas supply.

Proper lubrication minimizes machine breakdown and the resulting down-time while increasing savings in labor, production time, and costly repairs. Savings from proper lubrication can amount to 10 to 100 times the cost of lubricants purchased per year.

Liquid Lubricants

A *liquid lubricant* is a lubricant that uses a liquid, such as oil, to separate two surfaces. Liquid lubricants are the preferred lubricants because of their reliability, versatility, and flexibility. Besides reducing friction, liquid lubricants are used for heat removal (in combustion engines) or as a sealer (in hydraulic cylinders and pumps). Liquid lubricants include animal/vegetable oils, petroleum fluids, and synthetic fluids.

Animal/Vegetable Oils. Until 1860, animal and vegetable oils were the primary substances used for lubrication. Fish oils, animal tallow, and sperm oil were, and in some cases, are still used for lubrication because their fatty oils provide superior slipperiness (oiliness) and smoothness. Animal and vegetable oils are used mostly in the food industry. Food grade lubricants are lubricants approved for use on food machinery. Food grade lubricants can contact food being processed without being detrimental to human health.

Animal and vegetable fatty oils are applied to other lubricants to increase their load-carrying capabilities because of their great slipperiness. These lubricant compounds are used in automatic transmission fluids, industrial gear oils, and marine engine oils. However, due to their organic origin, animal and vegetable oils can support bacteria and require sterilization. Also, germicides and antiseptic agents must be added to reduce rusting and bad odors when used in water-soluble lubricants such as cutting oils.

Exxon Company

Synesstic synthetic lubricants from Exxon are used in refinery hot liquid pumps to reduce pump failures.

Animal and vegetable oils contain fatty acids which tend to form more fatty acids and gums as they oxidize. Care must be taken to prevent an increase in the fluid's oxygen content. Also, these lubricants must be replenished regularly because they break down.

Petroleum Fluids. A *petroleum fluid* is a fluid consisting of hydrocarbons. A *hydrocarbon* is any substance that is composed mostly of hydrogen and carbon. Petroleum is composed of 12% hydrogen and 85% carbon, with a small amount of other elements such as oxygen, sulfur, nitrogen, etc. Petroleum fluids make up approximately 90% of the total lubricants used. Petroleum is not as susceptible to oxidation, bacteria, and the formation of acid and gummy residues as animal or vegetable oils.

Petroleum is formed by an evolutionary process that takes many millions of years. This process begins with oil vapor given off into the atmosphere by plants. The oil vapor, sometimes seen as a blue haze over heavily vegetated areas, settles or is washed to the ground by rain or snow. The oil works its way deep into rock voids where it is concentrated and processed under high temperatures and pressures. See Figure 9-5.

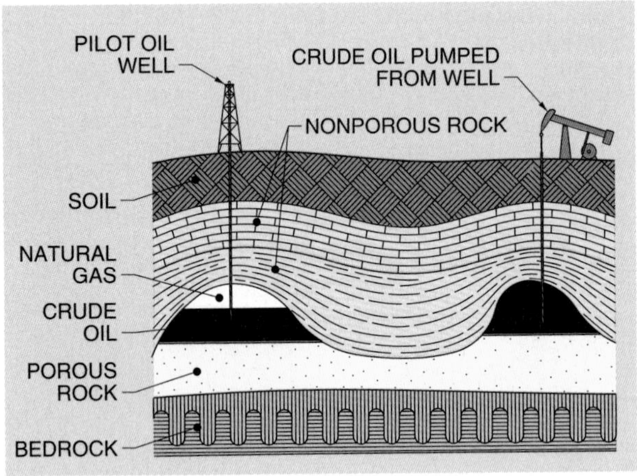

Figure 9-5. All animals and plants release hydrocarbons that, when absorbed into the earth's surface, combine to form pools of gas or crude oil.

> ***Never pressurize, cut, weld, drill, grind, or expose empty lubricant containers to heat. Empty lubricant containers retain residue and may explode and cause injury or death. Do not attempt to clean the residue because even a trace of remaining material constitutes an explosive hazard.***

The concentrated fluid body (crude oil) works its way back to the earth's surface through natural oil and gas seeps or through drilling. The earth's plants release approximately 175 million tons of hydrocarbons into the air each year. The evolution of the hydrocarbons into petroleum takes from approximately 50 million to 500 million years.

Crude oil is found in various physical forms ranging from a light gas such as methane to a heavy tar. To maintain a consistent, stable, uniform, and reliable lubricating liquid, the crude oil is processed in steps of heating, distilling, and filtering. The final step of processing a petroleum lubricant is the application of certain additives for individual and special applications. New compounds are being developed daily to handle average consumer needs. Recent submarine, subterranean, and outer space equipment applications have added to the research and development process. See Figure 9-6.

Lubricant Additives. Additives are used to intensify and improve certain characteristics of a base oil for specific applications. Additives protect a machine from harm, maintain the integrity of the lubricant, and improve the physical properties of the lubricant, such as odor or color control. Additives are included by the manufacturer after performing considerable tests.

Additives include oxidation inhibitors (to provide long bearing or gear life), rust inhibitors (to prevent rust), fatty materials (to improve film strength), powdered lead or graphite (to prevent galling), viscosity index improvers (to ease machine movement in cold weather), and demulsifiers (to separate out water). A technician must understand the various choices and characteristics of the different lubricants before specifying a certain lubricant and must also understand the damage that can occur to a machine if its lubricant is arbitrarily changed.

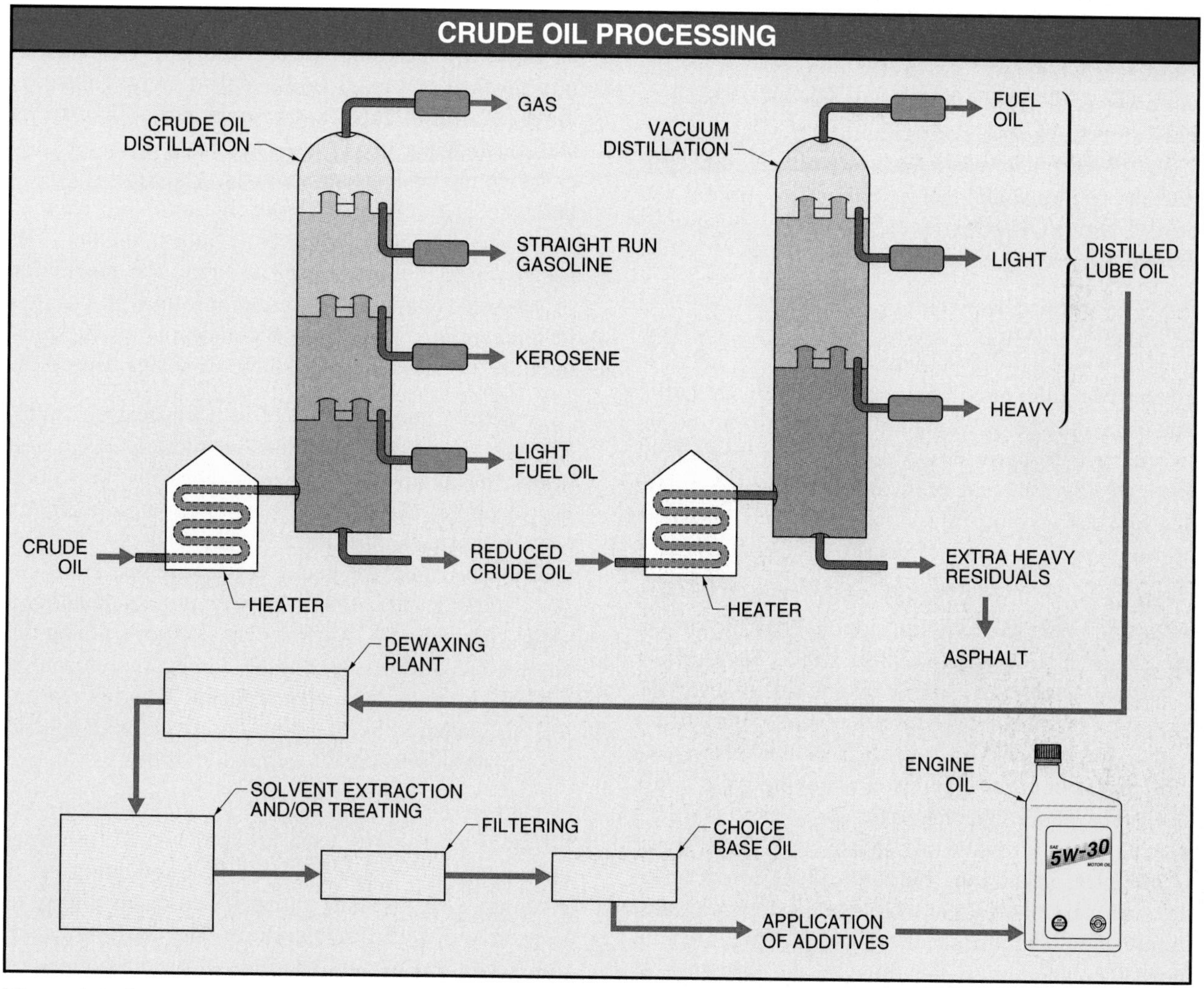

Figure 9-6. Crude oil must be processed to become a uniform and dependable product.

Exxon Company

UNIVIS Special oils from Exxon are widely used in aluminum cold rolling applications because they were designed to minimize aluminum staining during the annealing process.

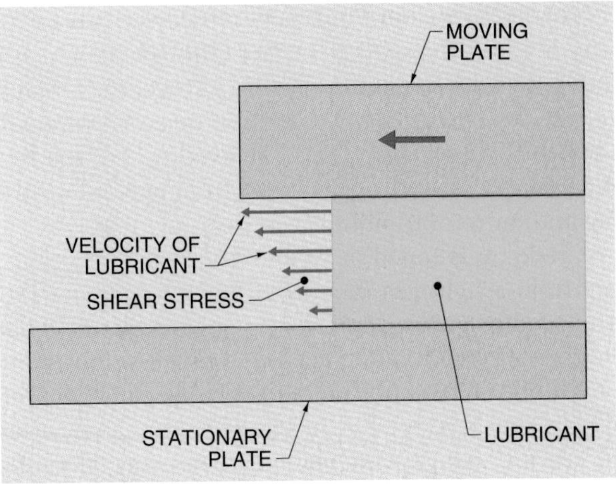

Figure 9-7. Shear strength, a liquid's ability to remain as a separator between solids in motion, relies greatly on the liquid's viscosity.

Viscosity. *Viscosity* is the measurement of the resistance of a fluid's molecules to move past each other. The flow rate is the most important property of a lubricant. Flow rate is directly proportional to the viscosity of a fluid. In general, the viscosity of a fluid is its thickness. For example, syrup is more viscous (thicker) than water. Viscosity affects fluid flow rate and frictional and thermal properties. Higher viscosity fluids offer greater frictional resistance, thereby increasing thermal activity.

Gear oil viscosity should be high enough to protect tooth surfaces and low enough to offer a good heat transfer. Generally, high-viscosity lubricants result in greater film thickness. Low-viscosity lubricants offer thin film thickness. Maintaining any separation between moving parts ensures equipment protection. However, as loads increase or viscosities decrease due to a rise in temperature, the protective layer can be ruptured as a result of decreased shear strength.

Shear strength is a liquid's ability to remain as a separator between solids in motion. It is the ability of a material to withstand shear stress. *Shear stress* is stress in which the material on one side of a surface pushes on the material on the other side of the surface with a force parallel to the surface. For example, a lubricant is contained between a moving plate and a stationary plate. The lubricant in contact with the moving plate moves with, and has the same speed as, the plate in motion. The lubricant in contact with the stationary plate attempts to remain stationary. See Figure 9-7. The influence of the moving plate to move the lubricant is the same as the influence of the stationary plate to stop the lubricant movement.

Shear strength relies greatly on the liquid's viscosity. As the viscosity of a lubricant decreases or as forces increase, the shear strength of a lubricant is weakened or broken and metal to metal contact is made. A liquid lubricant's shear strength may be weakened when the liquid's viscosity decreases because of increasing temperatures. The viscosity/temperature characteristic for each liquid lubricant is cataloged and given a viscosity index number. *Viscosity index* is a scale used to show the magnitude of viscosity changes in lubrication oils with changes in temperature. Under basic conditions, as the temperature of oils increases, their viscosity decreases.

Additives must be applied to a lubricant to stabilize the viscosity/temperature characteristic because one of the main functions of a lubricant is cooling. For example, automobile engine oils must maintain even viscosity/temperature characteristics. Low-viscosity lubricants are better for low-temperature engine starting. The low viscosity is also beneficial while the engine is warming up. However, once the engine and oil heat up, engine damage may occur if the oil viscosity is greatly reduced. For this reason, oils are prepared by blending improvers with special base materials to create thermal stability.

Commercial lubricating oil is categorized by six different groups with each group designed for a specific application. See Figure 9-8. Lubricating oil is given an SAE viscosity rating based on its ability to flow at a specific temperature. The SAE viscosity rating is a number assigned based on the volume of a base oil that flows through a specific orifice at a specified temperature, atmospheric pressure, and time period. A high viscosity rating results from a small volume of oil flowing through the orifice caused by high resistance to flow. A low viscosity rating results from a large volume of oil flowing through the orifice caused by low resistance to flow. The higher the viscosity rating number, the thicker the oil. For example, a 40 weight oil is thicker than a 10 weight oil. The viscosity rating number assigned to an oil does not change, but oil viscosity can change with temperature and ambient pressure.

OIL GROUPS/APPLICATION
Group A: Automotive
SAE 10W SAE 20W SAE 30 SAE 40 SAE 50
Group B: Gear Trains and Transmissions
General Purpose Oils
Group C: Machine Tools
SUS 75 SUS 80 SUS 90 SUS 140 SUS 250
Group D: Marine Propulsions and Stationary Power Turbines
Turbine Oils
Group E: Turbojet Engines
Aviation Oils
Group F: Reciprocating Engines
Aviation Oils

Figure 9-8. Commercial lubricating oil is categorized by six different groups with each group designed for a specific application.

During startup, oil is cool and does not flow easily. As the machine and oil warms, it flows more easily. For this reason, most machine wear occurs during startups when the cool oil provides less lubrication. Some oils may require a longer machine operating time before proper oil flow to bearing surfaces occurs. This situation can cause premature wear during startup when lubrication may be deficient.

Oil film thickness decreases with an increase in oil temperature and can be completely depleted in high operating temperatures. Oil specified for an application should flow at low temperatures but still protect the engine at high ambient and/or operating temperatures. Oil manufacturer's recommendations use standards provided by the Society of Automotive Engineers (SAE), the American Society for Testing and Materials (ASTM), and the American Petroleum Institute (API). These organizations provide standards in both viscosity and additive packages for the majority of lubricants manufactured worldwide.

Always follow the recommendations of the machine manufacturer or the recommendations of the oil manufacturer when selecting an oil for an application. In general, operating temperature, frequency of stopping and starting, shock producing actions, and any unusual conditions must be considered when selecting a lubricant.

Synthetic Fluids. A *synthetic fluid* is a lubricant, having a petroleum base, which has improved heat, chemical resistance, and other characteristics than straight petroleum products. Synthetic lubricants are higher priced than petroleum lubricants but have certain advantages over petroleum lubricants. One major advantage is that synthetic lubricant viscosity changes less with temperature changes. Another advantage is that the oxidation rate of a synthetic lubricant is more stable at higher temperatures.

Synthetic lubricants evolved out of petroleum product shortcomings, such as a relatively low temperature stability, high oxidation rate, and short shelf life. Many of the additives designed for petroleum lubricant enhancement are now used in synthetic lubricants.

The various synthetic lubricant compositions behave differently and have improved qualities over petroleum lubricants. However, some are not compatible when mixed. A mixture of certain synthetic lubricants with petroleum lubricants, or one synthetic composition with another, may cause rapid deterioration of seals or an increase in oxidation. During oil changes, old lubricant should be replaced with the same type of lubricant.

In some cases, a recommended change may have to be made. Great care must be taken when switching lubricants. During replacement of a petroleum product with a synthetic product, any residual amounts of petroleum product left in a machine may be enough to cause a chemical reaction and deteriorate seal material. In most cases, a flushing fluid is required. Consult both lubricant manufacturers for the correct replacement method.

Synthetic lubricants, due to their increased efficiency, high performance, and long life, will be used in an increasing number of applications despite the initial cost savings of petroleum lubricants. In addition, if properly maintained, machine life can be extended by using synthetic lubricants while petroleum supplies are decreasing.

Oil-Rite® Corporation

The new grease dispenser developed by Oil-Rite® Corporation uses an air signal to pulse the meter, causing a predetermined amount of grease to be ejected. The dispenser is available with single feed or multiple feed outlets. The grease injectors are precisely adjustable for fluid outputs from 0 cu in. to .012 cu in. per cycle.

Semisolid Lubricants

A *semisolid lubricant* is a lubricant that combines low-viscosity oils with thickeners, such as soap or other finely dispersed solids. A *dispersed solid* is a solid that is finely ground in order to be spread. Dispersed solids such as soap, clay, lead, etc. must be able to adsorb or trap lubricating oils. A *grease* is a semisolid lubricant created by combining low-viscosity oils with thickeners, such as soap or other finely dispersed solids. The oil and thickeners in grease act as a whole with only slight bleeding of the oil.

Semisolid lubricant base oils may consist of mineral oils, silicones, diesters, esters, or fluorocarbons. Thickeners may consist of soaps from aluminum, sodium, lithium, or calcium. Complex soaps may be used with other solids such as clay, graphite, Teflon, or lead. When thickeners are mixed, one thickener is predominant, such as soap, and the other thickeners are included as an additive. Each grease base has its own specific characteristics. For example, aluminum soap offers clarity, calcium soap is water-resistant, lithium soap allows high-temperature use, clay is used for extreme temperatures, and fiber is added to resist being thrown off.

Grease is classified by thickener grade. The National Lubricating Grease Institute (NLGI) has established a series of nine consistency grades. The higher the NLGI number, the stiffer the grease and the less penetration it has. See Figure 9-9. Each group is designed for a specific temperature range and purpose. For example, grades NLGI 00 and 0 can be used at temperature as low as −30°F. Grades NLGI 1 and 2 are recommended for applications operating in the temperature range of 0°F to 350°F. Grades NLGI 00, 0, and 1 are recommended for centralized lubrication systems because of their relatively high flow characteristics. The increasing NLGI grade corresponds to an increase in the percentage of thickener. Grades 0, 1, and 2 are the most widely used in industry. The more fluid grades, such as 000 and 00 are used where thickened oil is desired, such as in gearboxes, where leakage may occur when using oil lubricants.

NLGI GREASE GRADES

NLGI Grade	Penetration*	Stiffness
000	1.75–1.87	VERY SOFT
00	1.57–1.69	↑
0	1.32–2.30	
1	1.22–1.33	
2	1.04–1.16	
3	.86–.98	
4	.68–.80	
5	.51–.62	↓
6	.33–.45	VERY HARD

* in in.

Figure 9-9. The National Lubricating Grease Institute (NLGI) has established a series of nine consistency grades for grease.

Effects of Temperature

Grease consistency varies with a change in temperature. Grease typically softens as temperatures increase. As temperatures increase, greases become soft enough to separate the oil from the thickener. This is known as the grease dropping point. The *grease dropping point* is the maximum temperature a grease withstands before it softens enough to flow through a laboratory testing orifice. In practical applications, as machine speeds or loads increase the temperature of a grease, the oil within the grease is generally lightened enough to run off or bleed away from the grease's thickener, leaving behind a hardened, dark-colored substance. The dropping point of greases vary according to type. Some special greases do not exhibit a dropping point. See Figure 9-10.

GREASE CHARACTERISTICS			
Grease Thickener	Oil Viscosity*	Dropping Point**	Maximum Usable Temperature**
Soap Base			
Aluminum	275	195	175
Calcium	300	180	175
Lithium	300	340	300
Nonsoap Base			
Colloidal silica	400	†	†
Bentonite	400	†	†

* SSU @ 100°F

** in °F

† these products bleed oil but do not exhibit a melting point

Figure 9-10. Grease base thickeners are chosen according to the temperature and viscosity demands placed on the lubricating product.

Prolonged or repeated contact of oils and greases with the skin can result in plugging of sweat glands and hair follicles. This may lead to skin irritation or dermatitis. Oils and greases should be removed from the skin promptly. In case of skin contact, wash skin thoroughly with soap and warm water. A waterless hand cleaner and a soft skin brush can also be used to remove oils and greases.

The viscosity of grease also changes with a change in its shear rate (applied pressure). As shear rates increase, the viscosity of the thickener decreases to the point of the base oil. This is common in many high-speed machines. A grease with an extreme pressure additive is used in applications where high speeds are combined with high loads.

PLI Incorporated

MEMOLUB® is a self-contained electromechanical positive displacement lubricating system containing a motor, gearbox, piston pump, microprocessor, and battery pack. The system ejects lubricant at over 200 psi and is used in applications where frequent, reliable lubrication is required.

Solid Lubricants

A *solid lubricant* is a material such as graphite, molybdenum disulfide, or polytetrafluoroethylene (PTFE) that shears easily between sliding surfaces. Most solid lubricants are used to provide a dry film in a high-load, low-speed application. Solid lubricants may be used with semisolid lubricants (greases), which fuse or bond the lubricant to the substrate, or as a solid block.

Graphite, produced from coal, is milled to obtain grayish black crystalline flakes. Graphite has low shearing forces, especially between each graphite flake layer. The graphite flake layers run parallel to the bearing surface. Although graphite slides easily over itself, it also adheres well to bearing surfaces. Graphite requires moisture, such as atmospheric humidity, to be effective. This makes graphite a poor lubricator where humidity is absent, such as aerospace applications. Graphite has a low coefficient of friction and an ability for service in temperatures up to 500°C. Graphite may be used dry, mixed with oil or grease, or as a solid block.

Molybdenum disulfide is a common substance that is similar in appearance to graphite. Molybdenum disulfide is found as an ore in molybdenite. Molybdenum disulfide does not require the presence of moisture. This allows it to be used in dry atmospheres, high vacuums, and in high temperatures. This makes molybdenum disulfide perfect for aerospace applications. Molybdenum disulfide becomes a bonded film when used with the binder combination of ceramic and resin. Its film thickness is generally .001″ or .002″ and this thickness is typically its limitation for life.

Polytetrafluoroethylene (PTFE) is a long-chain polymer produced from ethylene, which is a coal by-product. A *polymer* is the result of a chemical reaction in which two or more small molecules combine to form larger molecules. PTFE is reinforced by being combined with other substances, such as glass fibers, rayon, or other synthetics. This offers outstanding strength capabilities with low wear and a low coefficient of friction. PTFE is used in pharmaceutical and food plants as well as chemical industries because it is non-toxic and has excellent chemical stability. PTFE is commonly used for non-stick surfaces of household cooking utensils.

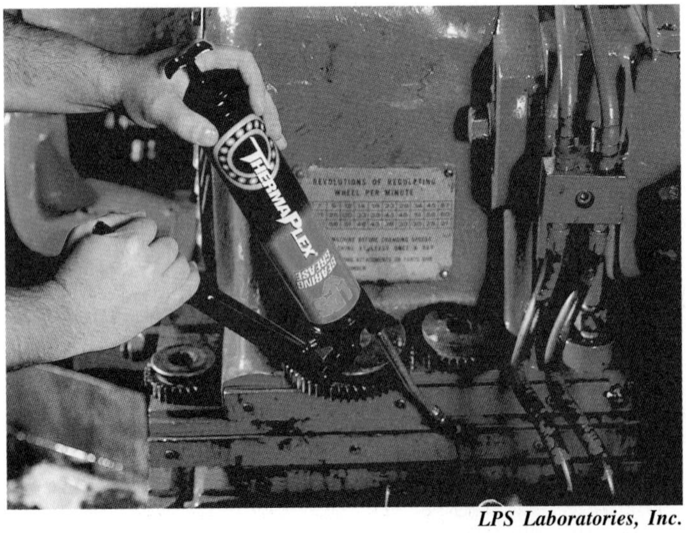

LPS Laboratories, Inc.

ThermaPlex Lo-Temp Bearing Grease from LPS Laboratories is effective to −58°F and is used in refrigeration equipment, mechanical bridges, automatic safety gates, and starter motors.

LUBRICANT APPLICATION

Lubricants must be distributed within a mechanical apparatus so all parts requiring lubrication receive the proper amount. Except in sealed units where there is a void of oxygen, lubricants require replenishment or replacement through occasional topping or by means of a feed system.

Oil Application

Oil must be replenished because oil, being a liquid, does not cling to and remain on the sides of moving parts. In addition, splashing or sloshing vaporizes the oil, thus reducing the amount of oil present for an application. Oil lubrication may be applied to bearing elements by submersion, wick, drip, or centralized systems. See Figure 9-11.

Submersion Systems. A *submersion system* is a lubrication system in which the bearings are submerged below oil for lubrication. Oil submersion systems allow oil to be carried throughout load bearing surfaces. The level of oil in most systems is critical to prevent churning or drag when the level is too high or inadequate lubrication accompanied by high temperatures when the level is too low. Proper oil levels require that ball or roller bearings be immersed halfway up the lowest ball or roller, and gears be immersed to twice the tooth height. Where gears are designed in a vertical train, the oil level should be just below the shaft of the lowest gear. When oil transfer is inadequate using these methods, ring, chain, or splash devices may be added to assist oil movement.

Wick Systems. A *wick system* is a lubrication system that uses capillary action to convey oil to a bearing surface. *Capillary action* is the action by which the surface of a liquid is elevated on a material due to its relative molecular attraction. This is seen in the raising of a fuel through the wick in an oil lamp. In a wick system, wicks or felt pads consisting of pores or spaces allow oil to penetrate and spread from an oil source to bearing surfaces. Oil feed rate is controlled by the thickness of felt pads, number of wick strands, or the length of material immersed in the oil.

In applications where repeated skin contact with lubricants occurs, use protective skin creams, such as silicone-base creams, applied to clean hands prior to contact.

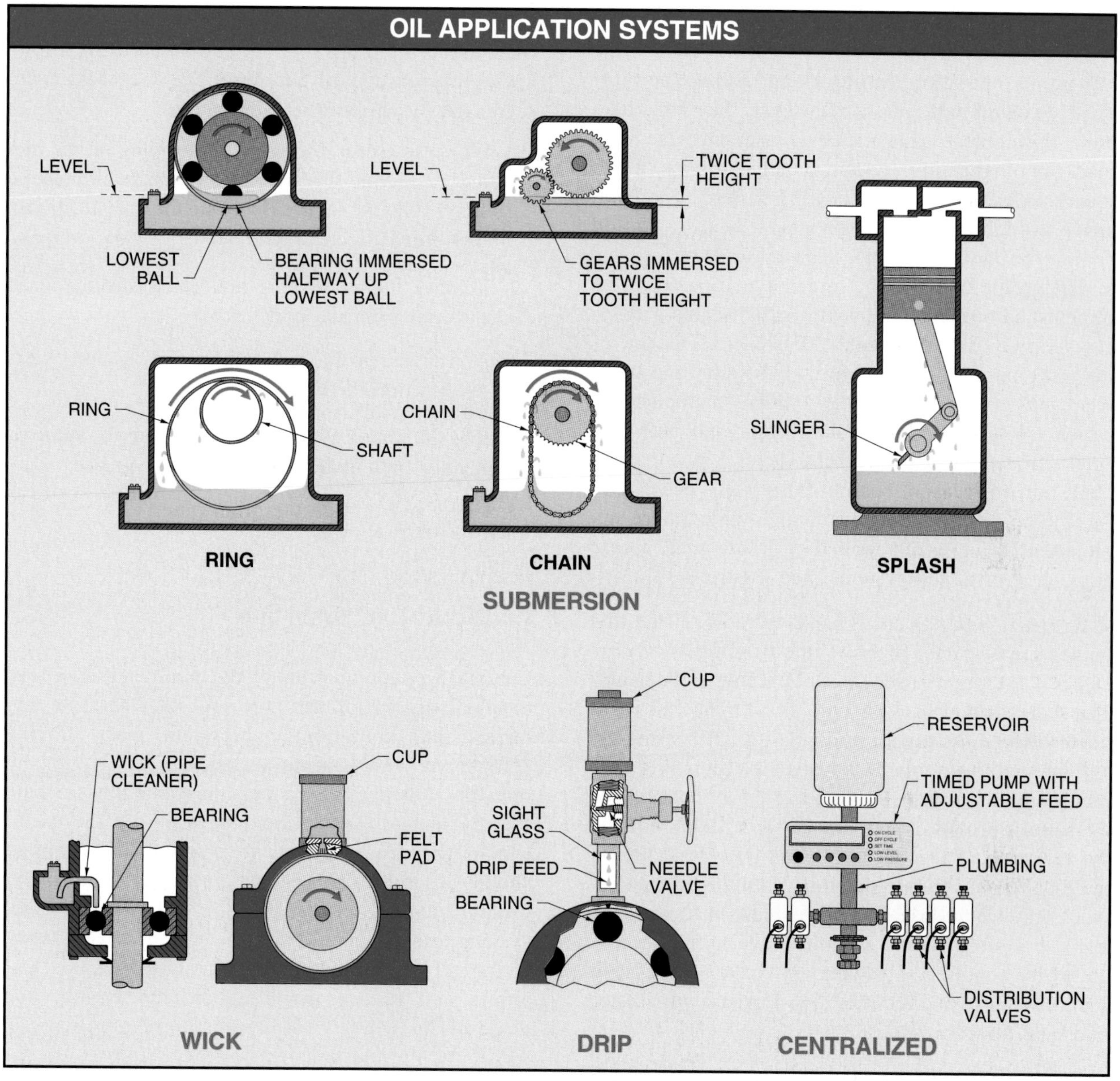

Figure 9-11. Lubricant delivery methods vary according to the design, speed, and accessibility of the machinery.

Drip Systems. A *drip system* is a gravity-flow lubrication system that provides drop-by-drop lubrication from a manifold or manually-filled cup through a needle valve. The needle valve is adjusted for flow regulation. A drip system offers flow regulation in addition to a higher rate of liquid flow than that of wick systems. Most drip systems are equipped with a sight glass for liquid level and dripping motion observation.

Centralized Systems. A *centralized system* is a lubrication system that contains permanently installed plumbing, distribution valves, reservoir, and pump to provide lubrication. These systems, although initially high in cost, simplify lubrication processes. The initial cost is offset by longer running, dependable operating equipment with more reliable lubrication periods and less possibility of contamination.

Empty lubricant drums should be completely drained, properly bunged, and promptly returned to a drum reconditioner. All other empty lubricant containers should be disposed of in an environmentally safe manner and in accordance with local, state, and federal regulations for petroleum distillates.

Grease Application

For proper operation, rolling elements must be thoroughly coated with grease. However, they must also have the correct quantity of grease. Overgreasing leads to overheating, aerating, and churning of the grease, resulting in early bearing failure. The total space available in a rolling element bearing should contain no more than 50% grease lubricant.

The grease of all open, unsealed bearings must be replenished because the oil from all unsealed greases eventually bleeds off from the thickeners. This leaves the thickeners, having no value, to solidify and burn. Also, all greases eventually oxidize and corrode, which is evident by their dark color and burnt oil smell. In sealed bearings, relubrication is not practical. For this reason, bearing failure will occur over time, especially when used in high-temperature applications. Greases are applied by grease guns, grease cups, or centralized systems. See Figure 9-12.

A *grease gun* is a small hand-operated device that pumps grease under pressure into bearings. A *grease cup* is a receptacle used to apply grease to bearings. The receptacle is packed with grease. The cap is rotated to force the grease into bearings. A centralized system contains permanently installed plumbing, reservoir, and air supply to provide the lubrication. The air supply provides pressure above a diaphragm in the reservoir. The pressure forces the grease below the diaphragm into a pipe which is routed to the devices requiring lubrication. The grease used in a centralized system should be one grade softer than is otherwise required. For example, a No. 0 grease is used instead of the required No. 1 grease when used in a centralized system.

Motor Regreasing. Motors equipped with grease fittings and drain plugs must be regreased using a low-pressure grease gun. See Figure 9-13. Motors are regreased by applying the procedure:

1. Wipe the grease fitting and drain plug on the motor. Also wipe the grease gun nozzle and expel a small amount of grease from the gun to ensure that grease is uncontaminated.
2. Remove the drain plug and clean any hardened material from the port.
3. Add grease to the motor until new grease is expelled from the drain plug port.
4. Run the motor without the drain plug for approximately 10 minutes to expel excess grease.
5. Clean and replace the drain plug.

Lubricant Contamination

Lubricant contamination is the main cause of mechanical system failure. Dirt and other abrasive materials that contaminate lubricant wear moving components and bearing surfaces as the equipment operates. Mechanical system lubricants must remain free of damaging contaminants to ensure the useful life of the drive system. A maintenance program should be used to provide a routine schedule for lubricant changes and/or filtering. Some mechanical systems use oil purifiers to clean and recycle lubricating oil while the equipment is operating. Oil purifiers can reduce maintenance, lubrication, and disposal costs.

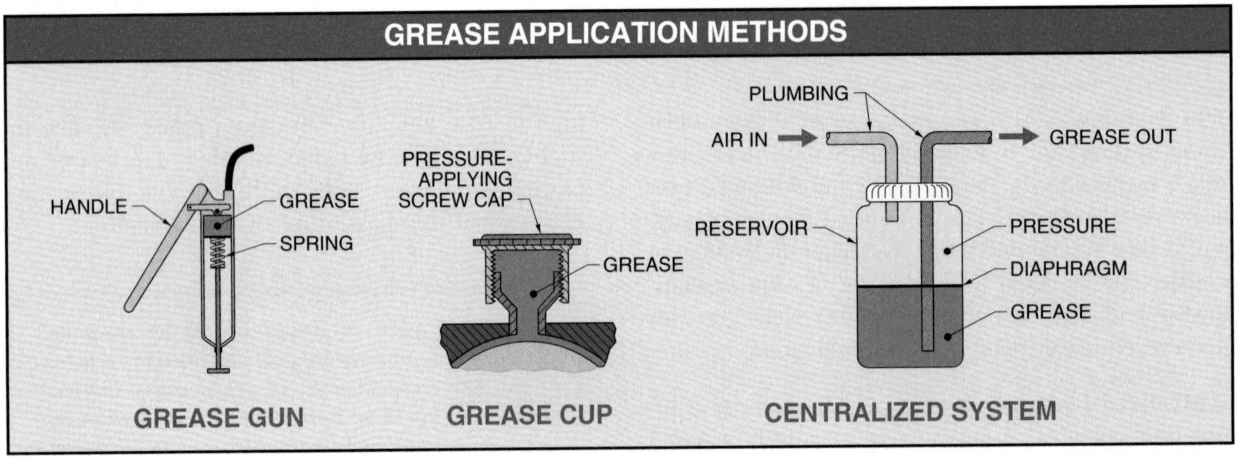

Figure 9-12. Applying grease may be accomplished through hand-operated pumping methods or by centralized systems serving whole facilities.

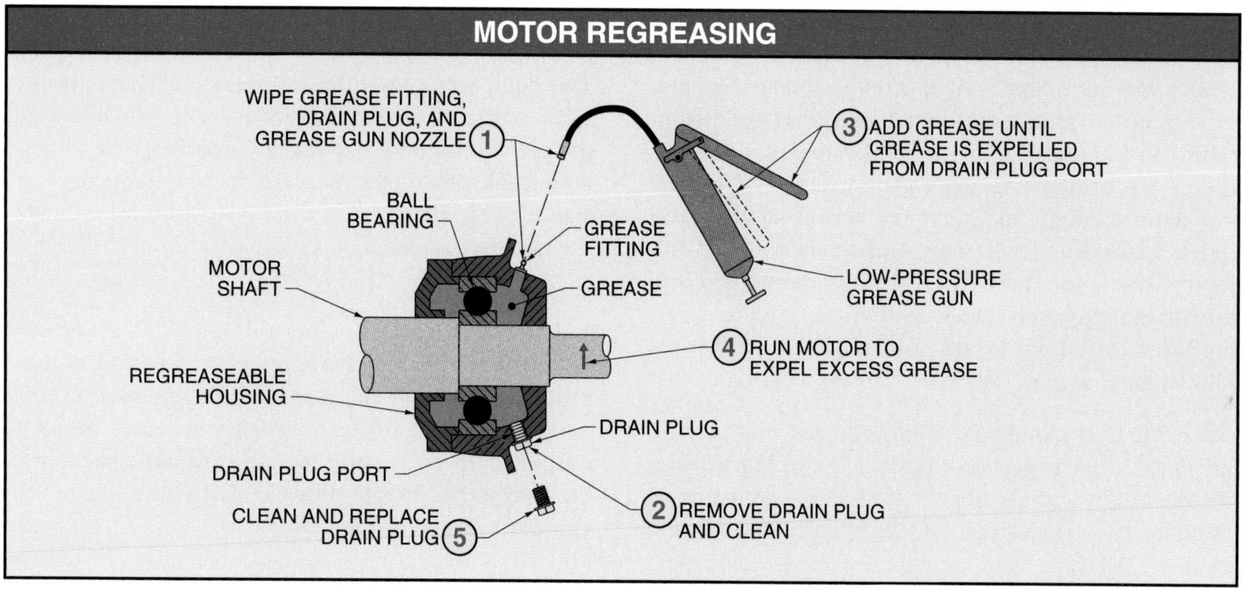

Figure 9-13. Regreasing a motor may include purging the old grease from between the shaft and the housing.

Contamination of the lubricant can also reduce the effectiveness of the lubricant. Common contaminants include dirt and water. Lubricant contaminated with dirt subjects moving components to a constant flow of abrasives. Water that mixes with the lubricant also reduces the effectiveness of the lubricant. The water causes bearing components to rust, increasing friction and eventually causing bearing failure. Sources of water can include condensation and environments with high humidity. Periodic oil changes are necessary to remove water from the lubricant. Oil that is contaminated with water has a milky appearance.

LUBRICATION PROGRAMS

A lubrication program should be established within an organization to ensure that the criteria needed for dependable operation are met. The lubrication program should establish the parameters of each lubricant used and include training of personnel in the methods of application of each lubricant. The program should establish the scheduling and frequency requirements needed by each machine. Finally, recorded results of equipment failure should be reviewed frequently to determine possible faulty lubrication or lubrication practices.

Minimize exposure to petroleum and synthetic-base hydrocarbons because these liquids and vapors pose potential human health risks that vary from person to person.

Oil Analysis

Oil analysis is a predictive maintenance technique that detects and analyzes the presence of acids, dirt, fuel, and wear particles in lubricating oil to predict equipment failure. Lubricating oil analysis is performed on a scheduled basis. An oil sample is taken from a machine to determine the condition of the lubricant and moving parts. Oil samples are commonly sent to a company specializing in lubricating oil analysis. See Figure 9-14.

Equipment commonly used for oil analysis is a spectrometer. A *spectrometer* is a device that vaporizes elements in the oil sample into light. The light is separated into a spectrum and then converted into electrical signals, which are processed and displayed by a computer.

Data management software for lubricant analysis allows the user to be connected (via modem) directly to an analysis laboratory where oil samples are tested and results are sent to the user providing a quick, comprehensive analysis and trending. The analysis commonly includes oil viscosity, particle count, wear particle concentration analysis, and wear particle analysis. Analysis information allows the user to determine if wear is occurring, what component is affected, what is causing the wear, and how far damage has progressed.

The lubricant condition rating is identified as normal, marginal, or critical based on the results of the sample, comparison with previous data, and the machine condition analyst's experience with the particular type of equipment. A normal condition rating indicates the lubricant is within expected levels and require no corrective action. A marginal condition rating indicates that critical physical properties and/or trace elements are outside expected levels and require minor maintenance action such as increased sampling frequency. A critical condition indicates that the majority of physical properties are outside the expected levels. A lubricant and/or wear condition problem exists that requires definitive maintenance action.

Wear Particle Analysis. *Wear particle analysis* is the study of wear particles present in the lubricating oil. While lubricating oil analysis focuses on the condition of the lubricating oil, wear particle analysis concentrates on the size, frequency, shape, and composition of the particles produced from worn parts. The equipment condition is assessed by monitoring wear particles in the lubricating oil. Normal wear occurs as equipment parts are routinely in contact with each other. An increase in the frequency and size of wear particles in the lubricating oil indicates a worn part or predicts possible failure.

For example, lubricating oil samples having consistent wear particle readings over a period of time provide a baseline measurement. An increase in wear particles may indicate premature wearing of parts. Large, sharp wear particles indicate parts sheared in the equipment. Fractured wear particles indicate broken parts in the equipment.

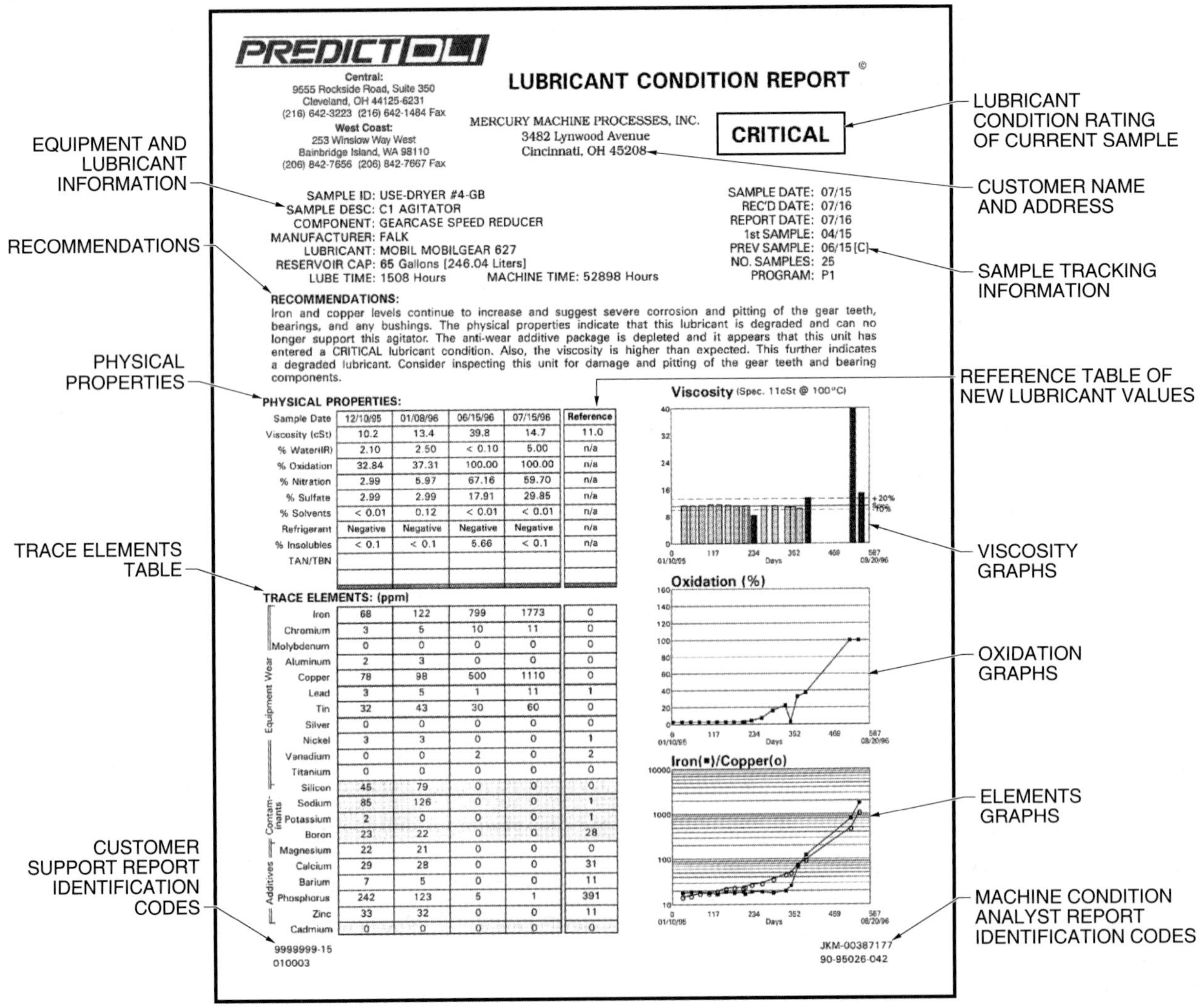

PREDICT DLI

Central:
9555 Rockside Road, Suite 350
Cleveland, OH 44125-6231
(216) 642-3223 (216) 642-1484 Fax
West Coast:
253 Winslow Way West
Bainbridge Island, WA 98110
(206) 842-7656 (206) 842-7667 Fax

LUBRICANT CONDITION REPORT ©

MERCURY MACHINE PROCESSES, INC.
3482 Lynwood Avenue
Cincinnati, OH 45208

CRITICAL

SAMPLE ID: USE-DRYER #4-GB
SAMPLE DESC: C1 AGITATOR
COMPONENT: GEARCASE SPEED REDUCER
MANUFACTURER: FALK
LUBRICANT: MOBIL MOBILGEAR 627
RESERVOIR CAP: 65 Gallons [246.04 Liters]
LUBE TIME: 1508 Hours MACHINE TIME: 52898 Hours

SAMPLE DATE: 07/15
REC'D DATE: 07/16
REPORT DATE: 07/16
1st SAMPLE: 04/15
PREV SAMPLE: 06/15 [C]
NO. SAMPLES: 25
PROGRAM: P1

RECOMMENDATIONS:
Iron and copper levels continue to increase and suggest severe corrosion and pitting of the gear teeth, bearings, and any bushings. The physical properties indicate that this lubricant is degraded and can no longer support this agitator. The anti-wear additive package is depleted and it appears that this unit has entered a CRITICAL lubricant condition. Also, the viscosity is higher than expected. This further indicates a degraded lubricant. Consider inspecting this unit for damage and pitting of the gear teeth and bearing components.

PHYSICAL PROPERTIES:

Sample Date	12/10/95	01/08/96	06/15/96	07/15/96	Reference
Viscosity (cSt)	10.2	13.4	39.8	14.7	11.0
% Water(IR)	2.10	2.50	< 0.10	5.00	n/a
% Oxidation	32.84	37.31	100.00	100.00	n/a
% Nitration	2.99	5.97	67.16	59.70	n/a
% Sulfate	2.99	2.99	17.91	29.85	n/a
% Solvents	< 0.01	0.12	< 0.01	< 0.01	n/a
Refrigerant	Negative	Negative	Negative	Negative	n/a
% Insolubles	< 0.1	< 0.1	5.66	< 0.1	n/a
TAN/TBN					

TRACE ELEMENTS: (ppm)

Equipment Wear	Iron	68	122	799	1773	0
	Chromium	3	5	10	11	0
	Molybdenum	0	0	0	0	0
	Aluminum	2	3	0	0	0
	Copper	78	98	500	1110	0
	Lead	3	5	1	11	1
	Tin	32	43	30	60	0
	Silver	0	0	0	0	0
	Nickel	3	3	0	0	1
	Vanadium	0	0	2	0	2
	Titanium	0	0	0	0	0
Contaminants	Silicon	45	79	0	0	0
	Sodium	85	126	0	0	1
	Potassium	2	0	0	0	1
	Boron	23	22	0	0	28
Additives	Magnesium	22	21	0	0	0
	Calcium	29	28	0	0	31
	Barium	7	5	0	0	11
	Phosphorus	242	123	5	1	391
	Zinc	33	32	0	0	11
	Cadmium	0	0	0	0	0

9999999-15
010003

JKM-00387177
90-95026-042

Predict\DLI

Figure 9-14. Lubricating oil analysis is used to predict potential equipment malfunction or failure.

Bearings

10
Chapter

NTN Bearing Corporation of America

Bearings guide and position moving parts to reduce friction, vibration, and temperature. Machine efficiency and accuracy depends on proper bearing selection, installation and handling, and maintenance procedures. Bearings are classified as rolling-contact or plain bearings. Rolling-contact bearings include ball, roller, and needle bearings. Successful bearing installation requires cleanliness, correct bearing selection, mounting methods, tool use, and tolerance specifications.

BEARINGS

A *bearing* is a machine part that supports another part, such as a shaft, which rotates or slides in or on it. A bearing guides and positions moving parts to reduce friction, vibration, and temperature. The length of time a machine retains proper operating efficiency and accuracy depends on proper bearing selection, installation and handling, and maintenance procedures. Bearings are available with many special features, but all incorporate the same basic functioning parts.

Bearings are designed to support radial, axial, and radial and axial loads. A *radial load* is a load in which the applied force is perpendicular to the axis of rotation. For example, a rotating shaft resting horizontally on, or being supported by, a bearing surface at each end has a radial load due to the weight of the shaft itself. See Figure 10-1. In a radial load, the shaft should have negligible end-to-end movement. An *axial load* is a load in which the applied force is parallel to the axis of rotation. For example, a rotating vertical shaft has an axial load due to the weight of the shaft itself.

Radial and axial loads occur when a combination of the two loads are present. For example, the shaft of a fan blade is supported horizontally (radial load) and is pulled or pushed (axial load) by the fan blade. Bearings are classified as rolling-contact (anti-friction) or plain bearings.

Rolling-Contact Bearings

A *rolling-contact (anti-friction) bearing* is a bearing composed of rolling elements between an outer and inner ring. Rolling-contact bearings are referred to as anti-friction bearings because they are designed to roll on a film of lubricant, which separates the metal components. All rolling-contact bearings are given a life expectancy (fatigue life). *Fatigue life* is the maximum useful life of a bearing. The fatigue life of a bearing is determined by expected speed, temperature, lubrication, and load rate standards. These standards are recommended by the Anti-Friction Bearings Manufacturers Association (AFBMA).

Standard bearings never wear out or reach their fatigue life. Average bearings normally exceed the life of the machine where they are installed. Bearing failure is generally the result of a deviation from bearing fatigue life standards. Properly maintained bearings do not wear out. They are meant to eventually, if run long enough, fail due to fatigue. Close examination of a failed bearing often provides evidence to the cause of failure. For example, dark discolored metals indicate high temperatures, rusting surfaces indicate high moisture and/or improper lubrication, and split or fractured rings indicate an improper fit or assembly.

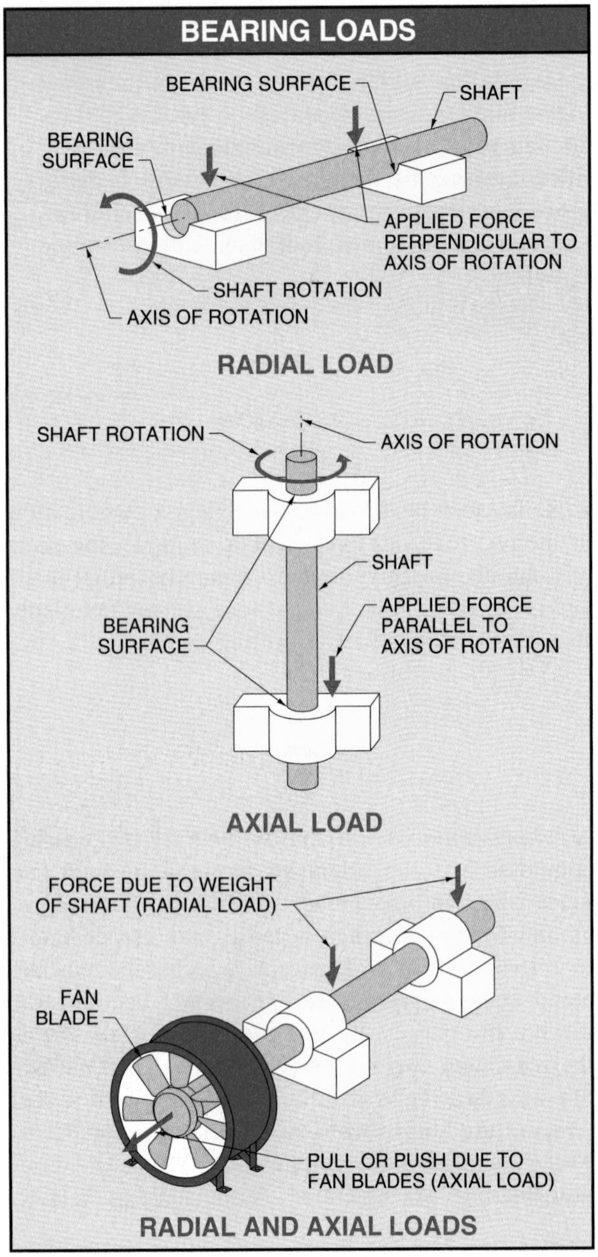

Figure 10-1. Bearings are designed to support radial loads, axial loads, or radial and axial loads.

Early bearing failure due to a minor deviation from bearing fatigue life standards is regarded as bearing service life. *Service life* is the length of service received from a bearing. Service life is generally shorter than fatigue life. This is due to less-than-optimal operating conditions. For example, doubling the speed of a bearing reduces its service life to one half its fatigue life. Doubling the load on a bearing reduces its service life by 6 to 8 times.

Rolling-contact bearings include ball, roller, and needle bearings. A *ball bearing* is an anti-friction bearing that permits free motion between a moving part and a fixed part by means of balls confined between inner and outer rings. A *roller bearing* is an anti-friction bearing that has parallel or tapered steel rollers confined between inner and outer rings. A *needle bearing* is an anti-friction roller-type bearing with long rollers of small diameter. See Figure 10-2.

The rolling-contact bearing categories may be further divided into more specific designs or configurations. For example, ball bearing races may be made deeper for supporting radial and axial loads or additional rolling elements (balls or rollers) may be installed to support heavier loads.

Rolling-Contact Bearing Construction. Ball and roller bearings are constructed of an outer ring (cup), balls or rollers, and an inner ring (cone). The outer ring is generally slid or pressed easily into a housing. The inner ring is generally pressed on a shaft with a tighter fit than the outer ring. Ball bearing rings are designed with a groove known as a race. A *race* is the track on which the balls of a bearing move.

Needle bearings contain an outer ring (cup) and rollers. The rollers are retained in a cage and bear directly on the rotating shaft. Bearing precision and cost is determined by the smoothness of the ground surfaces (grade of finish) and the quality of tolerances. Better finishes produce less friction, lower temperatures, smoother movements, and longer bearing life, but usually cost more.

Additional bearing components include cages, separators, seals, and snap rings. Cages are devices used to hold the balls or rollers in place. Bearings are designed as open for lubrication injection, or as sealed which hold lubricant for the life of the bearing. Seals are used to retain lubrication as well as prevent contamination from dust, dirt, or other solids. Snap rings allow the bearing to be inserted into a housing and held at a certain depth.

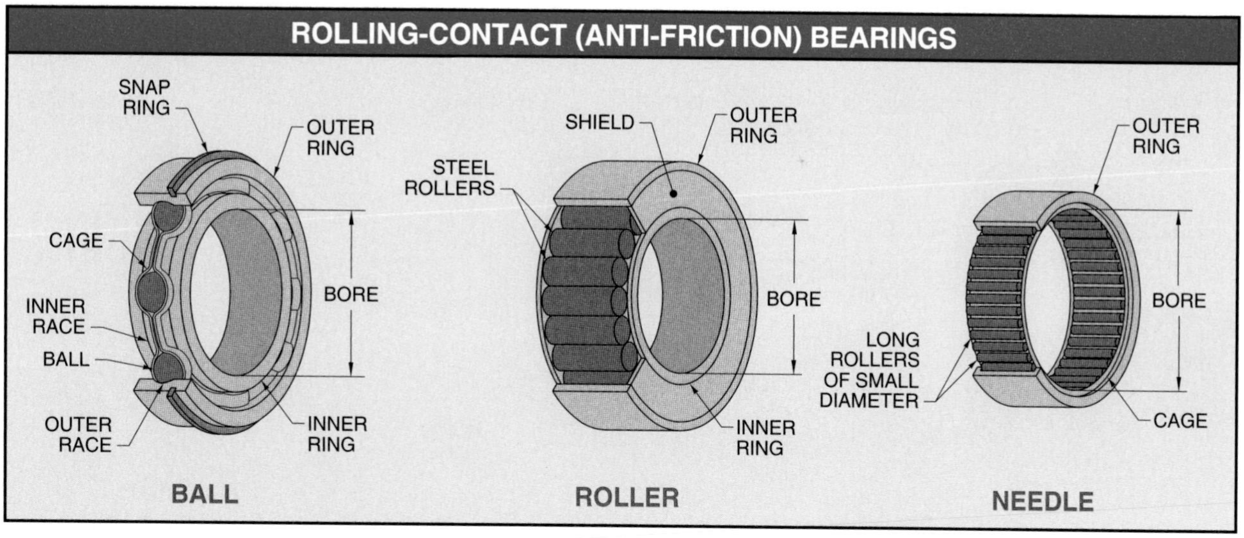

Figure 10-2. Rolling-contact (anti-friction) bearings include ball, roller, and needle bearings.

Ball Bearings. Ball bearings are anti-friction bearings that permit free motion between a moving part and a fixed part by means of balls confined between inner and outer rings. Ball bearings are selected based on the application of the bearing. Ball bearings may be designed for light or heavy loads, radial or axial loads (or combination of each), or harsh or clean environments.

General-use ball bearings are designed as single-row radial, single-row angular-contact (axial), or double-row radial or axial based on the direction of applied force. See Figure 10-3.

A *radial bearing* is a rolling-contact bearing in which the load is transmitted perpendicular to the axis of shaft rotation. Single-row radial bearings have a single row of balls and may be designed with or without loading slots. A *loading slot* is a groove or notch on the inside wall of each bearing ring to allow insertion of balls. Bearings with loading slots are referred to as maximum capacity bearings due to the ability to add the maximum number of balls. Applying axial (thrust) loads to maximum capacity bearings causes rapid damage. A *Conrad bearing* is a single-row ball bearing without a loading slot that has deeper-than-normal races. Conrad bearings allow for axial and radial loads. Installing Conrad bearings in the wrong direction results in immediate damage.

An *angular-contact bearing* is a rolling-contact bearing designed to carry both heavy axial (thrust) loads and radial loads. The ability to provide axial (thrust) support is due to the high shoulder on one side of the outer race. This high-shouldered race, acting as a seat for the balls, provides high thrust load capacities in one direction only.

Bearings that are designed for thrust loads must be installed in only one direction to prevent the load from separating the bearing components. These bearings have a face and back side for ease in identifying the thrust direction. The back side receives the thrust and is marked with the bearing number, tolerance, manufacturer, and, in some cases, the word "thrust".

Emerson Power Transmission

Self aligning tapered roller bearings handle a combination of radial and thrust loads and are used in a wide variety of applications such as chemical processing, sawmills, and pulp and paper mills.

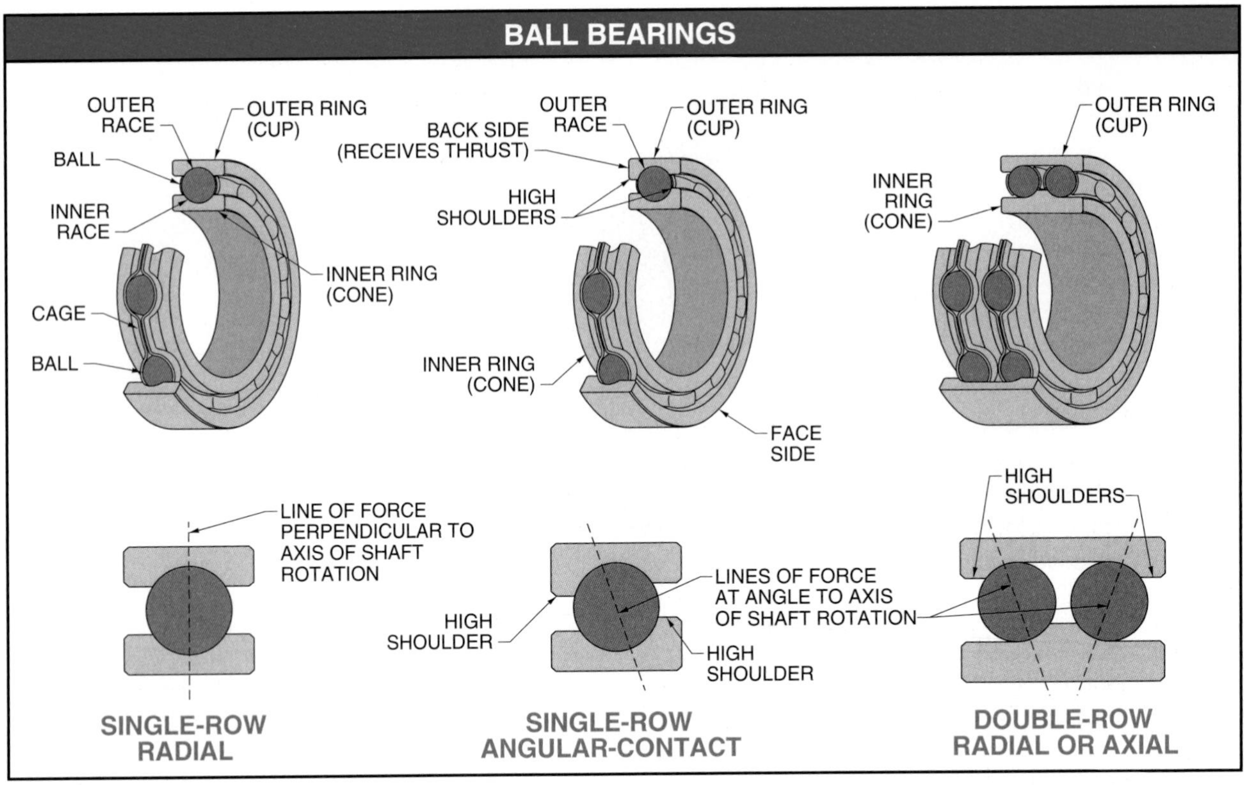

Figure 10-3. General-use ball bearings are designed as single-row radial, single-row angular-contact (axial), or double-row radial or axial.

Double-row bearings, also known as duplex bearings, are matched pairs of angular-contact bearings. They are capable of heavy radial and thrust loads in both directions. Double-row bearings are designed as matched sets and are identified based on their configuration, such as back-to-back, face-to-face, etc. Never use two single-row bearings when replacement requires the use of double-row bearings. Pairing unmatched single-row bearings adversely affects shaft rotation.

Angular-contact bearings are always used in pairs, sometimes at opposite ends of a spindle shaft. Combined sets are installed back-to-back, face-to-face, or back-to-face. See Figure 10-4. Bearing numbers normally indicate that each set has been ground to mount as a pair. Mixing different manufacturer's bearings is not recommended because each angular-contact ball bearing has a ground finish on the back and face.

Ball bearings are installed with one ring being a press fit and the other a push fit. A press fit requires the force of an arbor or hydraulic press to install the ring. A push fit allows the ring to be slid into place by hand. Generally, the press-fit ring is pressed onto or into the rotating part, and the push-fit ring is installed onto or into the stationary component.

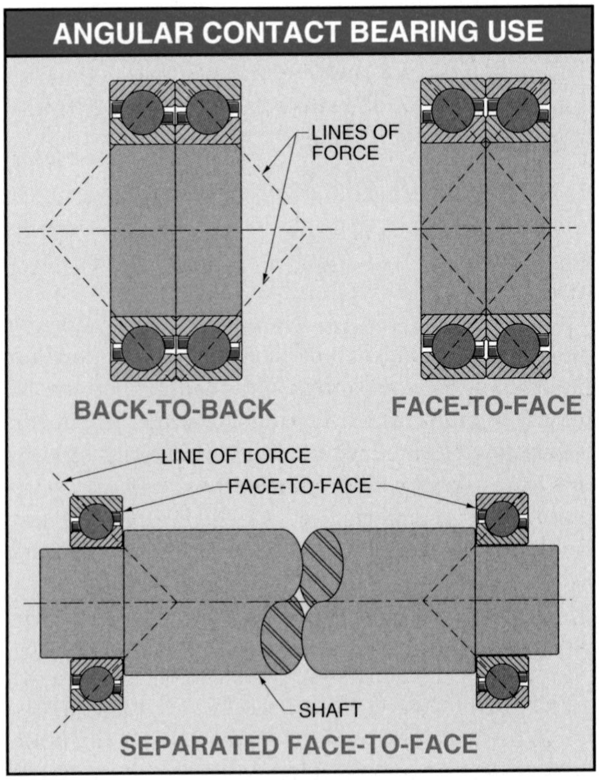

Figure 10-4. Angular-contact bearings are used in sets.

Both inner and outer rings may be press fit where large bearings encounter high speeds or high loads. Under normal load conditions, ball bearings generally have .00025″ interference per inch of shaft when the inner race is press fit. *Interference fit* is fit in which the internal member is larger than the external member so that there is always an actual interference of metal. Less interference is required when the outer race is press fit.

Roller Bearings. Roller bearings are anti-friction bearings that have cylinder-shaped or tapered steel rollers confined between an outer ring (cup) and an inner ring (cone). See Figure 10-5. Roller bearings are designed for loads and applications similar to those of ball bearings. Roller bearings are designed for heavy radial and axial loads. Cylindrical rollers are used for radial loads and tapered rollers are used for radial and axial loads. Roller bearings are precision devices and must be kept clean and handled with care.

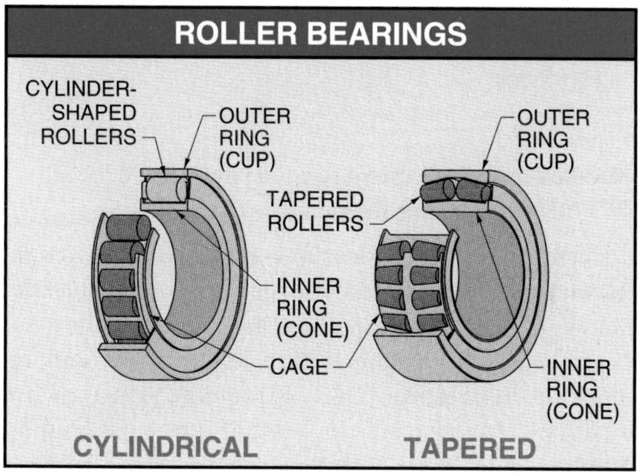

Figure 10-5. Cylindrical and tapered roller bearings are used for heavy loads. Tapered roller bearings are also used for axial loads.

A *cylindrical roller bearing* is a roller bearing having cylinder-shaped rollers. Cylindrical roller bearings are also known as radial roller bearings. Cylindrical roller bearings are used in high-speed, high-load applications and may contain as many as four rows of rollers. These bearings have a high radial load capacity but are not designed for axial loads or misalignment.

A *tapered roller bearing* is a roller bearing having tapered rollers. Tapered roller bearings are normally used for heavy radial and adjustable axial loads. Tapered roller bearings are available in more than 20 configurations from many different manufacturers. The different tapered roller bearings are generally interchangeable and may be cross-referenced because most manufacturers subscribe to the International Standards Organization (ISO).

Needle Bearings. Needle bearings are anti-friction roller-type bearings with long rollers of small diameter. Needle bearings, similar to cylindrical roller bearings, are chosen for relatively high radial loads. Needle bearings are characterized by their rollers being of small diameter compared to their length. The ratio of length to diameter may be as much as 10 : 1. Needle bearings are often used in applications with limited space. Needle bearings normally have tightly packed rollers without separators or inner rings. Special needle bearings that are used for oscillating motion in aircraft elements may have separators with inner rings. Needle bearing cases may have machined surfaces or be drawn and formed for roller retention. Needle bearings are generally press fit. A firm, even, and square press during installation prevents damage to the bearing case.

Depending on the material used, plain bearing load capacity can greatly exceed the load capacity of rolling-contact bearings because of the increased area of surface contact with a shaft.

Plain Bearings

A *plain bearing* is a bearing in which the shaft turns and is lubricated by a sleeve. Plain bearings are used in areas of heavy loads where space is limited. Plain bearings are quieter, less costly, and if kept lubricated, have little metal fatigue compared to other bearings. See Figure 10-6.

Plain bearings may support radial and axial (thrust) loads. In addition, plain bearings can conform to the part in contact with the bearing because of the sliding rather than rolling action. This allows the plain bearing material to yield to any abnormal operating condition rather than distort or damage the shaft or journal. A *journal* is the part of a shaft, such as an axle or spindle, that moves in a plain bearing. The sliding motion of a shaft or journal, whether rotating or reciprocating, is generally against a softer, lower friction bearing material.

The service life of a plain bearing depends on the surface condition of the shaft. A shaft with nicks, gouges, scratches, or rough machine marks wears a plain bearing rapidly. In addition, a shaft that is ground too fine does not allow lubricant retention and wears once the lubricant is squeezed out.

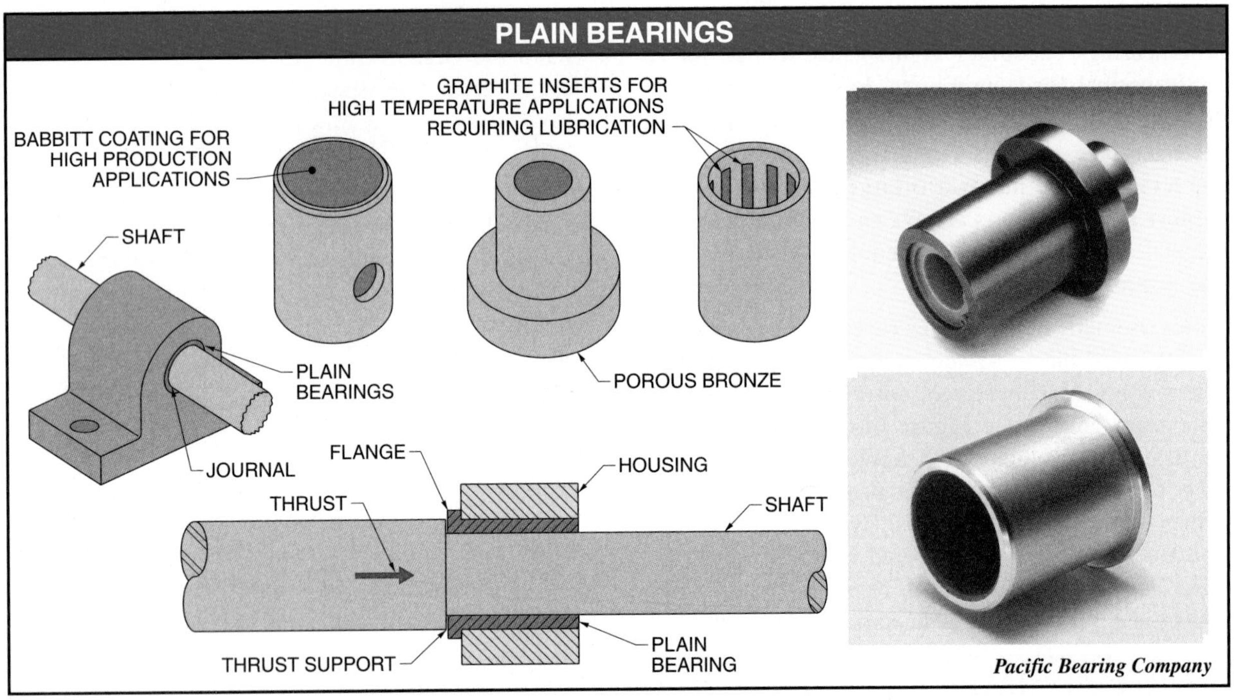

Figure 10-6. Plain bearings provide sliding contact between mating surfaces.

Plain Bearing Materials. Special materials, or combinations of materials, must be selected for plain bearings because of the momentary metal-to-metal contact that occurs during shaft stopping and starting. Plain bearing material must be corrosion- and fatigue-resistant, able to handle running loads and thermal activity, and compatible with other materials used.

Tin-base and lead-base babbitt metals are the best metals for plain bearing loads. *Babbitt metals* are alloys of soft metals such as copper, tin, and lead, and a hard material such as antimony. Copper-leads, bronze, and aluminum base metals are used for plain bearings requiring increased load-carrying capacities. Babbitt metals are used in a thin layer over a steel support for heavy commercial applications, such as armature bearings used in hand drills.

The various metals used for plain bearings are chosen because of their ability to perform under specific conditions. Copper-lead bearings are designed for their ability to withstand high temperatures and high loads, such as engine connecting rod bearings. Porous bronze bearings become self-lubricating when impregnated with oil. These bearings are capable of absorbing oil equal to 30% of their total volume. Porous bronze bearings must be relubricated even though they are used in confined locations or where supplying lubricant is difficult.

Nylon and Teflon bearings are designed for light loads because these materials readily deform under heavy loads. However, they are ideally suited for chemical or high-temperature applications and require no lubrication. Carbon-graphite bearings are designed to operate in temperatures exceeding 700°F and withstand 300 psi of load force without a lubricant. Carbon-graphite bearings are ideally suited for oven applications. Metal bearings with machined grooves containing graphite inserts may be used with high temperature applications requiring lubrication.

Bearing hardness must also be considered in addition to operating conditions. Normal wear and scoring must take place on the less costly bearing surface, not on the surface of the shaft or journal. For this to occur, plain bearings must be at least 100 Brinell points softer than the shaft or journal. A Brinell hardness test measures the hardness of a metal or alloy by hydraulically pressing a hardened steel ball into the metal to be tested and then measuring the area of indentation. The Brinell hardness number is found by measuring the diameter of the indentation and finding the corresponding hardness number on a calibrated chart.

BEARING REMOVAL

Proper tools and maintenance procedures are required when working with precision elements such as bearings. A clean working environment prevents dust, dirt, or other solids from contaminating the bearing, shaft, or housing.

Many bearing failures are due to contaminants that have worked their way into or around a bearing before it has been placed in operation. Internal abrasive particles permanently indent balls, rollers, and raceways. This alters the shape of the surface and begins bearing erosion. Bearing tolerances are such that a solid particle of a few thousands of an inch (.003″) lodged between the housing and the outer ring can distort raceways enough to reduce critical clearances.

Work benches, tools, clothing, wiping cloths, and hands must be clean and free from dust, dirt, and other contaminants. For example, the rolling elements and races of high precision aircraft bearings can be contaminated the moment they are touched by bare hands. The acid from the skin surface is enough to begin corrosion of the metallic surfaces.

Bearing removal is more difficult than bearing installation. A firm, solid contact must be made for bearing removal. Bearings should always be removed from a shaft with even pressure against the ring. The removing pressure should be applied to the inner ring when the bearing is press fit on a shaft. The removing pressure should be applied to the outer ring when the bearing is press fit in a housing. Bearings are removed from shafts or housings using bearing pullers, gear pullers, arbor presses, or manual impact. These methods enable easy bearing removal and reduce the damage to the bearing. See Figure 10-7.

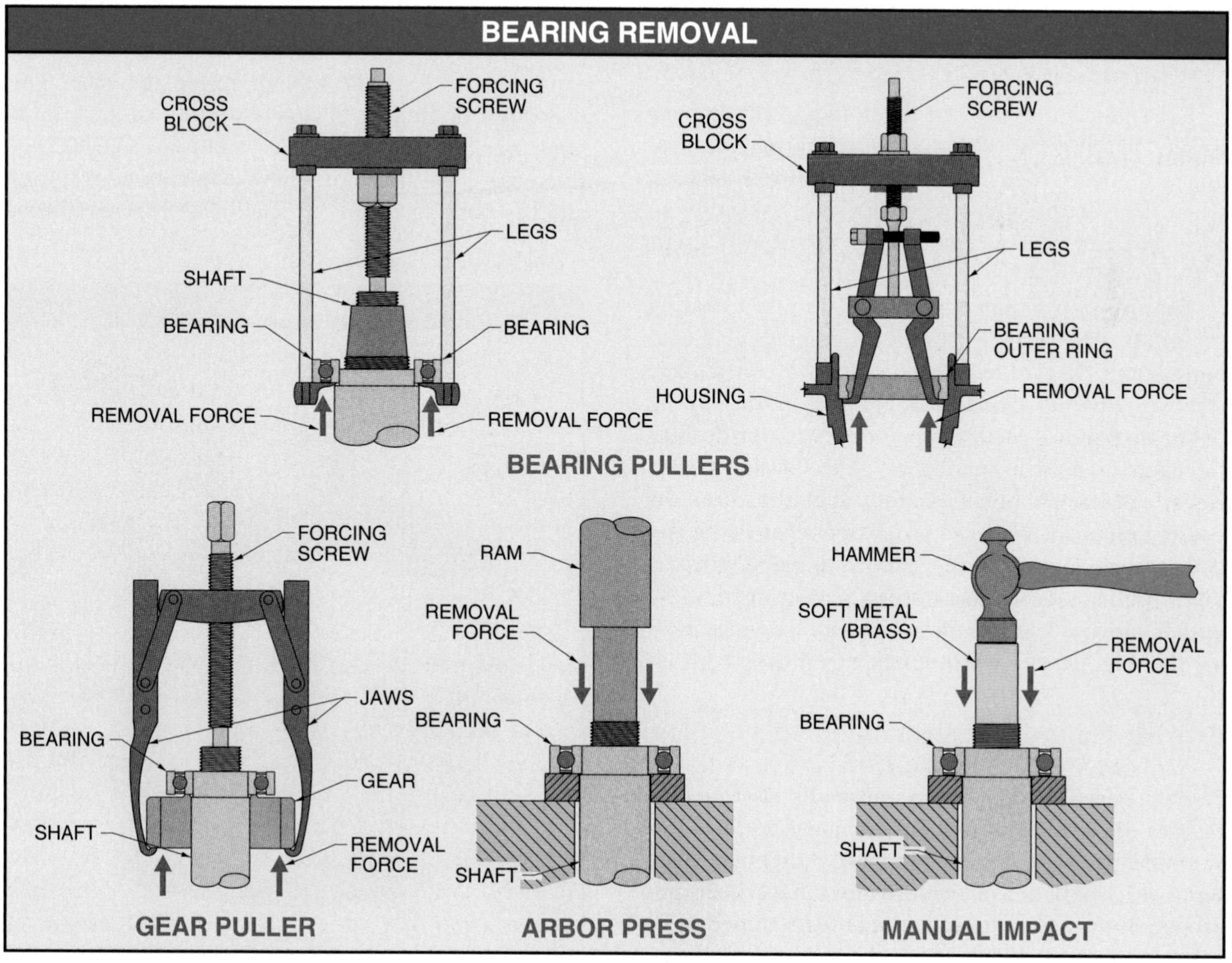

Figure 10-7. To prevent bearing damage, bearing removal forces should be applied to the back side of the ring that is pressed in place.

Power Team, Division of SPX Corporation

Hydraulically powered bearing pullers from Power Team are available with pumps that develop pressures up to 10,000 psi for effortless bearing removal.

Extreme caution must be taken to prevent damage to any bearing part. Most damage during removal goes unnoticed. Also, mark each part as to its location when disassembling a bearing housing assembly so each part can be reassembled the way it was originally positioned.

The use of a hammer and chisel to pry a bearing off of its shaft usually results in damage and contamination. Discard any bearing that was difficult to remove. Bearing damage is possible if the wrong puller or removal method is used. Check for designs intended to hold a bearing in place, such as snap rings, set screws, or pins before applying force for bearing removal. Wear eye protection when using any device that applies force, such as pullers, presses, vises, or hammers. Never strike a bearing directly with a hammer because these are both hardened metals that can shatter when struck together.

Bearing Failure Investigation

Bearing service life, which is normally shorter than fatigue life, may be shortened for many reasons. For example, the load may be too heavy, alignment may be poor, installation procedures may have been improper, or the environment around the machinery may be excessively dirty. Whatever the reason for a bearing failure, each bearing provides a indication as to the cause of its damage. In many cases, it is possible to examine a damaged bearing to determine the cause of failure. This information is useful in taking corrective action to prevent recurring failure. Examining a damaged bearing is most reliable when damage or wear is at an early stage.

Analyzing a bearing to determine the cause of failure is sometimes difficult due to the stages of problems and symptoms. For example, a bearing may have a dark discolored appearance of failure due to high temperatures. However, this bearing may have gone through a series of other failures first, such as spalling or fretting corrosion.

Conditions providing clues when analyzing bearing failure include spalling, false Brinell damage, excessive temperature damage, fretting corrosion, misalignment wear, thrust damage, or electrical pitting and fluting.

Spalling. *Spalling* is the flaking away of metal pieces due to metal fatigue. *Metal fatigue* is the fracturing of worked metal due to normal operating conditions or overload situations. A ball or roller that is under a load and rolls over a bearing race momentarily distorts the metal of the ball or roller and race. This distortion or flexing occurs 4 million times in a 40 hour week if the bearing is rotating at 1700 rpm. Eventually, millions of microscopic fractures form and the bearing begins to spall. See Figure 10-8.

SPALLING

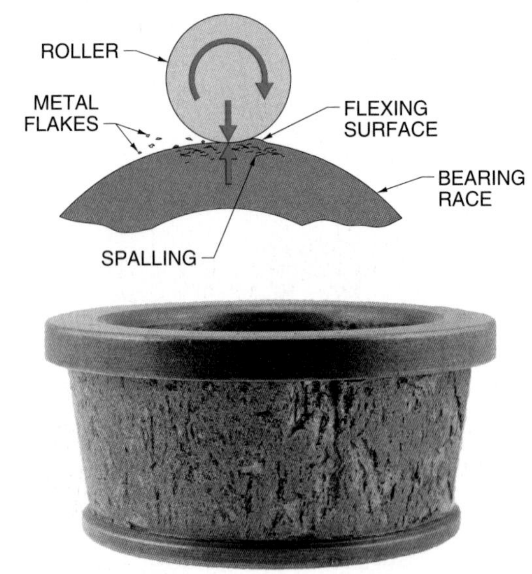

The Timken Company

Figure 10-8. Spalling is the flaking away of metal pieces due to metal fatigue.

Even though there are conditions of advanced destruction that produce little evidence to the initial problem, there are also conditions where clues may be used to correct an ongoing problem. For example, in a bearing failure due to spalling, the flaked metal created excess friction, overheated the lubricant, broke down the lubricant to an acidic condition, and burned up the bearing. Little evidence is given to determine the cause of the failure because the bearing was totally destroyed. Total destruction of the bearing and machine downtime could have been prevented if the bearing was analyzed at the spalling stage. A determination would still be needed to establish whether the flaking was due to overload or if the bearing had reached its life expectancy.

False Brinell Damage. *False Brinell damage* is bearing damage caused by forces passing from one ring to the other through the balls or rollers. False Brinell damage occurs on poorly installed bearings and bearings such as motor bearings or wheel bearings that sit on shelves that vibrate or are roughly transported over distances without rotation. False Brinell damage is also caused by pressure applied to the ring that has a loose fit during bearing removal. The vibration and hammering on a non-rotating bearing causes marks or indentations on the race that are spaced exactly the same distance apart as the balls or rollers. See Figure 10-9.

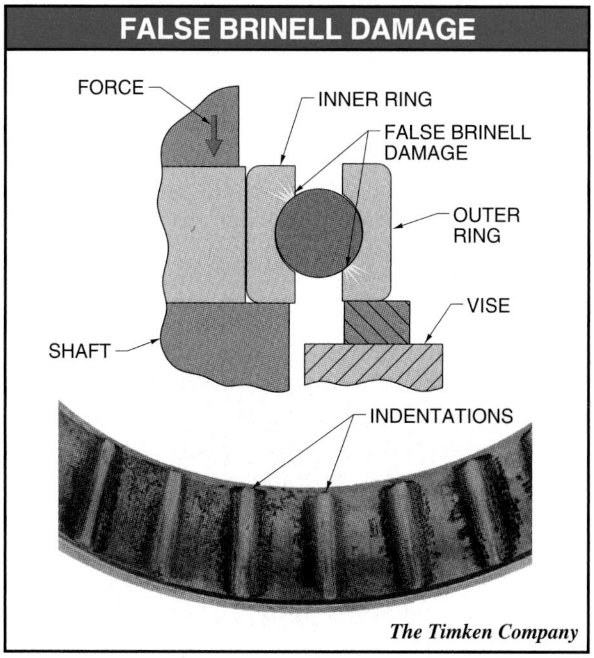

Figure 10-9. False Brinell damage is the indentations on the inner and outer ring races caused by the balls or rollers during rough handling or improper removal.

Excessive Temperature Damage. As the temperature of steel increases, it discolors, turning from silver to blue to black. In addition, the hardness of steel decreases with an increase in temperature. Bearings at excessive temperatures deform more than normal which creates greater resistance and friction.

Another sign that bearings have overheated is the presence of solid or caked lubricant. The darkened, brittle grease has gone through the stages of being heated to its dropping point, allowing the thickener to be baked and burned. The *dropping point of grease* is the temperature at which the oil in the grease separates from the thickener and runs out, leaving just the thickener. This condition is created due to poor alignment, contaminated lubricant, overloading, or high speeds. See Figure 10-10.

The Timken Company

Figure 10-10. Excessive temperature damage is the result of melted and deformed bearing metal.

Fretting Corrosion. *Fretting corrosion* is the rusty appearance that results when two metals in contact are vibrated, rubbing loose minute metal particles that become oxidized. In many cases, fretting is a normal condition that appears as the discoloration on the outer surface of the outer ring between the outer ring and the housing. This happens as moisture from the air settles between the two contacting and unprotected metal surfaces.

Fretting corrosion becomes harmful when the oxidation breaks down supporting wall surfaces, creating looseness. Fretting corrosion is also harmful as its oxidation particles (oxides) mix with and break down the bearing lubricant. See Figure 10-11.

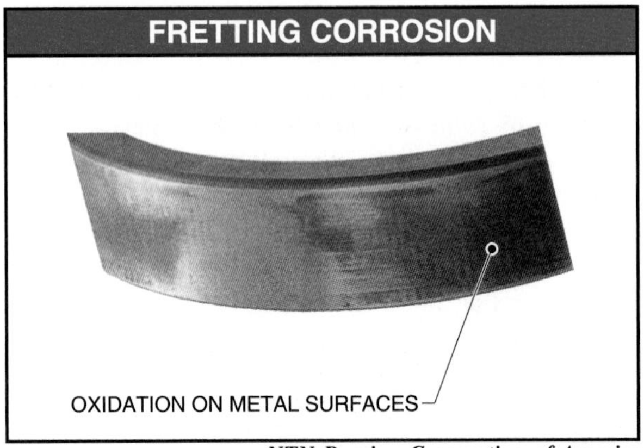

NTN Bearing Corporation of America

Figure 10-11. Fretting corrosion is the rusty appearance that results when two metals in contact are vibrated, rubbing loose minute metal particles that become oxidized.

Misalignment Wear. Bearing surfaces that are misaligned appear as worn surfaces on one side or opposing sides of a bearing. Rollers in roller bearings can leave wear marks on one side of the bearing inner race. The roller may also show high and low trails on the inside of the outer race. Misalignment wear may also appear as lack of fretting on two sides of the outer surface of the outer ring. See Figure 10-12.

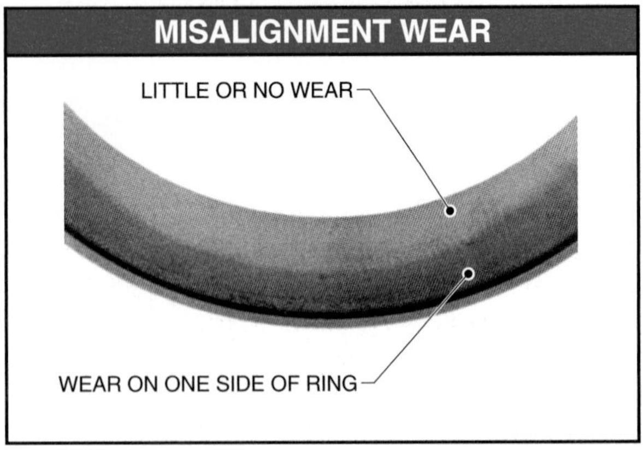

The Timken Company

Figure 10-12. Bearing surfaces that are misaligned appear as worn surfaces on one side or opposing sides of a bearing.

Contact local lubricant suppliers for lubricant recommendations. Failure to maintain proper lubrication of a bearing can result in equipment failure, creating a risk of serious bodily harm.

Thrust Damage. *Thrust damage* is bearing damage due to axial force. Thrust damage on ball bearings appears as marks on the shoulder or upper portion of the inner and outer race and will be anywhere from a slight discoloration to heavy galling. *Galling* (adhesive wear) is a bonding, shearing, and tearing away of material from two contacting, sliding metals. The amount of galling is proportional to the applied load forces. Thrust damage on plain bearings appears as heavy wear at the bearing ends. See Figure 10-13.

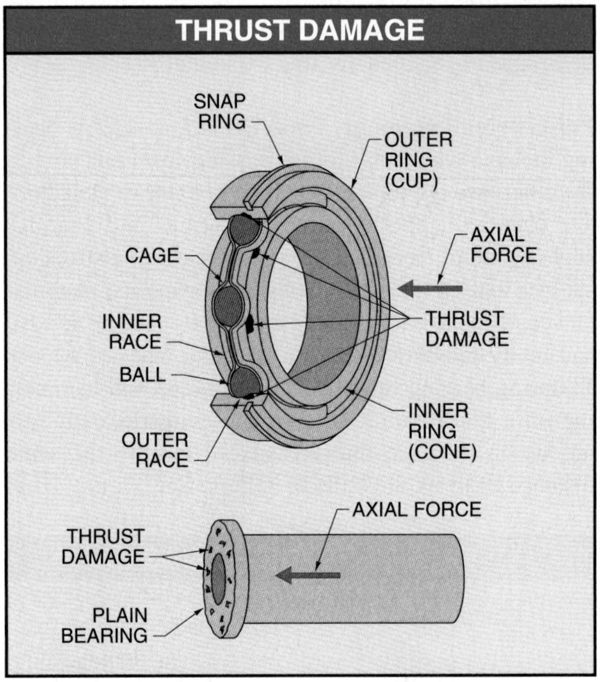

Figure 10-13. Thrust damage in ball bearings appears as marks on the shoulder or upper portion of the inner and outer races or as heavy wear at the bearing ends of plain bearings.

Electrical Pitting and Fluting. Electrical current passing through a machine can be harmful to humans, other electrical components, and machine bearings. Current may pass through machine parts without harm to humans. This happens as current passes from its introduction, such as static electricity created by the manufacturing process, electrical system feedback, or welding currents, to a grounded connection. When it passes through bearings, the current can etch or pit bearing surfaces. Mild electrical currents may not etch the metal, but can create high enough temperatures as current transfers through the bearing to burn and break down lubricants.

Welding current damage is observed as short pitted lines on balls or rollers that were stationary when the current was present. See Figure 10-14. The race has corresponding damage, but this is not normally observed unless the bearing is destroyed. Electrical feedback created by certain forces throughout plant electrical usage, faulty wiring, and static electricity can be prevented from flowing through a machine if extra grounding is provided. Extra grounding of a machine can be as simple as running a wire from the machine to a pneumatic or water line.

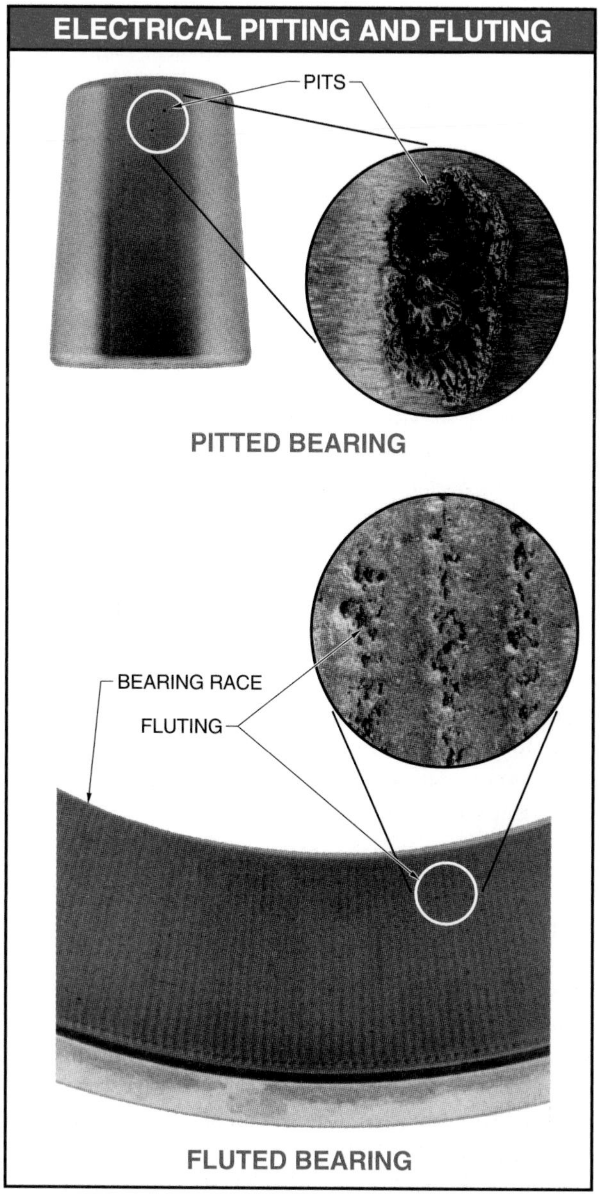

The Timken Company

Figure 10-14. Welding on a machine that is not grounded properly can permanently damage the machine bearings.

Warning: Never ground a machine by connecting a wire from the machine to a gas or oil pipe.

Fluting is observed in roller bearings that were rotating while welding currents passed through them. *Fluting* is the elongated and rounded grooves or tracks left by the etching of each roller on the rings of an improperly grounded roller bearing during welding. In roller bearings, fluting is caused by electrical arcing and pitting the length of each roller in the bearing. Damage from welding can be prevented by attaching the welding ground clamp in a location where no bearings are between the ground and the weld.

The Timken Company

Bearings are inspected after they are cleaned and dried by removing one roller to enable the inner race, cage, and rollers to be examined for damage.

PARTS PREPARATION

After all bearing parts have been dismantled, they should be cleaned and spread out on a clean surface for inspection. Cleaning is accomplished by dipping or washing the housing, shaft, bearing, spacers, and other parts in a clean, nonflammable cleaning solvent. Remove all traces of dirt, grease, oil, rust, or any other foreign matter. Dry all parts with a lint-free cloth.

Caution should be taken when using part cleaning solutions. Seals, O-rings, and other soft materials may deteriorate due to incompatibility. Check the cleaning solution label or contact the supplier before use. If in doubt, a lightweight, warm mineral oil is a good cleaning and flushing fluid. Care should be taken to clean housing and shaft bearing seats, corners, and keyways.

Bearings may be blown dry using clean, dry compressed air. Do not allow the air pressure to spin the bearings because this scratches the surfaces and the bearings may fly apart. Wipe all clean and dry parts with a lightweight oil and wrap or cover them to protect from dust and dirt. Inspect all parts carefully for nicks, burrs, or corrosion on shaft seats, shoulders, or faces. Closely inspect bearing components for indication of abnormality or obvious defects. Check both inner and outer races for cracks and the balls or rollers for wear or breaks.

Remove nicks, corrosion, rust, and scuffs on shaft or housing surfaces that may make assembly unsatisfactory. Check that corners and bearing seats are square and that all diameters are round, in tolerance, and without runout. Replace any worn spacers, shafts, bearings, or housings. Check the housing shoulder to ensure that it is square enough to clear the bearing corner.

Once lightly oiled, inspection of a bearing is accomplished by holding the inner ring while rotating the outer ring. This inspection may show uneven tolerances, particle contamination, or chipped elements. Replace any bearing when there is doubt about the condition of the bearing.

SPM Instrument, Inc.

Bearings must be handled and installed with care to prevent damage because a greater number of bearings are damaged during installation than during use.

BEARING INSTALLATION

Successful bearing installation requires cleanliness, correct bearing selection, mounting methods, tool use, and tolerance specifications. Proper bearing assembly is required for proper bearing performance, durability, and reliability.

Caution: More bearings are destroyed, damaged, and abused during the installation stage than from malfunction during their life expectancy. Often, mechanics handle bearings as though it is impossible to damage them by force, dirt, or misalignment. Within a few short seconds, carelessness can destroy the protective measures of any bearing.

Proper Bearing Selection

Bearings are often selected without the use of manufacturer's specifications. Certain factors other than dimensions must be observed when a replacement bearing is chosen by comparison of a removed bearing instead of from an equipment manual or parts book. Factors to be considered include the exact replacement part number, the type and position of any seal, the direction of force and positioning of a required high shoulder, and whether a retaining ring is required. Early failure is possible if a replaced bearing, even though it is dimensionally correct, lacks any other requirement. The factors in bearing selection include the load type and orientation of the bearing being replaced. Close attention must also be paid to any special features of a removed bearing, such as special seal configurations, spherical OD's, non-circular bores, etc.

All types of bearings, especially rolling-contact bearings, are available in many design variations and vary greatly in internal design. Always ensure that a bearing is a direct replacement for the one required by the manufacturer when replacing the bearing. Check the equipment manual for the proper bearing number. A previous bearing replacement may have been incorrect. Bearing numbers on the box should be checked before opening the box and bearings in the box should be checked to ensure the number on the bearing corresponds to the number on the box. Duplex mounted bearings must not be mismatched and must be from the same manufacturer.

Know and identify angular contact bearings and their position within an assembly. For example, replacing a Conrad bearing with a general-purpose bearing may produce rapid deterioration of the bearing. This is because the Conrad bearing may be used for axial support and general-purpose bearings cannot provide axial support.

Bearing Mounting

Bearing mounting procedures have a great effect on performance, durability, and reliability. Precautions should be taken to allow a bearing to perform without excessive temperature rise, noise from misalignment or vibration, and shaft movements. During installation, force must be applied uniformly on the face or ring that is to be press fit. Any method that presses the bearing on squarely without damage may be used. Press fits may be accomplished by using a piece of tubing, steel plate, and hammer; an arbor press; or a hydraulic ram. Wood should not be used if there is any possibility of contaminating a bearing with wood splinters or fibers.

When reinstalling bearings, insert the lightly oiled shaft in the bearing and line up the marks made at disassembly. Start the new or used bearing assembly by hand. If pipe is used as an installation tool, press the bearing onto the shaft by placing the pipe only on the press-fit ring. This pipe must be the proper diameter, clean, and have both ends cut square. See Figure 10-15. Bearings that can be mounted in either direction should be mounted with the part number facing out for ease of future identification.

Bearings must be mounted so their inner and outer race is not distorted, their rolling elements do not become bound, and each machine part remains in its proper relationship. This is accomplished by proper mounting alignment (trueness) of machined parts, using proper machine tolerances (internal clearances), and allowing for additional tolerances due to thermal movement.

Damage can easily occur internally to a bearing if removal or installation force is passed from one bearing ring to the other through the ball or roller set.

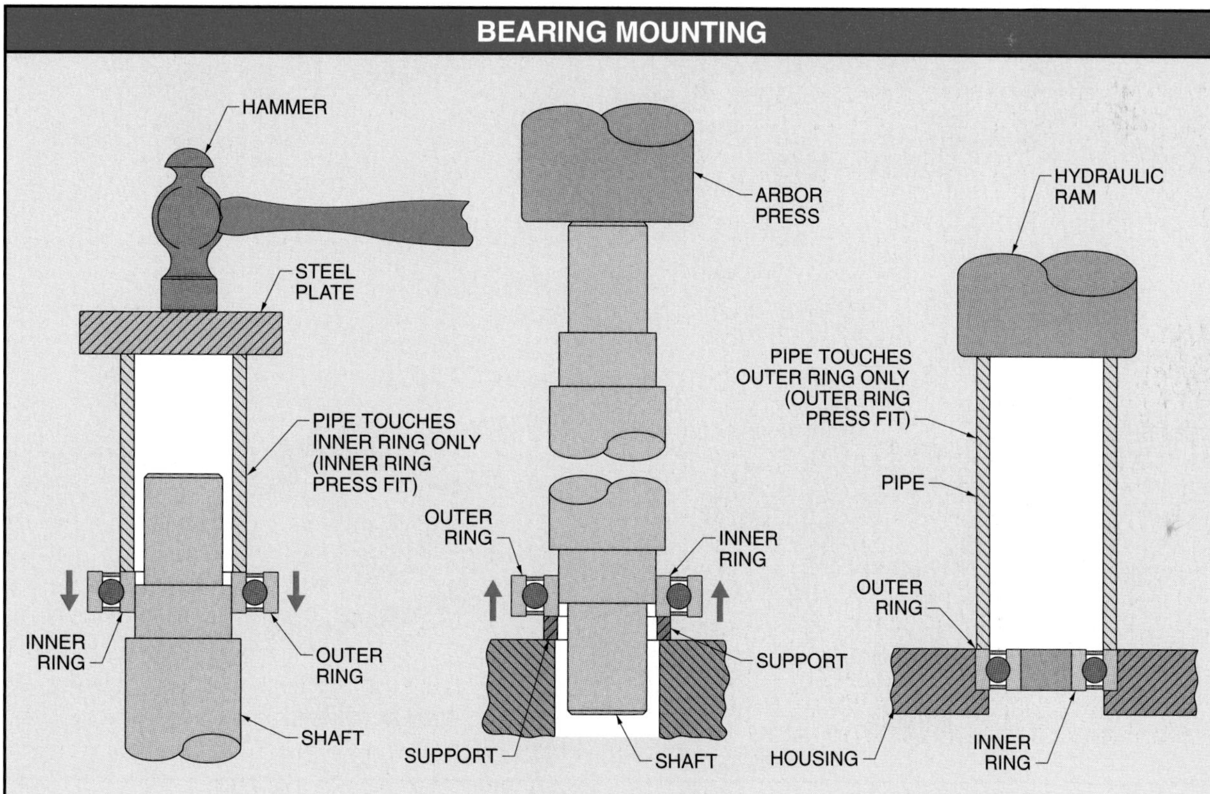

Figure 10-15. Pressure applied when mounting must be applied squarely to the ring being pressed.

NTN Bearing Corporation of America

Uneven wear between a bearing inner ring, outer ring, and rollers is an indication of poor bearing mounting.

Bearings are mounted with the rotating ring pressed on a shaft or in a housing. The non-rotating ring is pushed or slid on a shaft or in a housing and may be held in place by a snap ring. Generally, the rotating element of a machine is the shaft with the inner ring pressed on to the shaft. However, if the outer part of the machine is the rotating element, the outer ring is press fit. The exception to this rule is with the use of heavy-duty cylindrical roller bearings where the extra loads require that both rings be press fit. The press fit ring is then clamped to prevent axial movement. The inner ring of a bearing mounted on a shaft is generally held in place by a locknut, clamp plate, or castle-nut and cotter pin. See Figure 10-16.

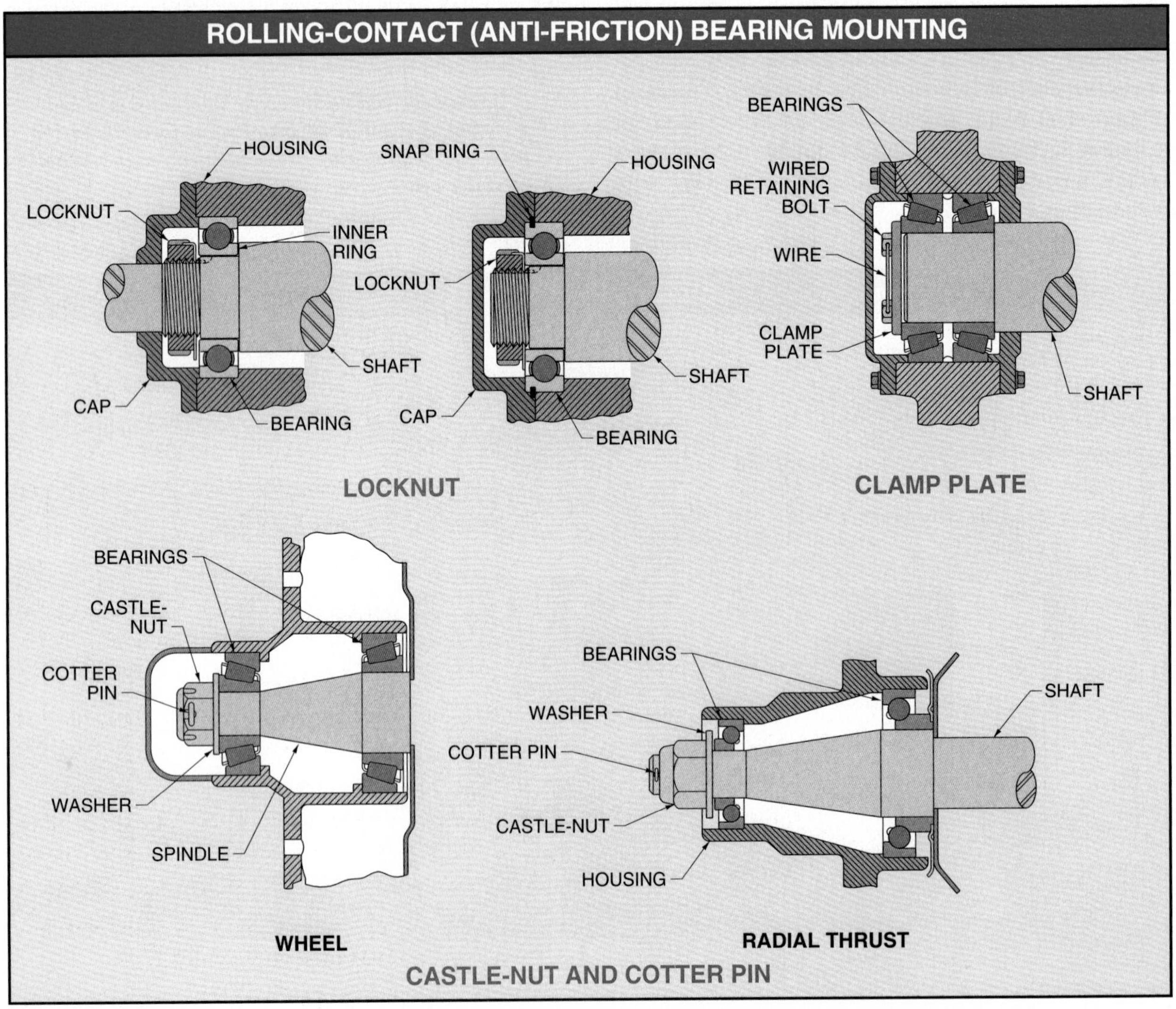

Figure 10-16. Bearing mounting requires the rotating ring to be mounted securely using locknuts, clamp plates, or castle-nuts and cotter pins.

Precision Class Bearings. Precision class bearings require quality mounting and control of shaft, housing, and bearing deflection. Deviations of more than .0005″ in precision class assemblies can cause significant vibration. For example, deviations of guided missile bearing applications must be limited to less than .00002″. Precision class bearings are generally marked with their high points of runout. Outer rings may be marked with a "V" on each ring and each ring should be aligned with the other. Check for the bearing manufacturer's runout marking. Inner race runout may be identified by a dot, a copper dot, or some other mark at the high point. After identifying the high point on the shaft, place the bearing runout mark 180° from the shaft high point.

Bearing Mounting Using Temperature. During bearing installation, it may be necessary to increase the inner ring diameter by heating the bearing. Regardless of bearing size, increasing the diameter with heat is the simplest way to mount a bearing that must be press fit on a shaft. Prelubricated bearings must not be heated for installation. Heating bearings may be accomplished using a light bulb, an oven, clean hot oil that has a high flash point, a hot plate, or induction heat. The light bulb, oven, and induction heating methods are most reliable because their temperature is easy to control. The hot oil method is less reliable because the oil temperature is harder to control and the oil is difficult to keep clean. The light bulb heating method is the best and most economical. This method uses a light bulb placed in the bore of a bearing to provide heat to increase the diameter of the bearing. The temperature is controlled by the length of time the bulb is placed in the bearing. This method works well because the inner ring is the only component to be brought to high heat, allowing handling and assembly using the outer ring. See Figure 10-17.

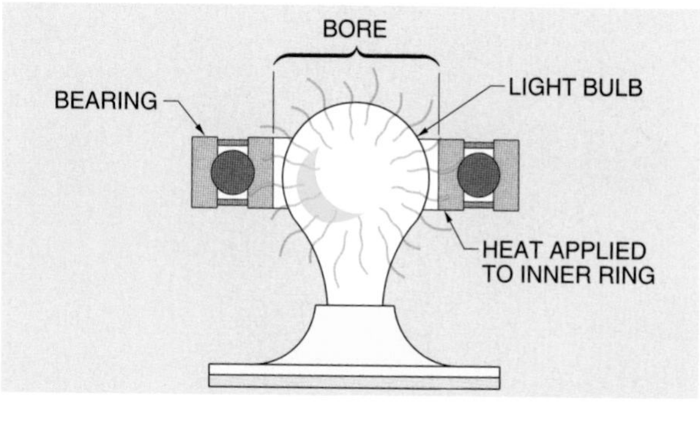

Figure 10-17. The light bulb heating method enlarges the diameter of a bearing inner ring through an increase in the ring temperature.

Temperatures must be even throughout the inner bearing ring and controlled to between 175°F and 200°F. Torches heat unevenly, distort diameters, and must not be used. Temperatures over 250°F may reduce the hardness of bearing metals, resulting in early failure. Bearing temperatures may be determined using thermal crayons, which melt at specific temperatures. Induction heaters, operating electrically and rapidly, leave metals magnetized. All small metal particles are drawn to the assembly if bearings are not demagnetized.

Freezing may be required to reduce the outside diameter of a bearing to allow installation into a housing when heating methods are not possible. Freezing causes shaft sizes to be reduced to allow installation of prelubricated bearings. A mixture of dry ice and alcohol can be used to lower the temperature of a bearing to −20°F. Condensation forms on bearings in areas where ambient conditions are humid. Corrosion is prevented if the condensate is wiped or blown off after assembly, followed by thorough lubrication. Corrosion should not be a problem if components were lightly oiled before freezing. Precautions that must be taken when mounting bearings include:

- Know the bearing function in a machine.
- Keep all bearings wrapped or in their sealed container until ready to use. Reusable bearings should be treated as new.
- Maintain clean tools, hands, and work surface, and work in a clean environment.
- Use clean, lint-free cloths when wiping bearings.
- Never attempt to remove the rust preventive compound used by the manufacturer unless specifically recommended.
- Use the best bearings available within reason. The life and reliability of a bearing is generally related to its cost.
- Always follow the instructions of the heating equipment manufacturer when bearings are to be heated for assembly.
- Use rings, sleeves, or adaptors that provide uniform, square, and even movements.
- Prevent cocking during bearing installation by starting races evenly on the shaft without pressure devices.

- Never strike the bearing with a wooden mallet or wooden block.
- Never apply pressure on the outer ring if the inner ring is press fit and never apply pressure on the inner ring if the outer ring is press fit.
- Be careful not to abuse, strike, force, press on, scratch, or nick bearing seals or shields.

Understanding Bearing Function. No universal bearing exists that can do all of the functions and applications required in industry. In many cases, a review of the machine function and its bearing requirements may indicate if proper bearings are being used.

Provisions made for thermal expansion within a machine are generally published by the machine manufacturer and are listed as space tolerances between housing, bearing components, and shaft. Greater space tolerances are allowed for plain bearings than for rolling-contact bearings because plain bearings are more susceptible to damage from higher temperatures.

Cone Mounter Company

Proper bearing installation is required for maximum bearing life. Heating a bearing before installation expands the inner race, allowing easy bearing mounting on a shaft.

Rolling-contact bearings must be firmly mounted so end play and shaft expansion and contraction due to thermal activity is minimized. *End play* is the total amount of axial movement of a shaft. To work properly, all bearings require axial and radial operating clearance, but are not intended to move or flex under load. Tapered roller bearings are adjusted to a specified end play or end lateral movement with the use of a dial indicator. Bearing movement is erratic, noisy, and damaging if end play is excessive. See Figure 10-18.

In some cases, zero clearance or preloading of tapered roller bearings is necessary. *Preloading* is an initial pressure placed on a bearing when axial load forces are expected to be great enough to overcome preload force, thereby resulting in proper clearances. Preloading is accomplished by the use of shims or adjustable nut settings. If preloading is too tight, lubrication is squeezed out, metal-to-metal contact occurs, and the increase in temperature damages the bearing. If preloading is too loose, bearing movement is sloppy and damaging. In some cases, preloading is measured with a torque wrench to determine the force required to rotate the bearing assembly. Proper bearing adjustment specifications are generally provided by equipment manufacturers.

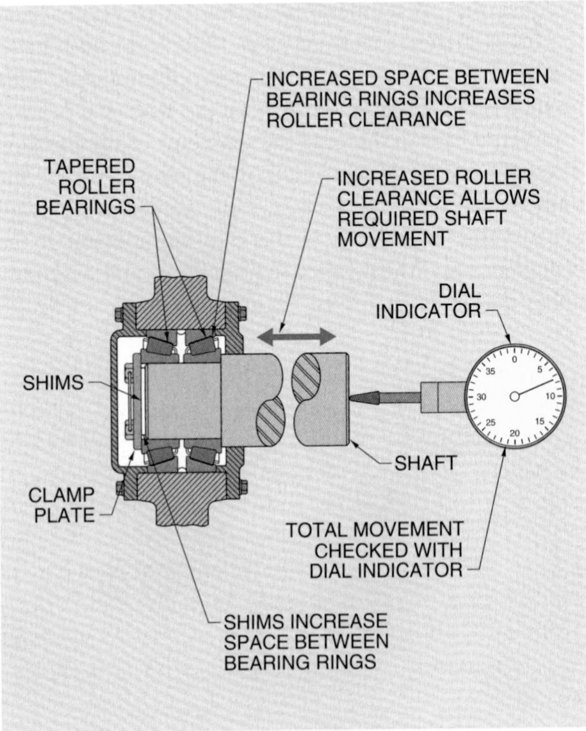

Figure 10-18. Tapered roller bearings are adjusted to a specified end play with the use of a dial indicator.

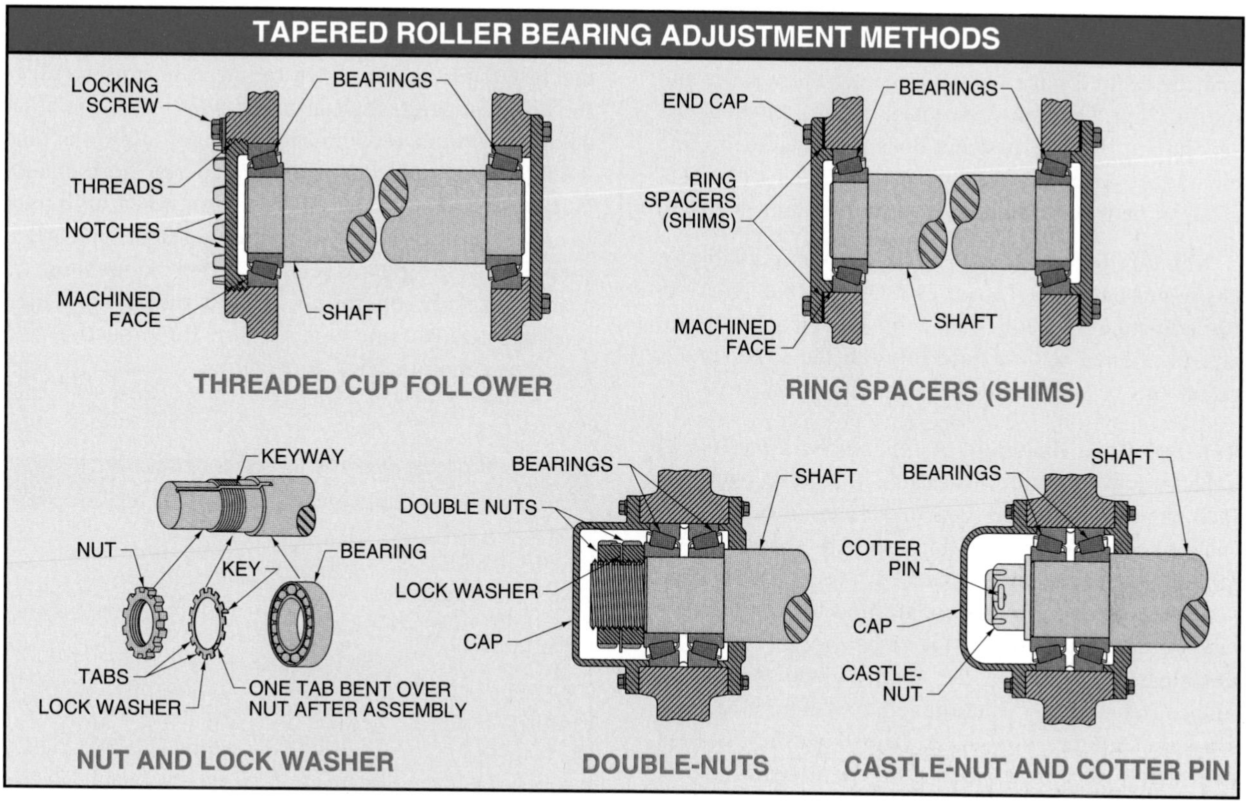

Figure 10-19. Tapered roller bearings are adjusted using a threaded cup follower, ring spacers (shims), nut and lock washer, double-nuts, or castle-nut and cotter pin.

Tapered Roller Bearing Adjustment. Tapered roller bearings are generally arranged in pairs for opposed mounting. This ensures that thrust force in either direction is taken by one of the bearings. Careful control of the clearance between bearing faces is required to prevent excessive shaft movement. Adjustments are accomplished using methods such as a threaded cup follower, ring spacers (shims), nut and lock washer, double-nuts, or castle-nut and cotter pin. See Figure 10-19.

A *threaded cup follower* is a tapered bearing gap adjusting device that is used to adjust shaft endplay by controlling the amount of clearance between the bearings. A threaded cup follower is a dish-shaped cap that has a threaded OD and a machined face. This adjusting device is screwed into a housing until its machined face makes contact with the bearing cup (outer ring). The threaded cup follower is then screwed in until the proper shaft end play is achieved. Once proper adjustment has been met, a locking screw is secured to prevent cup follower movement.

Special attention must be given to the arrangement of axial/radial bearing pairs. Assembly design may require back-to-back, face-to-face, or tandem arrangement. Ring spacer (shim) assemblies are generally used where a non-adjustable (NA) application exists and the ring spacer thickness is determined by the equipment manufacturer. Ring spacers are placed beneath end-cap assemblies to provide adjustment between the tapered roller bearing and the machined face of the end cap. In most cases, clearance is increased as ring spacers are added, and clearance is decreased as ring spacers are removed.

Nut and lock washer adjustment is made using a keyed washer, keyed shaft, and a nut with a group of notches on the outside diameter. The more notches designed into the nut, the finer the assembly may be adjusted. Upon final adjustment, a tab on the washer is bent over the corresponding notch on the nut. This assembly is held firm by a tab on the washer engaged in the shaft keyway.

The double-nut method of adjusting the running clearance of a tapered bearing assembly is accomplished using a keyed shaft, two notched nuts, and a keyed washer. With the bearing assembled onto the shaft and into the housing, the first nut is screwed in until proper bearing clearance has been established. The keyed washer is then slid onto the shaft and the second nut tightened against the washer and the first nut. Care must be taken when tightening the second nut that the first nut does not turn, losing the clearance. After the second nut has been tightened, a tab is bent over both nuts, locking them in place.

Similar to the nut and lock washer assembly, a castle-nut and cotter pin uses a notched (slotted) nut for adjustment. Upon final adjustment, a castle-nut slot is aligned with a hole through the shaft, and a cotter pin is inserted to hold the assembly firm.

Tapered Bore Bearings. A *tapered bore bearing* is a bearing whose bore varies in diameter from the face to the back of the bearing. Tapered bore bearings are used directly on tapered shafts or on straight shafts using tapered sleeves. Tapered sleeves (split adapter sleeves) allow firm mounting of tapered bore ball and roller bearings. The internal clearance between the shaft and the bearing is reduced as a locknut is tightened, forcing the bearing onto the tapered sleeve. A tabbed washer is bent over a flat on a locknut to hold the assembly firm. Tapered bore bearings using a removable sleeve must always be firm against the shaft shoulder. See Figure 10-20.

Pack bearing housings half full of grease because too much grease causes excessive churning and extremely high temperatures.

MACHINE RUN-IN

A machine run-in check should be made after bearing assembly is complete. A run-in check starts with a hand check of the torque of the shaft.

Warning: Ensure power is locked out when manually rotating a shaft. Unusually high torques normally indicate a problem with a tight fit, misalignment, or improper assembly of machine parts. Listen carefully for clicks, squeals, or thumps and investigate. Restore machine power and listen for unusual noises. High noise levels may indicate excessive loading or cocked or damaged bearings. Correct the problem before continuing.

Final checks are accomplished by measuring machine temperatures. High initial temperatures are common due to bearings being packed with grease, which produces excess friction. Generally, turning a machine OFF to cool down before restarting resolves the high temperature. Run-in temperatures should decrease to within recommended ranges. Any machine with temperatures that continue to run high should be corrected before proceeding. Continued high temperatures are generally a sign of tight fit, misalignment, or improper assembly. After a machine is placed back in operation, record motor amperage readings and bearing temperature for proactive and predictive maintenance procedures.

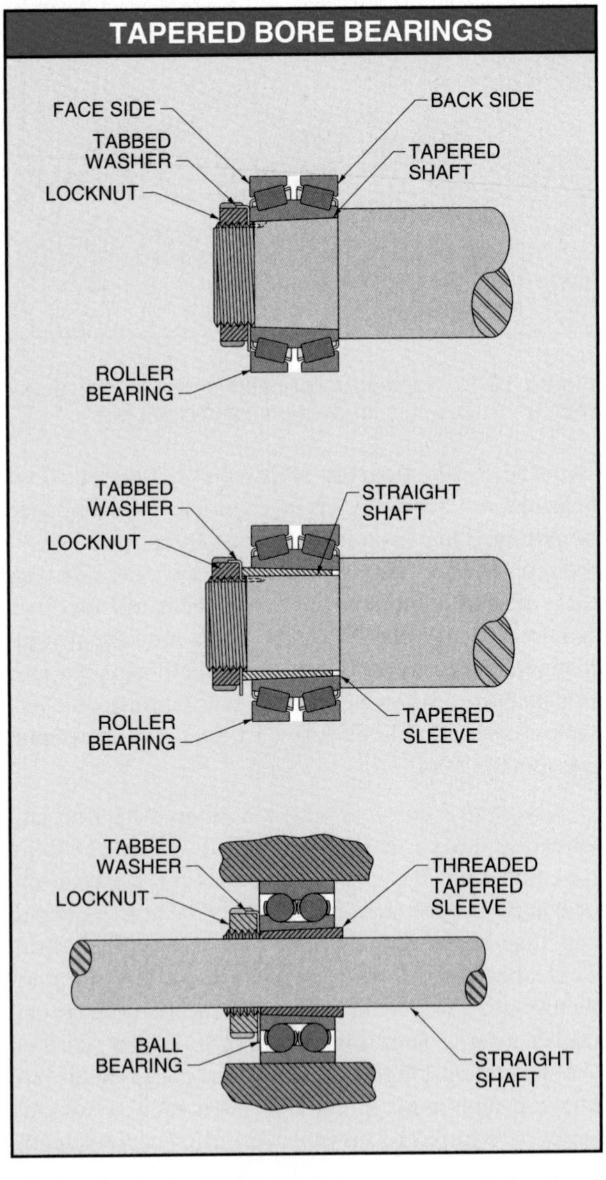

Figure 10-20. Tapered bore bearings are used directly on tapered shafts or on straight shafts using tapered sleeves.

Flexible Belt Drives

11 Chapter

The Gates Rubber Company

Flexible belt drives are systems in which resilient flexible belts are used to drive one or more shafts. Belts used for power transmission in flexible belt drives include V-belts, double V-belts, and timing belts. V-belts that are not aligned properly are destroyed prematurely due to excessive side wear, broken or stretched tension members, or rolling over in the pulley. The main concerns when working with flexible belt drives are the safety of the technician working on the machine, co-workers working in the area of the machine, and the machine itself.

FLEXIBLE BELT DRIVES

A *flexible belt drive* is a system in which a resilient flexible belt is used to drive one or more shafts. Flexible belt drives offer convenient power transmission between two or more shafts on a machine. A *machine* is a group of mechanical devices that transfer force, motion, or energy input at one device into a force, motion, or energy output at another device. Belts used for power transmission in flexible belt drives include V-belts, double V-belts, and timing belts.

V-belts

A *V-belt* is an endless power transmission belt with a trapezoidal cross section. V-belts are made of molded fabric and rubber for body and bending action. V-belts contain fiber or steel cord reinforcement (tension members) as their major pulling strength material. The *tension member* is the load-carrying element of a belt which prevents stretching. V-belts become thinner and weaker as tension members break.

V-belts are resilient, quiet, and able to absorb many shocks because they are made of cloth and rubber. V-belts are generally classified as standard or high-capacity. Standard V-belts are designated as A, B, C, D, or E. High-capacity V-belts are designated as 3V, 5V, or 8V. The letter or number designation also indicates the cross-sectional dimension and thickness of the belt. See Figure 11-1.

V-belts run in a pulley (sheave) with a V-shaped groove. V-belts transmit power through the wedging action of the tapered sides of the belt in the pulley groove. The wedging action results in an increased coefficient of friction. See Figure 11-2. V-belts do not normally contact the bottom of the pulley. A pulley or belt that has worn enough so that the belt touches the bottom of the pulley should be replaced. The belt becomes shiny and slips and burns if allowed to bottom out. More than one belt may be used if additional power transmission is required. However, each belt must be of the same type and size.

V-belt Sizes. V-belt replacement, whether for preventive maintenance or equipment breakdown, starts with proper identification and sizing of the belt being replaced. The technician can prevent many premature belt failures by selecting the proper belt, belt size, and installation procedure.

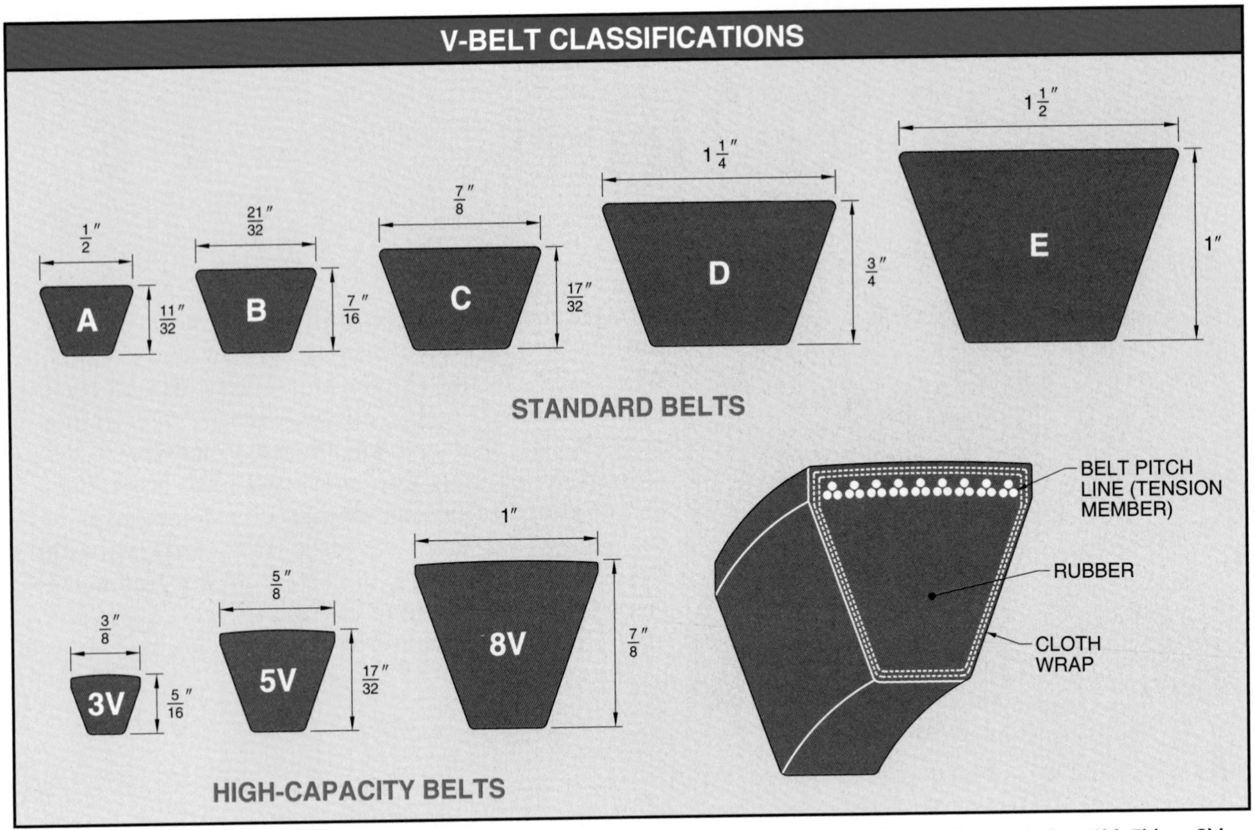

Figure 11-1. Standard V-belts are designated as A, B, C, D, or E and high-capacity V-belts are designated as 3V, 5V, or 8V.

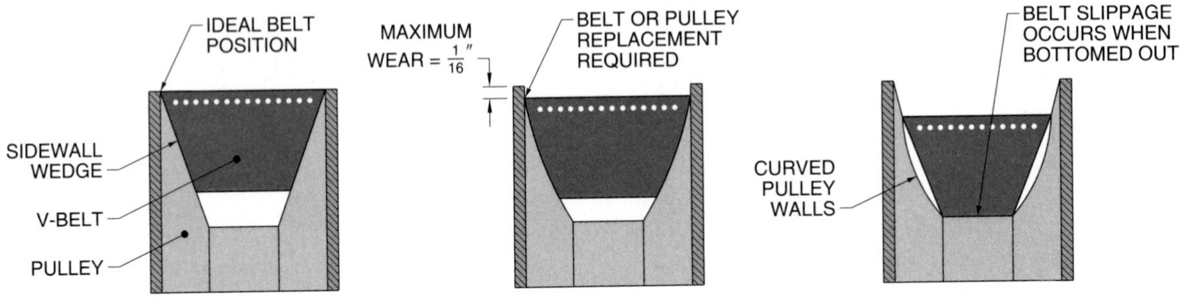

Figure 11-2. V-belts transmit power through the wedging action of the tapered sides of the belt in a pulley groove.

Standard V-belt sizes are identified by letter/number combinations. Letters identify the cross-sectional dimensions and numbers identify the nominal length. A *nominal value* is a designated or theoretical value that may vary from the actual value. For example, a flexible belt with a nominal length of 24″ may be significantly different when compared to another flexible belt with a nominal length of 24″. An X following the letter indicates that the belt is a molded notch belt. A *molded notch belt* is a belt that has notches molded into its cross-section along the full length of the belt. These notches provide for cooler operation along with greater flexibility.

Manufacturers measure each belt under tension and mark the belt with an additional code number because belt fibers and rubber allow some stretching. Typically, the code numbers following the manufacturer's name indicate the effective (working) belt length in tenths of an inch.

The additional code numbers begin with 50, indicating that the belt is within nominal size tolerance range. A number is added or subtracted from 50 for each 1⁄10″ over or under the nominal length. For example, a BX60 55 belt is a 21⁄32″ notched belt, 60″ long, manufactured 5⁄10″ longer than the nominal length. See Figure 11-3.

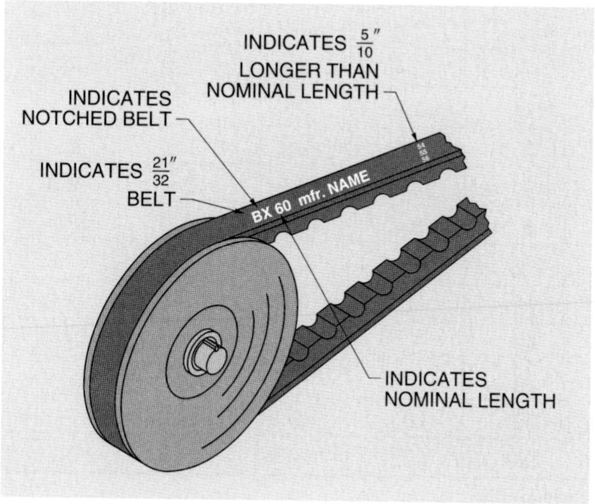

Figure 11-3. Standard V-belt sizes are identified by letter/number combinations.

Even though a belt may be classified as being the same nominal length, code numbers indicate that belts can vary greatly between actual and nominal length. Code numbers must match or be within a matching number range when more than one belt is used as a set. In general, sets should be made of belts with the same code range numbers. See Figure 11-4. For example, if a set of three belts are 300″ long and one belt reads D300 50, the other two belts must be the same or within a range of 4⁄10″ of the first belt. These belts may read anywhere from D300 48 to D300 52.

V-belts are selected based on the pulley shape, pulley size, and the outside circumference around the pulleys. The easiest method of determining a belt size is to locate the belt letter/number code stamped on the top surface of the belt. A belt and sheave groove gauge may be used to determine the proper belt cross section if the number is not legible. A *belt and sheave groove gauge* is a gauge that has a male form to determine the size of a pulley and a female form to determine the size of a belt. A set of gauges, generally on a chain or ring, consists of various belt and pulley sizes.

RECOMMENDED BELT SET LENGTH VARIATIONS

Code Range**	Nominal Lengths*				
	A	B	C	D	E
2	26 – 180	35 – 180	51 – 180	—	—
3	—	195 – 315	195 – 255	120 – 255	144 – 240
4	—	—	270 – 360	270 – 360	270 – 360
6	—	—	390 – 420	390 – 660	390 – 660

* in in.
** for matching numbers

Figure 11-4. Variations in length for belt sets allow only a few tenths of an inch over or under the belt's nominal length.

V-belt length may be determined by measuring the outside circumference using flexible tape while the belt is still on the pulleys. Generally, the correct belt length is the next standard belt shorter than the measurement. For example, an A belt having a 27″ outside circumference is replaced with an A26 belt. The reduced length is closer to the actual belt pitch line. *Belt pitch line* is a line located on the same plane as the belt tension member. Belt pitch line is the effective length of a belt. A more accurate belt length may be determined using the pulley diameters. See Figure 11-5. Belt length is found by applying the formula:

$$L = 2 \times C + 1.57 \times (D + d) + \frac{(D - d)^2}{4 \times C}$$

where

L = belt length (in in.)

2 = constant

C = distance between pulley centers (in in.)

1.57 = constant

D = large pulley diameter (in in.)

d = small pulley diameter (in in.)

4 = constant

Belts should be replaced if there are signs of cracking, fraying, unusual wear, or loss of teeth in a timing belt.

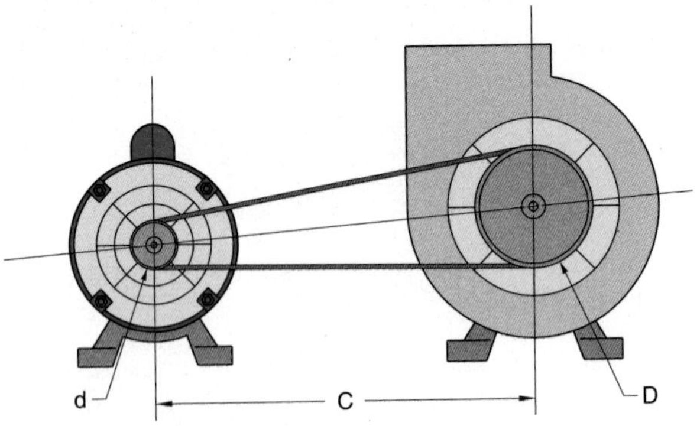

Figure 11-5. Belt length is calculated using the distance between pulley centers and the pulley diameters.

For example, what is the belt length for two pulleys 5″ and 10″ in diameter that are 30″ apart at their centers?

$$L = 2 \times C + 1.57 \times (D + d) + \frac{(D - d)^2}{4 \times C}$$

$$L = 2 \times 30 + 1.57 \times (10 + 5) + \frac{(10 - 5)^2}{4 \times 30}$$

$$L = 60 + 1.57 \times 15 + \frac{5^2}{120}$$

$$L = 60 + 23.55 + \frac{25}{100}$$

$$L = 83.55 + .208$$

$$L = \mathbf{83.758''}$$

V-belt Forces. V-belt forces are constantly changing as a belt bends around a pulley. Tension forces develop at the top of the belt and compressive forces build at the bottom of the belt. The amount of tension depends on the belt construction and the pulley diameter. Pulley diameters that are too small greatly reduce the life of the belt. Recommended minimum pulley diameters are used to reduce bearing loads and provide the longest possible belt life. For example, the minimum diameter for a pulley used on a system containing a 5 HP drive motor operating at 1750 rpm is 3.0″. See Figure 11-6.

The Gates Rubber Company

Abnormal wear on the bottom surface of a belt is caused by a belt too small for the pulleys or by worn or dirty pulleys.

RECOMMENDED MINIMUM PULLEY DIAMETERS*				
Motor HP	Motor Speed**			
	870	1160	1750	3500
½	2.2	—	—	—
¾	2.4	2.2	—	—
1	2.4	2.4	2.2	—
1½	2.4	2.4	2.4	2.2
2	3.0	2.4	2.4	2.4
3	3.0	3.0	2.4	2.4
5	3.8	3.0	3.0	2.4
7½	4.4	3.8	3.0	3.0
10	4.4	4.4	3.8	3.0
15	5.2	4.4	4.4	3.8
30	6.8	6.8	5.2	—
75	10.0	10.0	8.6	—
100	12.0	10.0	8.6	—

* in in.
** in rpm

Figure 11-6. Recommended minimum pulley diameters are used to reduce bearing loads and provide the longest possible belt life.

V-belt Pulleys

A *V-belt pulley* is a pulley with a V-shaped groove. V-belt pulleys are manufactured to exact dimensions. All pulley dimensions, regardless of the manufacturer, should be machined to the same standard sizes. This allows various manufacturer's parts to be interchanged. See Figure 11-7.

V-belt pulleys may be fixed bore or tapered bore pulleys. A *fixed bore pulley* is a machine-bored one-piece pulley. Fixed bore pulleys have an integral hub cast into the pulley. Fixed bore pulleys are chosen by specific bore, OD, and groove dimensions. A *tapered bore pulley* is a two-piece pulley that consists of a tapered pulley bolted to a tapered hub (bushing). The assembly becomes as sound as a press fit when the two pieces are bolted together. The close fit is due to the extreme pressure created from the angled force of the tapered mating surfaces. The angled force prevents fretting corrosion and freezing of parts. The tapered mating surfaces also allow easy removal. For this reason, no lubricant is required. Also, using a lubricant for assembly may cause pulley breakage due to hydraulic (grease or oil) forces created when tightening pulley bolts. See Figure 11-8.

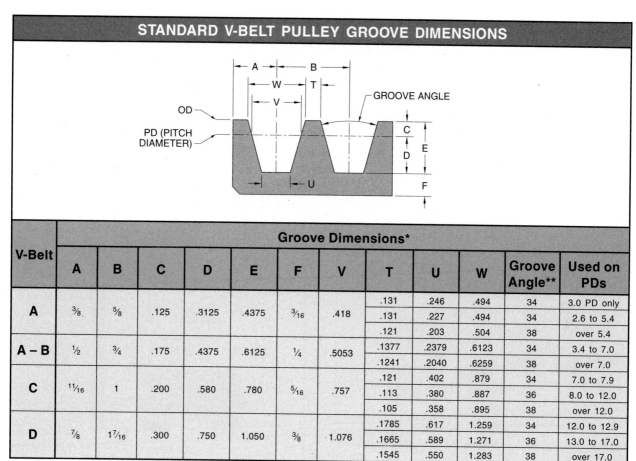

STANDARD V-BELT PULLEY GROOVE DIMENSIONS

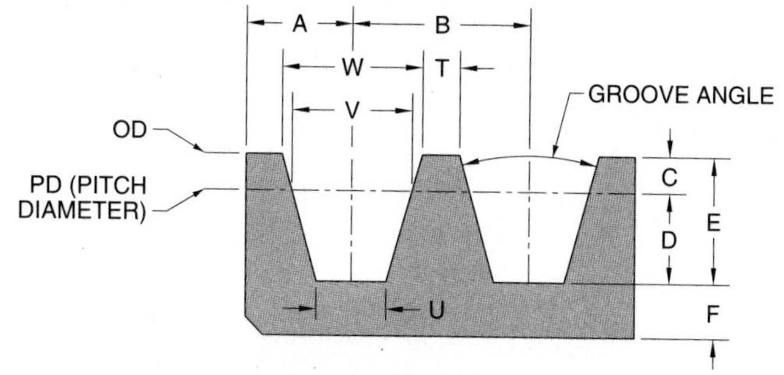

V-Belt	Groove Dimensions*										
	A	B	C	D	E	F	V	T	U	W	Groove Angle**	Used on PDs
A	3/8	5/8	.125	.3125	.4375	3/16	.418	.131	.246	.494	34	3.0 PD only
								.131	.227	.494	34	2.6 to 5.4
								.121	.203	.504	38	over 5.4
A – B	1/2	3/4	.175	.4375	.6125	1/4	.5053	.1377	.2379	.6123	34	3.4 to 7.0
								.1241	.2040	.6259	38	over 7.0
C	11/16	1	.200	.580	.780	5/16	.757	.121	.402	.879	34	7.0 to 7.9
								.113	.380	.887	36	8.0 to 12.0
								.105	.358	.895	38	over 12.0
D	7/8	1 7/16	.300	.750	1.050	3/8	1.076	.1785	.617	1.259	34	12.0 to 12.9
								.1665	.589	1.271	36	13.0 to 17.0
								.1545	.550	1.283	38	over 17.0

* in in.
** in °

Figure 11-7. Standard dimensions used by pulley manufacturers allow parts to be interchanged.

Tapered bore pulleys are installed by sliding the flange end of the hub on a shaft and inserting the key. The hub is positioned on the shaft and the set screw is lightly snugged over the key. The large end of the tapered bore is slid onto the hub and the unthreaded bolt holes on the pulley are lined up with the threaded bolt holes in the hub. The cap screws are screwed into the hub and tightened alternately and evenly. Never tighten enough to close the gap between the hub flange and the pulley. To ensure proper alignment and squareness to prevent wobble, pull up screws evenly and progressively to approximately one-half the required torque values and check alignment before completing torque settings. Finally, apply the proper torque to the screws and tighten the set screw on the key. Torque values are given for safely assembling the pulley to each hub size. See Figure 11-9. Different size tapered bore hubs require particular size keyseats and fit on a particular size bore.

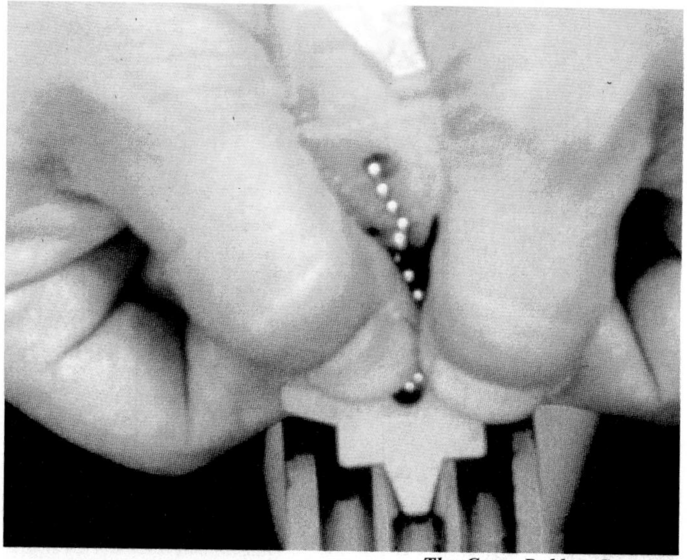

The Gates Rubber Company

A pulley gauge is used to check a pulley for wear. A pulley with 1/16" wear on either side of the gauge should be replaced.

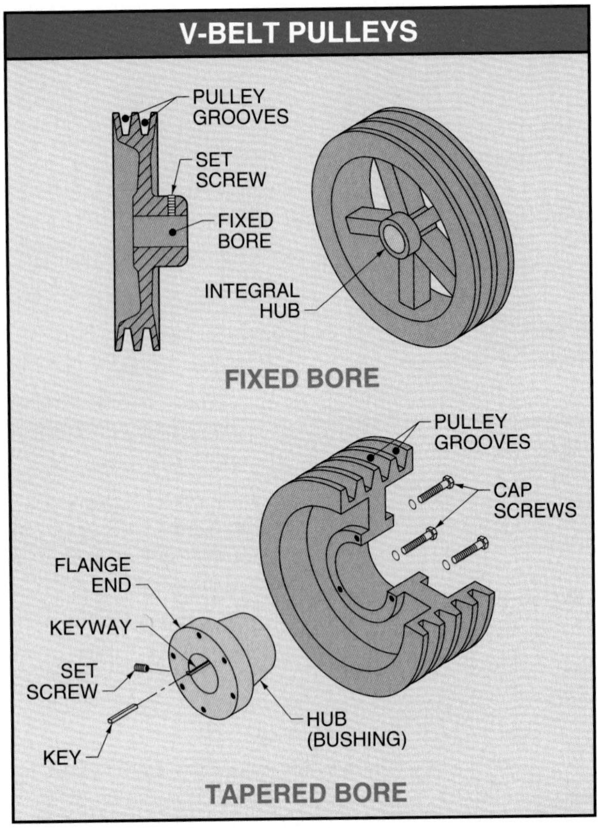

Figure 11-8. V-belt pulleys may be fixed bore or tapered bore pulleys.

RECOMMENDED TAPERED BORE HUB TORQUE VALUES

Hub Size	Cap Screw Size and Thread	Torque*
JA	No. 10 – 24	3
SH – SDS – SD	1⁄4 – 20	6
SK	5⁄16 – 18	10
SF	3⁄8 – 16	20
E	1⁄2 – 13	40
F	9⁄16 – 12	50
J	5⁄8 – 11	90
M	3⁄4 – 10	150
N	7⁄8 – 9	200
P	1 – 8	300
W	1 1⁄8 – 7	400
S	1 1⁄4 – 7	500

* in lb/ft

Figure 11-9. Overtightening cap screws can result in pulley or bushing breakage.

Some applications require the flange to be on the outside of the pulley assembly. In these applications, always leave room between the pulley and the motor or gear drive for tool movement. Pulleys should be placed as close as possible to the shaft bearing to prevent overhung loads. An *overhung load* is a force exerted radially on a shaft that may cause bending of the shaft or early bearing and belt failure. See Figure 11-10.

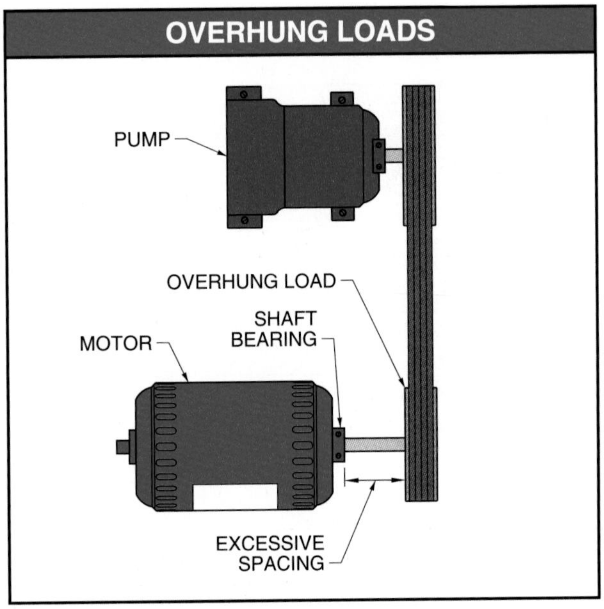

Figure 11-10. Bearings, pulleys, and belts are damaged from excess vibrations when the distance between the shaft bearing and the pulley is excessive.

Pulley Alignment. V-belts that are not aligned properly are destroyed prematurely due to excessive side wear, broken or stretched tension members, or rolling over in the pulley. Pulley misalignment may be offset, nonparallel, or angular. See Figure 11-11.

Offset misalignment is a condition where two shafts are parallel but the pulleys are not on the same axis. Offset misalignment may be corrected using a straightedge along the pulley faces. The straightedge may be solid or a string. Offset misalignment must be within 1⁄10″ per foot of drive center distance. *Nonparallel misalignment* is misalignment where two pulleys or shafts are not parallel. Nonparallel misalignment is also corrected using a string or straightedge. The device connected to the pulley that touches the straightedge at one point is rotated to bring it parallel with the other pulley so that the two pulleys touch the straightedge at four points. *Angular mis-*

alignment is a condition where two shafts are parallel but at different angles with the horizontal plane. Angular misalignment is corrected using a level placed on top of the pulley parallel with the pulley shaft. Angular misalignment must not exceed ½°.

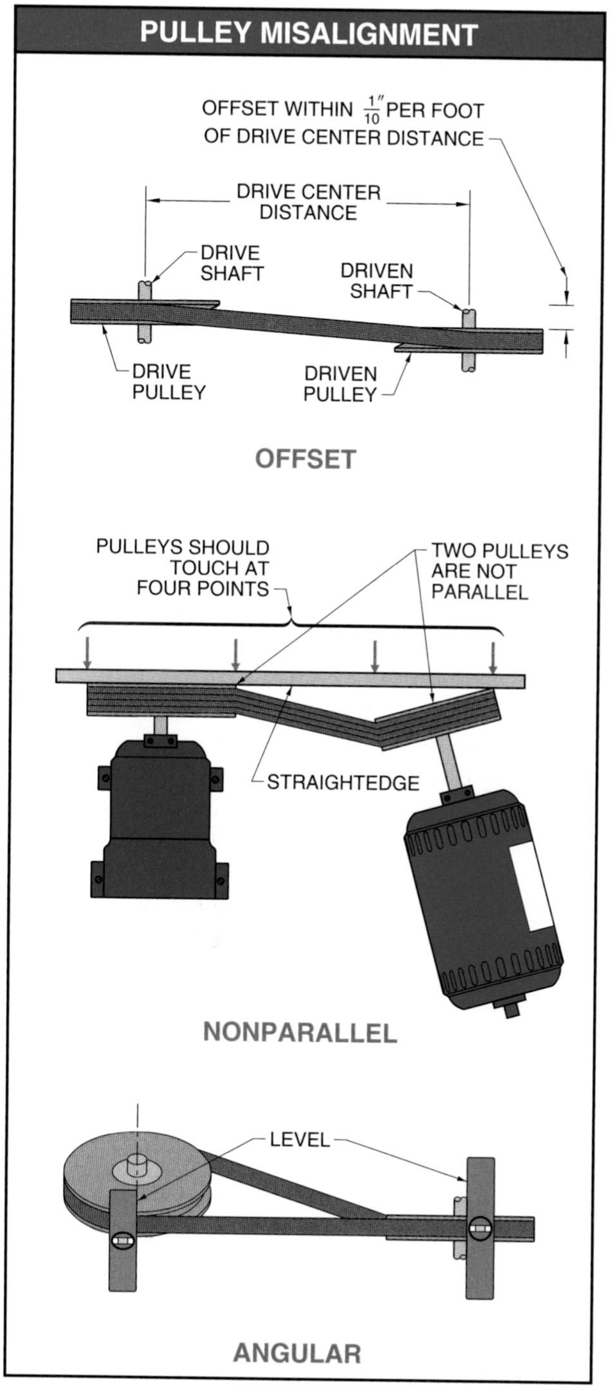

Figure 11-11. Pulley misalignment is corrected by placing a straightedge across the pulleys and adjusting the position of the equipment so the pulleys touch the straightedge at four points.

An excessive number of belts or belts that are too large can severely stress motor or driven shafts. This can happen when load requirements are reduced on a drive, but the belts are not redesigned accordingly. This can also happen when a drive is greatly overdesigned and forces created from belt tensioning are too great for the shafts.

V-belt Tensioning. Proper alignment and tension of V-belts produces long and trouble-free belt operation. Excessive tension produces excessive strain on belts, bearings, and shafts, causing premature wear. Too little tension causes belt slippage. Belt slippage causes excessive heat and premature belt and pulley wear. The best tension for a V-belt is the lowest tension at which the belt does not slip under peak loads. Do not use belt dressing as a remedy for pulley wear or belt slippage. Two methods used for tensioning a belt include the visual adjustment and belt deflection methods. See Figure 11-12.

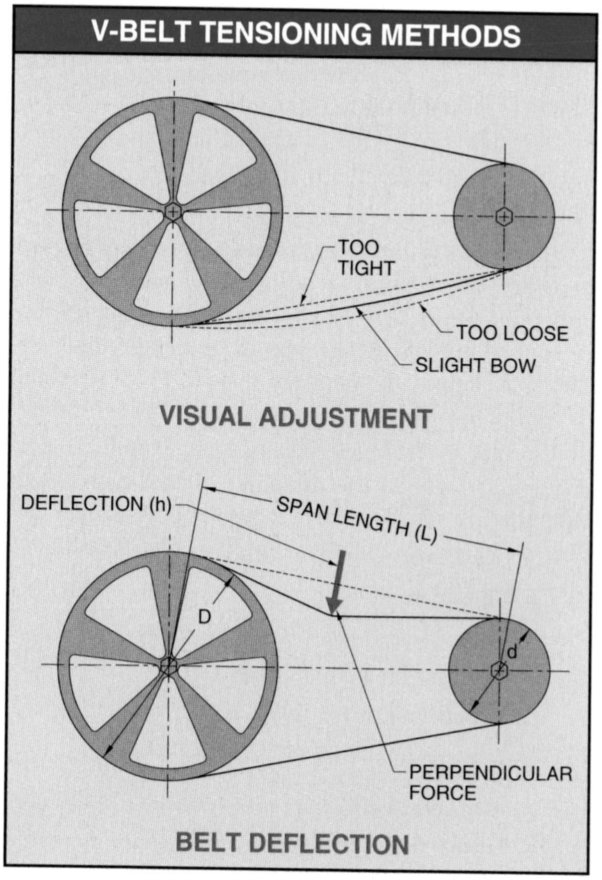

Figure 11-12. Belt tensioning methods include the visual adjustment and belt deflection methods.

The Gates Rubber Company

Alignment is checked visually after pulleys are installed by placing a straightedge along the outside face of both pulleys.

The *visual adjustment method* is a belt tension method in which the tension is adjusted by observing the slight sag at the slack side of the belt. The visual adjustment method is used to roughly adjust belt tension. The belt is placed on the pulleys without forcing the belt over the pulley flange. Never force or pry a belt over the pulley flange. Apply tension to the belt by increasing the drive center distance until the belt is snug. Run the machine for approximately 5 min to seat the belt. Apply load to the belt and observe the slack side of the belt. A slight sag is normal. A squeal or slip indicates that the belt is too loose.

The *belt deflection method* is a belt tension method in which the tension is adjusted by measuring the deflection of the belt. The belt is installed and tension is applied by increasing the drive center distance. The span length (L) between the pulley centers and the deflection height (h) is measured. Proper deflection height is $\frac{1}{64}''$ per inch of span length. A perpendicular force at the midpoint of the span length is applied to measure for proper tension. Proper deflection height is found by applying the formula:

$$h = L \times \frac{1}{64}''$$

where

h = deflection height (in in.)

L = span length (in in.)

$\frac{1}{64}''$ = constant (.0156″)

Timing belts do not generally require run-in and retension because of their positive engagement characteristics.

For example, what is the proper belt deflection of an assembly using a 10″ pulley and a 5″ pulley having a span length of 36″?

$$h = L \times \frac{1}{64}''$$

$$h = 36 \times .0156$$

$$h = \mathbf{.562''}$$

New V-belts seat rapidly during the first few hours. Check and retension new belts following the first 24 hr and 72 hr of operation. During retensioning, determine the deflection height from the normal position by using a straightedge or stretched string across the pulley tops. Belt tension tools are also available. Belt tension tools allow for faster, simpler, and more accurate tension checks.

Double V-belts

A *double V-belt* is a belt designed to transmit power from the top and bottom of the belt. Power must be able to be transmitted in the usual and reverse bend position. Double V-belts are used where the pulleys in a system are required to rotate in opposite directions. An example of a double V-belt system is that of the serpentine belt drive used in the automotive industry. See Figure 11-13.

Double V-belts, like standard V-belts, are identified by a letter-number combination, where the letter designates width and the number identifies nominal length. Like standard V-belts, double V-belts are rated as A, B, C, or D belts, but due to their double sides are designated as AA, BB, CC, or DD belts. For example, a double V-belt identified as DD180 has a 1¼″ width and a 180″ length.

Double V-belt tensioning is determined by the belt's width, speed in feet per minute (fpm), and the diameter of the pulley used. The tension is measured on the tight side of the drive system because every belt drive has a slack or loose side and a tight side. Basic recommended tensioning for double V-belts may be taken from double V-belt tension tables. See Appendix.

Timing Belts

A *timing (synchronous) belt* is a belt designed for positive transmission and synchronization between the drive shaft and the driven shaft. Timing belts consist of tension members, neoprene backing and teeth, and nylon facing. See Figure 11-14.

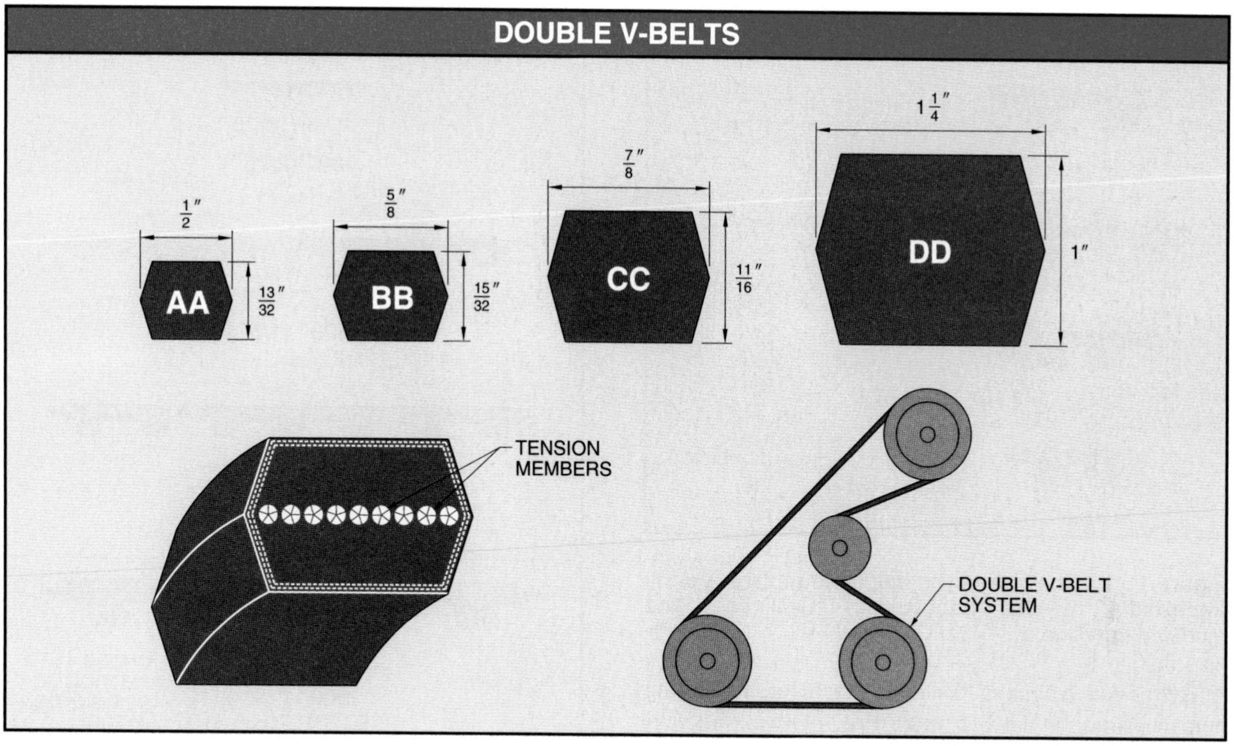

Figure 11-13. Double V-belts are designed with tension members in the center, which allow the belt to flex in both directions.

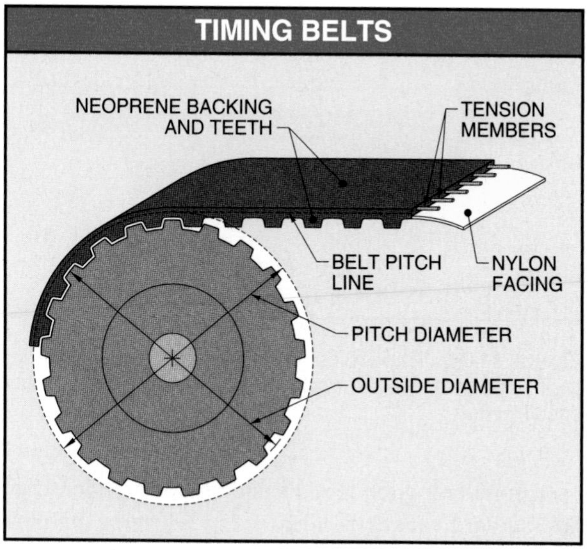

Figure 11-14. Timing belts are constructed of neoprene rubber for flexibility, tension members for strength, and nylon facing to offer low coefficient of friction.

Timing belts and their matching pulleys (sprockets) are toothed for force driving similar to chain drives. Many equipment manufacturers are replacing timing belts in power transmission systems where chains and sprockets were used previously. This is because timing belts are quieter than chain drives, require no lubrication, are able to operate efficiently in drive ranges up to 600 HP, and run at speeds over 10,000 fpm. Similar to gear drive operation, timing belts enter and leave pulleys in a smooth rolling motion with very low friction. Timing belts are identified by tooth profile, pitch length, circular pitch, and nominal width.

The four timing belt tooth profiles are the trapezoidal, double trapezoidal, curvilinear, and modified curvilinear. See Figure 11-15. A *trapezoidal belt* is a timing belt containing trapezoidal-shaped teeth. Trapezoidal belts are the most common timing belt used in industrial applications. A *double trapezoidal belt* is a timing belt containing two trapezoidal-shaped sets of teeth. Double trapezoidal belts are designed for serpentine drive applications where shaft rotations must run in opposite directions. A *curvilinear belt* is a timing belt containing circular-shaped teeth. Curvilinear belts were designed to provide increased capacity over a trapezoidal belt. A *modified curvilinear belt* is a timing belt containing modified circular-shaped teeth. Modified curvilinear belts were designed to maximize load life capacities while optimizing materials and tooth shape.

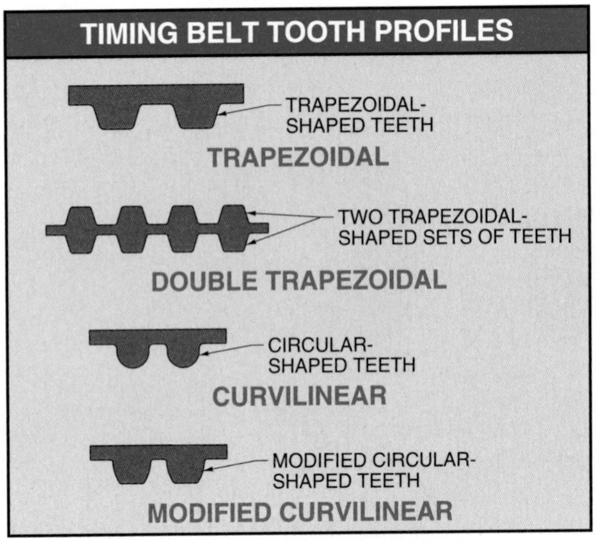

Figure 11-15. The four timing belt tooth profiles are the trapezoidal, double trapezoidal, curvilinear, and modified curvilinear.

Belt *pitch length* is the total length of the timing belt measured at the belt pitch line. *Circular pitch* is the distance from the center of one tooth on a timing belt to the center of the next tooth measured along the belt pitch line. Belt circular pitch is directly related to the cross section of the belt. The six basic cross sections of timing belts are mini-extra light (MXL), extra light (XL), light (L), heavy (H), extra heavy (XH), and double extra heavy (XXH). See Figure 11-16.

Fenner Drives

Product transfer lines use timing belts to synchronize rollers for the efficient transfer of product.

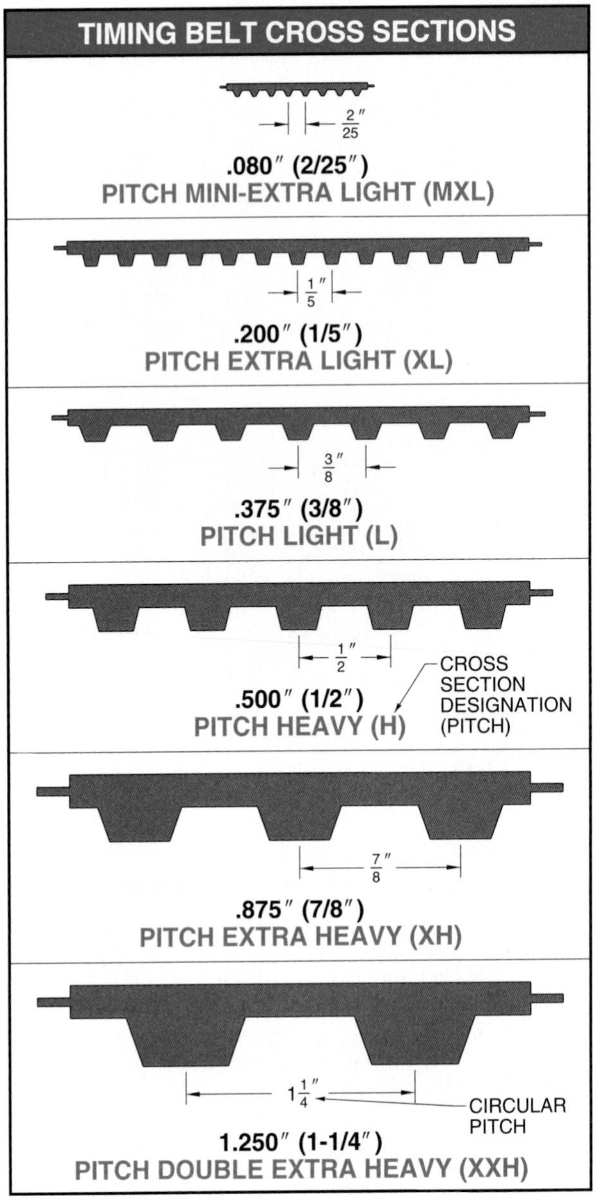

Figure 11-16. Trapezoidal timing belts are classified by their standard length, cross section designation, and circular pitch.

Timing belt pitch lengths and widths are identified by standard numbers. Identifying a timing belt includes pitch length, section, and width designations (nominal width times 100). The nominal pulley width is the maximum standard belt width the pulley accommodates. For example, an XL section belt with a pitch length of 8.000 and a standard width of 0.38 is specified as an 80 XL038 synchronous belt. Belt pitch lengths, number of teeth, length tolerances, and size designations may be taken from American National Standards Institute (ANSI) tables. See Appendix.

VARIABLE-SPEED BELT DRIVES

A *variable-speed belt drive* is a mechanism that transmits motion from one shaft to another and allows the speed of the shafts to be varied. Variable-speed belt drives use wide V-belts and spring-loaded adjustable cone-faced pulleys. Variable-speed belt drives operate by adjusting the pulley width, which changes the pulley pitch diameter. This increases or decreases the shaft speed without changing the drive mechanism (motor) speed. See Figure 11-17.

Variable-speed belt drives are used where speed variations can be obtained using conventional V-belts. V-belts used for variable-speed drives are generally thin, wide, rigid, and are capable of operating in adjustable pulley assemblies.

Changing Speeds

Changing the speed of a belt and pulley is accomplished using an adjustable (sliding) motor base. The motor base is adjusted to increase the pulley center-to-center distance, which increases the belt tension. The increased belt tension forces the flanges apart, which forces the belt to ride lower in the pulley groove. The motor base may also be adjusted to decrease the pulley center-to-center distance, which decreases the belt tension. The decreased belt tension allows the spring to force the flanges together, which forces the belt to ride higher in the pulley groove. Speed control is accomplished by the belt riding high or low in the pulley groove. Most variable-speed belt drives require the drive to be running for adjustment of the pulley width.

Variable-speed belt drive horsepower ratings range from fractional to over 100 HP and are available in speed ratios of 1.15 : 1 to over 9 : 1. Variable-speed drives are categorized as stationary-control or motion-control drives. Stationary-control drives require the drive to be stopped in order to manually adjust the pulley width. This adjustment usually requires drive component disassembly. Motion-control drives are distinguished by one or two movable flanges. The stationary flange is fixed to a central sleeve in drives where only one flange moves during adjustment. The movable flange maintains pressure on the belt by a spring. See Figure 11-18.

> ***The correct belt tension is the least amount of tension that enables the belt to run without slipping when a full load is applied.***

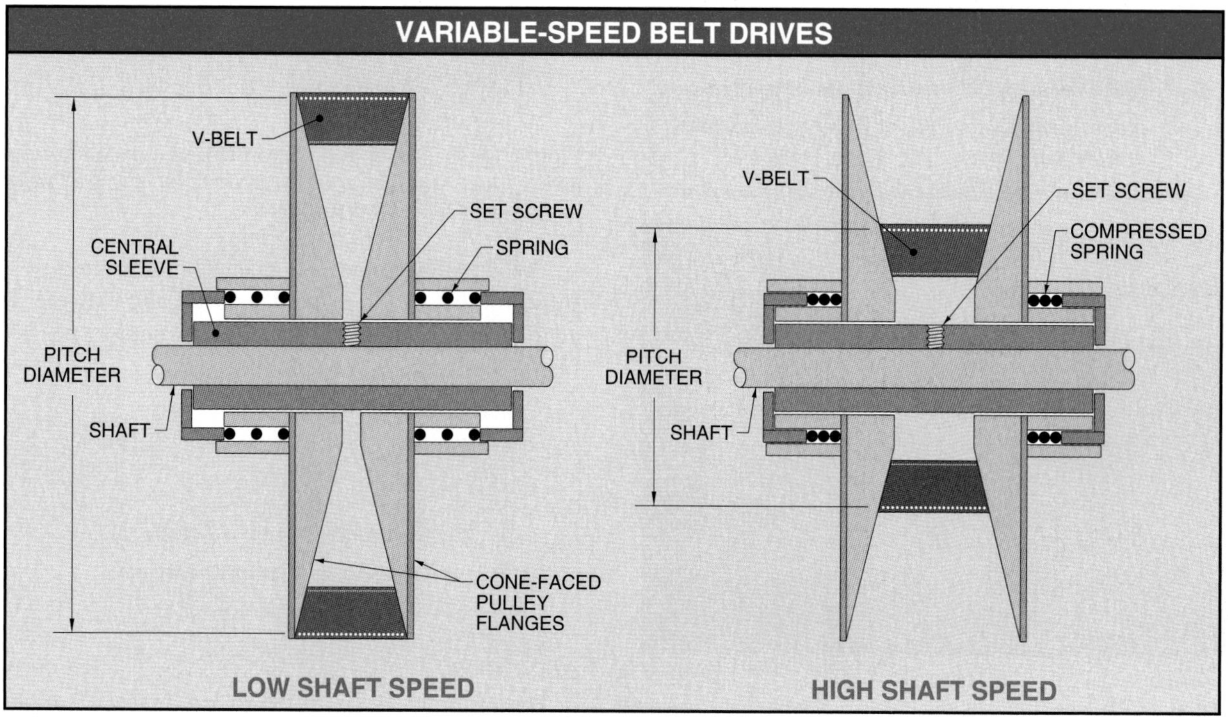

Figure 11-17. In a variable-speed belt drive, if the width of the pulley is increased, the pitch diameter is decreased, thus increasing the speed of the shaft.

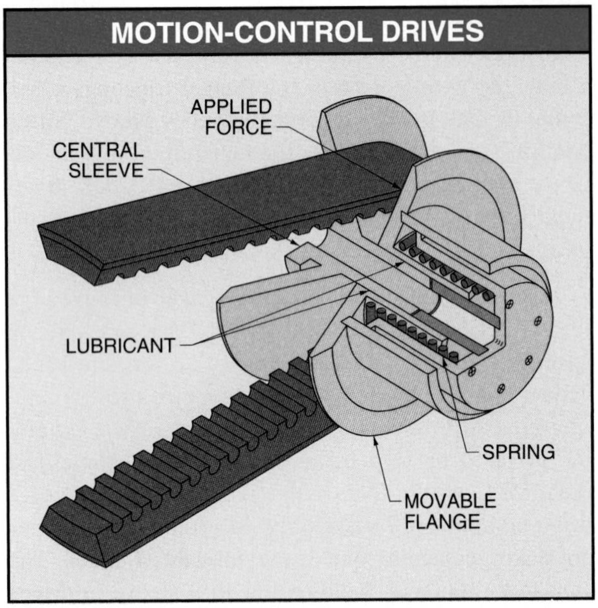

Figure 11-18. In a variable-speed drive, a layer of lubricant must be maintained for separation of parts and prevention of fretting corrosion due to the high forces applied on the central sleeve by the movable flange.

In motion-control drives where two flanges move, both flanges ride on a central sleeve. Each flange is backed up by a spring that exerts equal force on the belt. Extreme caution is required when attempting to disassemble any variable-speed drive assembly due to spring compression of the pulleys.

Never wear neckties, loose sleeves, or lab coats around belt drives.

The Gates Rubber Company

Change all belts on a multiple belt drive even when only one belt requires replacement. New belts that are run with old belts carry more of the load, which shortens belt life.

Calculating Pulley Speed and Size

A change in pulley size may be required for an increase or decrease in pulley output speed. In addition, the diameter of destroyed or missing pulleys must be determined for proper replacement. Pulley size is determined using calculations based on the speed and size of the driven or drive pulley.

The speed or diameter of a driven or drive pulley may be determined by using the appropriate formula. The fourth value (speed or size) can be determined by using the correct formula for calculating speed or diameter when the other three factors (speed and size) are known. Any one value may be determined by rearranging the formula when the other three values are known. Driven pulley speed is based on the diameter and speed of the drive pulley and the diameter of the driven pulley. See Figure 11-19.

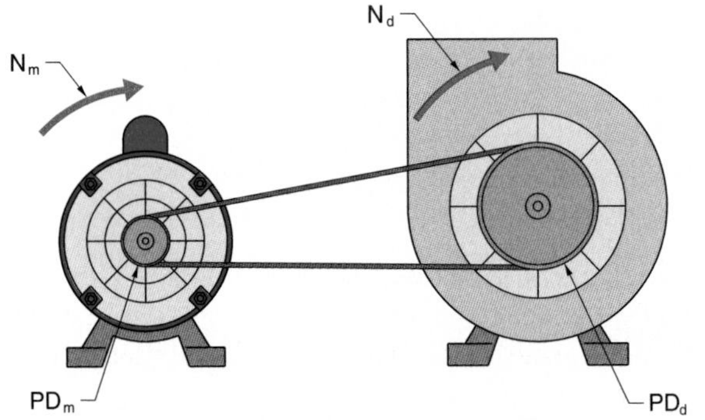

Figure 11-19. The speed or diameter of a driven or drive pulley may be determined by solving for any one value when the other three are known.

Drive pulley speed and diameter is denoted by a subscript m ($_m$). Driven pulley speed and diameter is denoted by a subscript d ($_d$). Driven pulley speed is found by applying the formula:

$$N_d = \frac{PD_m \times N_m}{PD_d}$$

where

N_d = driven pulley speed (in rpm)

PD_m = drive pulley diameter (in in.)

N_m = drive pulley speed (in rpm)

PD_d = driven pulley diameter (in in.)

For example, what is the speed of an 8″ driven pulley if the diameter of the drive pulley is 12″ and its speed is 150 rpm?

$$N_d = \frac{PD_m \times N_m}{PD_d}$$

$$N_d = \frac{12 \times 150}{8}$$

$$N_d = \frac{1800}{8}$$

$$N_d = \mathbf{225\ rpm}$$

Calculating driven pulley diameter is based on knowing the speed of the driven pulley and the diameter and speed of the drive pulley. Driven pulley diameter is found by applying the formula:

$$PD_d = \frac{PD_m \times N_m}{N_d}$$

For example, what is the diameter of a driven pulley rotating at 225 rpm if the diameter of the drive pulley is 12″ and its speed is 150 rpm?

$$PD_d = \frac{PD_m \times N_m}{N_d}$$

$$PD_d = \frac{12 \times 150}{225}$$

$$PD_d = \frac{1800}{225}$$

$$PD_d = \mathbf{8''}$$

Calculating drive pulley speed is based on knowing the diameters of the drive and driven pulleys and the driven pulley speed. Drive pulley speed is found by applying the formula:

$$N_m = \frac{PD_d \times N_d}{PD_m}$$

For example, what is the speed of a 12″ drive pulley if the diameter of the driven pulley is 8″ and its speed is 225 rpm?

$$N_m = \frac{PD_d \times N_d}{PD_m}$$

$$N_m = \frac{8 \times 225}{12}$$

$$N_m = \frac{1800}{12}$$

$$N_m = \mathbf{150\ rpm}$$

A rapid drop in belt tension normally occurs during the run-in period. Tension new drives with a ½ greater deflection force than the maximum recommended force. Check tension frequently during the first day of operation.

Calculating drive pulley diameter is based on knowing the speed of the drive pulley and the speed and diameter of the driven pulley. Drive pulley diameter is found by applying the formula:

$$PD_m = \frac{PD_d \times N_d}{N_m}$$

For example, what is the diameter of a drive pulley rotating at 150 rpm if the diameter of the driven pulley is 8″ and its speed is 225 rpm?

$$PD_m = \frac{PD_d \times N_d}{N_m}$$

$$PD_m = \frac{8 \times 225}{150}$$

$$PD_m = \frac{1800}{150}$$

$$PD_m = \mathbf{12''}$$

FLEXIBLE BELT DRIVE SAFETY

The main concerns when working with flexible belt drives, or any machine, are the safety of the technician working on the machine, co-workers working in the area of the machine, and the machine itself. Unsafe acts may jeopardize the health and welfare of those around the machine but can also damage or destroy the equipment that supports the technicians' employment. Safe work habits related to working on any mechanism in motion requires wearing proper clothing, maintaining a clean environment, and removing and locking out any energy supply.

Proper Clothing

Loose or bulky clothes, loose sleeves, and untucked shirts can cause serious injury while working on flexible belt drives and rotating machinery. Lab coats, neckties, loose belts, and loose long hair should never be worn when working around drive systems.

A technician is properly dressed for safety when wearing safety glasses for eye protection, safety toe shoes for protecting the toes against dropped heavy objects, and gloves to avoid being cut by sharp or nicked pulley edges. A technician must recognize any condition that may do harm and dress accordingly.

Clean Environment

Access to and around machinery must be safe. Floors must be free of oil, clutter, or obstructions. Good footing and balance is necessary while working on any machinery. Also, clean environments make it possible for technicians to devote full attention to the assigned task while not being obstructed by irrelevant objects.

The Gates Rubber Company

Never pry belts off their drives. Belt tension is adjusted by moving the motor and tightening the motor mount to the correct torque. Personal injury or machine damage may result if the pry bar slips.

Removing and Locking Out Energy Supplies

Any person involved in the service or repair of drive systems shall be responsible for placing the equipment inoperable. Assurance must be made that any energy source that has been placed inoperable cannot be inadvertently made operable by any other person.

On January 2, 1990, OSHA 29 CFR 1910.147, *Control of Hazardous Energy Sources (Lockout/Tagout)* became effective. This standard provides rules designed to protect industrial workers who operate, service, and repair power equipment and machinery. Controlling energy sources include lockout, tagout, and blockout procedures. Failure to place equipment energy sources inoperable before working on them is a major cause of serious injury and death. See Figure 11-20.

Placing energy sources in a position of total safety may require either lockout, tagout, or blockout of equipment or possibly all three. *Lockout* is the process of preventing the flow of energy from a power source to a piece of equipment. *Tagout* is the process of placing a tag on a power source that warns others not to restore energy. *Blockout* is the process of placing a solid object in the path of a power source to prevent accidental energy flow.

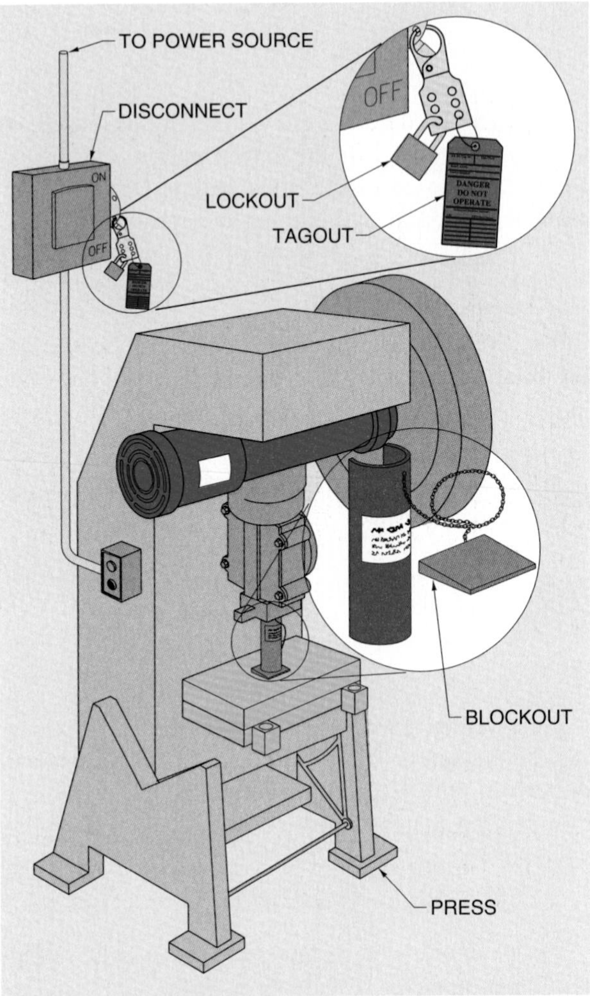

Figure 11-20. Controlling energy sources include lockout, tagout, and blockout procedures.

Mechanical lockout is required of any machine that is operated by any energy source other than electricity. Machines that are activated by compressed air or steam have valves that control movement. These valves need to be locked out and the system may also need to be bled to release any back pressure.

Additionally, coiled springs, spring-loaded devices, or suspended loads may need to be released so that their stored energy will not result in inadvertent movement. Chain or clamp-off large fan blades or turbines while changing belts or aligning. Any gust of wind or air flow may turn the blades enough to injure a worker.

Mechanical Drives

Chapter 12

SEW-Eurodrive, Inc.

Mechanical drives transmit power from one point to another. Mechanical drives use gears for the conversion of energy or the transmission of power. Gears may be spur, helical, rack, herringbone, bevel, miter, worm, or hypoid gears. Gear drives are designed for high speed, low speed, high thrust or radial loads, changing the angle of power, or compactness. Gear drives lose some of their efficiency to friction, but typically remain as high as 95% efficient.

MECHANICAL DRIVES

The law of conservation of energy states that energy can be neither created nor destroyed, but it can be converted from one form to another by appropriate mechanical means. *Mechanical* is pertaining to or concerned with machinery or tools. A *mechanical drive* is a system by which power is transmitted from one point to another. A *gear* is a toothed machine element used to transmit motion between rotating shafts. Gears are used in mechanical drives for the conversion of energy or the transmission of power. In addition to transmitting power, gears convert energy through changing shaft directions, reducing or increasing speed, and changing output torque.

Gear-driven mechanical drives are also used to transmit positive mechanical energy. Gear-driven mechanical drives are often chosen over belt drives because the efficiency of a gear drive is greater than that of a belt drive or friction disc. A *friction disc* is a device that transmits power through contact between two discs or plates. One disc or plate consists of a high frictional material, while the other disc or plate is fabricated from a soft metal. For example, one friction disc may be fabricated of fiberglass strands embedded in an epoxy base and the other friction disc may be fabricated from mild steel or brass. Belt drives or friction discs lose most of their efficiency through slippage. Gear drives lose some of their efficiency to friction, but typically remain as high as 95% efficient. See Figure 12-1. Energy and efficiency is the principle of work where work input equals useful work output plus work done against friction.

Transformation of Energy

A gear or gear drive is the rotational mechanism within a machine. A *machine* is any device by which the magnitude or method of application of a force is changed in order to achieve work. The force applied by a rotational mechanism is measured as its torque.

Torque. *Torque* is the twisting (rotational) force of a shaft. Torque is the amount of force at a distance from the center of a shaft required to achieve work. Torque is produced when turning something, such as removing a bottle cap or turning a doorknob. Torque is equal to force times the distance from the point of rotation to the point the force is applied. Torque is found by applying the formula:

$T = F \times D$

where

T = torque (in lb-ft)

F = force (in lb)

D = distance (in in. or ft)

For example, what is the torque developed if a 60 lb force is applied at the end of a 2′ lever arm?

$T = F \times D$

$T = 60 \times 2$

T = **120 lb-ft**

MACHINE ELEMENT EFFICIENCY	
Mechanical Component	**Efficiency***
Common bearings (single)	97
Ball bearings	99
Roller bearings	98
Belting	96 – 98
Bevel gears, including bearings	
Cast teeth	92
Cut teeth	95
Hydraulic couplings	98
Hydraulic jacks	80 – 90
Overhead cranes	30 – 50
Transmission chains (high grade)	98

* in %

Figure 12-1. Friction, slippage, or a combination of both significantly reduces the efficiency of a mechanical component.

In many cases, torque is expressed as a measurement of pound-inches (lb-in.). For example, a 50 HP hydraulic motor may have 20,000 lb-in. of torque. The inch value may be converted to a foot value by dividing the inch value by 12. For example, 20,000 lb-in. of torque equals 1666.667 lb-ft (20,000 ÷ 12 = 1666.667 lb-ft).

This formula can be used to determine the amount of torque required to rotate an object such as a doorknob or screw. However, the torque developed or given by a rotating machine such as a motor is determined using the machine's energy output rating (horsepower) and its speed in revolutions per minute (rpm). The torque (in lb-ft) of a rotating machine is found by applying the formula:

$$T = \frac{5252 \times HP}{rpm}$$

where

T = torque (in lb-ft)

HP = horsepower

rpm = revolutions per minute

5252 = constant (33,000 lb-ft ÷ π × 2)

For example, what is the available torque supplied by a 1 HP, 1750 rpm motor?

$$T = \frac{5252 \times HP}{rpm}$$

$$T = \frac{5252 \times 1}{1750}$$

$$T = \frac{5252}{1750}$$

T = **3 lb-ft**

Torque can be measured at the different components of a machine, but it is not a measure of work performance. Torque is the measurement of overcoming a resistance. This is because torque is not dependent on time. For example, it is possible to have a specific torque valve that offers high speed, low speed, or no movement at all. Torque with an added time element, such as revolutions per minute, becomes horsepower. *Horsepower* is a unit of power equal to 746 W or 33,000 lb-ft per minute (550 lb-ft per second). See Figure 12-2.

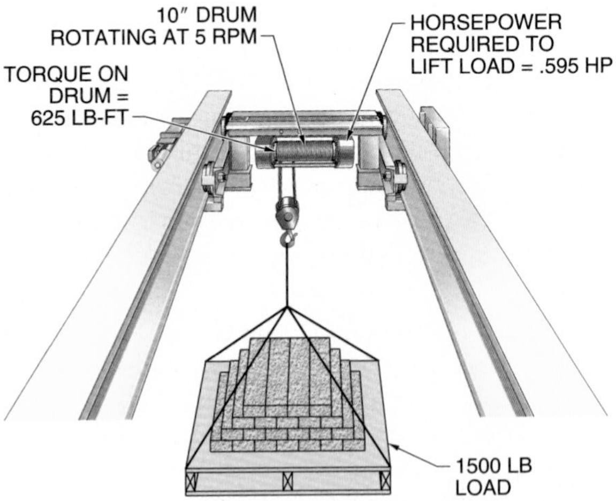

Figure 12-2. Horsepower required to lift a load is determined by speed and torque.

Horsepower is used to measure the energy produced by an electric motor while doing work. Horsepower required to overcome a load is found by applying the formula:

$$HP = \frac{T \times rpm}{5252}$$

For example, what is the horsepower required to turn a winch containing a 10″ drum at 5 rpm if a 1500 lb load is placed on the winch? *Note:* The torque at the drum equals 625 lb-ft (5 × 1500 ÷ 12 = 625 lb-ft).

$$HP = \frac{T \times rpm}{5252}$$

$$HP = \frac{625 \times 5}{5252}$$

$$HP = \frac{3125}{5252}$$

$$HP = \mathbf{.595\ HP}$$

Gear Speed. A *gear train* is a combination of two or more gears in mesh used to transmit motion between two rotating shafts. A *drive gear* is any gear that turns or drives another gear. A *driven gear* is any gear that is driven by another gear. For example, on a two-gear machine, one gear is the drive gear and the other gear is the driven gear. On a three-gear machine, two gears are drive gears (first and second) and two gears are driven gears (second and last). The second gear is the driven gear when referenced to the first gear and is the drive gear when referenced to the third gear. The speeds of the drive and driven gears are inversely proportional to the number of teeth on each gear. For example, a gear with a large number of teeth rotates at a slow speed and a gear with a small number of teeth rotates at a fast speed.

Although some gear drives are designed to deliver an increase in speed (rpm) over the input (drive) gear, the majority of gear drives are designed to produce a speed reduction. In gear transmission, the speed of the driven gear depends on the speed of the drive gear and the number of teeth of the drive and driven gears. See Figure 12-3. The speed of a driven gear is found by applying the formula:

$$N_2 = \frac{T_1 \times N_1}{T_2}$$

where

N_2 = speed of driven gear (in rpm)

T_1 = number of teeth on drive gear

N_1 = speed of drive gear (in rpm)

T_2 = number of teeth on driven gear

For example, what is the speed of a 50 tooth driven gear if the drive gear has 18 teeth and rotates at 100 rpm?

$$N_2 = \frac{T_1 \times N_1}{T_2}$$

$$N_2 = \frac{18 \times 100}{50}$$

$$N_2 = \frac{1800}{50}$$

$$N_2 = \mathbf{36\ rpm}$$

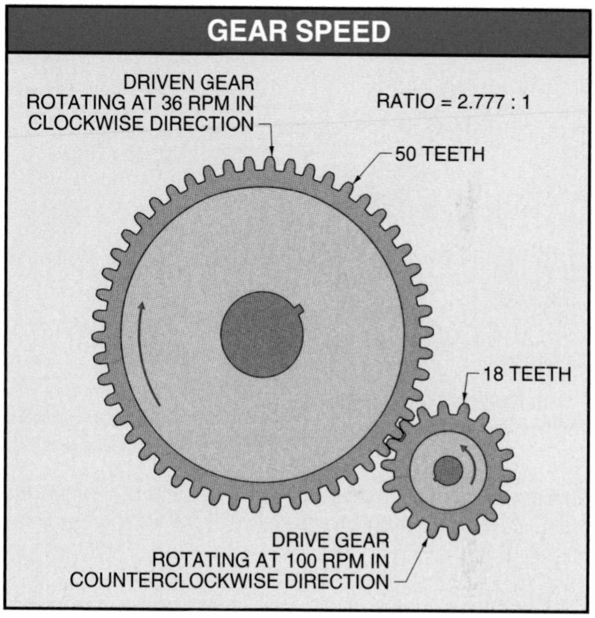

Figure 12-3. The speed of the driven gear depends on the speed of the drive gear and the number of teeth of the drive and driven gears.

This formula may be rearranged to find any one value when the other three are known. For example, the number of teeth required on a driven gear may be found by multiplying the number of teeth on the drive gear by the speed of the drive gear and dividing by the required speed of the driven gear. The number of teeth required on a driven gear to produce a required output speed is found by applying the formula:

$$T_2 = \frac{T_1 \times N_1}{N_2}$$

where

T_2 = number of teeth on driven gear

T_1 = number of teeth on drive gear

N_1 = speed of drive gear (in rpm)

N_2 = speed of driven gear (in rpm)

For example, what is the number of teeth on a driven gear required to produce 25 rpm if the drive gear having 40 teeth rotates at 75 rpm?

$$T_2 = \frac{T_1 \times N_1}{N_2}$$

$$T_2 = \frac{40 \times 75}{25}$$

$$T_2 = \frac{3000}{25}$$

$$T_2 = \mathbf{120\ teeth}$$

Exxon Company

Gear-driven mechanical drives are often chosen over belt drives for use in industrial machines because their efficiency ranges between 92% and 95%.

The speed of the gears in a gear train is inversely proportional to the ratio of the number of teeth of the gears. A *ratio* is the relationship between two quantities or terms. The ratio of one quantity to another is the first divided by the second. The colon (:) is the symbol used to indicate a relation between terms. For example, the ratio of 10 : 4 is $^{10}/_{4}$, which equals 2.5 (10 ÷ 4 = 2.5).

An *inverse ratio* is the ratio that results when the second term is divided by the first. An inverse ratio is the reciprocal of a given ratio. The inverse ratio of 10 : 4 is 4 : 10, or $^{4}/_{10}$, which equals .4 (4 ÷ 10 = .4). A *proportion* is an expression of equality between two ratios. Proportions are either direct proportions or inverse proportions. A *direct proportion* is a statement of equality between two ratios in which the first of four terms divided by the second equals the third divided by the fourth. For example, in the proportion 8 : 4 = 12 : 6, both ratios equal 2 (8 ÷ 4 = 2 and 12 ÷ 6 = 2). An increase in one term results in a proportional increase in the other related term.

An *inverse proportion* is a proportion in which an increase in one quantity results in a proportional decrease in the other related quantity. For example, two gears with 50 and 18 teeth respectively are meshed. The 18 tooth gear rotates at 100 rpm and the 50 tooth gear rotates at 36 rpm. The ratio of the number of teeth is 50 : 18, or $^{50}/_{18}$ and the inverse ratio of speeds is 100 : 36, or $^{100}/_{36}$. Using the inverse ratio, the two ratios are equal at 2.777 : 1.

Idler Gears. An *idler gear* is a gear that transfers motion and direction in a gear train, but does not change speeds. Two external gears in mesh operate in opposite directions unless an idler gear is used. Idler gears are drive and driven gears. Idler gears are generally used when the shafts of the driven gear and the drive gear are too far apart and the use of large gears is impractical. See Figure 12-4.

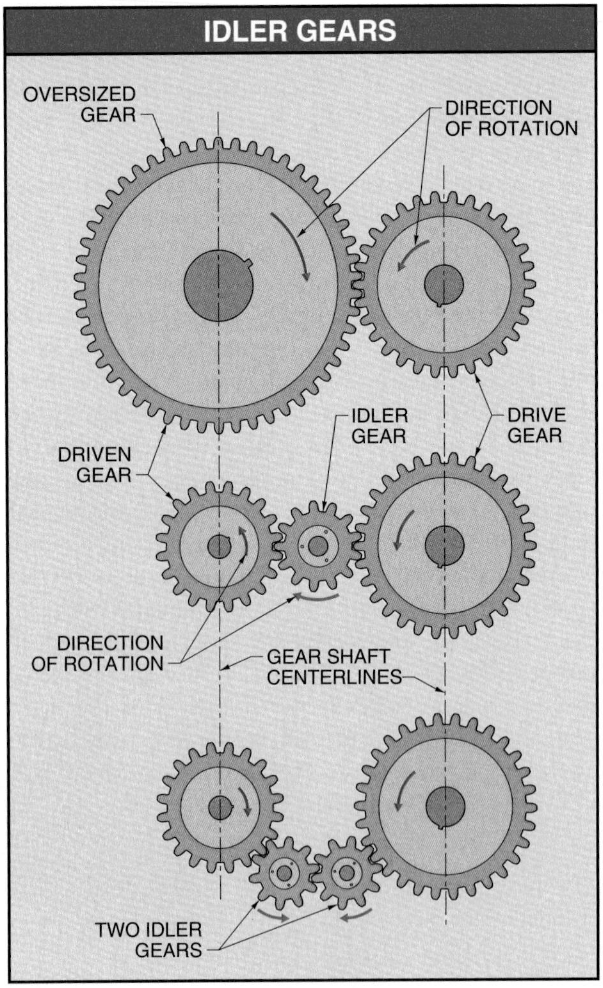

Figure 12-4. The use of idler gears in a gear train allows proper rotation of the output gear and offers compactness between gear shafts within the gear housing.

Two idler gears may be placed between the drive and driven gears when opposite rotation between the drive and driven gears is required. This configuration of the gear train, barring extra losses from friction, provides opposite rotations without affecting speed.

Gear drives that are shut down for longer than one week should be run at least 10 min each week to keep the gears coated with oil and help prevent rusting due to moisture condensation.

Compound Gear Trains

A *compound gear train* is two or more sets of gears where two gears are keyed and rotate on one common shaft. Compound gear trains produce higher speeds in less space than gear trains using simple gearing. See Figure 12-5.

In a four-gear compound gear train, gear 1 is the first drive gear. Gear 1 drives gear 2, which is the first driven gear. Gear 2 is keyed to the same shaft as gear 3, which is the second drive gear. Gear 2 and gear 3 are different sizes and contain a different number of teeth, but rotate at the same speed. Gear 3 drives gear 4, which is the second driven and output gear. Drive and driven gear determination is required when calculating compound gear train output speed. The formula for calculating the output speed of a compound gear train is based on the number of teeth of the drive gears multiplied by the speed of the first drive gear divided by the number of teeth of the driven gears. The output speed of a compound gear train is found by applying the formula:

$$N_4 = \frac{T_1 \times T_3 \times N_1}{T_2 \times T_4}$$

where

N_4 = speed of output gear (in rpm)
T_1 = number of teeth on first drive gear
T_3 = number of teeth on second drive gear
N_1 = speed of first drive gear (in rpm)
T_2 = number of teeth on first driven gear
T_4 = number of teeth on output gear

For example, what is the output speed of a 20 tooth output gear in a compound gear train in which gear 1 (first drive gear) contains 50 teeth and rotates at 25 rpm, gear 2 (first driven gear) contains 25 teeth and rotates at 50 rpm, and gear 3 (second drive gear) contains 75 teeth and rotates at 50 rpm?

$$N_4 = \frac{T_1 \times T_3 \times N_1}{T_2 \times T_4}$$

$$N_4 = \frac{50 \times 75 \times 25}{25 \times 20}$$

$$N_4 = \frac{93{,}750}{500}$$

$$N_4 = \mathbf{187.5\ rpm}$$

COMPOUND GEAR TRAIN—OUTPUT SPEED DETERMINATION

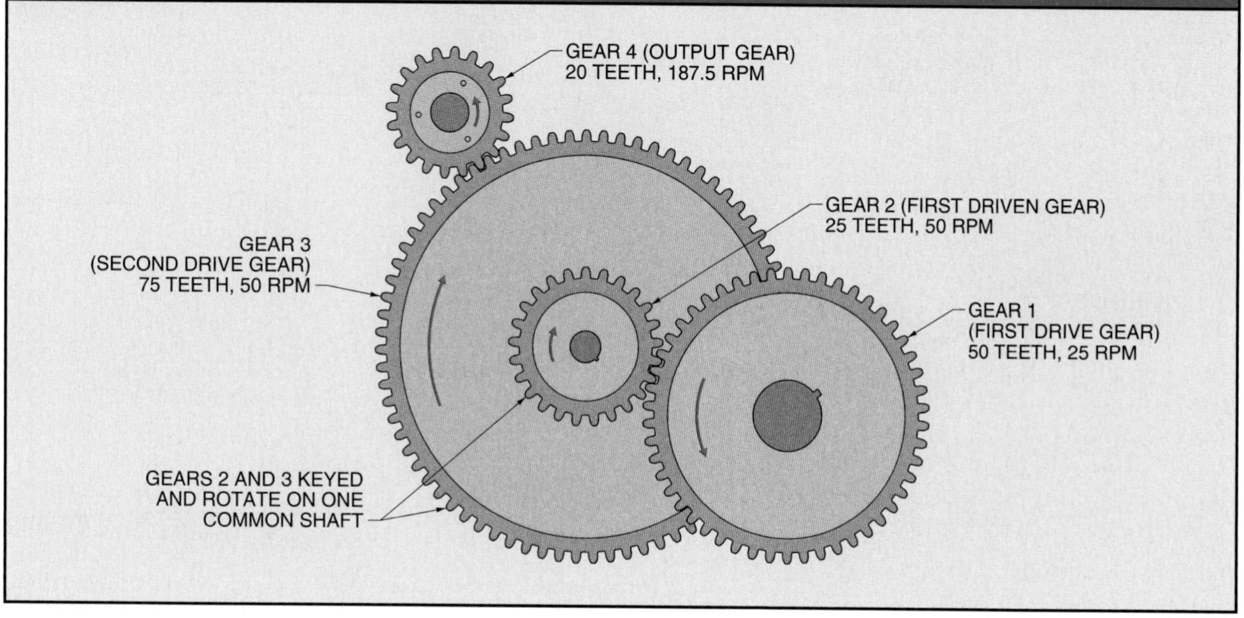

Figure 12-5. Compound gear trains produce higher speeds in less space than gear trains using simple gearing.

In compound gear trains, the number of teeth on the output gear is based on the number of teeth on the first and second drive gears and the speed of the first drive gear divided by the number of teeth of the first driven gear and the speed of the output gear. See Figure 12-6. The number of teeth on the output gear in a compound gear train is found by applying the formula:

$$T_5 = \frac{T_1 \times T_4 \times N_1}{T_3 \times N_5}$$

where

T_5 = number of teeth on output gear

T_1 = number of teeth on first drive gear

T_4 = number of teeth on second drive gear

N_1 = speed of first drive gear (in rpm)

T_3 = number of teeth on first driven gear

N_5 = speed of output gear (in rpm)

For example, how many teeth does an output gear rotating at 100 rpm have if connected to 5 gears in a compound gear train where gear 1 (first drive gear) contains 30 teeth and rotates at 300 rpm, gear 2 (idler gear) has 20 teeth, gear 3 (first driven gear) and gear 4 (second drive gear) have the same shaft and contain 50 and 35 teeth respectively?

$$T_5 = \frac{T_1 \times T_4 \times N_1}{T_3 \times N_5}$$

$$T_5 = \frac{30 \times 35 \times 300}{50 \times 100}$$

$$T_5 = \frac{315{,}000}{5000}$$

$$T_5 = \mathbf{63\ teeth}$$

GEAR FORM AND TERMINOLOGY

To transmit power smoothly from one gear to another, a special tooth form is used to allow sliding without damage or jerky motion. A *tooth form* is the shape or geometric form of a tooth in a gear when seen as its side profile. For example, the tooth form of a rack gear consists of three flat surfaces. The tooth form of a spur gear has a flat surface and two curved (involute) surfaces. *Rack teeth* are gear teeth used to produce linear motion. Smooth power transmission is accomplished using an involute tooth form. An *involute form* is a tooth form that is curled or curved. This tooth form is used for gear teeth to provide a uniform motion and straight line of action. See Figure 12-7.

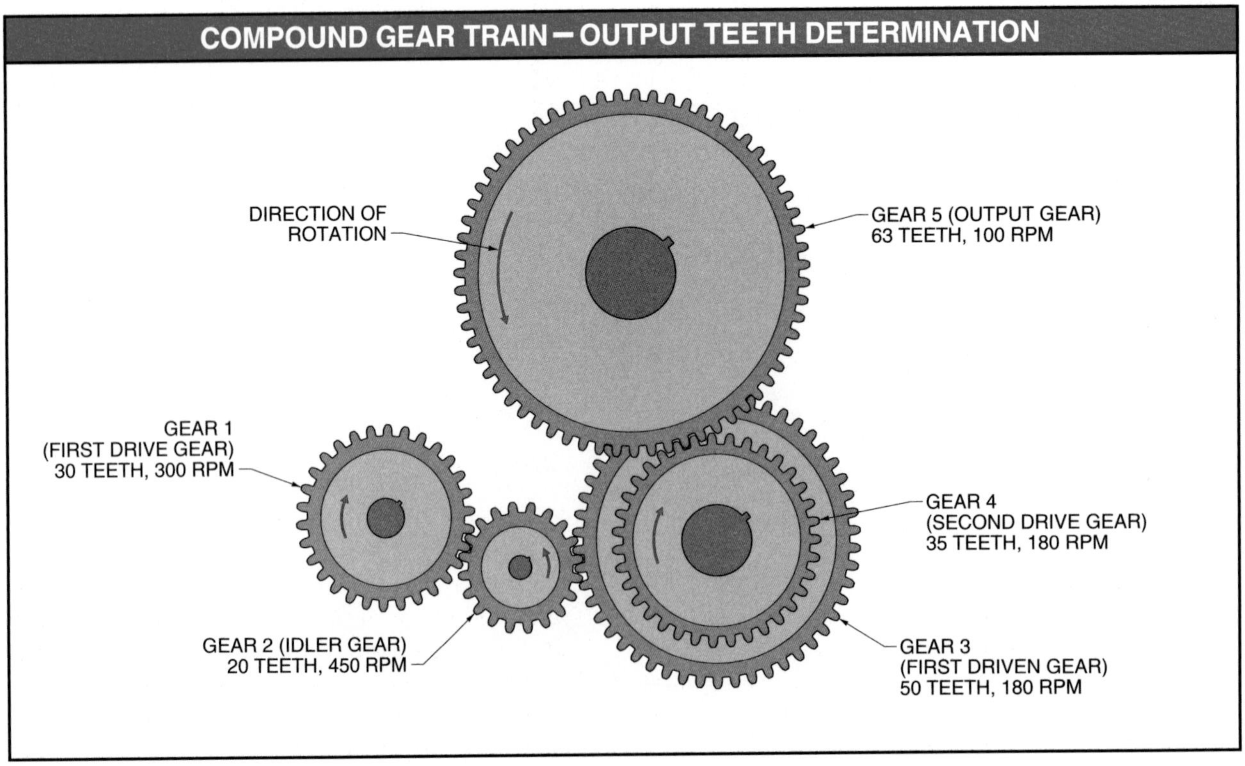

Figure 12-6. The idler gear is disregarded when determining the number of teeth or speed calculation in a compound gear train.

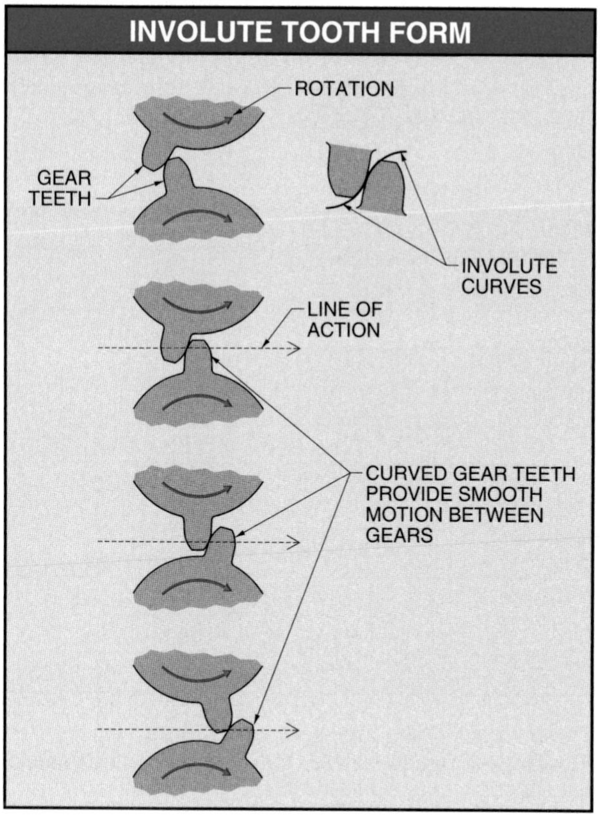

Figure 12-7. The involute tooth form on a gear allows a smooth and uniform motion between gears.

The involute form is a basic profile for gear teeth and is used on most gears. The involute form and gear tooth terminology are defined and standardized by ANSI and the American Gear Manufacturers Association (AGMA). See Figure 12-8.

Certain terms are used when fabricating, specifying, or defining a gear's function. Gear terminology includes pitch circle, pinion, diametral pitch, circular pitch, base diameter, clearance, working depth, and face width. *Pitch circle* is the circle that contains the operational pitch point. An *operational pitch point* is the tangent point of two pitch circles at which gears operate. A *pinion* is the smaller gear of a pair of gears, especially when engaging rack teeth. *Diametral pitch* is the ratio of the number of teeth in a gear to the diameter of the gear's pitch circle. *Circular pitch* is the distance from a point on a gear tooth to the corresponding point on the next gear tooth, measured along the pitch circle. The diametral pitch of a gear may be found if the pitch diameter and number of teeth are known. Diametral pitch is found by applying the formula:

$$DP = \frac{T}{D}$$

where

DP = diametral pitch

T = number of teeth

D = pitch diameter (in in.)

This formula may be rearranged to find any one value when the other two are known.

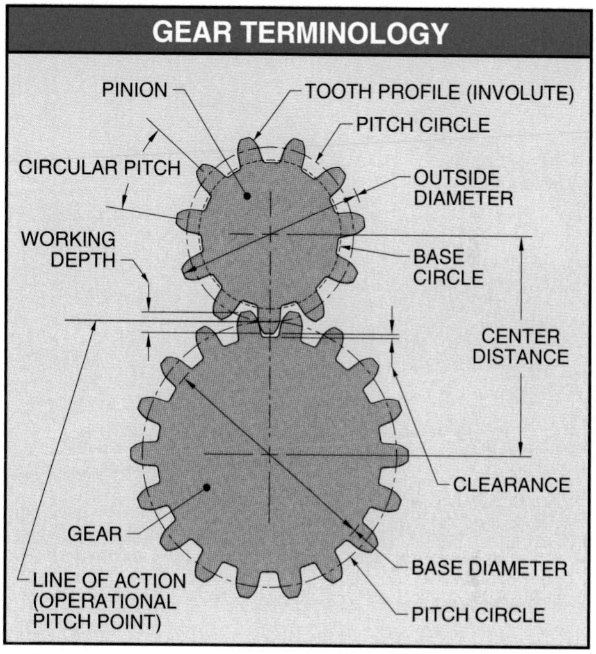

Figure 12-8. Gear terminology is standardized by ANSI and AGMA.

For example, what is the diametral pitch of a gear that has 30 teeth with a pitch diameter of 6″?

$$DP = \frac{T}{D}$$

$$DP = \frac{30}{6}$$

$$DP = \mathbf{5}$$

Base diameter is the diameter from which the involute portion of a tooth profile is generated. *Clearance* is the radial distance between the top of a tooth and the bottom of the mating tooth space when fully mated. Clearance is directly related to a gear's working depth. *Working depth* is the depth of engagement of two gears. Clearance is required because gears bind without clearance. Clearance is the gap between the base diameter of one gear and the top of the mating tooth. Proper clearance is generally accomplished when backlash measurements are met.

The *face width* is the length of the teeth in an axial plane. This is generally the width of any spur gear. However, the length or face width of angled gears, such as helix or spiral angled gears, are somewhat longer because the measurement is made along the cone distance of these gears.

Cone Drive Operations Inc./Subsidiary of Textron Inc.

Cone Drive manufactures worm gear drives with a double-enveloping design in which the worm and gear wrap around each other. This provides high shock resistance, low backlash, and increased load carrying capacity as more tooth area are in contact and more teeth are in mesh than other worm gear designs.

BACKLASH

Backlash is the play between mating gear teeth. Backlash allows for any errors in tooth profile, gear mounting, or shaft or gear runout. Backlash prevents gears from making contact on both sides of the teeth, causing seizing and failure, and allows for lubrication.

The amount of backlash does not affect the involute action of the gear, but too little backlash can cause overheating or overloading and too much backlash can cause damage from tooth slamming, especially when drives reverse frequently. Backlash measurement is made at the pitch diameter and is checked after rotating the gears in mesh to the point of closest engagement. *Pitch diameter* is the diameter of the pitch circle. Determining the proper amount of backlash is based on the diametral pitch. A chart may be used to determine backlash tolerances when the diametral pitch is known. See Figure 12-9.

RECOMMENDED GEAR BACKLASH	
Diametral Pitch	**Backlash***
3	.009 – .014
4	.007 – .011
5	.006 – .009
6	.005 – .008
8	.004 – .006
10	.003 – .005
12	.003 – .005
16	.002 – .004
20	.002 – .004
24	.002 – .004

* in in.

Figure 12-9. Recommended backlash for mating gears is based on the gear diametral pitch.

Measuring Backlash

Backlash is generally measured by holding one gear stationary while rocking the engaged gear back and forth. The movement is measured with the stem of a dial indicator in the plane of rotation on the driven gear and at the pitch diameter. Other methods of measuring backlash include using thickness gauges or a lead wire. Thickness (feeler) gauges are slid into the spacing between meshing teeth without using force or rotation. The gauge thickness is read for backlash tolerance. Lead wire may be placed between meshing teeth and the gears rotated once. The lead is squeezed and sized to the spacing between the meshed teeth. A micrometer is then used to measure the formed (flattened) lead to determine backlash tolerance. See Figure 12-10.

Gear-driven mechanical drives should be checked daily for oil leaks and unusual noises. If oil leaks are present, the drive should be shut down, the cause of the leakage corrected, and the oil level checked. If any unusual noises occur, the unit should be shut down until the cause of the noise has been determined and corrected.

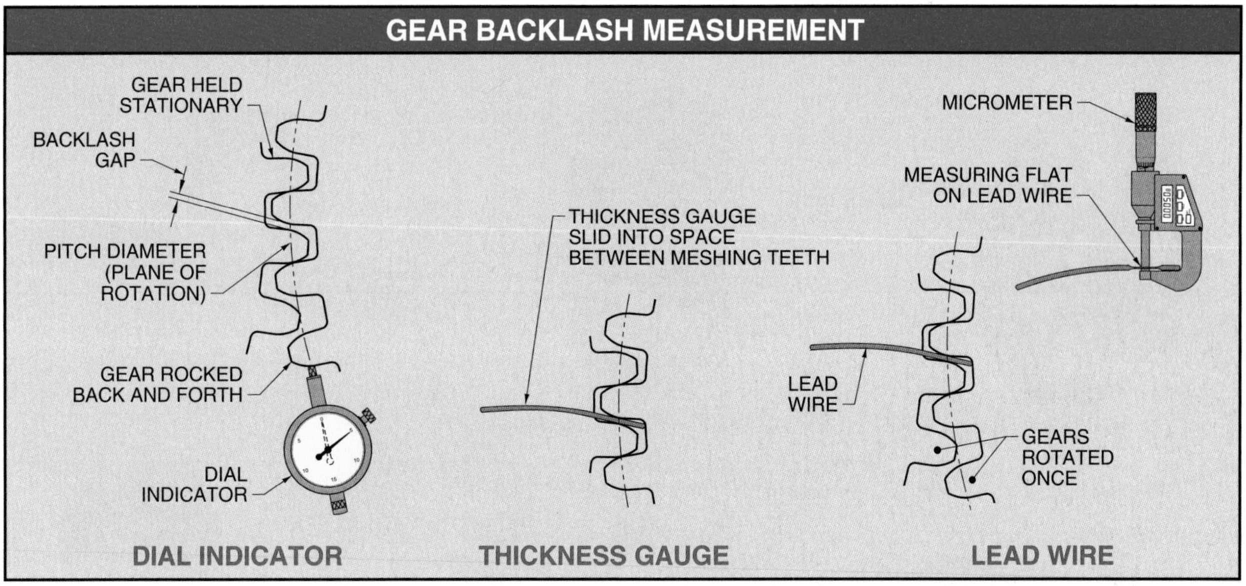

Figure 12-10. Backlash between two gears in mesh is measured using a dial indicator, thickness gauge, or lead wire.

GEARS

Gear drives are designed for high speed, low speed, high thrust or radial loads, changing the angle of power, or compactness. The transmission of power, torque, or angle is accomplished by the use of spur, helical, rack, herringbone, bevel, miter, worm, or hypoid gears. See Figure 12-11.

Spur

A *spur gear* is a gear that has straight teeth that are parallel to the shaft axes. Spur gears are the most commonly used gear and were originally designed for internal clock works. Spur gears are included in most mechanical drives and large industrial machinery. Spur gears are used to transmit power from one parallel shaft to another shaft where there is no end thrust or axial displacement.

Spur gears are excellent transmission gears due to their ability to slide and mesh from one gear size into another to change speeds. Spur gears of different diameters and numbers of teeth are interchangeable as long as they are of the same pitch. Spur gears are rougher running and noisier than other gears and usually run at slower speeds to reduce vibration and noise.

Helical

A *helical gear* is a gear with teeth that are cut at an angle to its axis of rotation. The steeper the angle, the quieter the gear. Side movement (thrust) is developed when a helical gear is used. Thrust energy caused by the tooth load must be supported by thrust bearings. The direction and amount of thrust depends on gear rotation and the direction of the helix. Helical gear drive angles may be anywhere from 0° to 90°.

Helical gears are quieter and smoother running than spur gears because pressure is transferred gradually and uniformly as successive teeth are meshed. Also, power is more widely distributed because it is placed on several teeth in mesh. The increased area of distribution allows for finer tooth sizes and equalized tooth wear.

Rack

A *rack gear* is a gear with teeth spaced along a straight line. Rack gears are used with pinion gears (spur gears) to convert rotary motion to linear motion or linear motion to rotary motion depending on which gear is the drive gear and which gear is the driven gear. Unlike the pinion gear, rack gear teeth do not have to be involute because the rack gear does not rotate. Rack and pinion gears may be either spur or helical in tooth form.

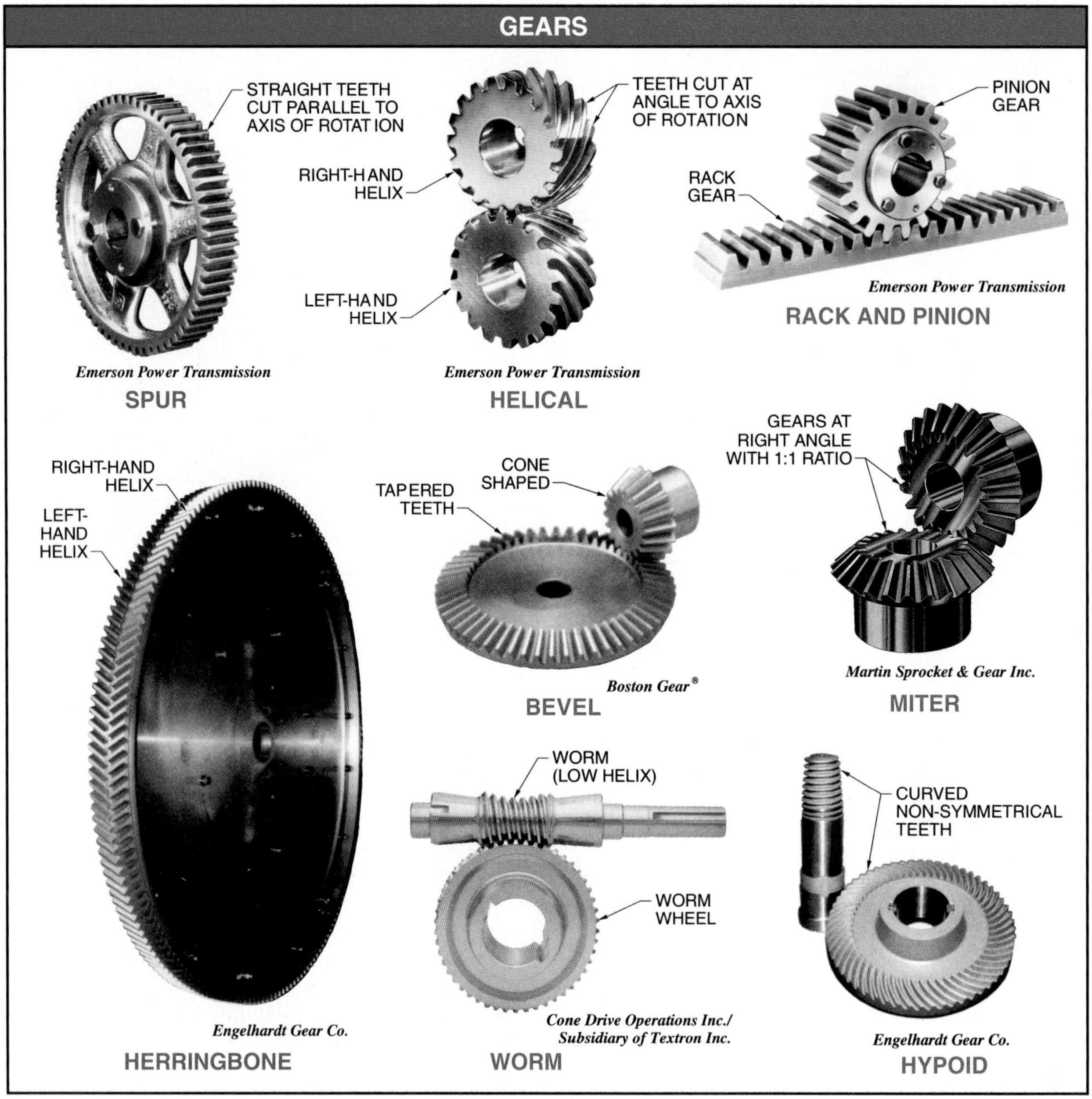

Figure 12-11. The gear used in a mechanical drive is determined by the required gear speed, load placed on the gear, angular requirements, and space constraints.

Herringbone

A *herringbone gear* is a double helical gear that contains a right- and left-hand helix. End thrust is avoided by the teeth being at opposite angles. Herringbone gears are used in parallel shaft transmission, especially where smooth, high speeds are required. Smooth transmission is possible due to the gradual overlapping mesh of the teeth and is made even smoother when the right- and left-hand helix is offset by one half pitch. Herringbone gears are often used for connecting steam turbines to electrical generators, pumps, propeller shafts, etc. because they are able to run at high speeds.

The operating temperature of a gear drive is the temperature of the oil inside the gear case. Under normal conditions, the maximum operating temperature should not exceed 180°F.

Bevel

A *bevel gear* is a gear that connects shafts at an angle in the same plane. The involute tooth form is used on bevel gears when a drive requires large ratios using high torque, high speed, and non-parallel shafts. Bevel gears are cone-shaped with teeth that are tapered. The tapered teeth allow uniform clearance along the length of the teeth even though the gears are meshed on an angle. Mounting of bevel gears should be rigid enough to allow only a maximum separation of .006″. Proper meshing and backlash of bevel gears is accomplished using shims to position each gear axially.

Extra power transmission and noise reduction is possible on bevel gears by using spiral tooth bevel gears. The tooth spiral allows the teeth to engage with one another gradually. The continuous pitch line contact of spiral bevel gears makes it possible to obtain superior performance with high speeds and silent operations.

Miter

A *miter gear* is a gear used at right angles to transmit horsepower between two intersecting shafts at a 1 : 1 ratio. Only miter gears with the same pitch, pressure angle, and number of teeth can be operated together. However, more than two miter gears may be used in sets, such as in automotive differentials. Similar to bevel gears, miter gears should also have thrust bearings to absorb axial thrust.

Worm

A *worm* is a shank having at least one complete tooth around the pitch surface. A *worm gear* is a set of gears consisting of a worm (drive gear) and a wheel (driven gear) that are used extensively as a speed reducer. The worm is of such a low helix angle that it cannot be reversed. The driven gear cannot drive the worm because the gearing automatically locks itself against backward motion. Also, because of the low helix angle, a proportionate increase in torque is offered between the drive and driven gears.

Worm gears may be coarse-pitch or fine-pitch. Coarse-pitch worm gears are the most commonly used worm gears. Coarse-pitch worm gears are primarily used in industrial applications because they can transmit power efficiently, provide considerable mechanical advantage, and transmit power at a significant reduction in velocity. Fine-pitch gears are used to transmit motion rather than power. For this reason, tooth strength is seldom an important factor with fine-pitch worm gears.

Emerson Power Transmission

Bevel or miter gears are used in right-angle gearboxes to change the angle of power for use in applications where space is limited.

Multiple threads are added to the worm when additional power is required from a coarse-pitch worm gear. More threads added to the worm produce greater tooth contact, thereby allowing increased power. Adding more teeth reduces the input/output ratio. The number of helical threads on a worm can be as high as seven. The ratio of a worm gear is found by applying the formula:

$$R = \frac{T}{T_W}$$

where

R = input/output ratio

T = number of teeth on worm wheel

T_W = number of threads on worm

For example, what is the input/output ratio of a 50 tooth speed reducer with a 3 thread worm?

$$R = \frac{T}{T_W}$$

$$R = \frac{50}{3}$$

$$R = \mathbf{16.667}$$

Worm gears are unique in that wear does not destroy the tooth form. The worm continues to keep a proper form even through wear. Unlike other gears, worm gears tend to wear in rather than wear out. Due to the sliding motion on the worm's helix, high heat is generated and must be kept to a minimum through proper lubrication.

Hypoid

A *hypoid gear* is a spiral bevel gear with curved, non-symmetrical teeth that are used to connect shafts at right angles. The hypoid gear has the tooth angle of a helical gear, the base angle of a bevel gear, and the straight tooth of a rack gear. The hypoid gear is related to the worm gear due to the extreme pressure angle of the pinion. Hypoid gears, used widely in automotive differentials, provide greater sliding angles than spiral bevel gears for smoother and quieter operation.

GEAR WEAR

The sliding and meshing of gear teeth under load causes gear failure due to wear. Gear manufacturers design certain parts of a gear train to wear out or break sooner than others. This is done because some gears are physically easier to replace than others or one gear may be cheaper than another in a set. For example, worm gears are made of hardened steel while their matching worm wheels are made of bronze. At times, cast iron or nonmetallic gears are placed in a train of steel gears to wear out first.

Hunting Teeth

A *hunting tooth* is a tooth added to mesh with every tooth on the mating gear to produce even tooth wear. Hunting tooth choice is based on gear train ratio and the teeth per gear. For example, a pair of gears in a 3 : 1 ratio set have 60 and 20 teeth. In this case, wear is uneven because every 3 rotations of the pinion (20 teeth) produces one rotation of the driven gear (60 teeth). However, gears with 61 and 20 teeth (3.05 : 1 ratio) produce evenly distributed wear. Hunting teeth ratios should be used when gears in a pair or train are subject to uneven or cyclic loads such as in indexing or crankshaft use.

Gear Wear Identification

Proper identification of gear wear and its causes can prevent many hours of equipment downtime. Identification is also a useful indicator for upgrading a preventive maintenance program. For example, lubrication may need to be changed if a gear is scuffing. In addition, lubricant may require more frequent changing if there are signs of abrasive wear. Early recognition and correction of gear wear may also prevent extensive equipment damage.

The American Gear Manufacturers Association has compiled gear wear identification standards as a guide to provide a common language on gear wear and to provide a means to document gear appearance as gears wear or fail. Gear wear may be abrasive wear, corrosive wear, electrical pitting, rolling and scuffing, or fatigue wear.

Abrasive Wear. *Abrasion* is the removal or displacement of material due to the pressure of hard particles. *Abrasive wear* is wear caused by small, hard particles. Abrasive wear is caused by particle-contaminated oil, where particles of metal, sand, scale, or other abrading material grind and scratch gear teeth as they make contact. The scratches appear as parallel furrows oriented in the direction of sliding. See Figure 12-12.

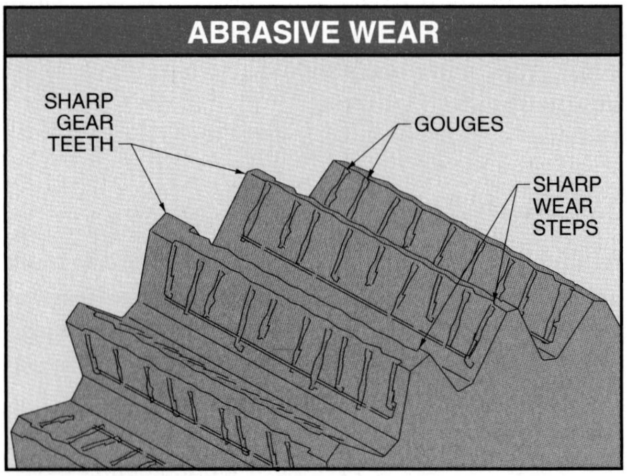

Figure 12-12. Abrasive wear is identified by sharp wear steps from the original surfaces, gouges in the direction of sliding, and tooth tips with sharp edges.

Abrasive material may enter a gear housing due to a harsh environment or may have been left in the gear housing as residual casting scale when the component was manufactured. Abrasive wear corrective action includes:

- Drain and flush residual oil.
- Scrape, flush, and wipe the internal surfaces of the gear housing.
- Clean out and flush any oil passages.
- Refill the housing with a light flushing grade oil and run without load for approximately 10 min.
- Clean breathers and replace seals and filters if suspected contamination was from the environment.
- Drain the flushing oil and refill with correct oil.

Corrosive Wear. *Corrosion* is the action or process of eating or wearing away gradually by chemical action. *Corrosive wear* is wear resulting from metal being attacked by acid. Acid is usually formed when the oil temperature becomes high enough to boil and separate acid-forming resins from the oil. High temperatures may be the result of an overloaded system, the wrong lubricant, or old lubricant that has broken down.

Corrosive wear may be initially identified by a stained or rusty appearance. More advanced corrosive wear is identified by rust-colored deposits along with extensive acid-etched pits. See Figure 12-13.

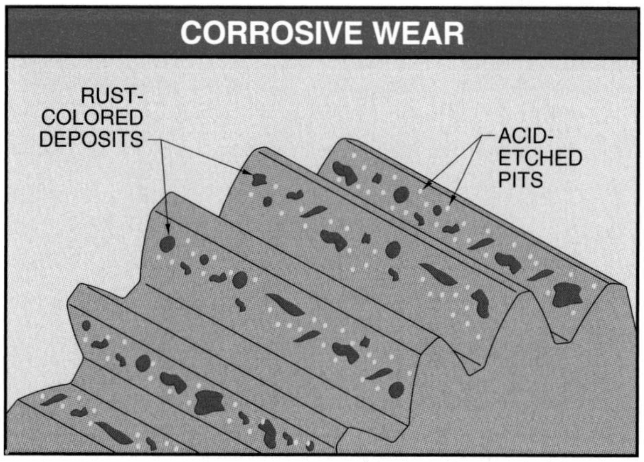

Figure 12-13. Corrosive wear is identified by rust-colored deposits along with acid-etched pits.

Corrosion attacks the entire gear, but wear is greatest on working surfaces because the build-up of corrosion itself becomes a partial insulator to the non-working surfaces. Corrosive wear corrective action includes:

- Reduce the load if the system is overloaded.
- Upgrade the system if the system is overloaded and the load cannot be reduced.
- Use an extreme-pressure lubricant if a system is overloaded and the load cannot be reduced nor the system upgraded.
- Check to see if the wrong grade of lubricant is being used. Contact the machine manufacturer or an oil company representative for proper lubricant specifications.
- Check the frequency of oil changes. It may be necessary to increase the oil change frequency.

Electrical Pitting. *Electrical pitting* is an electric arc discharge across the film of oil between mating gear teeth. The temperatures produced are high enough to locally melt gear tooth surfaces. Damage from electrical pitting may be caused by improperly grounded electrical connections, high static charges, or improper welding connections. Electrical pitting is identified by many small craters surrounded by burned or fused metal. See Figure 12-14.

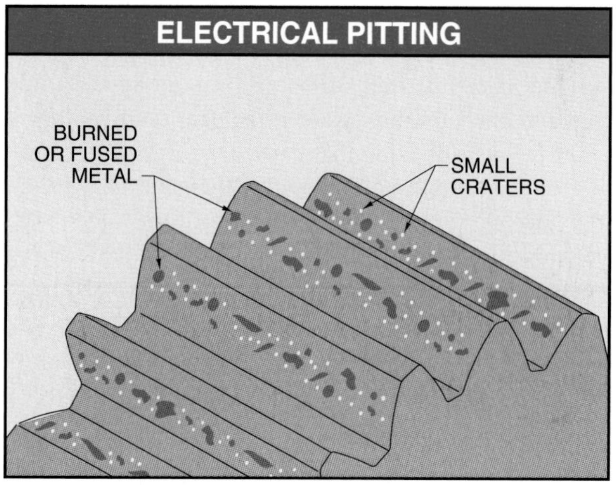

Figure 12-14. Electrical pitting appears as small craters surrounded by burned or fused metal.

Electric current on gears may break down lubricant if the current is not high enough to etch metal but is high enough to locally burn and break down lubricants. Electrical pitting corrective action includes:

- Place a ground clamp on the same side of a gear box when welding.
- Run grounding straps from a machine to rigid electrical or pneumatic piping to reduce static electricity created by manufacturing processes.
- Check the electrical system for proper installation and grounding.

Exxon Company

Gear wear and machine downtime is minimized by ensuring the machine is not overloaded, the proper lubricant is used, and the lubricant is changed at the correct frequency.

Rolling and Scuffing. *Rolling* is the deforming of metal on the active portion of gear teeth caused by high contact stresses. Rolling is a displacement of surface materials that forms grooves along the pitch line and burrs on the tips of drive gear teeth. *Scuffing* is the severe adhesion that causes the transfer of metal from one tooth surface to another due to welding and tearing. Scuffing generally occurs in localized patches due to the surface area of meshed teeth being mismatched or misaligned. Rolling and scuffing is created when gear teeth do not mesh properly and is progressive, meaning that wear continues and worsens until total damage has occurred. See Figure 12-15.

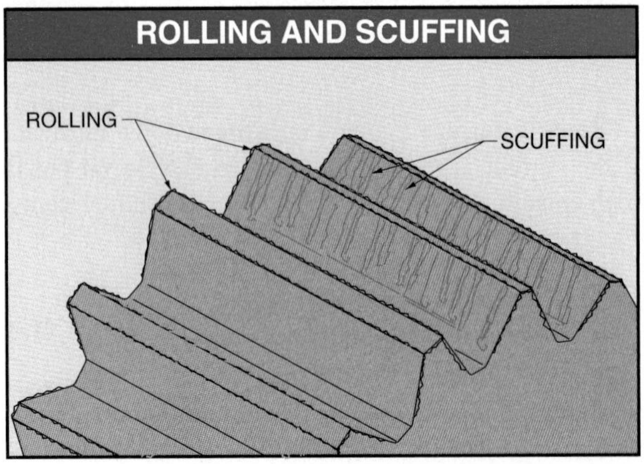

Figure 12-15. Rolling and scuffing is created when gear teeth do not mesh properly and continues until total damage has occurred.

Improper adjustments include radial/axial misalignment, improper end play, out of tolerance backlash, and manufacturer's defect. In gear teeth that do not mesh correctly due to misalignment, the gear wears at high points and removes metal until a mating profile is established. In cases such as manufacturer's defect, once a mating profile has been established, wear lightens or ceases. However, all other misalignments, if not corrected, continue to wear to gear destruction.

Fatigue Wear. *Fatigue wear* is gear wear created by repeated stresses below the tensile strength of the material. Fatigue may be identified as cracks or fractures. A *fatigue crack* is a crack in a gear that occurs due to bending, mechanical stress, thermal stress, or material flaws. A *fatigue fracture* is a breaking or tearing of gear teeth. Fatigue cracks usually culminate in a fracture when the fatigue crack grows to a point where the remaining tooth section can no longer support the load. Fatigue wear begins at the first moment a gear is used. Fatigue wear is repeated minute deformations under normal stress (normally unseen and immeasurable) that eventually produce cracks or fractures. See Figure 12-16.

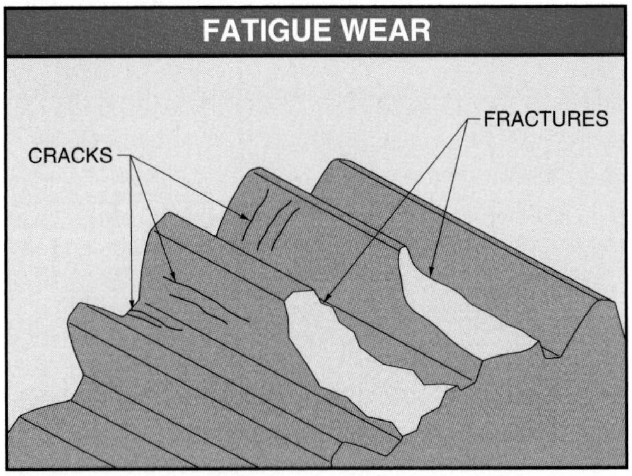

Figure 12-16. Fatigue wear is repeated minute deformations under normal stress that eventually produce cracks or fractures.

Sufficient alternating stresses (vibrations) produce rapid fatigue fracture in industrial gear trains. Other destructive noises, vibrations, overloading, and grinding of gears and gear surfaces must be identified early to be effectively corrected. Proper installation procedures, effective lubrication techniques, and regular periodic inspections with service can produce a successful, profit-oriented mechanical installation.

Vibration

13
Chapter

SPM Instrument, Inc.

More than 80% of all rotary equipment failures are related to vibration. Vibration characteristics include cycle, displacement, frequency, phase, velocity, and acceleration. Vibration characteristics are used to determine machine misalignment, unbalance, mechanical looseness, resonance, bearing wear, or gear defect. A successful vibration monitoring program compares initial vibration readings with present vibration readings to see if an increase in readings is developing.

VIBRATION

Vibration is a continuous periodic change in displacement with respect to a fixed reference. All objects on earth are constantly experiencing vibration. A vibration is a force that starts from a neutral position, travels a displaced distance to a positive upper limit (peak), reverses its direction to return to neutral, travels a displaced distance to a positive lower limit (peak), and returns to neutral. See Figure 13-1.

More than 80% of all rotary equipment failures are related to vibration. Vibration can break down the resiliency of seals and increase bearing and equipment temperatures. Vibration in one location may react with and add to vibration from another location. This vibration may be magnified, resulting in equipment damage.

Vibration also significantly reduces the expected life of bearing and rotating shaft seals. With recent motor designs, the effects of vibration become more critical because bearing and bearing fit tolerances have decreased, motor speeds have increased, and motor support frames have become lighter.

An estimate of vibration problems shows that 50% to 60% of damaging machinery vibrations are the result of shaft misalignment, 30% to 40% of damaging vibrations are the result of equipment unbalance, and 20% are the result of resonance. *Resonance* is the magnification of vibration and its noise by 20% or more. Resonance is often coupled with vibration from other sources. Less common vibration sources include bent shafts, loose parts, oil whirl, and defective bearings.

Industrial motors are rugged and able to handle heavy or continuous loads. For this reason, motors are manufactured with larger rotor shafts, which require larger bearings even though the rotors remain lightweight. The combination of lightweight rotors and large bearings should lengthen bearing service life considerably. In reality, previously allowable misalignment causes vibration that prematurely wears shaft seals, contaminates bearings, and shortens bearing life to less than ten years.

Vibration limits can be determined by comparison to standards developed by engineering standards organizations, manufacturer associations, or governmental bodies such as the American Petroleum Institute, American Gear Manufacturers Association, or American National Standards Institute.

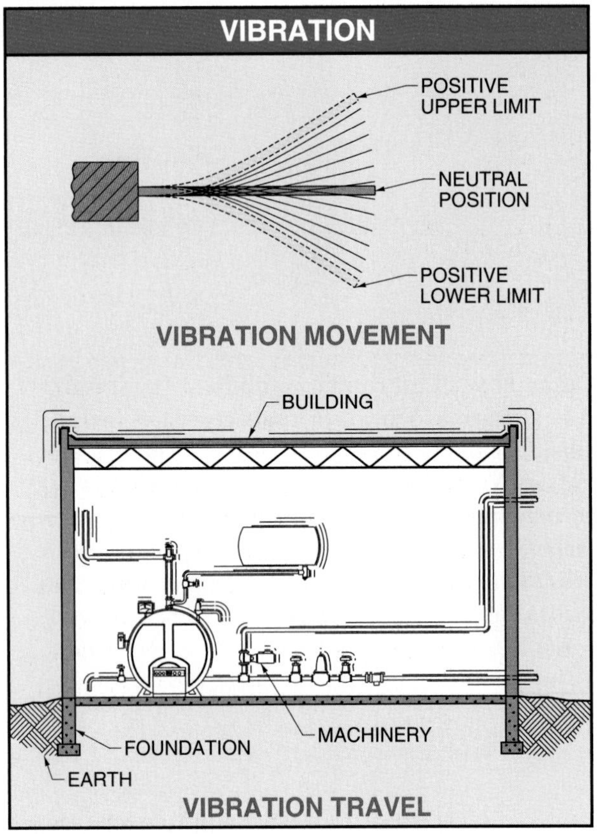

Figure 13-1. Vibration is a continuous periodic change in a displacement with respect to a fixed reference.

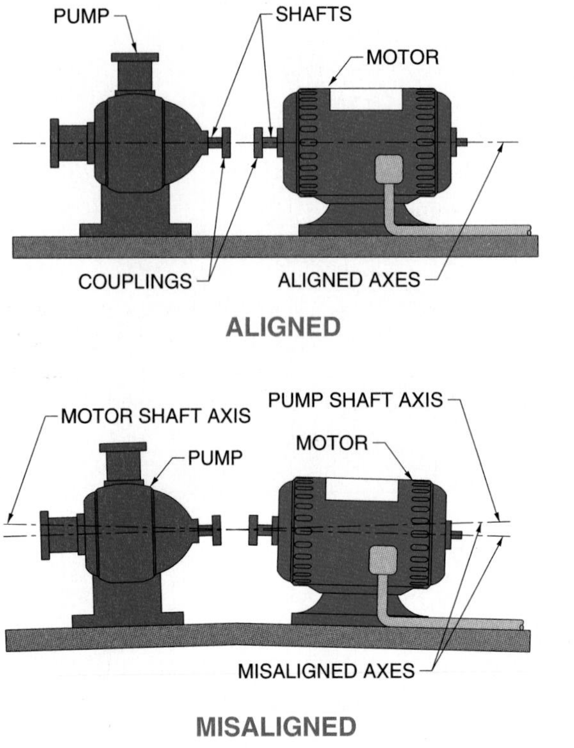

Figure 13-2. Properly aligned rotating shafts reduce vibration and add many years of service to pump/motor seals and bearings.

Alignment is the location (within tolerance) of one axis of a coupled machine shaft relative to another. *Misalignment* is the condition where the axes of two machine shafts are not aligned within tolerances. Properly aligned rotating shafts reduce vibration and add many years of service to pump/motor seals and bearings. See Figure 13-2.

Vibration Effects

Understanding vibration effects and establishing a testing, data collection, and analysis program goes beyond preventive maintenance and is a major tool in predictive maintenance programs. A complete vibration analysis program is one of the more complex and expensive predictive techniques. The cost of sophisticated electronic instruments able to collect, analyze, and store data is added to the cost of training personnel to interpret the data. However, the investment of total company commitment, power, and resources can provide a considerable payback through reduced equipment costs and less frequent machine downtime.

Many vibrations are unbearable for human comfort. These include loud noises and the shaking of buildings or vehicles. The magnitude of vibrations felt by humans is extremely small. The most damaging effect of vibration, especially in the case of machinery, occurs in ranges outside of human perception. These vibrations can produce fatigue failure in machine and structural elements, increase general wear of parts, and develop vibration through building foundations that are great enough to cause destruction or annoyance in another location.

Machinery Vibration

Machines vibrate even when in the best operating condition. This vibration and noise is generally due to minor defects or matching parts that are out of tolerance, such as clearances in bearings. Machinery vibration is a complex combination of signals created by a variety of internal vibration sources. Monitoring these signals is only the detection part of a maintenance program. A complete maintenance vibration program requires technicians to have a basic knowledge of how machines work, their common problems and how to repair them, the ability to recognize and pinpoint mechanical problems early and accurately, and the ability to understand and use applied technology diagnostics in determining a specific problem, its severity, and the machine part being affected.

The combined vibration and noise from every machine is different. The component producing the vibration can be identified when certain vibration signals are separated from the others. The measurements are monitored, recorded, and compared to previous measurements. The sign of developing mechanical problems may be determined when the vibration or noise readings continue to rise.

VIBRATION CHARACTERISTICS

Characteristics offered in vibration analysis become the clues toward describing and detecting unwanted motions in a machine. They are the symptoms used in determining any significant variation and reflect the true mechanical condition of a machine. Vibration characteristics such as cycle, displacement, frequency, phase, velocity, and acceleration define the dynamic properties of machine misalignment, unbalance, mechanical looseness, resonance, bearing wear, or gear defect.

Vibration characteristics become valuable in determining machine condition because an unwanted vibration is caused by either a change in direction or amount. The resulting characteristic is determined by the manner in which its forces are generated, with each cause of vibration having its own peculiar characteristic. For example, misalignment vibration is generally characterized by a 2x running speed vibration frequency with high axial levels. Unbalance is usually characterized by a sinusoidal frequency of 1x running speed with an increase in amplitude with an increase in speed. Gear or bearing problems may be characterized as having a vibration frequency equal to the number of teeth on a gear or balls in a bearing multiplied by their rotational frequency.

Most vibration characteristics do not stand alone. When there is a change in one, there is generally a corresponding change in another. Changing the rotational frequency of a machine also changes the displacement, phase, and velocity characteristics of the machine. For this reason, all running parameters such as load, pressure, speed, etc. must be, within reason, the same each time a condition-monitoring measurement is taken.

Vibration Cycle

Vibration, when measured, is referred to by its cycle or amplitude. A *vibration cycle* is the complete movement from beginning to end of a vibration. *Vibration amplitude* is the extent of vibration movement measured from a starting point to an extreme point. Amplitude may be measured as peak or peak-to-peak. *Peak* is the absolute value from a zero point (neutral) to the maximum travel on a waveform. *Peak-to-peak* is the absolute value from the maximum positive travel to the maximum negative travel on a waveform. A *waveform* is a graphic presentation of an amplitude as a function of time. The waveform shows a spectrum of a vibration. A *spectrum* is a representative combination of the amplitude (total movement) and frequency (time span) of a waveform. Vibration cycles continue as long as the object is disturbed. See Figure 13-3.

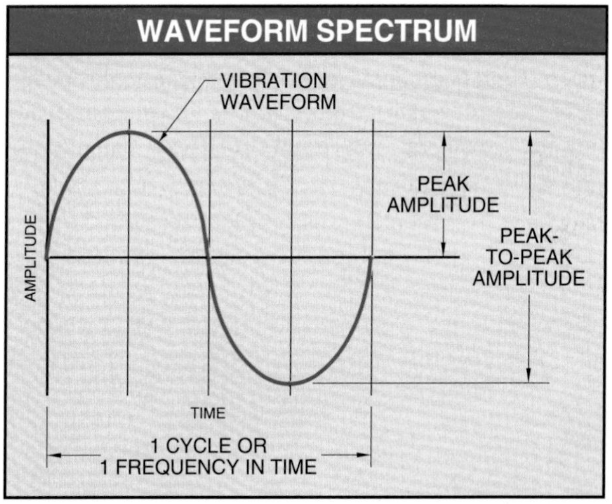

Figure 13-3. A waveform shows the frequency and amplitude of a vibration.

Displacement

Displacement is the measurement of the distance (amplitude) an object is vibrating. Peak-to-peak displacement is the distance from the upper limit to the lower limit of a vibration. Peak-to-peak displacement is measured to determine the severity of a vibration. Peak-to-peak displacement is expressed in mils, where 1 mil equals one-thousandth of an inch (.001″). See Figure 13-4. Displacement damage is similar to bending or flexing a twig. Increasing the amount of flex (displacement) increases its likeliness to snap.

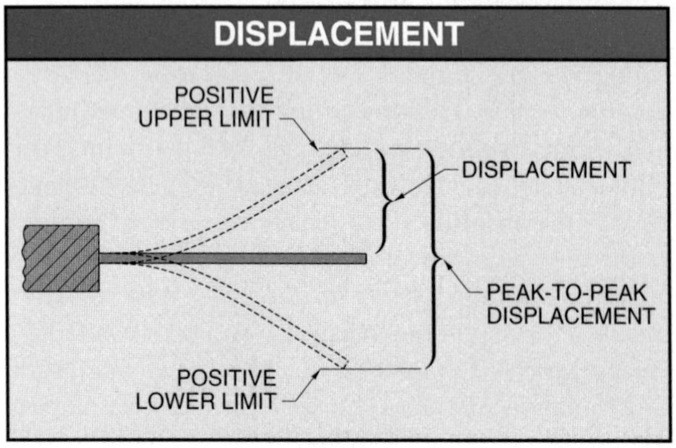

Figure 13-4. Displacement is the measurement of the distance (amplitude) an object is vibrating.

Frequency

Frequency is the number of cycles per minute (cpm), cycles per second (cps), or multiples of rotational speed (orders). An *order* is a multiple of a running speed (rpm) frequency. Orders are commonly referred to as the order of rotation, such as 1x (one time) for running speed, 2x (two times) for twice the running speed, etc.

Heidelberg Harris, Inc.

Complex automatic machinery must be kept running to justify its cost. A vibration monitoring program requires reliable, detailed, and readily available information to enable maintenance personnel to replace failing components with the least downtime to the production process.

Orders are used for vibration analysis because most vibration problems are related to the running speed of a machine. For example, a vibration reading of 32x 1800 is identified as a problem gear having 32 teeth and rotating at 1800 rpm. This reading has a frequency of 57,600 (32 × 1800 = 57,600) or an order of 32x running speed. Frequency damage is similar to bending a wire back and forth until it breaks. Increasing the frequency (rate of bending) reduces the time it takes the wire to break.

Machine-related frequencies correlate with the rotating speeds of the machine order and are expressed as either cycles per minute (cpm) or Hertz. *Hertz (Hz)* is a measurement of frequency equal to 1 cps. For example, 1 Hz is equal to 1 cps and 50 Hz is equal to 50 cps. Frequencies read as cycles per minute are divided by 60 sec to be converted to Hertz. For example, a motor operating at 1740 rpm is rotating at 29 revolutions per second (1740 rpm ÷ 60 = 29 Hz). Vibration frequencies within the range of human hearing fall between 15 Hz and 20,000 Hz. Machinery frequencies generally fall below 15,000 Hz.

Vibration frequencies within a machine are significant when analyzing the condition of the machine. Knowing and identifying the various frequencies within a machine enables a technician to pinpoint the part that may be at fault and determine the problem. Different machine problems create different vibration frequencies. Vibration due to gear problems is easily determined because the vibration generally occurs at a frequency equal to the meshing of gear teeth.

For example, a motor/gear unit indicates a vibration frequency of 16,000 cpm. The motor rotates at 800 rpm and has 4 gears with 15, 20, 40, and 28 teeth respectively. The problem component is found by dividing the frequency reading by the number of teeth in each gear. As a general rule, this frequency reading is indicating tooth wear on the 20 toothed gear (16,000 ÷ 20 = 800). See Figure 13-5.

Rolling-contact bearings also have high frequency readings. The frequency readings are often equal to the number of balls or rollers times the shaft speed. A bearing produces high frequency vibration even when only one of the rollers in a bearing is defective.

Worn or loose plain bearings produce a vibration frequency equal to twice the rotation of a shaft (order of 2x speed). The frequency increases as the bearing or shaft wears. The frequency may increase to an order as high as 10x speed, but does not increase consistently.

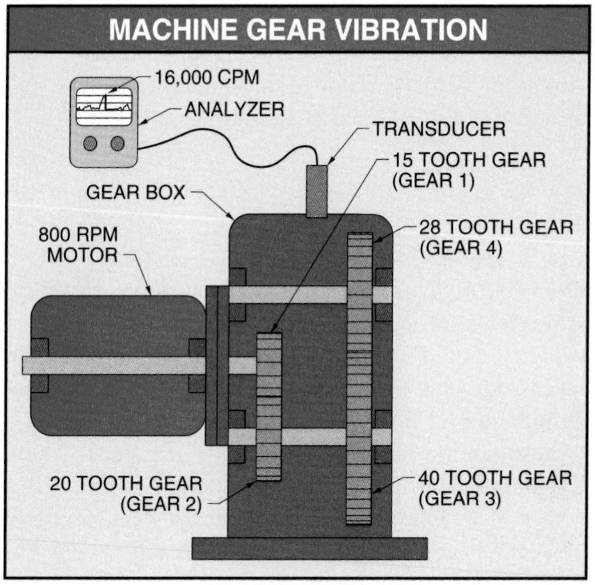

Figure 13-5. Vibration due to gear problems generally occurs at a frequency equal to the meshing of gear teeth.

Phase

Phase is the position of a vibrating part at a given moment with reference to another vibrating part at a fixed reference point. Phase readings are expressed in degrees from 0° to 360°, with one complete vibration cycle equaling 360°. Phase readings are a convenient method of comparing one vibration to another on a machine. *Note:* The end opposite the shaft is the front of the motor. When viewed from the front, forward rotation is clockwise and reverse rotation is counterclockwise. See Figure 13-6.

Phase readings are most commonly used in determining rotor unbalance where a rotor may be out of balance in one direction at shaft end and out of balance in the other direction at the opposite shaft end. Unbalance vibration has an order of 1x speed and may be equal rotor unbalance, opposing forces rotor unbalance, or coupling unbalance.

Equal rotor unbalance is the unbalance of weighted force across one side of a rotor or armature. Equal rotor unbalance produces a measurable vibration in only one direction. *Opposing forces rotor unbalance* is the unbalance of weighted forces on opposing ends and sides of a rotor or armature. Opposing forces rotor unbalance produces a measurable vibration in two directions.

Coupling unbalance is an unequal radial weight distribution where the mass and coupling geometric lines do not coincide. Coupling unbalance is an unbalance that occurs in different radial planes at opposite ends of a machine, similar to opposing forces rotor unbalance. See Figure 13-7. The problem is normally due to unbalance when the frequency of vibration is equal to the rotation of a shaft or an order of 1x speed. Unbalance always gives a radial reading of 1x speed.

Use permanently-installed transducers and remote measuring terminals for applications in which the bearings cannot be reached directly.

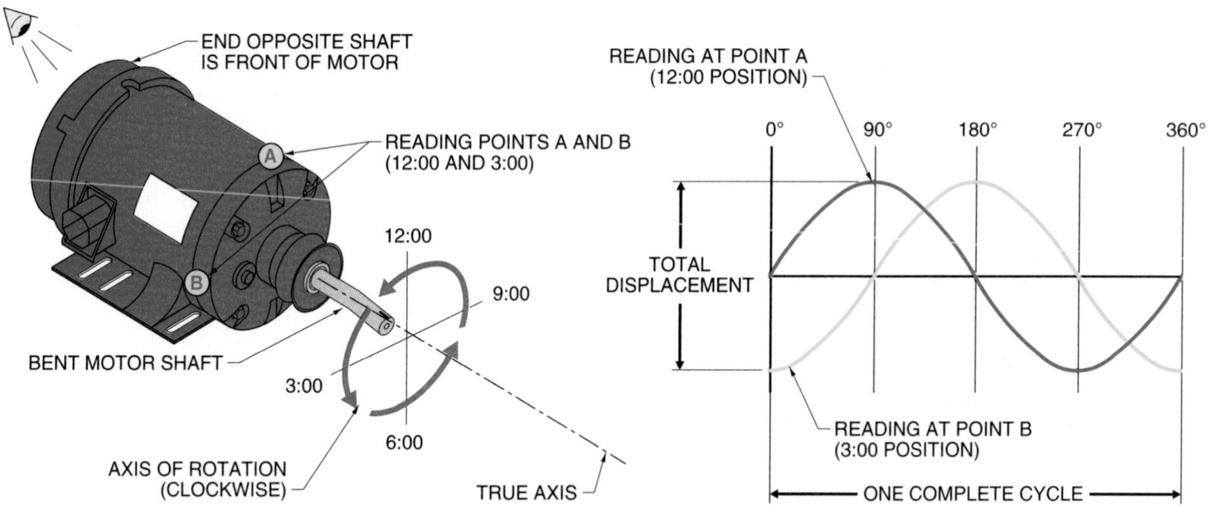

Figure 13-6. Phase readings between two signals of identical frequencies are helpful in determining unbalance or bent motor shafts.

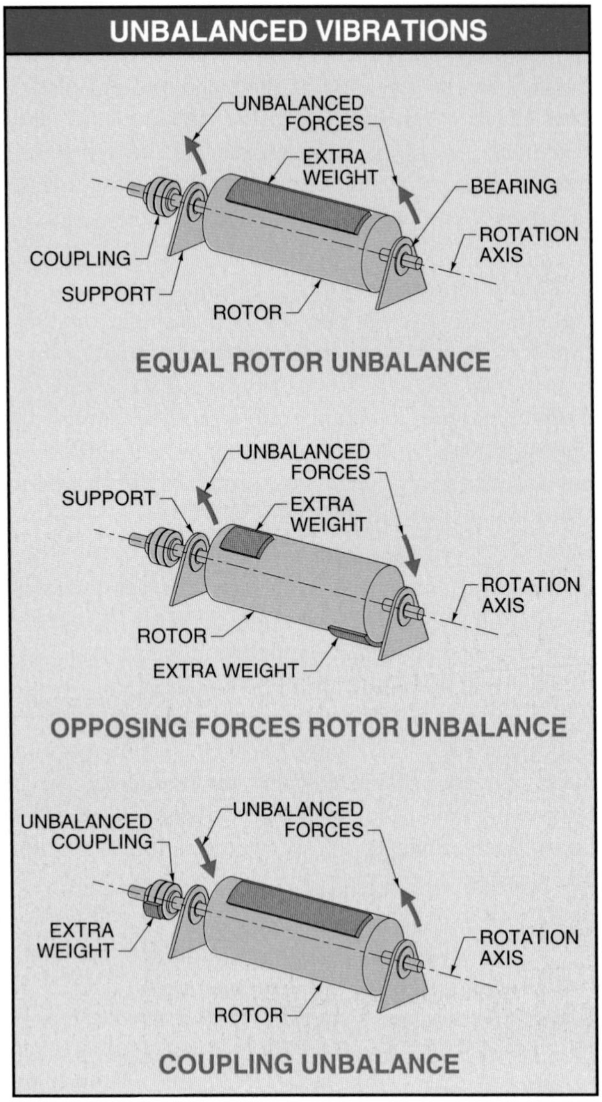

Figure 13-7. Unbalance forces have an order of 1x speed and vary according to phase.

Vibration Velocity

Vibration velocity is the rate of change of displacement of a vibrating object. The amplitude of a vibration is measured as the maximum value of its distance moved in relation to the time of movement. Vibration velocity is measured in inches per second peak (in./sec). The value recorded is its maximum value (peak velocity) when traveling through the neutral position.

Velocity is an excellent indicator of damage because it is proportional to the extent of component damage and not the speed of the machine. The reason velocity does not rely on the machine speed is because it is proportional to the energy content of the vibration. Velocity damage occurs from the repeated forceful cycles of flexing or fatigue.

Vibration Acceleration

Vibration acceleration is the increasing of vibration movement speed. It is the time rate of change of velocity. See Figure 13-8. The rate of acceleration reaches its maximum value as an object goes beyond its maximum limits of displacement. The peak value of acceleration is measured in units of g peak, where 1 g is equal to 386 inches per second squared (1 g = 386 ips^2). One g is also the international standard of acceleration produced by the force of gravity at the earth's surface and is used to indicate the force an object is subjected to when accelerated. Acceleration parameters are useful, especially with high frequencies, because accelerated forces at high frequency are extreme forces that ultimately cause a bearing to fail or lubricants to break down.

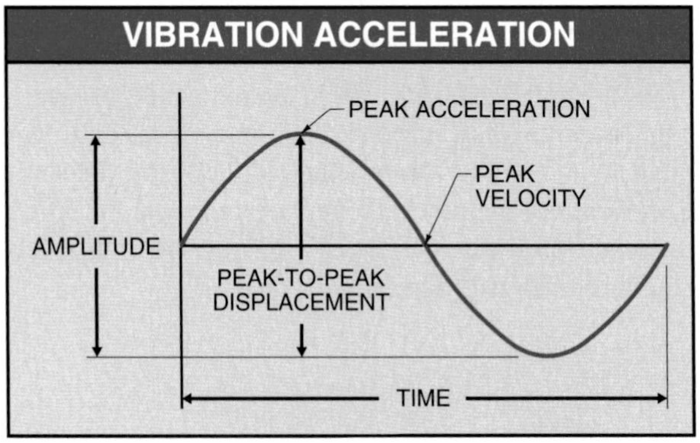

Figure 13-8. Vibration acceleration is the increasing of vibration movement speed.

Oil Whirl

Oil whirl is the buildup and resistance of a lubricant in a rolling-contact bearing that is rotating at excessive speeds. Oil whirl has a frequency of less than one-half the speed (rpm). Oil whirl occurs when a shaft is turning so fast that it attempts to roll over the lubricant rather than squeeze it out of the way. Oil whirl vibration may have a frequency rate of less than 4500 cpm while the shaft is rotating at 8600 rpm. See Figure 13-9.

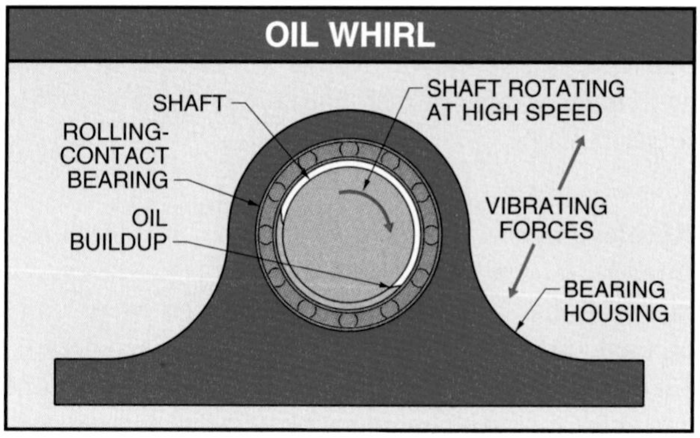

Figure 13-9. A shaft vibrates at less than 1x speed when it is turning so fast that it tries to roll over its lubricant.

Vibration Severity

Displacement, velocity, and acceleration are all direct measures of the severity of machine vibration. In most cases, taking only one of the three measurements is required to sufficiently describe machine vibration condition. Although displacement, velocity, and acceleration are directly related, the measurement taken is generally determined by the frequency (cpm) of the vibration. See Figure 13-10.

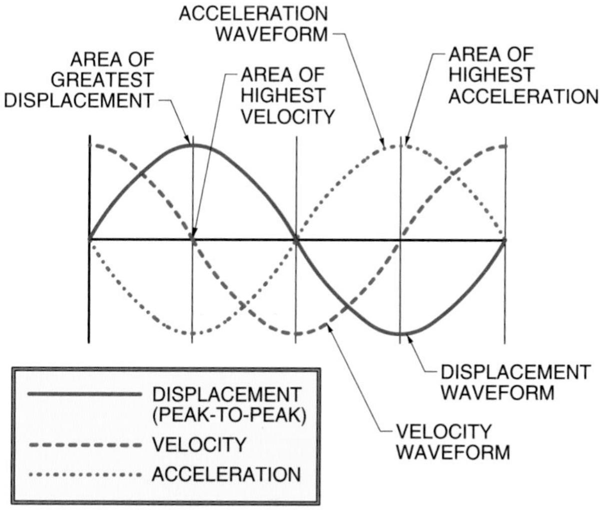

Figure 13-10. The phases of displacement, velocity, and acceleration are always 90° apart from each other.

Displacement is difficult to determine at high frequencies, but is a good choice for determining amplitudes of low-speed equipment. Displacement is best suited for frequencies between 1 cpm and 60,000 cpm. Velocity amplitudes are mostly used because the velocity range of 600 cpm to 60,000 cpm is typical of most industrial facility equipment. Acceleration parameters are useful in the higher frequency ranges between 18,000 cpm and 600,000 cpm.

VIBRATION MEASUREMENT METHODS

The internal mechanical condition of a machine is determined by the vibration measurement method. Vibration measurement methods are developed by choosing the vibration transducer (pickup), the proper placement of the transducer for the measurement to be taken, and analysis of the measurement readings.

SPM Instrument, Inc.

The hand-held instruments manufactured by SPM Instrument, Inc. are used for comprehensive machine condition monitoring.

Vibration Transducers

Vibration measurement takes the variety of internal vibrations and their complex signals and converts them into readings through the use of a vibration transducer and an analyzer. A *transducer* is a device that converts a physical quantity into another quantity, such as an electrical signal or a graphic display. A *vibration analyzer* is a meter that pinpoints a specific machine problem by identifying its unique vibration or noise characteristics. Various analyzers are available for gathering the required data. Each can transform complex signals into an understandable display useful for diagnosis. Analyzers can range from simple walk-around vibration meters to the more sophisticated tracking analyzers having a permanent link between transducer and analyzer which provides 24 hr inspection, detection, and protection.

Transducers are similar to a microphone in that a physical movement, such as sound, is converted into an electrical signal. The choice of transducer is generally determined by the frequency of the vibration. The correct transducer must be selected to obtain a signal that represents a vibration accurately. Vibration transducers include velocity, accelerometer, and displacement transducers. See Figure 13-11. Vibration severity charts are used to determine if the readings refer to a smooth or rough condition. See Appendix.

Velocity Transducers. A *velocity transducer* is an electromechanical device that is constructed of a coil of wire supported by light springs. Velocity transducers are the most common transducer used because they operate in the frequency range of most industrial applications. The coil surrounds a permanent magnet that moves following the motion of vibration. The cut of magnetic flux between the coil and the magnet creates a voltage in the coil. This voltage, expressed in millivolts per inch per second, is the output to the analyzer. See Figure 13-12. Voltage output is proportional to relative movement (vibration). The faster the movement, the higher the voltage output. Internal moving parts of a velocity transducer require recalibration because they wear out or change over a period of time. Velocity transducers give reliable results in a hand-held device at frequency rates of 600 cpm to 60,000 cpm (10 Hz to 1000 Hz).

Accelerometer Transducers. An *accelerometer transducer* is a device constructed of quartz crystal material that produces electric current when compressed. Quartz crystals are strongly piezoelectric, becoming polarized with a negative charge on one end and a positive charge on the other when subjected to pressure. *Piezoelectric* is the production of electricity by applying pressure to a crystal. Forces on piezoelectric crystals produce a voltage output proportional to vibration accelerations producing a display in gravitational (g) forces. See Figure 13-13.

Accelerometer transducers are becoming more popular than velocity transducers because they can operate at frequencies between 120 cpm and 600,000 cpm (2 Hz to 10,000 Hz), have no moving parts, and are more rugged than velocity transducers. They are not greatly affected by stray magnetic fields, which makes them useful around AC motors.

Modern electronic accelerometer transducers include protection circuits which prevent electronic damage due to use beyond specified range or accidental dropping.

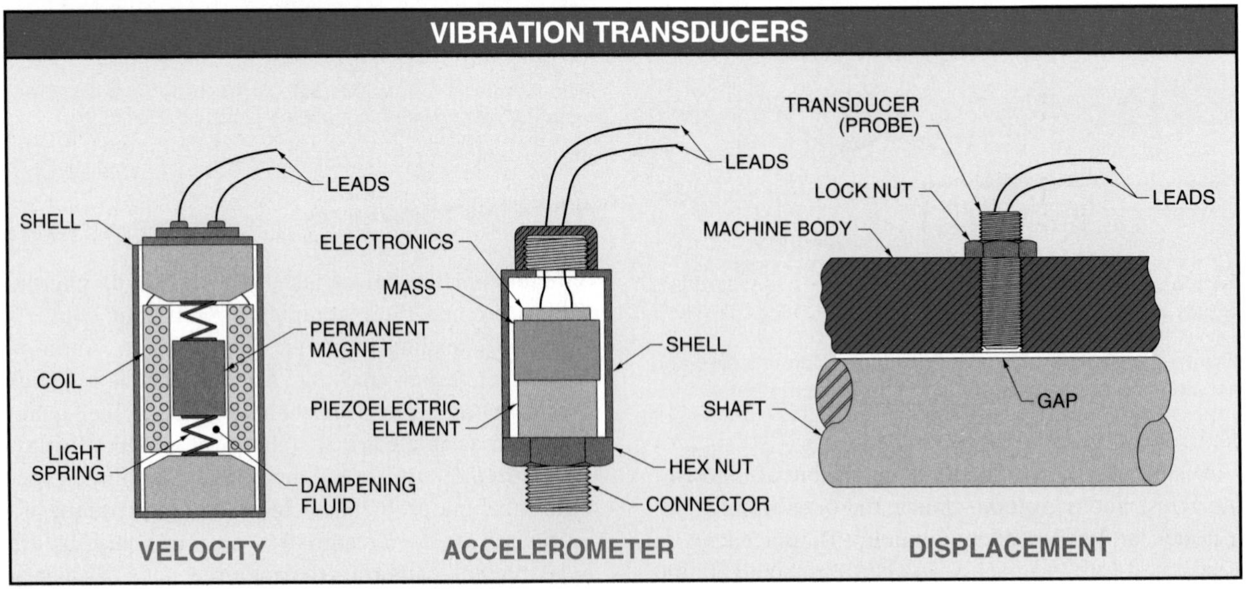

Figure 13-11. Transducers are similar to a microphone in that a physical movement, such as sound, is converted into an electrical signal.

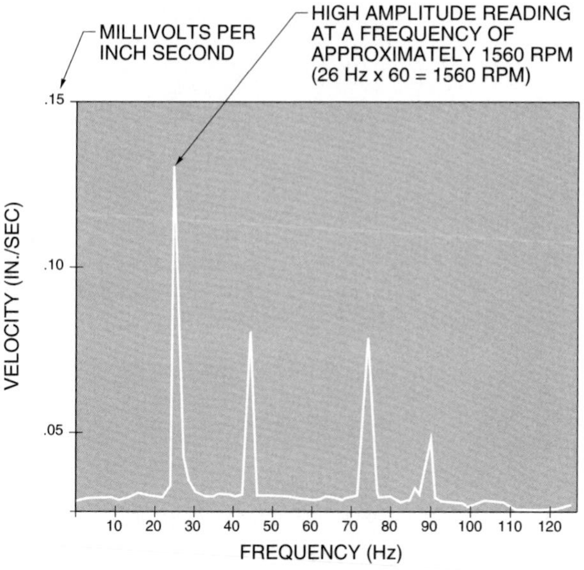

Figure 13-12. Velocity transducers relay a voltage output signal in millivolts, relative to vibration movement.

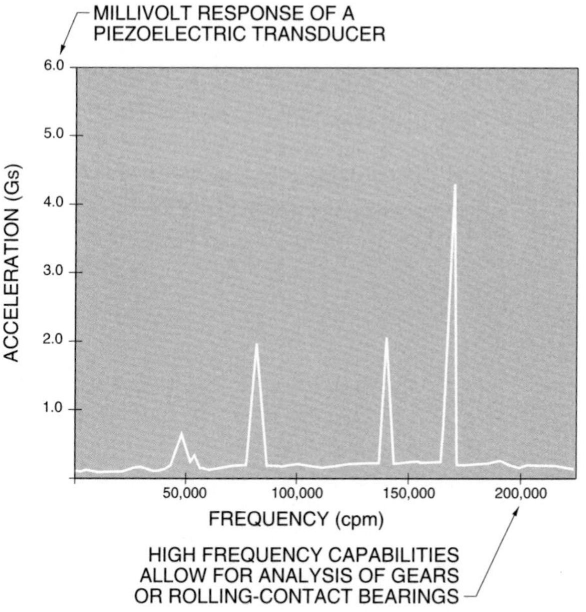

Figure 13-13. Accelerometer transducers can produce currents indicating up to 70 g of acceleration and measure frequencies up to 600,000 cpm.

Accelerometer transducers are also popular because they are small, lightweight, and are able to withstand high temperatures. Accelerometer transducers cannot be hand-held like velocity transducers because they can pick up interference. They must be firmly connected to the machine through the use of a mounting fixture that is glued, screwed, or clamped to the machine.

Baldor Electric Co.

Minimizing machine vibration reduces the damaging vibration transmitted to other nearby machinery.

Displacement Transducers. A *displacement transducer* is a mechanical sensor whose gap-to-voltage output is proportional to the distance between it and the measured object (usually a shaft). Displacement transducers are used mostly as permanent installations to continuously monitor machine displacement vibration. Also known as proximity pickups, displacement transducers detect displacement of a vibrating shaft as the shaft opens and closes a gap between the probe tip and shaft as it rotates. Gap settings vary according to transducer size and shaft material, but are typically .020″, .030″, .050″, .060″, or .100″. Gap readings may be taken whether the shaft is rotating or not. The ability to measure shaft vibration relative to the bearing housing offers an inside view of the condition of bearings or seals.

Gap readings are taken by means of a magnetic field (flux) being set up at the transducer (probe) tip. Magnetic field energy changes as the gap changes during vibration movements. Magnetic field energy is converted to a signal amplitude (displacement signal) at the transducer. A displacement measuring system senses the distance between the probe tip and a conductive surface using the transducer, an oscillator/amplifier, and a meter/scope. See Figure 13-14.

Displacement transducers are designed to measure the actual movement of a machine shaft relative to the transducer and therefore, must be rigidly mounted to a stationary structure to gain accurate, repeatable data.

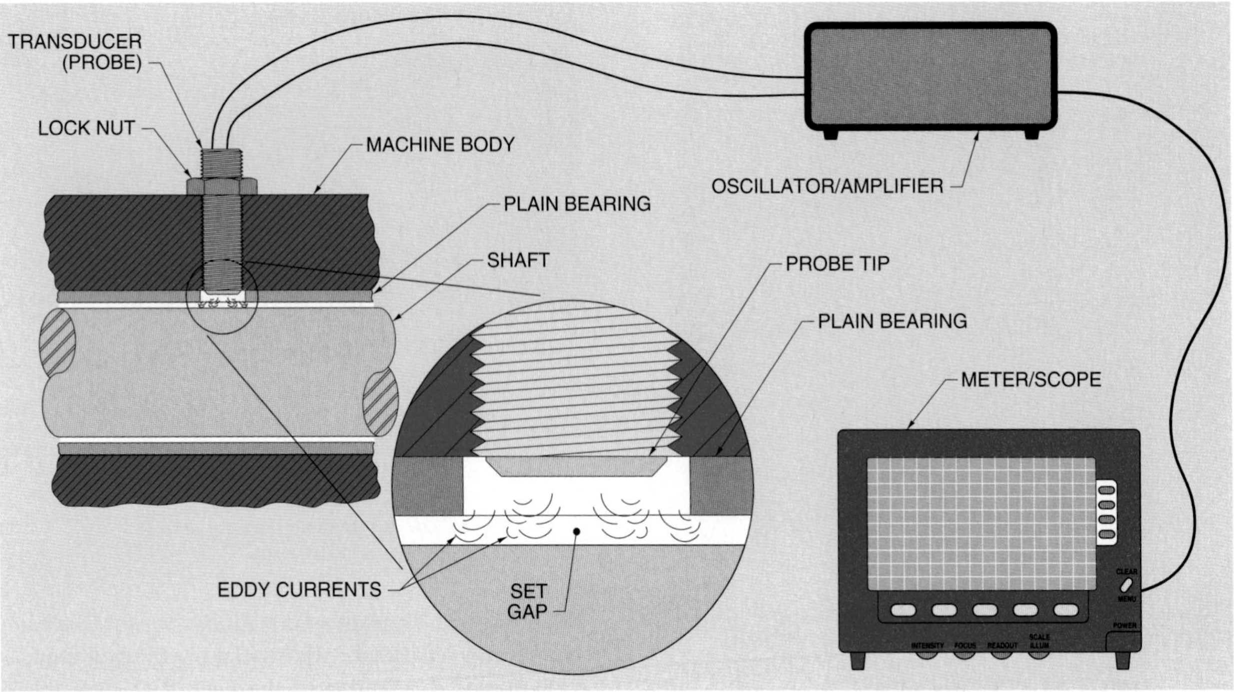

Figure 13-14. Displacement transducers are sent an electrical current to set up eddy currents at the probe tip. Eddy current voltages sent back to the meter as an output voltage vary according to the gap distance.

An *oscillator* is a device that generates a radio frequency (RF) field that, when sent to the transducer tip, creates eddy currents. An *eddy current* is an electric current that is generated and dissipated in a conductive material in the presence of an electromagnetic field. As the shaft vibrates relative to the transducer, energy changes modulate the oscillator voltage. An amplifier receives the feedback signal and amplifies the output signal to a meter, analyzer, or monitor. See Figure 13-15.

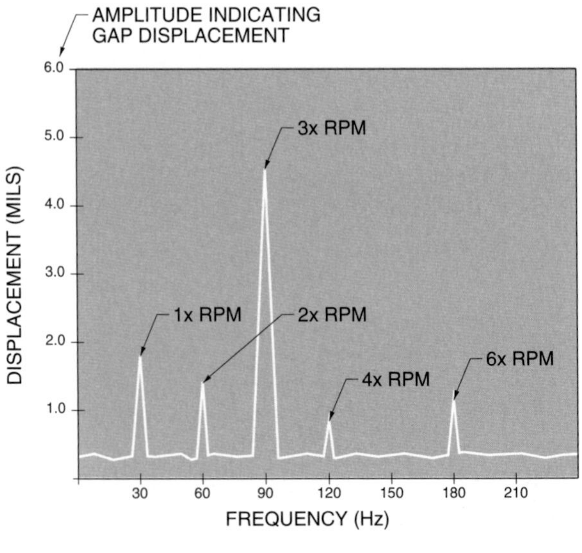

Figure 13-15. Displacement transducers produce an amplitude that represents the total peak-to-peak displacement of a vibrating shaft.

Transducer Selection

Selecting the proper transducer (velocity, accelerometer, or displacement) is determined by the design of the machine being monitored and the severity of the vibration. Velocity transducers detect vibration from defective components, loose parts, and rolling-contact bearings in the low-frequency range. Accelerometer transducers are helpful in detecting high-frequency defects in bearings, gears, or fan or turbine blades. These measurements are also helpful in detecting structural movements. Displacement transducers are used where dynamic stresses are found or where clearance within motors, gear boxes, or any mechanism using fluid film bearings can be measured. Displacement transducers may also be used to measure machine unbalance.

When mounting transducers with cables, secure the cables to the vibrating structure to minimize cable or connector fatigue failures and loss of data.

Transducer Placement

Vibration may occur in any direction. Transducers must be placed in a position to directly receive a vibration. Transducers must also be placed at the exact same location on a machine to ensure accurate readings. The three major directions vibration travels within a machine are horizontally (radial), vertically (radial), or axially.

Horizontal measurements are taken with transducers placed in the horizontal plane (X axis) with the axis of rotation. Vertical measurements are taken with transducers placed in the vertical plane (Y axis) with the axis of rotation. Axial measurements are taken with transducers placed at the centerline (A axis) with the axis of rotation. See Figure 13-16.

Transducer placement depends on the direction of the vibration to be detected. Radial in the vertical plane (Y axis) is generally chosen over radial in the horizontal plane (X axis) or axial if only one measurement is chosen. This is because most vibration is heightened by gravitational pull and also because it is normally easier to probe from the top. A complete measurement uses all three positions. Transducers used to measure radial vibration must be attached within 3″ of the bearing.

Transducers should be placed on or as near as possible to bearings because vibration forces are transmitted through the bearings.

SPM Instrument, Inc.

The early detection of bearing damage reduces the risk of breakdowns and enables maintenance personnel to plan the replacement and reduce the required downtime.

BASIC VIBRATION ANALYSIS

To analyze individual component condition, a vibration signature is read from an analyzer. A *vibration signature* is a set of vibration readings resulting from tolerances and movement within a new machine. Machine wear may then be plotted through the use of its vibration signature. Periodic analysis begins after a signature is established. The vibration signature of a machine in good operating condition provides a baseline measurement against which future measurements may be compared. A change in the vibration signature of a machine indicates the beginning of a defect.

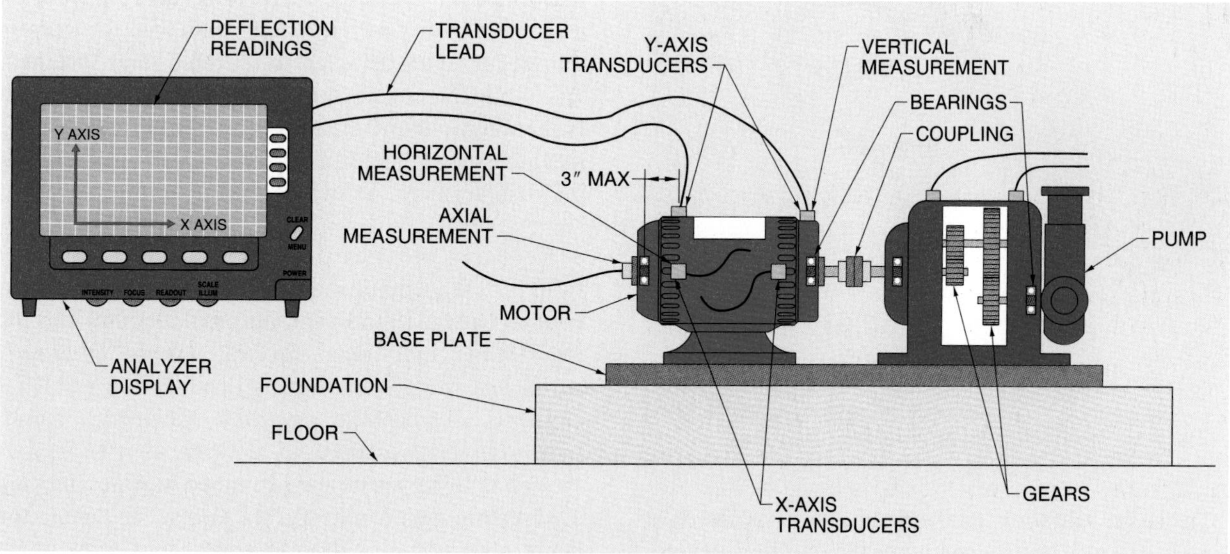

Figure 13-16. Transducers used to measure radial vibration must be placed within 3″ of bearings for transducer measurements and analyzer displays to be representative of machine condition.

Analysis of vibration signals allows the specific nature of problems to be found and an assessment for repair to be made while the machine is operating. A steady, continuous vibration signature change enables a technician to project machine condition and allow scheduled machine repairs in advance of machine failure.

Limits must be set to provide a basis for determining the condition of a machine. These limits must be close enough to normal values to allow corrective action before operating conditions begin to cause damage. Original limits are normally recommended limits supplied by charts or the equipment manufacturer. As data is developed or when necessary limits have been reached often, the limits should be adjusted and refined to more realistic values.

Limits must also be set according to the load. Radial loads wear gradually and continually until failure. This gradual wear is detected as a displacement change. Early corrective action is possible even though a radial position change due to bearing wear may not produce an increase in vibration amplitudes.

Thrust loads generally fail without much notice, where the only detectable movement prior to total failure is in the loss of lubrication space and some metal compression (about 2 mils). For this reason, limits must be set closer to the original limits for axial forces than for radial forces.

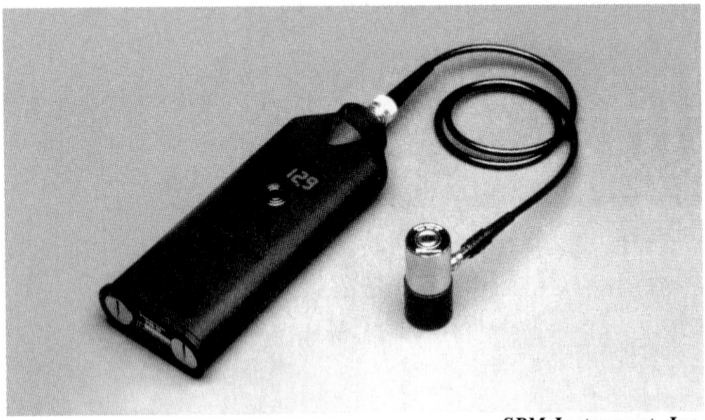

SPM Instrument, Inc.

The Vibrameter VIB-10 from SPM Instrument, Inc. is a hand-held instrument used for periodic measurements to detect unbalance, misalignment, and other mechanical faults in rotating machines.

A basic vibration analysis system consists of a signal pickup (transducer), signal recording device, signal analyzer, analysis software, and a computer for data storage. Relatively inexpensive units are available consisting of a programmable data collector and its analysis software. Any analysis system requires the use of a computer.

Analyzers include Fast Fourier Transform (FFT) or dynamic signal analyzers (DSAs). A *Fast Fourier Transform analyzer* is a microprocessor capable of displaying the FFT of an input signal. An *FFT* is a calculation method for converting a time waveform into a series of frequency vs. amplitude components. A *dynamic signal analyzer* is an analyzer that uses digital signal processing and the FFT to display a dynamic vibration signal as a series of frequency components. The DSA is an analyzer that uses both amplitude and frequency for display and has the capability to display both low frequencies and high frequencies on the same screen using a logarithmic scale. A *logarithmic scale* is an amplitude or frequency displayed in powers of ten.

Reading Amplitudes

A three-step vibration detection and analysis process is required before the characteristics presented by an analyzer showing unbalance, defective bearings or gears, etc. are placed into a maintenance record. See Figure 13-17. The three steps are: conversion of vibrations to electrical signals, reduction of electrical signals into component form, and identification of individual defect frequencies from component signals.

Converting vibration into an electrical signal is performed by the transducer. However, to achieve an accurate account of vibration condition, the correct transducer must be selected, located, and installed. An important consideration for transducer selection is the amplitude of the vibration parameters and the machine speed. Transducer selection is made regarding displacement, velocity, or acceleration.

Reducing the electrical signals into component form is best accomplished with a display using a conversion between the linear amplitude spectra and the logarithmic amplitude spectra. *Linear amplitude spectra* are amplitude signals displayed in equal increments. *Logarithmic amplitude spectra* are amplitude signals displayed in powers of 10. Signals viewed in linear spectra are summed together, making a waveform next to impossible to read. Reducing the many electrical signals into component form is accomplished by switching the linear spectra to the logarithmic spectra. See Figure 13-18.

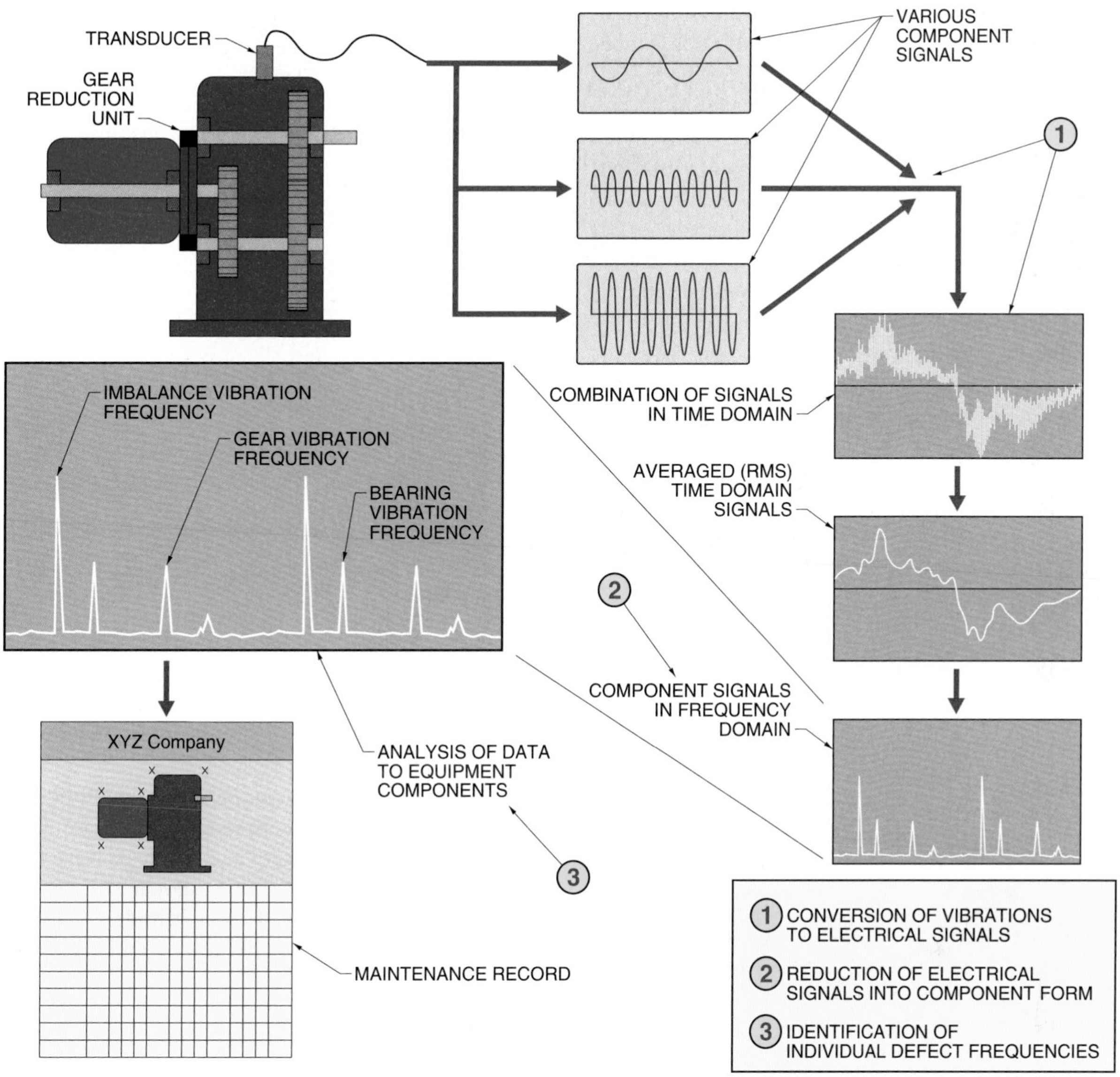

Figure 13-17. A three-step vibration detection and analysis process is required before the characteristics presented by an analyzer are placed into a maintenance record.

Linear spectra, although easier to read, may not show the value of a high scale and give an indication to the value of small signals. Logarithmic spectra allow all frequencies to be visible by compressing the large signal amplitudes and expanding the small ones. Most analyzers have the ability to shift between the linear and logarithmic amplitude scales. This offers the choice of viewing the amplitude of a single component on linear spectra or viewing logarithmic spectra for a full range of vibration data.

Identifying individual defect frequencies from component signals is the key to a good analysis. Each machine offers a unique set of component signals because the vibration measured is a response to a defect force, not the force itself. Amplitudes may be read as peak-to-peak, zero-to-peak, or root mean square (rms). See Figure 13-19.

Root mean square (rms) is the square root of the sum of a set of squared instantaneous values. Rms averages and smooths a signal containing high peaks, making the output more representative of unbalance or misalignment problems. An rms vibration signal produces a time-averaged amplitude proportional to the area within a time domain waveform. The extent of vibration may be read in either the time domain or the frequency domain. Rms amplitude of a true peak value may be determined by multiplying .707 by the peak value. For example, the rms value of a true peak of 3 is equal to 2.121 (3 × .707 = 2.121). When a true peak value is required, as in the case of measuring shaft vibration, a total rms peak-to-peak sine wave is multiplied by 1.414. For example, when a shaft vibration using an rms peak-to-peak signal shows a 4 mil waveform, the true peak-to-peak shaft displacement is equal to 5.656 mil (4 mil × 1.414 = 5.656 mil). The true peak value is one half of 5.656 mil or 2.828 mil (5.656 mil ÷ 2 = 2.828 mil). Dynamic signal analyzers perform rms averaging digitally on successive vibration spectra.

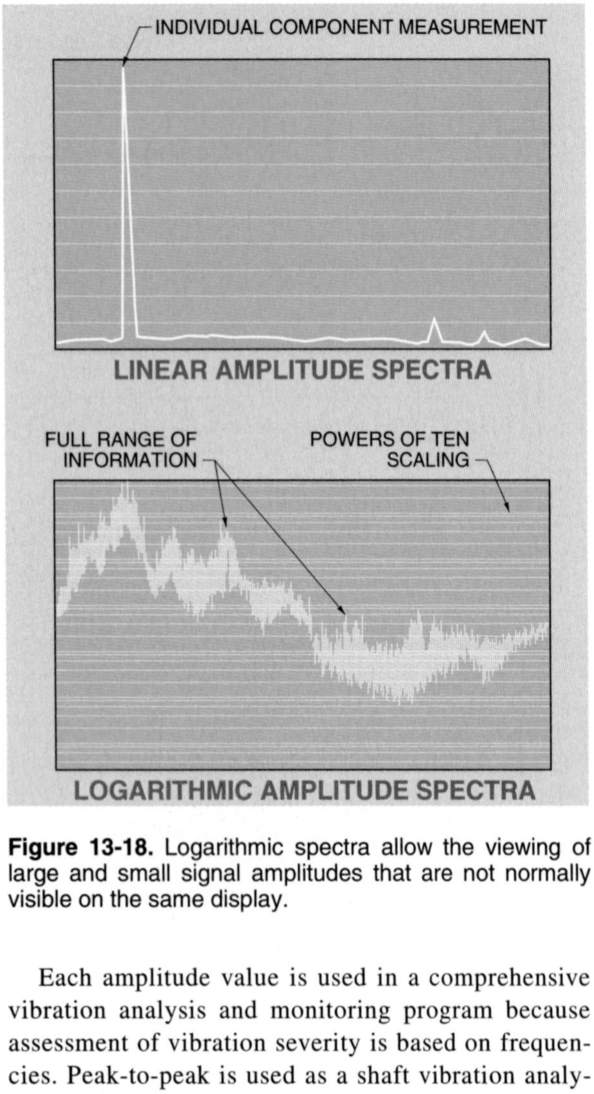

Figure 13-18. Logarithmic spectra allow the viewing of large and small signal amplitudes that are not normally visible on the same display.

Each amplitude value is used in a comprehensive vibration analysis and monitoring program because assessment of vibration severity is based on frequencies. Peak-to-peak is used as a shaft vibration analysis based on displacement characteristics. Peak is used with acceleration characteristics offering the greatest sensitivity required for defect detection of bearings or gears. Rms velocity characteristics are used when component levels vary significantly, such as with unbalance or misalignment. Vibration (noise) levels that vary greatly, such as in misalignment, are not statistically accurate with one measurement. More than one measurement is required to obtain accurate results.

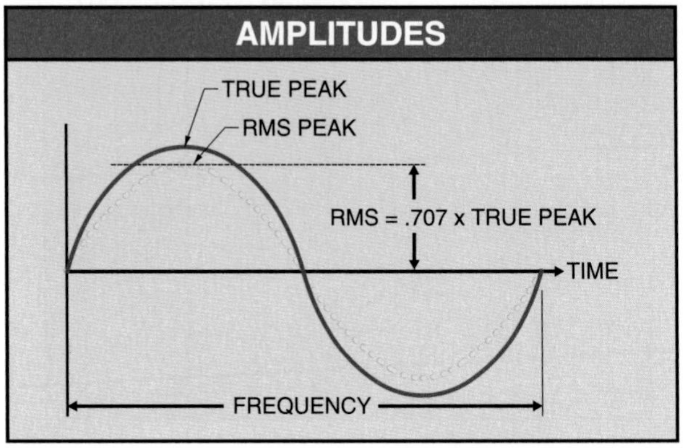

Figure 13-19. Smoothing a mix of many amplitudes and frequencies is accomplished by rms averaging.

Time Domain. *Time domain* is the amplitude as a function of time. Time domain signals are signals that appear as cycles where each cycle or wave occurs in a certain time period (generally in milliseconds). Time domain is displayed as amplitude versus time, with amplitude on the vertical axis and time on the horizontal axis.

The time domain is useful when a single source vibration, such as rotor unbalance or component looseness, is displayed. Although the time domain is used when a single vibration source is suspected, most machine vibration is a complex mix of many vibration frequencies, where each vibration requires an individual identification. The overall peak-to-peak vibration (displacement) of a machine is the sum (total) of the various individual vibrations of the machine.

For example, a gear transmission may have 2 mils of vibration occurring at 2x speed because of looseness, 2 mils of vibration occurring at 1x speed because of unbalance, and 1 mil of vibration at a high frequency due to bearing displacement. The total vibrating energy amplitude is approximately 5 mils peak-to-peak. See Figure 13-20.

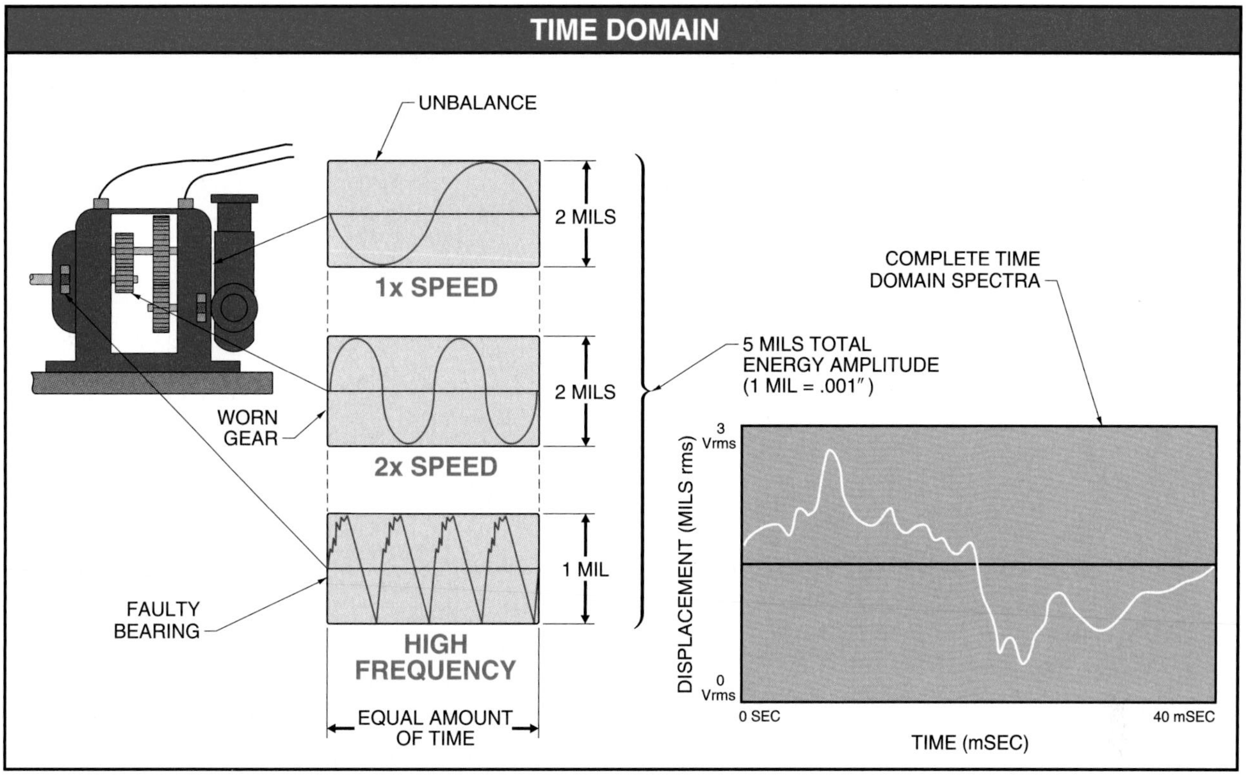

Figure 13-20. The time domain is useful when a single source vibration, such as rotor unbalance or component looseness, is displayed.

More than one vibration displayed in the time domain appears as a complex sum of amplitudes and is difficult to read. A particular frequency range that is not affected by other frequencies may be read by filtering out the other frequencies. A *filter* is a device that limits vibration signals so only a single frequency or group of frequencies can pass.

Frequency Domain. *Frequency domain* is the amplitude versus frequency spectrum observed on an FFT analyzer. Frequency domain is best understood by viewing a time domain cycle from a three-dimensional perspective. The frequency domain takes a combined multiple source vibration viewed in the time domain and separates the individual vibrations based on their frequencies. Each amplitude is read from left to right as an increase in frequency. See Figure 13-21.

A *frequency spectrum* is a representation of the frequency and content of a dynamic signal. The frequency spectrum is displayed with the vertical axis consisting of the amplitude (calibrated in convenient units) and the horizontal axis graduated in multiples of running frequencies. Abnormal characteristics related to balance, bearings, gears, and alignments are generally read within their separate frequencies. For example, basic unbalance displayed on a frequency domain analyzer is seen as an amplitude at the frequency of 1x speed.

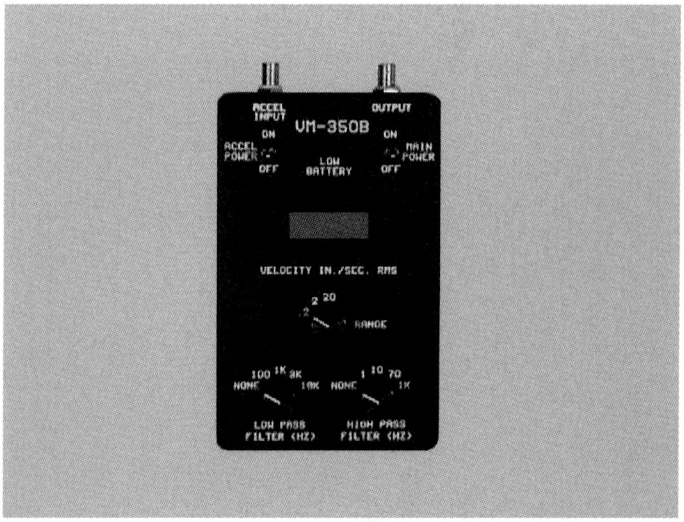

Vibration Monitoring Systems, Inc.

The Model VM-350B Vibration Meter measures vibration velocity and contains internal high pass and low pass frequency filters that allow the operator to pinpoint the vibration source.

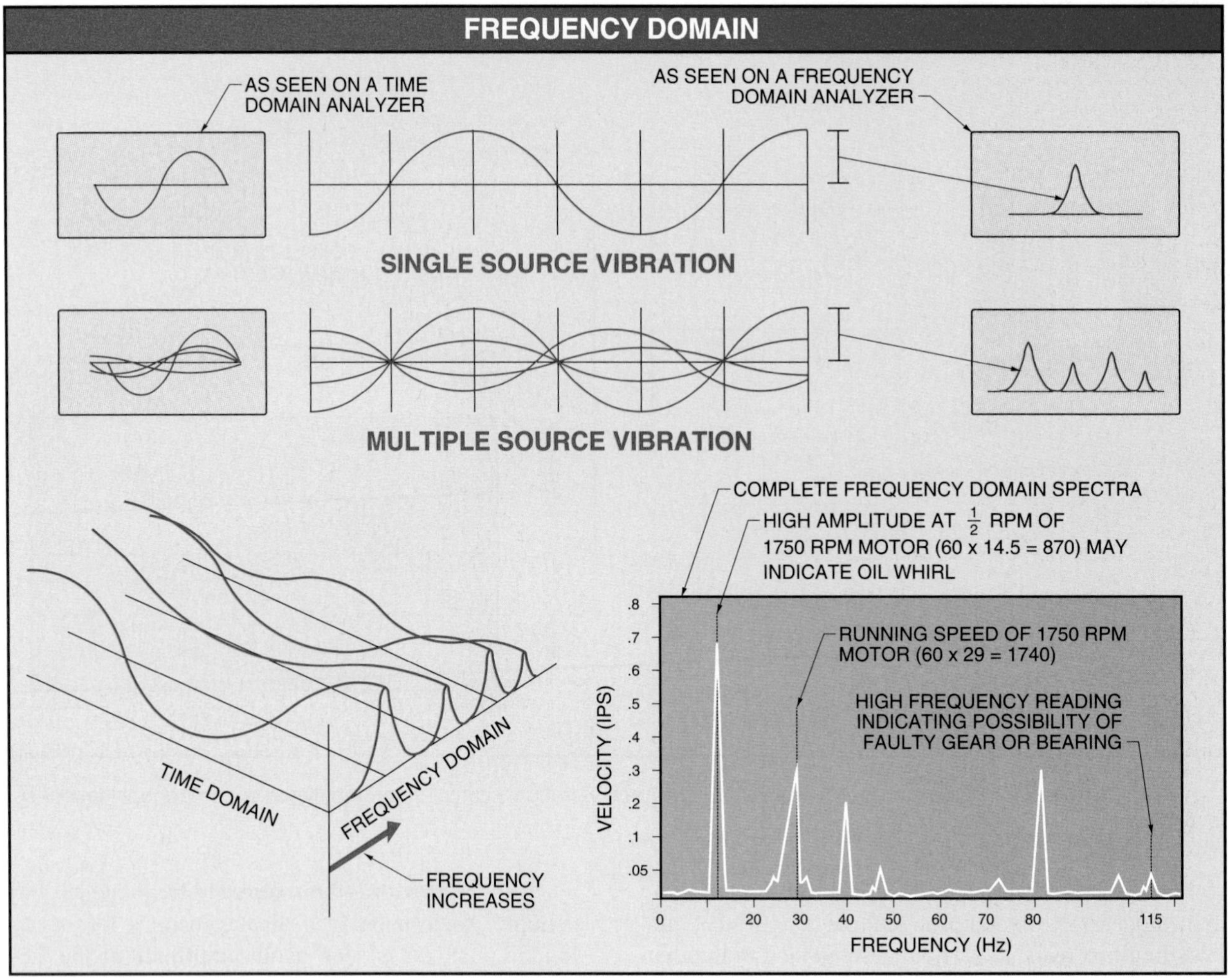

Figure 13-21. The frequency domain takes a combined multiple source vibration viewed in the time domain and separates the individual vibrations based on their frequencies.

GE Motors & Industrial Systems

Machine alignment is critical to long equipment life because small amplitude vibration caused by motor shaft deflection may drive out lubrication from contacting parts resulting in accelerated wear.

Current vibration analysis equipment is able to scan through a range of frequencies using digital filtering and display amplitudes at various frequencies. Complex vibration signals can be broken down using filters into simpler vibration components, resulting in clarification of the vibration.

Dynamic signal analyzers are able to amplify and display signals that are so small they would not show up on a time domain spectra. A DSA can simultaneously display a vibration that is 1000 times greater than another vibration. *Dynamic range* is the ratio between the smallest and largest signals that can be analyzed simultaneously.

When taking vibration readings, ensure that the transducer is pointed accurately because transducers are most sensitive along their central axis.

VIBRATION MONITORING PROGRAMS

The key to establishing a successful vibration monitoring program is to understand that the only objective for taking and recording readings is to compare previous readings with present readings and to see if an increase in readings is developing. Mechanical trouble is generally the reason why machine vibration increases.

To maintain consistency, every effort should be made to take readings under the same conditions. For example, consecutive readings should be taken as a machine is under a load if the first readings were taken while the machine was under a load (during production).

Vibration monitoring programs may be short-term or long-term. A short-term program includes permanently installing sensors on critical equipment such as turbines or high-speed rollers to give an alarm at the first sign of trouble. A long-term monitoring program includes manually monitoring critical areas of plant equipment to provide information on gradual or impending problems. Regardless of the program used, there must be recognized trends and alarms to issue appropriate warnings.

Technicians should be able to accurately identify specific gears within a failing gearbox, interpret resonance problems creating damage to bearings, shafts, and couplings, and identify improper bearing installation and shaft alignment. Skilled personnel are able to diagnose unbalance, angular misalignment, eccentric gears, or defects to inner or outer bearing races. Vibration characteristics charts are beneficial towards pinpointing a specific component defect. See Appendix.

Diagnosing machinery problems can be complex, requiring extensive training and experience. However, smaller scale programs may be instituted and should be used only as a progressive development towards a complete vibration monitoring program. Developmental stages consist of personnel training, facility layout, machinery layout, machinery component layout, and establishing files and data retrieval systems. Forms are also required for generating machine component specifications such as motor sizes, types, dimensions, speed, and coupled devices such as pumps, gearboxes, belt drives, etc. This includes specifications such as the type of gear, number of teeth, etc. Other forms include those for scheduling vibration checks and recording machinery maintenance history and vibration checks.

Exxon Company

Vibration monitoring programs help reduce wear and maintenance costs on vital machines and provide increased operating life.

Recordkeeping

Historical data is recorded and kept on a vibration check sheet. A vibration check sheet contains the basic machine configuration to be checked and its checkpoints, equipment type and location, check interval (weekly, monthly, etc.), check made (displacement, velocity, or acceleration), individual making the checks, date checked, initial horizontal, vertical, and axial readings, test equipment used, and a place to record the periodic checks. See Figure 13-22.

An interpretation may be made from a vibration check sheet as recordings stay the same or change. Recordings may be graphed to develop a trend. *Trending* is a graphic display used for interpretation of machine characteristics. Picture or graphic plotting provides a visual display of recordings. The visual display, when updated at each recording, provides an indication as to whether the component check is showing a gradual life expectancy decline or a rapid deterioration.

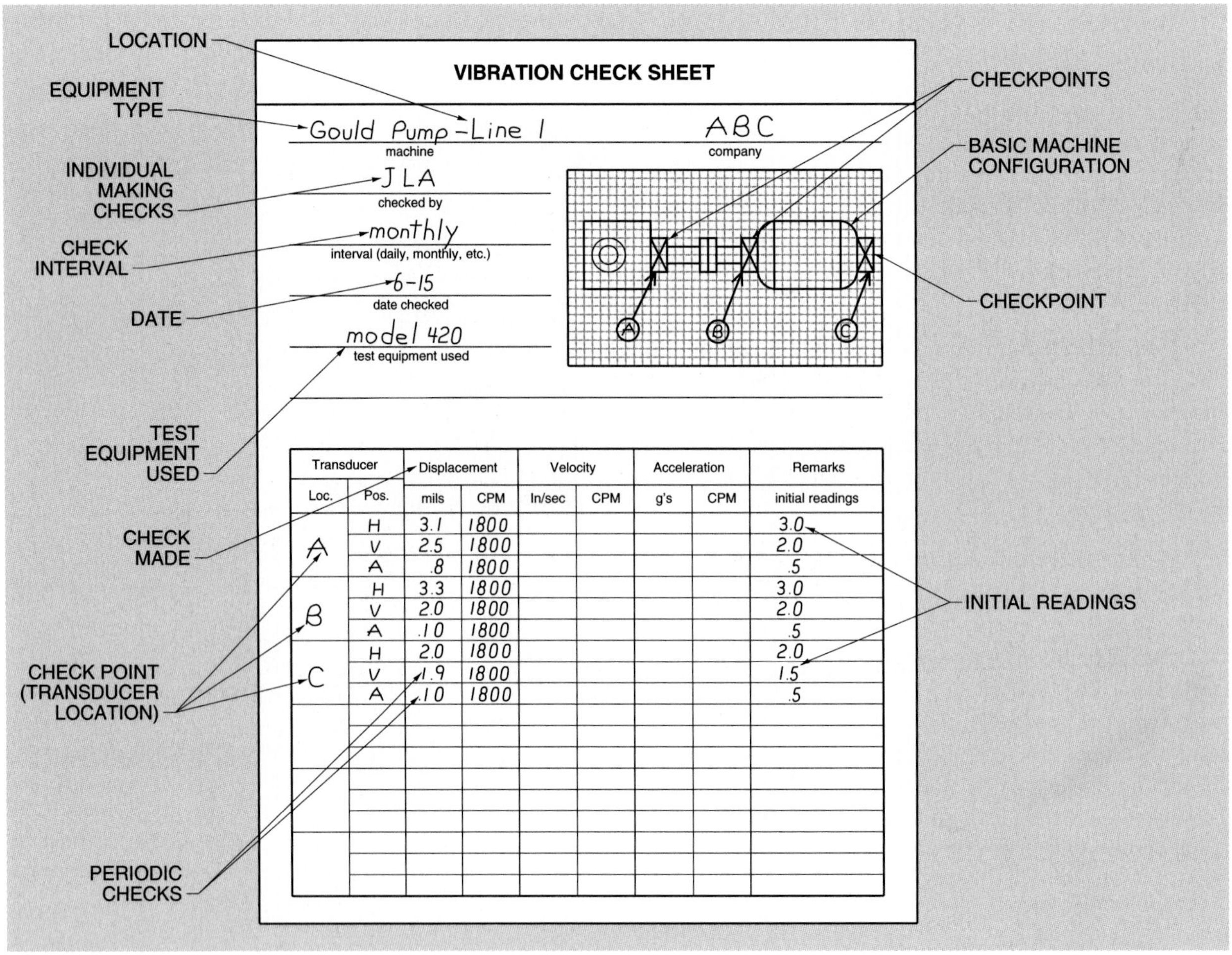

VIBRATION CHECK SHEET

Gould Pump – Line 1 (machine) ABC (company)

JLA (checked by)

monthly (interval (daily, monthly, etc.))

6-15 (date checked)

model 420 (test equipment used)

Transducer		Displacement		Velocity		Acceleration		Remarks
Loc.	Pos.	mils	CPM	In/sec	CPM	g's	CPM	initial readings
A	H	3.1	1800					3.0
	V	2.5	1800					2.0
	A	.8	1800					.5
B	H	3.3	1800					3.0
	V	2.0	1800					2.0
	A	.10	1800					.5
C	H	2.0	1800					2.0
	V	1.9	1800					1.5
	A	.10	1800					.5

Figure 13-22. The numerical data entered on a chart indicates if machine conditions are staying the same or are worsening.

Alignment

14
Chapter

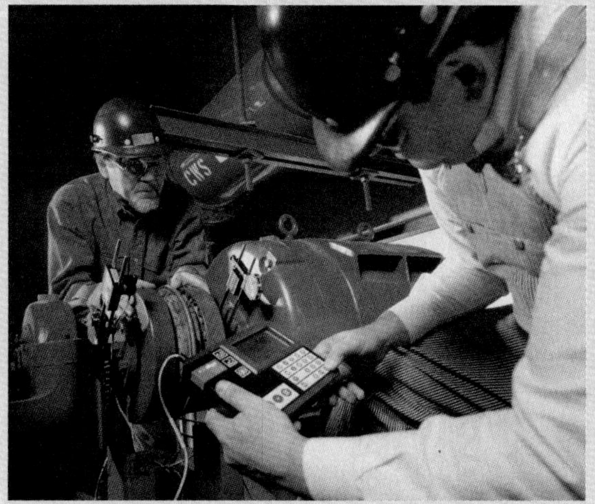

Ludeca Inc., representative of PRUEFTECHNIK AG.

More than 50% of vibration problems are caused by misaligned machinery. Vibration from misaligned shafts has a direct impact on the operating costs of a facility. Misaligned shafts require more power, create premature seal damage, and cause excessive forces on bearings. This leads to early bearing, seal, or coupling failure. Alignment specifications must be much more accurate now than in the past because operating speeds have increased, material weights have been reduced, and bearing tolerances have increased. Various methods, such as dial indicator fixtures and electronic and laser measuring devices, have been developed to achieve running condition accuracy.

MISALIGNMENT

Alignment is the location (within tolerance) of one axis of a coupled machine shaft relative to that of another. *Misalignment* is the condition where the centerlines of two machine shafts are not aligned within tolerances. Properly aligned rotating shafts reduce vibration and add many years of service to equipment seals and bearings. Misalignment of a coupling by .004″ can shorten its life by 50%. See Figure 14-1.

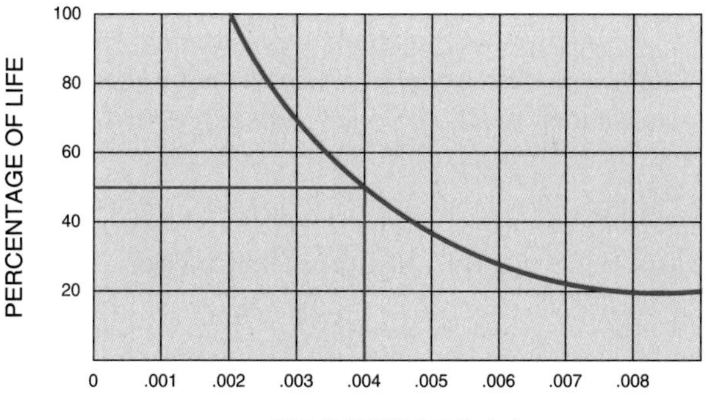

Figure 14-1. Misalignment of a coupling by .004″ can shorten its life by 50%.

Poor condition of equipment, such as worn bearings, bent shafts, stripped mounting bolts, bad gear teeth, or insufficient foundations or base plates, can create enough vibration to render any alignment effort useless. Once vibration starts, rapid wear of other components begins.

Misalignment exists when two shafts are not aligned within specific tolerances. Misalignment may be offset or angular. *Offset misalignment* is a condition where two shafts are parallel but are not on the same axis. *Angular misalignment* is a condition where one shaft is at an angle to the other shaft. Shaft misalignment is usually a combination of offset and angular misalignment. See Figure 14-2.

Offset and angular misalignment may be in the vertical or horizontal planes or both. Misalignment may be in the vertical offset, horizontal offset, vertical angularity, or horizontal angularity. Most misalignments are a combination of each. Many offsets in alignment are caused by improper machine foundations, weak supports, forces received from machine piping, soft foot, or thermal expansion. *Soft foot* is a condition that occurs when one or more feet of a machine do not make complete contact with its base. *Thermal expansion* is the dimensional change in a substance due to a change in temperature.

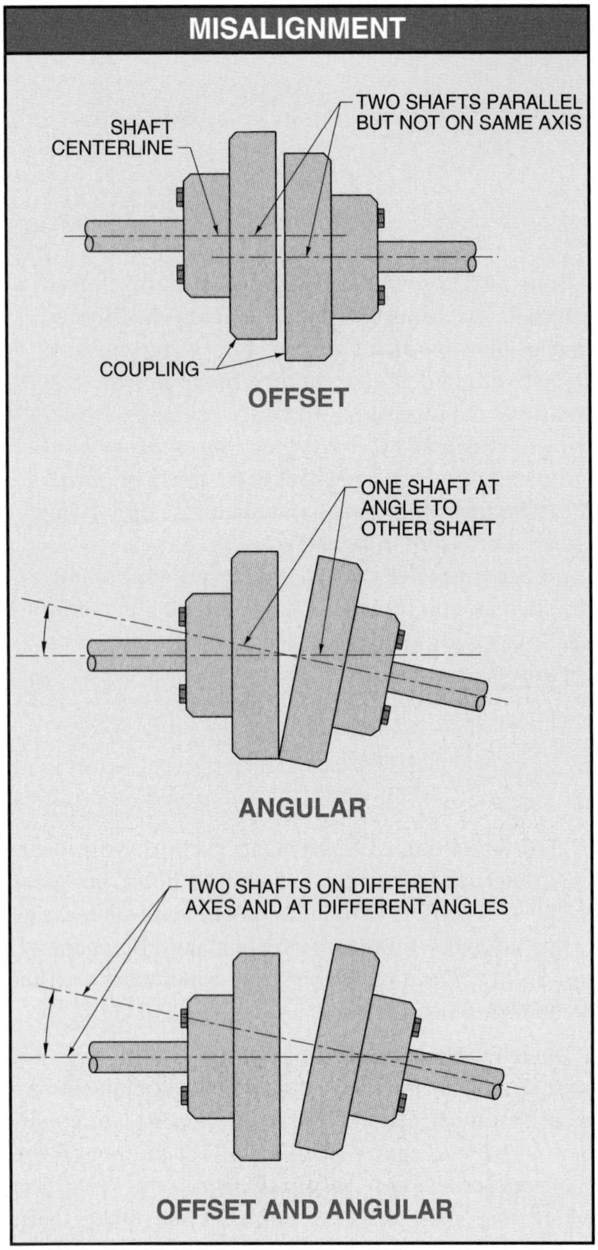

Figure 14-2. Misalignment may be offset or angular, but is generally a combination of the two.

Rotating machines are generally connected by couplings. A *coupling* is a device that connects the ends of rotating shafts. A *flexible coupling* is a coupling with a resilient center, such as rubber or oil, that flexes under temporary torque or misalignment due to thermal expansion. Flexible couplings can allow enough vibration to cause excessive wear to seals and bearings. Where flexible couplings are used, shaft alignment should be as accurate as it would be if solid couplings are used.

ALIGNMENT SEQUENCE

Vibration is now recognized as detrimental and industry is making many moves toward correction. Fundamental steps to reducing vibration include purchasing quality equipment, setting up good preventive maintenance procedures, and maintaining proper equipment alignment techniques and tolerances.

The objective of proper alignment is to perfectly couple two shafts under operating conditions so all forces that cause damaging vibration between the two shafts and their bearings are removed. The objective is to align the shafts, not the couplings. Aligning the couplings may result in misaligned shafts with aligned but irregularly-shaped couplings.

Each component that directly or indirectly affects the proper alignment of machinery must be identified and considered before actual alignment begins. Considerations for good alignment include:

- Proper preparation of foundations, foundation base plates, and machinery
- Proper machinery anchoring
- Proper machinery movement during alignment
- Soft foot
- Proper use of alignment methods or procedures

EQUIPMENT PREPARATION

Machinery to be aligned that is connected electrically must be locked out first. Before working on the equipment, challenge the electrical functions by testing the start switch. *Challenging* is the process of pressing the start switch of a machine to determine if the machine starts when it is not supposed to start. On completion of the lockout challenge, place all switches in OFF position. Also, any product pumps to be aligned must be blocked out to prevent product flow in the piping. See Figure 14-3.

Machine Foundation and Base Plates

Aligning any equipment begins with the foundation and base plate to which the equipment is anchored. A *foundation* is an underlying base or support. A *base plate* is a rigid steel support for firmly coupling and aligning two or more rotating devices. Foundations must be level and strong enough to provide support without movement. Base plates must be rigid enough to firmly support the equipment without stress and be securely anchored to the foundation.

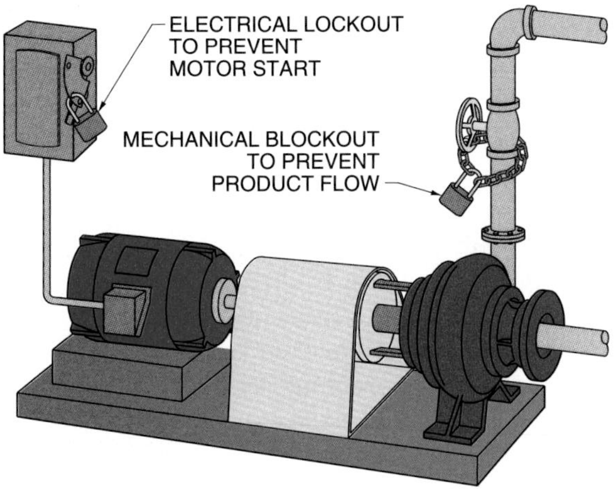

Figure 14-3. Lockout all electrical energy and blockout all mechanical energy before beginning alignment.

Originally, equipment base plates were made of thick cast iron that was strong enough to support the equipment in all operating conditions. The mounting surfaces were machined level. Currently, many equipment base plates are only sheet metal or plate metal welded or bolted to angle iron or I-beams. Flexing base plates must not be used. Good alignment is wasted when a flexing base plate is used. See Figure 14-4.

The feet on machines such as motors, gearboxes, pumps, etc., must be checked for cracks, breaks, rust, corrosion, or paint. This equipment should be bolted to a base plate, not anchored to concrete. The contacting surfaces between the motor, pump, gearbox, etc., and the base plate must be smooth, flat, and free of paint, rust, or foreign materials. Inspect all areas for burrs, rust, cracks, breaks, or any other damage. Finally, inspect the couplings, shafts, and bearings for damage, contamination, or inaccurate sizes.

ANCHORING MACHINERY

Each element related to anchoring machinery has a direct effect on the alignment forces of the equipment. Machine anchoring must consider piping and plumbing and the condition of anchoring components such as bolts and washers.

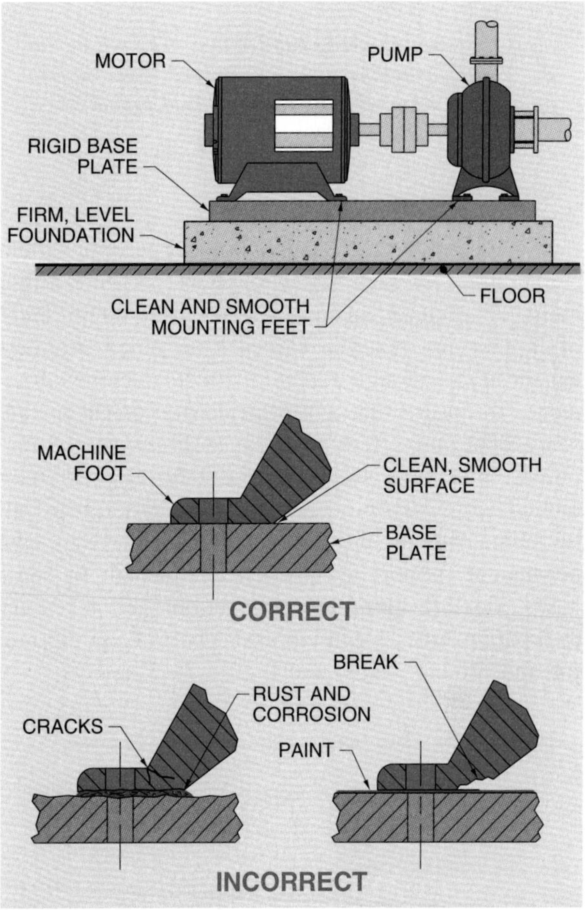

Figure 14-4. A clean, firm, and level base plate and foundation is required for proper alignment to ensure minimal flexing between machines.

Fluke Corporation

Ensure power to a motor is locked out and flow in piping is blocked out to prevent the devices from rotating during the alignment procedure.

Flexible plumbing connections are used on rotating machine piping to reduce stress caused by thermal expansion due to hot fluid processes.

Piping Strain

Pipe and conduit connections, if improperly installed, can produce enough force to affect machine alignment. Thermal expansion created by the temperature of liquids and reaction forces from piped products can produce enough force to affect machine alignment. To ensure that any transmission of an outside force does not affect the proper alignment of machines, machines should initially be aligned unattached from any piping if possible. Therefore, all plumbing must be properly aligned and have its own permanent support even when unattached. In some cases, flexible plumbing connections are necessary to separate stresses and vibrations between pump/motor and product lines. See Figure 14-5.

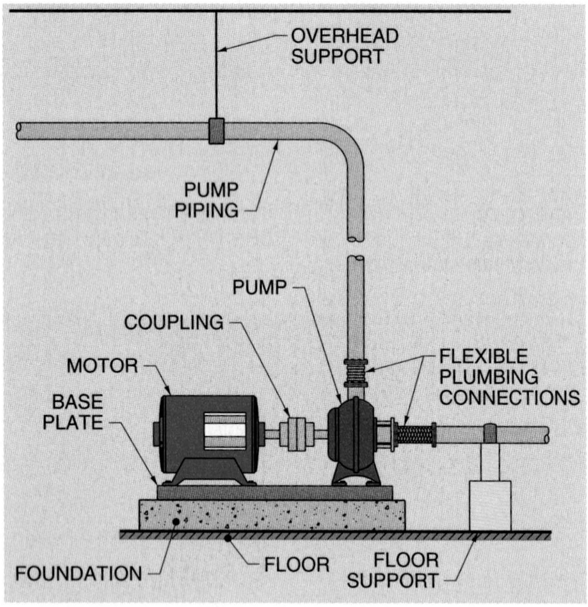

Figure 14-5. Pump piping must be independently supported to prevent angular forces from working against bearing and alignment tolerances.

Anchoring

Anchoring is any means of fastening a mechanism securely to a base or foundation. Firm but adjustable anchoring of mechanisms on a base plate is accomplished using the proper mechanical fasteners (bolts, screws, nuts, etc.). Adverse anchoring includes bolt-bound machines, bolts or bolt holes not of the proper size to allow for sufficient movement, and improper washers that create a dowel effect. Improper use, type, or fit of the anchoring bolts can make alignment of any machine impossible. See Figure 14-6.

Bolt Bound. *Bolt bound* is the prevention of the horizontal movement of a machine due to the contacting of the machine anchor bolts to the sides of the machine anchor holes. Bolt bound bolts prevent horizontal movement of a machine in any needed horizontal direction. Work is not wasted or duplicated if this condition is checked before any alignment checks begin. In many cases, bolt bound conditions may be checked using a straightedge placed along the machine shafts to determine if enough horizontal movement is available for proper alignment.

Loosening all anchoring bolts and shifting both machines to align the base plate mounting holes usually works if it is detected that sufficient movement is not available at the machine to be shimmed. The anchoring bolts must be turned down or the machine mounting holes enlarged if repositioning both machines does not work.

Anchoring bolts may be turned down without decreasing the tensile strength as long as the cut does not go beyond the root diameter (bottom portion) of the threads. Also, mounting holes may be enlarged, but this usually leads to the dowel effect. In some cases, a combination of undercutting the bolt and enlarging the holes may be necessary.

Dowel Effect. *Dowel effect* is a condition that exists when the bolt hole of a machine is so large that the bolt head forces the washer into the hole opening on an angle. Angled washers force the bolt to the center of the hole, making any horizontal movement impossible. Dowel effect is corrected by using machined washers 2 to 3 times thicker than the original washer. This prevents any deformation of the washer by bolt forces.

Always use the correct torque on an anchoring bolt to prevent excessive bolt stretch and reduce the possibility of distorting the base plate or machine frame.

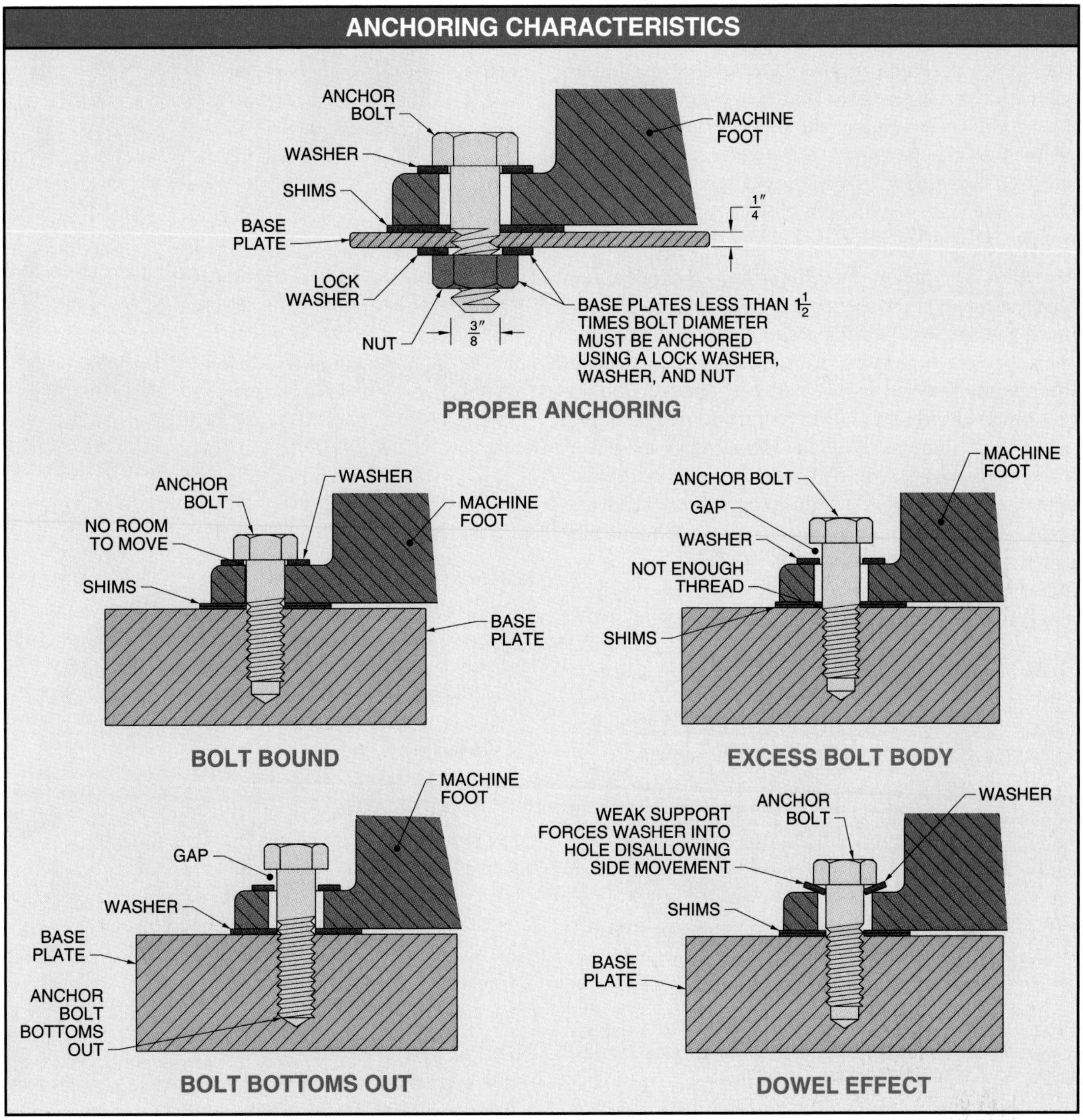

Figure 14-6. Improper bolt diameter, excessive bolt body or length, and weak washers can make alignment of any machine impossible.

Proper Bolt Installation. A base plate that is drilled and tapped to anchor a machine must be a minimum thickness of 1½ times the root diameter of the anchoring bolts. The threaded depth must be a minimum of 1½ times the root diameter of the bolt when a base plate is thicker than 1½ times the root diameter of the bolt. Mounting bolts require the use of nuts and lock washers if the base plate is less than 1½ times the bolt diameter. Always select anchor bolts that have the correct length of the unthreaded portion of the bolts. The bolt may run out of thread and leave an incomplete and loose anchor if the unthreaded portion is too long. Also, the bolt may bottom out in the base plate hole leaving a loose anchor if the threaded portion of the bolt is too long. Always use the correct grade and size anchor bolts to properly secure the machine frame to the base plate.

Controlled Machine Movements

The choice of which machine to move during alignment, the proper tools to use to prevent equipment damage, and the components used for precise movements must be known for an accurate, fast, and damage-free alignment. This knowledge includes the proper use and installation of jack screws and the proper use and selection of spacers and shim stock.

Machine Movement Specification. Generally, the heaviest machine or the machine attached to plumbing is the anchored machine and is referred to as the stationary machine (SM). The motor is generally the machine moved and is referred to as the machine to be shimmed (MTBS). Either machine may be the moved and shimmed machine. Regardless of which machine is moved, the SM must initially be higher than the MTBS to allow for proper vertical alignment. A good practice is to initially install the SM using .125″ shims under each foot. This practice requires raising the MTBS, but prevents any vertical movement requirements of the SM. See Figure 14-7.

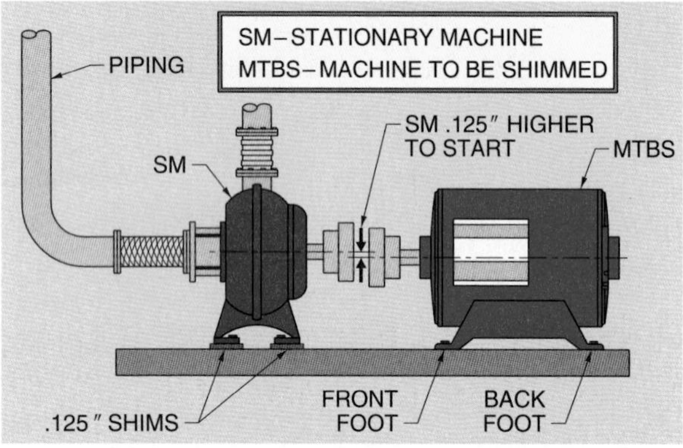

Figure 14-7. A machine is chosen as the MTBS because it is easier to move than the other machine, which may be larger or connected to piping.

Jack Screws. A *jack screw* is a screw inserted through a block that is attached to a machine base plate allowing for ease in machine movement. Jack screws should be installed on the base plate to easily and accurately control the horizontal movement of a machine. Typical jack screw blocks are 1½″ × 2½″ rectangular blocks made from ½″ thick steel (larger or thicker depending on the size of the machine) and are bolted or welded to the sides of the base plate with the jack screws directly in-line with each mounting hole. Each jack screw block is drilled and tapped to allow for a ½″ bolt assembly. See Figure 14-8.

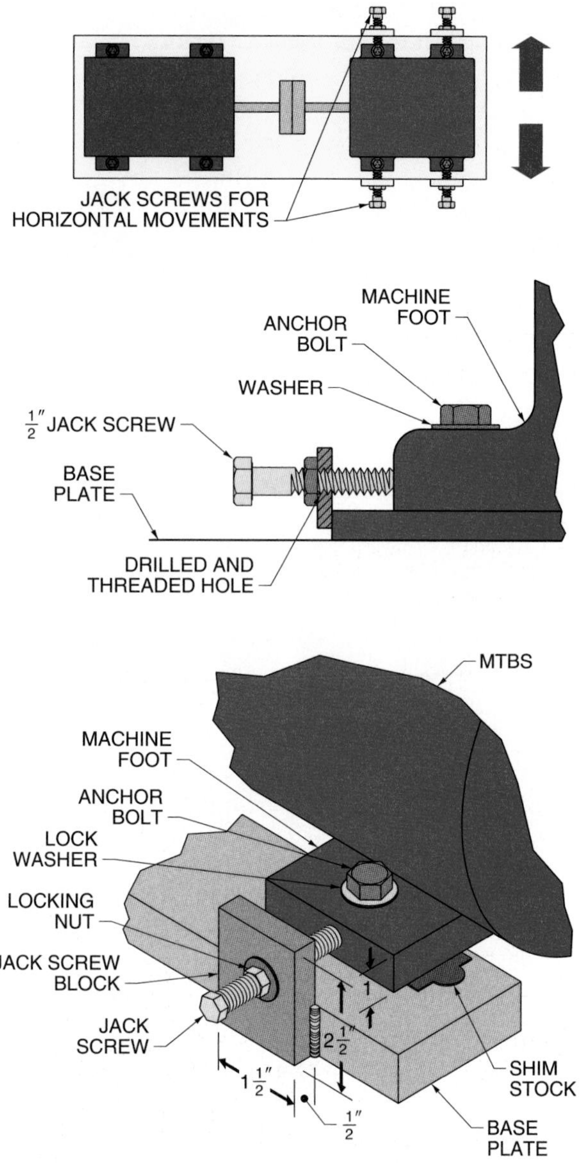

Figure 14-8. A jack screw is a screw attached to a block that is bolted or welded to a machine base plate to allow for ease in machine movement.

Jack screws are used for machine movement only. To prevent additional forces from being applied, jack screw pressure must be backed off after each tightening of the anchor bolts. Jack screws are invaluable when a machine is to be moved just a few thousandths of an inch.

A screwdriver or crowbar can be used for machine movement if it is not possible to install jack screws or if they are not available. An easy, steady prying force is safer and less damaging than a blow from a hammer. Use only a soft-blow hammer if a hammer is necessary.

Checking for Shaft Runout. *Runout* is a radial variation from a true circle. Any shaft that runs eccentric to the true centerline of a machine by more than .002″ makes achieving tolerance impossible and should be corrected. *Eccentric* is out-of-round or that which deviates from a circular path. An eccentric shaft produces high vibrations similar to those caused by a bent shaft. An eccentric shaft may be determined by the use of a dial indicator. A *dial indicator* is a device that measures the deviation from a true circular path. Correction may be accomplished by changing machines or by having the shaft recut. See Figure 14-9.

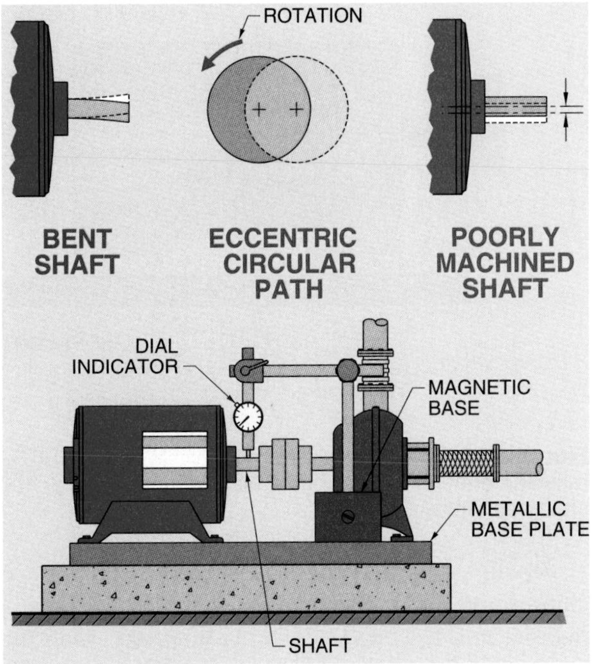

Figure 14-9. An eccentric shaft produces high vibrations similar to those caused by a bent shaft. An eccentric shaft may be determined by the use of a dial indicator.

Shim Stock. It is rare for any machine to have total contact of all of its feet with the base plate and also be within tolerance. Shims and spacers are used to adjust the height of a machine. *Shim stock* is steel material manufactured in various thicknesses, ranging from .0005″ to .125″. Shim stock can be purchased as a sheet, roll, or in precut shapes. A *spacer* is steel material used for filling spaces ¼″ or greater. The feet of a machine must be firmly anchored to the base plate without creating excessive forces or movements between mating shafts. To prevent stacking inaccuracies, limit the amount of shims or spacers on each foot to five or less. Always choose the combination that uses the least amount of shims or spacers when different shim or spacer combinations can be chosen. Any spacer .250″ (¼″) and over should equal 1 piece and may be mild steel or stainless steel. Any spacer .125″ (⅛″) or under is considered a shim and must be stainless steel.

For example, a spacer and shim combination is required to raise a motor .683″. To keep the spacer to 1 piece and the shim stack to 5 or less, the spacer selected should be ⅝″ (.625″) with the remaining .058″ taken up by shims. Shim stacks used are sometimes based on the sizes available in a shim pack set. Variations for the .058″ shim stack include one .050″ and two .004″ or two .025″, one .005″, one .002″, and one .001″. Extra spring is added each time a shim is added to a stack. The best combination is the one that uses the fewest shims.

Precut stainless steel shims are recommended for alignment purposes. Cutting shim stock for use can create rough edges and burrs, making a successful alignment improbable. Also, any material other than stainless steel is smashed under load and vibration or rusts and corrodes.

Good shim packs are laser cut with each size printed (not stamped) on the shim. Stack 4 or 5 shims with the printed size reversed on every other shim when checking for thickness accuracy. This arrangement, when checked with a micrometer, produces a true size and condition of the stack. See Figure 14-10.

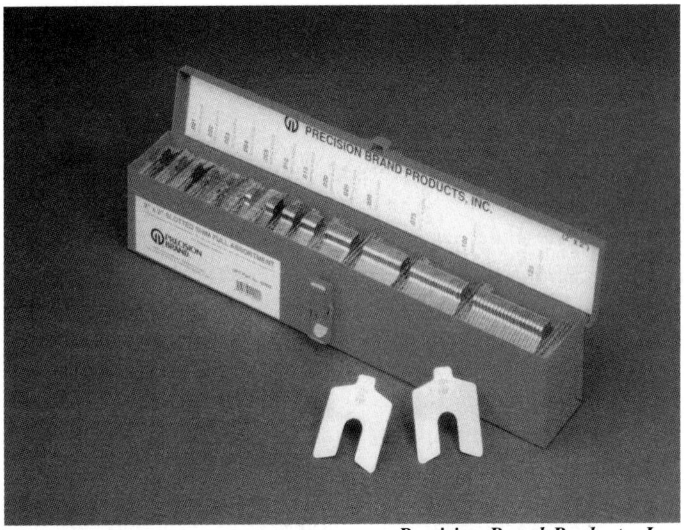

Precision Brand Products, Inc.

Precision Brand slotted shim assortment kits are economical, safe, and accurate, and reduce costs by eliminating hand cutting, material waste, and shim preparation.

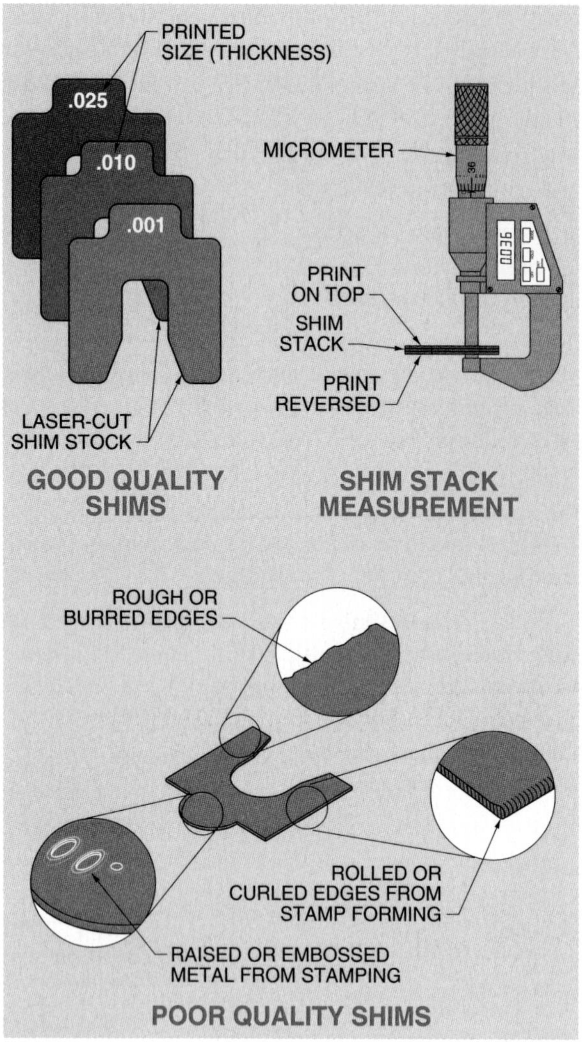

Figure 14-10. Good shim quality ensures a proper and firm machine-to-base contact.

Soft Foot

Soft foot is a condition that occurs when one or more machine feet do not make complete contact with the base plate. Distorted frames creating internal misalignment due to soft foot is a major reason for bearing failure. This internal misalignment and distortion loads the bearings and deflects the shaft. See Figure 14-11.

> ***All four corners of a machine foot should be checked with feeler gauges when a dial indicator check indicates soft foot. Angular soft foot is present if the total gap measured for any corner is different from the other corners. Any angularity greater than .001" per inch of foot must be corrected before alignment is started.***

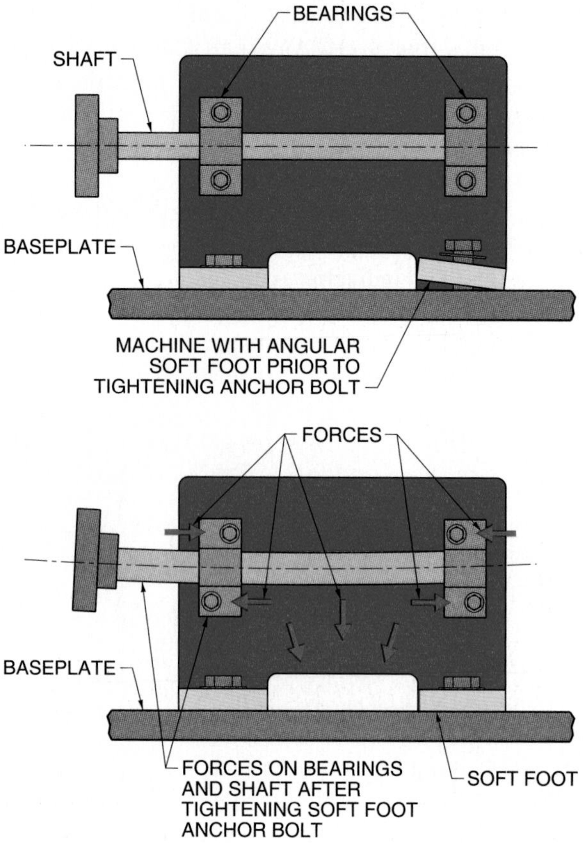

Figure 14-11. Uncorrected soft foot twists and distorts a machine enough to cause destructive forces on rotor and shaft bearings.

An angled or raised foot twists and distorts a machine frame when anchored. Distorted frames cause internal misalignment to bearing housings, shaft deflection, and distorted bearings, resulting in premature bearing and coupling failure. This condition also creates great difficulty in shaft alignment. Before aligning, any machine that has soft foot must be shimmed for equal and parallel support on all feet.

As a soft foot bolt is tightened, the shaft of the machine is deflected, which loads the bearings. This deflection can easily cause enough back-and-forth vibrations and pressures to damage the bearings, seals, and shaft. A shaft that rotates at 1800 rpm in a machine with a soft foot condition deflects 30 times each second or 108,000 times per hour. Soft foot should be checked on the MTBS and SM because soft foot can occur on any machine. Soft foot tolerance must be within .002″ of shaft movement. Soft foot may be parallel, angular, springing, or induced. Each is independent of the others. All four conditions may exist on the same machine. See Figure 14-12.

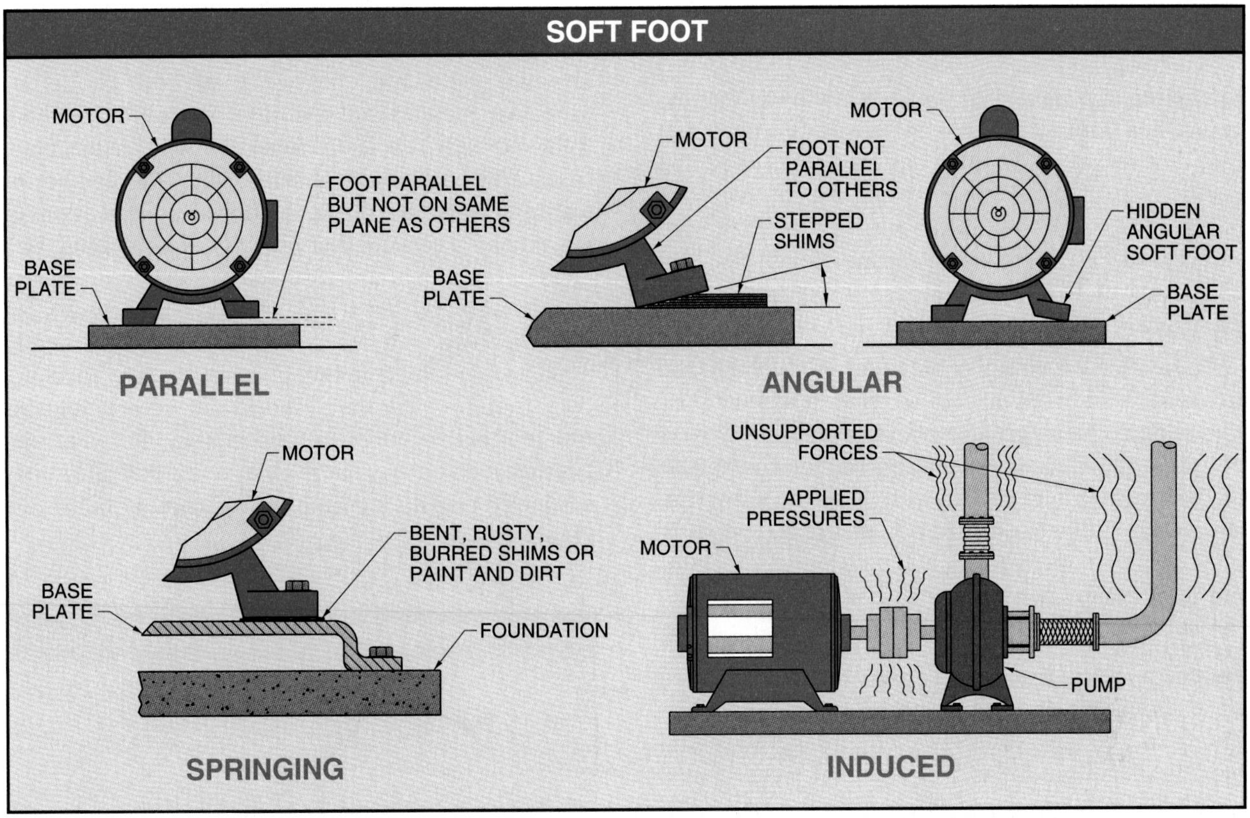

Figure 14-12. Soft foot may be parallel, angular, springing, or induced.

Parallel. *Parallel soft foot* is a condition that exists when one or two machine feet are higher than the others and parallel to the base plate. This condition occurs when a machine leg is short or when spacers of different thicknesses are used.

Correcting parallel soft foot is accomplished by first rocking the machine from side to side and determining the gap under the high foot using feeler gauges. A *feeler gauge* (thickness gauge) is a steel leaf at a specific thickness. Feeler gauges determine the air gap between two solids within thousandths of an inch.

Shim stock equal to the thickness of the soft foot gap is placed under the high foot and all four machine feet are rechecked. The shims should be moved from under the first foot to the other rocking foot if soft foot is noticed under a foot that was not shimmed. Check again for soft foot. The shim stock should be divided between the feet that were rocking if soft foot is noticed under two opposing feet. Checking more than once for proper parallel spacing is done as an attempt to find true, or close to true, parallel spacing without creating angular soft foot.

Angular. *Angular soft foot* is a condition that exists when one machine foot is bent and not on the same plane as the other feet. Generally, one corner of the angular foot is touching the base plate. Angular soft foot is usually the result of the machine being roughly handled or dropped or having uneven mounting pads due to poor machining or welding.

The machine may be sitting high on one side and low on the other if all four feet appear to have angular soft foot in the same direction. Angular soft foot may be determined when a .002″ feeler gauge can be placed under one side of a foot but not under the other side of the same foot.

Correction of angular soft foot is accomplished by machining the foot to be on the same plane as the other feet or by step shimming the foot to fill the gap. Step shimming begins by determining the direction and amount of slope and filling the sloping void with a series of steps (usually a maximum of 5 or 6). This is done by measuring the largest portion of the gap and dividing by 5 or 6, giving the thickness of each step. Finally, place each shim by hand, in steps, to fill the gap without lifting the machine. Check shaft movement using a dial indicator while tightening the machine bolts and correct as indicated from dial movement.

Springing. *Springing soft foot* is a condition that occurs when a dial indicator at the shaft shows soft foot, but feeler gauges show no gaps. This condition occurs from shims that are burred or bent, corroded bases or feet, dirt, grease, or rust between the feet, shims, and base plate, or too many shims. The machine acts as if it were mounted on springs due to each imperfection.

Prevention of springing soft foot is accomplished by using solid bases that are cleaned to the metal by removing all paint, grime, rust, and corrosion. Layers of grease can even act as a spring. The top and bottom of the machine feet must also be free of rust, paint, and grime. Shims used must be flat, clean, and without stamping imperfections. After cleaning, check for out-of-tolerance movement using a dial indicator at the shaft as the bolts are tightened.

Induced. *Induced soft foot* is soft foot that is created by external forces such as coupling misalignment, piping strain, tight jack screws, or improper structural bracing. Coupling forces from vertical or horizontal misalignment are noticed when couplings are difficult to bolt up or a spring or snap is noticed as couplings are disconnected.

Any external force in any direction to coupled pipe flanges on a pump strains the machine. This condition may be seen when checking for soft foot before, during, and after piping connections or when checking for structural bracing strain.

Lovejoy, Inc.

Shaft couplings allow drive and driven equipment connection and provide protection against misalignment, vibration, and shock.

Measuring Soft Foot

The two methods of measuring soft foot are the at-each-foot method and the shaft deflection method. Both methods use a dial indicator with a magnetic base to determine movement. More than one set of readings must be made, in the same direction of movement, to ensure that readings are constant. Two identical readings are assumed correct if three sets of readings are taken and one set is different from the other two. All other feet must remain securely tightened and the coupling uncoupled when checking each foot for soft foot. Check and correct angular soft foot before either method is used. The shaft deflection method is easier because the indicator does not interfere with loosening the anchoring bolts. See Figure 14-13.

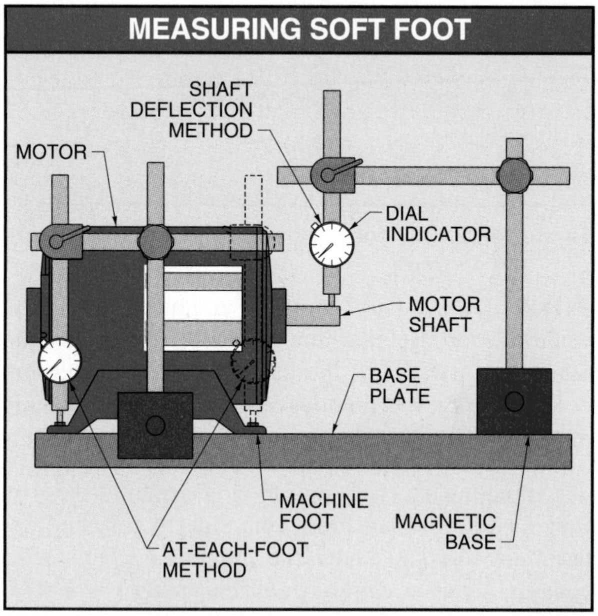

Figure 14-13. Soft foot may be measured at the shaft or each foot of a machine.

At-Each-Foot Method. In the at-each-foot method, soft foot is checked at each foot of the machine. A dial indicator with a magnetic base is secured to the base plate at the foot to be tested. The dial indicator is adjusted so the stem is above and perpendicular to the top of the foot. Ensure that all four feet are anchored firmly. With the dial adjusted to zero, watch the dial movement as the bolt for that foot is slowly but steadily loosened. The required shim thickness can be determined from the indicator movement if the foot rises and the dial moves more than .002″.

Place shims beneath the foot equal to the amount of the indicator movement and re-tighten the bolt. Repeat this process until the movement is less than .002″. Relocate the dial indicator to another foot and repeat the above procedure until all four feet have been checked and corrected.

Shaft Deflection Method. In the shaft deflection method, the shaft is checked for deflection when anchoring bolts are tightened or loosened. Critical distortion has not occurred if there is no measured deflection even though a foot has movement. Correction is necessary if there is movement. This method is quicker and more accurate than the at-each-foot method.

A magnetic base dial indicator is secured to the base plate and adjusted so the stem is above and perpendicular to the top of the shaft or coupling, whichever is the farthest from the MTBS. The farther the dial indicator is from the first set of feet, the greater the dial movement, which increases accuracy. Ensure that all four feet are anchored firmly. Zero the dial on the indicator and slowly loosen the bolt on the first foot. A dial that indicates .003″ rise requires the placement of shim stock beneath that foot totaling .003″. Tighten the bolt and check the foot again. Continue checking and shimming each foot individually until shaft movement is within the .002″ tolerance.

Check all feet for hidden angular soft foot when the fourth or final foot has been corrected and one foot rises when double-checking. Once soft foot conditions have been corrected, shaft alignment may begin. Shaft alignment is difficult or impossible if soft foot conditions have not been corrected to within tolerance.

Thermal Expansion

For proper alignment, two coupled shafts must be on the same horizontal and vertical plane under operating conditions. However, there could be a significant change in physical dimensions when there is a change in operating condition temperature and thermal expansion results. Thermal expansion is the growing or shrinking of metal molecules under hot or cold conditions. A temperature change between startup and running conditions can influence machine alignment because metal expands when heated and contracts when chilled. See Figure 14-14.

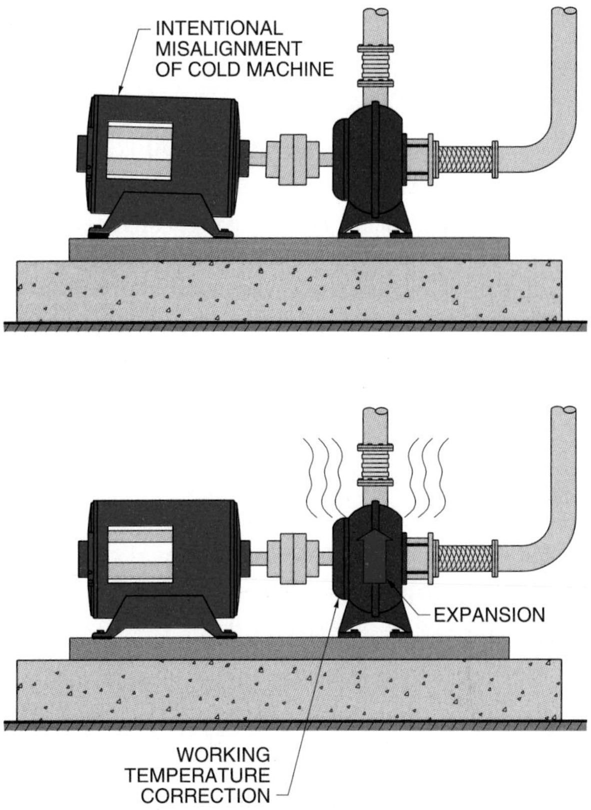

Figure 14-14. Temperature differences from cold startup to working temperature influence the position of one machine relative to another.

Temperature conditions that can change enough to affect critical alignment measurements may be caused by the temperature of product being pumped, excessive room or ambient temperature, or loaded motor temperatures. Most materials expand when heated and contract when chilled. For example, a piece of steel exactly 12″ long and 2″ in diameter in a room temperature of 72°F grows to 12.008″ long and 2.001″ in diameter when heated to 172°F. This change is affected by the material and how much the temperature has changed.

Change in material length is calculated to determine specific tolerances to absorb thermal expansion and accommodate different materials. A thermal expansion constant is given based on the material used. Constants include .0000063 for cast iron, .000009 for stainless steel, and .0001 for plastic. Thermal expansion is found by applying the formula:

$\Delta L = L \times \Delta T \times C$

where

ΔL = change in length (in in.)

L = original length (in in.)

ΔT = change in temperature (in °F)

C = material constant

For example, what is the change in length of a pump and cast iron frame motor combination when the operating temperature of the motor increases from 75°F to 140°F and the motor measures 15″ from its base to the shaft center? *Note:* The temperature change equals 65°F (140°F − 75°F = 65°F).

$\Delta L = L \times \Delta T \times C$

$\Delta L = 15 \times 65 \times .0000063$

$\Delta L = \mathbf{.006''}$

The vertical plane of the motor should be reduced by .006″ because the motor shaft rises as it gets warmer (operating temperature). At times, the SM, which may be a pump that pumps hot liquid, is the machine that rises. Compensating shims may be added under the MTBS when this is the case.

Manufacturing & Maintenance Systems, Inc.

Dial indicator probes must be mounted perpendicular to the contact surface when used for machine alignment.

DIAL INDICATORS AND ALIGNMENT

A dial indicator is a precise, jeweled movement instrument similar to a watch and must be treated as such if it is to be an accurate, useful tool. Dial indicators are required to be mounted perpendicular to their contacted surface. A dial indicator 10° out of perpendicular results in an immediate error of 2%. Blows from mallets or hammers on a machine into an indicator damage the indicator or throw off readings because indicators have jeweled movements. Always use slow, forceful movements when adjusting into an indicator.

To prevent incorrect readings, position a dial indicator on the circumference of a shaft so that the centerline of the indicator probe runs through the centerline of the shaft.

Dial Indicator Use

Dial indicators are read using the total movement of the dial needle. Readings do not have to begin at zero to determine total movement. Total indicator readings (TIR) are used as indicator readings regardless of whether the needle begins before, at, or after zero. Total dial indicator readings are determined from readings above (positive) or below (negative) zero, or a combination of both.

Total indicator readings are found by subtracting the low reading from the high reading when all indicator readings are positive. For example, total indicator readings on an indicator that has a high reading of +.022″ and a low reading of +.006″ is .016″ (.022″ − .006″ = .016″). The same subtraction rule holds true for dial readings that are negative. An indicator reading of −.006″ is subtracted from a greater reading of −.022″ to give a TIR of .016″. The exception to this rule is when a high reading is positive and the low reading is negative. In this case, the TIR is the sum of the two values. For example, an indicator reading of +.022″ and −.006″ gives a TIR of .028″ [.022″ + (−.006″) = .028″]. See Figure 14-15.

Verifying Dial Readings. Indicators must be run through their movement as many as three times to ensure a proper reading. Dial movement must be in the same direction. False readings are given if shaft rotation is reversed. This is due to the play or tolerances of indicator moving parts.

Aligning shafts that rotate must be accomplished using an indicator to check both vertical (up and down) and horizontal (back and forth) positioning. Straight movements (not angular) are known as offset movements. Dial indicator offset measurements are twice their actual offset. For example, a total indicator reading of −.020″ is indicating a shaft centerline offset of −.010″.

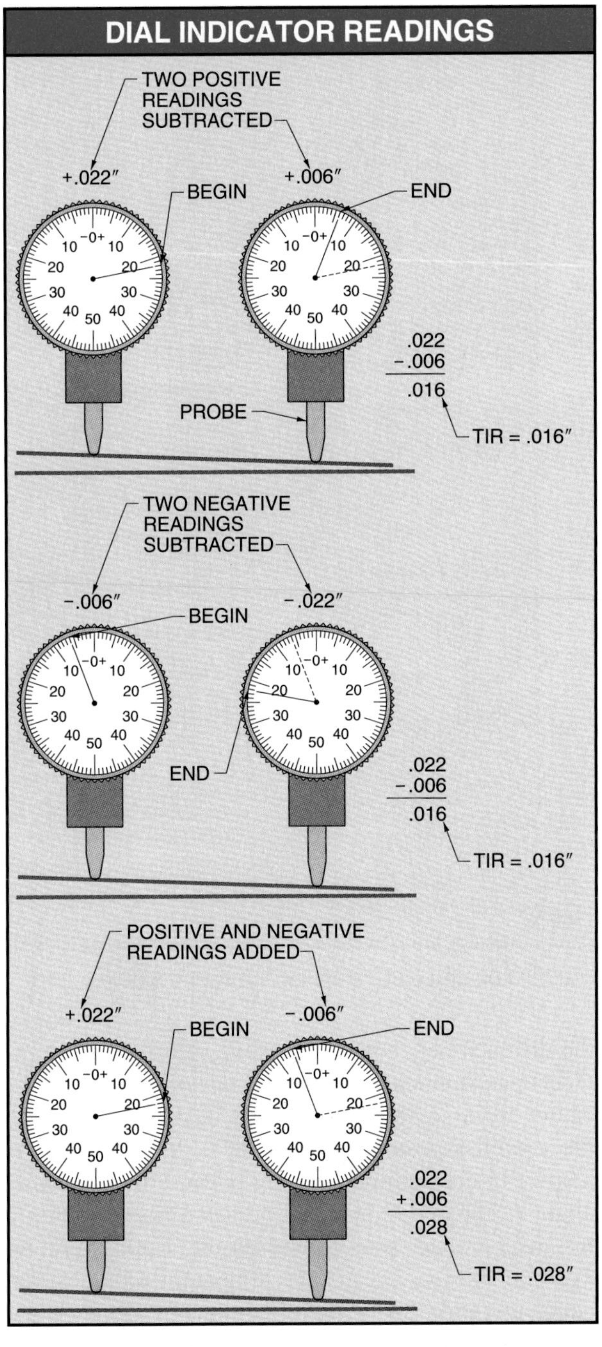

Figure 14-15. Dial indicators are read using the total movement of the dial needle and do not have to begin at zero.

An offset is checked by placing the indicator tip at the top of the shaft or coupling and zeroing the indicator. The top position is known as the 12:00 position. The indicator and shaft that it is attached to are rotated to the farthest position, known as the 3:00 position. Both shafts must be rotated with the coupling unencumbered.

Both coupling halves must be unattached from each other and must be turned together. Turning the shafts together gives a true condition of the shaft centerlines and is not affected by coupling faces that may not be square with the plane of the shaft or coupling rims that may not be machined round or concentric with their inside dimension. Any alignment using improper methods results in having good readings on poorly aligned machinery.

An easy way to verify whether 12:00, 3:00, 6:00, and 9:00 readings are correct is to add the 3:00 and 9:00 readings and compare the sum to the 6:00 and 12:00 readings. The sum of 3:00 and 9:00 readings should equal the 6:00 reading if the 12:00 reading started at zero. Readings may be taken at only three positions when complete rotation is not possible. See Figure 14-16.

The shaft being checked is above the centerline of the shaft on which the indicator is mounted if the indicator reading at 6:00 is negative. The shaft being checked is below the centerline of the shaft on which the indicator is mounted if the indicator reading at 6:00 is positive. The shaft being checked is below the centerline of the shaft on which the indicator is mounted if the angular reading at 6:00 is negative. The shaft being checked is above the centerline of the shaft on which the indicator is mounted if the angular reading at 6:00 is positive.

Manufacturing & Maintenance Systems, Inc.

Dial indicators are used during machine alignment procedures to indicate the amount of misalignment between two machine shafts.

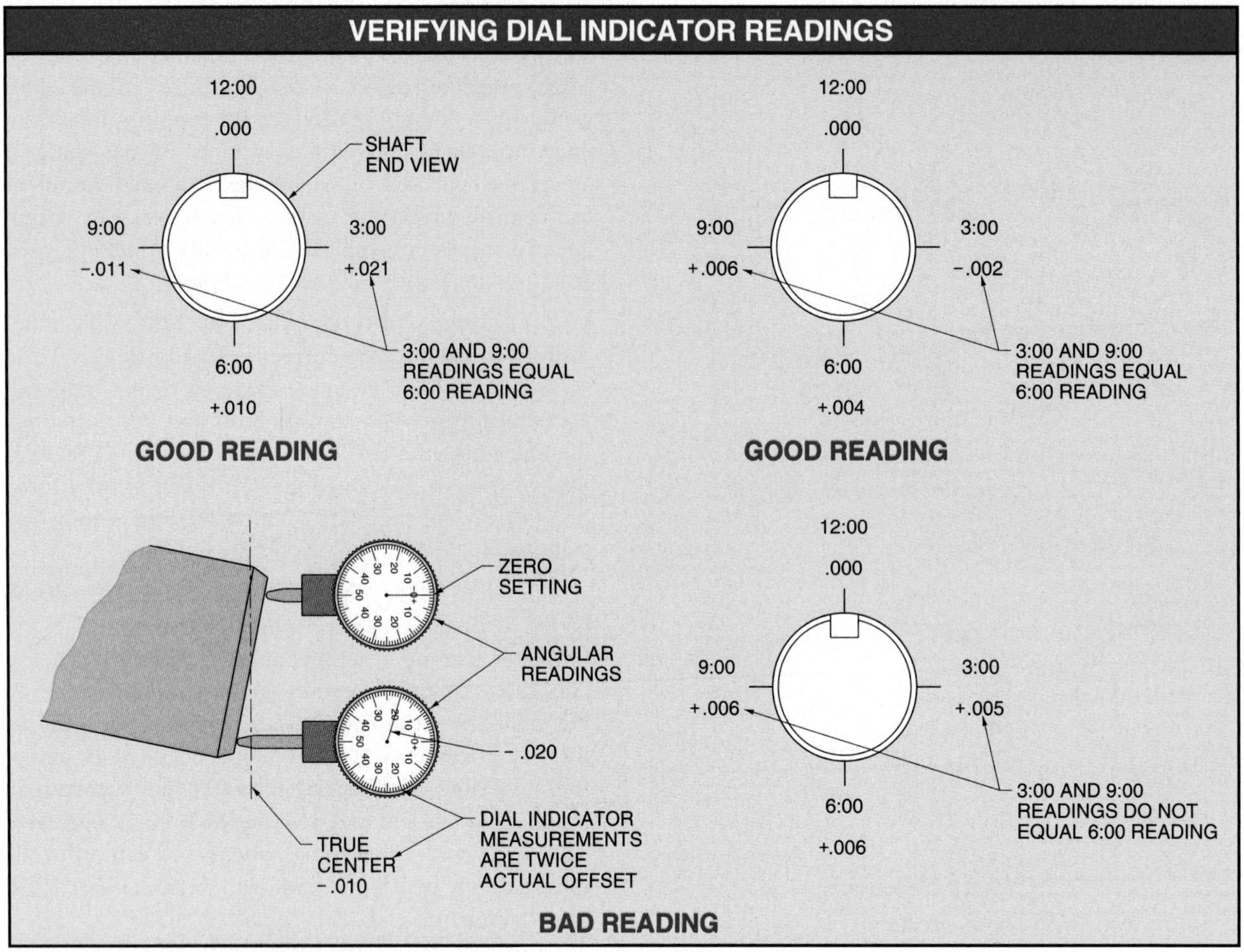

Figure 14-16. Good indicator readings are those where the sum of the 3:00 and 9:00 readings equals the 6:00 reading.

Indicator Rod Sag

To indicate the alignment of one shaft to another, the indicating device that indicates the alignment of one shaft must be clamped or strapped to the opposing shaft. The entire assembly generally consists of the clamp or strap, a riser rod, two 90° rod couplings, a spanning rod, and an indicating device. The indicating device may be a dial indicator or an electronic indicator. The greater the distance between the rod couplings, the more the weight of the indicating device creates a sag in the spanning rod. See Figure 14-17. This sag, if not accounted for, can throw readings off. A 3/8″ diameter spanning rod at a distance of 8″ between couplings can throw readings off by .010″.

Electronic indicators calculate rod sag when measurements are keyed in. Rod sag from a dial indicator must be determined by the technician. Accurate indicator rod sag is determined by first establishing the distance between rod couplings during the alignment measurement. All parts (rods, couplings, indicator) are assembled on a solid shaft or pipe using the established coupling distance. A solid shaft is helpful in determining rod sag because it is not misaligned. The dial indicator is adjusted to get a reading off the bar and then zeroed at the 12:00 position. The dial shows a negative reading at the 6:00 position equal to twice the actual amount of rod sag when the bar is rotated 180°. Record half of the total reading so it may be subtracted from the actual alignment readings. The actual sag is –.005″ if the readings on the bar indicated a total sag of .010″.

For example, an actual alignment reading indicates a rotational misalignment of .024″. This is divided by two, giving a vertical reading of .012″. The indicator rod sag of .005″ is then subtracted from the vertical reading, giving an actual vertical offset reading of .007″ (.012″ – .005″ = .007″).

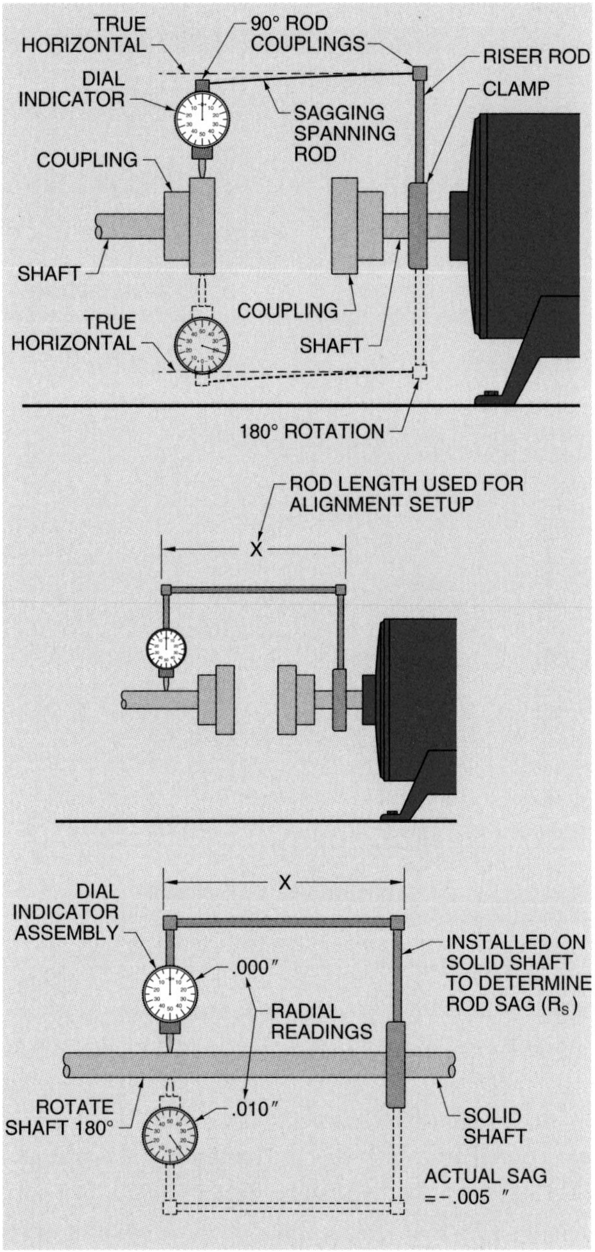

Figure 14-17. Gravity and the weight of the dial indicator produce a change in indicator readings from rod sag toward and away from the point of reading.

ALIGNMENT METHODS

Five methods are available to align machinery, each having its own degree of accuracy. The five methods include straightedge, rim-and-face, reverse dial, electronic reverse dial, and laser rim-and-face methods. See Figure 14-18.

All alignment techniques require that a specific order of adjustment be made. Any attempt to align a machine outside of the specific order is considered trial and error adjustment, which can only lead to frustration. The specific order of shaft alignment is: angular in the vertical plane (up and down angle), parallel in the vertical plane (up and down offset), angular in the horizontal plane (side to side angle), and parallel in the horizontal plane (side to side offset).

Once angular in the vertical plane and parallel in the vertical plane have been corrected, they generally are not lost when angular in the horizontal plane and parallel in the horizontal plane are in the process of being corrected. This step-by-step process is used regardless of the alignment method. Always double check each corrective move.

Accuracy Expectations

The choice of alignment method is based on cost, accuracy required, ease of use, and time required to perform the alignment. The accuracy of any alignment is based on the skill level of the individual doing the alignment and the alignment method used. For example, straight-edge measurements are usually made without the knowledge of coupling irregularities and require the feel of thickness gauge measurements. Therefore, the accuracy of straight-edge alignment generally is no better than 1⁄64″.

Dial indicators and electronic measuring devices (except laser) measure in the thousandths of an inch, which allows for an accuracy of alignment within .001″. Laser alignment methods are generally exact and quick with a possible accuracy of .0002″.

Alignment Tolerance. Alignment tolerance requirements of two or more shafts are based on the speed (rpm) of the motor or drive unit. At times, a manufacturer may indicate the alignment tolerance for its machine. A shaft alignment tolerances chart may be used if manufacturer tolerances are not available. See Figure 14-19. A shaft alignment tolerances chart indicates suggested tolerances by speed in thousandths per inch. Some shaft alignment tolerances charts show the tolerance for angularity in degrees, minutes, and seconds rather than mils or inches. This means that before and after each adjustment, an indicator reading must be taken and converted into angles of degrees, minutes, and seconds.

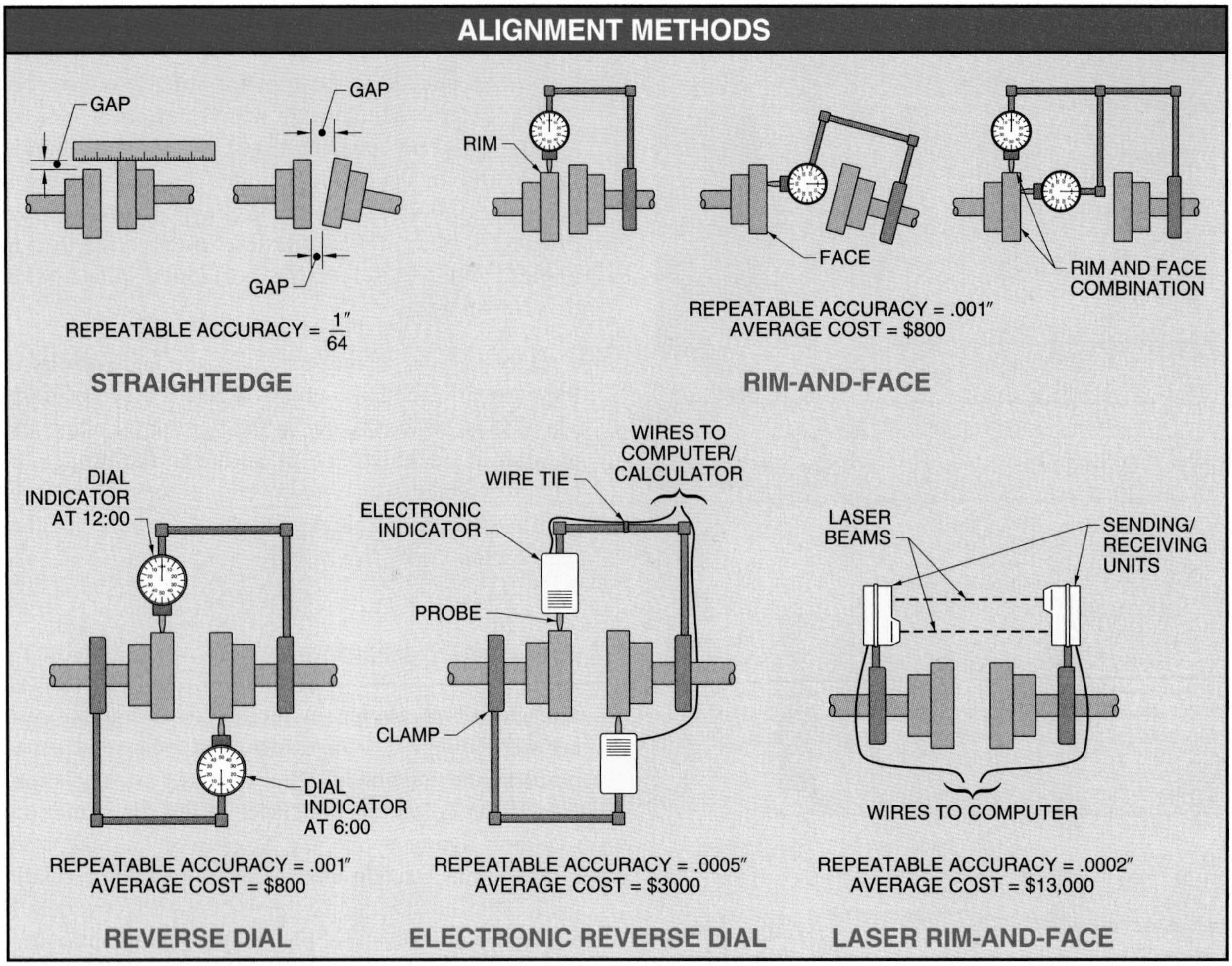

Figure 14-18. The five methods available to align machinery include straightedge, rim-and-face, reverse dial, electronic reverse dial, and laser rim-and-face methods.

SPM Instrument, Inc.

The MAC-10 from SPM Instrument, Inc. is an alignment computer that uses electronic indicators and the electronic reverse dial method to correct machine misalignment.

For example, a technician is aligning a pump/motor combination that has a 5″ coupling and operates at 1300 rpm. The vertical offset is .005″, the horizontal offset is .003″, the vertical angularity is .006″, and the horizontal angularity is .005″. Using the shaft alignment tolerances chart indicates that at 1300 rpm, the acceptable offset tolerance is .0038″. The chart also indicates that the angularity tolerance of a 10″ coupling operating at 1300 rpm is acceptable at .010″.

The angular measurements must be doubled to equate with a 10″ coupling because the equipment coupling is 5″. This gives a recorded reading of .012″ for vertical angularity and .010″ for horizontal angularity. The results of the recorded readings indicate that both horizontal readings are in tolerance and both vertical readings are out of tolerance.

SHAFT ALIGNMENT TOLERANCES*		
Offset (Thousandths/Inch)		
Speed**	**Excellent**	**Acceptable**
0 – 999	.003	.005
1000 – 1999	.002	.0038
2000 – 2999	.0015	.0025
3000 – 3999	.0008	.0015
4000 – 4999	.0005	.0010
5000 – 5999	.0004	.0008
Angularity (Measured as Gap Size at 10″)		
Speed**	**Excellent**	**Acceptable**
0 – 999	.007	.020
1000 – 1999	.003	.010
2000 – 2999	.0025	.005
3000 – 3999	.002	.004
4000 – 4999	.0015	.003
5000 – 5999	.001	.002

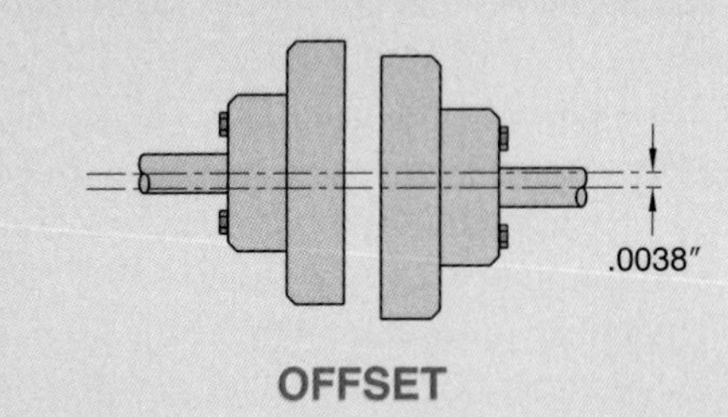

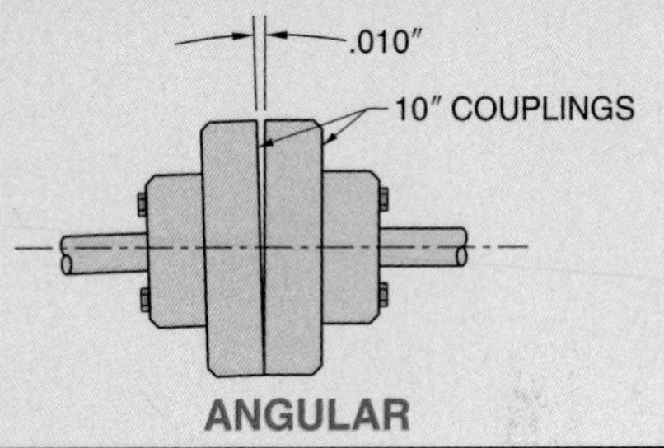

* in in.

** in rpm

Figure 14-19. A shaft alignment tolerances chart indicates suggested tolerances by speed in thousandths per inch.

Horizontal Movements

Angular and offset movements in the horizontal plane are normally made as jack screws are screwed in as dial indicator movement is observed. To move the MTBS away .020″, a dial indicator is placed at the back edge of the machine base plate and directly in-line with the adjusting jack screw. The front jack screw is screwed in until the indicator registers a .020″ movement. See Figure 14-20. When only one indicator is used, a movement at one end of a machine changes the machine position at the other end, complicating accurate movement at both ends. To overcome possible confusion during horizontal movement, two indicators, one at each foot, are used to display exact front and back motor/pump unit movement.

Angular Movements

Angular movements are generally made by adjustments at two machine feet. However, it is necessary to adjust only one set of feet such as the machine front feet or back feet only. For example, if calculations call for the front feet to be raised .050″ and the back feet .025″, the total angularity position is corrected by raising the front feet .025″ (.050″ – .025″ = .025″).

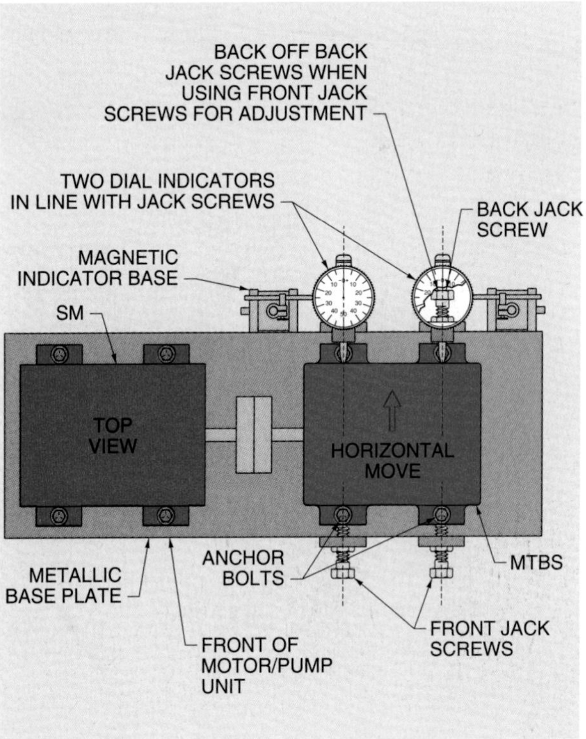

Figure 14-20. Angular and offset movements in the horizontal plane are normally made as jack screws are screwed in as dial indicator movement is observed.

To prevent compounding errors, new readings must be taken after each adjustment to the MTBS. To reduce the chances for error, always rotate dial indicators in one direction. Start back at the zero setting if movement direction has been reversed. Clock position movements (clockwise/counterclockwise) are those viewed from the MTBS toward the SM. Repeat dial indicator movements and recheck readings at least three times.

Angular measurements are easier to interpret when it is realized that each misalignment angle, in its own plane (vertical or horizontal), is the same whether it is measured off of the coupling face or the misalignment at the feet of the machine. Known readings of distance and gap may be used to determine angularity and gap at any other distance. This principle can be used to determine the shim thickness to eliminate any angle. The angle reading (gap) must always start at zero for this to occur. Indicator readings that begin at a number greater than zero must have the initial reading subtracted from the total needle movement to give the proper gap. See Figure 14-21.

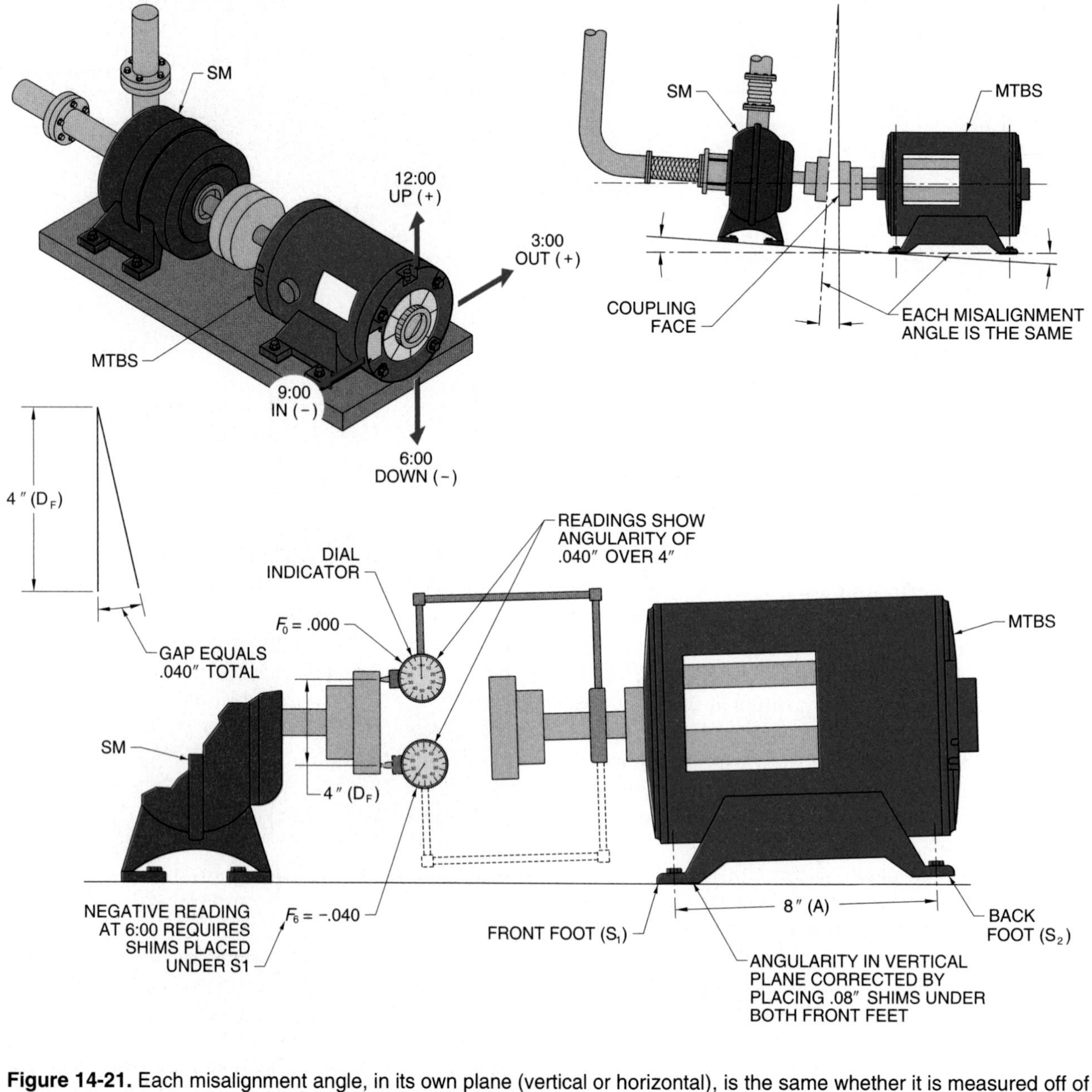

Figure 14-21. Each misalignment angle, in its own plane (vertical or horizontal), is the same whether it is measured off of the coupling face or the misalignment at the feet of the machine.

The gap at the desired distance is found by applying the formula:

$$G = \frac{g}{d} \times D$$

where

G = gap at desired distance (in in.)

g = known gap (in in.)

d = known distance from zero (in in.)

D = distance desired from zero (in in.)

For example, what is the gap at 8″ if the gap at 4″ from zero is .04″?

$$G = \frac{g}{d} \times D$$

$$G = \frac{.04}{4} \times 8$$

$$G = .01 \times 8$$

$$G = \mathbf{.08''}$$

Straightedge Method

The *straightedge alignment method* is a method of coupling alignment in which an item with an edge that is straight and smooth, such as a steel rule, feeler gauge, or taper gauge, is used to align couplings. A *taper gauge* (sometimes referred to as a gap gauge) is a flat, tapered strip of metal with graduations in thousandths of an inch or millimeters marked along its length. As the gauge is placed in a hole or gap, the reading on the gauge at the hole or gap edge is the diameter of the hole or gap at that point. A taper gauge used for this purpose is more reliable and less likely to give false trial-and-error readings like those of feeler gauges. See Figure 14-22.

Straightedge alignment is the oldest method used for measuring misalignment at couplings. This method, although easy to understand and perform, is highly inaccurate, produces no proper resolutions, and does not offer any repeatability. Any proper measuring with tolerances within a few thousandths of an inch must be read in thousandths of an inch and not fractions. Also, proper measurement is of no value if the readings are not consistent or repeatable. Another drawback is that the alignment is being measured on machined surfaces and the OD of many coupling hubs are not true with their bore, nor is the face of a hub perpendicular with the shaft. With the straight-edge method of alignment, coupling hub face and OD runout must be checked and compensated for in making corrective calculations. The straight-edge method is, however, an excellent method for getting shafts roughly close for other alignment methods.

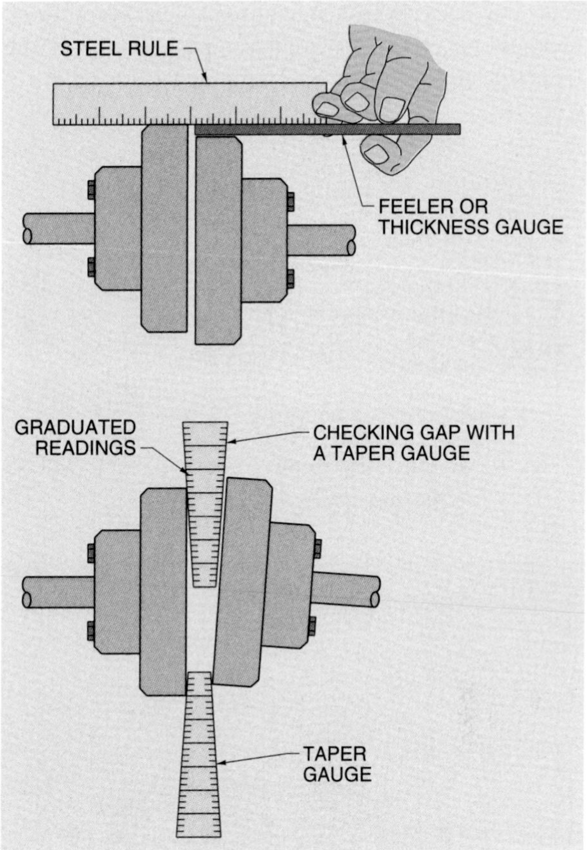

Figure 14-22. The straightedge alignment method uses a steel rule, feeler gauge, or taper gauge to align couplings.

To begin alignment, ensure that the MTBS is not bolt bound and the SM is firmly anchored. Check angular differences in the vertical plane (up and down angle) using a taper gauge. This measurement is made at the 12:00 and 6:00 positions on the shafts or couplings. Vertical or horizontal angularity corrections using shim stock must be performed at one location only, such as at the MTBS front feet or back feet. The opposing feet are used as the angle pivot point. The selection of the adjusted feet and the pivot feet is determined by the 6:00 reading. The front feet (S_1) are shimmed if the 6:00 reading is positive (greater than the 12:00 reading). The back feet are shimmed if the 6:00 reading is negative. The difference between the 12:00 and 6:00 readings is used to determine the thickness of the shim stock placed under the feet to eliminate the angle. See Figure 14-23. Shim stock thickness to eliminate angular misalignment in the vertical plane is based on the vertical angular gap (gap at 12:00 minus gap at 6:00), the diameter of the coupling, and the distance between front and back MTBS mounting holes. Shim stock thickness to eliminate angular misalignment in the vertical plane is found by applying the formula:

$$S = \frac{V_A}{2} \div \frac{D}{2} \times A$$

where

S = shim stock thickness (in in.)

V_A = vertical angular gap (in in.)

2 = constant

D = diameter of coupling (in in.)

A = distance between front and back MTBS mounting holes (in in.)

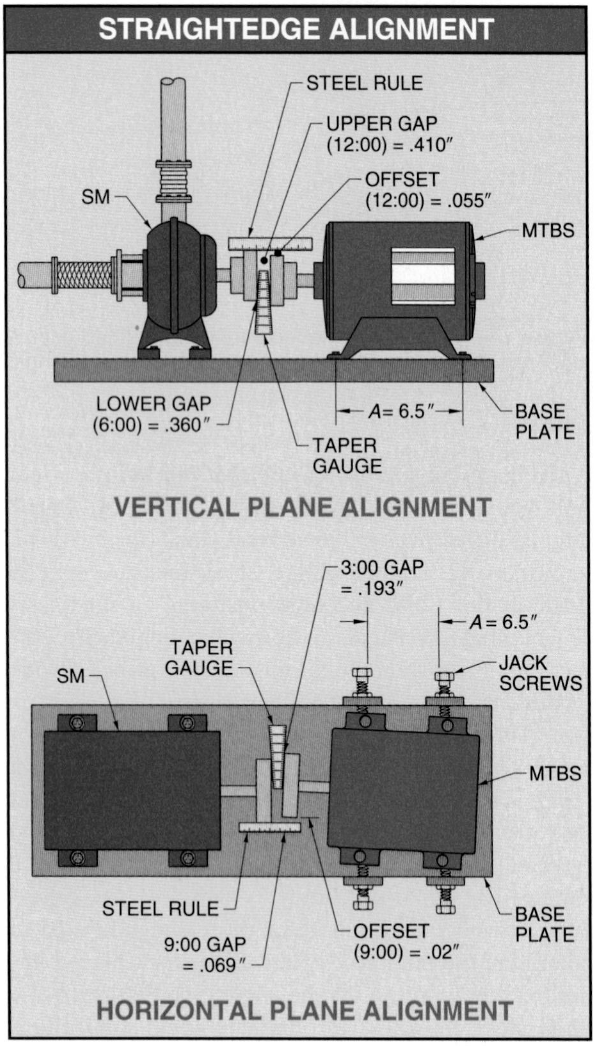

Figure 14-23. Straightedge alignment is generally used for rough alignment prior to using a more precise alignment method.

For example, what is the shim stock thickness required to correct the angular misalignment in the vertical plane of a pump and motor assembly having a 5″ diameter coupling, a vertical angular gap of .050″ (.410″ − .360″ = .050″), and an MTBS mounting hole distance of 6.5″?

$$S = \frac{V_A}{2} \div \frac{D}{2} \times A$$

$$S = \frac{.050}{2} \div \frac{5}{2} \times 6.5$$

$$S = \frac{.025}{2.5} \times 6.5$$

$$S = .01 \times 6.5$$

$$S = \mathbf{.065''}$$

The misalignment in the vertical plane is adjusted by placing shims equal to .065″ under both back feet of the MTBS because the gap between the couplings is wider at 12:00 (.410″) than at 6:00 (.360″).

After the angular misalignment in the vertical plane has been corrected, in this case by placing .065″ shims under both back feet, the MTBS is checked for offset misalignment in the vertical plane (vertical offset). This is accomplished by laying a straightedge across the top of the shafts or couplings. With the straightedge held firmly and parallel against the highest shaft or coupling, feeler gauges are slid into the offset (if there is one) to determine the distance that the MTBS must be raised or lowered. Always check for shafts or couplings of unequal diameters.

For example, a straightedge placed on top of a 5″ coupling shows a .055″ offset between the straightedge and the coupling half connected to the MTBS. The MTBS must be raised .055″. After inserting shims equaling .055″ under all four feet and tightening mounting bolts, recheck the angular position in the vertical plane and the offset position in the vertical plane.

The angular position in the horizontal plane (side to side angle) and the offset position in the horizontal plane (side to side offset) are checked and corrected similar to the vertical plane adjustments. These movements are accomplished with the use of jack screws. The angular position in the horizontal plane is checked with the jack screws fingertight against the feet of the MTBS and the mounting bolts tight. The angular gap is checked with the taper gauge at the 3:00 and the 9:00 positions.

The adjustment to eliminate angular misalignment in the horizontal plane is based on the horizontal angular gap (gap at 3:00 minus gap at 9:00 or gap at 9:00 minus gap at 3:00), the diameter of the coupling, and the distance between the front and back MTBS mounting holes. Horizontal adjustment to eliminate angular misalignment in the horizontal plane is found by applying the formula:

$$S = \frac{H_A}{2} \div \frac{D}{2} \times A$$

where

S = shim stock thickness (in in.)

H_A = horizontal angular gap (in in.)

2 = constant

D = diameter of coupling (in in.)

A = distance between front and back MTBS mounting holes (in in.)

For example, what is the adjustment required to correct the angular misalignment in the horizontal plane of a pump and motor assembly having a 5″ diameter coupling, a horizontal angular gap of .124″ (.193″ – .069″ = .124″), and a MTBS mounting hole distance of 6.5″?

$$S = \frac{H_A}{2} \div \frac{D}{2} \times A$$

$$S = \frac{.124}{2} \div \frac{5}{2} \times 6.5$$

$$S = \frac{.062}{2.5} \times 6.5$$

$$S = .0248 \times 6.5$$

$$S = .161''$$

The horizontal angular adjustment of .161″ is at the back foot (S_2) because the gap is wider at the 3:00 position (.193″) than the 9:00 position (.069″).

Offset misalignment in the horizontal plane is corrected similar to correcting offset misalignment in the vertical plane. A straightedge and feeler gauge is placed at the 9:00 position and the offset is measured.

For example, a straightedge placed at the 9:00 position of a 5″ coupling shows a .020″ offset between the straightedge and the coupling half connected to the MTBS. The gap being between the straightedge and the MTBS coupling half at the 9:00 position requires the motor to be moved to the front a distance of .020″ by the jack screws. Recheck the angular position in the horizontal plane and the offset position in the horizontal plane after making any adjustments.

Straightedge alignment is not necessarily true to the shaft axis because straightedge alignment measures the condition of a coupling and shaft assembly along with machined surfaces of the coupling. Final results may easily be far from tolerance if this method is used as the total alignment method. Upon completion of alignment, release (unscrew) any pressure from jack screws.

When using feeler gauges, start with the smallest thickness feeler and keep adding thicknesses until a slight interference is felt with the feeler between the two parts.

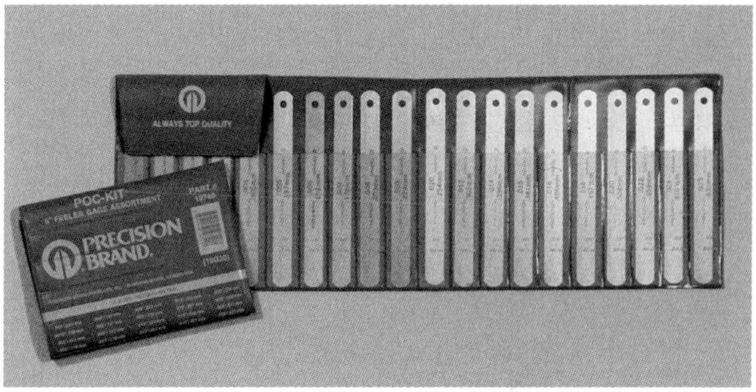

Precision Brand Products, Inc.

Feeler gauges are used to determine the distances that machines must be moved for correct alignment.

Rim-And-Face Method

The *rim-and-face alignment method* is an alignment method in which the offset and angular gap of two shafts is determined using two dial indicators that measure the rim and face of a coupling. A *coupling rim* is the outside diameter surface of a coupling. A *coupling face* is the flat surface of a coupling half, facing the flat surface of the connecting coupling half.

A dial indicator measuring at the rim position measures offset directly under the indicator stem. Also, the difference in offset over the distance between two indicators is the angularity in thousandths per inch. Before shim size adjustments are determined, offset and angular gaps must be calculated and used to determine if a machine must be moved up or down or back or forth.

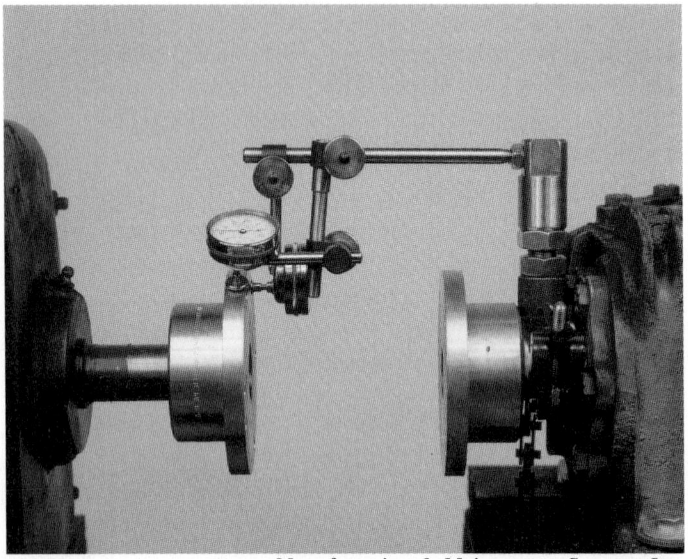

Manufacturing & Maintenance Systems, Inc.

The combination rim-and-face alignment method uses one dial indicator to read the rim to measure offset and the other dial indicator to read the face to measure angularity.

The rim-and-face method is the most widely used, most widely misused, and most troublesome of the precision alignment methods. It is misused by the technician who turns only one shaft to check alignment and troublesome because misuse creates a never ending search in trying to be within tolerance. This method also has additional error sources such as axial float and irregular coupling shapes. *Axial float* is the axial movement of a shaft due to bearing and bearing housing clearances.

Rim-and-face alignment may be accomplished by using the individual rim-and-face method or the combination rim-and-face method. The combination method is considerably faster and more accurate than the individual method.

Individual Rim-and-Face Alignment. The individual rim-and-face alignment method uses an indicator that is attached and used to measure the coupling face (angularity) and then repositioned to measure the coupling rim (offset). Rim and face readings must be taken with the coupling disconnected and both coupling halves rotated together. This is best accomplished when a mark is made on both coupling halves and kept in-line as the couplings are rotated. By rotating both couplings, shaft centerlines are measured, whereas rotating only one shaft measures one shaft in relation to the opposing coupling face or diameter.

Ensure that neither the driven nor drive shafts move axially when taking face readings because this distorts the face readings.

Combination Rim-and-Face Alignment. The combination rim-and-face alignment method uses two dial indicators. One indicator reads the rim offset, and the other reads the face to measure the angularity. Both indicators are assembled using rods and couplings (hardware) and are assembled at the same end of the spanning rod. The two indicators measure the vertical and horizontal planes of the same shaft simultaneously.

Angular gap and offset information must be obtained before shim thickness and location can be determined. Angular gap and offset information used for shim placement is found by checking angular misalignment in the vertical plane (up and down), offset misalignment in the vertical plane (vertical offset), angular misalignment in the horizontal plane (side to side), and offset misalignment in the horizontal plane (side to side offset). The misalignment values are found by measuring with dial indicators. See Figure 14-24. Combination rim-and-face alignment is performed by applying the procedure:

1. Check for angular misalignment in the vertical plane (up and down).

Angular misalignment in the vertical plane is checked by measuring the face of the coupling at the 12:00 and 6:00 positions. The vertical angular gap equals the 6:00 reading minus the 12:00 reading if both total indicator readings are either positive or negative. The vertical angular gap equals the 6:00 reading plus the 12:00 reading if one reading is positive and the other reading is negative.

The shim stock thickness to adjust for angular misalignment in the vertical plane can be found once the vertical angular gap is determined. The shim stock thickness is based on the diameter traveled by the face indicator tip, the vertical angle gap, and the distance between front and back MTBS mounting holes. Shim stock thickness to eliminate angular misalignment in the vertical plane is found by applying the formula:

$$S = \frac{V_A}{2} \div \frac{D_F}{2} \times A$$

where

S = shim stock thickness (in in.)

V_A = vertical angular gap (in in.)

2 = constant

D_F = diameter traveled by the face indicator tip (in in.)

A = distance between front and back MTBS mounting holes (in in.)

2. Check for offset misalignment in vertical plane.

Offset misalignment in the vertical plane is checked by measuring the rim of the coupling at the 12:00 and 6:00 positions.

The shim stock thickness to adjust for offset misalignment in the vertical plane can be found once the vertical offset is determined. The shim stock thickness is based on the difference between the 12:00 and 6:00 rim readings minus the rod sag. Shim stock thickness to eliminate offset misalignment in the vertical plane is found by applying the formula:

$$V_O = \frac{R_0 - R_6 - RS}{2}$$

where

V_O = offset in vertical plane (in in.)

R_0 = reading of rim at 12:00 (in in.)

R_6 = reading of rim at 6:00 (in in.)

RS = rod sag (in in.)

2 = constant

3. Check for angular misalignment in the horizontal plane (side to side).

Angular misalignment in the horizontal plane is checked by measuring the face of the coupling at the 3:00 and 9:00 positions. The horizontal angular gap equals the 9:00 reading minus the 3:00 reading if both indicator readings are either positive or negative. The horizontal angular gap equals the 9:00 reading plus the 3:00 reading if one reading is positive and the other negative.

The horizontal adjustment to eliminate angular misalignment in the horizontal plane is based on the horizontal angular gap (gap at 3:00 minus gap at 9:00 or the gap at 9:00 minus the gap at 3:00), the diameter traveled by the face indicator tip, and the distance between the front and back MTBS mounting holes. Horizontal adjustment to eliminate angular misalignment in the horizontal plane is found by applying the formula:

$$S = \frac{H_A}{2} \div \frac{D_F}{2} \times A$$

where

S = shim stock thickness (in in.)

H_A = horizontal angular gap (in in.)

2 = constant

D_F = diameter traveled by the face indicator tip (in in.)

A = distance between the front and back MTBS mounting holes (in in.)

4. Check for offset misalignment in the horizontal plane (horizontal offset). Offset misalignment in the horizontal plane is checked by measuring the rim of the coupling at the 3:00 and 9:00 positions.

The shim stock thickness to adjust for offset misalignment in the horizontal plane can be found once the horizontal offset is determined. The shim stock thickness is based on the difference between the 3:00 and 9:00 rim readings. Rod sag deviation has no effect in the horizontal plane. Shim stock thickness to eliminate offset misalignment in the horizontal plane is found by applying the formula:

$$H_O = \frac{R_3 - R_9}{2}$$

where

H_O = offset in horizontal plane (in in.)

R_3 = reading of the rim at 3:00 (in in.)

R_9 = reading of the rim at 9:00 (in in.)

2 = constant

Manufacturing & Maintenance Systems, Inc.

Many years of service to equipment seals and bearings may be added by keeping machine shafts within alignment tolerances.

RIM-AND-FACE ALIGNMENT

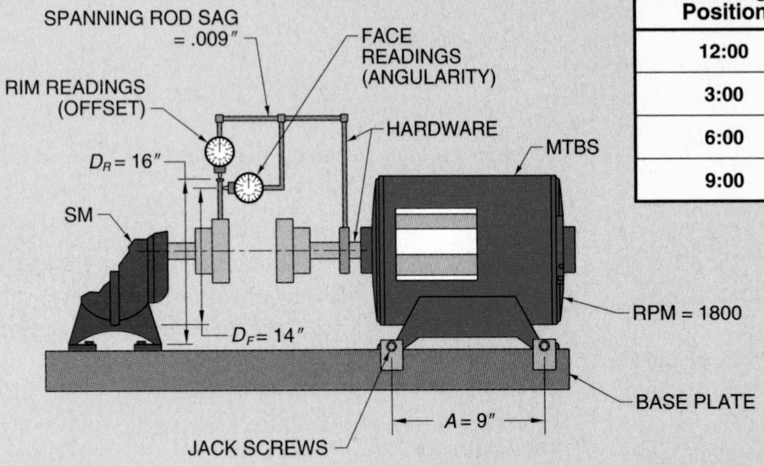

DIAL INDICATOR DATA		
Reading Position	**Face Reading**	**Rim Reading**
12:00	.000	.000
3:00	−.006	−.022
6:00	−.016	−.021
9:00	−.010	+.001

What are the proper shims to be placed under the feet of a pump/motor combination with readings as shown in Dial Indicator Data?

1. Check for angular misalignment in vertical plane.

Measure the face of the coupling at the 12:00 and 6:00 positions (−.016″ + .000″ = −.016″). Shim stock is placed beneath the front feet (S_1) because the 6:00 reading is negative.

$$S = \frac{V_A}{2} \div \frac{D_F}{2} \times A$$

$$S = \frac{-.016}{2} \div \frac{14}{2} \times 9$$

$$S = \frac{-.008}{7} \times 9$$

$$S = -.001 \times 9$$

$$S = \mathbf{-.009''}$$

2. Check for offset misalignment in the vertical plane.

Measure the rim of the coupling at the 12:00 and 6:00 positions (.000″ − .021″ = −.021″).

$$V_O = \frac{R_0 - R_6 + RS}{2}$$

$$V_O = \frac{.000 - .021 - .009}{2}$$

$$V_O = \frac{-.030}{2}$$

$$V_O = \mathbf{-.015''}$$

Shims are placed beneath the feet of the MTBS to raise the MTBS because the 6:00 reading is negative (R_6 = −.021″). To compensate for vertical offset, shims equaling .015″ are placed under all four feet.

3. Check for angular misalignment in the horizontal plane.

Measure the face of the coupling at the 3:00 and 9:00 positions [−.010″ − (−.006″) = −.004″].

$$S = \frac{H_A}{2} \div \frac{D_F}{2} \times A$$

$$S = \frac{.004}{2} \div \frac{14}{2} \times 9$$

$$S = \frac{.002}{7} \times 9$$

$$S = .00028 \times 9$$

$$S = \mathbf{.0025''}$$

The angular gap of .004″ is adjusted at the front feet of the MTBS because the lesser gap reading is at 3:00. Adjust the front jack screws for an indicator movement of .0025″ and anchor the MTBS.

4. Check for offset misalignment in the horizontal plane.

Measure the rim of the coupling at the 3:00 and 9:00 positions (−.022″ + .001″ = −.021″).

$$H_O = \frac{R_3 - R_9}{2}$$

$$H_O = \frac{.022 - .001}{2}$$

$$H_O = \frac{.021}{2}$$

$$H_O = \mathbf{.0105''}$$

Lightly snug the anchor bolts and make the .0105″ adjustment. At this stage, the machine should be rechecked and corrected until a minimum acceptable tolerance for the 1800 rpm motor is met (offset = .002″ to .003″ and angularity = .003″ to .010″).

Figure 14-24. Rim and face readings determine the offset and angle respectively of one shaft relative to another.

Reverse Dial Method

The reverse dial method is an alignment method that uses two dial indicators to take readings off of opposing sides of coupling rims, giving two sets of shaft runout readings. Since it is faster and more accurate than the rim-and-face method, the reverse dial method is not affected by axial float. Each indicator shows both angle and offset. Reverse dial indicator readings can be illustrated by a plotted layout. See Figure 14-25.

Plots are laid out using graph paper. Each square represents a horizontal and vertical measurement. Horizontal measurements are in inches, providing a representative view of the overall machine dimensions. Vertical measurements are total indicator readings plotted in thousandths of an inch. Plotting may be done to view the top and side relative positions.

Plotting a graph of alignment measurements allows easy experimentation with various shim moves, produces an exact hard copy description of the amount of shim movement needed that can be stored in the plant maintenance files for future reference, and is less expensive than calculators and computers.

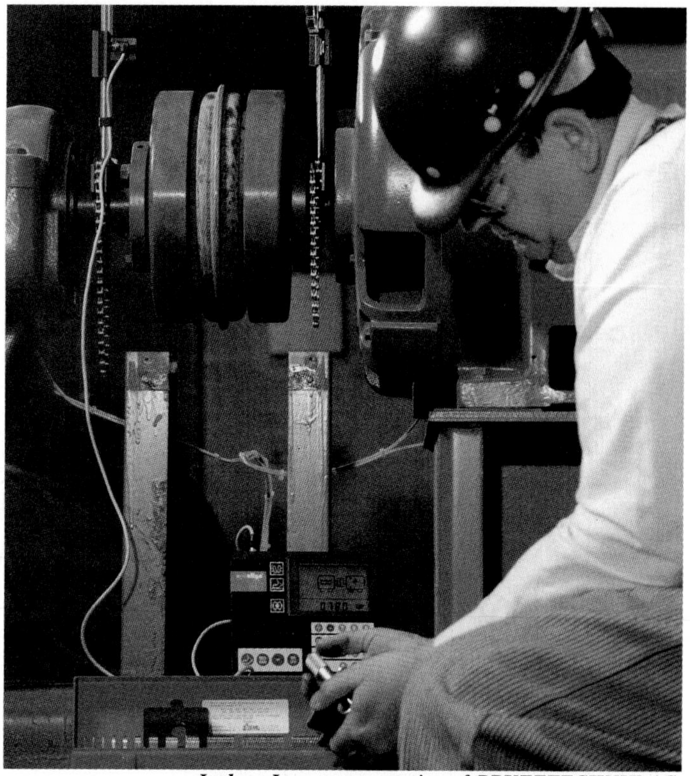

Ludeca Inc., representative of PRUEFTECHNIK AG.

Vertical MTBS adjustments are made by selecting the proper shim thickness to bring both machines into proper alignment.

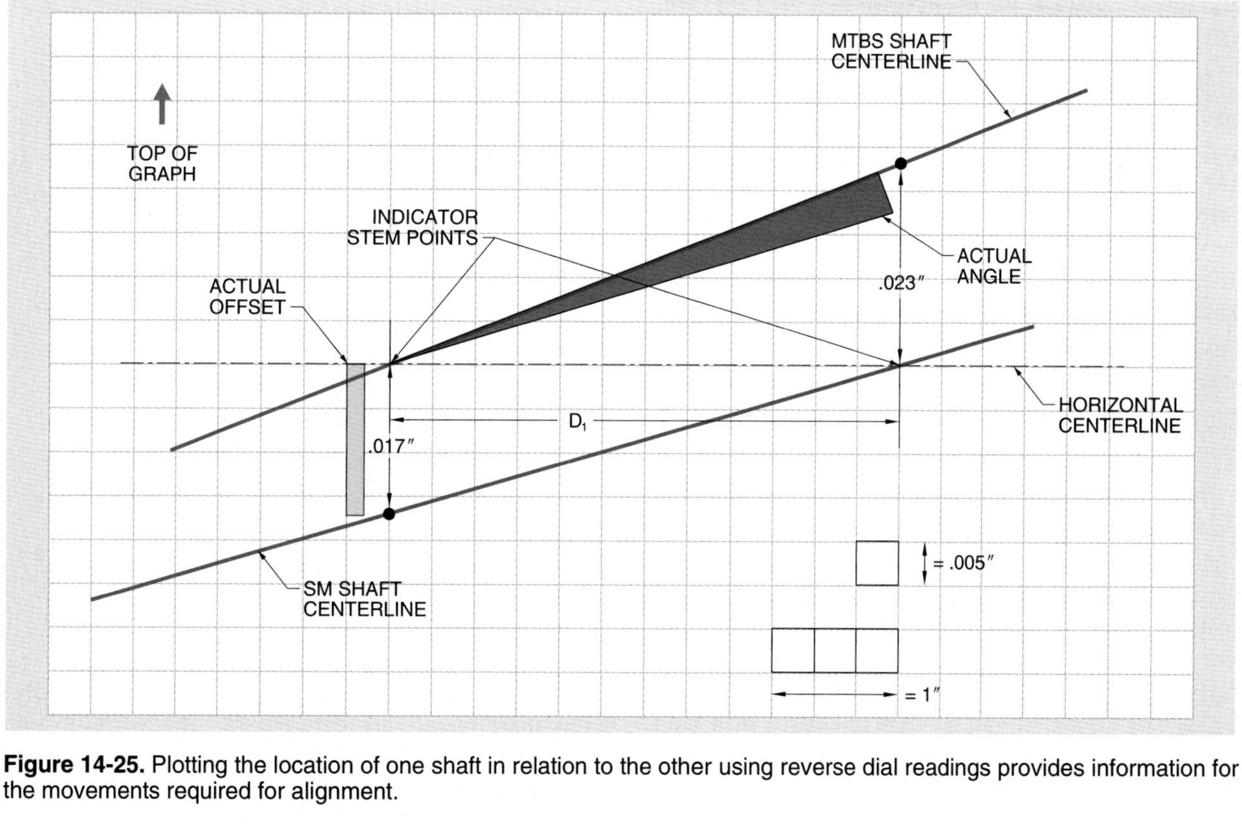

Figure 14-25. Plotting the location of one shaft in relation to the other using reverse dial readings provides information for the movements required for alignment.

Manufacturing & Maintenance Systems, Inc.

In the reverse dial method, two dial indicators are used to take readings off of opposing couplings, giving two sets of shaft runout readings.

The plot shows the relative position of shaft centerlines and the indicator reading dimensions. The horizontal squares are 1″ per three-square division and the vertical squares are .005″ per division.

Plotting begins by drawing a horizontal centerline and placing a mark on the centerline that represents an indicator stem point. Count the appropriate number of squares and place a mark at the second indicator stem point. This distance represents the distance between the indicator stem points on the two shafts. To plot the MTBS shaft centerline, plot half of the MTBS reading toward the top of the graph from one indicator point and make a mark. For example, if the total MTBS reading is .046″, a mark is made .023″ toward the top of the graph from one of the indicator stem points ($4\frac{1}{2}$ squares).

A line is drawn from this mark through the opposing indicator stem point. To plot the SM shaft centerline, plot half of the SM reading from the indicator point toward the bottom of the graph and make a mark. Draw a line from this mark through the opposing indicator point. These lines indicate that reverse dial readings give offset and angular positions. The alignment objective is to end up with both lines parallel. For example, if the total SM reading is .034″, a mark is made .017″ toward the bottom of the graph from the opposite indicator stem point. The difference between these lines should be adjusted for proper alignment.

Plotting provides a graphic illustration as well as an indication of the movements required for alignment. In this case, it is shown that any corrective move must be made by raising the SM or lowering the MTBS.

Exact movements and shim thicknesses are determined by calculating the angular and offset dimensions in the vertical and horizontal planes. See Figure 14-26. Reverse dial alignment is obtained by applying the procedure:

1. Correct vertical offsets.

Angular and offset conditions are corrected using the SM and MTBS offset readings. Corrections are first determined by adjusting the TIR to represent the shaft offset. The vertical shaft offset at the SM is found by applying the formula:

$$ST_V = \frac{S_0 - S_6 - RS_1}{2}$$

where

ST_V = vertical shaft offset at SM (in in.)

S_0 = SM indicator reading at 12:00 (in in.)

S_6 = SM indicator reading at 6:00 (in in.)

RS_1 = SM rod sag (in in.)

2 = constant

The vertical shaft offset at the MTBS is found by subtracting the MTBS rod sag and indicator reading at 6:00 from the MTBS indicator reading at 12:00 and dividing by 2. The vertical shaft offset at the MTBS is found by applying the formula:

$$M_V = \frac{M_0 - M_6 - RS_2}{2}$$

where

M_V = vertical shaft offset at MTBS (in in.)

M_0 = MTBS indicator reading at 12:00 (in in.)

M_6 = MTBS indicator reading at 6:00 (in in.)

RS_2 = MTBS rod sag (in in.)

2 = constant

The vertical shim corrections are calculated after the shaft offsets are determined. The vertical shim correction under both MTBS front feet (S_1) is found by applying the formula:

$$VS_1 = \left\{(ST_V + M_V) \times \left(\frac{D_2}{D_1}\right)\right\} - ST_V$$

where

VS_1 = vertical shim correction under both MTBS front feet (in in.)

ST_V = vertical shaft offset at SM (in in.)

M_V = vertical shaft offset at MTBS (in in.)

D_2 = distance between SM indicator and MTBS front feet (in in.)

D_1 = distance between indicators (in in.)

The vertical shim correction under both MTBS back feet (S_2) is found by applying the formula:

$$VS_2 = \left\{(ST_V + M_V) \times \left(\frac{D_3}{D_1}\right)\right\} - ST_V$$

where

VS_2 = vertical shim correction under both MTBS back feet (in in.)

ST_V = vertical shaft offset at SM (in in.)

M_V = vertical shaft offset at MTBS (in in.)

D_3 = distance between SM indicator and MTBS back feet (in in.)

D_1 = distance between indicators (in in.)

2. Correct horizontal offsets.

Angular and offset conditions are corrected using reverse dial offset readings. The horizontal offset TIR of both machines is adjusted to represent true offsets divided by 2. The horizontal shaft offset at the MTBS is found by applying the formula:

$$ST_H = \frac{S_9 - S_3 - RS_1}{2}$$

where

ST_H = horizontal shaft offset at MTBS (in in.)

S_9 = SM indicator reading at 9:00 (in in.)

S_3 = SM indicator reading at 3:00 (in in.)

RS_1 = SM rod sag (in in.)

2 = constant

The horizontal shaft offset at the SM is found by applying the formula:

$$M_H = \frac{M_9 - M_3 - RS_2}{2}$$

where

M_H = horizontal shaft offset at SM (in in.)

M_9 = MTBS indicator reading at 9:00 (in in.)

M_3 = MTBS indicator reading at 3:00 (in in.)

RS_2 = MTBS rod sag (in in.)

2 = constant

Horizontal shaft offsets are then used to calculate the side movement of the MTBS. The horizontal corrective movement of the MTBS front feet (S_1) is found by applying the formula:

$$HS_1 = \left\{(ST_H + M_H) \times \left(\frac{D_2}{D_1}\right)\right\} - ST_H$$

where

HS_1 = horizontal corrective movement at MTBS front feet (in in.)

ST_H = horizontal shaft offset at MTBS (in in.)

M_H = horizontal shaft offset at SM (in in.)

D_2 = distance between SM indicator and MTBS front feet (in in.)

D_1 = distance between indicators (in in.)

The horizontal corrective movement of the MTBS back feet (S_2) is found by applying the formula:

$$HS_2 = \left\{(ST_H + M_H) \times \left(\frac{D_3}{D_1}\right)\right\} - ST_H$$

where

HS_2 = horizontal corrective movement at MTBS back feet (in in.)

ST_H = horizontal shaft offset at MTBS (in in.)

M_H = horizontal shaft offset at SM (in in.)

D_3 = distance between SM indicator and MTBS back feet (in in.)

D_1 = distance between indicators (in in.)

Sprecher + Schuh

Alignment performed to the correct motor/pump tolerance increases the life of pump seals and bearings.

REVERSE DIAL ALIGNMENT

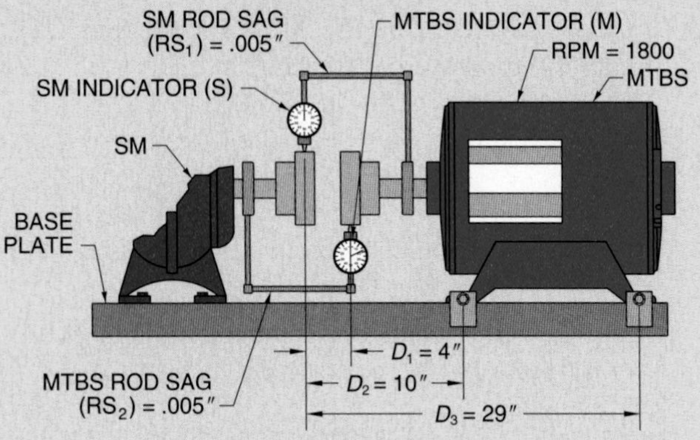

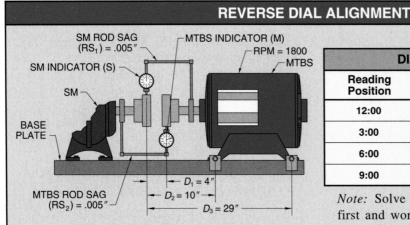

DIAL INDICATOR DATA		
Reading Position	SM Reading	MTBS Reading
12:00	.000	.000
3:00	+ .005	- .006
6:00	+ .035	- .046
9:00	+ .030	- .040

Note: Solve for values inside the parentheses first and work from the inside out.

What adjustments must be made to the MTBS using the reverse dial alignment method for proper alignment?

1. Correct vertical offsets.

$$ST_V = \frac{S_0 - S_6 - RS_1}{2}$$

$$ST_V = \frac{.000 - .035 - .005}{2}$$

$$ST_V = \frac{.040}{2}$$

$$ST_V = .020''$$

$$M_V = \frac{M_0 - M_6 - RS_2}{2}$$

$$M_V = \frac{.000 - .046 - .005}{2}$$

$$M_V = \frac{-.051}{2}$$

$$M_V = -.025''$$

$$VS_1 = \{(ST_V + M_V) \times \left(\frac{D_2}{D_1}\right)\} - ST_V$$

$$VS_1 = \{(.020 - .025) \times \left(\frac{10}{4}\right)\} - .020$$

$$VS_1 = (-.005 \times 2.5) - .020$$

$$VS_1 = -.0125 - .020$$

$$VS_1 = \mathbf{-.032''}$$

$$VS_2 = \{(ST_V + M_V) \times \left(\frac{D_3}{D_1}\right)\} - ST_V$$

$$VS_2 = \{(.020 - .0255) \times \left(\frac{29}{4}\right)\} - .020$$

$$VS_2 = (-.005 \times 7.25) - .020$$

$$VS_2 = -.036 - .020$$

$$VS_2 = \mathbf{-.056''}$$

VS_1 and VS_2 results indicate that the MTBS front feet (S_1) are lowered .032″ and the back feet (S_2) are lowered .056″.

Vertical correction of the MTBS is not possible without raising the SM. Raise the SM by placing .100″ shims under all feet and start over.

2. Correct horizontal offsets.

$$ST_H = \frac{S_9 - S_3 - RS_1}{2}$$

$$ST_H = \frac{.030 - .005 - .005}{2}$$

$$ST_H = \frac{.020}{2}$$

$$ST_H = .010''$$

$$M_H = \frac{M_9 - M_3 - RS_2}{2}$$

$$M_H = \frac{-.040 - .006 - .005}{2}$$

$$M_H = \frac{-.051}{2}$$

$$M_H = -.0255''$$

$$HS_1 = \{(ST_H + M_H) \times \left(\frac{D_2}{D_1}\right)\} - ST_H$$

$$HS_1 = \{(.010 - .0255) \times \left(\frac{10}{4}\right)\} - .010$$

$$HS_1 = (-.0155 \times 2.5) - .010$$

$$HS_1 = -.0387 - .010$$

$$HS_1 = \mathbf{-.049''}$$

$$HS_2 = \{(ST_H + M_H) \times \left(\frac{D_3}{D_1}\right)\} - ST_H$$

$$HS_2 = \{(.010 - .0255) \times \left(\frac{29}{4}\right)\} - .020$$

$$HS_2 = (-.0155 \times 7.25) - .020$$

$$HS_2 = -.112 - .020$$

$$HS_2 = \mathbf{-.092''}$$

Using two zeroed indicators at the back of the MTBS (one on the same plane as S_1 and one on the same plane as S_2, adjust the front feet (S_1) forward (toward the technician) by .049″ and the back feet (S_2) forward (toward the technician) by .092″. The large numbers indicate a very coarse adjustment.

Figure 14-26. Reverse dial alignment procedures offer net misalignment values at one setting from which shim placement can be determined.

Electronic Reverse Dial Method

The *electronic reverse dial method* is an alignment method that uses the reverse dial as a base method with the dial indicators replaced with electromechanical sensing devices. The electronic reverse dial method is supported by computer-aided electronic instrumentation. The sensing devices detect physical movement and convert the movement to an electrical signal, which is sent to a calculator. Before the calculator can process an adjustment response, it must have the electronic movement signal and certain physical dimensions, which must be entered by the technician. The calculator computes MTBS movements as its response. See Figure 14-27.

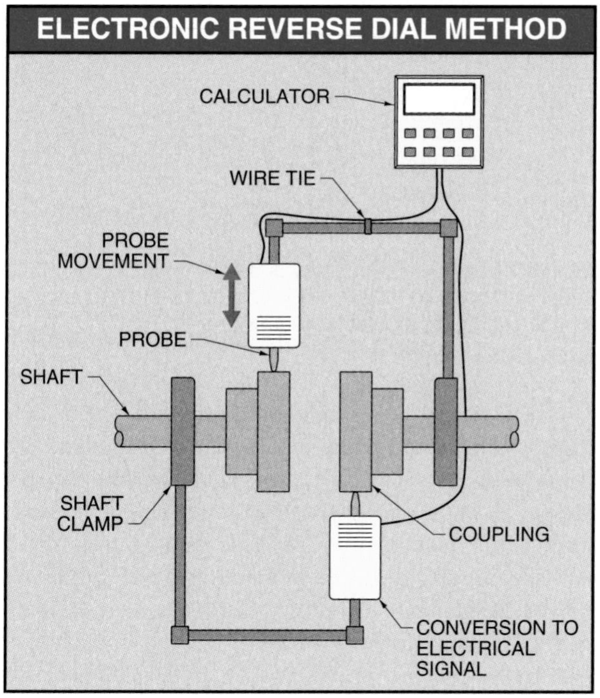

Figure 14-27. In the electronic reverse dial method, electromechanical aligning devices convert a mechanical movement into an electrical signal.

Inspect equipment condition before beginning any alignment procedure. Check motor and pump bases for breaks or cracks. All contact surfaces must be clean, smooth, and flat. Foundation and base plates must be in good condition. Check shaft, coupling, and bearing condition for any irregularities. Finally, organize tools, materials, and workplace. Anchor bolt condition, eccentric shafts, and soft foot must be checked and corrected before the alignment procedure begins.

The electronic reverse dial method requires that two sensing devices are used, each being assembled using rods and couplings and attached to a coupling or shaft. Installation is similar to using dial indicators where one sensor is opposite the other. Sensor wires must be secured with slack to prevent unwanted tugging forces.

Some manufacturers recommend that the coupling be disconnected while others suggest that they be connected. Always follow manufacturer's recommendations. However, any force, including coupling force, can create enough resistance to produce adverse responses.

The shafts are rotated together and readings are entered with the press of a button at the 12:00, 3:00, 6:00, and 9:00 positions. Some manufacturers recommend entering readings at 12:00, then forward (clockwise) to 3:00, back (counterclockwise) to 9:00, forward (clockwise) to 6:00, and back (counterclockwise) to 12:00. Check manufacturer's requirements. Also, add any necessary thermal expansion compensation information to the calculation.

In preparation for corrective horizontal movements, place two dial indicators against the farthest front and back feet of the MTBS and zero both indicators as a starting reference. Vertical and horizontal movements are made according to the calculator's indication.

Finally, repeat all measurements until shaft runout is within tolerance. Upon startup, vibration readings can be taken to compare previous readings with present readings to determine alignment condition and establish data in order to determine future progressing conditions.

Laser Rim-and-Face Method

The *laser rim-and-face alignment method* is an alignment method in which laser devices are placed opposite each other to measure alignment. The laser rim-and-face method is used when extreme accuracy and fast alignment is required. Even though the initial cost of laser equipment is higher than that of other methods, known and hidden payback costs are generally worth the extra expense. Known paybacks due to extreme accuracy are those items that are measurable, such as less emergency downtime, less need for extra or spare replacement inventories, and less utility costs. Energy savings on an accurately aligned machine are 7% to 12% per alignment over a marginally acceptable aligned machine. Hidden savings are those that are not measurable, such as the availability and utilization of power elsewhere or the increased morale within a smoother running facility.

Laser alignment devices operate using the rim-and-face method, with the dial indicator being replaced with a laser beam. The beam is directed to a 90° prism reflector, which is directed back to the sending unit where a receiving transducer (photo position detector) accepts the signal and converts it into an impulse for the calculator. See Figure 14-28. A benefit of using a laser beam is that there is no rod sag to calculate and the alignment is not affected by distance or axial float. The beam sensor is able to detect up and down movement. Offsets and angles are read and measured accurately when the return beam is detected at one position (12:00) and then detected at another position (6:00). The calculator determines offset as a direct send/receive reading and determines the angularity by measuring right triangles.

Ludeca Inc., representative of PRUEFTECHNIK AG.

Turbalign® is a laser alignment device that uses a visible laser combined with a 4-axis position detector to measure machine misalignment.

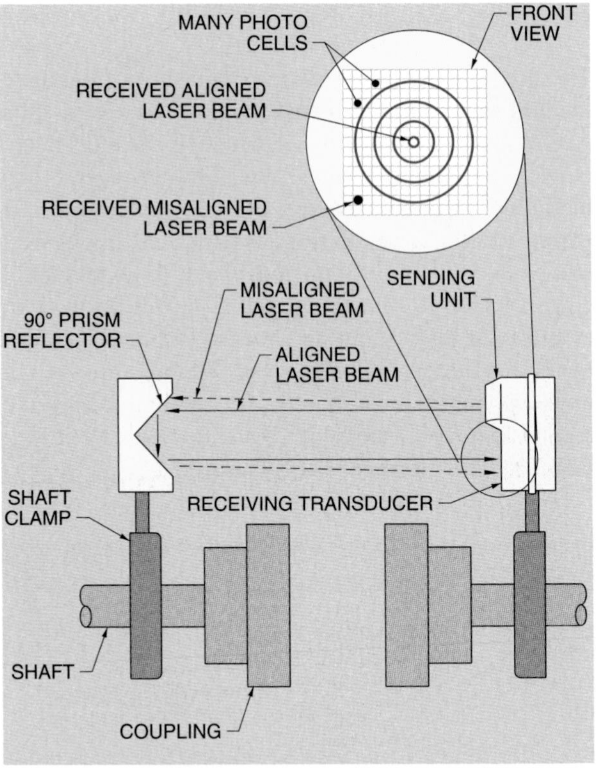

Figure 14-28. Laser accuracy is based upon being able to send a pinpoint light beam, reflect it, and send it to another location without deflection.

Once set up, laser alignment devices check soft foot, compensate for thermal expansion, indicate the movement of a machine during alignment, rapidly couple machines with multiple couplings, and determine shim placement. The ease and simplicity of making alignment moves is noted when all adjusting moves are being observed on a screen as they happen. A graphic display presents the condition of each foot when soft foot or angular soft foot is checked. Finally, all rechecking is completed and corrected in a matter of seconds.

An advantage of a laser alignment device is that measurements are not required to be read, recorded, and calculated for proper movements to be made. Also, the corrective values for the machine feet appear automatically in the computer display.

A laser alignment device requires care in handling to maintain its high level of calibration. Dropping or bumping may result in loss of calibration and alignment integrity. Care must also be used with lasers because steam, dust, and sunlight can adversely affect the laser beam.

Electricity

15
Chapter

Fluke Corporation

Electricity is the main energy source used to provide power for equipment. The National Electrical Code® safeguards persons and property form hazards arising from the use of electricity. Components normally encountered by a mechanical technician include transformers, contactors, motor starters, fuses, circuit breakers, GFCIs, switches, and solenoids. Electrical safety rules should be practiced by all personnel working with electricity.

ELECTRICITY

Electricity is a physical occurrence involving electric charges and their effects when in motion and at rest. Electricity may be static (stored) or dynamic (flowing). *Static electricity* is the accumulation of charge. *Dynamic electricity* (electric current) is electron flow from one atom to another atom.

An *atom* is the smallest building block of matter that cannot be divided into smaller units without changing its basic character. The three fundamental particles contained in atoms are protons, neutrons, and electrons. Protons and neutrons make up the nucleus, and electrons whirl about the nucleus in orbits or shells. The *nucleus* is the heavy, dense center of an atom. The nucleus has a positive electrical charge. A *proton* is a particle with a positive electrical charge of one unit. A *neutron* is a particle with no electrical charge. The nucleus is surrounded by one or more electrons. An *electron* is a negatively charged particle whirling around the nucleus at great speeds in a shell. See Figure 15-1.

Atoms are combined to form molecules. A *molecule* is the smallest particle of a substance that retains all the properties of the substance. Atoms that make up various molecules each have a different number of electrons and a different number of electron shells. A *shell* is an orbiting layer of electrons in an atom. The electrons travel at such a high rate of speed in their orbits that they form a shell around the nucleus. Each shell has a specific number of electrons, which becomes greater with each consecutive shell from the nucleus with the exception of the last (valence) shell. A *valence shell* is the outermost shell of an atom. A *valence electron* is an electron located in the outermost shell of an atom. For example, a hydrogen atom has only 1 electron, which makes it the valence electron. Silver atoms have 47 electrons with 1 valence electron, and lead atoms have 82 electrons with 4 valence electrons. See Appendix.

Each shell can hold a specific number of electrons. The innermost shell can hold two electrons. The second shell can hold eight electrons. The third shell can hold 18 electrons, etc. The shells are filled starting with the inner shell and working outward, so when the inner shells are filled with as many electrons as they can hold, the next shell is started. Electrons and protons have equal amounts of opposite charges. There are as many electrons as there are protons in an atom, which leaves the atom electrically neutral.

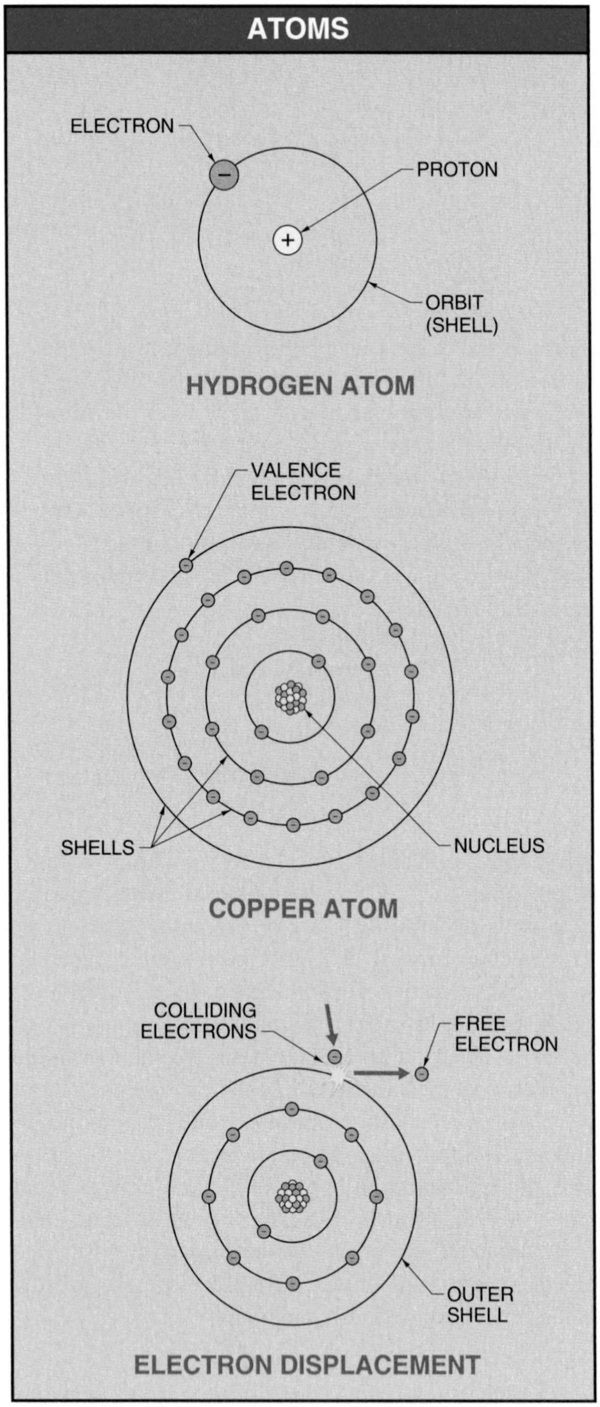

Figure 15-1. An atom is the smallest particle of an element that cannot be divided into smaller units without changing its basic character.

The total number of electrons is not as important to electricity as is the number of valence electrons. *Electron flow* (electrical current) is the traveling of a displaced (free) valence electron from one atom to another. The amount or number of electrons in the valence shell and the ease at which electrons are pulled out of their orbit is the difference between a conductor and an insulator. A *conductor* is a material that has very little resistance and permits electrons to move through it easily. Valence shells carrying one or two electrons make up materials that are conductors, like copper, which contains only one valence electron. One or two valence electrons are unstable and held loosely, allowing the atom to give them up with little effort. The best conductor materials are silver, copper, and aluminum.

An *insulator* is a material that has a very high resistance and resists the flow of electrons. Atoms containing a large number of valence electrons make good insulating material. Common insulators include rubber, plastic, air, glass, and paper. Atoms with a large number of valence electrons have a strong force of attraction with the nucleus. This makes it difficult to force an electron out of orbit or allow another electron into the orbit.

Electron Flow

Current (electron flow) is the amount of electrons flowing through an electrical circuit. Current is measured in amperes. Electron flow may be compared to the amount and force of liquid flow in a pipe. An *amp* (short for amperage) is a measure of a quantity of electron movement per second, commonly referred to as electric current. This, however, is only a measurement of speed. The quantity moved at this speed is the coulomb. A *coulomb* is a quantity of electrons equal to 6.25×10^{18} (6.25 quintillion). The coulomb and amp can be compared to a hydraulic system's liquid flow in gallons per minute. An amp (A) is a flow of 6,250,000,000,000,000,000 (one coulomb) electrons per second.

Voltage (V) is the amount of electrical pressure in a circuit. Voltage is measured in volts. A volt is a measure of electromotive force (electrical pressure). Voltage is the force that pushes the electrons through a conductor. Voltage pushes current through a wire, but voltage cannot flow through a wire. This is similar to the pressure of a hydraulic system. There may be pressure potential, but there is no flow. There is voltage (potential) at a wall outlet, but no current flow until an electrical device is connected and turned ON.

Resistance is the opposition to electron flow. Resistance is measured in ohms (Ω). An *ohm* is the resistance of a conductor in which an electrical pressure of 1 V causes an electrical current of 1 A to flow.

Just as water pressure in a long hose is reduced by friction (resistance), voltage is reduced by the zigzag activity, detouring, and collisions of countless electrons in motion within a long wire. Resistance to the flow of electrons varies according to the physical dimensions and the material through which electric current passes. A large wire allows a greater number of free electrons to flow with less resistance than a smaller wire. See Figure 15-2.

It takes 1 V (pressure) to push 1 A (current) through 1 Ω (resistance). Even though the theoretical speed of an electron is greater than the speed of light (186,000 miles per second), the actual forward motion of an electron is about 2″ to 3″ per minute. Increasing the electrical pressure enables a higher rate of electrical flow (current) through a larger conductor.

Hard steel is difficult to magnetize and demagnetize, making it a good permanent magnet. Soft iron is ideal for temporary magnets used in control devices because it does not retain residual magnetism very easily.

OHM'S LAW

Ohm's law is the relationship between voltage (E), current (I), and resistance (R) in a circuit. Ohm's law states that current in a circuit is proportional to the voltage and inversely proportional to the resistance. If the resistance in a circuit remains constant, a change in current is directly proportional to a change in voltage. If voltage in a circuit remains constant, current in a circuit decreases with an increase in resistance, and current in the circuit increases with a decrease in resistance. Using Ohm's law, any value in this relationship can be found when the other two are known. The relationship between voltage, current, and resistance may be visualized by presenting Ohm's law in a pie chart form. See Figure 15-3.

Using Ohm's Law

Ohm's law can be used for determining voltage, current, or resistance requirements during circuit design and for predicting circuit characteristics before power is applied. For example, in a heating element (resistive load) circuit, a fixed load resistance of 4 Ω is

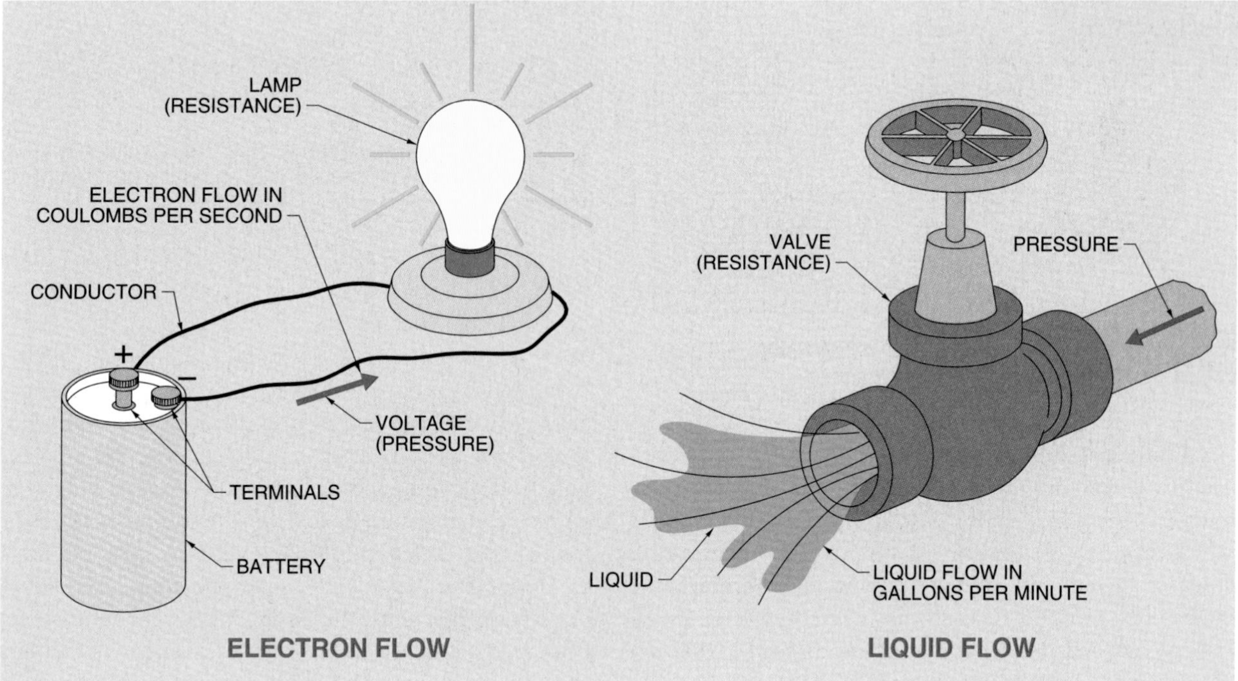

Figure 15-2. Current (electron flow), voltage (pressure), and electrical resistance are similar to the flow, pressure, and resistance of liquid in a pipe.

connected to a variable power supply which supplies 0 V to 24 V. See Figure 15-4. The current in the circuit may be found for any voltage by applying Ohm's law.

For example, what is the current in a circuit if the voltage is set at 8 V and the resistance is 4 Ω?

$$I = \frac{E}{R}$$

$$I = \frac{8}{4}$$

$$I = \mathbf{2\ A}$$

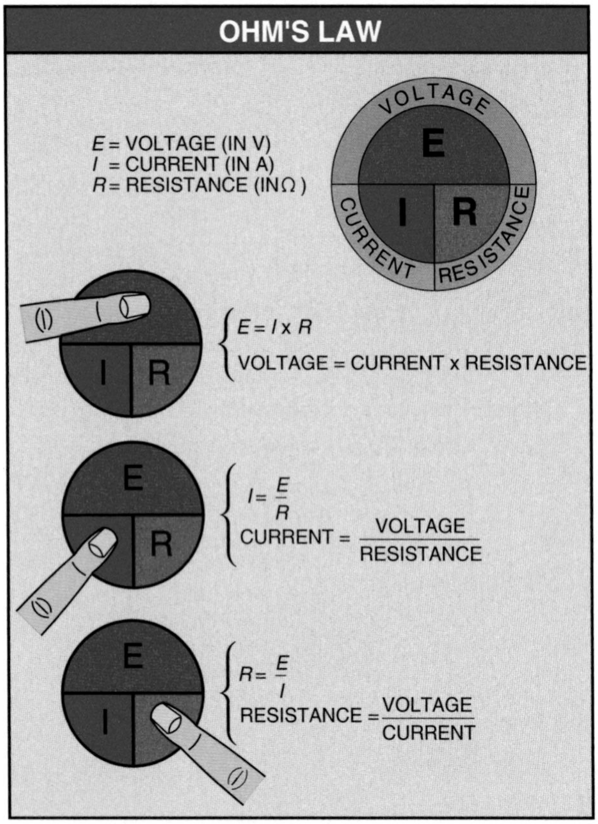

Figure 15-3. Ohm's law is the relationship between voltage, current, and resistance in a circuit.

In troubleshooting applications, Ohm's law can be used to determine how a circuit should operate and how it is operating under power. For example, when the insulation of equipment or conductors breaks down, a current measurement higher than normal indicates that circuit resistance has decreased or circuit voltage has increased. This information is used when identifying potential problems such as insulation breakdown or a high-voltage condition. In the same circuit, a current measurement lower than normal indicates that the circuit resistance has increased or the circuit voltage has decreased. An increase in circuit resistance is usually caused by poor connections, loose connections, corrosion, and/or damaged components.

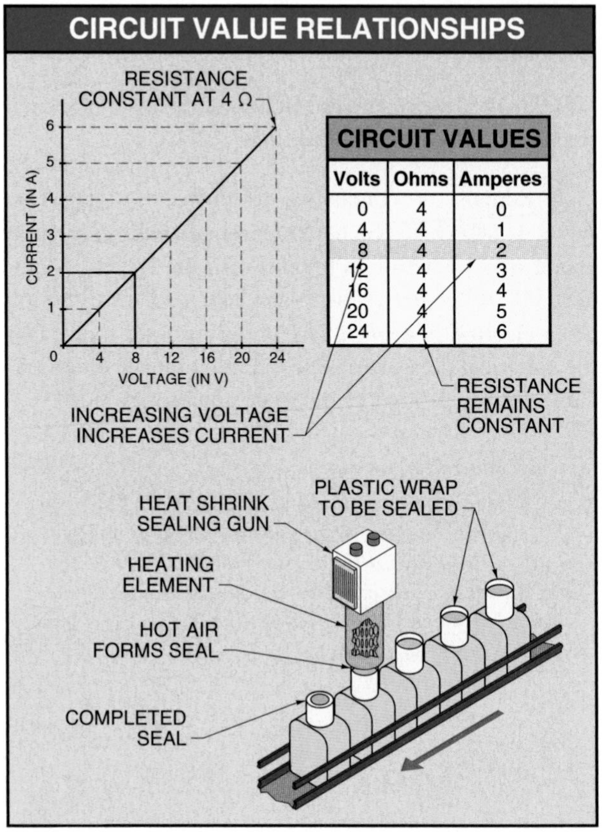

Figure 15-4. If resistance in a circuit remains constant, a change in current is directly proportional to a change in voltage.

MAGNETISM

A *magnet* is a device that attracts iron and steel because of the molecular alignment of its material. *Magnetism* is a force that interacts with other magnets and ferromagnetic materials. A *ferromagnetic material* is a material, such as soft iron, that is easily magnetized. *Magnetic flux lines* are the invisible lines of force that make up a magnetic field. The more dense the flux lines, the stronger the magnetic force. Flux is most dense at the ends of a magnet. For this reason, the magnetic force is strongest at the ends of a magnet. Many flux lines surround an electron or magnetic bar. See Figure 15-5.

The flux lines leave the north pole and enter the south pole of a magnet or magnetic field. *Polarity* is the positive (+) or negative (–) state of an object. The fundamental law of magnetism is that unlike poles attract each other and like poles repel each other. For example, when two magnetic forces are aligned, the north or south poles face each other and repel each other. The force of repulsion increases as the two poles are moved closer together.

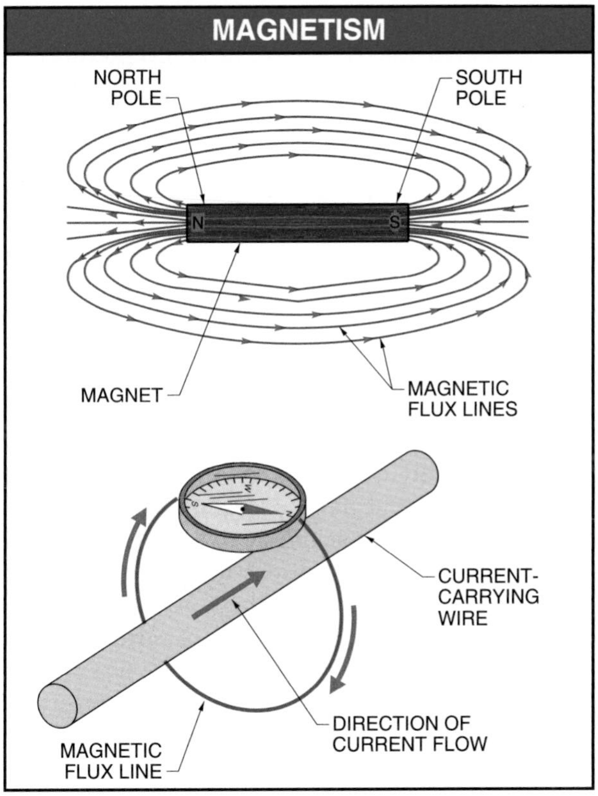

Figure 15-5. Magnetic flux lines are the invisible lines of force that make up a magnetic field.

An iron bar may become a magnet when its electrons are aligned and uniform in force. The electrons in iron combine so they share valence electrons in such a way that they spin in the same direction, causing their magnetic fields to combine. Ordinary iron can be converted into a magnet by induction. *Induction* is the process of causing electrons to align or uniformly join to create a magnetic or electrical force. Placing an iron bar in a strong magnetic field causes induction. Electrons join forces and the lines of force in the field combine to pass through the iron bar, causing the bar to become a magnet.

Magnetism is also a product of electric current due to each electron within the current having its own line of magnetic force. Therefore, as current flows through a straight wire, each electron, grouped with other electrons, sets up a circular field of magnetism that surrounds the wire. The flux lines circling the wire are magnetic, which sets up magnetic fields the length of the wire, creating an electromagnet. An *electromagnet* is a magnet created when electricity passes through a wire.

A coiled wire with an electric current passing through it has a magnetic field set up around the entire coil. Placing an iron core within the coil concentrates the flux lines and transfers the magnetism of the wire to the iron core by induction. The iron core added to a coil of wire increases the strength of the magnetic field. The iron core remains magnetized as long as current continues to flow through the coiled wire and ceases when the current is turned OFF. See Figure 15-6.

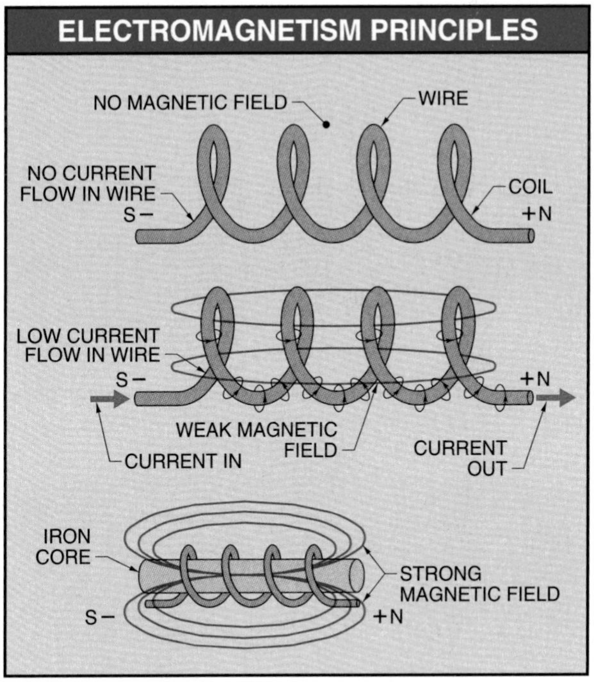

Figure 15-6. The magnetism created as electric current is introduced into a coil of wire may be strengthened by placing an iron core in the coil.

Voltage measurements are normally taken to establish that there is voltage at a given point in a circuit and to determine if the voltage is at the proper level.

Electromagnetic Induction

Electromagnetic induction is the process by which voltage is induced in a wire by a magnetic field when lines of force cut across the wire. See Figure 15-7. Voltage is induced in a wire by electromagnetic induction when the wire is moved across a magnetic field. Increasing the speed of the wire movement through the magnetic field increases the voltage.

A *generator* is a device that converts mechanical energy into electrical energy. A generator consists of a loop of wire (armature), which rotates between north and south magnetic poles. An *armature* is the movable part of a generator or motor. A flow of current is induced as the loop of wire cuts through the magnetic field between the poles. The direction of current flow is determined by the magnetic north pole and the direction of the wire movement through the magnetic field.

Ruud Lighting, Inc.

Approximately 40% of the electrical energy generated in the United States is used to provide power for equipment in industry.

Alternating Current. *Alternating current* (AC) is a flow of electrons that reverses its direction of flow at regular intervals. AC is generated as a loop of wire (armature) enters and leaves a magnetic field. See Figure 15-8. In position 1, the rotor is just about to rotate in the clockwise direction. There is no current flow at this point because the rotor is not cutting any magnetic flux lines. As the rotor rotates from position 1 to position 2, the rotor begins to cut across the magnetic flux lines. The voltage in segments AB and CD increases as the rotor rotates. The maximum number of magnetic flux lines are cut when the rotor is in position 2. The induced voltage is greatest in this position.

From position 2 to position 3, the voltage decreases to zero because the rotor is cutting less and less magnetic flux lines, and finally none at position 3. As the rotor continues to rotate from position 3 to position 4, the voltage increases, but in the opposite direction. The voltage reaches a maximum negative value at position 4, then returns to zero at position 1. Alternations of current (reversing polarity) continue as long as the loop rotates. Slip rings are used to connect the external load to the armature without interfering with its rotation.

Direct Current. *Direct current* (DC) is a flow of electrons in only one direction. DC is generated using the principle of electromagnetic induction. A DC generator produces direct current using a commutator with brushes instead of slip rings. A *commutator* is a device for reversing the direction of current flowing through the wires of a rotating armature that ensures a single-direction flow of current from the generator. The commutator is split into two parts with an insulated gap between each part.

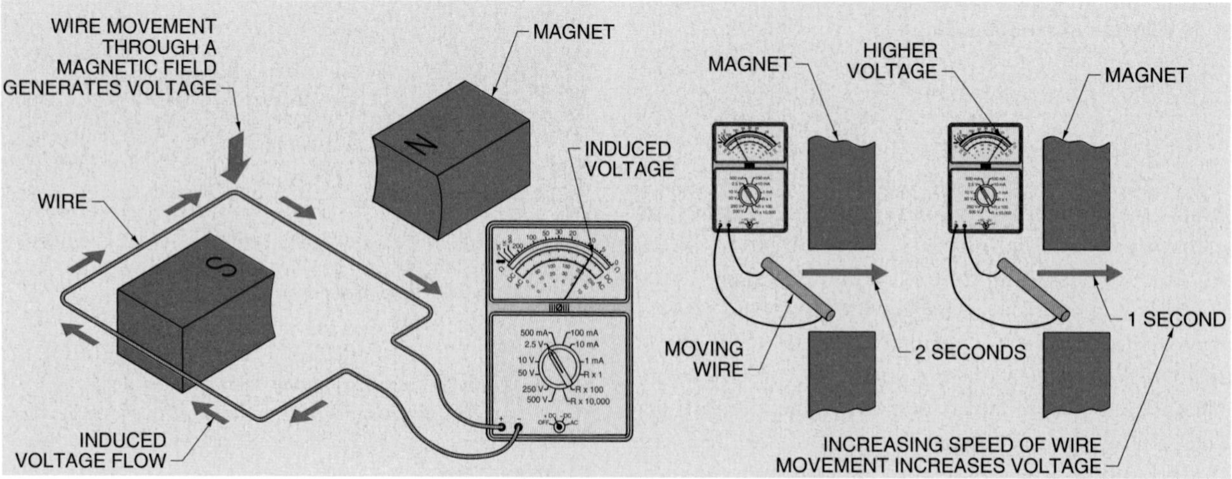

Figure 15-7. A voltage is induced in a wire by electromagnetic induction when the wire is moved across a magnetic field.

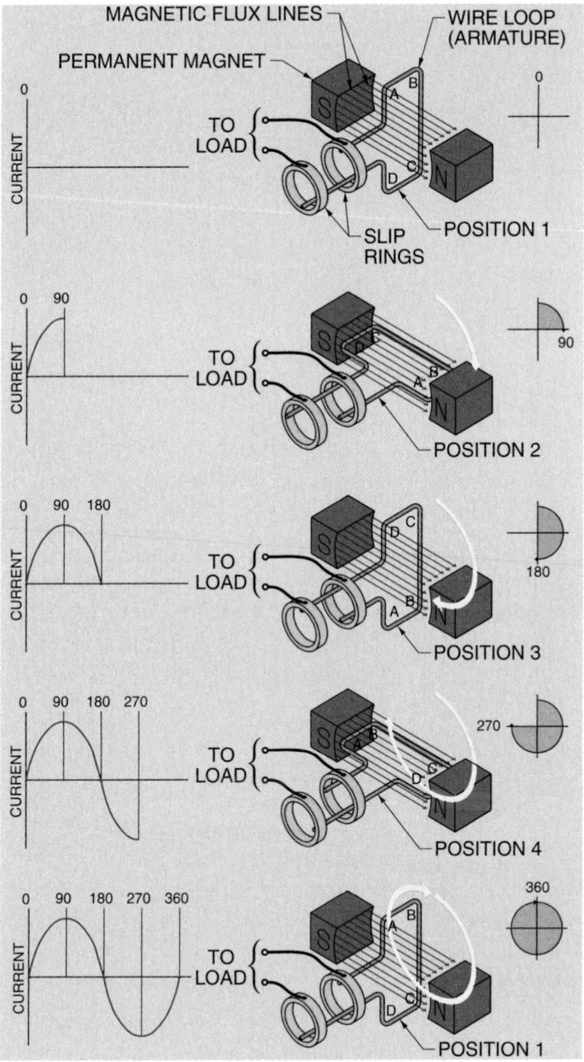

Figure 15-8. Voltage changes direction in a wire as it alternately passes through north and south magnetic poles.

Positive and negative brushes riding against the commutator segments carry the power to the load circuit. The open or insulated areas of the split commutator match the brushes and the neutral area of the armature. As the commutator turns and the brushes pass the neutral area, the other half of the commutator contacts the brushes, reversing the current flow. See Figure 15-9.

The vast majority of electricity is produced by converting potential (stored) energy into a force that can turn a generator. Forms of energy used to produce electricity include fossil fuels, nuclear power, and hydroelectric (water) power.

The direction of current is reversed at the same time as the rotating armature reverses its polarity. The commutator maintains the correct polarity to each brush. DC generation is made possible because the wire cutting through the north pole magnetic field is always connected to the negative brush, and the wire cutting through the south pole magnetic field is always connected to the positive brush.

In an actual device, the armature consists of many wires wound around a core of laminated iron sections. Similarly, the commutator must have sections to match and connect each wire of the armature. Each section is separated by an insulating material. The DC voltage is generated by the relative motion of the armature wires through the magnetic field. The induced voltage is proportional to the number of flux lines cut per second. The output voltage is determined by the speed of the armature if the number of turns of wire used to make the magnetic field remains constant. Increasing the speed of the armature and its cutting action increases the output voltage. Decreasing the rotational speed decreases the output voltage. DC voltages of 6 V, 12 V, 24 V, 36 V, 125 V, 250 V, and 600 V are typically used to drive loads.

Transformers enable AC voltage to be stepped up for transmission over power lines and then stepped down for final use.

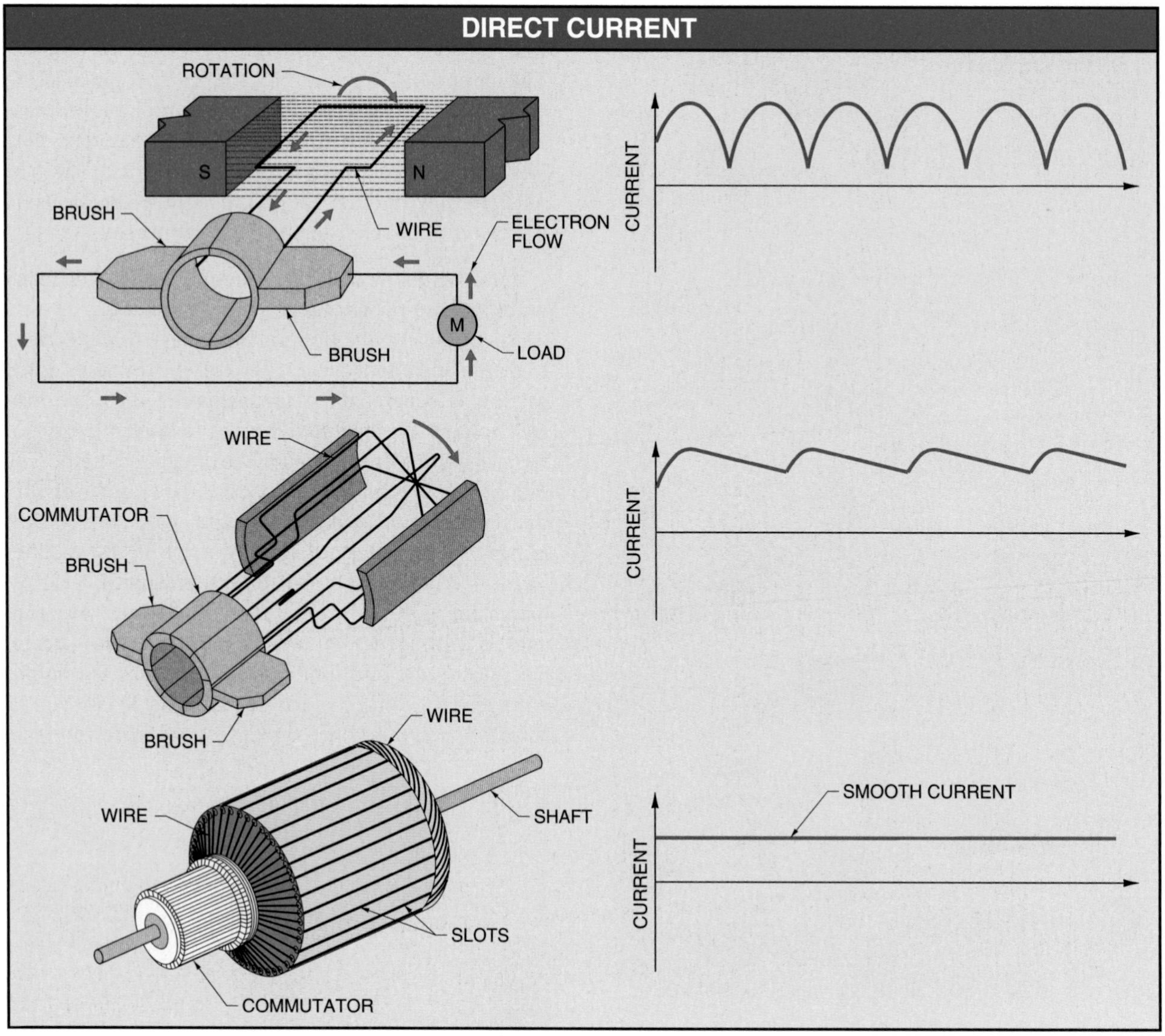

Figure 15-9. Increasing the number of wires that pass through the flux field creates a smooth generated current.

Power Distribution. *Power distribution* is the process of delivering electrical power to where it is needed. Power distribution includes all parts of the electrical utility system from the power generating plant to the customer's service-entrance equipment. Power control, protection, transformation, and regulation must take place before any power is delivered. See Figure 15-10. The distribution system includes the following:

- *Step-up transformers.* The generated voltage is stepped up to a transmission voltage level. The transmission voltage level is usually between 12.47 kV and 245 kV.
- *Power plant transmission lines.* The 12.47 kV to 245 kV power plant transmission lines deliver power to the transmission substations.
- *Transmission substations.* The transmission substations transform the voltage to a lower primary (feeder) voltage. The primary voltage level is usually between 4.16 kV and 34.5 kV.
- *Primary transmission lines.* The 4.16 kV to 34.5 kV primary transmission lines deliver power to the distribution substations and heavy industry.
- *Distribution substations.* The distribution substations transform the voltage down to utilization voltages. Utilization voltage levels range from 480 V to 4.16 kV.
- *Distribution lines.* Distribution lines carry the power from the distribution substation along the street or rear-lot lines to the final step-down transformers.

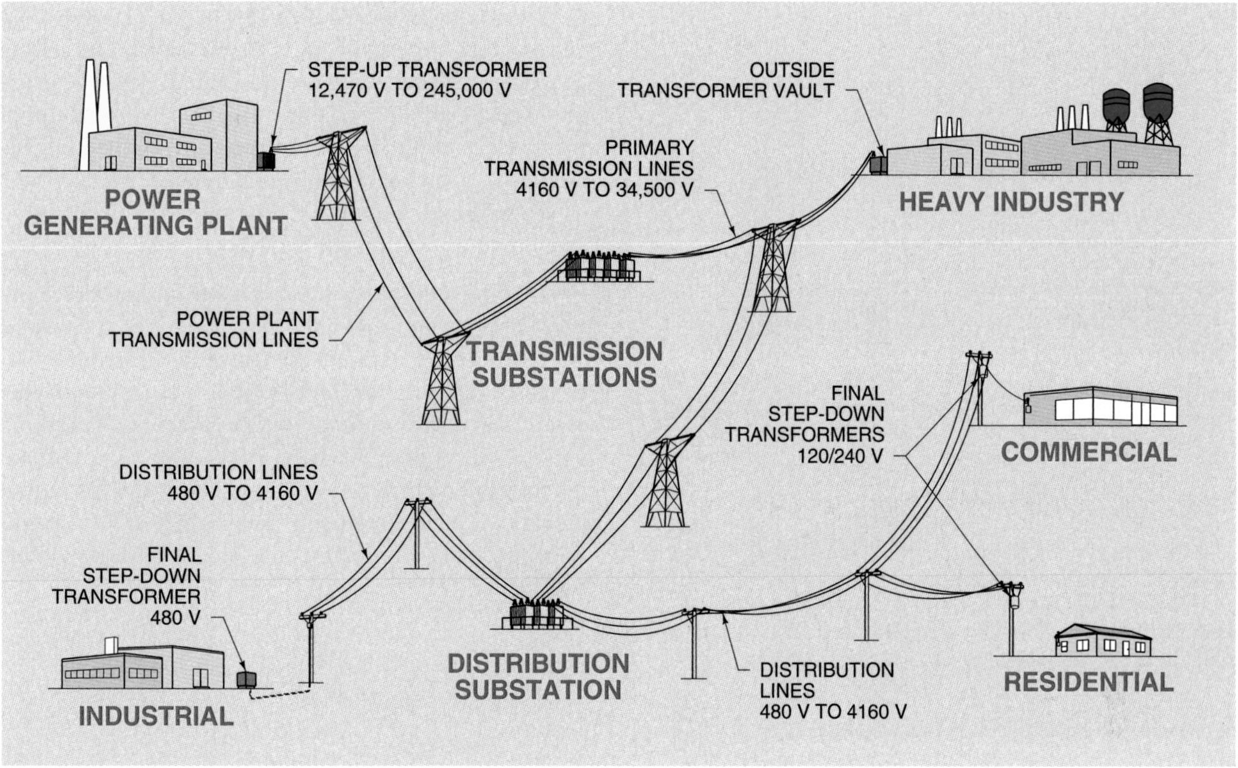

Figure 15-10. High-voltage electricity from power plants is transmitted through electrical lines and substations.

- *Final step-down transformers.* The final step-down transformer transforms the voltage to 480 V. The final step-down transformers may be installed on poles, grade-level pads, or in underground vaults. The secondary of the final step-down transformers is connected to service drops that deliver the power to the customer's service-entrance equipment.

A typical power distribution system delivers power to industrial, commercial, and residential customers. In a power distribution system, standard voltage levels at fixed current ratings are delivered to set points, such as receptacles. These voltage levels are typically 110 V, 115 V, 120 V, 208 V, 220 V, 240 V, 277 V, 430 V, 440 V, 460 V, and 480 V. However, there is no such thing as a standard voltage or current level in an electrical circuit, because these levels are continuously being changed to meet the circuit requirements. In a typical heavy industrial facility, the electricity is delivered directly from a transmission substation to an outside transformer vault.

AC generators normally produce voltages of up to 22,000 V which are stepped up to a higher voltage for economical transmission.

NATIONAL ELECTRICAL CODE®

The *National Electrical Code® (NEC®)* is published by the National Fire Protection Association, Inc. The purpose of the NEC® is the practical safeguarding of persons and property from hazards arising from the use of electricity. The NEC® is updated on a three-year cycle.

The NEC® is adopted by governmental bodies that have legal jurisdiction over electrical installations and for use by insurance inspectors. The authority having jurisdiction (AHJ) is responsible for enforcing the NEC®. See Figure 15-11.

The NEC's scope of coverage includes:

- Electrical conductors and equipment in public and private buildings, structures, mobile homes, RVs, floating buildings, and yards, carnivals, parking lots, and industrial substations.
- Installations of conductors and equipment connected to the electrical supply.
- Installations of other outside conductors and equipment.
- Installations of optical fiber cable and raceways.

Figure 15-11. The authority having jurisdiction (AHJ) is responsible for enforcing the NEC®.

- Installations in buildings used by the electric utility that are not part of the generating plant, substation, or control center.

The NEC® does not cover:

- Ships, trains, aircraft, or automotive vehicles other than mobile homes and RVs.
- Installations in mines.
- Installations of communication equipment controlled by communications utilities and located outdoors.
- Installations controlled by electric utilities for communications, metering, generation, control, transformation, transmission, or distribution of electrical energy.

Grounding

Electrical circuits are grounded to safeguard equipment and personnel against the hazards of electrical shock. Proper grounding of electrical tools, machines, equipment, and delivery systems is one of the most important factors in preventing hazardous conditions.

Grounding is the connection of all exposed noncurrent-carrying metal parts to the earth. Grounding provides a direct path for unwanted (fault) current to the earth without causing harm to persons or equipment. Grounding is accomplished by connecting the circuit to a metal underground pipe, a metal frame of a building, a concrete-encased electrode, or a ground ring. See Figure 15-12.

Noncurrent-carrying metal parts that are connected to ground include all metal boxes, raceways, enclosures, and equipment. Unwanted current exists because of insulation failure or if a current-carrying conductor makes contact with a noncurrent-carrying part of the system. In a properly grounded system, the unwanted current flow blows fuses or trips circuit breakers. Once the fuse is blown or circuit breaker is tripped, the circuit is open and no additional current flows.

Hazardous Locations

A *hazardous location* is a location where there is an increased risk of fire or explosion due to the presence of flammable gases, vapors, liquids, combustible dusts, or easily-ignitable fibers or flyings. The use of electrical equipment in areas where explosion hazards are present can lead to an explosion and fire. This danger exists in the form of escaped flammable gases such as naphtha, benzine, propane, and others. Coal, grain, and other dust suspended in air can also cause an explosion. Article 500 of the NEC® covers hazardous locations. Any hazardous location requires the maximum in safety and adherence to local, state, and federal guidelines and laws, as well as in-plant safety rules. Hazardous locations are indicated by Class, Division, and Group. See Figure 15-13.

ELECTRICAL MEASURING DEVICES

A mechanical technician must often determine the electrical condition of a component within a system. A mechanical technician must determine if the component is receiving an electric current when required, if the conductors are able to carry an electric current, and if an electrical component operates when electricity is present. These questions may be answered by using a continuity tester, a voltage tester, or a multimeter.

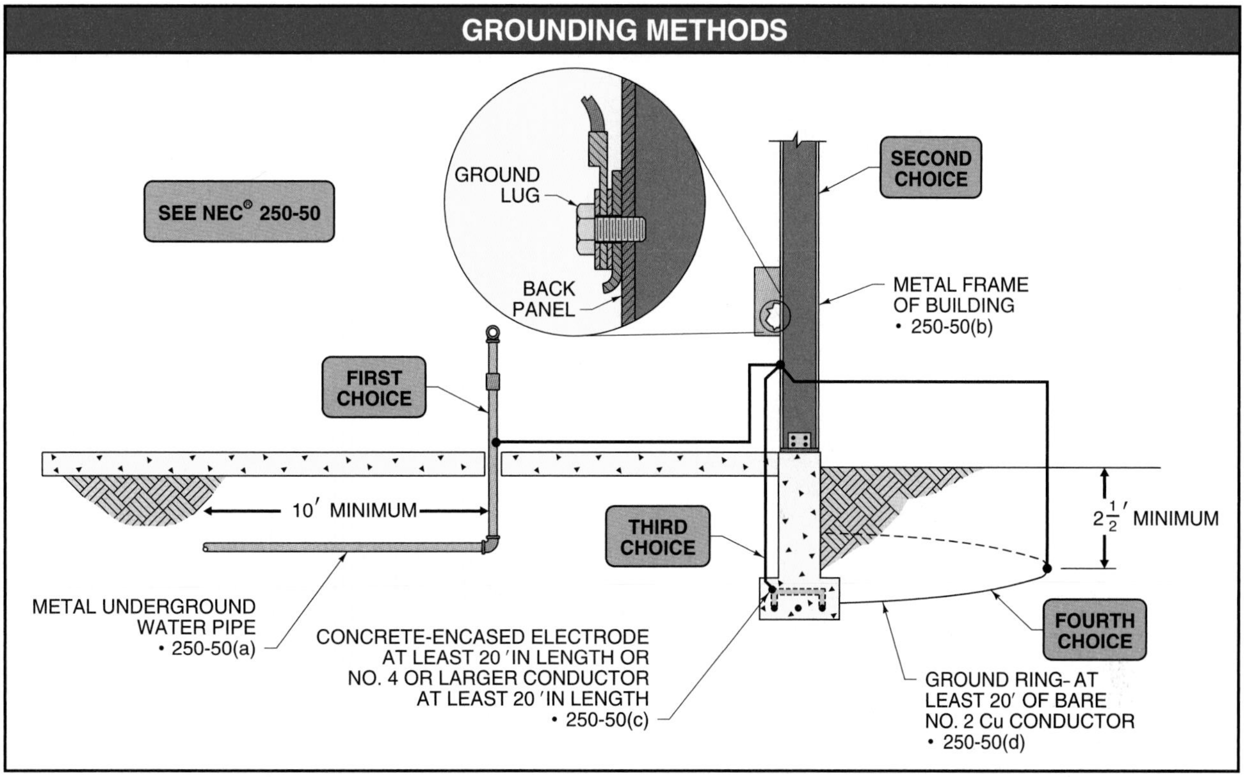

Figure 15-12. Grounding is accomplished by connecting the circuit to a metal underground pipe, a metal frame of a building, a concrete-encased electrode, or a ground ring.

Continuity Testers

A *continuity tester* is a device that indicates if a circuit is open or closed. An *open circuit* is an electrical circuit that has a gap or opening that does not allow current flow. A break in a conductor or an open switch causes an open circuit, preventing the flow of electricity. A *closed circuit* is an electrical circuit with a completed path that allows current flow. Continuity is generally determined using a continuity tester or a multimeter. A continuity tester supplies its own voltage and current by the use of batteries. A continuity tester must be used only in an inoperative circuit with the power OFF. See Figure 15-14.

Continuity Tester Use. A continuity tester may be used on an electrical component to check a wire to determine if it is continuous or broken. The circuit is complete and the continuity tester lights when the tester leads are touching both ends of a continuous (unbroken) wire. The wire may be broken if the tester does not light.

Continuity testers are also used to check for short circuits between component wiring and the body of the component. This is accomplished by touching one tester lead to one of the wire ends and the other tester lead to the component body. A short circuited component, meaning a bare spot in the wire insulation, allows the wire to come in contact with the component body, giving an indication of a short from the tester.

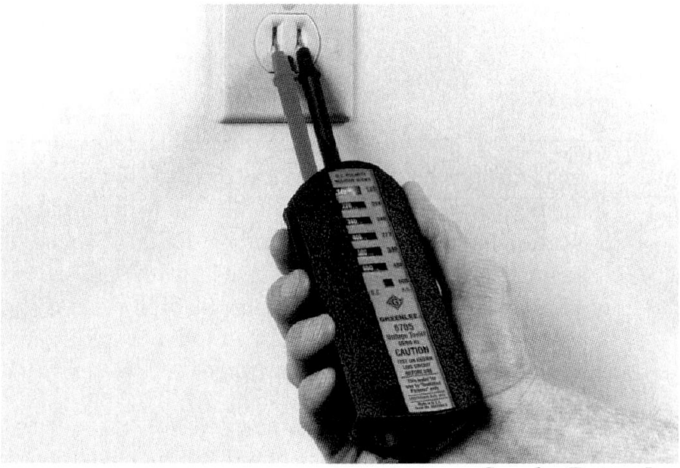

Greenlee Textron Inc.

A voltage tester is used to check if voltage is present and gives an indication of its approximate level.

HAZARDOUS LOCATIONS—500

Classes	Likelihood that a flammable or combustible concentration is present
I	Sufficient quantities of flammable gases and vapors present in air to cause an explosion or ignite hazardous materials
II	Sufficient quantities of combustible dust are present in air to cause an explosion or ignite hazardous materials
III	Easily-ignitable fibers or flyings are present in air, but not in a sufficient quantity to cause an explosion or ignite hazardous materials

Divisions	Location containing hazardous substances
1	Hazardous location in which hazardous substance is normally present in air in sufficient quantities to cause an explosion or ignite hazardous materials
2	Hazardous location in which hazardous substance is not normally present in air in sufficient quantities to cause an explosion or ignite hazardous materials

Groups	Atmosphere containing flammable gases or vapors or combustible dust	
Class I	Class II	Class III
A B C D	E F G	none

DIVISION I EXAMPLES

Class I:

- Spray booth interiors
- Areas adjacent to spraying or painting operations using volatile flammable solvents
- Open tanks or vats of volatile flammable liquids
- Drying or evaporation rooms for flammable solvents
- Areas where fats and oil extraction equipment using flammable solvents are operated
- Cleaning and dyeing plant rooms that use flammable liquids
- Gas generator rooms
- Pump rooms for flammable gases or volatile flammable liquids that do not contain adequate ventilation
- Refrigeration or freezer interiors that store flammable materials
- All other locations where sufficient ignitable quantities of flammable gases or vapors are likely to occur during routine operations

Class II:

- Grain and grain products
- Pulverized sugar and cocoa
- Dried egg and milk powders
- Pulverized spices
- Starch and pastes
- Potato and woodflour
- Oil meal from beans and seeds
- Dried hay
- Any other organic materials that may produce combustible dusts during their use or handling

Class III:

- Portions of rayon, cotton, or other textile mills
- Manufacturing and processing plants for combustible fibers, cotton gins, and cotton seed mills
- Flax processing plants
- Clothing manufacturing plants
- Woodworking plants
- Other establishments involving similar hazardous processes or conditions

Figure 15-13. Hazardous locations are indicated by Class, Division, and Group.

Voltage Testers

A *voltage tester* is a device that indicates approximate voltage level and type (AC or DC) by the movement of a pointer on a scale. Basic voltage testers do not have meter movement or digital display. A voltage tester should not be used on low-power circuits because a voltage tester draws high current and no voltage indication is given on low-power circuits. The range for most voltage testers is between 90 V and 600 V.

Most voltage testers contain a solenoid that vibrates when connected to AC and does not vibrate when connected to DC, even though the pointer indicates the tested voltage. Most voltage testers also include neon lights that determine if the voltage is AC or DC. Voltage testers are designed for intermittent use and should not be connected to a power supply for more than 15 sec. Simple voltage testers that are compact and rugged enough for mechanical technicians generally have the capability of being a voltage tester and a continuity tester. See Figure 15-15.

Voltage Tester Use. Caution must be taken when testing any voltage over 24 V. **Warning:** Ensure no part of the body contacts any part of a live circuit, including the metal contact points at the tip of a tester. Always assume that circuits are powered (hot) until they have been checked.

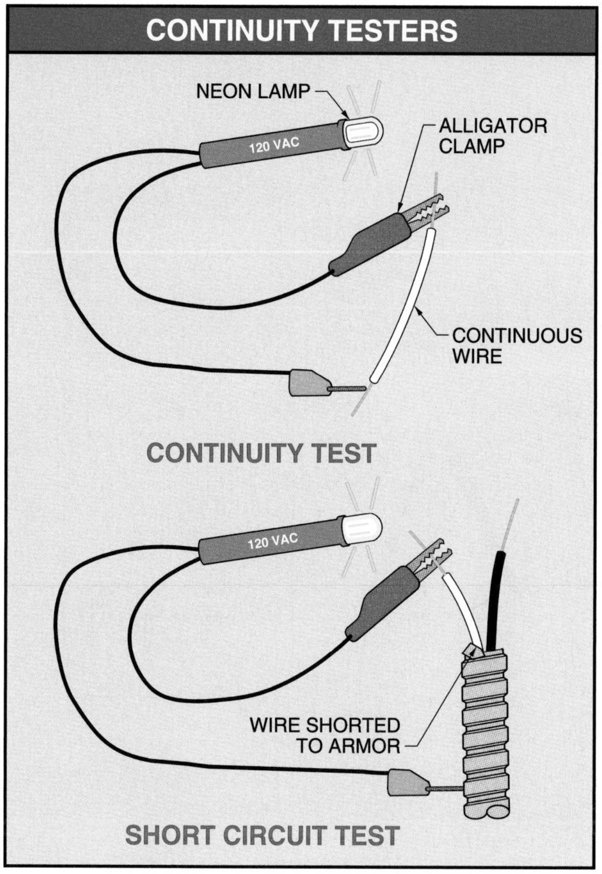

Figure 15-14. A continuity tester uses its own power circuit to determine if a wire is continuous or broken or if a short exists between a wire and its housing.

The procedure for using a voltage tester is to connect one test probe to one side of the circuit and then connect the other test probe to the other side of the circuit. A reading is taken from the pointer. The white wire, when present, should be checked first to determine that it is the neutral conductor. A *neutral conductor* is a wire that carries current from one side of the load to ground. Neutral conductors are connected directly to loads and never connected through fuses, circuit breakers, or switches. Connecting a voltage tester between the neutral wire and a grounded surface should give no voltage indication. All other hot wires are then checked to ground or neutral and should show a voltage reading (voltage present) on the voltage tester.

Before using any electrical test equipment, always refer to the user's manual for proper operating procedures, safety precautions, and limits.

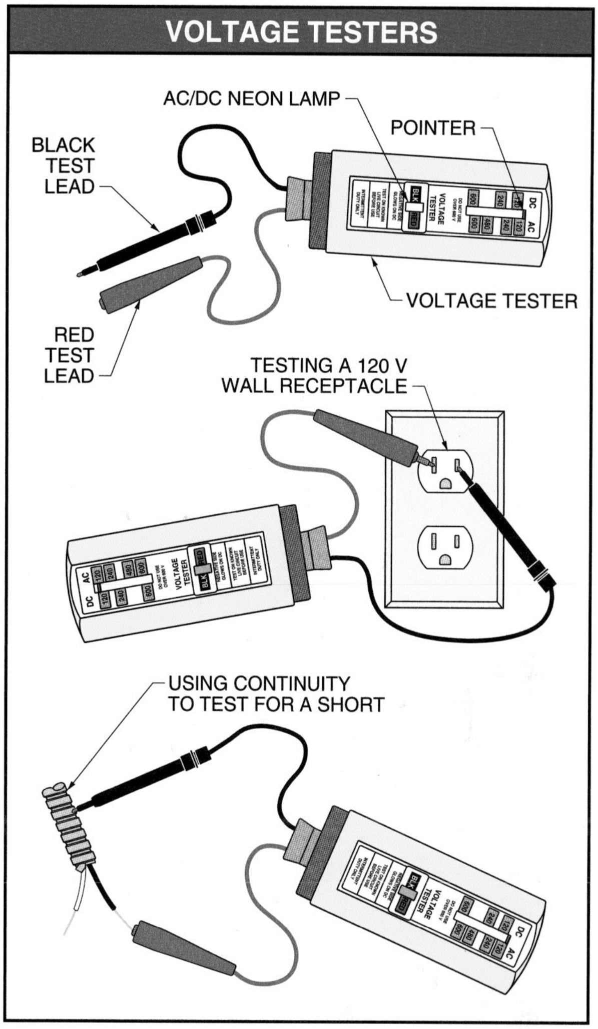

Figure 15-15. A voltage tester is effective in determining voltages up to 600 V.

Multimeters

A *multimeter* is a test tool used to measure two or more electrical values. Multimeters may be analog or digital. An analog multimeter indicates readings by the mechanical motion of a pointer. The value is read from right to left on the scale. Extremely high readings on the scale are to the far left and are regarded as infinity (∞). A digital multimeter indicates readings as numerical values. Digital multimeters help eliminate human error when taking readings by displaying exact values measured. Multimeters have a function switch that enables the testing of various electrical values such as current, voltage, resistance, etc. The function switch is set on AC when measuring alternating voltage (VAC) and is set on DC when measuring direct voltage or current (DC). See Figure 15-16.

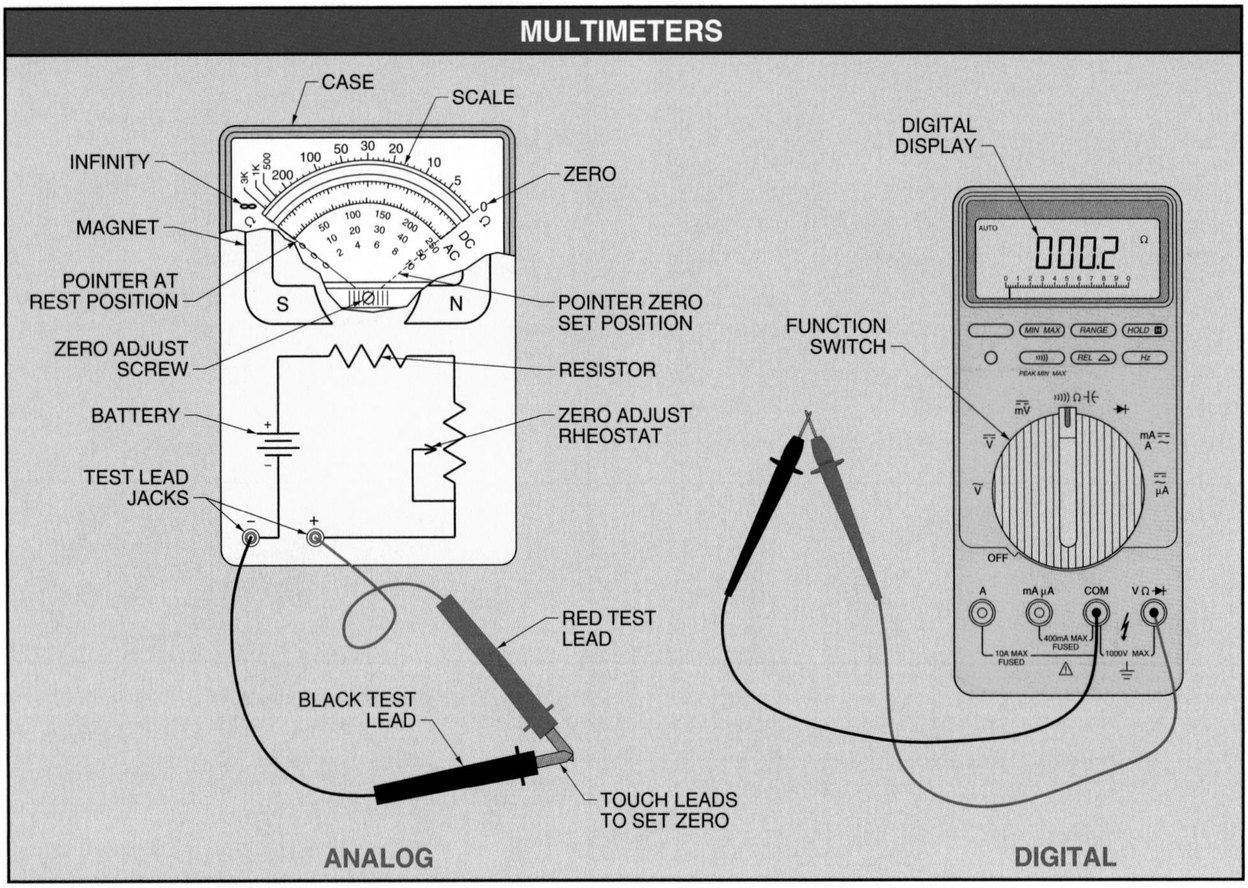

Figure 15-16. A multimeter is a test tool used to measure two or more electrical values.

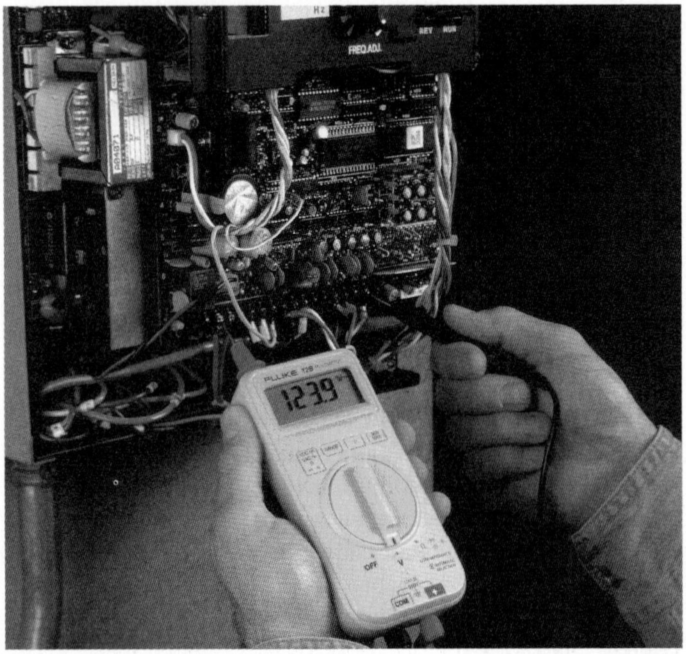

Fluke Corporation

A multimeter enables the testing of several different electrical values.

If a multimeter does not have a separate setting for resistance, the positive DC setting is used. The function and range switches must be set at the correct quantity and the correct scale must be read when taking electrical measurements using a multimeter.

Multimeter Use. Multimeters measure many different electrical quantities. Care must be taken to ensure that a multimeter is set on the correct settings, connected to a circuit correctly, and the correct scale is read accurately. Ensure that the multimeter is properly set before connecting it to a circuit.

Resistance is measured by first checking that all power is OFF in the circuit being tested. Resistance is checked only in an inoperative circuit with the power turned OFF and the component removed from the circuit. Multimeters do not require an external power source because they use a battery to supply their own current and voltage. Resistance measurements are so sensitive that normal circuit voltages burn out a multimeter. However, many multimeters are protected from incorrect use and damage by fuses in their circuitry. Ensure that the multimeter batteries are in good condition. Plug the black test lead into the negative jack (–) and plug the red test lead into the positive jack (+). In an analog multimeter, adjust for zero (far right) by touching the test leads together and, using the adjustment knob, set the zero adjustment by placing the pointer on the 0 Ω mark. Connect the meter leads across the component under test and read the displayed resistance. Turn the meter OFF after measurements are taken to save battery life.

ELECTRICAL DEVICES

Electrical circuits are used to produce work. To produce work, an electrical circuit must include a component (load), a source of electricity (battery, generator, etc.), and a method of controlling the flow of electricity (switch). In addition, a circuit should also include a protection device (fuse, circuit breaker, etc.) to ensure that the circuit operates safely and within its electrical limits. Components encountered by an mechanical technician include transformers, contactors, motor starters, fuses, circuit breakers, GFCIs, switches, and solenoids.

Transformers

A *transformer* is an electric device that uses electromagnetism to change AC voltage from one level to another. A transformer consists of two looped conductors (windings) and a laminated iron core. The iron core is used as a magnetic coupling between the windings and is laminated to reduce power loss due to eddy current. *Eddy current* is an electric current that is generated and dissipated in a conductive material in the presence of an electromagnetic field. A solid iron core has many unwanted eddy currents flowing in two directions as AC electric current alternates through the core. Eddy currents create an excessive amount of heat. Laminating (layering) iron reduces the amount of eddy currents. The core, like that in an electromagnet, concentrates the flux lines of the conductors. See Figure 15-17.

A transformer is overloaded when it is required to deliver more power than its rating. Transformers are not damaged when overloaded for a short period of time.

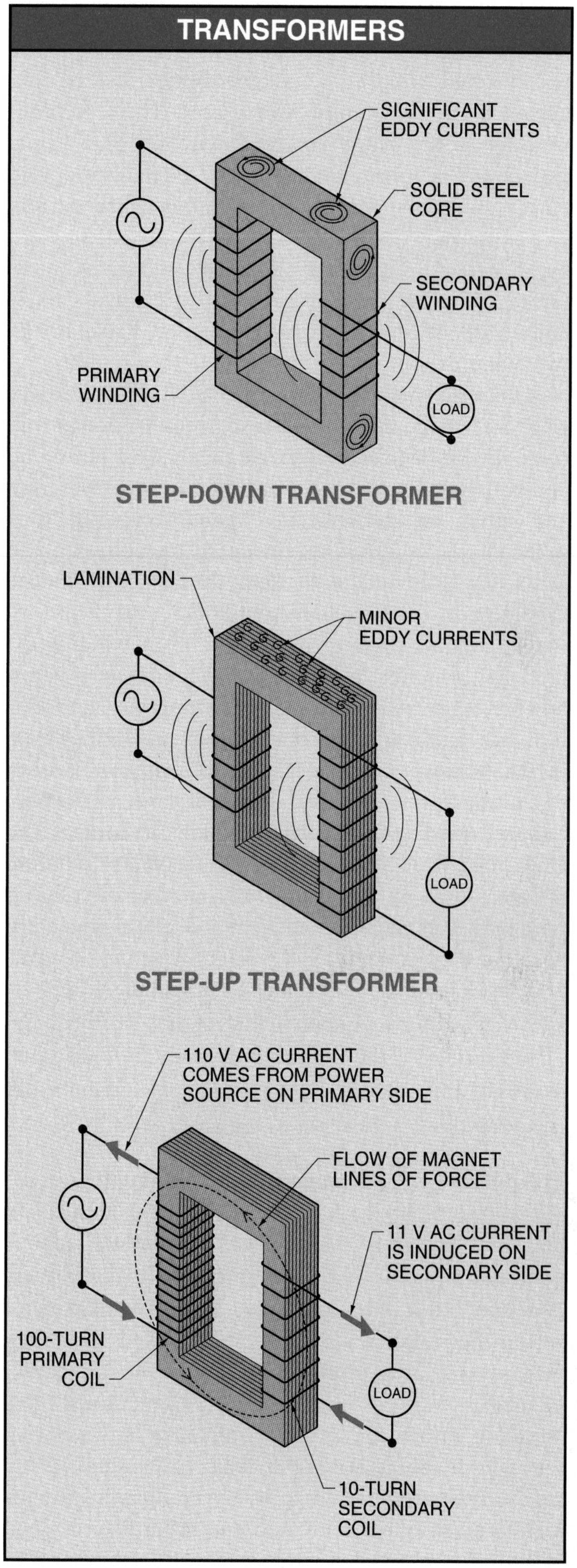

Figure 15-17. A transformer consists of two looped conductors (windings) and a laminated iron core.

The two windings are referred to as the primary and secondary windings. The *primary winding* is the power input winding of a transformer and is connected to the incoming power supply. The *secondary winding* is the output or load winding of a transformer and is connected to the load. A magnetic field builds up around the primary winding when current passes through its conductor. As the magnetic field builds around the primary winding, current is transferred (induced) in the secondary winding. This is done without any physical connection between the two windings.

Alternating current is connected to the primary winding when an alternating current is transferred through a transformer. The magnetic field builds up and collapses in one direction, then reverses direction and builds up and collapses again, completing one cycle of the alternating current. Concentrating the primary winding magnetic field, the iron core induces a voltage in the secondary winding.

The number of turns in the secondary winding establishes whether the transformer steps up or steps down voltage. A transformer is a step-up transformer if it has more turns in the secondary winding than in the primary winding. The transformer is a step-down transformer if there are fewer turns in the secondary winding than the primary winding. The amount of voltage induced in the secondary winding is determined by the ratio of the number of turns of wire in the secondary winding to those in the primary winding. For example, if a primary winding contains 100 turns and is connected to an input supply of 120 V, and the secondary winding has 50 turns ($\frac{1}{2}$ of the primary), the output voltage to the load is 60 V ($\frac{1}{2}$ the input voltage). In this case, the transformer is a step-down transformer with a 2 : 1 ratio.

Troubleshooting Transformers. The input and output voltages should be checked if there appears to be a problem with a transformer. The transformer is good as far as voltages are concerned if the voltage is within ±10% of the nameplate rating. A multimeter set on the resistance setting may be used if a break in the wires or a short is suspected. This is accomplished by first zeroing the meter. Then, check for a break in either the primary or secondary winding wires by touching the meter leads to the ends of the wires of the circuit. Normally, the reading is approximately midway between zero and infinity for a good transformer. Exact resistance values for each winding may be found by comparing a good transformer with one of the same type and size. See Figure 15-18.

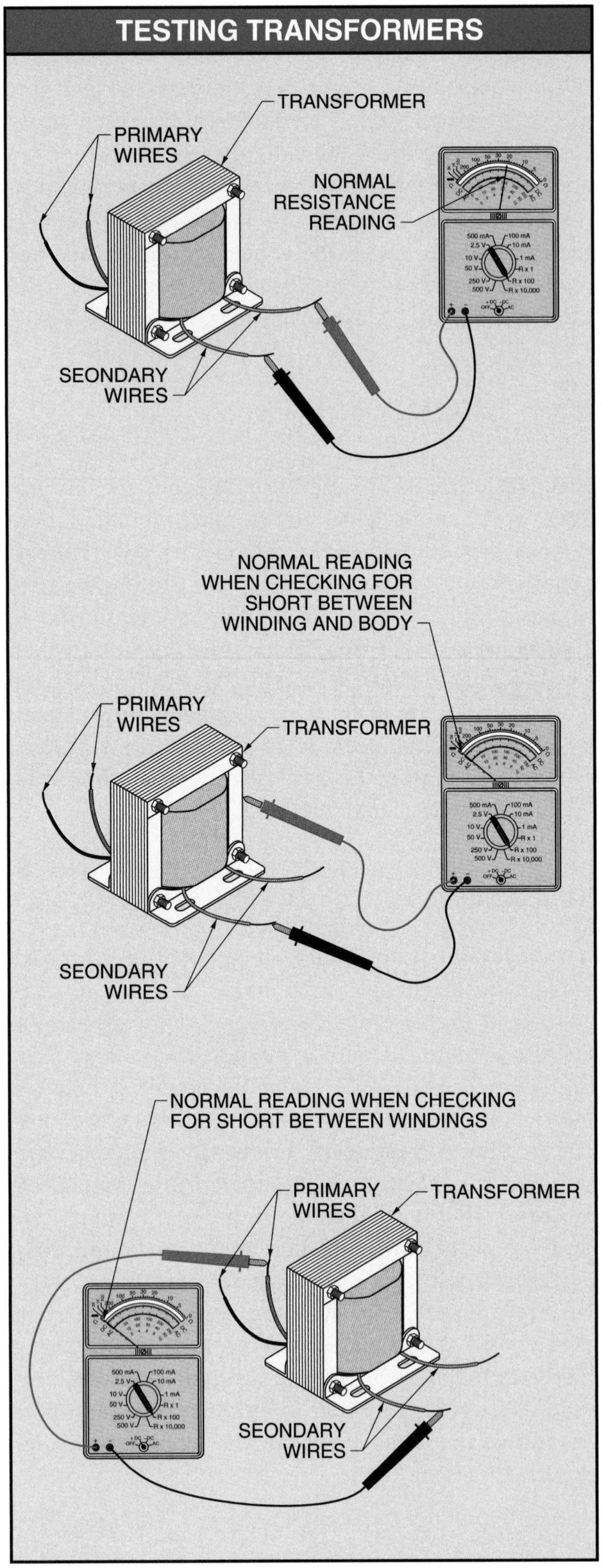

Figure 15-18. Transformers are tested by checking the resistance of the primary and secondary windings and checking for shorts or breaks between each wire and between each wire and the core.

To check a transformer for a short, the meter leads should be touched between each of the secondary and primary wires, or between each of the secondary and primary wires and the core. An unshorted transformer produces a reading of infinity. Connect the test lead to the metal of the core, not to the paint or varnish, when checking between the wires and the core. During a test for shorts, any reading other than infinity indicates a short.

Contactors and Motor Starters

A *contactor* is a control device that uses a small control current to energize or de-energize the load connected to it. A *motor starter* is an electrically-operated switch (contactor) that includes motor overload protection. Manual contactors and motor starters use pushbuttons to energize or de-energize the load connected to them. Magnetic contactors and motor starters use a coil which magnetically closes a set of contacts to energize and de-energize the load connected to them. Magnetic contactors and motor starters contain an armature and coil. The armature is connected to a set of contacts which, when activated, make contact and close a circuit. See Figure 15-19.

Contactors and motor starters consist of completely separate power circuits and control circuits. A *power circuit* is the part of an electrical circuit that connects the load to the main power lines. A *control circuit* is a circuit that allows electrical devices to be controlled from remote locations. A control circuit requires the use of a pilot device, such as a pushbutton, float switch, flow switch, etc. The voltage of a control circuit may be lower, higher, or the same as the power circuit voltage because the control circuit is independent of the power circuit.

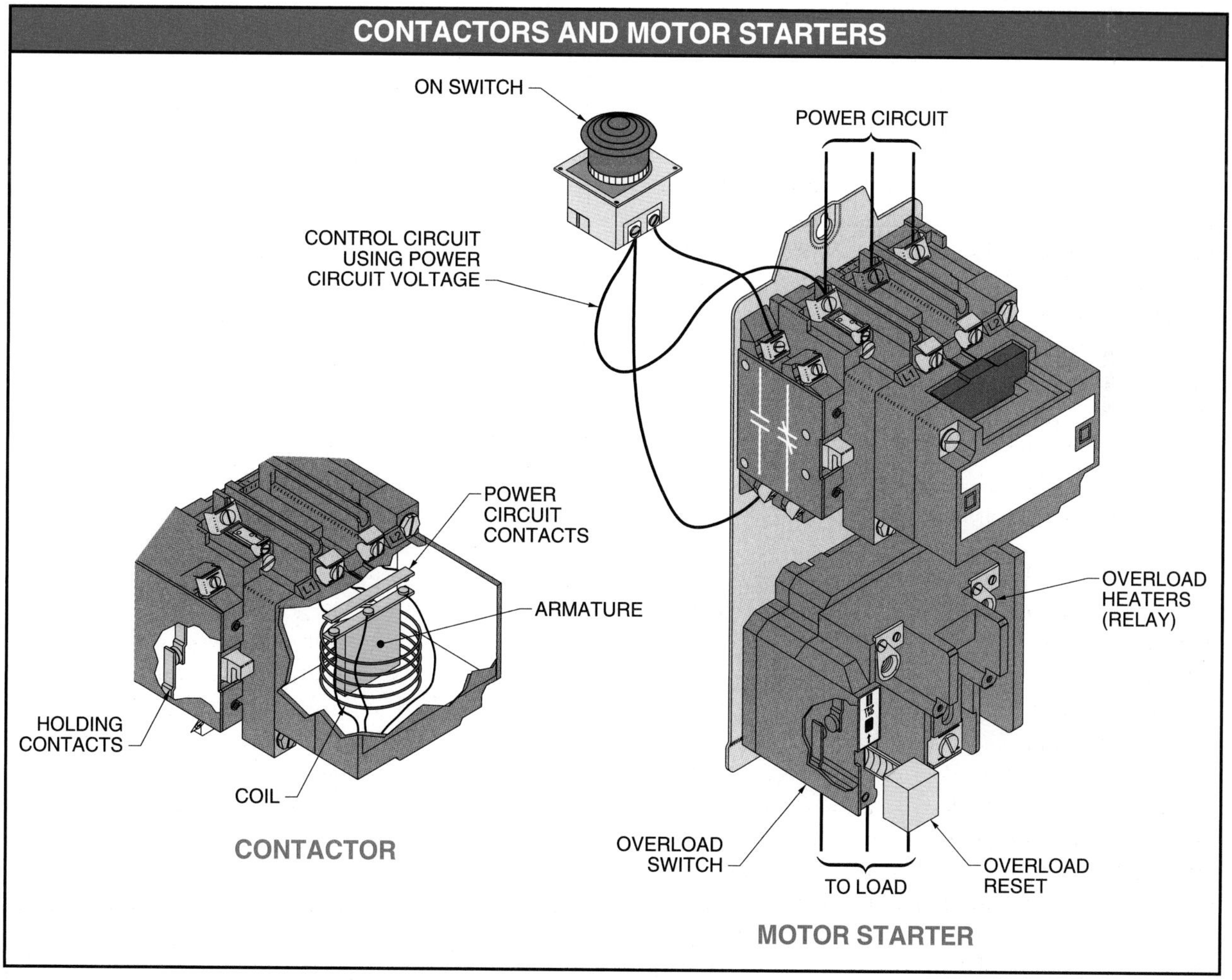

Figure 15-19. A magnetic contactor or motor starter is operated when a control current is sent to a coil, which magnetically closes a set of power circuit contacts.

The two circuits are always at different current levels. The control circuit may be constructed of a light-gauge wire because its required current is only the amount needed to produce a magnetic force for contactor or motor starter operation. Many control circuits operate on less than 1 A.

Contactors and motor starters (relays) allow a pilot operator in one location to close a switch in another location. A *relay* is an interface that controls one electrical circuit by opening and closing contacts in another circuit. A relay sends a signal (mechanical input) to a decision maker (contactor or motor starter), which creates the action by turning ON the load (motor, heater, light, etc.). A relay consists of a wire coil wound on an iron bar to produce a magnet. The magnet, produced by a small current from the pilot operator, moves and connects contacts, closing a circuit of another (usually higher) voltage. A *contact* is a conducting part of a relay that acts as a switch to connect or disconnect a circuit or component.

Contactors or motor starters may relay a signal to the load through a continual push of a button or operator input switch. In a switch requiring a continual push, power to the load is constant as long as the input switch is held closed. This may not always be the best situation. Most pushbutton switches are pressed and released, leaving the load operational and requiring only a momentary pilot signal to operate a load. The load continues to operate until the control circuit is signaled to open. Opening the control circuit is accomplished by using a stop pushbutton switch, which requires only a momentary signal to open the circuit.

The control voltage of a motor starter is checked by touching the voltage tester leads to the control circuit terminals.

Continuous operation using a momentary pilot signal is accomplished through use of a holding contact. A *holding contact* is an auxiliary contact used to maintain current flow to the coil of a relay. Holding contacts are physically attached to the side of the contactor or motor starter and open and close with the power contacts. Once an ON pushbutton is pressed, holding contacts may use the current of the circuit that it is a component of to maintain the closed condition of the circuit. Once an OFF pushbutton is pressed, the circuit is opened and current flow to the holding contact coil is removed, signaling the holding contacts and power contacts to open.

Troubleshooting Contactors and Motor Starters. Motor starters usually have built-in motor overload protection. Overload protection is accomplished through the use of overload relays. An *overload relay* is a time-delay device that senses motor current temperatures and disconnects the motor from the power supply if the current is excessive for a certain length of time. Overload relays are the main difference between a motor starter and a contactor. Contactors or motor starters should be checked first when there is a contactor or motor starter in a circuit and the load is inoperative because these components are the point where the incoming power, load, and control circuits are connected. See Figure 15-20.

Testing contactors and motor starters is done by visually inspecting the contactor or starter for physical damage. Check for incoming voltage of the power circuit. Each hot line must be within ±10% of the load voltage rating. Also, check each incoming power wire to ground for a blown fuse upstream. Check the voltage used by the control circuit and the control device, which may be a start/stop switch or pilot operation (pressure switch, float switch, flow switch, etc.).

On motor starters, check if the overload has tripped by testing each side of the overloads for the presence of voltage. The overloads may have tripped if there is supply voltage and power voltage but no output voltage. Do not attempt to reset the overloads unless authorized to do so. Overloads protect a motor and normally trip when there is a high current condition. Other checks must be made by an authorized technician to determine the overload condition before the resetting of overloads.

To meet motor protection needs, overload relays are designed to have a time delay to allow harmless temporary overloads without disrupting the circuit.

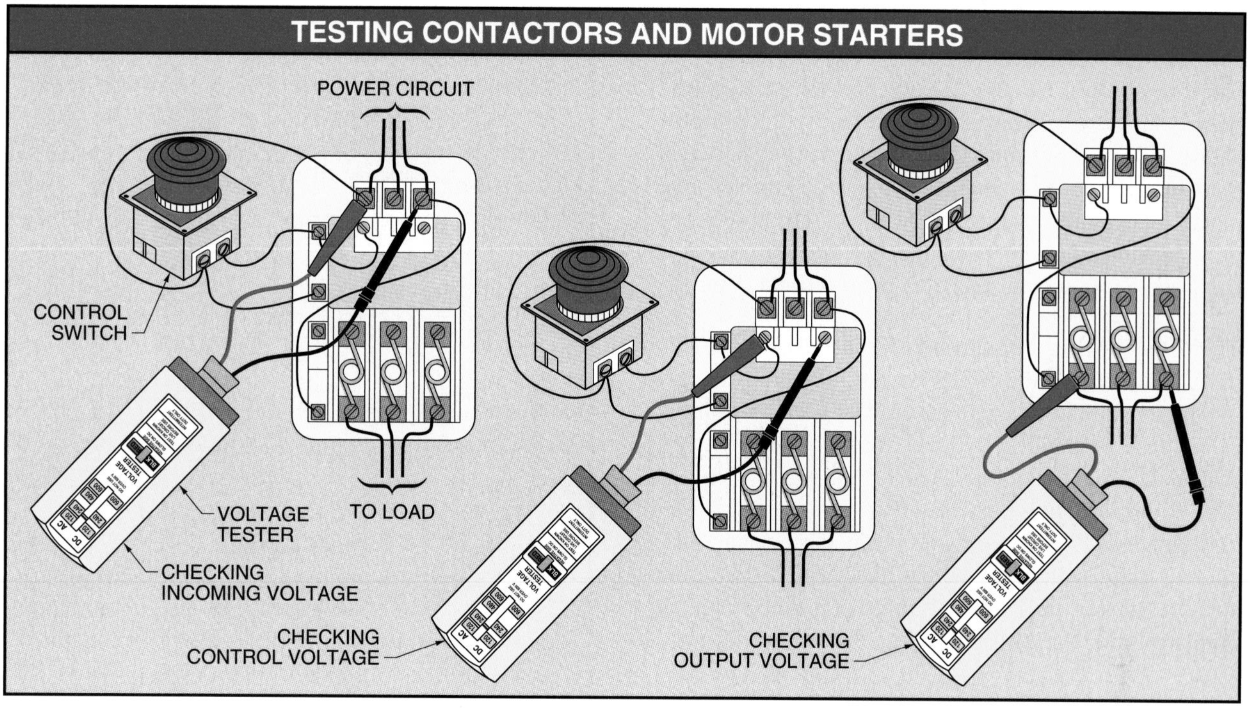

Figure 15-20. The incoming voltage, control voltage, and output voltage should be checked when testing a contactor or motor starter.

Fuses and Circuit Breakers

A *fuse* is an overcurrent protection device with a fusible link that melts and opens the circuit on an overcurrent condition. A *circuit breaker* is a device with a mechanical mechanism that may manually or automatically open a circuit when an overload condition or short circuit occurs. Fuses and circuit breakers automatically interrupt the flow of current in a circuit if the current exceeds the amount the circuit was designed to handle. Overcurrent conditions may be caused by an overloaded circuit that uses too many lights, appliances, or other loads, or by a short circuit where a bare hot wire touches another hot wire or any metal that is grounded.

Excessive heat is generated in a conductor that has more current than the conductor can handle because of the activity of the free electrons in the conductor. This principle is used in safety devices such as fuses and circuit breakers. Fuses and circuit breakers are an intentional weak spot in a circuit. Fuses and circuit breakers have a higher resistance than the rest of the circuit. Fuses are designed to have high resistance by the use of a material other than copper (usually zinc alloy), which melts at a lower temperature than the copper conductor. The melted alloy opens the electrical circuit before damage is done to the copper conductor or other electrical devices.

Fuses are checked using a voltage tester by touching one probe to one supply side terminal and the other probe to the load side terminal of another line.

Fuses or circuit breakers may be bimetallic. A *bimetallic device* is a device that consists of two dissimilar metals. One metal, being a lesser conductor than the other, offers greater resistance and allows less electron activity. The other metal, being a better conductor, offers little resistance and allows greater electron activity. During a high flow of current, the great electron excitement in the better conductor causes its temperature to rise. The difference between the electron activity in the two metals causes one metal to expand more than the other. The difference in expansion rates between the two metals creates a bending action, which separates the electrical contacts.

Troubleshooting Fuses and Circuit Breakers. To troubleshoot fuses and circuit breakers, begin by checking the incoming voltage to ensure that power is coming to the fuse or breaker. An electrical problem may be as simple as a disconnect switch that was not turned ON. A disconnect switch (disconnect) is a switch that disconnects electrical circuits from motors and machines. Also, within a problem electrical distribution system, power may be fed through other fuses or breakers with the problem being further upstream.

Incoming power to fuses or circuit breakers is tested using a voltage tester by checking between each incoming power lead and ground or by checking between each pair of power leads. See Figure 15-21. Incoming voltage should be within ±10% of the nameplate voltage. A high or low voltage situation indicates a supply problem. In a 240 V circuit, each leg should have 120 V.

A check should be made for proper grounding of the enclosure, conduit, boxes, etc., if proper voltage is present. To test for proper grounding, connect one probe of the voltage tester to an unpainted metal part of the equipment (enclosure) and touch the other probe to each of the line terminals. Properly grounded equipment produces a voltage indication.

Fuses or circuit breakers are also tested for an open circuit. This may be accomplished using a voltage tester and line power, or with the fuses or circuit breakers removed using a continuity tester. Always turn OFF the power and remove the fuses or breakers when checking with a continuity tester. To prevent accidental dead shorts, remove fuses using a fuse puller. A *fuse puller* is a device made of a non-conductive material such as nylon that is used to grasp and remove cartridge fuses. See Figure 15-22. With the fuse or breaker removed, place continuity tester leads across the fuse or breaker. The fuse or breaker is good if the continuity tester lights. The fuse or breaker is bad if the continuity tester does not light. The circuit breaker must be in the ON position when testing using a continuity tester.

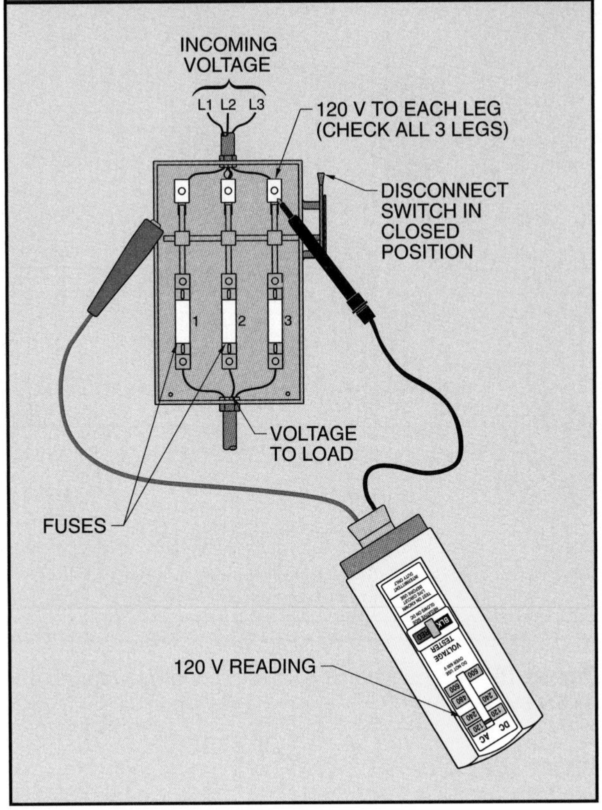

Figure 15-21. Always check the incoming voltage supply when fuses or circuit breakers are suspected to be faulty.

Testing a fuse or breaker using line power and a voltage tester is accomplished with one probe touching one supply side terminal and the other probe touching a terminal on the load side of another line. Never check between terminals of the same line when checking fuse or breaker condition using a voltage tester. For example, a voltage tester can be used to check the fuse condition of a 3ϕ circuit. A *3ϕ circuit* is a circuit that has three incoming wires, with each wire having the same voltage. Fuse 3 may be checked with one probe contacting the L1 or L2 supply terminals and the other probe on the F3 load side terminal. Testing Fuse 2 is accomplished by placing one probe of the voltage tester on the L1 or L3 supply terminals and the other probe contacting the F2 load terminal.

Ensure the switch for a circuit is open or disconnected before removing a fuse from a circuit. Always use an approved puller and break contact on the hot side of the circuit first. Install the fuse into the load side of the fuse clip, then into the line side.

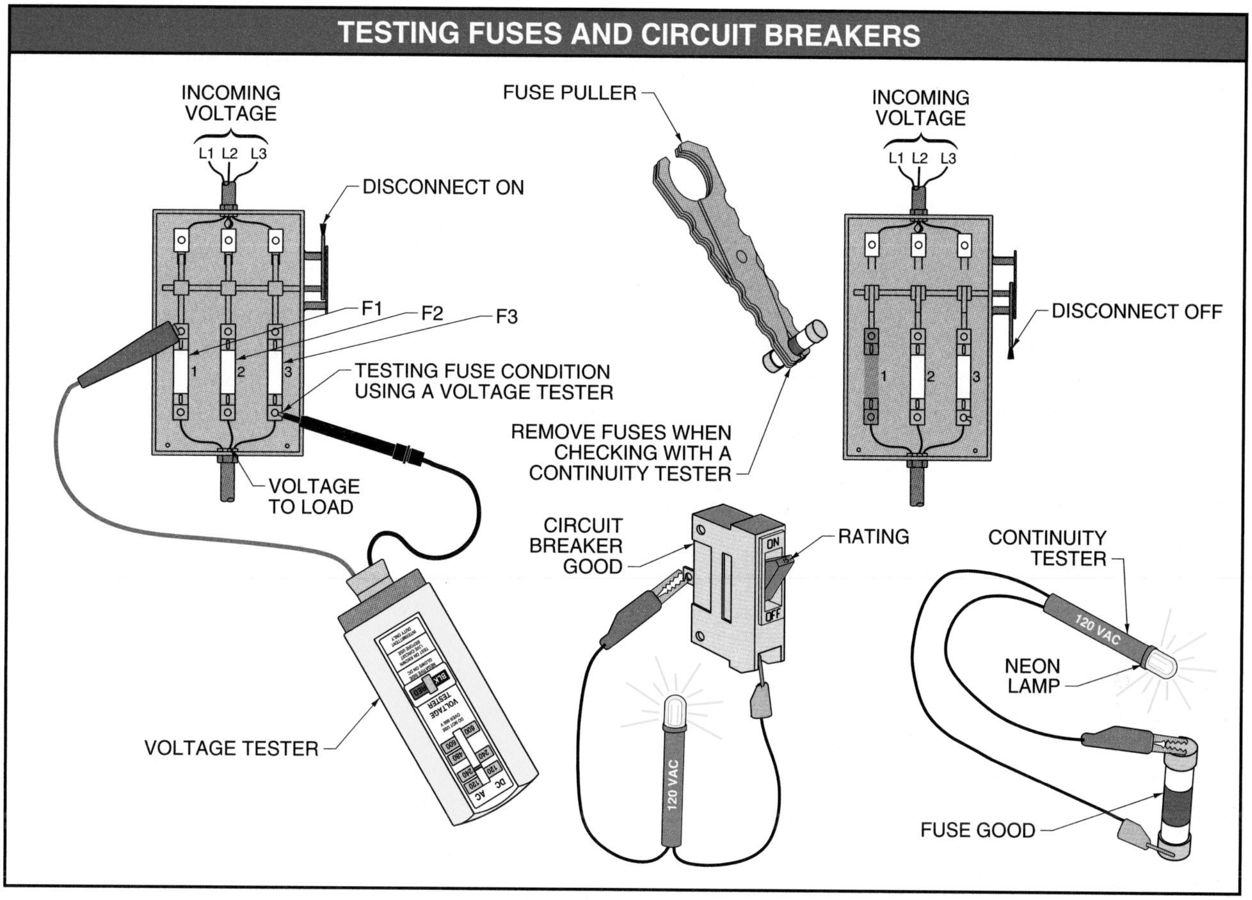

Figure 15-22. Fuse or circuit breaker condition may be checked using a voltage tester with the disconnect ON or with a continuity tester after the disconnect has been opened and the fuses or breakers removed.

Ground Fault Circuit Interrupter (GFCI)

A ground fault exists when an unintended current path is established between ground and an ungrounded (hot) conductor. A ground fault may result from defective electrical equipment, improperly installed equipment, or through misuse of good equipment. Ground faults may damage equipment. More importantly, however, a ground fault may cause an electrical shock resulting in injury or death to any person who becomes part of the ground fault circuit. See NEC® 210-8, 215-9, 215-10, and 230-95. See Figure 15-23.

A *ground fault circuit interrupter (GFCI)* is an electrical device that protects personnel by detecting potentially hazardous ground faults and quickly disconnecting power from the circuit. A potentially dangerous ground fault is any amount of current above the level that may deliver a dangerous shock. Any current over 8 mA is considered potentially dangerous depending upon the path the current takes, the amount of time exposed to the shock, and the physical condition of the person receiving the shock. Therefore, GFCIs are required in such places as dwellings, hotels, motels, construction sites, marinas, receptacles near swimming pools and hot tubs, underwater lighting, fountains, and other areas in which a person may experience a ground fault.

A GFCI compares the amount of current in the ungrounded (hot) conductor with the amount of current in the neutral conductor. If the current in the neutral conductor becomes less than the current in the hot conductor, a ground fault condition exists. The amount of current that is missing is returned to the source by some path other than the intended path (fault current). A fault current as low as 4 mA to 6 mA will activate the GFCI and interrupt the circuit. Once activated, the fault condition is cleared and the GFCI is manually reset before power may be restored to the circuit. See Figure 15-24.

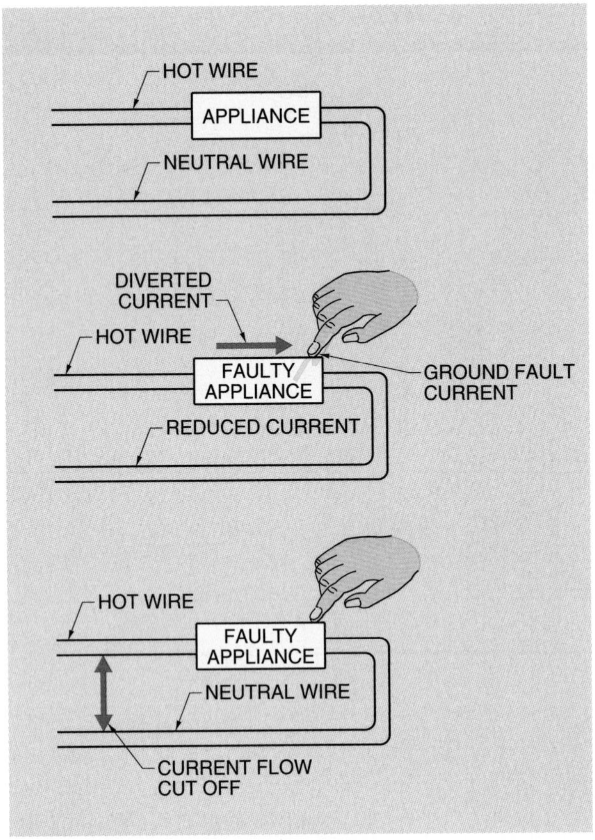

Figure 15-23. A ground fault exists when an unintended current path is established to ground.

GFCI protection may be installed at different locations within a circuit. Direct-wired GFCI receptacles provide a ground fault protection at the point of installation.

GFCI receptacles may also be connected to provide GFCI protection at all other receptacles installed downstream on the same circuit. GFCI CBs, when installed in a load center or panelboard, provide GFCI protection and conventional circuit overcurrent protection for all branch circuit components connected to the CB. Plug-in GFCIs provide ground fault protection for devices plugged into them. These plug-in devices are often used by personnel working with power tools in an area that does not include GFCI receptacles.

Switches

A *switch* is a device that starts or stops the flow of electrical energy. Regardless of the source being used, power must be under control at all times. Control of power means that a system must contain devices that permit it to be safely turned ON or OFF, limited in strength, or controlled in direction. Controlling the ON/OFF function of power is accomplished using a switch. A switch is closed when it allows current to flow in a circuit. A switch is open when it prevents flow of current in a circuit. See Figure 15-25.

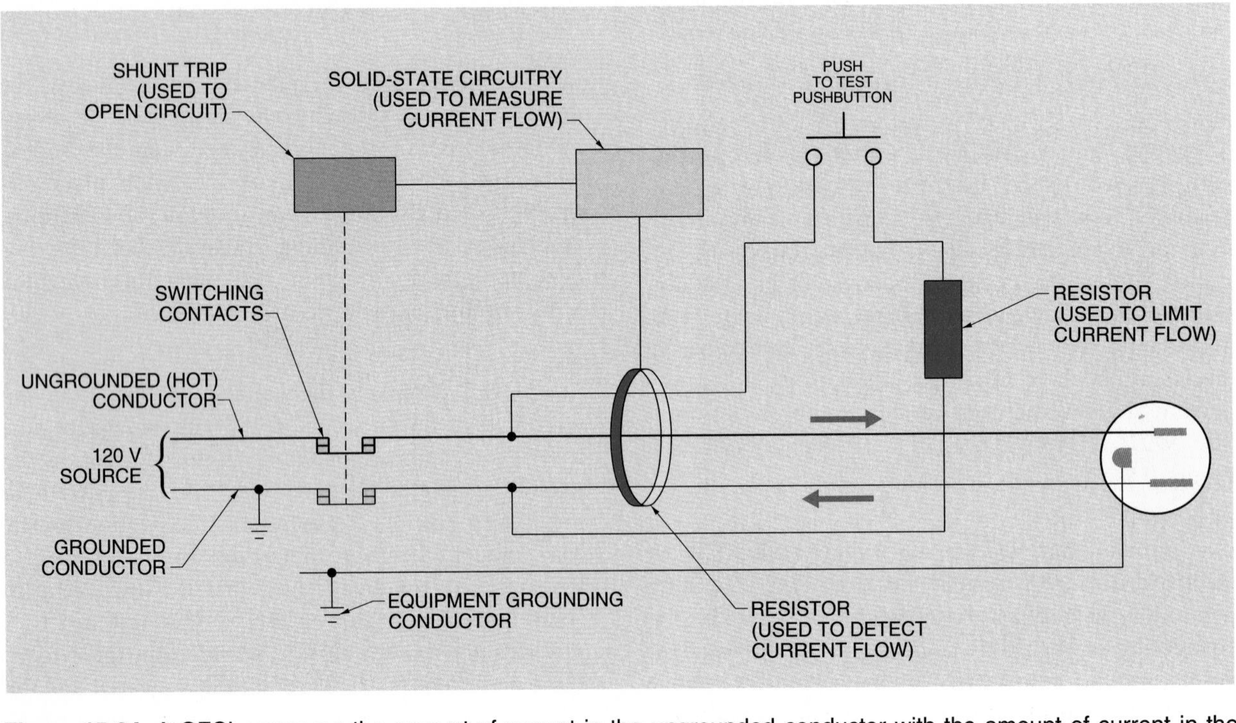

Figure 15-24. A GFCI compares the amount of current in the ungrounded conductor with the amount of current in the grounded conductor.

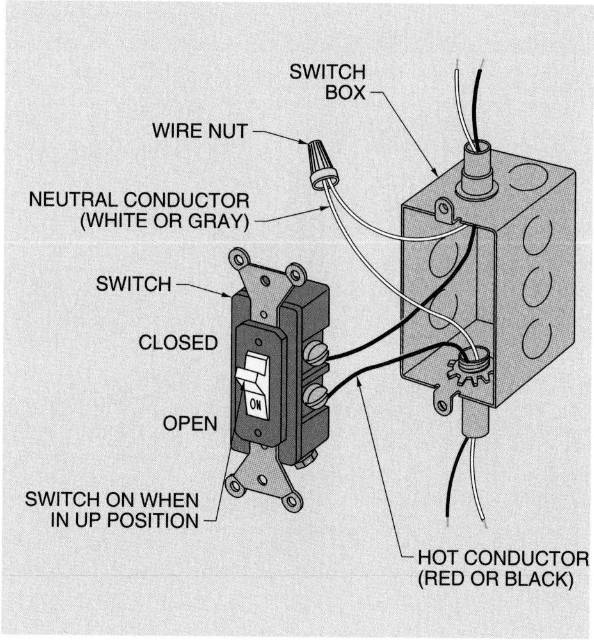

Figure 15-25. A switch is used to allow or prevent the flow of current in a circuit by switching the hot (live) conductor.

Conductors to and from a switch must be identified. The neutral conductor is grounded to earth and is always white or gray. Hot wires are generally red or black, although colors other than red or black are sometimes used. Switching neutral wires should never be used to open or close a circuit unless all the wires in the circuit are used for switching. Generally, only the hot wire is used for switching purposes.

Troubleshooting Switches. Testing a switch may be accomplished by using a voltage tester when the switch is connected to a voltage supply or by using a continuity tester when the switch is removed from the circuit. Checking a switch to determine that it is working properly is done by first checking the supply voltage to verify that the voltage is present at the switch terminals using a voltage tester. This is accomplished by checking between the supply line terminal of the switch and ground. Ground may be found as a separate wire (generally bare) within the conduit, cable, or raceway, or as the metal enclosure and conduit itself. A voltage should be indicated as one lead is connected to the supply side of a switch (hot side) and the other lead is connected to ground. A voltage reading should also be indicated when the switch is in the ON position and one tester lead is connected to the load side of the switch and the other lead is connected to ground. A voltage reading should not be indicated when the switch is in the OFF position and the tester leads are connected between the load side and ground. See Figure 15-26.

Testing a switch with a continuity tester is accomplished with the switch removed from electrical energy because a continuity tester uses its own power supply. A continuity tester is used for switch testing to ensure that the mechanisms within the switch are properly making or breaking contacts and allowing or preventing continuity of current flow. This is accomplished by connecting one of the tester leads to either of the switch terminals and the other lead to the remaining terminal. A switch in good operating condition indicates a light condition when the switch is in the ON position and a no light condition when the switch is in the OFF position. Continuity testers may also be used to give an indication of current paths on 3-way light switches.

Solenoids

A *solenoid* is a device that converts electrical energy to a linear, mechanical force. Solenoid use is common in the operation of valves, clutches, contactors, and motor starters. A solenoid consists of a plunger (armature) and a coil. A magnetic field is set up around the coil when the coil is energized by an electric current. The magnetic field causes the plunger to move into the coil. See Figure 15-27.

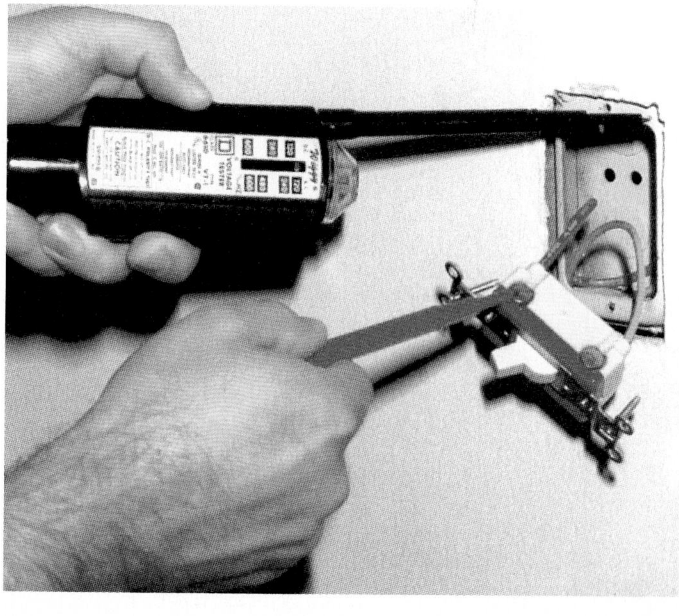

Voltage into a switch is tested by connecting one lead to the supply side of the switch and the other lead to ground.

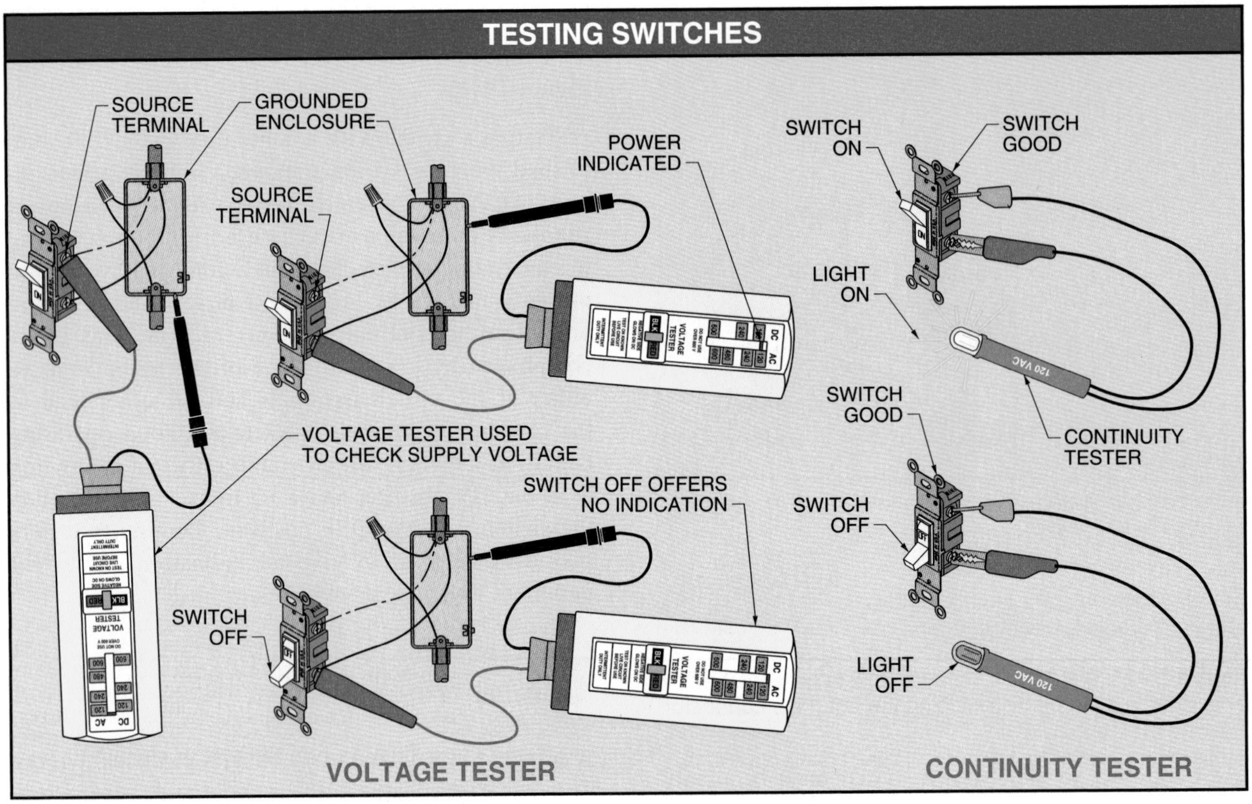

Figure 15-26. Testing a switch may be accomplished using a voltage tester when the switch is connected to a voltage supply or a continuity tester when the switch is removed from the circuit.

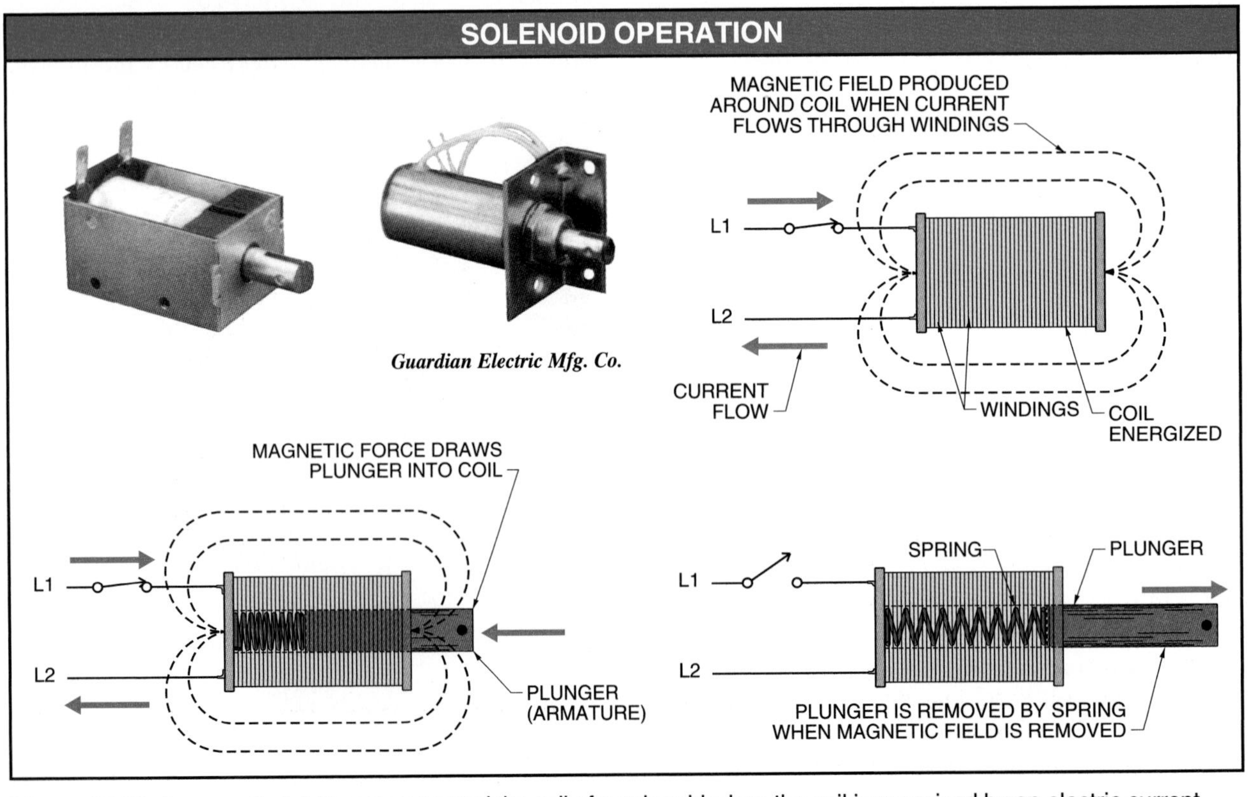

Figure 15-27. A magnetic field is set up around the coil of a solenoid when the coil is energized by an electric current.

The force of the plunger movement is transmitted into a useful function by being combined with or attached to some other device. For example, a plunger attached to a set of electrical contacts becomes a relay switch. A plunger attached to a friction pad becomes a brake or clutch. A plunger attached to a spool in a housing becomes a valve.

Troubleshooting Solenoids. To troubleshoot a solenoid, first check that the solenoid is capable of receiving the proper voltage when required. Then, lock out all electrical power to the solenoid or solenoid circuit. Double-check the electrical lockout by challenging the solenoid circuit. *Challenging* is the process of pressing or selecting the start switch of a machine to determine if the machine starts when it is not supposed to start. At this time, check for any voltages that may be coming from another source. Remove the solenoid cover and visually inspect for burnt, broken, or frozen parts. Verify that the plunger is free to move and is not jammed. Most solenoids have ways to be manually shifted.

Electrical troubleshooting begins by disconnecting the solenoid wires from the electrical circuit to check the coil resistance. Connect the leads of a multimeter set on the resistance setting to the solenoid wires. The condition of the solenoid is based on the meter reading. The solenoid is good if the meter is within ±15% of the normal coil or rated value. Unknown values may be found by comparing a good solenoid with that of the same type and size. A low or zero reading indicates a short in the solenoid coil windings. No reading or a reading of infinity indicates that the coil has a broken wire and is open. See Figure 15-28.

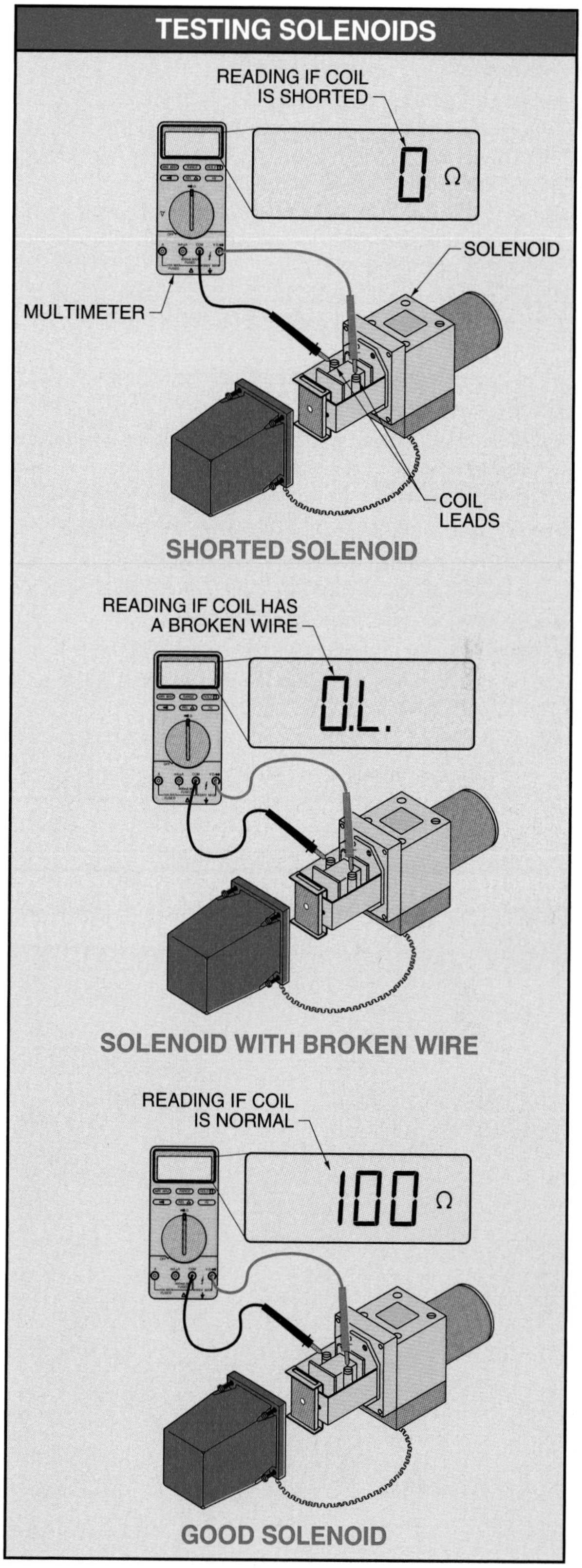

Figure 15-28. Solenoids are tested using a multimeter set on the resistance setting to check for shorts, broken wires, or normal resistance.

ELECTRICAL SAFETY

Improper electrical wiring or misuse of electricity causes destruction of equipment and fire damage to property. Electricity is the number one cause of fires. More than 100,000 people are killed in electrical fires each year. Safe working habits are required when troubleshooting an electrical circuit or component because the electric parts that are normally enclosed are exposed. Electrical safety rules should be practiced by all personnel working with electricity. See Figure 15-29.

ELECTRICAL SAFETY

- Always comply with the NEC®.
- Use UL® approved appliances, components, and equipment.
- Keep electrical grounding circuits in good condition. Ground any conductive component or element that does not have to be energized. The grounding connection must be a low-resistance conductor heavy enough to carry the largest fault current that may occur.
- Turn OFF, lockout, and tag disconnect switches when working on any electrical circuit or equipment. Test all circuits after they are turned OFF. Insulators may not insulate, grounding circuits may not ground, and switches may not open the circuit.
- Use double-insulated power tools or power tools that include a third conductor grounding terminal which provides a path for fault current. Never use a power tool that has the third conductor grounding terminal removed.
- Always use protective and safety equipment.
- Know what to do in an emergency.
- Check conductors, cords, components, and equipment for signs of wear or damage. Replace any equipment that is not safe.
- Never throw water on an electrical fire. Turn OFF the power and use a Class C rated fire extinguisher.
- Work with another individual when working in a dangerous area or with dangerous equipment.
- Learn CPR and first aid.
- Do not work when tired or taking medication that causes drowsiness.
- Do not work in poorly lighted areas.
- Always use nonconductive ladders. Never use a metal ladder when working around electrical equipment.
- Ensure there are no atmospheric hazards such as flammable dust or vapor in the area. A live electrical circuit may emit a spark at any time.
- Use one hand when working on a live circuit to reduce the chance of an electrical shock passing through the heart and lungs.
- Never bypass or disable fuses or circuit breakers.
- Extra care must be taken in an electrical fire because burning insulation produces toxic fumes.
- Always fill out accident forms and report any electrical shock.

Figure 15-29. Electrical safety rules should be practiced by all personnel working with electricity.

Fluke Corporation

Proper test and personal safety equipment must be used when testing any electrical circuit.

Electrical current always searches for the path of least resistance (direction back to ground). *Ground* is an electrical connection to the earth. A dead short (short circuit) occurs if an uncontrolled flow of current goes back to ground. A short circuit could overheat wires and burn up motors or switches and even start a fire. Also, electricity does not require wire to form a circuit or to complete its path. Electricity flows through gas, water, air, and even human flesh when given the opportunity.

The technician must make a conscious effort to remain safe when working around electric current. The severity at which an individual is affected by electrical shock depends on the rate of current flow through the body and the resistance to electricity of an individual's body. See Figure 15-30.

Over 1000 people are killed each year in the United States from electrical shock and over 65,000 injuries occur due to failure to properly control hazardous energy sources during maintenance.

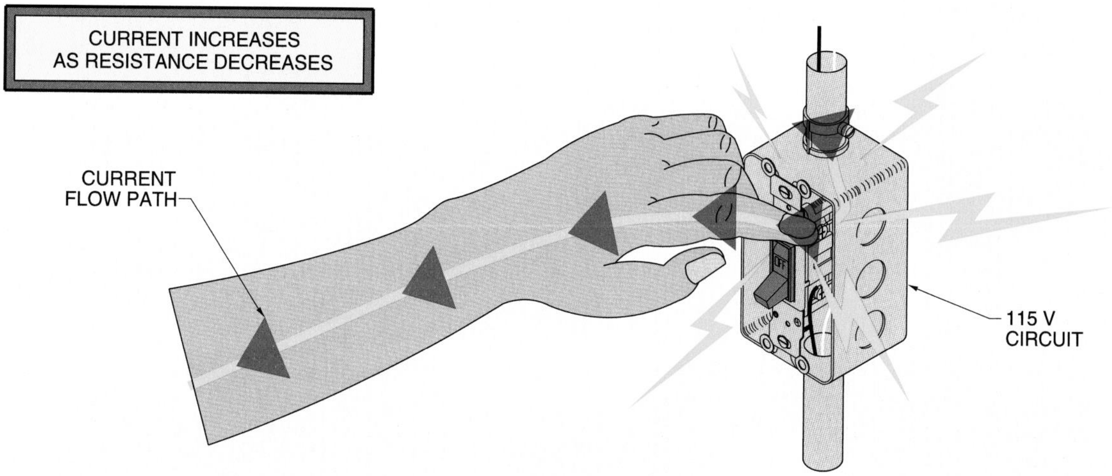

EFFECT OF ELECTRIC CURRENT	
Current (in mA)	**Effect On Body**
8 or less	Sensation of shock but probably not painful
8 to 15	Painful shock Removal from contact point by natural reflexes
15 to 20	Painful shock May be frozen or locked to point of electric contact until circuit is de-energized
over 20	Causes severe muscular contractions, paralysis of breathing, heart convulsions

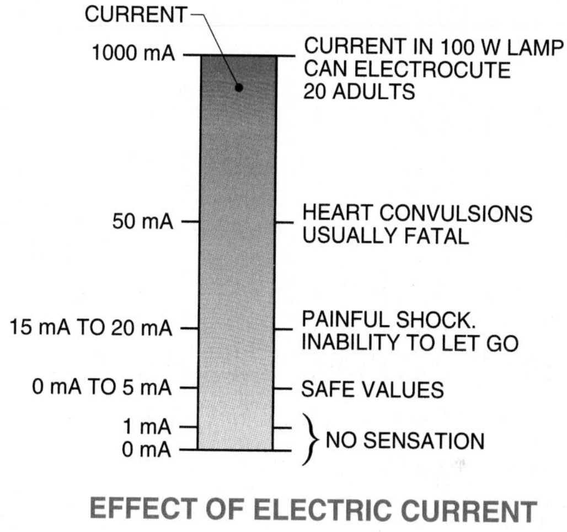

Figure 15-30. The severity at which an individual is affected by electrical shock depends on the rate of current flow through the body and the resistance to electricity of an individual's body.

A shock may not even be felt if the hands are dry and the floor is dry. However, an individual may be killed when perspiring and standing on a wet surface. Severe internal injury may occur if an electrical shock causes a person to freeze to the conductor. The longer the electrical contact, the greater the current flow and shock. The most dangerous path a current can take is a path through vital organs. Current flow from hand-to-hand, hand-to-head, or foot-to-hand or head can cause severe internal damage and paralysis, especially to the heart and lungs. Whenever possible, work with electricity using one hand. This lessens the chance of an accidental dead short through the body.

Only an authorized individual is to work on electrical systems or equipment because of the danger of electrical shock. An *authorized individual* is a knowledgeable individual to whom the authority and responsibility to perform a specific assignment has been given. Also, any individual who works with electricity must know, understand, and use all safety regulations, including lockout and tagout procedures. See Figure 15-31. Causes of the most severe injuries and deaths in industry include failure to shut OFF and stop a machine while working on it, failure to lockout or tagout energy sources, failure to release residual energy, such as accumulators or capacitors, accidental restarting of machinery, and failure to clear the work area and notify coworkers before reactivating a machine.

Panduit Corp.

Figure 15-31. Equipment must be locked out and tagged out before maintenance or servicing is performed.

Advanced Assembly Automation Inc.

Electrical devices should be wired in an orderly manner to promote efficiency and safety.

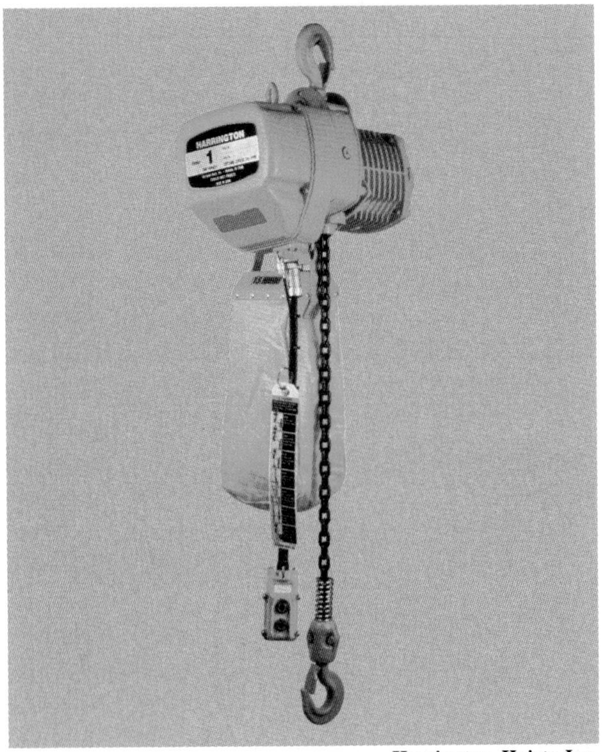

Harrington Hoists Inc.

The Peerless ES electric chain hoist has an electric limit-switch mechanism to automatically stop the hoist instantly in case of overlifting or overloading.

Appendix

English System . . . 348
Metric System . . . 349
Prefixes . . . 350
Conversion Table . . . 350
English to Metric Equivalents . . . 351
Metric to English Equivalents . . . 352
Stock Material Weight . . . 353
Hoisting Equipment Checklist . . . 354
AISI-SAE Designation System . . . 355
Unified Numbering System (UNS) for Metals and Alloys . . . 355
Standard Series Threads – Graded Pitches . . . 356
Metric Drill Sizes . . . 356
Inch – Millimeter Equivalents . . . 357
Twist Drill Fractional, Number, and Letter Sizes . . . 358
Metric Screw Threads . . . 358
Drilled Hole Tolerances . . . 358
Regular Nut Eyebolts . . . 359
Shoulder Nut Eyebolts . . . 360
Machinery Eyebolts . . . 361
Wire Rope Strength . . . 362
Sling Angle Loss Factors . . . 362
Choker Sling Loss Factors . . . 362
Filter Pressure Loss Constants . . . 363
Pipe Pressure Loss Constants (100 ft of Schedule 40 Pipe) . . . 363
Pipe Fitting Pressure Loss Constants . . . 363
Pipe . . . 364
Fluid Weights/Temperature Standards . . . 364
Fluid Power Graphic Symbols . . . 365
Logic Symbols . . . 368
Fluid Power Abbreviations/Acronyms . . . 369
Timing Belt Standard Pitch Lengths and Tolerances . . . 370
Timing Belt Standard Widths and Tolerances . . . 371
Allowable Tight Side Tension For AA Section V-Belts . . . 372
Allowable Tight Side Tension For BB Section V-Belts . . . 373
Allowable Tight Side Tension For CC Section V-Belts . . . 374
Allowable Tight Side Tension For DD Section V-Belts . . . 375
Vibration Characteristics . . . 376
Vibration Severity . . . 377
Three-Phase Voltage Values . . . 378
Power Formula Abbreviations and Symbols . . . 378
Ohm's Law and Power Formula . . . 378
Power Formulas – 1ϕ, 3ϕ . . . 378
Horsepower to Torque Conversion . . . 379
Motor Horsepower . . . 380
Common Nails . . . 381
Chemical Elements . . . 382

ENGLISH SYSTEM

		Unit	Abbr	Equivalents
LENGTH 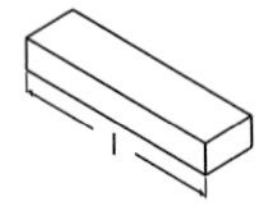		mile	mi	5280 , 320 rd, 1760 yd
		rod	rd	5.50 yd, 16.5
		yard	yd	3′, 36
		foot	ft *or*	12″, .333 yd
		inch	in. *or*	.083′, .028 yd
AREA $A = l \times w$ 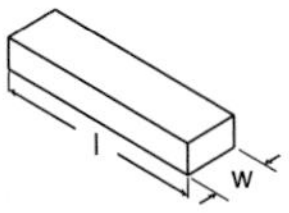		square mile	sq mi *or* mi^2	640 A, 102,400 sq rd
		acre	A	4840 sq yd, 43,560 sq ft
		square rod	sq rd *or* rd^2	30.25 sq yd, .00625 A
		square yard	sq yd *or* yd^2	1296 sq in., 9 sq ft
		square foot	sq ft *or* ft^2	144 sq in., .111 sq yd
		square inch	sq in. *or* in^2	.0069 sq ft, .00077 sq yd
VOLUME $V = l \times w \times t$ 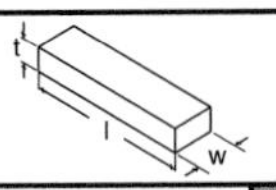		cubic yard	cu yd *or* yd^3	27 cu ft, 46,656 cu in.
		cubic foot	cu ft *or* ft^3	1728 cu in., .0370 cu yd
		cubic inch	cu in. *or* in^3	.00058 cu ft, .000021 cu yd
CAPACITY: WATER, FUEL, ETC. 	*U.S. liquid measure*	gallon	gal.	4 qt (231 cu in.)
		quart	qt	2 pt (57.75 cu in.)
		pint	pt	4 gi (28.875 cu in.)
		gill	gi	4 fl oz (7.219 cu in.)
		fluidounce	fl oz	8 fl dr (1.805 cu in.)
		fluidram	fl dr	60 min (.226 cu in.)
		minim	min	1/6 fl dr (.003760 cu in.)
VEGETABLES, GRAIN, ETC. 	*U.S. dry measure*	bushel	bu	4 pk (2150.42 cu in.)
		peck	pk	8 qt (537.605 cu in.)
		quart	qt	2 pt (67.201 cu in.)
		pint	pt	1/2 qt (33.600 cu in.)
DRUGS 	*British imperial liquid and dry measure*	bushel	bu	4 pk (2219.36 cu in.)
		peck	pk	2 gal. (554.84 cu in.)
		gallon	gal.	4 qt (277.420 cu in.)
		quart	qt	2 pt (69.355 cu in.)
		pint	pt	4 gi (34.678 cu in.)
		gill	gi	5 fl oz (8.669 cu in.)
		fluidounce	fl oz	8 fl dr (1.7339 cu in.)
		fluidram	fl dr	60 min (.216734 cu in.)
		minim	min	1/60 fl dr (.003612 cu in.)
MASS AND WEIGHT: COAL, GRAIN, ETC. 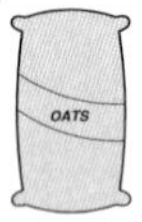	*avoirdupois*	ton		2000 lb
		short ton	t	2000 lb
		long ton		2240 lb
		pound	lb *or* #	16 oz, 7000 gr
		ounce	oz	16 dr, 437.5 gr
		dram	dr	27.344 gr, .0625 oz
		grain	gr	.037 dr, .002286 oz
GOLD, SILVER, ETC. 	*troy*	pound	lb	12 oz, 240 dwt, 5760 gr
		ounce	oz	20 dwt, 480 gr
		pennyweight	dwt *or* pwt	24 gr, .05 oz
		grain	gr	.042 dwt, .002083 oz
DRUGS 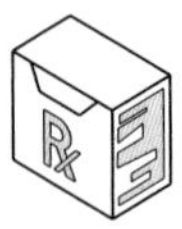	*apothecaries'*	pound	lb ap	12 oz, 5760 gr
		ounce	oz ap	8 dr ap, 480 gr
		dram	dr ap	3 s ap, 60 gr
		scruple	s ap	20 gr, .333 dr ap
		grain	gr	.05 s, .002083 oz, .0166 dr ap

METRIC SYSTEM

	Unit	Abbreviation	Number of Base Units
LENGTH 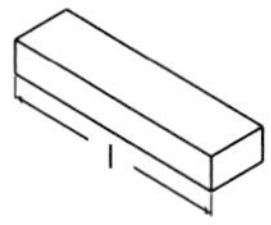	kilometer	km	1000
	hectometer	hm	100
	dekameter	dam	10
	meter*	m	1
	decimeter	dm	.1
	centimeter	cm	.01
	millimeter	mm	.001
AREA A = l x w 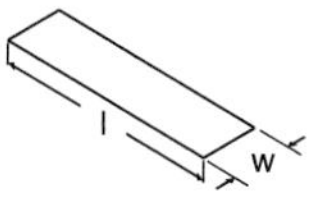	square kilometer	sq km *or* km^2	1,000,000
	hectare	ha	10,000
	are	a	100
	square centimeter	sq cm *or* cm^2	.0001
VOLUME V = l x w x t 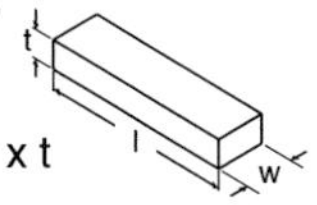	cubic centimeter	cu cm, cm^3, *or* cc	.000001
	cubic decimeter	dm^3	.001
	cubic meter*	m^3	1
CAPACITY	kiloliter	kl	1000
WATER, FUEL, ETC. 	hectoliter	hl	100
	dekaliter	dal	10
	liter*	l	1
VEGETABLES, GRAIN, ETC. 	cubic decimeter	dm^3	1
	deciliter	dl	.10
DRUGS 	centiliter	cl	.01
	milliliter	ml	.001
MASS AND WEIGHT	metric ton	t	1,000,000
COAL, GRAIN, ETC. 	kilogram	kg	1000
	hectogram	hg	100
GOLD, SILVER, ETC. 	dekagram	dag	10
	gram*	g	1
DRUGS 	decigram	dg	.10
	centigram	cg	.01
	milligram	mg	.001

* base units

PREFIXES			
Multiples and Submultiples	**Prefixes**	**Symbols**	**Meaning**
$1,000,000,000,000 = 10^{12}$	tera	T	trillion
$1,000,000,000 = 10^{9}$	giga	G	billion
$1,000,000 = 10^{6}$	mega	M	million
$1000 = 10^{3}$	kilo	k	thousand
$100 = 10^{2}$	hecto	h	hundred
$10 = 10^{1}$	deka	d	ten
Unit $1 = 10^{0}$			
$.1 = 10^{-1}$	deci	d	tenth
$.01 = 10^{-2}$	centi	c	hundredth
$.001 = 10^{-3}$	milli	m	thousandth
$.000001 = 10^{-6}$	micro	μ	millionth
$.000000001 = 10^{-9}$	nano	n	billionth
$.000000000001 = 10^{-12}$	pico	p	trillionth

CONVERSION TABLE												
Initial Units	**Final Units**											
	giga	**mega**	**kilo**	**hecto**	**deka**	**base**	**deci**	**centi**	**milli**	**micro**	**nano**	**pico**
giga		3R	6R	7R	8R	9R	10R	11R	12R	15R	18R	21R
mega	3L		3R	4R	5R	6R	7R	8R	9R	12R	15R	18R
kilo	6L	3L		1R	2R	3R	4R	5R	6R	9R	12R	15R
hecto	7L	4L	1L		1R	2R	3R	4R	5R	8R	11R	14R
deka	8L	5L	2L	1L		1R	2R	3R	4R	7R	10R	13R
base	9L	6L	3L	2L	1L		1R	2R	3R	6R	9R	12R
deci	10L	7L	4L	3L	2L	1L		1R	2R	5R	8R	11R
centi	11L	8L	5L	4L	3L	2L	1L		1R	4R	7R	10R
milli	12L	9L	6L	5L	4L	3L	2L	1L		3R	6R	9R
micro	15L	12L	9L	8L	7L	6L	5L	4L	3L		3R	6R
nano	18L	15L	12L	11L	10L	9L	8L	7L	6L	3L		3R
pico	21L	18L	15L	14L	13L	12L	11L	10L	6L	6L	3L	

R = move the decimal point to the right
L = move the decimal point to the left

ENGLISH TO METRIC EQUIVALENTS

		Unit	Metric Equivalent
LENGTH 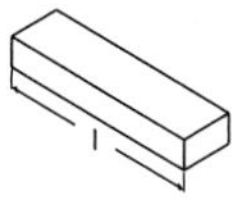		mile	1.609 km
		rod	5.029 m
		yard	.9144 m
		foot	30.48 cm
		inch	2.54 cm
AREA A = l x w 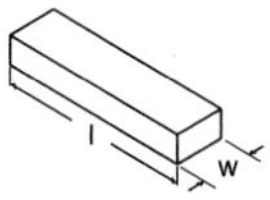		square mile	2.590 k^2
		acre	.405 hectacre, 4047 m^2
		square rod	25.293 m^2
		square yard	.836 m^2
		square foot	.093 m^2
		square inch	6.452 cm^2
VOLUME V = l x w x t 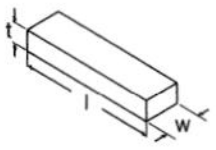		cubic yard	.765 m^3
		cubic foot	.028 m^3
		cubic inch	16.387 cm^3
CAPACITY WATER, FUEL, ETC. 	*U.S. liquid measure*	gallon	3.785 l
		quart	.946 l
		pint	.473 l
		gill	118.294 ml
		fluidounce	29.573 ml
		fluidram	3.697 ml
		minim	.061610 ml
VEGETABLES, GRAIN, ETC. 	*U.S. dry measure*	bushel	35.239 l
		peck	8.810
		quart	1.101 l
		pint	.551 l
DRUGS 	*British imperial liquid and dry measure*	bushel	.036 m^3
		peck	.0091 m^3
		gallon	4.546 l
		quart	1.136 l
		pint	568.26 cm^3
		gill	142.066 cm^3
		fluidounce	28.412 cm^3
		fluidram	3.5516 cm^3
		minim	.059194 cm^3
MASS AND WEIGHT COAL, GRAIN, ETC. 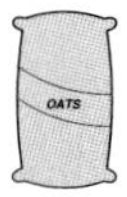	*avoirdupois*	short ton	.907 t
		long ton	1.016 t
		pound	.454 kg
		ounce	28.350 g
		dram	1.772 g
		grain	.0648 g
GOLD, SILVER, ETC. 	*troy*	pound	.373 kg
		ounce	31.103 g
		pennyweight	1.555 g
		grain	.0648 g
DRUGS 	*apothecaries'*	pound	.373 kg
		ounce	31.103 g
		dram	3.888 g
		scruple	1.296 g
		grain	.0648 g

METRIC TO ENGLISH EQUIVALENTS

	Unit	English Equivalent		
LENGTH 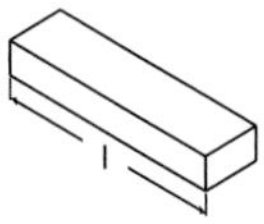	kilometer	.62 mi		
	hectometer	109.36 yd		
	dekameter	32.81′		
	meter	39.37″		
	decimeter	3.94″		
	centimeter	.39″		
	millimeter	.039″		
AREA A = l x w 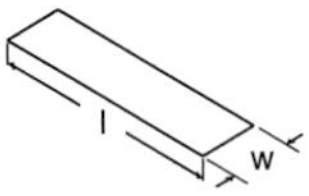	square kilometer	.3861 sq mi		
	hectacre	2.47 A		
	are	119.60 sq yd		
	square centimeter	.155 sq in.		
VOLUME V = l x w x t 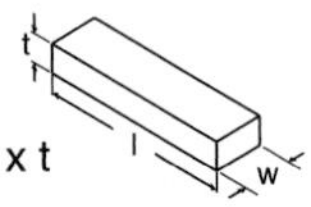	cubic centimeter	.061 cu in.		
	cubic decimeter	61.023 cu in.		
	cubic meter	1.307 cu yd		
CAPACITY		***cubic***	***dry***	***liquid***
WATER, FUEL, ETC. 	kiloliter	1.31 cu yd		
	hectoliter	3.53 cu ft	2.84 bu	
VEGETABLES, GRAIN, ETC. 	dekaliter	.35 cu ft	1.14 pk	2.64 gal.
	liter	61.02 cu in.	.908 qt	1.057 qt
DRUGS 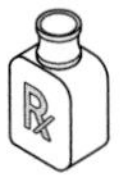	cubic decimeter	61.02 cu in.	.908 qt	1.057 qt
	deciliter	6.1 cu in.	.18 pt	.21 pt
	centiliter	.61 cu in.		338 fl oz
	milliliter	.061 cu in.		.27 fl dr
MASS AND WEIGHT	metric ton	1.102 t		
COAL, GRAIN, ETC. 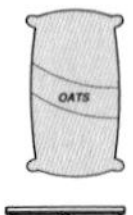	kilogram	2.2046 lb		
	hectogram	3.527 oz		
GOLD, SILVER, ETC. 	dekagram	.353 oz		
	gram	.035 oz		
DRUGS 	decigram	1.543 gr		
	centigram	.154 gr		
	milligram	.015 gr		

STOCK MATERIAL WEIGHT*

Material	Weight	Material	Weight	Material	Weight
METALS		Chestnut	30	Granite	172
Aluminum, cast hammered	165	Cypress, southern	32	Greenstone, trap	187
Aluminum, bronze	481	Douglas fir	34	Gypsum, alabaster	159
Antimony	416	Elm, American	35	Limestone	160
Arsenic	358	Hemlock, eastern, western	28	Magnesite	187
Bismuth	608	Hickory	53	Marble	168
Brass, cast-rolled	534	Larch, western	36	Phosphate rock, apatite	200
Chromium	428	Maple, red, black	38 – 40	Pumice, natural	40
Cobalt	552	Oak	51	Quartz, flint	165
Copper, cast-rolled	556	Pine, white, yellow, western	27 – 28	Sandstone, bluestone	147
Gold, cast-hammered	1205	Poplar, yellow	28	Slate, shale	172
Iron, cast, pig	450	Redwood	30	Soapstone, talc	169
Iron, wrought	485	Spruce	28	**BITUMINOUS SUBSTANCES**	
Iron, slag	172	Tamarack	37	Asphaltum	81
Lead	706	Walnut	39 – 40	Coal, anthracite	97
Magnesium	109	**LIQUIDS**		Coal, bituminous	84
Manganese	456	Alcohol, 100%	49	Coal, lignite	78
Mercury	848	Acids, muriatic, 40%	75	Coal, coke	75
Molybdenum	562	Acids, nitric, 91%	94	Graphite	131
Nickel	545	Acids, sulphuric, 87%	112	Paraffin	56
Platinum, cast-hammered	1330	Lye, soda, 66%	106	Petroleum, crude	55
Silver, cast-hammered	656	Oils	58	Petroleum, refined	50
Steel	490	Petroleum	55	Pitch	69
Tin, cast-hammered	459	Gasoline	42	Tar, bituminous	75
Tungsten	1180	Water, 4°C	62	**BRICK MASONRY**	
Vanadium	350	Water, ice	56	Pressed brick	140
Zinc, cast-rolled	440	Water, snow, fresh fallen	8	Common brick	120
SOLIDS		Water, sea water	64	Soft brick	100
Carbon, amorphous, graphitic	129	**GASES**		CONCRETE	
Cork	15	Air, 0°C	.08071	Cement, stone, sand	144
Ebony	76	Ammonia	.0478	Cement, slag, etc.	130
Fats	58	Carbon dioxide	.1234	Cement, cinder, etc.	100
Glass, common, plate	160	Carbon monoxide	.0781	**BUILDING MATERIAL**	
Glass, crystal	184	Gas, natural	.038 – .039	Ashes, cinders	40 – 45
Phosphorous, white	114	Hydrogen	.00559	Cement, Portland, loose	90
Resins, rosin, amber	67	Nitrogen	.0784	Cement, Portland, set	183
Rubber	58	Oxygen	.0892	Lime, gypsum, loose	65 – 75
Silicon	155	**MINERALS**		Mortar, set	103
Sulphur, Amorphous	128	Asbestos	153	Slags, bank slag	67 – 72
Wax	60	Basalt	184	Slags, bank, screenings	98 – 117
TIMBER, U.S. SEASONED		Bauxite	159	Slags, machine slag	96
Ash, white	41	Borax	109	**EARTH**	
Beech	44	Chalk	137	Clay, damp, plastic	110
Birch, yellow	43	Clay	137	Dry, packed	95
Cedar, white, red	22 – 23	Dolomite	181	Mud, packed	115

* in lb/cu ft

HOISTING EQUIPMENT CHECKLIST

1. Prior to Installation:

- ❑ Check for any possible damage during shipment. Do not install a damaged hoist.
- ❑ Check all lubricant levels.
- ❑ Check wire rope for damage if hoist wire-rope type. Be sure wire rope is properly seated in drum grooves and sheaves.
- ❑ Check chain for damage if hoist is chain type. Be sure chain properly enters sprockets and chain guiding points.
- ❑ Check to be sure that power supply shown on serial plate of hoist is the same as the power supply planned for connection to the hoist.

2. Installation:

- ❑ Install stationary mounting or trolley mounting to monorail beam exactly as instructed by the manufacturer's instructions.
- ❑ Check supporting structure, including monorail, to make sure it has a load rating equal to that of the hoist installed.

3. Power Supply:

- ❑ Make sure all electrical connections are made in accordance with manufacturer's wiring diagram, which is usually found inside the cover of the control enclosure.
- ❑ Make sure electrical supply system is in compliance with the National Electrical Code®.

4. Phase Connections:

- ❑ Depress the UP button on the pendant control to determine the direction of hook travel. If hook travel is upward, the hoist is properly phased. If it is downward, discontinue operation until phasing is corrected.
- ❑ Correct power connections if hoist is improperly phased by changing any two power line leads to the hoist. Never change internal wiring connections in the hoist or pendant control.
- ❑ Recheck operation of hoist after interchanging power line leads to confirm proper direction of motion.

5. Upper Limit Switch:

- ❑ Raise unloaded hook until it is approximately 1′ below the upper limit switch trip point. Slowly jog hook upward until hook can be raised no further. Lower block about 2′ and raise without jogging until limit switch trips and hook can be raised no further.
- ❑ Disconnect power supply and check all electrical connections if upper limit switch does not operate, or trip point is too close to hoist.
- ❑ Make any necessary adjustments.
- ❑ Reconnect power supply and recheck hoist operation after checking connections or making adjustments.

6. Lower Limit Switch:

- ❑ Check operation of hoist having a lower limit switch in same manner as for one with an upper limit switch. Never adjust lower limit switch to a point where less than one wrap of wire rope remains on the drum.

7. Lower Hook Travel (when hoist does not have lower limit switch):

- ❑ Lower the unloaded hook to its lowest possible operating point, or, for wire rope hoists, until two full wraps of wire remain on the drum.
- ❑ If it appears that less than two wraps of wire rope will be on the drum at the lowest possible operating point, the hoist cannot be installed or used unless it is equipped with a lower limit device.

8. Trolley Operation:

- ❑ Operate a trolley-mounted hoist over its entire travel distance on a monorail beam while the hoist is unloaded to check all clearances and verify that no interference occurs.

9. Braking System:

- ❑ Raise and lower hook, without load, stopping the motion at several points to test the operation of the brakes.
- ❑ Raise hook with capacity load several inches and stop to check that brake holds the load and that the load does not drift downward. If drift does not occur, raise and lower hook with capacity load, stopping the motion at several points to test the operation of the brakes.

10. Load Test:

- ❑ Load test the hoist with a load equal to 125% of the rated capacity load. If the hoist is equipped with a load limiting device that prevents the lifting of 125% of the rated load, testing should be accomplished with a load equal to 100% of the rated capacity load, followed by a test to check the function of the load limiting device.

11. Filing the Report:

- ❑ Prepare written report outlining installation procedures, problems encountered, and results of all checks and tests conducted. This report should indicate the approval or certification of the equipment for plant use, and should be signed by the responsible individual and filed in the equipment folder.

12. Operating Instructions:

- ❑ Issue instructions for hoist operators based on instructions and warning in hoist manufacturer's manual.
- ❑ Check warning tag or label on the hoist and make sure it stays there. Warning tag is a recent code requirement for new equipment. It is highly recommended for existing equipment. The warning tag should contain the following message:

WARNING:

To avoid injury, do not:

- lift more than rated load
- lift people or load over people
- operate with twisted, kinked, or damaged rope or chain
- operate damaged or malfunctioning hoist
- make side pulls that misalign rope or chain with hoist
- operate if rope is not seated in groove or chain in pockets
- operate unless travel devices limit function; test each shift
- operate hand-powered hoist except with hand power

AISI-SAE DESIGNATION SYSTEM					
Type of Steel	**Numbers and Digits**	**Nominal Alloy Content**	**Type of Steel**	**Numbers and Digits**	**Nominal Alloy Content**
Carbon	10xx	Plain carbon (1% Mn max)	**Nickel-chromium-molybdenum cont.**	87xx	.55% Ni; .50% Cr; .25% Mo
				88xx	.55% Ni; .50% Cr; .35% Mo
				93xx	3.25% Ni; 1.20% Cr; .12% Mo
	11xx	Resulfurized		94xx	.45% Ni; .40% Cr; .12% Mo
	12xx	Resulfurized and rephosphorized		97xx	.55% Ni; .20% Cr; .20% Mo
	15xx	Plain carbon (1.00% Mn — 1.65% Mn max)		98xx	1.00% Ni; .80% Cr; .25% Mo
Manganese	13xx	1.75% Mn	**Nickel-molybdenum**	46xx	.85% Ni and 1.82% Ni; .20% Mo and .25% Mo
Nickel	23xx	3.5% Ni		48xx	3.50% Ni; .25% Mo
	25xx	5% Ni	**Chromium**	50xx	.27% Cr, .40% Cr, .50% Cr, and .65% Cr
Nickel-chromium	31xx	1.25% Ni; .65% Cr and .80% Cr		51xx	.80% Cr, .87% Cr, .92% Cr, .95% Cr, 1.00% Cr, and 1.05% Cr
	32xx	1.75% Ni; 1.07% Cr	**Chromium**	50xxx	.50% Cr (C 1.00% min)
	33xx	3.50% Ni; 1.50% Cr and 1.57% Cr		51xxx	1.02% Cr (C 1.00% min)
	34xx	3.00% Ni; .77% Cr		52xxx	1.45% Cr (C 1.00% min)
Molybdenum	40xx	.20% Mo and .25% Mo	**Chromium-vanadium**	61xx	.60% Cr, .80% Cr, and .95% Cr; .10% V and .15% V min
	44xx	.40% Mo and .52% Mo	**Tungsten-chromium**	72xx	1.75% W; .75% Cr
Chromium-molybdenum	41xx	.50% Cr, .80% Cr, and .95% Cr; .12% Mo, .20% Mo, .25% Mo, and .30% Mo	**Silicon-manganese**	92xx	1.40% Si and 2.00% Si; .65% Mn, .82% Mn, and .85% Mn; 0% Cr and .65% Cr
Nickel-chromium-molybdenum	43xx	1.82% Ni; .50% Cr and .80% Cr; .25% Mo	**High-strength low-alloy**	9xx	Various SAE grades
	43BVxx	1.82% Ni; .50% Cr; .12% Mo and .25% Mo; .03% V min	**Boron**	xxBxx	B denotes Boron steel
	47xx	1.05%Ni; .45%Cr; .20% Mo and .35% Mo	**Leaded**	xxLxx	L denotes Leaded steel
	81xx	.30% Ni; .40% Cr; .12% Mo			
	86xx	.55% Ni; .50% Cr; .20% Mo			

UNIFIED NUMBERING SYSTEM (UNS) FOR METALS AND ALLOYS			
UNS Series	**Metal**	**UNS Series**	**Metal**
Nonferrous Metals and Alloys		**Ferrous Metals and Alloys**	
A00001 to A99999	Aluminum and aluminum alloys	D00001 to D99999	Specified mechanical property steels
C00001 to C99999	Copper and copper alloys	F00001 to F99999	Cast irons
E00001 to E99999	Rare earth and rare earth-like metals and alloys	G00001 to G99999	AISI and SAE carbon and alloy steels (except tool steels)
L00001 to L99999	Low melting metals and alloys	H00001 to H99999	AISI H-steels
M00001 to M99999	Miscellaneous nonferrous metals and alloys	J00001 to J99999	Cast steels (except tool steels)
P00001 to P99999	Precious metals and alloys	K00001 to K99999	Miscellaneous steels and ferrous alloys
R00001 to R99999	Reactive and refractory metals and alloys	S00001 to S99999	Heat and corrosion resistant (stainless) steels
Z00001 to Z99999	Zinc and zinc alloys	T00001 to T99999	Tool steels

STANDARD SERIES THREADS – GRADED PITCHES

NOMINAL DIAMETER	UNC		UNF		UNEF	
	TPI	TAP DRILL	TPI	TAP DRILL	TPI	TAP DRILL
0 (.0600)			80	3/64		
1 (.0730)	64	No. 53	72	No. 53		
2 (.0860)	56	No. 50	64	No. 50		
3 (.0990)	48	No. 47	56	No. 45		
4 (.1120)	40	No. 43	48	No. 42		
5 (.1250)	40	No. 38	44	No. 37		
6 (.1380)	32	No. 36	40	No. 33		
8 (.1640)	32	No. 29	36	No. 29		
10 (.1900)	24	No. 25	32	No. 21		
12 (.2160)	24	No. 16	28	No. 14	32	No.13
1/4 (.2500)	20	No. 7	28	No. 3	32	7/32
5/16 (.3125)	18	F	24	I	32	9/32
3/8 (.3750)	16	5/16	24	Q	32	11/32
7/16 (.4375)	14	U	20	25/64	28	13/32
1/2 (.5000)	13	27/64	20	29/64	28	15/32
9/16 (.5625)	12	31/64	18	33/64	24	33/64
5/8 (.6250)	11	17/32	18	37/64	24	37/64
11/16 (.6875)					24	41/64
3/4 (.7500)	10	21/32	16	11/16	20	45/64
13/16 (.8125)					20	49/64
7/8 (.8750)	9	49/64	14	13/16	20	53/64
15/16 (.9375)					20	57/64
1 (1.000)	8	7/8	12	59/64	20	61/64

METRIC DRILL SIZES

DRILL DIAMETER						DRILL DIAMETER					
mm	in.	mm	in.	mm	in.	mm	in.	mm	in.	mm	in.
.40	.0157	1.95	.0768	4.70	.1850	8.00	.3150	13.20	.5197	25.50	1.0039
.42	.0165	2.00	.0787	4.80	.1890	8.10	.3189	13.50	.5315	26.00	1.0236
.45	.0177	2.05	.0807	4.90	.1929	8.20	.3228	13.80	.5433	26.50	1.0433
.48	.0189	2.10	.0827	5.00	.1969	8.30	.3268	14.00	.5512	27.00	1.0630
.50	.0197	2.15	.0846	5.10	.2008	8.40	.3307	14.25	.5610	27.50	1.0827
.55	.0217	2.20	.0866	5.20	.2047	8.50	.3346	14.50	.5709	28.00	1.1024
.60	.0236	2.25	.0886	5.30	.2087	8.60	.3386	14.75	.5807	28.50	1.1220
.65	.0256	2.30	.0906	5.40	.2126	8.70	.3425	15.00	.5906	29.00	1.1417
.70	.0276	2.35	.0925	5.50	.2165	8.80	.3465	15.25	.6004	29.50	1.1614
.75	.0295	2.40	.0945	5.60	.2205	8.90	.3504	15.50	.6102	30.00	1.1811
.80	.0315	2.45	.0965	5.70	.2244	9.00	.3543	15.75	.6201	30.50	1.2008
.85	.0335	2.50	.0984	5.80	.2283	9.10	.3583	16.00	.6299	31.00	1.2205
.90	.0354	2.60	.1024	5.90	.2323	9.20	.3622	16.25	.6398	31.50	1.2402
.95	.0374	2.70	.1063	6.00	.2362	9.30	.3661	16.50	.6496	32.00	1.2598
1.00	.0394	2.80	.1102	6.10	.2402	9.40	.3701	16.75	.6594	32.50	1.2795
1.05	.0413	2.90	.1142	6.20	.2441	9.50	.3740	17.00	.6693	33.00	1.2992
1.10	.0433	3.00	.1181	6.30	.2480	9.60	.3780	17.25	.6791	33.50	1.3189
1.15	.0453	3.10	.1220	6.40	.2520	9.70	.3819	17.50	.6890	34.00	1.3386
1.20	.0472	3.20	.1260	6.50	.2559	9.80	.3858	18.00	.7087	34.50	1.3583
1.25	.0492	3.30	.1299	6.60	.2598	9.90	.3898	18.50	.7283	35.00	1.3780
1.30	.0512	3.40	.1339	6.70	.2638	10.00	.3937	19.00	.7480	35.50	1.3976
1.35	.0531	3.50	.1378	6.80	.2677	10.20	.4016	19.50	.7677	36.00	1.4173
1.40	.0551	3.60	.1417	6.90	.2717	10.50	.4134	20.00	.7874	36.50	1.4370
1.45	.0571	3.70	.1457	7.00	.2756	10.80	.4252	20.50	.8071	37.00	1.4567
1.50	.0591	3.80	.1496	7.10	.2795	11.00	.4331	21.00	.8268	37.50	1.4764
1.55	.0610	3.90	.1535	7.20	.2835	11.20	.4409	21.50	.8465	38.00	1.4961
1.60	.0630	4.00	.1575	7.30	.2874	11.50	.4528	22.00	.8661	40.00	1.5748
1.65	.0650	4.10	.1614	7.40	.2913	11.80	.4646	22.50	.8858	42.00	1.6535
1.70	.0669	4.20	.1654	7.50	.2953	12.00	.4724	23.00	.9055	44.00	1.7323
1.75	.0689	4.30	.1693	7.60	.2992	12.20	.4803	23.50	.9252	46.00	1.8110
1.80	.0709	4.40	.1732	7.70	.3031	12.50	.4921	24.00	.9449	48.00	1.8898
1.85	.0728	4.50	.1772	7.80	.3071	12.80	.5039	24.50	.9646	50.00	1.9685
1.90	.0748	4.60	.1811	7.90	.3110	13.00	.5118	25.00	.9843		

INCH — MILLIMETER EQUIVALENTS*

Inches		MM	Inches		MM	Inches		MM
	.00004	.001		.11811	3		.550	13.970
	.00039	.01	1/8	.1250	3.175		.55118	14
	.00079	.02		.13780	3.5	9/16	.56250	14.2875
	.001	.025	9/64	.14063	3.5719		.57087	14.5
	.00118	.03		.150	3.810	37/64	.57813	14.6844
	.00157	.04	5/32	.15625	3.9688		.59055	15
	.00197	.05		.15748	4	19/32	.59375	15.0812
	.002	.051	11/64	.17188	4.3656		.600	15.24
	.00236	.06		.1750	4.445	39/64	.60938	15.4781
	.00276	.07		.17717	4.5		.61024	15.5
	.003	.0762	3/16	.18750	4.7625	5/8	.6250	15.875
	.00315	.08		.19685	5		.62992	16
	.00354	.09		.20	5.08	41/64	.64063	16.2719
	.00394	.1	13/64	.20313	5.1594		.64961	16.5
	.004	.1016		.21654	5.5		.650	16.51
	.005	.1270	7/32	.21875	5.5562	21/32	.65625	16.6688
	.006	.1524		.2250	5.715		.66929	17
	.007	.1778	15/64	.23438	5.9531	43/64	.67188	17.0656
	.00787	.2		.23622	6	11/16	.68750	17.4625
	.008	.2032	1/4	.250	6.35		.68898	17.5
	.009	.2286					.700	17.78
	.00984	.25		.25591	6.5	45/64	.70313	17.8594
	.01	.254	17/64	.26563	6.7469		.70866	18
	.01181	.3		.275	6.985	23/32	.71875	18.2562
1/64	.01563	.3969		.27559	7		.72835	18.5
	.01575	.4	9/32	.28125	7.1438	47/64	.73438	18.6531
	.01969	.5		.29528	7.5		.74803	19
	.02	.508	19/64	.29688	7.5406	3/4	.750	19.050
	.02362	.6		.30	7.62			
	.025	.635	5/16	.3125	7.9375	49/64	.76563	19.4469
	.02756	.7		.31496	8		.76772	19.5
	.0295	.75	21/64	.32813	8.3344	25/32	.78125	19.8438
	.03	.762		.33465	8.5		.78740	20
1/32	.03125	.7938	11/32	.34375	8.7375	51/64	.79688	20.2406
	.0315	.8		.350	8.89		.800	20.320
	.03543	.9		.35433	9		.80709	20.5
	.03937	1	23/64	.35938	9.1281	13/16	.81250	20.6375
	.04	1.016		.37402	9.5		.82677	21
3/64	.04687	1.191	3/8	.375	9.525	53/64	.82813	21.0344
	.04724	1.2	25/64	.39063	9.9219	27/32	.84375	21.4312
	.05	1.27		.39370	10		.84646	21.5
	.05512	1.4		.400	10.16		.850	21.590
	.05906	1.5	13/32	.40625	10.3188	55/64	.85938	21.8281
	.06	1.524		.41339	10.5		.86614	22
1/16	.06250	1.5875	27/64	.42188	10.7156	7/8	.875	22.225
	.06299	1.6		.43307	11		.88583	22.5
	.06693	1.7	7/16	.43750	11.1125	57/64	.89063	22.6219
	.07	1.778		.450	11.430		.900	22.860
	.07087	1.8		.45276	11.5		.90551	23
	.075	1.905	29/64	.45313	11.5094	29/32	.90625	23.0188
5/64	.07813	1.9844	15/32	.46875	11.9062	59/64	.92188	23.4156
	.07874	2		.47244	12		.92520	23.5
	.08	2.032	31/64	.48438	12.3031	15/16	.93750	23.8125
	.08661	2.2		.49213	12.5		.94488	24
	.09	2.286	1/2	.50	12.7		.950	24.130
	.09055	2.3				61/64	.95313	24.2094
3/32	.09375	2.3812		.51181	13		.96457	24.5
	0.9843	2.5	33/64	.51563	13.0969	31/32	.96875	24.6062
	.1	2.54	17/32	.53125	13.4938		.98425	25
	.10236	2.6		.53150	13.5	63/64	.98438	25.0031
7/64	.10937	2.7781	35/64	.54688	13.8906	1	1.0000	25.4

TWIST DRILL FRACTIONAL, NUMBER, AND LETTER SIZES

Drill No.	Frac	Deci	Drill No.	Frac	Deci	Drill No.	Frac	Deci	Drill No.	Frac	Deci
80	—	.0135	42	—	.0935	7	—	.201	X	—	.397
79	—	.0145	—	3/32	.0938	—	13/64	.203	Y	—	.404
—	1/64	.0156				6	—	.204			
78	—	.0160	41	—	.0960	5	—	.206	—	13/32	.406
77	—	.0180	40	—	.0980	4	—	.209	Z	—	.413
			39	—	.0995				—	27/64	.422
76	—	.0200	38	—	.1015	3	—	.213	—	7/16	.438
75	—	.0210	37	—	.1040	—	7/32	.219	—	29/64	.453
74	—	.0225				2	—	.221			
73	—	.0240	36	—	.1065	1	—	.228	—	15/32	.469
72	—	.0250	—	7/64	.1094	A	—	.234	—	31/64	.484
			35	—	.1100				—	1/2	.500
71	—	.0260	34	—	.1110	—	15/64	.234	—	33/64	.516
70	—	.0280	33	—	.1130	B	—	.238	—	17/32	.531
69	—	.0292				C	—	.242			
68	—	.0310	32	—	.116	D	—	.246	—	35/64	.547
—	1/32	.0313	31	—	.120	—	1/4	.250	—	9/16	.562
			—	1/8	.125				—	37/64	.578
67	—	.0320	30	—	.129	E	—	.250	—	19/32	.594
66	—	.0330	29	—	.136	F	—	.257	—	39/64	.609
65	—	.0350				G	—	.261			
64	—	.0360	—	9/64	.140	—	17/64	.266	—	5/8	.625
63	—	.0370	28	—	.141	H	—	.266	—	41/64	.641
			27	—	.144				—	21/32	.656
62	—	.0380	26	—	.147	I	—	.272	—	43/64	.672
61	—	.0390	25	—	.150	J	—	.277	—	11/16	.688
60	—	.0400				—	9/32	.281			
59	—	.0410	24	—	.152	K	—	.281	—	45/64	.703
58	—	.0420	23	—	.154	L	—	.290	—	23/32	.719
			—	5/32	.156				—	47/64	.734
57	—	.0430	22	—	.157	M	—	.295	—	3/4	.750
56	—	.0465	21	—	.159	—	19/64	.2297	—	49/64	.766
—	3/64	.0469				N	—	.302			
55	—	.0520	20	—	.161	—	5/16	.313	—	25/32	.781
54	—	.0550	19	—	.166	O	—	.316	—	51/64	.797
			18	—	.170				—	13/16	.813
53	—	.0595	—	11/64	.172	P	—	.323	—	53/64	.828
—	1/16	.0625	17	—	.173	—	21/64	.328	—	27/32	.844
52	—	.0635				Q	—	.332			
51	—	.0670				R	—	.339			
50	—	.0700	16	—	.177	—	11/32	.344	—	55/64	.859
			15	—	.180				—	7/8	.875
49	—	.0730	14	—	.182	S	—	.348	—	57/64	.891
48	—	.0760	13	—	.185	T	—	.358	—	29/32	.906
—	5/64	.0781	—	3/16	.188	—	23/64	.359	—	59/64	.922
47	—	.0785				U	—	.368			
46	—	.0810	12	—	.189	—	3/8	.375	—	15/16	.938
			11	—	.191				—	61/64	.953
45	—	.0820	10	—	.194	V	—	.377	—	31/32	.969
44	—	.0860	9	—	.196	W	—	.386	—	63/64	.984
43	—	.0890	8	—	.199	—	25/64	.391	—	1	1.000

METRIC SCREW THREADS

Coarse (general purpose)		Fine	
Nom Size & Thd Pitch	Tap Drill Dia (mm)	Nom Size & Thd Pitch	Tap Drill Dia (mm)
M1.6 × 0.35	1.25	—	—
M1.8 × 0.35	1.45	—	—
M2 × 0.4	1.60	—	—
M2.2 × 0.45	1.75	—	—
M2.5 × 0.45	2.05	—	—
M3 × 0.5	2.50	—	—
M3.5 × 0.6	2.90	—	—
M4 × 0.7	3.30	—	—
M4.5 × 0.75	3.75	—	—
M5 × .8	4.20	—	—
M6.3 × 1	5.30	—	—
M7 × 1	6.00	—	—
M8 × 1.25	6.80	M8 ö 1	7.00
M9 × 1.25	7.75	—	—
M10 × 1.5	8.50	M10 ö 1.25	8.75
M11 × 1.5	9.50	—	—
M12 × 1.75	10.30	M12 ö .25	10.50
M14 × 2	12.00	M14 ö 1.5	12.50
M16 × 2	14.00	M16 ö 1.5	14.50
M18 × 2.5	15.50	M18 ö 1.5	16.50
M20 × 2.5	17.50	M20 ö 1.5	18.50
M22 × 2.5	19.50	M22 ö 1.5	20.50
M24 × 3	21.00	M24 ö 2	22.00
M27 × 3	24.00	M27 ö 2	25.00
M30 × 3.5	26.50	M30 ö 2	28.00
M33 × 3.5	29.50	M30 ö 2	31.00
M36 × 4	32.00	M36 ö 3	33.00
M39 × 4	35.00	M39 ö 3	36.00
M42 × 4.5	37.50	M42 ö 3	39.00
M45 × 4.5	40.50	M45 ö 3	42.00
M48 × 5	43.00	M48 ö 3	45.00
M52 × 5	47.00	M52 ö 3	49.00
M56 × 5.5	50.50	M56 ö 4	52.00
M60 × 5.5	54.50	M60 ö 4	56.00
M64 × 6	58.00	M64 ö 4	60.00
M68 × 6	62.00	M68 ö 4	64.00
M72 × 6	66.00	—	—
M80 × 6	74.00	—	—
M90 × 6	84.00	—	—
M100 × 6	94.00	—	—

DRILLED HOLE TOLERANCES

Drill Size	Tolerance*	
	Plus	Minus
.0135 (No. 80) – .185 (No. 13)	.003	.002
.1875 – .246 (D)	.004	.002
.250 (E) – .750	.005	.002
.756 – 1.000	.007	.003
1.0156 – 2.000	.010	.004
2.0312 – 3.500	.015	.005

* generally accepted tolerances for good practice

REGULAR NUT EYEBOLTS

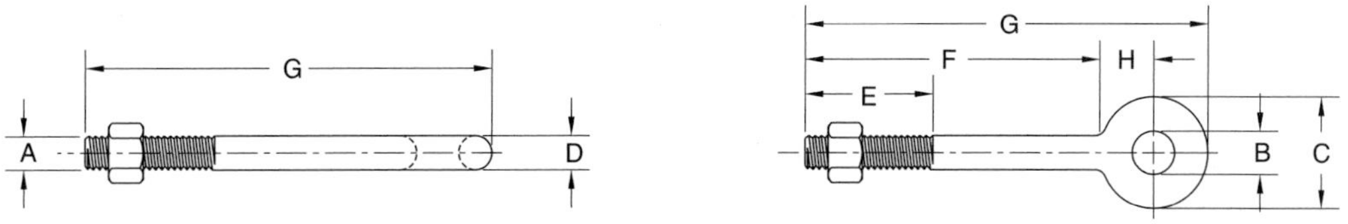

Shank Diameter and Length*	G-291 Stock No. Galv	Working Load Limit**	Weight per 100**	Dimensions*							
				A	B	C	D	E	F	G	H
¼ × 2	1043230	650	6.00	.25	.50	1.00	.25	1.50	2.00	3.06	.56
¼ × 4	1043258	650	13.50	.25	.50	1.00	.25	2.50	4.00	5.06	.56
5⁄16 × 2¼	1043276	1200	18.75	.31	.62	1.25	.31	1.50	2.25	3.56	.69
5⁄16 × 4¼	1043294	1200	25.00	.31	.62	1.25	.31	2.50	4.25	5.56	.69
⅜ × 2½	1043310	1550	24.33	.38	.75	1.50	.38	1.50	2.50	4.12	.88
⅜ × 4½	1043338	1550	37.50	.38	.75	1.50	.38	2.50	4.50	6.12	.88
⅜ × 6	1043356	1550	43.75	.38	.75	1.50	.38	2.50	6.00	7.62	.88
½ × 3¼	1043374	2600	50.00	.50	1.00	2.00	.50	1.50	3.25	5.38	1.12
½ × 6	1043392	2600	62.50	.50	1.00	2.00	.50	3.00	6.00	8.12	1.12
½ × 8	1043418	2600	75.00	.50	1.00	2.00	.50	3.00	8.00	10.12	1.12
½ × 10	1043436	2600	88.00	.50	1.00	2.00	.50	3.00	10.00	12.12	1.12
½ × 12	1043454	2600	100.00	.50	1.00	2.00	.50	3.00	12.00	14.12	1.12
⅝ × 4	1043472	5200	101.25	.62	1.25	2.50	.62	2.00	4.00	6.69	1.44
⅝ × 6	1043490	5200	120.00	.62	1.25	2.50	.62	3.00	6.00	8.69	1.44
⅝ × 8	1043515	5200	131.00	.62	1.25	2.50	.62	3.00	8.00	10.69	1.44
⅝ × 10	1043533	5200	162.50	.62	1.25	2.50	.62	3.00	10.00	12.69	1.44
⅝ × 12	1043551	5200	175.00	.62	1.25	2.50	.62	4.00	12.00	14.69	1.44
¾ × 4½	1043579	7200	185.90	.75	1.50	3.00	.75	2.00	4.50	7.69	1.69
¾ × 6	1043597	7200	180.00	.75	1.50	3.00	.75	3.00	6.00	9.19	1.69
¾ × 8	1043613	7200	200.00	.75	1.50	3.00	.75	3.00	8.00	11.19	1.69
¾ × 10	1043631	7200	237.50	.75	1.50	3.00	.75	3.00	10.00	13.19	1.69
¾ × 12	1043659	7200	251.94	.75	1.50	3.00	.75	4.00	12.00	15.19	1.69
¾ × 15	1043677	7200	300.00	.75	1.50	3.00	.75	5.00	15.00	18.19	1.69
⅞ × 5	1043695	10,600	275.00	.88	1.75	3.50	.88	2.50	5.00	8.75	2.00
⅞ × 8	1043711	10,600	325.00	.88	1.75	3.50	.88	4.00	8.00	11.75	2.00
⅞ × 12	1043739	10,600	400.00	.88	1.75	3.50	.88	4.00	12.00	15.75	2.00
1 × 6	1043757	13,300	425.00	1.00	2.00	4.00	1.00	3.00	6.00	10.31	2.31
1 × 9	1043775	13,300	452.00	1.00	2.00	4.00	1.00	4.00	9.00	13.31	2.31
1 × 12	1043793	13,300	550.00	1.00	2.00	4.00	1.00	4.00	12.00	16.31	2.31
1 × 18	1043819	13,300	650.00	1.00	2.00	4.00	1.00	7.00	18.00	22.31	2.31
1¼ × 8	1043837	21,000	750.00	1.25	2.50	5.00	1.25	4.00	8.00	13.38	2.88
1¼ × 12	1043855	21,000	900.00	1.25	2.50	5.00	1.25	4.00	12.00	17.38	2.88
1¼ × 20	1043873	21,000	1150.00	1.25	2.50	5.00	1.25	6.00	20.00	25.38	2.88

* in in.

** in lb

SHOULDER NUT EYEBOLTS

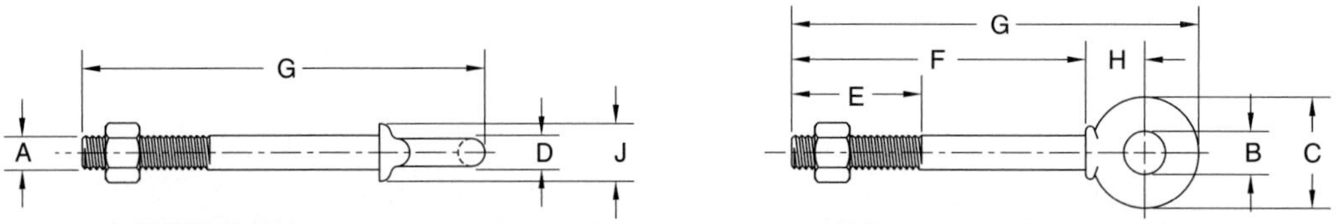

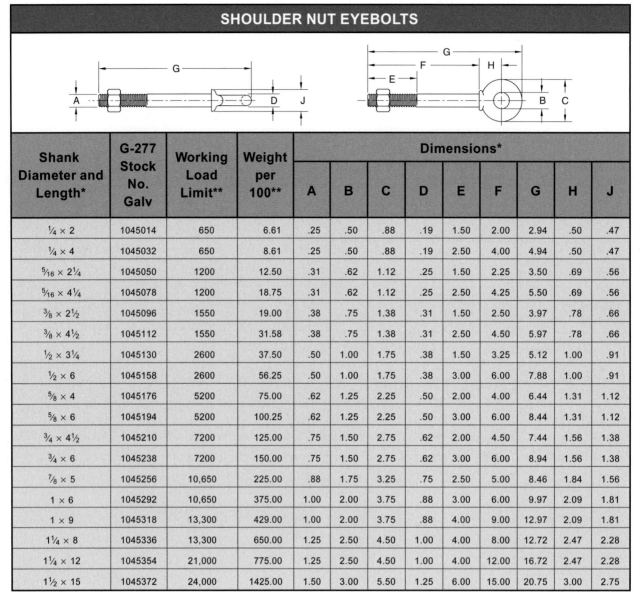

Shank Diameter and Length*	G-277 Stock No. Galv	Working Load Limit**	Weight per 100**	Dimensions*								
				A	B	C	D	E	F	G	H	J
1/4 × 2	1045014	650	6.61	.25	.50	.88	.19	1.50	2.00	2.94	.50	.47
1/4 × 4	1045032	650	8.61	.25	.50	.88	.19	2.50	4.00	4.94	.50	.47
5/16 × 2 1/4	1045050	1200	12.50	.31	.62	1.12	.25	1.50	2.25	3.50	.69	.56
5/16 × 4 1/4	1045078	1200	18.75	.31	.62	1.12	.25	2.50	4.25	5.50	.69	.56
3/8 × 2 1/2	1045096	1550	19.00	.38	.75	1.38	.31	1.50	2.50	3.97	.78	.66
3/8 × 4 1/2	1045112	1550	31.58	.38	.75	1.38	.31	2.50	4.50	5.97	.78	.66
1/2 × 3 1/4	1045130	2600	37.50	.50	1.00	1.75	.38	1.50	3.25	5.12	1.00	.91
1/2 × 6	1045158	2600	56.25	.50	1.00	1.75	.38	3.00	6.00	7.88	1.00	.91
5/8 × 4	1045176	5200	75.00	.62	1.25	2.25	.50	2.00	4.00	6.44	1.31	1.12
5/8 × 6	1045194	5200	100.25	.62	1.25	2.25	.50	3.00	6.00	8.44	1.31	1.12
3/4 × 4 1/2	1045210	7200	125.00	.75	1.50	2.75	.62	2.00	4.50	7.44	1.56	1.38
3/4 × 6	1045238	7200	150.00	.75	1.50	2.75	.62	3.00	6.00	8.94	1.56	1.38
7/8 × 5	1045256	10,650	225.00	.88	1.75	3.25	.75	2.50	5.00	8.46	1.84	1.56
1 × 6	1045292	10,650	375.00	1.00	2.00	3.75	.88	3.00	6.00	9.97	2.09	1.81
1 × 9	1045318	13,300	429.00	1.00	2.00	3.75	.88	4.00	9.00	12.97	2.09	1.81
1 1/4 × 8	1045336	13,300	650.00	1.25	2.50	4.50	1.00	4.00	8.00	12.72	2.47	2.28
1 1/4 × 12	1045354	21,000	775.00	1.25	2.50	4.50	1.00	4.00	12.00	16.72	2.47	2.28
1 1/2 × 15	1045372	24,000	1425.00	1.50	3.00	5.50	1.25	6.00	15.00	20.75	3.00	2.75

*in in.

** in lb

MACHINERY EYEBOLTS

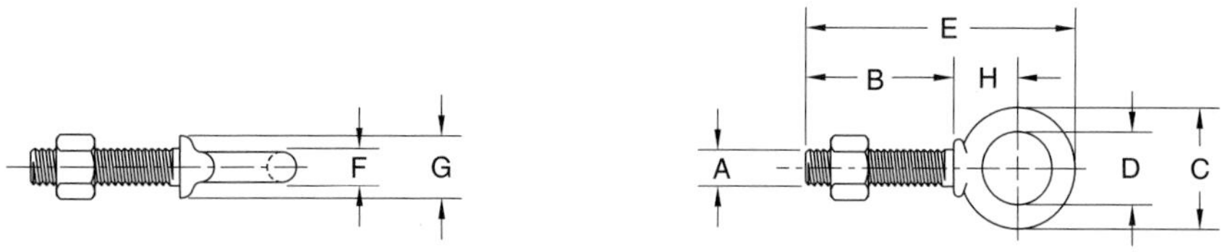

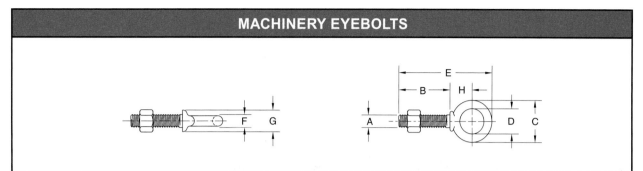

Shank Diameter and Length*	Stock No.	Working Load Limit**	Weight per 100**	Dimensions*							
				A	B	C	D	E	F	G	H
1/4 × 1	9900182	650	3.20	.25	1.00	.88	.50	1.94	.19	.47	.50
5/16 × 1 1/8	9900191	1200	6.20	.31	1.13	1.12	.62	2.38	.25	.56	.69
3/8 × 1 1/4	9900208	1550	12.50	.38	1.25	1.38	.75	2.72	.31	.66	.78
1/2 × 1 1/2	9900217	2600	25.00	.50	1.50	1.75	1.00	3.38	.38	.91	1.00
5/8 × 1 3/4	9900226	5200	50.00	.63	1.75	2.25	1.25	4.19	.50	1.12	1.31
3/4 × 2	9900235	7200	87.50	.75	2.00	2.75	1.50	4.94	.62	1.38	1.56
7/8 × 2 1/4	9900244	10,600	150.00	.88	2.25	3.25	1.75	5.72	.75	1.56	1.84
1 × 2 1/2	9900253	13,300	218.00	1.00	2.50	3.75	2.00	6.47	.88	1.81	2.09
1 1/4 × 3	9900262	21,000	380.00	1.25	3.00	4.50	2.50	7.72	1.00	2.28	2.47
1 1/2 × 3 1/2	9900271	24,000	700.00	1.50	3.50	5.00	3.00	9.25	1.25	2.75	3.00

* in in.

** in lb

WIRE ROPE STRENGTH

Nominal Diameter*	Classification	Nominal Breaking Strength per 2000 lb Ton	
		IPS**	EIPS†
1/4	6 x 19 STANDARD HOISTING FIBER CORE	2.74	–
	6 x 19 STANDARD HOISTING IWRC	–	3.40
	8 x 19 SPECIAL FLEXIBLE HOISTING FIBER CORE	2.35	–
	18 x 7 NONROTATING	2.51	–
3/8	6 x 19 STANDARD HOISTING FIBER CORE	6.10	–
	6 x 19 STANDARD HOISTING IWRC	–	7.55
	8 x 19 SPECIAL FLEXIBLE HOISTING FIBER CORE	5.24	–
	18 x 7 NONROTATING	5.59	–
1/2	6 x 19 STANDARD HOISTING FIBER CORE	10.7	–
	6 x 19 STANDARD HOISTING IWRC	–	13.3
	8 x 19 SPECIAL FLEXIBLE HOISTING FIBER CORE	9.23	–
	18 x 7 NONROTATING	9.85	–

STANDARD HOISTING
6 x 19 SEALE
WITH FIBER CORE

SPECIAL FLEXIBLE HOISTING
8 x 19 WARRINGTON
WITH FIBER CORE

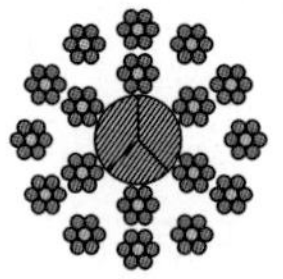

NONROTATING WIRE ROPE
18 x 7
WITH FIBER CORE

* in in.
** IPS - improved plow steel
† EIPS - extra improved plow steel

SLING ANGLE LOSS FACTORS

Angle from Horizontal*	Loss Factor
90	1.000
85	.996
80	.985
75	.966
70	.940
65	.906
60	.866
55	.819
50	.766
45	.707
40	.643
35	.574
30	.500

* in degrees

CHOKER SLING LOSS FACTORS

Choke Angle*	Sling Angle Loss Factor
120 – 180	.75
90 – 119	.65
60 – 89	.55
30 – 59	.40

* in degrees

FILTER PRESSURE LOSS CONSTANTS								
Filter Length*	Micron Size	Pipe Size*						
		1/8	1/4	3/4	1	1 1/4	1 1/2	2
3.5	5	115.0	55.0	—	—	—	—	—
	25	112.0	49.0	—	—	—	—	—
	100	92.0	41.0	—	—	—	—	—
14	5	—	—	0.47	0.34	0.34	0.34	—
	25	—	—	0.34	0.23	0.20	0.20	—
	50	—	—	0.32	0.20	0.19	0.19	—
	75	—	—	0.32	0.20	0.19	0.19	—
17	25	—	—	—	—	—	0.05	0.028
	50	—	—	—	—	—	0.036	0.020
	75	—	—	—	—	—	0.032	0.018

* in in.

PIPE PRESSURE LOSS CONSTANTS (100′ OF SCHEDULE 40 PIPE)	
Pipe Size*	Constant (C)
1/8	2300.00
1/4	450.00
3/8	91.00
1/2	26.40
3/4	5.93
1	1.66
1 1/4	.40
1 1/2	.174
2	.046
2 1/2	.018
3	.006

* in in.

PIPE FITTING PRESSURE LOSS CONSTANTS									
Pipe Fitting	Pipe Size*								
	1/8	1/4	3/8	1/2	3/4	1	1 1/4	1 1/2	2
45 Elbow*	8.30	2.20	.53	.21	.059	.021	.007	.004	.001
90 Elbow	15.40	4.09	1.09	.42	.12	.043	.014	.007	.002
Gate Valve	6.7	1.76	.47	.18	.05	.018	.006	.003	.001
Globe Valve	175.3	46.40	12.70	4.75	1.36	.48	.16	.08	.03
Tee-Side Flow	31.0	8.14	2.37	.81	.24	.08	.03	.014	.005
Tee-Run Flow	10.4	2.74	.80	.26	.08	.03	.009	.004	.002

* in in.

PIPE							
Nominal ID*	OD (BW Gauge)	Inside Diameter (BW Gauge)			Nominal Wall Thickness		
		Standard	Extra-Heavy	Double Extra-Heavy	Schedule 40	Schedule 60	Schedule 80
1/8	0.405	0.269	0.215	—	0.068	0.095	—
1/4	0.540	0.364	0.302	—	0.088	0.119	—
3/8	0.675	0.493	0.423	—	0.091	0.126	—
1/2	0.840	0.622	0.546	0.252	0.109	0.147	0.294
3/4	1.050	0.824	0.742	0.434	0.113	0.154	0.308
1	1.315	1.049	0.957	0.599	0.133	0.179	0.358
1 1/4	1.660	1.380	1.278	0.896	0.140	0.191	0.382
1 1/2	1.900	1.610	1.500	1.100	0.145	0.200	0.400
2	2.375	2.067	1.939	1.503	0.154	0.218	0.436
2 1/2	2.875	2.469	2.323	1.771	0.203	0.276	0.552
3	3.500	3.068	2.900	2.300	0.216	0.300	0.600
3 1/2	4.000	3.548	3.364	2.728	0.226	0.318	—
4	4.500	4.026	3.826	3.152	0.237	0.337	0.674
5	5.563	5.047	4.813	4.063	0.258	0.375	0.750
6	6.625	6.065	5.761	4.897	0.280	0.432	0.864
8	8.625	7.981	7.625	6.875	0.322	0.500	0.875
10	10.750	10.020	9.750	8.750	0.365	0.500	—
12	12.750	12.000	11.750	10.750	0.406	0.500	—

* in in.

FLUID WEIGHTS/TEMPERATURE STANDARDS		
Fluid	Weight*	Temperature**
Air	4.33×10^{-5}	20°C/68°F @ 29.92 in. Hg
Gasoline	.0237 – .0249	20°C/68°F
Kerosene	.0296	20°C/68°F
Mercury	.49116	0°C/32°F
Oil, lubricating	.0307 – .0318	15°C/59°F
Oil, fuel	.0336 – .0353	15°C/59°F
Water	.0361	4°C/39°F
Sea water	.0370	15°C/59°F

* in lb/cu in.

** laboratory temperature under which numerical values are defined

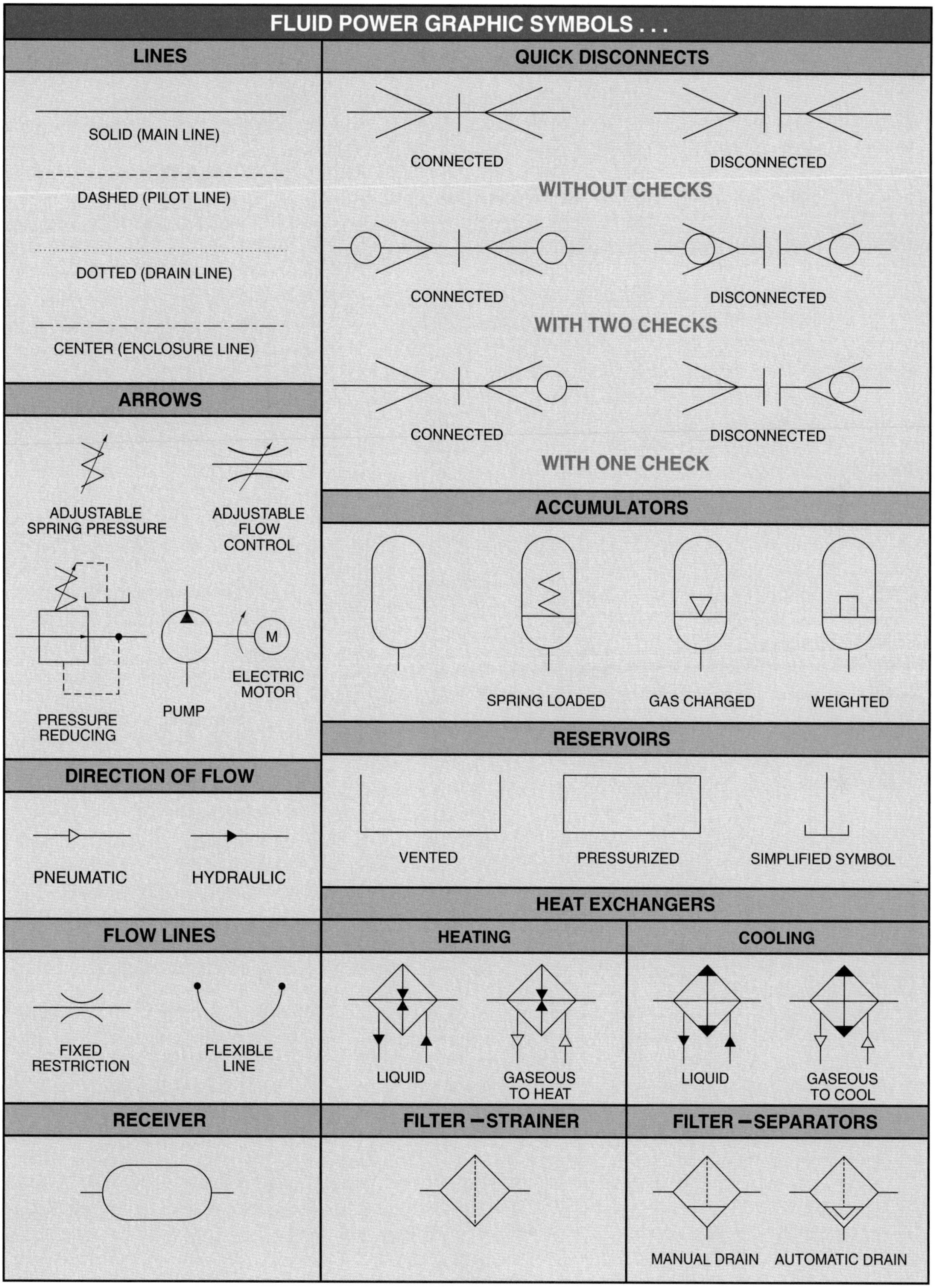
FLUID POWER GRAPHIC SYMBOLS . . .
LINES
SOLID (MAIN LINE)
DASHED (PILOT LINE)
DOTTED (DRAIN LINE)
CENTER (ENCLOSURE LINE)
QUICK DISCONNECTS
CONNECTED
DISCONNECTED
WITHOUT CHECKS
CONNECTED
DISCONNECTED
WITH TWO CHECKS
CONNECTED
DISCONNECTED
WITH ONE CHECK
ARROWS
ADJUSTABLE SPRING PRESSURE
ADJUSTABLE FLOW CONTROL
M
PRESSURE REDUCING
PUMP
ELECTRIC MOTOR
ACCUMULATORS
SPRING LOADED
GAS CHARGED
WEIGHTED
RESERVOIRS
VENTED
PRESSURIZED
SIMPLIFIED SYMBOL
DIRECTION OF FLOW
PNEUMATIC
HYDRAULIC
HEAT EXCHANGERS
HEATING
COOLING
FLOW LINES
FIXED RESTRICTION
FLEXIBLE LINE
LIQUID
GASEOUS TO HEAT
LIQUID
GASEOUS TO COOL
RECEIVER
FILTER — STRAINER
FILTER — SEPARATORS
MANUAL DRAIN
AUTOMATIC DRAIN

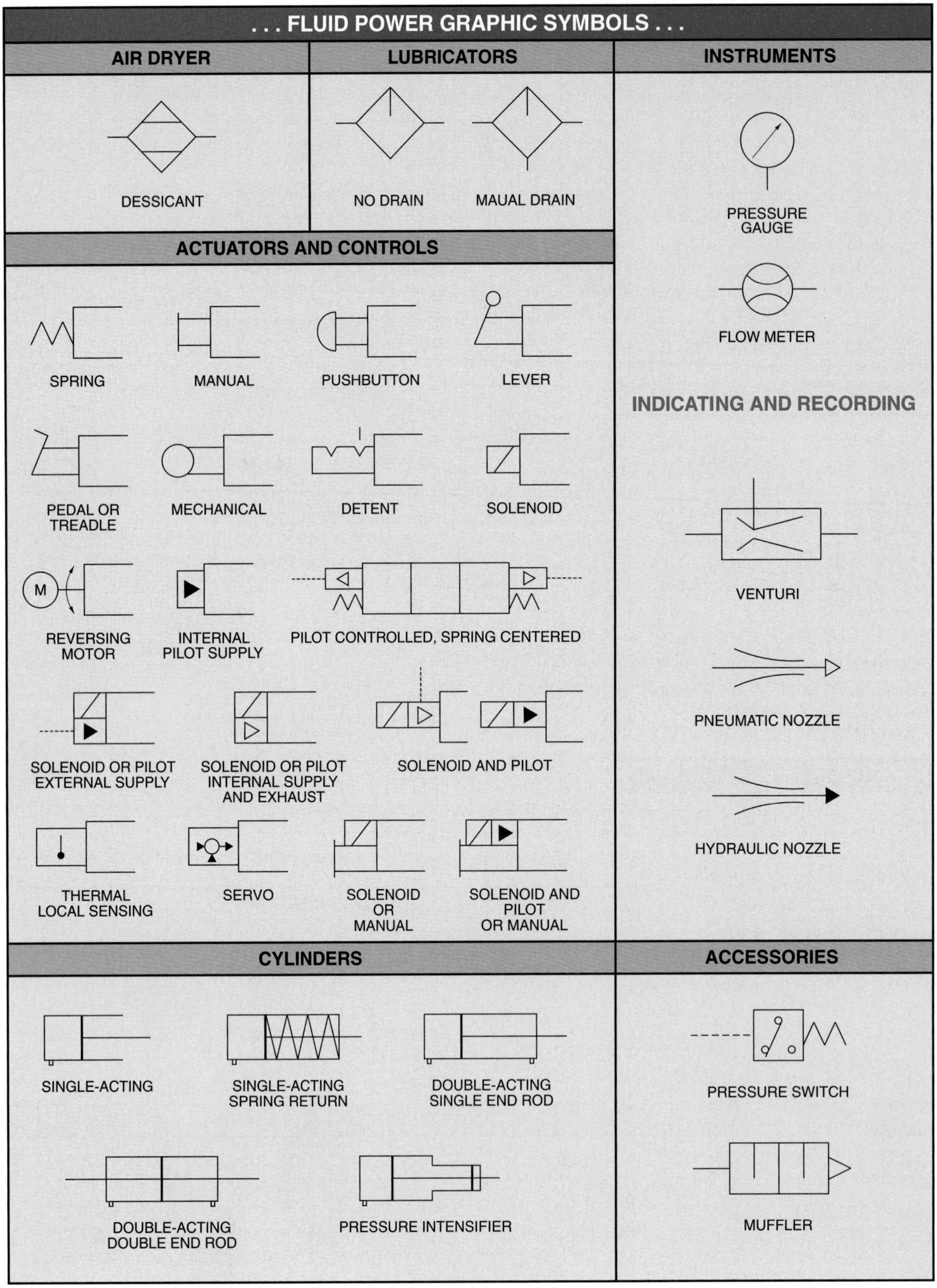
. . . FLUID POWER GRAPHIC SYMBOLS . . .
AIR DRYER
DESSICANT
LUBRICATORS
NO DRAIN
MAUAL DRAIN
INSTRUMENTS
PRESSURE GAUGE
FLOW METER
INDICATING AND RECORDING
VENTURI
PNEUMATIC NOZZLE
HYDRAULIC NOZZLE
ACTUATORS AND CONTROLS
SPRING
MANUAL
PUSHBUTTON
LEVER
PEDAL OR TREADLE
MECHANICAL
DETENT
SOLENOID
M
REVERSING MOTOR
INTERNAL PILOT SUPPLY
PILOT CONTROLLED, SPRING CENTERED
SOLENOID OR PILOT EXTERNAL SUPPLY
SOLENOID OR PILOT INTERNAL SUPPLY AND EXHAUST
SOLENOID AND PILOT
THERMAL LOCAL SENSING
SERVO
SOLENOID OR MANUAL
SOLENOID AND PILOT OR MANUAL
CYLINDERS
SINGLE-ACTING
SINGLE-ACTING SPRING RETURN
DOUBLE-ACTING SINGLE END ROD
DOUBLE-ACTING DOUBLE END ROD
PRESSURE INTENSIFIER
ACCESSORIES
PRESSURE SWITCH
MUFFLER

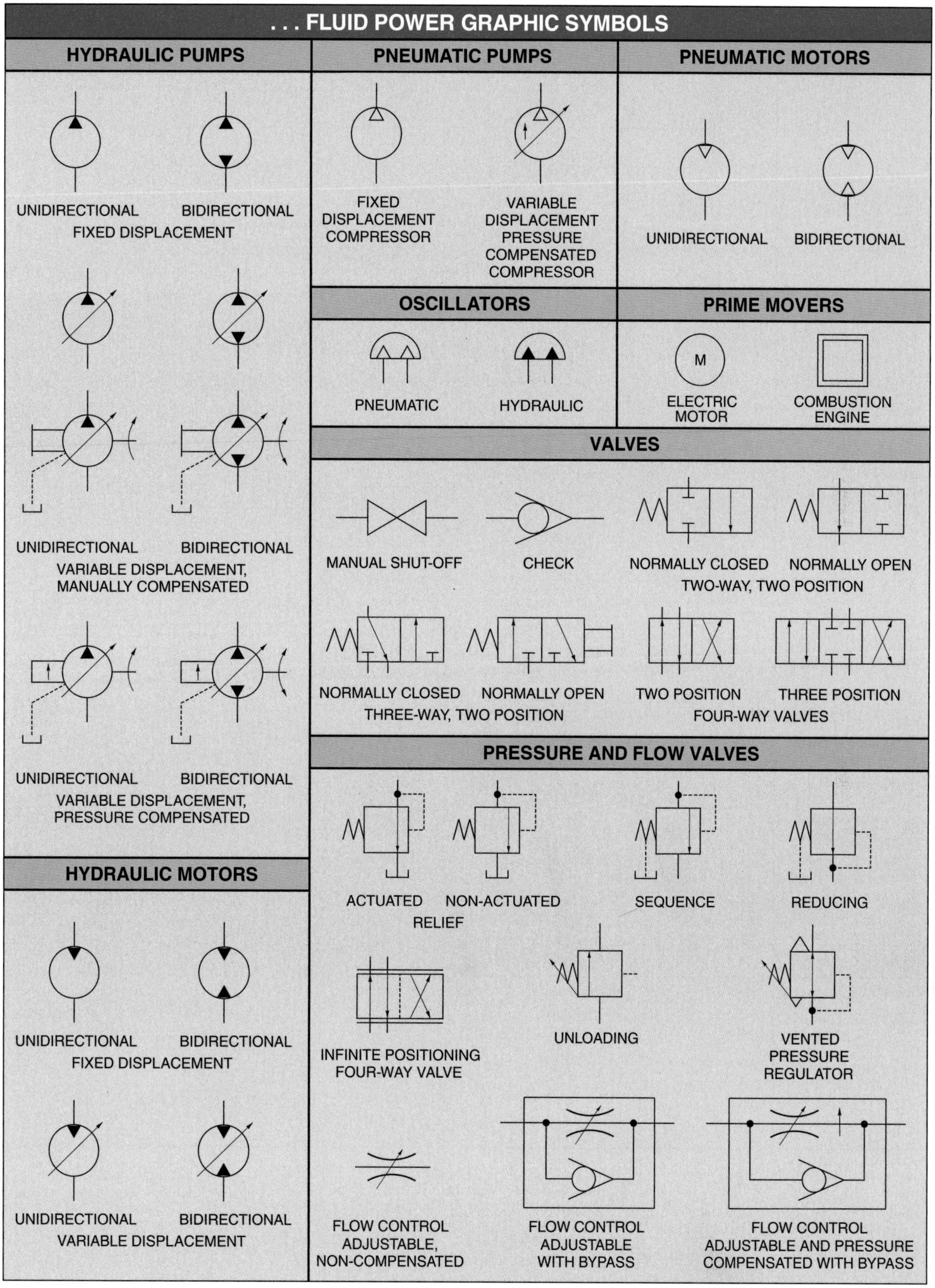
. . . FLUID POWER GRAPHIC SYMBOLS
HYDRAULIC PUMPS
UNIDIRECTIONAL
BIDIRECTIONAL
FIXED DISPLACEMENT
UNIDIRECTIONAL
BIDIRECTIONAL
VARIABLE DISPLACEMENT, MANUALLY COMPENSATED
UNIDIRECTIONAL
BIDIRECTIONAL
VARIABLE DISPLACEMENT, PRESSURE COMPENSATED
HYDRAULIC MOTORS
UNIDIRECTIONAL
BIDIRECTIONAL
FIXED DISPLACEMENT
UNIDIRECTIONAL
BIDIRECTIONAL
VARIABLE DISPLACEMENT
PNEUMATIC PUMPS
FIXED DISPLACEMENT COMPRESSOR
VARIABLE DISPLACEMENT PRESSURE COMPENSATED COMPRESSOR
PNEUMATIC MOTORS
UNIDIRECTIONAL
BIDIRECTIONAL
OSCILLATORS
PNEUMATIC
HYDRAULIC
PRIME MOVERS
M
ELECTRIC MOTOR
COMBUSTION ENGINE
VALVES
MANUAL SHUT-OFF
CHECK
NORMALLY CLOSED
NORMALLY OPEN
TWO-WAY, TWO POSITION
NORMALLY CLOSED
NORMALLY OPEN
THREE-WAY, TWO POSITION
TWO POSITION
THREE POSITION
FOUR-WAY VALVES
PRESSURE AND FLOW VALVES
ACTUATED
NON-ACTUATED
RELIEF
SEQUENCE
REDUCING
INFINITE POSITIONING FOUR-WAY VALVE
UNLOADING
VENTED PRESSURE REGULATOR
FLOW CONTROL ADJUSTABLE, NON-COMPENSATED
FLOW CONTROL ADJUSTABLE WITH BYPASS
FLOW CONTROL ADJUSTABLE AND PRESSURE COMPENSATED WITH BYPASS

LOGIC SYMBOLS					
LOGIC ELEMENT	**AND**	**OR**	**NOT**	**NAND**	**NOR**
LOGIC ELEMENT FUNCTION	OUTPUT IF ALL CONTROL INPUT SIGNALS ARE ON	OUTPUT IF ANY ONE OF THE CONTROL INPUTS IS ON	OUTPUT IF SINGLE CONTROL INPUT SIGNAL IS OFF	OUTPUT IF ALL CONTROL INPUT SIGNALS ARE ON	OUTPUT IF ANY OF THE CONTROL INPUTS ARE ON
MIL-STD-806B AND ELECTRONIC LOGIC SYMBOL	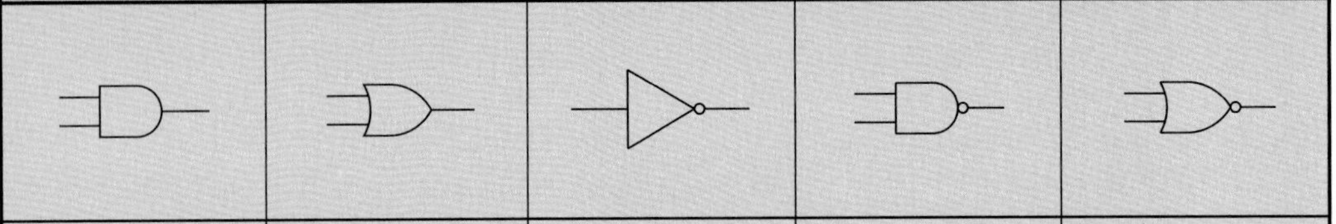				
ELECTRICAL RELAY LOGIC SYMBOL	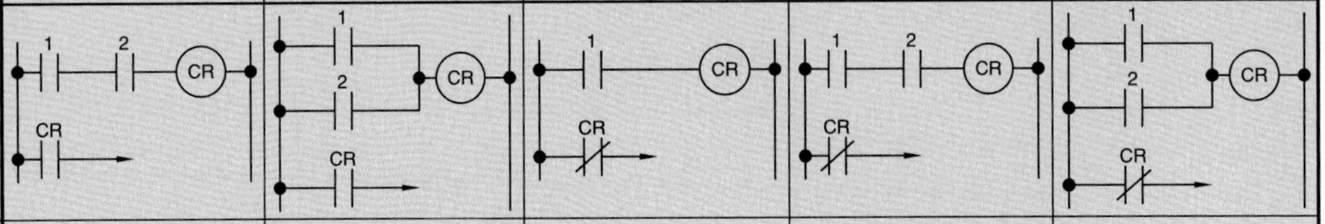				
ELECTRICAL SWITCH LOGIC SYMBOL	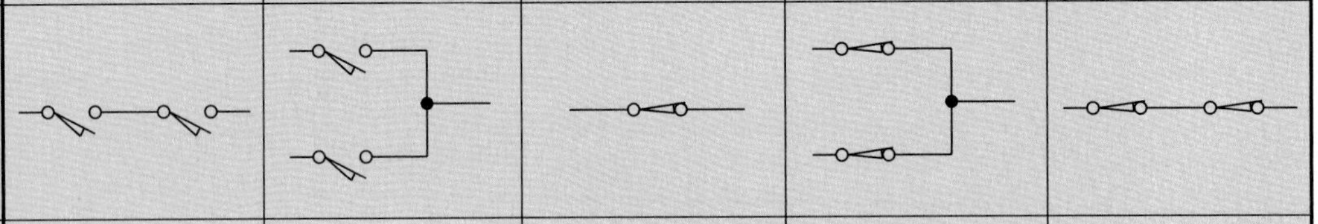				
ASA (JIC) VALVING SYMBOL	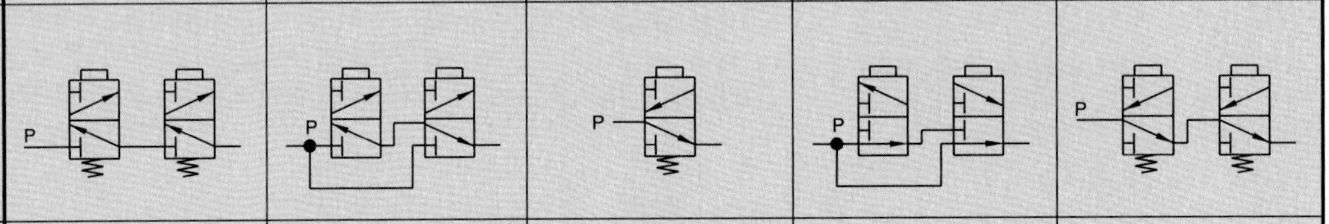				
ARO PNEUMATIC LOGIC SYMBOL	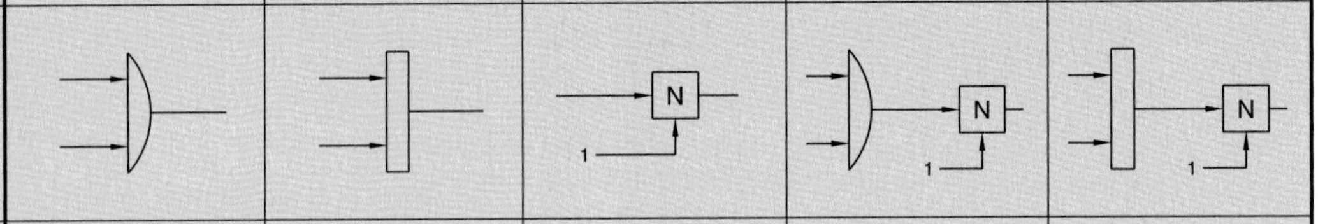				
NFPA STANDARD	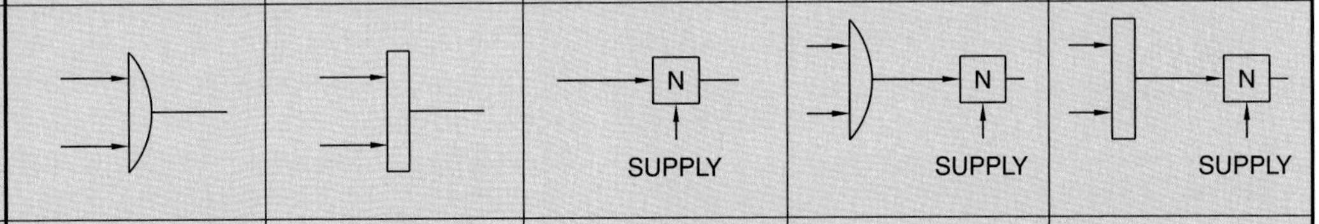				
BOOLEAN ALGEBRA SYMBOL	$(\ )\cdot(\ )$	$(\ )+(\ )$	$\overline{(\ )}$	$\overline{(\ )\cdot(\ )}$	$\overline{(\ )+(\ )}$
FLUIDIC DEVICE TURBULENCE AMPLIFIER	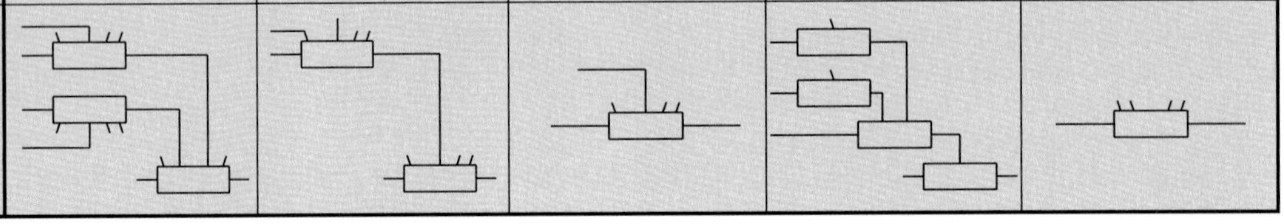				

FLUID POWER ABBREVIATIONS/ACRONYMS			
Abbr/Acronym	**Meaning**	**Abbr/Acronym**	**Meaning**
BTU	British Thermal Unit	IN-LB	Inch Pound
C	Degrees Celsius	INT	Internal
CC	Closed Center (valves)	I/O	Input/Output
CCW	Counterclockwise	IPM	Inches Per Minute
CFM	Cubic Feet Per Minute	IPS	Inches Per Second
CFS	Cubic Feet Per Second	LB	Pound
CIM	Cubic Inches Per Minute	MAX	Maximum
COM	Common	MIN	Minimum
CPM	Cycles Per Minute	NC	Normally Closed
CPS	Cycles Per Second	NO	Normally Open
CW	Clockwise	NPT	National Pipe Thread
CYL	Cylinder	OC	Open Center (valves)
D	Drain	OZ	Ounce
DIA	Diameter	PO	Pilot Operated
EXT	External	PRES or P	Pressure
F	Degrees Fahrenheit	PSI	Pounds Per Square Inch
FL	Fluid	PSIA	PSI Absolute
FPM	Feet Per Minute	PSIG	PSI Gauge
FS	Full Scale	PT	Pint
FT	Foot	QT	Quart
FT-LB	Foot Pound	R	Radius
GA	Gauge	RMS	Root Mean Square
GAL	Gallon	RPM	Revolutions Per Minute
GPM	Gallons Per Minute	RPS	Revolutions Per Second
HP	Horsepower	SOL	Solenoid
Hz	Hertz	T	Torque; Thrust; Tank
ID	Inside Diameter	VAC	Vacuum
IN	Inches	VI	Viscosity Index
		VISC	Viscosity

TIMING BELT STANDARD PITCH LENGTHS AND TOLERANCES*

Belt Length Designation	Pitch Length	Number of Teeth for Standard Lengths			Belt Length Designation	Pitch Length	Number of Teeth for Standard Lengths				
		MXL (.080)	XL (.200)	L (.375)			XL (.200)	L (.375)	H (.500)	XH (.875)	XXH (1.250)
36	3.600	45	—	—	230	23.000	115	—	—	—	—
40	4.000	50	—	—	240	24.000	120	64	48	—	—
44	4.400	55	—	—	250	25.000	125	—	—	—	—
48	4.800	60	—	—	255	25.500	—	68	—	—	—
56	5.600	70	—	—	260	26.000	130	—	—	—	—
60	6.000	75	30	—	270	27.000	—	72	54	—	—
64	6.400	80	—	—	285	28.500	—	76	—	—	—
70	7.000	—	35	—	300	30.000	—	80	60	—	—
72	7.200	90	—	—	322	32.250	—	86	—	—	—
80	8.000	100	40	—	330	33.000	—	—	66	—	—
88	8.800	110	—	—	345	34.500	—	92	—	—	—
90	9.000	—	45	—	360	36.000	—	—	72	—	—
100	10.000	125	50	—	367	36.750	—	98	—	—	—
110	11.000	—	55	—	390	39.000	—	104	78	—	—
112	11.200	140	—	—	420	42.000	—	112	84	—	—
120	12.000	—	60	—	450	45.000	—	120	90	—	—
124	12.375	—	—	33	480	48.000	—	128	96	—	—
124	12.400	155	—	—	507	50.750	—	—	—	58	—
130	13.000	—	65	—	510	51.000	—	136	102	—	—
140	14.000	175	70	—	540	54.000	—	144	108	—	—
150	15.000	—	75	40	560	56.000	—	—	—	64	—
160	16.000	200	80	—	570	57.000	—	—	114	—	—
170	17.000	—	85	—	600	60.000	—	160	120	—	—
180	18.000	225	90	—	630	63.000	—	—	126	72	—
187	18.750	—	—	50	660	66.000	—	—	132	—	—
190	19.000	—	95	—	700	70.000	—	—	140	80	56
200	20.000	250	100	—	750	75.000	—	—	150	—	—
210	21.000	—	105	56	770	77.000	—	—	—	88	—
220	22.000	—	110	—	800	80.000	—	—	160	—	64
225	22.500	—	—	60	840	84.000	—	—	—	96	—

* in in.

TIMING BELT STANDARD WIDTHS AND TOLERANCES*					
Belt Section	Standard Belt Widths		Tolerances on Width for Pitch Lengths		
	Designation	Dimensions	Up to and Including 33″	Over 33″ up to and Including 66″	Over 66″
MXL (.080)	012	.12	+.02	—	—
	019	.19	−.03		
	025	.25	+.02		
XL (.200)	025	.25	−.03	—	—
	037	.38	+.03		
L (.375)	050	.50	−.03	+.03 −.05	—
	075	.75	+.03		
	100	1.00	−.03		
H (.500)	075	.75	+.03 −.03	+.03 −.05	+.03 −.05
	100	1.00			
	150	1.50			
	200	2.00	+.03	+.05	+.05
			−.05	−.05	−.06
	300	3.00	+.05	+.06	+.06
			−.06	−.06	−.08
XH (.875)	200	2.00	—	+.19 −.19	+.19 −.19
	300	3.00			
	400	4.00			
XXH (1.250)	200	2.00	—	—	+.19 −.19
	300	3.00			
	400	4.00			
	500	5.00			

* in in.

ALLOWABLE TIGHT SIDE TENSION FOR AA SECTION V-BELTS*								
Belt Speed**	Pulley Effective Diameter†							
	3.0	3.5	4.0	4.5	5.0	5.5	6.0	6.5
200	30	46	57	66	73	79	83	88
400	23	38	49	58	65	71	76	80
600	18	33	44	53	60	66	71	75
800	14	30	41	50	57	63	67	72
1000	12	27	38	47	54	60	65	69
1200	9	24	36	45	52	57	62	66
1400	7	22	34	42	49	55	60	64
1600	5	20	32	40	47	53	58	62
1800	3	18	30	38	46	51	56	60
2000	1	16	28	37	44	50	54	58
2200	—	15	26	35	42	48	53	57
2400	—	13	24	33	40	46	51	55
2600	—	11	23	31	39	44	49	53
2800	—	9	21	30	37	43	47	51
3000	—	8	19	28	35	41	46	50
3200	—	6	17	26	33	39	44	48
3400	—	4	16	24	31	37	42	46
3600	—	2	14	23	30	35	40	44
3800	—	1	12	21	28	34	38	43
4000	—	—	10	19	26	32	37	41
4200	—	—	8	17	24	30	35	39
4400	—	—	6	15	22	28	33	37
4600	—	—	4	13	20	26	31	35
4800	—	—	2	11	18	24	29	33
5000	—	—	—	9	16	22	27	31
5200	—	—	—	7	14	20	24	28
5400	—	—	—	4	12	17	22	26
5600	—	—	—	2	9	15	20	24
5800	—	—	—	—	7	13	18	22

* in lb-ft

** in fpm

† in in.

ALLOWABLE TIGHT SIDE TENSION FOR BB SECTION V-BELTS*									
Belt Speed**	Pulley Effective Diameter†								
	5.0	5.5	6.0	6.5	7.0	7.5	8.0	8.5	9.0
200	81	93	103	111	119	125	130	135	140
400	69	81	91	99	107	113	118	123	128
600	61	74	84	92	99	106	111	116	121
800	56	68	78	87	94	101	106	111	115
1000	52	64	74	83	90	96	102	107	111
1200	48	60	71	79	86	93	98	103	107
1400	45	57	67	76	83	89	95	100	104
1600	42	54	64	73	80	86	92	97	101
1800	39	51	61	70	77	84	89	94	98
2000	36	49	59	67	74	81	86	91	96
2200	34	46	56	64	72	78	84	89	93
2400	31	43	53	62	69	75	81	86	90
2600	29	41	51	59	67	73	78	83	88
2800	26	38	48	57	64	70	76	81	85
3000	23	35	45	54	61	68	73	78	82
3200	21	33	43	51	59	65	70	75	80
3400	18	30	40	49	56	62	68	73	77
3600	15	27	37	46	53	59	65	70	74
3800	12	24	35	43	50	57	62	67	71
4000	9	22	32	40	47	54	59	64	69
4200	7	19	29	37	45	51	56	61	66
4400	4	16	26	34	42	48	53	58	63
4600	1	13	23	31	39	45	50	55	60
4800	—	10	20	28	35	42	47	52	57
5000	—	6	16	25	32	39	44	49	53
5200	—	3	13	22	29	35	41	46	50
5400	—	—	10	18	26	32	38	42	47
5600	—	—	6	15	22	29	34	39	43
5800	—	—	3	11	19	25	31	36	40

* in lb-ft
** in fpm
† in in.

ALLOWABLE TIGHT SIDE TENSION FOR CC SECTION V-BELTS*									
Belt Speed**	Pulley Effective Diameter†								
	7.0	8.0	9.0	10.0	11.0	12.0	13.0	14.0	15.0
200	121	158	186	207	228	244	257	268	278
400	99	135	164	187	206	221	234	246	256
600	85	122	151	173	192	208	221	232	242
800	75	112	141	164	182	198	211	222	232
1000	67	104	133	155	174	190	203	214	224
1200	60	97	126	149	167	183	196	207	217
1400	54	91	120	142	161	177	190	201	211
1600	48	85	114	137	155	171	184	196	205
1800	43	80	108	131	150	166	179	190	200
2000	38	75	103	126	145	160	174	185	195
2200	33	70	98	121	140	155	169	180	190
2400	28	65	93	116	135	150	164	175	185
2600	23	60	88	111	130	145	159	170	180
2800	18	55	83	106	125	140	154	165	175
3000	13	50	78	101	120	135	149	160	170
3200	8	45	73	96	115	130	144	155	165
3400	3	39	68	91	110	125	138	150	160
3600	—	34	63	86	104	120	133	145	154
3800	—	29	58	80	99	115	128	139	149
4000	—	24	52	75	94	109	123	134	144
4200	—	18	47	70	88	104	117	128	138
4400	—	12	41	64	83	98	112	123	133
4600	—	7	35	58	77	93	106	117	127
4800	—	1	29	52	71	87	100	111	121
5000	—	—	23	46	65	81	94	105	115
5200	—	—	17	40	59	75	88	99	109
5400	—	—	11	34	53	68	81	93	103
5600	—	—	5	27	46	62	75	86	96
5800	—	—	—	21	40	55	68	80	90

* in lb-ft
** in fpm
† in in.

ALLOWABLE TIGHT SIDE TENSION FOR DD SECTION V-BELTS*									
Belt Speed**	Pulley Effective Diameter†								
	12.0	13.0	14.0	15.0	16.0	17.0	18.0	19.0	20.0
200	243	293	336	373	405	434	459	482	503
400	195	245	288	325	358	386	412	434	455
600	167	217	259	297	329	358	383	406	426
800	146	196	239	276	308	337	362	385	405
1000	129	179	222	259	291	320	345	368	389
1200	114	164	207	244	277	305	331	353	374
1400	101	151	194	231	263	292	318	340	361
1600	89	139	182	219	251	280	305	328	349
1800	78	128	170	207	240	269	294	317	337
2000	67	117	159	196	229	258	283	306	326
2200	56	106	149	186	218	247	272	295	316
2400	45	95	138	175	208	236	262	284	305
2600	35	85	128	165	197	226	251	274	294
2800	24	74	117	154	187	215	241	263	284
3000	14	64	106	144	176	205	230	253	273
3200	3	53	96	133	165	194	219	242	263
3400	—	42	85	122	155	183	209	231	252
3600	—	31	74	111	144	172	198	220	241
3800	—	20	63	100	132	161	186	209	230
4000	—	9	51	89	121	150	175	198	218
4200	—	—	40	77	109	138	163	186	207
4400	—	—	28	65	97	126	152	174	195
4600	—	—	16	53	85	114	139	162	183
4800	—	—	3	40	73	102	127	150	170
5000	—	—	—	28	60	89	114	137	158
5200	—	—	—	15	47	76	101	124	145
5400	—	—	—	1	34	62	88	111	131
5600	—	—	—	—	20	49	74	97	118
5800	—	—	—	—	6	35	60	83	104

* in lb-ft

** in fpm

† in in.

VIBRATION CHARACTERISTICS			
Problem	**Frequency**	**Amplitude**	**Remarks/Comments**
Bent Shaft	Normally 1x rpm. Often 2x and 3x rpm.	High amplitude in axial direction	A bent shaft is indicated by a phase difference between two bearings of the same machine.
Angular Misalignment	Normally 1x and 2x rpm. Often 3x rpm.	High amplitude in axial direction	Angular misalignment is indicated by a large phase difference between two bearings of a direct-coupled machine.
Offset Misalignment	1x and 2x rpm	High amplitude in radial direction	Normally produces a radial vibration at a frequency of 2x rpm.
Misaligned bearing on shaft	Normally 1x and 2x rpm. Often 3x rpm.	High amplitude in axial direction	Vibration eliminated by correctly installing bearing.
Mechanical Looseness	2x rpm	Often erratic	May be accompanied by unbalance and/or misalignment. May result from loose mounting bolts or excessive bearing clearance.
Plain Bearings	1x rpm	High amplitude in horizontal direction. Often high amplitude in vertical direction	Vibration normally due to excessive bearing clearance, looseness, or lubrication problems.
Oil Whirl	Slightly less than ½x rpm	Often quite severe	Oil whirl vibration drives shaft into whirling path around bearing. Causes include excessive bearing wear, an increase in lube oil pressure, or a change of oil frequency.
Unbalance	1x rpm	High amplitude in radial direction	Unbalance is the most common cause of vibration.
Gear Tooth Wear	Very high. Gear teeth x rpm	High amplitude at gear mesh frequency	Tooth wear indicated by excitation of gear natural frequency and includes sidebands at running speed of bad gear. Also indicated by excitation at gear mesh frequency.
Gear Eccentricity/ Backlash	1x gear rpm	Moderately high amplitude sidebends at gear mesh frequency	Improper backlash normally excites gear natural frequency and gear mesh frequency.
Gear Misalignment	1x, 2x, and 3x gear mesh frequency	Low amplitude at 1x gear mesh frequency. High amplitude at 2x and 3x gear mesh frequency	Gear misalignment normally excites second order gear mesh frequency harmonics and includes sidebands at running speed.
Broken Gear Tooth	1x rpm of gear	High amplitude	Broken gear tooth excites gear natural frequency with sidebands at gear running speed.
Worn or Loose V-Belts	Less than motor or driven machine rpm	Normally unsteady amplitude readings	Worn or loose belts often produce high frequency vibration and noise such as a chirp or squeal.
V-Belt Pulley Misalignment	1x rpm	High amplitude vibration in axial direction	V-belt pulley misalignment normally produces highest axial vibrations at the fan rpm.
Eccentric V-Belt Pulleys	1x rpm of eccentric pulley	High amplitude vibration in-line with belts	Largest vibration occurs in the direction of belt tension.
Beat Vibration	Continually changing	Continually changing	Belt vibration occurs as beats or pulses occurring at regular interval.

VIBRATION SEVERITY

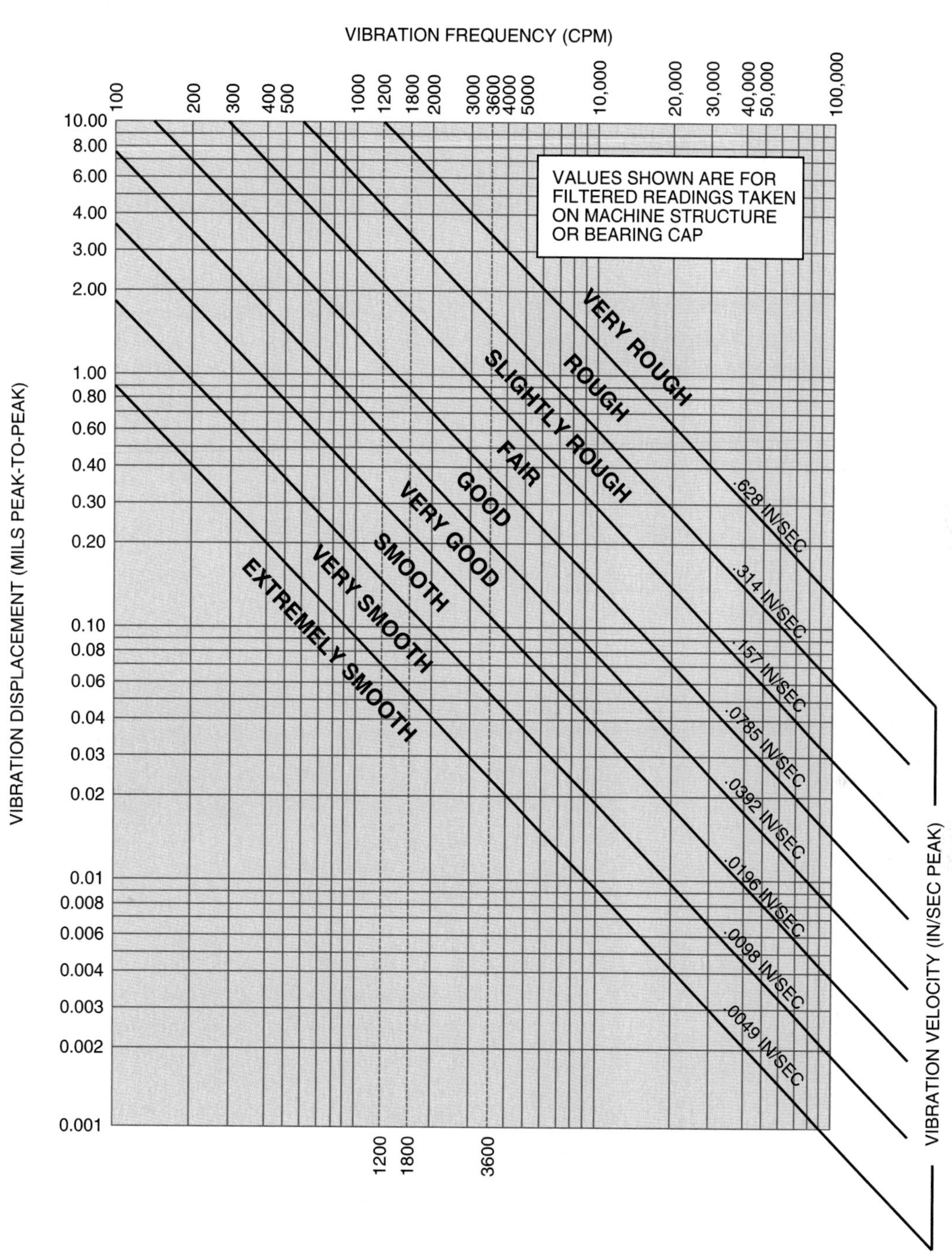

THREE-PHASE VOLTAGE VALUES
For 208 V × 1.732, use 360
For 230 V × 1.732, use 398
For 240 V × 1.732, use 416
For 440 V × 1.732, use 762
For 460 V × 1.732, use 797
For 480 V × 1.732, use 831

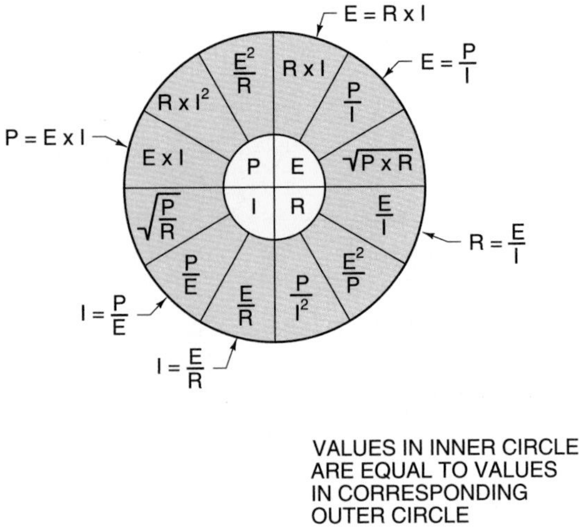

OHM'S LAW AND POWER FORMULA

POWER FORMULA ABBREVIATIONS AND SYMBOLS	
P = Watts	V = Volts
I = Amps	VA = Volt Amps
A = Amps	ϕ = Phase
R = Ohms	√ = Square Root
E = Volts	

POWER FORMULAS – 1ϕ, 3ϕ

Phase	To Find	Use Formula	Example		
			Given	Find	Solution
1ϕ	I	$I = \frac{VA}{V}$	32,000 VA, 240 V	I	$I = \frac{VA}{V}$ $I = \frac{32{,}000\text{ VA}}{240\text{ V}}$ I = **133 A**
1ϕ	VA	$VA = I \times V$	100 A, 240 V	VA	$VA = I \times V$ $VA = 100\text{ A} \times 240\text{ V}$ VA = **24,000 VA**
1ϕ	V	$V = \frac{VA}{I}$	42,000 VA, 350 A	V	$V = \frac{VA}{I}$ $V = \frac{42{,}000\text{ VA}}{350\text{ A}}$ V = **120 V**
3ϕ	I	$I = \frac{VA}{V \times \sqrt{3}}$	72,000 VA, 208 V	I	$I = \frac{VA}{V \times \sqrt{3}}$ $I = \frac{72{,}000\text{ VA}}{360}$ I = **200 A**
3ϕ	VA	$VA = I \times V \times \sqrt{3}$	2 A, 240 V	VA	$VA = I \times V \times \sqrt{3}$ $VA = 2 \times 416$ VA = **832 VA**

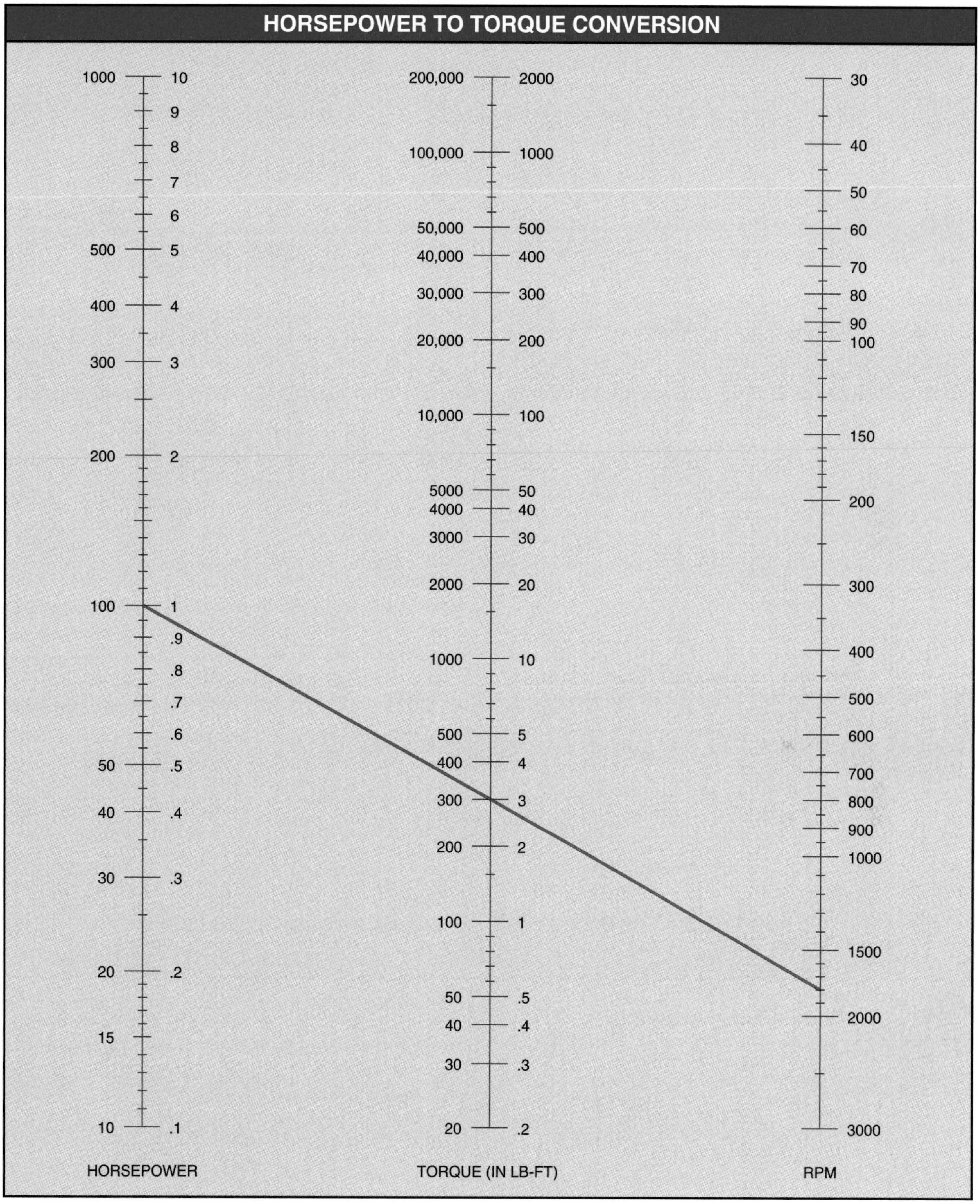
HORSEPOWER TO TORQUE CONVERSION
1000
10
9
8
7
6
500
5
400
4
300
3
200
2
100
1
.9
.8
.7
.6
50
.5
40
.4
30
.3
20
.2
15
10
.1
HORSEPOWER
200,000
2000
100,000
1000
50,000
500
40,000
400
30,000
300
20,000
200
10,000
100
5000
50
4000
40
3000
30
2000
20
1000
10
500
5
400
4
300
3
200
2
100
1
50
.5
40
.4
30
.3
20
.2
TORQUE (IN LB-FT)
30
40
50
60
70
80
90
100
150
200
300
400
500
600
700
800
900
1000
1500
2000
3000
RPM

MOTOR HORSEPOWER											
Pump Flow*	Pump Pressure**										
	100	250	500	750	1000	1250	1500	2000	3000	4000	5000
1	.07	.18	.36	.54	.72	.91	1.09	1.45	2.18	2.91	3.64
2	.14	.36	.72	1.09	1.45	1.82	2.18	2.91	4.37	5.83	7.29
3	.21	.54	1.09	1.64	2.18	2.73	3.28	4.37	6.56	8.75	10.93
4	.29	.72	1.45	2.18	2.91	3.64	4.37	5.83	8.75	11.66	14.58
5	.36	.91	1.82	2.73	3.64	4.55	5.46	7.29	10.93	14.58	18.23
8	.58	1.45	2.91	4.37	5.83	7.29	8.75	11.66	17.50	23.33	29.17
10	.72	1.82	3.64	5.46	7.29	9.11	10.93	14.58	21.87	29.17	36.46
12	.87	2.18	4.37	6.56	8.75	10.93	13.12	17.50	26.25	35.00	43.75
15	1.09	2.73	5.46	8.20	10.93	13.67	16.40	21.87	32.81	43.75	54.69
20	1.45	3.64	7.29	10.93	14.58	18.23	21.87	29.17	43.75	58.34	72.92
25	1.82	4.55	9.11	13.67	18.23	22.79	27.34	36.46	54.69	72.92	91.16
30	2.18	5.46	10.93	16.40	21.87	27.34	32.81	43.75	65.63	87.51	109.39
35	2.55	6.38	12.76	19.14	25.52	31.90	38.28	51.05	76.57	102.10	127.62
40	2.91	7.29	14.58	21.87	29.17	36.46	43.75	58.34	87.51	116.68	145.85
45	3.28	8.20	16.40	24.61	32.81	41.02	49.22	65.63	98.45	131.27	164.08
50	3.64	9.11	18.23	27.34	36.46	45.58	54.69	72.92	109.39	145.85	182.32
55	4.01	10.20	20.05	30.08	40.11	50.13	60.16	80.22	120.33	160.44	200.55
60	4.37	10.93	21.87	32.81	43.75	54.69	65.63	87.51	131.27	175.02	218.78
65	4.74	11.85	23.70	35.55	47.40	59.25	71.10	94.80	142.21	189.61	237.01
70	5.10	12.76	25.52	38.28	51.05	63.81	76.57	102.10	153.13	204.20	255.25
75	5.46	13.67	27.36	41.02	54.69	68.37	82.04	109.39	164.08	218.78	273.48
80	5.83	14.58	29.17	43.75	58.34	72.92	87.51	116.68	175.02	233.37	291.71
90	6.56	16.40	32.81	49.22	65.63	82.04	98.45	131.27	196.90	262.54	328.17
100	7.29	18.23	36.46	54.69	72.92	91.16	109.39	145.85	218.78	291.71	364.64

* in gpm

** pump pressure in psi (efficiency assumed to be 80%)

COMMON NAILS

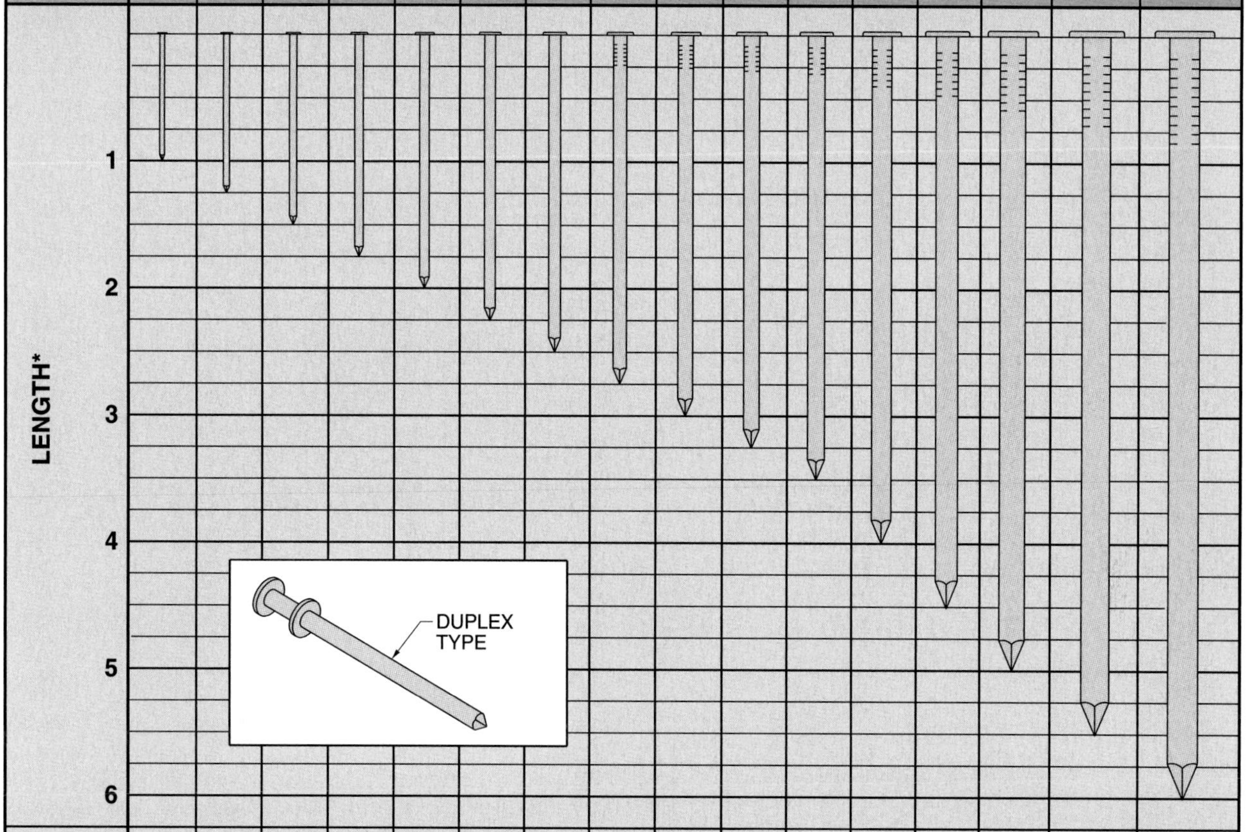

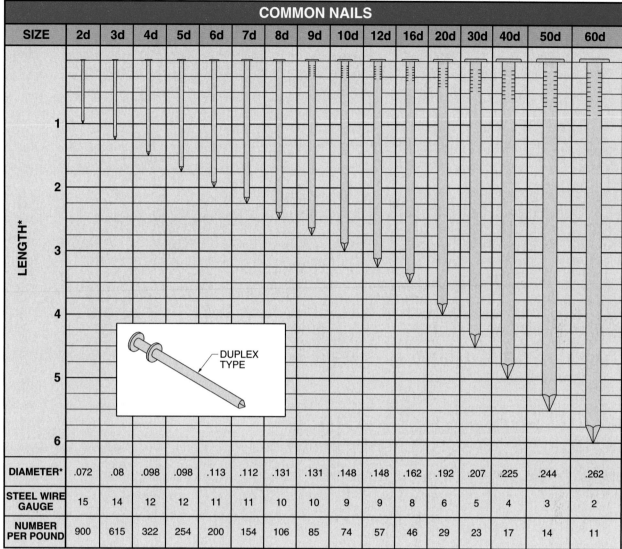

SIZE	2d	3d	4d	5d	6d	7d	8d	9d	10d	12d	16d	20d	30d	40d	50d	60d
DIAMETER*	.072	.08	.098	.098	.113	.112	.131	.131	.148	.148	.162	.192	.207	.225	.244	.262
STEEL WIRE GAUGE	15	14	12	12	11	11	10	10	9	9	8	6	5	4	3	2
NUMBER PER POUND	900	615	322	254	200	154	106	85	74	57	46	29	23	17	14	11

* in in.

CHEMICAL ELEMENTS

Name	Symbol	Valence Electrons	Atomic Weight*	Atomic Number	Name	Symbol	Valence Electrons	Atomic Weight*	Atomic Number
Actinium	Ac	2	[227]	89	Neon	Ne	8	20.183	10
Aluminum	Al	3	26.9815	13	Neptunium	Np	2	[237]	93
Americium	Am	2	[243]	95	Nickel	Ni	2	58.71	28
Antimony	Sb	5	121.75	51	Niobium	Nb	1	92.906	41
Argon	Ar	8	39.948	18	Nitrogen	N	5	14.0067	7
Arsenic	As	5	74.9216	33	Nobelium	No	2	[255]	102
Astatine	At	7	[210]	85	Osmium	Os	2	190.2	76
Barium	Ba	2	137.34	56	Oxygen	O	6	15.9994	8
Berkelium	Bk	2	[247]	97	Palladium	Pd	—	106.4	46
Beryllium	Be	2	9.0122	4	Phosphorus	P	5	30.9738	15
Bismuth	Bi	5	208.980	83	Platinum	Pt	1	195.09	78
Boron	B	3	10.811	5	Plutonium	Pu	2	[244]	94
Bromine	Br	7	79.909	35	Polonium	Po	6	[210]	84
Cadmium	Cd	2	112.40	48	Potassium	K	1	39.102	19
Calcium	Ca	2	40.08	20	Praseodymium	Pr	2	140.907	59
Californium	Cf	2	[251]	98	Promethium	Pm	2	[145]	61
Carbon	C	4	12.01115	6	Protactinium	Pa	2	[231]	91
Cerium	Ce	2	140.12	58	Radium	Ra	2	[226]	88
Cesium	Cs	1	132.905	55	Radon	Rn	8	[222]	86
Chlorine	Cl	7	35.453	17	Rhenium	Re	2	186.2	75
Chromium	Cr	1	51.996	24	Rhodium	Rh	1	102.905	45
Cobalt	Co	2	58.9332	27	Rubidium	Rb	1	85.47	37
Copper	Cu	1	63.54	29	Ruthenium	Ru	1	101.07	44
Curium	Cm	2	[247]	96	Samarium	Sm	2	150.35	62
Dysprosium	Dy	2	162.50	66	Scandium	Sc	2	44.956	21
Einsteinium	Es	2	[254]	99	Selenium	Se	6	78.96	34
Erbium	Er	2	167.26	68	Silicon	Si	4	28.086	14
Europium	Eu	2	151.96	63	Silver	Ag	1	107.870	47
Fermium	Fm	2	[257]	100	Sodium	Na	1	22.9898	11
Fluorine	F	7	18.9984	9	Strontium	Sr	2	87.62	38
Francium	Fr	1	[223]	87	Sulfur	S	6	32.064	16
Gadolinium	Gd	2	157.25	64	Tantalum	Ta	2	180.948	73
Gallium	Ga	3	69.72	31	Technetium	Tc	2	[97]	43
Germanium	Ge	4	72.59	32	Tellurium	Te	6	127.60	52
Gold	Au	1	196.967	79	Terbium	Tb	2	158.924	65
Hafnium	Hf	2	178.49	72	Thallium	Tl	3	204.37	81
Helium	He	2	4.0026	2	Thorium	Th	2	232.038	90
Holmium	Ho	2	164.930	67	Thulium	Tm	2	168.934	69
Hydrogen	H	1	1.00797	1	Tin	Sn	4	118.69	50
Indium	In	3	114.82	49	Titanium	Ti	2	47.90	22
Iodine	I	7	126.9044	53	Tungsten	W	2	183.85	74
Iridium	Ir	2	192.2	77	Unnilennium	Une	2	[266]	109
Iron	Fe	2	55.847	26	Unnilhexium	Unh	2	[263]	106
Krypton	Kr	8	83.80	36	Unniloctium	Uno	—	[265]	108
Lanthanum	La	2	138.91	57	Unnilpentium	Unp	2	[262]	105
Lawrencium	Lr	2	[256]	103	Unnilquadium	Unq	2	[261]	104
Lead	Pb	4	207.19	82	Unnilseptium	Uns	2	[262]	107
Lithium	Li	1	6.939	3	Uranium	U	2	238.03	92
Lutetium	Lu	2	174.97	71	Vanadium	V	2	50.942	23
Magnesium	Mg	2	24.312	12	Xenon	Xe	8	131.30	54
Manganese	Mn	2	54.9380	25	Ytterbium	Yb	2	173.04	70
Mendelevium	Md	2	[258]	101	Yttrium	Y	2	88.905	39
Mercury	Hg	2	200.59	80	Zinc	Zn	2	65.37	30
Molybdenum	Mo	1	95.94	42	Zirconium	Zr	2	91.22	40
Neodymium	Nd	2	144.24	60					

*a number in brackets indicates the mass number of the most stable isotope

Glossary

A

abrasion: The removal or displacement of material due to the pressure of hard particles.

abrasive wear: Wear caused by small, hard particles.

absolute pressure: Pressure above a perfect vacuum.

absolute rating: An indication of the largest opening in a strainer element.

absolute temperature: The temperature on a scale that begins with absolute zero.

absolute zero: The temperature at which substances possess no heat.

acceleration: An increase in speed.

accelerometer transducer: A device constructed of quartz crystal material that produces an electric current when compressed.

accumulator: A container in which fluid is stored under pressure.

action: The work of an actuator or a pilot operator, which becomes the input for another section of a control circuit.

actuator: A device that transforms fluid energy into linear or rotary mechanical force.

acute angle: An angle that contains less than 90°.

acute triangle: A scalene triangle with each angle less than 90°.

additive: A chemical compound added to a fluid to change its properties.

adjacent angles: Angles that have the same vertex and one side in common.

adsorption: The adhesion of a gas or liquid to the surface of a porous material.

aftercooler: A heat exchanger that cools air that has been compressed.

air compressor: A device that takes air from the atmosphere and compresses it to increase its pressure.

air cylinder: A device that converts compressed air energy into linear mechanical energy.

air motor: An air-driven device that converts fluid energy into rotary mechanical energy.

alignment: The location (within tolerance) of one axis of a coupled machine shaft relative to that of another.

alternating current (AC): A flow of electrons that reverses its direction of flow at regular intervals.

altitude: **1.** The perpendicular dimension from the vertex to the base of a triangle. **2.** The perpendicular distance between the two bases of a prism. **3.** The perpendicular distance from the vertex to the base of a pyramid.

ambient temperature: The temperature of the air surrounding a piece of equipment.

amp: A measure of a quantity of electron movement per second, commonly referred to as electric current.

anchoring: Any means of fastening a mechanism securely to a base or foundation.

AND logic element: A logic element that provides a logic level 1 only if all inputs are at logic level 1.

angle: The intersection of two lines or sides.

angular-contact bearing: A rolling-contact bearing designed to carry both heavy axial (thrust) loads and radial loads.

angular misalignment: **1.** In flexible belt drives, a condition where two shafts are parallel but at different angles with the horizontal plane. **2.** In motor couplings, a condition where one shaft is at an angle to the other shaft.

angular soft foot: A condition that exists when one machine foot is bent and not on the same plane as the other feet.

arc: A portion of the circumference.

area: The number of unit squares equal to the surface of an object.

armature: The moveable part of a generator or motor.

asymmetrical load: A load in which one-half of the load is not a mirror image of the other half.

atmospheric pressure: The force exerted by the weight of the atmosphere on the Earth's surface.

atom: The smallest building block of matter that cannot be divided into smaller units without changing its basic character.

authorized individual: A knowledgeable individual to whom the authority and responsibility to perform a specific assignment has been given.

auto-ignition: The temperature at which oil ignites by itself.

axial float: The axial movement of a shaft due to bearing and bearing housing clearances.

axial load: A load in which the applied force is parallel to the axis of rotation.

B

babbitt metals: Alloys of soft metals such as copper, tin, and lead, and a hard material such as antimony.
backlash: The play between mating gear teeth.
balanced-vane pump: A vane pump that has two sets of internal ports and contains an elliptical cam ring.
ball bearing: An anti-friction bearing that permits free motion between a moving part and a fixed part by means of balls confined between inner and outer rings.
base: The side upon which a triangle stands.
base diameter: The diameter from which the involute portion of a tooth profile is generated.
base plate: A rigid steel support for firmly coupling and aligning two or more rotating devices.
bases: The ends of a prism.
bearing: A machine part that supports another part, such as a shaft, which rotates or slides in or on it.
bed section: The lower section of an extension ladder.
bellows: A device that draws air in through a flapper valve when expanded and expels the air through a nozzle when contracted.
belt and sheave groove gauge: A gauge that has a male form to determine the size of a pulley and a female form to determine the size of a belt.
belt deflection method: A belt tension method in which the tension is adjusted by measuring the deflection of the belt.
belt pitch line: A line located on the same plane as the belt tension member.
bend: A permanent movement of a hook attempting to straighten.
bending strength: A metal's resistance to bending or deflection in the direction in which the load is applied.
bevel gear: A gear that connects shafts at an angle in the same plane.
bight: A loose or slack part of a rope between two fixed ends.
bimetallic device: A device that consists of two dissimilar metals.
binary system: A system that has two values such as pressure or no pressure.
bird caging: A damage condition of wire rope where the strands separate and open forming a shape similar to a bird cage.
blackwall hitch: A hitch made for securing a rigging rope to a hoisting hook.
bladder gas charged accumulator: An accumulator consisting of a seamless steel shell, a rubber bladder (bag) with a gas valve, and a poppet valve.
block: An assembly of hook(s), pulley(s), and frame suspended by hoisting ropes.
blockout: The process of placing a solid object in the path of a power source to prevent accidental energy flow.
bolt bound: The prevention of the horizontal movement of a machine due to the contacting of the machine anchor bolts to the sides of the machine anchor holes.
boundary lubrication: The condition of lubrication in which the friction between two surfaces in motion is determined by the properties of the surfaces and the properties of the lubricant other than viscosity.
Bourdon tube: A hollow metal tube made of brass or similar material.
bowline knot: A knot that forms a loop that is absolutely secure.
breakaway torque: The initial energy required to get a nonmoving load to turn.
bridge girder: The principal horizontal beam that supports a hoist trolley and is supported by end trucks.
butt spur: A notched, pointed, or spiked end of a ladder which helps prevent the ladder butt from slipping.

C

cab: A compartment or platform attached to a bridge girder from which an operator may ride while controlling a crane.
cabling: A rope's attempt to rotate and untwist its strand lays while under stress.
cage: A barrier or enclosure mounted on the siderails of a fixed ladder or fastened to a structure.
cam ring: A metal ring that provides an area for fluid flow and a surface against which vanes ride.
cantilever: A projecting beam or member supported at only one end.
capacity: The ability to hold or contain something.
capillary action: The action by which the surface of a liquid is elevated on a material due to its relative molecular attraction.
carrier: The track of a ladder safety system consisting of a flexible cable or rigid rail secured to a ladder or structure.
cats-paw hitch: A hitch used as a light-duty, quickly-formed eye for a hoisting hook.
caustic solution: A liquid that creates heat and corrosion.
cavitation: The process in which microscopic gas bubbles expand in a vacuum and suddenly implode when entering a pressure.
center of gravity: The balancing point of a load.
centerpoint: The point a circle or arc is drawn around.
centralized system: A lubrication system that contains permanently installed plumbing, distribution valves, reservoir, and pump to provide lubrication.
centrifugal force: The outward force produced by a rotating object.
chain: A series of metal rings connected to one another and used for support, restraint, or transmission of mechanical power.
challenging: The process of pressing or selecting the start switch of a machine to determine if the machine starts when it is not supposed to start.

check valve: A valve that allows flow in only one direction.
chemisorption: A chemical adsorption process in which weak chemical bonds are formed between liquid or gas molecules and solid surfaces.
choker hook: A sliding hook used in a choker sling and hooked to a sling eye.
chord: A line from circumference to circumference not through the centerpoint.
circle: **1.** A plane figure generated about a centerpoint. **2.** A plane figure formed by a cutting plane perpendicular to the axis of a cone.
circuit: A closed path through which hydraulic fluid flows.
circuit breaker: A device with a mechanical mechanism that may manually or automatically open a circuit when an overload condition or short circuit occurs.
circular pitch: **1.** In belts, the distance from the center of one tooth to the center of the next tooth, measured along the pitch line. **2.** In gears, the distance from a point on a gear tooth to the corresponding point on the next gear tooth, measured along the pitch circle.
circumference: The boundary of a circle.
clearance: The radial distance between the top of a tooth and the bottom of the mating tooth space when fully mated.
cleat: A narrow wood piece, nailed across another board or boards, to provide support or to prevent movement.
closed circuit: An electrical circuit with a completed path that allows current flow.
clove hitch: A quick hitch used to secure a rope temporarily to an object.
coalescing filter: A device that removes submicron solids and vapors of oil or water by uniting very small droplets into larger droplets.
coefficient of friction: The measure of the frictional force between two surfaces in contact.
commutator: A device for reversing the direction of current flowing through the wires of a rotating armature that ensures a single-direction flow of current from a generator.
competent person: A person capable of recognizing and evaluating employee exposure to hazardous substances or to other unsafe conditions and of specifying the necessary protection and precautions to be taken to ensure the safety of all employees.
complementary angles: Two angles formed by three lines in which the sum of the two angles equals 90°.
compound gear train: Two or more sets of gears where two gears are keyed and rotate on one common shaft.
concentric circles: Two or more circles with different diameters but the same centerpoint.
condensation: The change in state from a gas to a liquid.
conductor: A material that has very little resistance and permits electrons to move through it easily.
cone: A solid generated by a straight line moving in contact with a curve and passing through the vertex.
conic section: A curve produced by a plane intersecting a right circular cone.
connecting link: A three-part chain attachment used to assemble and connect the master link to a chain.
connecting rod: The rod that connects the crankshaft to a piston.
Conrad bearing: A single-row ball bearing without loading slots that has deeper-than-normal races.
contact: A conducting part of a relay that acts as a switch to connect or disconnect a circuit or component.
contactor: A control device that uses a small control current to energize or de-energize the load connected to it.
continuity tester: A device that indicates if a circuit is open or closed.
control circuit: A circuit that allows electrical devices to be controlled from remote locations.
controlled flow: The fluid flow after a flow control device has reduced the volume (gpm) of the fluid flow.
core protrusion: A damage condition of wire rope where compressive forces from within the rope force the strands apart.
corrosion: The action or process of eating or wearing away gradually by chemical action.
corrosive wear: Wear resulting from metal being attacked by acid.
coulomb: A quantity of electrons equal to 6.25×10^{18}.
coupling: A device that connects the ends of rotating shafts.
coupling face: The flat surface of a coupling half, facing the flat surface of the connecting coupling half.
coupling rim: The outside diameter surface of a coupling.
coupling unbalance: An unequal radial weight distribution where the mass and coupling geometric lines do not coincide.
cow hitch: A hitch used to secure a tag line to a load.
crankshaft: A shaft that has one or more eccentric surfaces that produce a reciprocating motion when the shaft is rotated.
crossover: One wrap winding on top of the preceding wrap.
crowning: A reverse strand splice that is used when an enlarged rope end is desired or not objectionable.
cubic foot: Contains 1728 cu in. (12″ × 12″ × 12″ = 1728 cu in.).
cubic inch: Measures 1″ × 1″ × 1″ or its equivalent.
cubic yard: Contains 27 cu ft (3′ × 3′ × 3′ = 27 cu ft).
cup seal: A lip seal whose lip forms the shape of a cup.
current (electron flow): The amount of electrons flowing through an electrical circuit.
curved line: A line that continually changes direction.
curve/rope (D/d) ratio: The ratio between the diameter of a curved component (D), such as a pulley, and the nominal diameter of the rope (d).
curvilinear belt: A timing belt containing circular-shaped teeth.
cutaway diagram: A diagram showing the internal details of components and the path of fluid flow.
cylinder: A solid generated by a straight line (genatrix) moving in contact with a curve and remaining parallel to the axis and its previous position.
cylindrical roller bearing: A roller bearing having cylinder-shaped rollers.

D

decision: A judgement or conclusion reached or given.
demulsification: The act of separating water and oil quickly.
desiccant dryer: A device that removes water vapor by adsorption.
dew point: The temperature to which air must be cooled in order for the moisture in the air to begin condensing.
dial indicator: A device that measures the deviation from a true circular path.
diameter: The distance from circumference to circumference through the centerpoint.
diametral pitch: The ratio of the number of teeth in a gear to the diameter of the gear's pitch circle.
diaphragm gas charged accumulator: An accumulator with a flexible diaphragm separating the gas and fluid.
direct-acting valve: A valve that is activated or directly moved by fluid pressure from the primary port.
direct current (DC): A flow of electrons in only one direction.
directional control valve: A valve whose primary function is to direct or prevent flow through selected passages.
direct proportion: A statement of equality between two ratios in which the first of four terms divided by the second equals the third divided by the fourth.
disconnect switch (disconnect): A switch that disconnects electrical circuits from motors and machines.
dispersed solid: A solid that is finely ground in order to be spread.
displacement: **1.** The volume of oil moved during each cycle of a pump. **2.** The measurement of the distance (amplitude) an object is vibrating.
displacement transducer: A mechanical sensor whose gap-to-voltage output is proportional to the distance between it and the measured object (usually a shaft).
dodecahedron: A regular solid of twelve pentagons.
double-acting cylinder: A cylinder that requires fluid flow for extending and retracting.
double hitch knot: A knot with two half hitch knots.
double-pole scaffold: A wood scaffold with both sides resting on the floor or ground and is not structurally anchored to a building or other structure.
double trapezoidal belt: A timing belt containing two trapezoidal-shaped sets of teeth.
dowel effect: A condition that exists when the bolt hole of a machine is so large that the bolt head forces the washer into the hole opening on an angle.
drift: The slippage of a load caused by insufficient braking.
drip system: A gravity-flow lubrication system that provides drop-by-drop lubrication from a manifold or manually-filled cup through a needle valve.
drive gear: Any gear that turns or drives another gear.
driven gear: Any gear that is driven by another gear.
dropping point of grease: The temperature at which the oil in grease separates from the thickener and runs out, leaving just the thickener.
drum wrap: The rope length required to make one complete turn around the drum of a hoist or crane.
dry air: Air free of water vapor or oil droplets.
dryer: A device that dries air through cooling and condensing.
dynamic electricity: Electron flow from one atom to another atom.
dynamic head: The head of fluid in motion.
dynamic head pressure: The pressure and velocity of a fluid produced by a liquid in motion.
dynamic lift: The lift of fluid in motion.
dynamic range: The ratio between the smallest and largest signals that can be analyzed simultaneously.
dynamic seal: A seal used between moving parts that prevents leakage or contamination.
dynamic signal analyzer: An analyzer that uses digital signal processing and the FFT to display a dynamic vibration signal as a series of frequency components.
dynamic total column: Dynamic head plus dynamic lift.

E

eccentric: Out-of-round or that which deviates from a circular path.
eccentric circles: Two or more circles with different diameters and different centerpoints.
eccentric surface: A surface that has a different center than the center of a crankshaft.
eddy current: An electric current that is generated and dissipated in a conductive material in the presence of an electromagnetic field.
efficiency: A measure of a component's or system's useful output energy compared to its input energy.
electrical pitting: An electric arc discharge across the film of oil between mating gear teeth.
electricity: A physical occurrence involving electric charges and their effects when in motion and at rest.
electromagnet: A magnet created when electricity passes through a wire.
electromagnetic induction: The process by which voltage is induced in a wire by a magnetic field when lines of force cut across the wire.
electron: A negatively charged particle whirling around the nucleus at great speeds in a shell.
electron flow (electrical current): The traveling of a displaced (free) valence electron from one atom to another.
electronic reverse dial method: An alignment method that uses the reverse dial as a base method with the dial indicators replaced with electromechanical sensing devices.

element: A logic device that is capable of making a 0 or 1 output decision based on its input.
ellipse: A plane figure formed by a cutting plane oblique to the axis of a cone, but at a greater angle with the axis than with the elements of the cone.
emulsification: The act of mixing oil and water.
end play: The total amount of axial movement of a shaft.
end truck: A roller assembly consisting of a frame, wheels, and bearings generally installed or removed as complete units.
energy: A measure of the ability to do work.
energy-isolating device: A device that prevents the transmission or release of energy.
equal rotor unbalance: The unbalance of weighted force across one side of a rotor or armature.
equation: A means of showing that two numbers or two groups of numbers are equal to the same amount.
equilateral triangle: A triangle that has three equal angles and three equal sides.
equilibrium: The condition when all forces and torques are balanced by equal and opposite forces and torques.
exhaust flow: The fluid flow from an actuator, back through a valve, to a reservoir.
extension ladder: An adjustable-height ladder with a fixed bed section and sliding, lockable fly section(s).
eyebolt: A bolt with a looped head.
eye loop: A rope splice containing a thimble.

F

face width: The length of gear teeth in an axial plane.
false Brinell damage: Bearing damage caused by forces passing from one ring to the other through the balls or rollers.
Fast Fourier Transform analyzer: A microprocessor capable of displaying the FFT of an input signal.
fatigue crack: A crack in a gear that occurs due to bending, mechanical stress, thermal stress, or material flaws.
fatigue fracture: A breaking or tearing of gear teeth.
fatigue life: The maximum useful life of a bearing.
fatigue wear: Gear wear created by repeated stresses below the tensile strength of a material.
feeler gauge (thickness gauge): A steel leaf at a specific thickness.
ferromagnetic material: A material, such as soft iron, that is easily magnetized.
ferrous metals: Metals containing iron.
ferrule: A metal sleeve used for joining one piece of tube to another.
FFT: A calculation method for converting a time waveform into a series of frequency vs. amplitude components.
fiberglass ladder: A ladder constructed of fiberglass.
filler wire: Wire rope that uses fine wires to fill the gaps between the major wires.
filter: 1. A device containing a porous substance through which a fluid can pass but particulate matter cannot. **2.** A device that limits vibration signals so only a single frequency or group of frequencies can pass.
fire point: The temperature at which oil ignites when touched with a flame.
fixed bore pulley: A machine-bored one-piece pulley.
fixed ladder: A ladder that is permanently attached to a structure.
flared fitting: A fitting that is connected to a tubing whose end is spread outward.
flareless (compression) fitting: A fitting that seals and grips by manual adjustable deformation.
flash point: The temperature at which oil gives off enough gas vapor to ignite briefly when touched with a flame.
flexible belt drive: A system in which a resilient flexible belt is used to drive one or more shafts.
flexible coupling: A coupling with a resilient center, such as rubber or oil, that flexes under temporary torque or misalignment due to thermal expansion.
flow: The movement of a fluid.
flow control valve: A valve whose primary function is to regulate the rate of fluid flow.
flow rate: The volume of fluid flow.
fluid: A substance that tends to flow or conform to the outline of its container (such as a liquid or a gas).
fluid flow: The movement of fluid caused by a difference in pressure between two points.
fluting: The elongated and rounded grooves or tracks left by the etching of each roller on the rings of an improperly grounded roller bearing during welding.
fly section: The upper section(s) of an extension ladder.
foaming: Excessive air in hydraulic fluid.
foot pad: A metal swivel attachment with rubber or rubber-like tread which helps prevent a ladder butt from slipping.
force: The energy that produces movement.
formula: A mathematical equation that contains a fact, rule, or principle.
foundry hook: A hook with a wide, deep throat that fits the handles of molds or castings.
four-way directional control valve: A valve that has four main ports that change fluid flow from one port to another.
fracture: A small crack in metal caused by the stress or fatigue of repeated pulling or bending forces.
free air: Air at atmospheric pressure and ambient temperature.
frequency: The number of cycles per minute (cpm), cycles per second (cps), or multiples of rotational speed (orders).
frequency domain: The amplitude versus frequency spectrum observed on an FFT analyzer.
frequency spectrum: A representation of the frequency and content of a dynamic signal.
fretting corrosion: The rusty appearance that results when two metals in contact are vibrated, rubbing loose minute metal particles that become oxidized.
friction disc: A device that transmits power through contact between two discs or plates.

frustum: The remaining portion of a pyramid or cone with a cutting plane passed parallel to the base.
fulcrum: A support on which a lever turns or pivots and is located somewhere between the effort force and the resistance force.
fuse: An overcurrent protection device with a fusible link that melts and opens the circuit on an overcurrent condition.
fuse puller: A device made of a non-conductive material such as nylon that is used to grasp and remove cartridge fuses.

G

galling (adhesive wear): A bonding, shearing, and tearing away of material from two contacting, sliding metals.
gantry crane: A crane with bridge beams supported on legs.
gas: A fluid that has neither independent shape nor volume and tends to expand indefinitely.
gas charged accumulator: An accumulator that uses compressed gas over hydraulic fluid to store energy.
gasket: A seal used between machined parts or around pipe joints to prevent the escape of fluids.
gas laws: The relationships between the volume, pressure, and temperature of a gas.
gas lubricant: A lubricant that uses pressurized air to separate two surfaces.
gate valve: A two-position valve that has an internal gate that slides over the opening through which fluid flows.
gauge pressure: Pressure above atmospheric pressure that is used to express pressures inside a closed system.
gear: A toothed machine element used to transmit motion between rotating shafts.
gear pump: A pump consisting of two meshing gears enclosed in a close-fitting housing.
gear train: A combination of two or more gears in mesh used to transmit motion between two rotating shafts.
generator: A device that converts mechanical energy into electrical energy.
globe valve: An infinite-position valve that has a disk that is raised or lowered over a port through which fluid flows.
grab hook: A hook used to adjust or shorten a sling leg through the use of two chains.
graphic diagram: A drawing that uses simple line shapes (symbols) with interconnecting lines to represent the function of each component in a circuit.
grease: A semisolid lubricant created by combining low-viscosity oils with thickeners, such as soap or other finely dispersed solids.
grease cup: A receptacle used to apply grease to bearings.
grease dropping point: The maximum temperature a grease withstands before it softens enough to flow through a laboratory testing orifice.
grease gun: A small hand-operated device that pumps grease under pressure into bearings.
great circle: The circle formed by passing a cutting plane through the center of a sphere.
ground: An electrical connection to the earth.
ground fault circuit interrupter (GFCI): An electrical device that protects personnel by detecting potentially hazardous ground faults and quickly disconnecting power from the circuit.
grounding: The connection of all exposed noncurrent-carrying metal parts to the earth.
guardrail: A rail secured to uprights and erected along the exposed sides and ends of a platform.
guyline: A rope, chain, rod, or wire attached to equipment as a brace or guide.

H

half hitch knot: A binding knot where the working end is laid over the standing part and stuck through the turn from the opposite side.
halyard: A rope used for hoisting or lowering objects.
hand chain: A continuous chain grasped by an operator to operate a pocket wheel.
hand-chain drop: The distance between the lower portion of a hand chain to the upper limit of the hoist hook travel.
hand-chain hoist: A manually-operated chain hoist used for moving a load.
hazardous location: A location where there is an increased risk of fire or explosion due to the presence of flammable gases, vapors, liquids, combustible dusts, or easily-ignitable fibers or flyings.
head: The difference in the level of a liquid (fluid) between two points.
head pressure: The pressure created by fluid stacked on top of itself.
headroom: The distance from the cup of the top hook to the cup of a hoist hook when the hoist hook is at its upper limit of travel.
heat energy: The ability to do work (usually destructive) using the heat stored or built up in a fluid.
heat exchanger: A device which transfers heat through a conducting wall from one fluid to another.
helical gear: A gear with teeth that are cut at an angle to its axis of rotation.
helical screw compressor: A compressor that contains meshing screw-like helical rotors that compress air as they turn.
helix: A spiral or screw shape form.
herringbone gear: A double helical gear that contains a right- and left-hand helix.

Hertz (Hz): A measurement of a frequency equal to one cycle per second.

hexahedron: A regular solid of six squares.

hitch: The interlacing of rope to temporarily secure it without knotting the rope.

hoist chain: The chain that raises a load.

hoisting apparatus chain: A precisely-measured chain calibrated to function in pocket-type wheels used in manual or powered chain hoists.

hoisting hook: A steel alloy hook used for overhead lifting and connected directly to the piece being lifted.

hoist trolley: The unit carrying the hoisting mechanism that travels on a bridge girder.

holding contact: An auxiliary contact used to maintain current flow to the coil of a relay.

hook: A curved or bent implement for holding, pulling, or connecting another implement.

hook drift: The slippage of a hook caused by insufficient braking.

horizontal line: A line that is parallel to the horizon.

horizontal weight center: A weight mass above a pivot point that causes a load to topple because it is top heavy.

horsepower: A unit of power equal to 746 W or 33,000 lb-ft per minute (550 lb-ft per second).

hose: A flexible tube for carrying fluids under pressure.

humidity: The amount of moisture in the air.

hunting tooth: A tooth added to mesh with every tooth on a mating gear to produce even tooth wear.

hydraulic actuator: A device that converts hydraulic energy into mechanical energy.

hydraulic cylinder: A device that converts hydraulic energy into straight-line (linear) mechanical energy.

hydraulic diagram: The layout, plan, or sketch of a hydraulic circuit that is designed to explain, demonstrate, or clarify the relationship or functions between hydraulic components.

hydraulic motor: A device that converts hydraulic energy into rotary mechanical energy.

hydraulics: The branch of science that deals with the practical application of water or other liquids at rest or in motion.

hydraulic scissor lift: A mobile hydraulically-operated platform controlled by remote switches attached at the platform.

hydrocarbon: Any substance that is composed mostly of hydrogen and carbon.

hydrodynamics: The study of the forces exerted on a solid body by the motion or pressure of a fluid.

hydrostatics: The study of liquids at rest and the forces exerted on them or by them.

hyperbola: A plane figure formed by a cutting plane that has a smaller angle with the axis than with the elements of a cone.

hypoid gear: A spiral bevel gear with curved, non-symmetrical teeth that are used to connect shafts at right angles.

hypotenuse: The side of a right triangle opposite the right angle.

I

icosahedron: A regular solid of twenty triangles.

idler gear: A gear that transfers motion and direction in a gear train, but does not change speeds.

imbalance: A lack of balance.

impact flaring method: A basic flaring method in which a flaring tool is inserted into the tubing end and hammered into the tubing until the tubing end is spread (flared) as required.

implosion: An inward bursting.

inching: Slow movement in small degrees.

inclined line: A line that is slanted.

induced soft foot: Soft foot that is created by external forces such as coupling misalignment, piping strain, tight jack screws, or improper structural bracing.

induction: The process of causing electrons to align or uniformly join to create a magnetic or electrical force.

industrial crane: A crane with structural beam supports for lifting equipment.

inert gases: Gases that lack active properties.

instrumentation: The area of industry that deals with the measurement, evaluation, and control of process variables.

insulator: A material that has a very high resistance and resists the flow of electrons.

intake filter: A filter that removes solids from free air at a compressor inlet port.

intake flow: The fluid flow from a reservoir, through filters, to a pump.

intensifier (booster): A device that converts low-pressure fluid power into high-pressure fluid power.

intercooler: A piped connection between the discharge of one compression stage and the inlet of the next compression stage.

intercooling: The process of removing a portion of the heat of compression as the air is fed from one compression stage to another.

interference fit: Fit in which the internal member is larger than the external member so that there is always an actual interference of metal.

inverse proportion: A proportion in which an increase in one quantity results in a proportional decrease in the other related quantity.

inverse ratio: The ratio that results when the second term is divided by the first.

involute form: A tooth form that is curled or curved.

irregular polygon: A polygon with unequal sides and unequal angles.

irregular polyhedra: Solids with faces that are irregular polygons (unequal sides).
isosceles triangle: A triangle that contains two equal angles and two equal sides.

J

jack screw: A screw inserted through a block that is attached to a machine base plate allowing for ease in machine movement.
jib crane: A crane that is mounted on a single structural leg.
journal: The part of a shaft, such as an axle or spindle, that moves in a sleeve bearing.

K

kinetic energy: The energy of motion.
kinking: A sharp permanent bending.
knot: The interlacing of rope to form a permanent connection.
knotting: Fastening a part of a rope to another part of the same rope by interlacing it and drawing it tight.

L

ladder: A structure consisting of two siderails joined at intervals by steps or rungs for climbing up and down.

ladder duty rating: The weight (in lb) a ladder is designed to support under normal use.

ladder jack: A ladder accessory that supports a plank to be used for scaffolding.

ladder safety system: An assembly of components whose function is to arrest the fall of a worker.

lang-lay rope: A rope in which the wires and strands are laid in the same direction.

lanyard: A rope or webbing device used to attach a worker's harness to a lifeline.

laser rim-and-face alignment method: An alignment method in which laser devices are placed opposite each other to measure alignment.

lay: A complete helical wrap of the strands of a rope.

lead line: The part of a rope to which force is applied to hold or move a load.
left lang-lay rope: A rope in which the wires are laid to the left and the strands are laid to the left.
left regular-lay rope: A rope in which the strands are laid to the left and wires are laid to the right.
lever-operated hoist: A lifting device that is operated manually by the movement of a lever.
lichens: Fungi normally seen as a growth on tree trunks or rocks.

lift: 1. In hoisting, the distance between the hoist's upper and lower limits of travel. **2.** In pumping, the height at which atmospheric pressure forces a fluid above the elevation of its supply source.

lifting: Hoisting equipment or machinery by mechanical means.

lifting lug: A thick metal loop (eyebolt) welded or screwed to a machine to allow balanced lifting.

limit switch: A device that cuts off the power automatically at or near the upper limit of hoist travel.

limit valve: A mechanically-actuated 3-way valve that is used to either monitor motion or measure position of an object.
line: The boundary of a surface.
linear amplitude spectra: Amplitude signals displayed in equal increments.

lip seal: A seal that is made of a resilient material that has a sealing edge formed into a lip.

liquid: A fluid that can flow readily and assume the shape of its container.

liquid lubricant: A lubricant that uses a liquid, such as oil, to separate two surfaces.
litmus paper: A color-changing, acid-sensitive paper that is impregnated with lichens.

loading slot: A groove or notch on the inside wall of each bearing ring to allow insertion of balls.

lobe: The screw helix of a rotor.
lockout: The process of preventing the flow of energy from a power source to a piece of equipment.
logarithmic amplitude spectra: Amplitude signals displayed in powers of 10.

logarithmic scale: An amplitude or frequency displayed in powers of ten.

logic: The science of correct reasoning.
loop: The folding or doubling of a line, leaving an opening through which another line may pass.
loop eye: A length of webbing folded back and spliced to a sling body, forming an opening.
lubricant: A substance placed between two solid surfaces to reduce their friction.
lubrication: The process of maintaining a fluid film between solid surfaces to prevent their physical contact.
lubricator: A device that injects atomized oil into the air sent to pneumatic components.

M

machine: A group of mechanical devices that transfer force, motion, or energy input at one device into a force, motion, or energy output at another device.

magnet: A device that attracts iron and steel because of the molecular alignment of its material.

magnetic flux lines: The invisible lines of force that make up a magnetic field.

magnetism: A force that interacts with other magnets and ferromagnetic materials.

main header: The main air supply line that runs between a receiver and the circuits in a pneumatic system.

manifold: A device that contains passageways that enable one input signal to be divided into several output signals.

master link: A chain attachment with a ring considerably larger than that of the chain to allow for the insertion of a hook.

maximum intended load: The total of all loads, including the working load, the weight of the scaffold, and any other loads that may be anticipated.

mechanical: Pertaining to or concerned with machinery or tools.

mechanical advantage: The ratio of the output force of a device to the input force.

mechanical drive: A system by which power is transmitted from one point to another.

mercury barometer: An instrument that measures atmospheric pressure using a column of mercury.

mesh: **1.** In rope, the size of the openings between the rope or twine of a net. **2.** In filters, the number of horizontal and vertical threads per square inch.

metal fatigue: The fracturing of worked metal due to normal operating conditions or overload situations.

metal ladder: A ladder constructed of metal.

metering: Regulating the amount or rate of fluid flow.

micron (μ): A unit of length equal to one millionth of a meter (.000039″).

midrail: A rail secured to uprights approximately midway between the guardrail and the platform.

misalignment: The condition where the axes of two machine shafts are not aligned within tolerances.

miter gear: A gear used at right angles to transmit horsepower between two intersecting shafts at a 1 : 1 ratio.

modified curvilinear belt: A timing belt containing modified circular-shaped teeth.

moisture separator: A device that separates a large percentage of water from cooled air through a series of plates or baffles.

molded notch belt: A belt that has notches molded into its cross-section along the full length of the belt.

molecule: The smallest division of matter that can be made and have a substance still retain its chemical identity.

motor starter: An electrically-operated switch (contactor) that includes motor overload protection.

multimeter: A test tool used to measure two or more electrical values.

multiple-point suspension scaffold: A suspension scaffold supported by four or more ropes.

multistage compressor: A compressor that uses two or three cylinders, each with a progressively smaller diameter, to produce progressively higher pressures.

N

needle bearing: An anti-friction roller-type bearing with long rollers of small diameter.

needle valve: An infinite-position valve that has a narrow tapered stem (needle) positioned in line with a tapered hole or orifice.

neutral conductor: A wire that carries current from one side of a load to ground.

neutron: A particle with no electrical charge.

nip: A pressure and friction point created when a rope crosses over itself after a turn around an object.

nominal value: A designated or theoretical value that may vary from the actual value.

nonparallel misalignment: Misalignment where two pulleys or shafts are not parallel.

nonpositive displacement pump: A pump that is not sealed between its inlet and outlet.

nonpositive seal: A seal that allows a minute amount of fluid through to provide lubrication between surfaces.

NOT logic element: A logic element that provides an output that is the opposite of the input.

nucleus: The heavy, dense center of an atom.

O

oblique cylinder: A cylinder with the axis not perpendicular to the base.

oblique prism: A prism with lateral faces not perpendicular to the bases.

obtuse angle: An angle that contains more than 90°.

obtuse triangle: A scalene triangle with one angle greater than 90°.

octahedron: A regular solid of eight triangles.

offset misalignment: **1.** In flexible belt drives, a condition where two shafts are parallel but the pulleys are not on

the same axis. **2.** In motor couplings, a condition where two shafts are parallel but are not on the same axis.
ohm: The resistance of a conductor in which an electrical pressure of 1 V causes an electrical current of 1 A to flow.
Ohm's law: The relationship between voltage (E), current (I), and resistance (R) in a circuit.
oil analysis: A predictive maintenance technique that detects and analyzes the presence of acids, dirt, fuel, and wear particles in lubricating oil to predict equipment failure.
oil carry-over: The released lubricating oil from the walls of a compressor cylinder and piston.
oil whirl: The buildup and resistance of a lubricant in a rolling-contact bearing that is rotating at excessive speeds.
open circuit: An electrical circuit that has a gap or opening that does not allow current flow.
operational pitch point: The tangent point of two pitch circles at which gears operate.
opposing forces rotor unbalance: The unbalance of weighted forces on opposing ends and sides of a rotor or armature.
order: A multiple of a running speed (rpm) frequency.
orifice: A precisely-sized hole through which fluid flows.
O-ring: A molded synthetic rubber seal having a round cross section.
OR logic element: A logic element that provides a logic level 1 if one or more inputs are at logic level 1.
oscillator: A device that generates a radio frequency (RF) field that, when sent to a probe tip, creates eddy currents.
overhead crane: A crane that is mounted between overhead runways.
overhung load: A force exerted radially on a shaft that may cause bending of the shaft or early bearing and belt failure.
overload relay: A time-delay device that senses motor current temperatures and disconnects the motor from the power supply if the current is excessive for a certain length of time.
oxidation: The combining of oxygen with elements in oil which break down the basic oil composition.

P

packing: A bulk deformable material or one or more mating deformable elements reshaped by manually adjustable compression.

parabola: A plane figure formed by a cutting plane oblique to the axis and parallel to the elements of a cone.

parallelepiped: A prism with bases that are parallelograms.

parallel lines: Two or more lines that remain the same distance apart.

parallelogram: A four-sided plane figure with opposite sides parallel and equal.

parallel soft foot: A condition that exists when one or two machine feet are higher than the others and parallel to the base plate.
part: A rope length between the lower (hook) block and the upper block or drum.
particulate: A fine solid particle which remains individually dispersed in a gas.
pawl: A mechanism used to prevent a ratchet wheel from turning backwards.
pawl lock: A pivoting hook mechanism attached to the fly section(s) of an extension ladder.
peak: The absolute value from a zero point (neutral) to the maximum travel on a waveform.
peak-to-peak: The absolute value from the maximum positive travel to the maximum negative travel on a waveform.
pendant: A pushbutton or lever control suspended from a crane or hoisting apparatus.
perpendicular line: A line that makes a 90° angle with another line.
petroleum fluid: A fluid consisting of hydrocarbons.
phase: The position of a vibrating part at a given moment with reference to another vibrating part at a fixed reference point.
pictorial diagram: A diagram that uses drawings or pictures to show the relationship of each component in a circuit.
piezoelectric: The production of electricity by applying pressure to a crystal.
pilot line: A passage used to carry fluid to control a valve.
pilot-operated valve: A valve that is actuated by fluid in the line that is otherwise sent back to the reservoir.
pilot operation: Controlling the function of a valve using system pressure or pressure supplied by an external (pilot) source.
pinion: The smaller gear of a pair of gears, especially when engaging rack teeth.
pipe: A hollow cylinder of metal or other material of substantial wall thickness.
piston compressor: A compressor in which air is compressed by reciprocating pistons.
piston gas charged accumulator: An accumulator with a floating piston acting as a barrier between the gas and fluid.
piston pump: A pump in which fluid flow is produced by reciprocating pistons.
pitch circle: The circle that contains the operational pitch point.
pitch diameter: The diameter of a pitch circle.
pitch length: The total length of the timing belt measured at the belt pitch line.
pitting: Localized corrosion that has the appearance of cavities (pits).
plane figure: A flat figure with no depth.
plank: A board 2″ to 4″ thick and at least 8″ wide.
platform: A landing surface which provides access/egress or rest from a fixed ladder.
plumb: An exact verticality (determined by a plumb bob and line) with the surface of the earth.

ply: A layer of a formed material.

pneumatic circuit: A combination of air-operated components that are connected to perform work.

pneumatic hoist: A power-operated hoist operated by a geared reduction air motor.

pneumatic logic element: A miniature air valve used as a switching device to provide decision making signals in a pneumatic circuit.

pneumatics: The branch of science that deals with the transmission of energy using a gas.

pneumatic system: A system that transmits and controls energy through the use of a pressurized gas within an enclosed circuit.

pocket wheel: A pulley-like wheel with chain link pockets that is connected to a hoist mechanism.

polarity: The positive (+) or negative (–) state of an object.

pole scaffold: A wood scaffold with one or two sides firmly resting on the floor or ground.

polygon: A many-sided plane figure.

polyhedra: Solids bound by plane surfaces (faces).

polymer: The result of a chemical reaction in which two or more small molecules combine to form larger molecules.

port plate: A device that contains ports that connect the pump internal inlet and discharge areas to the pump housing inlet and outlet ports.

position: A specific location of a spool within a valve which determines the direction of fluid flow through the valve.

positive displacement: The moving of a fixed amount of a substance with each cycle.

positive displacement compressor: A compressor that compresses a fixed quantity of air with each cycle.

positive-displacement pump: A pump that delivers a definite volume of fluid for each cycle of the pump at any resistance encountered.

positive seal: A seal that does not allow the slightest amount of fluid to pass.

potential energy: Stored energy a body has due to its position, chemical state, or condition.

power: The rate or speed of doing work.

power circuit: The part of an electrical circuit that connects the load to the main power lines.

power distribution: The process of delivering electrical power to where it is needed.

precharge pressure: The pressure of the compressed gas in an accumulator prior to the admission of hydraulic fluid.

preformed rope: Wire rope in which the strands are permanently formed into a helical shape during fabrication.

preloading: An initial pressure placed on a bearing when axial load forces are expected to be great enough to overcome preload force, thereby resulting in proper clearances.

pressure: The force per unit area.

pressure compensated flow control valve: A needle valve that makes allowances for pressure changes before or after an orifice through the use of a spring and spool.

pressure-compensated vane pump: A vane pump equipped with a spring on the low-displacement side of the cam ring.

pressure compensator: A displacement control that alters displacement in response to pressure changes in a system.

pressure drop: The pressure differential between upstream and downstream fluid flow caused by resistance.

pressure energy: The ability to do work by applying pressure to a fluid.

pressure filter: A very fine filter placed after a pump for protection of system components.

pressure gauge: A device that measures the intensity of a force applied to a fluid.

pressure-reducing valve: A valve that limits the maximum pressure at its outlet, regardless of the inlet pressure.

pressure regulator: A valve that restricts and/or blocks downstream air flow.

pressure-relief valve: A valve that sets a maximum operating pressure level for a circuit to protect the circuit from overpressure.

pressure switch: A device that senses a high- or low-pressure condition and relays an electrical signal to turn the compressor motor ON or OFF.

primary port: The source or inlet port.

primary winding: The power input winding of a transformer that is connected to the incoming power supply.

prime mover: An electric motor or engine that supplies rotational force at a constant speed.

prism: A solid with two bases that are parallel and identical polygons.

process variable: Any characteristic that changes its value during any operation within the process.

proportion: An expression of equality between two ratios.

proton: A particle with a positive electrical charge of one unit.

pseudocavitation: Artificial cavitation caused by air being allowed into a pump suction line.

pump: A mechanical device that causes fluid to flow.

pyramid: A solid with a base that is a polygon and sides that are triangles.

Pythagorean Theorem: States that the square of the hypotenuse of a right triangle is equal to the sum of the squares of the other two sides.

Q

quadrant: One-fourth of a circle containing 90°.

quadrilateral: A four-sided polygon with four interior angles.

quad-ring: A molded synthetic rubber seal having a basically square cross-sectional shape.

R

race: The track on which the balls of a bearing move.
rack gear: A gear with teeth spaced along a straight line.
racking: The ability to be forced out of shape or form.
rack teeth: Gear teeth used to produce linear motion.
radial bearing: A rolling-contact bearing in which the load is transmitted perpendicular to the axis of shaft rotation.
radial load: A load in which the applied force is perpendicular to the axis of rotation.
radius: The distance from the centerpoint to the circumference.
ratchet: A mechanism that consists of a toothed wheel and a spring-loaded pawl.
ratio: The relationship between two quantities or terms.
raveling: The unwinding or untwisting of rope.
reach: The distance between the cup of a top hook and the cup of a hoist hook when the hoist hook is at its lower limit of travel.
reciprocate: To move forward and backward alternately.
reciprocating compressor: A device that compresses gas by means of a piston(s) that moves back and forth in a cylinder.
rectangle: A quadrilateral with opposite sides equal and four 90° angles.
rectangular parallelepiped: A prism with bases and faces that are all rectangles.
reel: A wooden assembly on which wire rope is wound for shipping and storage.
reeving: Passing a rope through a hole or opening or around a series of pulleys.
refrigerant dryer: A device designed to lower the temperature of compressed air to 35°F.
regular-lay rope: A rope in which the wires in the strands are laid in the opposite direction to the lay of the strands.
regular polygon: A polygon with equal sides and equal angles.
regular pyramid: A base that is a regular polygon and a vertex that is perpendicular to the center of the base.
regular solids (polyhedra): Solids with faces that are regular polygons (equal sides).
relative humidity: The percentage of moisture contained in air compared to the maximum amount of moisture (saturation) it is capable of holding.
relay: An interface that controls one electrical circuit by opening and closing contacts in another circuit.
reservoir: A container for storing fluid in a hydraulic system.
resilience: The capability of a material to regain its original shape after being bent, stretched, or compressed.
resistance: The opposition to electron flow.
restrictive check valve: A check valve with a specific sized hole drilled through its center.
resonance: The magnification of vibrations and their noise by 20% or more.
return-line filter: A filter positioned in a circuit just before the reservoir.
reverse dial method: An alignment method that uses two dial indicators to take readings off of opposing sides of coupling rims, giving two sets of shaft runout readings.
rhomboid: A quadrilateral with opposite sides equal and no 90° angles.
rhombus: A quadrilateral with all sides equal and no 90° angles.
rigging: Securing equipment or machinery in preparation for lifting by means of rope, chain, or webbing.
right angle: Two lines that intersect perpendicular to each other.
right circular cone: A cone with the axis at a 90° angle to the circular base.
right cylinder: A cylinder with the axis perpendicular to the base.
right lang-lay rope: A rope in which the wires are laid to the right and the strands are laid to the right.
right parallelepiped: A prism with all edges perpendicular to the bases.
right prism: A prism with lateral faces perpendicular to the bases.
right regular-lay rope: A rope in which the strands are laid to the right and the wires are laid to the left.
right triangle: A triangle that contains one 90° angle and no equal sides.
rim-and-face alignment method: An alignment method in which the parallel and angular offset of two shafts is determined using two dial indicators that measure the rim and face of a coupling.
roller bearing: An anti-friction bearing that has parallel or tapered steel rollers confined between inner and outer rings.
rolling: The deforming of metal on the active portion of gear teeth caused by high contact stresses.
rolling-contact (anti-friction) bearing: A bearing composed of rolling elements between an outer and inner ring.
root mean square (rms): The square root of the sum of a set of squared instantaneous values.
rope grab: A device that clamps securely to a rope.
rope lay: The length of rope in which a strand makes a complete helical wrap around the core.
rounding off: The process of increasing or decreasing a number to the nearest acceptable number.
round sling: A sling consisting of one or more continuous polyester fiber yarns wound together to make a core.
running torque: The energy that a motor develops to keep a load turning.
runout: A radial variation from a true circle.
runway: The rail and beam on which a crane operates.
rust: A form of oxidation in which metal oxides are chemically combined with water to form a reddish-brown scale on metal.

S

safety net: A net made of rope or webbing for catching and protecting a falling worker.
safety relief valve: A device that prevents excessive pressure from building up by venting air to the atmosphere.
safety sleeve: A moving element with a locking mechanism that is connected between a carrier and the worker's body belt.
saturated air: Air that holds as much moisture as it is capable of holding.
scaffold: A temporary or movable platform and structure for workers to stand on when working at a height above the floor.
scaffold hitch: A hitch used to hold or support planks or beams.
scalene triangle: A triangle that has no equal angles or equal sides.
scuffing: The severe adhesion that causes the transfer of metal from one tooth surface to another due to welding and tearing.
seal: A device that creates positive contact between cylinder components to contain pressure and prevent leakage.
Seale wire: Wire rope that uses different size wire in different layers.
secant: A straight line touching the circumference at two points.
secondary port: An external passage that allows fluid flow to other components.
secondary winding: The output or load winding of a transformer that is connected to the load.
sectional metal-framed scaffold: A metal scaffold consisting of preformed tubes and components.
sector: A pie-shaped piece of a circle.
segment: The portion of a circle set off by a chord.
seizing: The wrapping placed around all strands of a rope near the area where the rope is cut.
seizing bar: A round bar ½″ to ⅝″ in diameter and about 18″ long used to seize rope.
selvedge: A knitted or woven edge of a webbing formed to prevent raveling.
semicircle: One-half of a circle containing 180°.
semisolid lubricant: A lubricant that combines low-viscosity oils with thickeners, such as soap or other finely dispersed solids.
sequence: The order of a series of operations or movements.
sequence valve: A pressure-operated valve that diverts flow to a secondary actuator while holding pressure on the primary actuator at a predetermined minimum value after the primary actuator completes its travel.
serpentine belt (double-V or hex belt): A belt designed to transmit power from the top and bottom of the belt.
service life: The length of service received from a bearing.
shackle: A U-shaped metal link with the ends drilled to receive a pin or bolt.
shear strength: **1.** A metal's resistance to a force applied parallel to its contacted plane. **2.** A liquid's ability to remain as a separator between solids in motion.
shear stress: Stress in which the material on one side of a surface pushes on the material on the other side of the surface with a force parallel to the surface.
shell: An orbiting layer of electrons in an atom.
shim stock: Steel material manufactured in various thicknesses, ranging from .0005″ to .125″.
signal: In pneumatic logic, a condition that initiates a start or stop of fluid flow by opening or closing a valve.
single-acting cylinder: A cylinder in which fluid pressure moves the piston in only one direction.
single ladder: A ladder of fixed length having only one section.
single-pole scaffold: A wood scaffold with one side resting on the floor or ground and the other side structurally anchored to the building.
single-stage compressor: A compressor that uses one piston to compress air in a single stroke before it is discharged.
slant height: The distance from the base to the vertex parallel to a side.
sleeve bearing: A bearing in which the shaft turns and is lubricated by a sleeve.
sling: A line consisting of a strap, chain, or rope used to lift, lower, or carry a load.
sling apex: The uppermost point where sling legs meet.
slip clutch: A spring-loaded, friction-held fiber disc that is adjusted to slip at 125% to 150% of the hoist-rated load.
slip knot: A knot that slips along the rope from which it is made.
slippage: The internal leaking of hydraulic fluid from a pump outlet to a pump inlet.
small circle: The circle formed by passing a cutting plane through a sphere but not through the center.
socket: A rope attachment through which a rope end is terminated.
soft foot: A condition that occurs when one or more machine feet do not make complete contact with its base.
solenoid: A device that converts electrical energy into a linear, mechanical force.
solid lubricant: A material such as graphite, molybdenum disulfide, or polytetrafluoroethylene (PTFE) that shears easily between sliding surfaces.
sorting hook: A hook with a tapered throat and a point designed to fit into holes.
spacer: Steel material used for filling spaces ¼″ or greater.
spalling: The flaking away of metal pieces due to metal fatigue.
spectrometer: A device that vaporizes elements in the oil sample into light.
spectrum: A representative combination of the amplitude (total movement) and frequency (time span) of a waveform.

speltered socket: A socket assembled by separating the wire rope ends after inserting the rope through the socket collar.
sphere: A solid generated by a circle revolving about one of its axes.
splice: **1.** The joining of two rope ends to form a permanent connection. **2.** The lapped and secured load-bearing part of a loop eye.
springing soft foot: A condition that occurs when a dial indicator at the shaft shows soft foot, but feeler gauges show no gaps.
spring-loaded accumulator: An accumulator that applies force to a fluid by means of a spring.
spur gear: A gear that has straight teeth that are parallel to the shaft axes.
square: A quadrilateral with all sides equal and four 90° angles.
square foot: Contains 144 sq in. (12″ × 12″ = 144 sq in.).
square inch: Measures 1″ × 1″ or its equivalent.
staging: The process of dividing the total pressure among two cylinders by feeding the outlet from the first large (low-pressure) cylinder into the inlet of a second small (high-pressure) cylinder.
standard: A guideline adopted by regulating authorities.
standing end: The end of the rope that is normally fixed to a permanent apparatus or drum, or is rolled into a coil.
standing part: The portion of the rope that is not active in the knot-making process.
standoff: A ladder accessory that holds a single or an extension ladder a fixed distance from a wall.
starting torque: The energy required to start a load turning after it has been broken away from a standstill.
static electricity: The accumulation of charge.
static energy (potential energy): The ability of a fluid to do work using the height and weight of the fluid above some reference point.
static head: The height of a fluid above a given point in a column at rest.
static head pressure: A force over an area created by the weight of the fluid itself.
static lift: The height to which atmospheric pressure causes a column of fluid to rise above the supply to restore equilibrium.
static load: A load that remains steady.
static seal: A seal used as a gasket to seal nonmoving parts.
static total column: Static head plus static lift.
steel alloy: Metallic material formulated from the fusing or combining of two or more metals.
stepladder: A folding ladder that stands independently of support.
straight angle: Two lines that intersect to form a straight line.
straightedge alignment method: A method of coupling alignment in which an item with an edge that is straight and smooth, such as a steel rule, feeler gauge, or taper gauge, is used to align couplings.
straight line: The shortest distance between two points.
strainer: A fine metal screen that blocks contaminant particles.
strand: Several pieces of yarn helically laid about an axis.
submersion system: A lubrication system in which the bearings are submerged below oil for lubrication.
suction strainer: A coarse filter attached to a pump inlet.
supplementary angles: Two angles formed by three lines in which the sum of the two angles equals 180°.
suspension scaffold: A scaffold supported by overhead wire ropes.
swaged socket: A compressed socket assembled to the wire rope under high pressure.
switch: A device that starts or stops the flow of electrical energy.
symbol: A graphic element which indicates a particular device, etc.
symmetrical load: A load in which one-half of the load is a mirror image of the other half.
synthetic fluid: A lubricant, often based on petroleum, which has improved heat, chemical resistance, and other characteristics than straight petroleum products.
synthetic yarn: Yarn made of twisted, manufactured fibers such as nylon or polyester.
system operating pressure: The pressure of a fluid after the pump until the flow is reduced, metered, or returned to the reservoir.

T

tackle: The combination of ropes and block assemblies arranged to gain mechanical advantage for lifting.
tag line: A rope, handled by an individual, to control rotational movement of a load.
tagout: The process of placing a tag on a power source that warns others not to restore energy.
tangent: A straight line touching the curve of the circumference at only one point.
tapered bore bearing: A bearing whose bore varies in diameter from the face to the back of the bearing.
tapered bore pulley: A two-piece pulley that consists of a tapered pulley bolted to a tapered hub (bushing).
tapered roller bearing: A roller bearing having tapered rollers.
taper gauge: A flat, tapered strip of metal with graduations in thousandths of an inch or millimeters marked along its length.
tempering: The process in which metal is brought to a temperature below its critical temperature and allowed to cool slowly.
tensile strength: A measure of the greatest amount of straight-pull stress metal can bear without tearing apart.

tension member: The load-carrying element of a belt which prevents stretching.

tetrahedron: A regular solid of four triangles.

thermal expansion: The dimensional change of a substance due to a change in temperature.

thimble: A curved piece of metal around which the rope is fitted to form a loop.

3ϕ circuit: A circuit that has three incoming wires, with each wire having the same voltage.

three-way directional control valve: A valve that has three main ports that allow or stop fluid flow or exhaust.

thrust damage: Bearing damage due to axial force.

throttling: Permitting the passing of a regulated flow.

timber hitch: A binding knot and hitch combination used to wrap and drag lengthy material.

timing (synchronous) belt: A belt designed for positive transmission and synchronization between the drive shaft and the driven shaft.

time domain: The amplitude as a function of time.

toeboard: A barrier to guard against the falling of tools or other objects.

tooth form: The shape or geometric form of a tooth in a gear when seen as its side profile.

top hook: The hook assembled to the top of a hoisting mechanism to allow for overhead suspension.

top support: The area of a ladder that makes contact with a structure.

torque: The twisting (rotational) force of a shaft.

total column: The fluid head plus lift.

total energy: A measure of a fluid's ability to do work.

transducer: A device that converts a physical quantity into another quantity, such as an electrical signal or a graphic display.

transformer: An electric device that uses electromagnetism to change AC voltage from one level to another.

trapezium: A quadrilateral with no sides parallel.

trapezoid: A quadrilateral with two sides parallel.

trapezoidal belt: A timing belt containing trapezoidal-shaped teeth.

trending: A graphic display used for interpretation of machine characteristics.

triangle: A three-sided polygon with three interior angles.

troubleshooting: The systematic elimination of the various parts of a system, circuit, or process to locate a malfunctioning part.

truth table: A table that lists the output condition of a logic element or combination of logic elements for every possible input condition.

tube: A thin-walled, seamless or seamed, hollow cylinder.

tuck set: Wedging a strand of rope into and between two other rope strands.

two-point suspension scaffold: A suspension scaffold supported by two overhead wire ropes.

two-way directional control valve: A valve that has two main ports that allow or stop the flow of fluid.

tying off: Securely connecting a harness directly or indirectly to an overhead anchor point.

U

union: A fitting used to connect or disconnect two tubes that cannot be turned.

unlay: The untwisting of the strands in a rope.

unloading valve: A device that senses a high-pressure condition and removes the compression energy.

unthread rotation: Counterclockwise rotation of an eyebolt having right-handed threads, or the clockwise rotation of an eyebolt having left-handed threads.

U-ring seal: A lip seal shaped like the letter U.

V

vacuum: A pressure lower than atmospheric pressure.

valence electron: An electron located in the outermost shell of an atom.

valence shell: The outermost shell of an atom.

valve: A device that controls the pressure, direction, or rate of fluid flow.

valve actuator: A device that changes the position of a valve spool.

vane air motor: An air motor that contains a rotor with vanes that are rotated by compressed air.

vane compressor: A positive-displacement compressor that has multiple vanes located in an offset rotor.

vane pump: A pump that contains vanes in an offset rotor.

variable displacement pump: A pump in which the displacement per cycle can be varied.

variable-speed belt drive: A mechanism that transmits motion from one shaft to another and allows the speed of the shafts to be varied.

V-belt: An endless power transmission belt with a trapezoidal cross section.

V-belt pulley: A pulley with a V-shaped groove.

vector: A quantity that has a magnitude and direction.

velocity: The distance a fluid travels in a specified time.

velocity transducer: An electromechanical device that is constructed of a coil of wire supported by light springs.

vertex: 1. The point of intersection of the sides of an angle. **2.** The common point of the triangular sides that forms a pyramid.

vertical line: A line that is perpendicular to the horizon.

vibration: A continuous periodic change in displacement with respect to a fixed reference.
vibration acceleration: The increasing of vibration movement speed.
vibration amplitude: The extent of vibration movement measured from a starting point to an extreme point.
vibration analyzer: A meter that pinpoints a specific machine problem by identifying its unique vibration or noise characteristics.
vibration cycle: The complete movement from beginning to end of a vibration.
vibration signature: A set of vibration readings resulting from tolerances and play within a new machine.
vibration velocity: The rate of change of displacement of a vibrating object.
viscosity: The measurement of the resistance of a fluid's molecules to move past each other.
viscosity index: A scale used to show the magnitude of viscosity changes in lubrication oils with changes in temperature.
visual adjustment method: A belt tension method in which the tension is adjusted by observing the slight sag at the slack side of the belt.
voltage: The amount of electrical pressure in a circuit.
voltage tester: A device that indicates approximate voltage level and type (AC or DC) by the movement of a pointer on a scale.
volume: The three-dimensional size of an object measured in cubic units.
volumetric efficiency: The percentage of actual pump output compared to the pump output if there were no slippage.
V-ring seal: A lip seal shaped like the letter V.

W

wagoneer's hitch knot: A knot that creates a load-securing loop from the standing part of the rope.
Warrington wire: Wire rope constructed of strands consisting of more than one size wire staggered in layers.
waveform: A graphic presentation of an amplitude as a function of time.
way: A flow path through a valve.
wear pad: A leather or webbed pad used to protect the web sling from damage.
wear particle analysis: The study of wear particles present in the lubricating oil.
webbing: A fabric of high-tenacity synthetic yarns woven into flat narrow straps.
web sling body: The part of the sling which is between the loop eyes or end fittings (if any).
web sling length: The distance between the extreme points of a web sling, including any fittings.
wedge socket: A socket with the rope looped within the socket body and secured by a wedging action.
weight-loaded accumulator: An accumulator that applies force to a fluid by means of heavy weights.
well (shaft): A walled enclosure around a fixed ladder.
whipping: Tightly binding the end of a rope with twine before it is cut.
wick system: A lubrication system that uses capillary action to convey oil to a bearing surface.
wiper: A seal designed to prevent foreign abrasive or corrosive material from entering a cylinder.
wood ladder: A ladder constructed of wood.
work: The energy used when a force is exerted over a distance.
working depth: The depth of engagement of two gears.
working end: The end of the working part of a rope.
working height: The distance from the ground to the top support.
working load limit (WLL): The maximum pull that should be applied to a vertical load.
working part: The portion of the rope where the knot is formed.
worm: A shank having at least one complete tooth around the pitch surface.
worm gear: A set of gears consisting of a worm (drive gear) and a wheel (driven gear) that are used extensively as a speed reducer.

Y

yarn: A continuous strand of two or more fibers twisted together.

Index

A

abrasion, 268
abrasive wear, *268* – 269
absolute pressure, 112, 170
absolute rating, 140
absolute temperature, 172
absolute zero, 172
acceleration, 124
accelerometer transducer, 278 – 279
accumulator, *164* – 165
  hydro-pneumatic, *164*, 165
    bladder, *164*, 165
    diaphragm, *164*, 165
    piston, *164*, 165
  spring-loaded, *164*
  weight-loaded, *164*, 165
acid, 27
action, 206
actuator, 161 – 164, 194, 201 – 205
acute angle, *3*
acute triangle, *6*, 7
additive, 138
adjacent angles, 4
adsorption, 182
aftercooler, 179
air compressor, 185 – 188
air contaminants, 178 – 182
air cylinder, 201 – *202*
air flow requirements, *69*
air motor, 204 – *205*
  vane, *205*
alignment, *272*, 289 – 318
  methods, 303 – 318
    electronic reverse dial, *317*
    laser rim-and-face, 317 – *318*
    reverse dial, 313 – 316
    rim-and-face, 309 – 312
    straightedge, 307 – 309
  sequence, 290
  tolerance, 303 – 305
alkali, 27
alternating current, 324
altitude, 6, 9, 11
ambient temperature, 69
American Gear Manufacturers Association, 263, 268
American National Standards Institute, 72, 135
American Petroleum Institute, 217
American Society for Testing and Materials, 217
American Society of Mechanical Engineers, 72, 135
amp, 320
amplitudes, *284*
anchoring, 292 – 293
  machinery, 291 – 300
AND logic element, *207*
angle, *3*
  acute, *3*
  obtuse, *3*
  right, *3*
  straight, *3*
angles
  adjacent, 4
  complimentary, *3*
  supplementary, *3* – 4
angle symbol, 3
angular-contact bearing, 227 – *228*
angular lifting, *77*
angular misalignment, 248, *249*, 289, *290*
angular sling load capacity, 25
angular soft foot, *297*
Anti-Friction Bearings Manufacturers Association, 225
arc, 4, *6*
Archimedes water-screw, *110*
area, 4 – *5*, *113*
armature, 324
asymmetrical load, *17*, 18
atmospheric pressure, 111 – 112, 169
atom, 168
authority having jurisdiction, 327 – 328
authorized individual, 345
auto-ignition, 138
axial float, 310
axial load, 225, *226*
axis, 36

B

babbitt metals, 230
backlash, 264
  measuring, 264 – *265*
ball bearing, 226 – 228
base, 6, 9
  diameter, *263*
  -mounted jib crane, 81 – *82*
  plate, 290 – 291
basket sling, 22 – *24*
bearing, 225 – 242
  angular-contact, 227 – *228*
  ball, 226 – 228
  Conrad, 227
  failure investigation, 232 – 235
  installation, 236 – 242
  mounting, 237 – 242
    using temperature, 239 – 240
  needle, 226, *227*, 229
  plain, 229 – *230*
  precision class, 239
  radial, 227 – *228*
  removal, 231 – 235
  roller, 226, *227*, *229*
    cylindrical, *229*
    tapered, *229*
  rolling-contact, 225 – 229
bed section, *92*
bellows, *167*
belt and sheave groove gauge, 245
belt deflection method, 250
belt pitch length, 252
belt pitch line, 245
bending strength, 53
bevel gear, 66 – *67*, *266* – 267
bight, *36*
bimetallic device, 337 – 338
binary system, 205
bird caging, 58, *60*
blackwall hitch, *46*
bladder gas charged accumulator, *164*, 165
block and tackle, 63 – 66
blockout, 256
body belt, *107*
bolt bound, 292, *293*
boundary lubrication, 212
Bourdon tube, *154*
bowline knot, *42*
Boyle's Law, 171 – 172
breakaway torque, 131
bridge girder, *83*
bridle sling, 22 – *24*
British measurement system, 2
butt spur, 87

C

cab, *83*
cabling, 32
cage, 91
calculating load weight, 19 – 21
calculating pulley speed and size, 254 – 255
calculations, 1 – 16

cam ring, *149*
cantilever, 81 – *82*
capacity, 121 – 123
  cylinder, 121 – 123
    when retracting, 121 – 123
capillary action, 220
carrier, 91, *105*
cat's-paw hitch, *45*
caustic solution, 27
cavitation, *150*
center of gravity, 18
centerpoint, 4 – *6*
centralized system, *221*
centrifugal force, 110
chain, *22*, 52 – 57
  construction, 53 – 54
  grade 43, 53, *54*
  grade 70, 54
  grade 80, 54
  hoisting apparatus, 54
  inspection, 61, 62
  working load limit, 53 – 54
challenging, 290, 343
Charles' Law, 172 – 173
check valve, *155*, 193, *195*
chemisorption, 213
choker hook, 57, *58*
choker sling, 22 – *24*
chord, 4, *6*
circle, 4 – *6*, *12*, 13
  arc, 4, *6*
  area of, 5 – 6
  centerpoint, 4 – *6*
  chord, 4, *6*
  circumference, 4 – 5
  diameter, 4 – *6*
  quadrant, 4, 6
  radius, 4 – *6*
  secant, 4, *6*
  sector, 4, *6*
  segment, 4, *6*
  semicircle, 4, *6*
  tangent, 4, *6*
circles
  concentric, 4, *6*
  eccentric, 4, *6*
circuit breaker, 337 – 339
  troubleshooting, 338 – 339
circular pitch, 252, *263*
circumference, 4 – 5
clearance, *263*
cleat, 100
climbing techniques, 90
clip, *23*
closed circuit, 329
clove hitch, *44*
coalescing filter, 180 – *181*
coefficient of friction, 212
combination rim-and-face alignment, 310
combined gas law, 174 – 175
common formulas, 2
commutator, 324, *326*
competent person, 98
complimentary angles, *3*
compound gear train, 261 – 262
compression, 175 – 177
  intercooling, *177*
  multistage, 176, *177*
compressor
  air, 185 – 188
  helical screw, *187*
  multistage, 185
  piston, 185 – 187
  positive displacement, 185
  reciprocating, 176
  single-stage, 185
  vane, *188*
concentric circles, 4, *6*
condensation, 179
conditioning compressed air, 194 – 196
conditions affecting rope, 27 – 29
  bending, *28*, 29
  chemical activity, 27, *28*
  moisture, 27, *28*
  temperature, 27, *28*
conductor, 320
cone, *12*
  altitude, *12*
  slant height, *12*
  vertex, *12*
  volume of, *14*, 15
conic sections, *12* – 13
  circle, *12*, 13
  ellipse, *12*, 13
  hyperbola, *12*, 13
  parabola, *12*, 13
connecting link, 56
connecting rod, *186*
Conrad bearing, 227
contact, 336
  holding, 336
contactor, 335 – 337
  troubleshooting, 336 – 337
continuity tester, 329, *331*
  use, 329, *331*
control circuit, 335
controlled flow, 134
core protrusion, 58, *60*
corrosion, 27, 269
corrosive wear, *269*
coulomb, 320
coupling, *290*
  face, 309
  flexible, 290
  rim, 309
  unbalance, 275 – *276*
cow hitch, *45*
crane, *80* – 86
  gantry, *81* – *82*
  jib, 81 – 83
  operation, 84 – 86
    hand signals, 85 – *86*
  overhead, *83* – 84
  safety, 85 – *86*
Crane Manufacturers Association of America, Inc., 72
crankshaft, *186*
crossover, *71*
crowning, 38, *39*
cubic foot, 14
cubic inch, 13
cup seal, 162, *163*, *203*
curve/rope ratio, 28 – 29
curvilinear belt, 251, *252*
cutaway diagram, *134*
cutting wire rope, 32 – 33
cylinder, *11*
  air, 201 – *202*
  double-acting, *161*, 201 – *202*
  hydraulic, 161 – 163
  oblique, *11*
  right, *11*
  single-acting, *161*, 201 – *202*
  volume of, *14*, 15

D

decision, 206
demulsification, 138
desiccant dryer, *182*
dew point, 179
diagrams
  cutaway, *134*
  graphic, 134, *135*
  pictorial, *133* – 134
dial indicator, *295*, 300 – 303
  rod sag, 302 – *303*
  use, 300 – 302
  verifying readings, 300 – *302*
diameter, 4 – *6*
  rope, 26 – *27*
diametral pitch, 263
diaphragm gas charged accumulator, *164*, 165
direct-acting valve, 151, *152*
direct current, 324 – 325, *326*
directional control valve, 154 – *156*, 197 – 200
  four-way, 155, *156*, *197*
  three-way, 155, *156*, *197*
  two-way, 155, *156*, *197*
direct proportion, 260
dispersed solid, 218
displacement, 120, 273 – *274*
  transducer, *278* – *280*
distribution line, 326, *327*
distribution substation, 326, *327*
dodecahedron, 9, *10*
double-acting cylinder, *161*
double hitch knot, 40, *41*
double trapezoidal belt, 251, *252*
double V-belts, 250, *251*
dowel effect, 292, *293*

drift, 85
drip system, *221*
drive gear, 259
driven gear, 259
drum wrap, *71*
  crossover, *71*
dry air, 182
dryer, 181 – 182
  desiccant, *182*
  refrigerant, *182*
duty rating, *89*
dynamic electricity, 319
dynamic head, 116
  pressure, 116
dynamic lift, *117*
dynamic range, 286
dynamic seal, 162, 202
dynamic signal analyzer, 282
dynamic total column, *117*

E

eccentric, 295
  circles, 4, *6*
  surface, 186
eddy current, 280, 333
efficiency, 128 – *129*
electrical devices, 333 – 343
electrical measuring devices, 328 – 333
  continuity tester, 329, *331*
  multimeter, 331 – 333
  voltage tester, 330 – *331*
electrical pitting, 234 – *235*, *269*
electrical safety, 343 – 346
electric hoist, *70*
  checklist, *73*
electricity, 319 – 346
  dynamic, 319
  static, 319
electromagnet, 323
electromagnetic induction, 324 – 327
electron, 319, *320*
  flow, 320 – *321*
  valence, 319
electronic reverse dial alignment, *317*
element, 205
ellipse, *12*, 13
emulsification, 138
end play, 240
end truck, 81 – *82*
energy, 127 – 132
  heat, *128*
  -isolating device, 165 – 166
  kinetic, *128*, 165
  pressure, *128*
  static, *128*
  total, *128*
equal rotor unbalance, 275 – *276*
equation, *1*
equilateral triangle, *6*, 7
equilibrium, 109
exhaust flow, 134
extension ladder, *92* – 96
  angle positioning, *95*
  bed section, *92*
  fly section, *92*
  overlap and height, 94 – *95*
  raising, *93* – *94*
  pawl lock, *92*
  section overlap, *95*
eyebolt, *23*, 76 – *80*
  angular lifting, *77*
  capacities, *78* – 79
  formed steel, *77*
  loads, 79 – 80
  machinery, *77* – 80
  regular nut, *77* – 80
  shoulder nut, *77* – *80*
    working load limits, *80*
eye loop, 39, *40*

F

face width, 264
fall-arrest sequence, 106
false Brinell damage, *233*
Fast Fourier Transform analyzer, 282
fatigue crack, 270
fatigue fracture, 270
fatigue life, 225
fatigue wear, *270*
feeler gauge, 297
ferromagnetic material, 322
ferrous metals, 138
ferrule, *147*
fiberglass ladder, 88 – 89
fiber rope, *22*, 34 – 46
  applications, 36
    hitch, 36
    knotting, 36
  construction, 35 – 36
  inspection, 59, *60*
  natural, 35
  strength, 36
  synthetic, 35
filter, 138, *195*, 285
  coalescing, 180 – *181*
  intake, 178
  pressure, 140
  return-line, 140
  suction, 140
final step-down transformer, *327*
fire point, 138
fixed bore pulley, 246, *248*
fixed ladders, *90* – *91*
  installation, *91*
flange seal, 162, *163*
flared fitting, *144*, 145
flared joint tightening, *145*
flareless fitting, *147*
flash point, 138
flexible belt drive, 243 – 256
  safety, 255
    clean environment, 256
    proper clothing, 255
    removing and locking out energy supplies, 256
flow, 118 – 125
  rate, *124*
flow control valve, 158 – *160*, 200, *201*
fluid, 111
  maintenance, 166
fluting, 234 – *235*
fly section, *92*
foaming, 137
foot pad, 87
force, 65 – 66, 111
formed steel eyebolt, *77*
formulas, 1 – *2*
foundation, 290 – 291
foundry hook, 57, *58*
four-part reeving, *66*
four-way directional control valve, 155 – *156*
fracture, 53
free air, 178
frequency, 274 – 275
  domain, 285 – *286*
  spectrum, 285 – *286*
fretting corrosion, 233 – *234*
friction, *118* – *119*
  disc, 257
frustum, *13*
fulcrum, 125
fuse, 337 – 339
  puller, 338
  troubleshooting, 338 – 339

G

galling, 234
gantry crane, *81* – *82*
gas, 111
  characteristics, 168 – 175
  laws, 171 – 175
    Boyle's Law, 171 – 172
    Charles' Law, 172 – 173
    combined gas law, 174 – 175
    Gay-Lussac's Law, 173 – 174
  lubricant, 213
gas charged accumulator, *164*, 165
gasket, 163
gate valve, 159, *160*
gauge pressure, 112, 169
Gay-Lussac's Law, 173 – 174
gear, 66 – 67, 257
  bevel, 66 – *67*, *266* – 267

drive, 259
driven, 259
helical, *148*, 265 – *266*
herringbone, *148*, *266*
hypoid, *266*, 268
idler, *260* – 261
miter, *266* – 267
pump, 120, *148*
rack, 265 – *266*
speed, *259* – 260
spur, *148*, 265 – *266*
tooth form, 262
train, 259
compound, 261 – 262
wear, 268 – 270
abrasive, *268* – 269
corrosive, *269*
electrical pitting, *269*
fatigue, *270*
rolling, *270*
scuffing, *270*
worm, 66 – *67*, *266* – 268
generator, 324
globe valve, 159, *160*
grab hook, 57, *58*
graphic diagram, 134, *135*, 183 – *184*
graphic symbols, 135, *136*, 183 – *184*
graphite, 219 – 220
grease, 218
application, 222
cup, *222*
dropping point, 219, 233
gun, *222*
great circle, *13*
ground, 344
ground fault circuit interrupter, 339 – 340
grounding, 328
methods, *329*
guardrail, *98*
guyline, *101*

H

half hitch knot, 40, *41*
halyard, *92*, *94*
hand chain, *67*
hand-chain drop, 67
hand-chain hoist, 67 – *68*
hand signals, 85 – *86*
harness, *107*
hazardous location, 328, 329
head, *114* – 116
dynamic, *115* – 116
pressure, *114* – 116
static, *115* – 116
headroom, 67
heat energy, *128*
heat exchanger, *142*
helical gear, *148*, 265 – *266*
helical screw compressor, *187*
helix, 26
heptagon, *8*
herringbone gear, *148*, *266*
Hertz, 274
hexagon, *8*
hexahedron, 9, *10*
hitch, 40
blackwall, *46*
cat's-paw, *45*
clove, *44*
cow, *45*
scaffold, *46*
timber, 43, *44*
hoist, 66 – 76
inspection, 72 – 76
brake lining, *76*
motor brake, 75 – 76
regularly-scheduled, 72 – 74
load tests, 76
manually-operated, 66 – *68*
hand-chain, 67 – *68*
lever-operated, 67 – *68*
power-operated, 68 – 71
electric, *70*
checklist, *73*
pneumatic, *69* – 70
safety, 72 – 76
trolley, *83*
hoist chain, *67*
hoisting apparatus chain, 54
hoisting hook, 57
capacity, 59
holding contact, 336
hook, *23*, *57*, *58*
bend, 57
choker, 57, *58*
drift, 75
foundry, 57, *58*
grab, 57, *58*
hoisting, 57
mousing, *57*
sorting, 57, *58*
horizontal lines, *3*
horizontal weight center, 18
horsepower, *130*, 258 – 259
hose, 142, *143*
installation, *143*
humidity, 177
relative, 178
hunting tooth, 268
hydraulic
actuators, 161 – 164
circuit components, 137 – 165
circuit maintenance, 165 – 170
circuitry, 133 – 136
cylinder, 161 – 164
diagram, 133
fluid, 137 – 141
maintenance, 166
motors, *163*, 164
scissor lift, *102*
system history, 110 – 111
hydraulics, 109 – 132
hydrocarbon, 214
hydrodynamics, *110*
hydrostatics, *109*
hyperbola, *12*, 13
hypoid gear, *266*, 268
hypotenuse, 7

I

icosahedron, 9, 10
idler gear, *260* – 261
imbalance, 18
impact flaring method, 146
implosion, *150*
inching, 76
inclined line, *3*
independent wire rope core, 30
individual rim-and-face alignment, 310
induced soft foot, *297*, 298
induction, 323
industrial crane, *80* – 86
gantry, *81* – *82*
jib, 81 – 83
operation, 84 – 86
hand signals, 85 – *86*
overhead, *83* – 84
safety, 85 – *86*
inert gases, 213
instrumentation, 181
insulator, 320
intake filter, 178
intake flow, 134
intensifier, *204*
intercooler, *177*
intercooling, *177*
interference fit, 229
International Organization for Standardization, 72
inverse proportion, 260
inverse ratio, 260
involute form, 262
irregular polygon, *8*
irregular polyhedra, 9, *10*
isosceles triangle, *6*, 7

J

jack screw, *294*
jib crane, 81 – 83
base-mounted, 81 – *82*
mast, 81 – *82*
journal, 229, *230*

K

kinetic energy, *128*, 165
kinking, 58, 60
knot, 40 – 43
  bowline, *42*
  double hitch, 40, *41*
  half hitch, 40, *41*
  slip, *41*
  wagoneer's hitch, *43*

L

ladder, 87 – 96
  climbing techniques, 90
  duty rating, *89*
  extension, *92* – 96
    overlap and height, 94 – *95*
    raising, *93* – 94
  fall protection, 105
  fiberglass, 88 – 89
  fixed, *90* – *91*
    installation, *91*
  jacks, *95*
  metal, 88
  position protection, 105 – *106*
  regulations and standards, 89
  safety, 105 – 107
  safety feet, *89*
  safety system, 91
  single, 91
  standoffs, *96*
  step, *96*
  wood, 87
lanyard, 106 – *107*
laser rim-and-face alignment, 317 – *318*
lay, 36
lead line, 63 – *64*
  factors, *66*
left lang-lay rope, *26*
left regular-lay rope, *26*
lever-operated hoist, 67 – *68*
lichens, 138
lift, 67, *116* – 117
  dynamic, *117*
  static, 116 – *117*
  total column, *117*
    dynamic, *117*
    static, *117*
lifting, 17, 63 – 86
  devices, 63 – 71
    block and tackle, 63 – 66
  lug, 18
limit switch, *70*
limit valve, 206
linear amplitude spectra, 282 – *283*
linear measure, 4
lines, 2 – *3*
  horizontal, *3*
  inclined, *3*
  parallel, *3*
  perpendicular, *3*
  straight, *3*
  vertical, *3*
lip seal, *162*, 163, 202 – *203*
  cup, *162*, 163
  flange, *162*, 163
  U-ring, *162*, 163
  V-ring, *162*, 163
liquid characteristics, 111 – 125
  flow, 118 – 125
    capacity, 121 – 123
    friction, 118 – *119*
    speed, 125
    velocity, *123* – 125
    viscosity, 119 – 120
    volume, *120* – 121
  pressure, 111 – 117
    head, *114* – 116
      dynamic, *115* – 116
      static, *115* – 116
    lift, *116* – *117*
      dynamic, *117*
        total column, *117*
      static, 116 – *117*
        total column, *117*
liquid lubricant, 214
  animal/vegetable oil, 214
  petroleum fluid, 214
litmus paper, 138
load
  asymmetrical, *17*, 18
  balance, 17
  static, 29
  symmetrical, *17*
  tests, 76
  tipping, 18
  toppling, 18
  weight calculation, 19 – 21
loading slot, 227
lobe, 187
lockout, *256*
logarithmic amplitude spectra, 282 – *283*
logarithmic scale, 282
logic, 205
  element combination, 209 – 210
loop, *36*
loop eye, *47*
lubricant, 211
  additives, 215
  application, 220
  contamination, 222 – 223
  gas, 213
  liquid, 214
  semisolid, 218 – 219
  solid, 219 – 220
lubrication, 211 – 224
  programs, 223 – 224
lubricator, 195, *196*

M

machine, 243, 257
  element efficiency, 258
  movement, 294 – 296
  run-in, 242
machinery eyebolt, *77* – 80
machinery vibration, 272 – 273
magnet, 322
magnetic flux lines, 322, *323*
magnetism, 322 – 327
main header, 192, *193*
main sling components, *22*
  chain, *22*
  fiber rope, *22*
  round sling, *22*
  webbing, *22*
  wire rope, *22*
manifold, 206
manually-operated hoists, 66 – *68*
  hand-chain, 67 – *68*
  lever-operated, 67 – *68*
master link, *23*, *56*
mast jib crane, 81 – *82*
material weight calculations, 21
maximum intended load, 98
measurement systems, 2
measuring soft foot, 298 – 299
  at-each-foot method, 298 – 299
  shaft deflection method, 299
mechanical, 257
  advantage, 63 – *65*, *125* – 127
  drive, 257 – 270
mercury barometer, 111
mesh, 104, 139
metal fatigue, 232
metal ladder, 88
metering, 159
metric measurement system, 2
micron, 139
midrail, *98*
misalignment, *272*, 289 – *290*
  angular, 289, *290*
  offset, 289, *290*
misalignment wear, *234*
miter gear, *266* – 267
mobile scaffold, 101 – *102*
modified curvilinear belt, 251, *252*
moisture separator, 179
molded notch belt, 244
molecule, 319
molybdenum disulfide, 220
motor
  air, 204 – *205*
  hydraulic, *163*, 164
  regreasing, 222 – *223*
  starter, 335 – 337
    troubleshooting, 336 – 337
mousing, 57
multimeter, 331 – 333

use, 332
multiple-point suspension scaffold, 102 – *103*
multiple-wire strand core, 30
multistage compression, 176, *177*
multistage compressor, 185

N

National Association of Chain Manufacturers, 53
National Electrical Code®, 72, 327 – 328
National Fire Protection Association, 72
National Fluid Power Association, 206
National Lubricating Grease Institute, 218
natural fiber rope, 35
needle bearing, 226, *227*, 229
needle valve, 159, *160*
neutral conductor, 331
neutron, 319, *320*
nip, 40
nominal value, 29, 244
nonparallel misalignment, 248, *249*
nonpositive displacement pump, 110
nonpositive seal, 162
NOT logic element, *208*
nucleus, 319, *320*

O

oblique cylinder, *11*
oblique prism, 9, *10*
obtuse angle, *3*
obtuse triangle, *6*, 7
Occupational Safety and Health Administration, 72
octagon, *8*
octahedron, 9, *10*
offset misalignment, 248, *249*, 289, *290*
ohm, 320 – 321
Ohm's law, 321 – 322
oil analysis, 223
oil application, 220 – 221
oil carry-over, 179
oil whirl, 276 – *277*
one-part reeving, *64*
open circuit, 329
operational pitch point, *263*
opposing forces rotor unbalance, 275 – *276*
order, 274
orifice, 159, *160*
O-ring, *162*, 202
OR logic element, 207 – *208*
oscillator, 280
overhead crane, *83* – 84
overhung load, *248*
overload relay, 336
oxidation, 137, 138

P

packing, 162, *163*, 203
parabola, *12*, 13
parallelepiped, 9, *10*
right, 9, *10*
parallel lines, *3*
parallelogram, *8*, 9
rectangle, *8*, 9
rhomboid, *8*, 9
rhombus, *8*, 9
square, *8*, 9
parallel soft foot, *297*
part, 63
particulate, 178
Pascal's law, 126, 169
pawl, *68*
pawl lock, *92* – *93*
peak, *273*
peak-to-peak, *273*
pendant, *70*
pentagon, *8*
perpendicular line, *3*
petroleum fluid, 214
phase, *275* – *276*
pictorial diagram, *133* – 134
piezoelectric, 278 – *279*
pilot line, 152
pilot-operated valve, *152*
pilot operation, 152
pinion, *263*
pipe, 143, *144*
piping strain, 292
piston compressor, 185 – 187
piston cushioning device, *203*
piston pump, *150*
pitch circle, *263*
pitch diameter, 264
pitting, 137
plain bearing, 229 – *230*
materials, 230
plane figures, 2 – 9
circle, 4 – *6*
plank, *95*
planking spans, *100*
platform, 91
plumbing, 142 – 147
ply, 46
pneumatic
circuit, 183
components, 194 – 205
circuitry, 183 – 184
hoist, *69* – 70
logic, 205 – 210
element, 206 – 208
principles, 167 – 182
system, 167, 183, 185 – 194
leaks, 192
liquid drainage, 192, *193*
pipe installation, 193
plumbing, 189
materials, 193
pressure loss, 191
pneumatics, 167 – 168
pocket wheel, *67*
polarity, 323
pole scaffold, *98* – 101
components, *99*
double-pole, 98 – 101
planking spans, *100*
single-pole, 98 – 101
polygon, *8*
heptagon, 8
hexagon, *8*
irregular, *8*
octagon, *8*
pentagon, *8*
regular, *8*
triangle, *8*
polyhedra, 9 – 13
polymer, 220
polytetrafluoroethylene, 220
port plate, *149*
position, *156*, 197 – *198*
position protection, 105 – *106*
positive-displacement, 147
compressor, 185
pump, 120
positive seal, 162
potential energy, 165
power, 129
circuit, 335
distribution, 326 – *327*
power-operated hoist, 68 – 71
electric, *70*
checklist, *73*
pneumatic, *69* – 70
power plant transmission lines, 326, *327*
precharge pressure, 165
precision class bearings, 239
preformed rope, 33
preloading, 240
pressure, 111 – 117, 168
absolute, 112, 170
atmospheric, 111, 169
compensated flow control valve, *160*
compensated vane pump, *149*
compensator, 149, 189
control, 189
drop, 119, 180
energy, *128*
equivalents, *112*
filter, 140
gauge, 112, 153, *154*, 169
-reducing valve, 153, *154*
regulator, *196*
-relief valve, *151*
switch, 189, *190*
primary port, *151*
primary transmission line, 326, *327*

prime mover, 147
prism, 9, *10*, *11*
  oblique, 9, 10, *11*
  parallelepiped, 9, *10*, *11*
  right, 9, *10*, *11*
process variable, 181
proportion, 260
  direct, 260
  inverse, 260
proton, 319, *320*
primary winding, 334
pseudocavitation, 151
pulley
  alignment, 248 – 249
  fixed bore, 246, *248*
  tapered bore, 246 – *248*
    recommended hub torque values, 248
pump, 147 – 151
  gear, *148*
  piston, 150
  vane, *148* – 150
    balanced, *149*
    pressure compensated, *149*
  variable displacement, 158
putlogs, 99
pyramid, *11* – 12
  altitude, 12
  regular, 12
  slant height, 12
  vertex, *11*
Pythagorean Theorem, 7

Q

quadrant, 4, *6*
quadrilateral, *8* – 9
  rectangle, *8*, 9
  rhomboid, *8*, 9
  rhombus, *8*, 9
  square, *8*, 9
  trapezium, *8*, 9
  trapezoid, *8*, 9
quad-ring, *162*

R

race, 226, *227*
rack gear, 265 – *266*
racking, 96
rack teeth, 262
radial bearing, 227 – *228*
radial load, 225, *226*
radius, 4 – *6*
raising ladders, *93* – *94*
ratchet, *68*
ratio, 260
  inverse, 260
reach, 67
reciprocate, 185
reciprocating compressor, 176
rectangle, *8*, 9
  area of, 9
rectangular parallelepiped, 9, *10*, *11*
reel, 71
reeving, 63 – *66*
  four-part, 66
  one-part, *66*
  three-part, *66*
  two-part, *66*
refrigerant dryer, *182*
regular-lay rope, *26*
regular nut eyebolt, *77* – 80
regular polygon, *8*
regular pyramid, 12
regular solids, 9, *10*
  dodecahedron, 9, *10*
  hexahedron, 9, *10*
  icosahedron, 9, *10*
  octahedron, 9, *10*
  tetrahedron, 9, *10*
  volume of, *14*
relative humidity, 178
relay, 336
  overload, 336
reservoir, *141*
resilience, 162
resonance, 271
restrictive check valve, 159, *160*
return-line filter, 140
reverse dial alignment, 313 – 316
rhomboid, *8*, 9
rhombus, *8*, 9
rigging, 17 – 62
  chain attachments, 55
    hook, *57*, *58*
    master link, 56
    shackle, 55, *56*
  chain strength, 54 – 55
  component inspection, 58 – 61
  component recordkeeping, 62
  equipment storage, 61, 62
  hardware attachments, *23*
    clip, *23*
    eyebolt, *23*
    hook, *23*
    master link, *23*
    shackle, *23*
    socket, *23*
    triangle choker fitting, *23*
    wedge socket, *23*
  techniques, *51*
right angle, *3*
right circular cone, *12*
right cylinder, *11*
right lang-lay rope, *26*
right parallelepiped, 9, *10*, *11*
right prism, 9, *10*
right regular-lay rope, *26*
right triangle, 6, 7
rim-and-face alignment, 309 – 312
  combination, 310
  individual, 310
roller bearing, 226, *227*, *229*
  cylindrical, *229*
  tapered, *229*
rolling, *270*
rolling-contact bearing, 225 – 229
root mean square, 283 – 284
rope, 25 – 46
  conditions affecting, 27 – 29
    bending, *28*, 29
      curve/rope ratio, 28 – 29
      efficiency, 29
    chemical activity, 27, *28*
    moisture, 27, *28*
    temperature, 27, *28*
  construction, *26*
  diameter, 26 – *27*
  fiber, 34 – 46
  grab, 106 – *107*
  lay, *26*
  left lang-lay, *26*
  left regular-lay, *26*
  preformed, 33
  regular-lay, *26*
  right lang-lay, *26*
  right regular-lay, *26*
  strength, 27 – 29
  wire, 29 – 34
rounding off, 21
round sling, *22*, 50 – *52*
  inspection, 60, *61*
  strength, 50
running torque, 131
runout, 295
runway, *83*
rust, 138

S

safety
  electrical, 343 – 346
safety net, *104* – 105
  maintenance, 104
  requirements, 104
  testing, 104 – 105
safety relief valve, 189, *191*
safety sleeve, 91, *105*
saturated air, 177
Saybolt viscometer, *119*
scaffold, *97* – 103
  fall protection, 106
  hitch, *46*
  hydraulic scissor lift, *102*
  mobile, 101 – *102*
  pole, *98* – 101

components, *99*
planking spans, *100*
single-pole, 98 – 100
anchoring, *100*
regulations and standards, 98
safety, 106, 108
sectional metal-framed, *101* – 102
suspension, 102 – *103*
multiple-point, 102 – *103*
two-point, 102 – *103*
scalene triangle, *6*, 7
scuffing, *270*
seal, 202
cup, *203*
dynamic, 202
lip, 202 – *203*
static, 202
V-ring, *203*
secant, 4, *6*
secondary port, *151*
secondary winding, 334
sectional metal-framed scaffold, *101* – 102
sector, 4, *6*
segment, 4, *6*
seizing, 32 – *33*
bar, 33
selvedge, 46, *47*
semicircle, 4, *6*
semisolid lubricant, 218 – 219
sequence, 152
sequence valve, 152, *153*
service life, 226
shackle, *23*, 55, *56*
shear strength, 53, 216
shear stress, 216
shell, 319, *320*
valence, 319, *320*
shim stock, 295, *296*
shim thickness for 90° rotation, *77*
shoulder nut eyebolt, *77* – *80*
working load limits, *80*
signal, 205
single-acting cylinder, *161*
single ladder, 91
single-stage compressor, 185
slant height, 12
sling, 22 – 25, *48*
angle, *25*
loss factors, 25
angular load capacity, 25
basket, 22
bridle, 22
choker, 22
combinations, 22 – *24*
material strength capacities, 32
round, 50 – *52*
U, 22
vertical, 22, *48*
sling apex, 18, *19*
sling eyebolt capacity loss, *78*
slip clutch, 68
slip knot, *41*
slippage, 120
small circle, *13*
Society of Automotive Engineers, 217
socket, *23*, *34*
speltered, *34*
swaged, *34*
wedge, *34*
soft foot, 289, 296 – 298
angular, *297*
induced, *297*, 298
measuring, 298 – 299
parallel, *297*
springing, *297*, 298
solenoid, 199, 341 – 343
troubleshooting, 343
valves, 200
solid lubricant, 219 – 220
solids, 9 – 13
sorting hook, 57, *58*
spacer, 295
spalling, *232* – 233
spectrometer, 223
spectrum, *273*
speed, 125
speltered socket, *34*
sphere, *13*
great circle, *13*
small circle, 13
volume of, *14*, 16
splice, 37, *47*
springing soft foot, *297*, 298
spur gear, *148*, 265 – *266*
square, *8*, 9
area of, 9
square foot, 4
square inch, 4
staging, 187
standard, 72
V-belt pulley groove dimensions, *247*
standards organizations, 72
standing end, *36*
standing part, *36*
standoff, *96*
starting torque, 131
static electricity, 319
static energy, *128*
static head, 115
pressure, 115
static lift, 116, *117*
static load, 29
static seal, 162, 202
static total column, *117*
steel alloy, 53
stepladder, *96*
step-up transformers, 326, *327*
stock material weight tables, 19, *20*
weight of steel and brass bar stock, 20
weight of steel plate, 20
straight angle, *3*
straightedge alignment, 307 – 309
straight line, *3*
strainer, 138
strand, *26*
submersion system, 220, *221*
suction strainer, 140
supplementary angles, *3*
suspension scaffold, 102 – *103*
multiple-point, 102 – *103*
two-point, 102 – *103*
swaged socket, *34*
switch, 340 – 341
troubleshooting, 341 – 342
symbol, 134, 183 – *184*
symmetrical load, *17*
synthetic fiber rope, 35
nylon, 35
polyester, 35
polypropylene, 35
synthetic fluid, 217 – 218
synthetic yarn, 46
system operating pressure, 134

T

tackle, 63
tag line, *45*
tagout, *256*
tangent, 4, *6*
tapered bore bearing, *242*
tapered bore pulley, 246 – *248*
recommended hub torque values, 248
tapered roller bearing adjustment, 240 – 242
taper gauge, *307*
tempering, 53
tensile strength, 53
tension member, 243
tetrahedron, 9, *10*
thermal expansion, 168, 289, 299 – 300
thimble, 29
threaded cup follower, 240
three-part reeving, *64*
three-phase circuit, 338
three-way directional control valve, 155 – *156*
throttling, 159
thrust damage, *234*
timber hitch, 43, *44*
time domain, 284 – *285*
timing belt, 250, *251*
toeboard, *98*
tooth form, 262
top hook, 67
top support, 95
torque, 66, *130* – 132, 257 – 258
breakaway, 131
running, 131
starting, 131

total column, *117*
dynamic, *117*
static, *117*
total energy, *128*
transducers, 277 – 281
accelerometer, *278 – 279*
displacement, *278 – 280*
placement, *281*
selection, 280
velocity, *278*
transformer, 333 – 335
troubleshooting, 334 – 335
transmission substation, 326, *327*
trapezium, *8*, 9
trapezoid, *8*, 9
trapezoidal belt, 251, *252*
triangle, *6 –* 7, *8*
acute, *6*, 7
altitude, 6
area of, 7
base, 6
choker fitting, *23*
equilateral, *6*, 7
isosceles, *6*, 7
obtuse, *6*, 7
right, *6*, 7
hypotenuse, 7
scalene, *6*, 7
symbol, 6
truth table, 207
tube, 144
installation, *146*
two-part reeving, *64*
two-point suspension scaffold, 102 – *103*
two-way directional control valve, 155 – *156*
tying off, 106

U

union, 144
units of measure, 2
unlay, 37
unloading valve, 189, *190*
unthread rotation, 78
U-ring seal, 162, *163*
U-sling, 22 – *24*

V

vacuum, 112, 169
valence electron, 319
valence shell, 319, *320*
valve, 151 – 161
actuator, 157
check, *155*, 193, *195*
direct-acting, 151, *152*
directional control, 154 – *156*, 197 – 200
flow control, 158 – *160*, 200, *201*
gate, 159, *160*
globe, 159, *160*
limit, 206
needle, 159, *160*
pilot-operated, *152*
position, 197 – *198*
pressure compensated flow control, *160*
pressure-reducing, 153, *154*
pressure-relief, *151*
safety relief, 189, *191*
sequence, 152, *153*
solenoid, 200
unloading, 189, *190*
vane air motor, *205*
vane compressor, *188*
vane pump, *148 –* 150
variable displacement pump, 158
variable-speed belt drive, *253 –* 255
changing speed, 253
V-belt, 243 – 246
classification, 244
double, 250, *251*
forces, 246
pulley, 246 – 250
fixed bore, 246, *248*
tapered bore, 246 – *248*
recommended hub torque values, 248
sizes, 243 – 246
tensioning, *249 –* 250
belt deflection method, 250
visual adjustment method, 250
vector, 123
velocity, *123 –* 125
transducer, *278*
vertex, 3, *11*
vertical line, *3*
vertical sling, 22, *24*
vibration, 271 – 288
acceleration, *276*
amplitude, *273*
analysis, 280 – 286
analyzers, 282
dynamic signal, 282
Fast Fourier Transform, 282
characteristics, 273 – 277
cycle, *273*
displacement, 273 – *274*
frequency, 274 – 275
oil whirl, 276 – *277*
phase, *275 – 276*
check sheet, *288*
effects, 272
machine gear, *275*
machinery, 272 – 273
measurement methods, 277 – 281
transducers, 277 – 281
accelerometer, *278 – 279*
displacement, *278 – 280*
placement, *281*
selection, 280
velocity, *278*
vibration analyzers, 277 – 278
monitoring programs, 287 – 288
recordkeeping, 287
trending, 287
severity, 277
signature, 281 – 282
velocity, 276
viscosity, 119 – 120, 138, 216
index, 120, 216
visual adjustment method, 250
voltage, 320 – 321
voltage tester, 330 – *331*
use, 330 – 331
volume, 13 – 16, *120 –* 121, 170
of cone, 15
of cylinder, 15, *120 –* 121
of rectangular solid, 14, 121
of sphere, 16
volumetric efficiency, 120
V-ring seal, 162, *163*, *203*

W

wagoneer's hitch knot, *43*
waveform, *273*
way, 155, *156*, *197*
wear pad, *47*
wear particle analysis, 224
webbing, *22*, 46 – 50
inspection, 60, *61*
strength, 49
web sling body, 47
web sling configurations, 47 – *48*
Type I, 47 – *48*
Type II, 47 – *48*
Type III, 47 – *48*
Type IV, *48*
Type V, *48*
Type VI, *48*
web sling length, 47
wedge socket, *23*, *34*, *35*
weight-loaded accumulator, *164*, 165
well (shaft), 91
Weston Brake, *74*
whipping, *37*
wick system, 220, *221*
wiper, *203*
wire rope, 22, 29 – 34
construction, *30*
cutting, 32
filler wire, 29, *30*
inspection, 58 – 59
rotation, 32
Seale, *30*
strength, 30 – 32

terminations, 33 – *34*
Warrington, *30*
Warrington-Seale, 29, *30*
wood ladder, 87
work, 127 – 132
working depth, *263*
working end, *36*
working height, 95
working load limit, 53 – 54
working part, *36*
worm, *266* – 267
worm gear, 66 – *67*, *266* – 268

Y

yarn, *26*